PATTERNS IN PHYSICS

Secondary Science Series

Physics: Consulting Editor
A W Trotter
North London Science Centre

Titles in the Secondary Science Series:

Pre-13
Explore and Discover—science for the middle years
J Darke and G Hughes

CSE
Chemistry—a new course for CSE
M Bowker

GCE Ordinary Level
Chemistry by Experiment and Understanding
J Gerrish and D Mansfield

GCE Advanced Level
Patterns in Physics
W Bolton

Question Books for the 13–16-year Age Group
Structured Questions in Chemistry
A D Gazard and E Wilkins

Objective Questions in Chemistry
K Valentine

Biology Question Book
P Holway

Other Titles:
Social Biology
Biology, Man and Society
A Cornwell

PATTERNS IN PHYSICS

W. Bolton

Former member, Nuffield Foundation Science Teaching Project

Advanced Physics Group

Physics Adviser, British Technical Aid Programme

Brazil

London · New York · St Louis · San Francisco · Düsseldorf
Johannesburg · Kuala Lumpur · Mexico · Montreal · New Delhi
Panama · Paris · São Paulo · Singapore · Sydney · Toronto

Published by

McGRAW-HILL Book Company (UK) Limited

MAIDENHEAD · BERKSHIRE · ENGLAND

---

07 094396 6

10 9 8 7 6 5 4 3 2

PRINTED IN GREAT BRITAIN

Preface

This book is addressed to all those who have an interest in learning about physics, something about a way of thinking. The level of the physics is intended to be that of pre-university courses in many countries.

The book owes much to the thoughts and writings of teachers in many countries. In particular my writing has been much influenced by the work of the Nuffield Foundation Advanced Physics Project, of which I was a team member. I hope the other members of the team will accept this brief note as an acknowledgement of their influence.

I have used quotations and data from many authors and apologize if in any way I have interpreted their writings in a manner different from that originally intended. Mrs D. Williams took a burden off my shoulders by undertaking the task of obtaining copyright clearance for the quotations and data, as well as tracking down the photographs used in this book.

Mr A. Trotter, Mrs A. Bradshaw, and Dr A. J. Walton read the entire manuscript and made many valuable suggestions for improvements. To them and all the others who gave advice I offer my thanks.

I also wish to thank the staff of McGraw-Hill, in particular Mr D. Hill, Mrs E. Woods, and Mrs A. Saunder whose problems were almost certainly increased by my being in Brazil during the entire production time.

W. Bolton

Acknowledgements

The following individuals and groups have generously given permission to reproduce extracts from publications and I am deeply indebted to them. If any have been omitted from the list, they have my apologies.

Academic Press Inc.: *Scientific uncertainty, and information.* Addison-Wesley Publishing Company, Inc.: *Development of concepts of physics; The Feynman lectures on physics,* Volume 1. The American Physical Society: *Physical Review. American Scientist:* 'Energy and Vision'. Johann Ambrosius Barth: *Annalen der Physik.* Basic Books Inc.: *The laws of physics* and *The world of the atom.* The British Interplanetary Society: *Journal of the British Interplanetary Society.* Prince Louis de Broglie and The Institute of Physics: 'Waves and Particles'. Cambridge University Press: 'The Development of the Theory of Atomic Structure', *Background to Modern Science, Proceedings of the Cambridge Philosophical Society,* and *Space, time and gravitation.* Frank Cass and Company Ltd: *A history of epidemics in Britain.* Clarendon Press, Oxford: *Concise Oxford Dictionary.* M. R. Cleland and *The Physical Review: Physical Review.* E. Condon and *The Physical Review:* ' Physics Today'. The Controller of Her Britannic Majesty's Stationery Office: CAP 127 (The Official Report on the Comet disaster). Dodd, Mead and Company: *No more War.* Doubleday & Company Inc. and Heinemann Educational Books Ltd: *The birth of a new physics* and *Heat engines.* Dover Publications Inc: *The restless universe.* Education Development Center Inc.: *PSSC Physics.* Dr L. Essen and The Royal Society: *Proceedings of the Royal Society.* Estate of Albert Einstein: *Einstein on peace.* R. P. Feynman and the British Broadcasting Corporation: *The character of physical law.* The Franklin Institute: *Journal of the Franklin Institute.* George Allen and Unwin Ltd: *An outline of philosophy* and *Science, theory and man.* George Allen and Unwin Ltd and Harper and Row: *Physics and philosophy.* George Allen and Unwin Ltd and Simon and Schuster: *History of western philosophy.* R. P. Hammond: 'Low Cost Energy: A New Dimension', *Science Journal* 1969. Harvard University Press: *The Copernican revolution* and *The nature of thermodynamics.* Hughes Massie Ltd and Dodd, Mead and Company: *The ABC murders.* Hutchinson Publishing Group: *Marconi—master of space.* Hutchinson Publishing Group and The World Publishing Corporation: *Quanta and reality.* The Institute of Electrical and Electronics Engineers, Inc.: *IEEE Spectrum* and *Proceedings of the Institute of Radio Engineers.* IPC Magazines Ltd: *New Scientist.* Sir Geoffrey Keynes and The Royal Society: Newton Tercentenary Celebrations. Alfred A. Knopf and Hodder & Stoughton Ltd: *Niels Bohr.* Macmillan Journals Limited: *Nature.* McGraw-Hill Book Company: *Fundamentals of optics.* W. H. Martin and the Institute of Physics: *Journal of Scientific Instruments.* National Academy of Sciences (USA): *Proceedings of the National Academy of Sciences.* A. O. Nier and the *American Scientist:* 'Mass Spectroscopy'. W. W. Norton and Company Inc. and Thomas Nelson and Sons Limited: *Special relativity.* Oxford and Cambridge Schools Examination Board: *Nuffield advanced physics* (1970)—*Short Answer Paper* (Questions 4, 6, 8, and 9), *Long Answer Paper* (Question 6), *Special Paper* (Question 7), *Physics Options Paper* (Questions 3 and 6); *Nuffield advanced physics* (1971)—*Short Answer Paper* (Questions 2, 5, and 7), *Long Answer Paper* (Question 5), *Special Paper* (Questions 2 and 10). Oxford University Press: *The second law.* Penguin Books Ltd: *Understanding weather; The invasion of the moon; On the nature of the universe;* and *Britain in figures.—A handbook of social statistics.* Pergamon Press Ltd: *Wave mechanics* and *The old quantum theory.* A. D. Peters & Co., The Macmillan Company and Hutchinson Publishing Group Ltd: *The act of creation* and *The sleepwalkers.* Princeton University Press: *Physics for the inquiring mind.* The Royal Society: *Philosophical Transactions* and *Proceedings.* Les Editions Seghers and Souvenir Press: *Enrico Fermi.* Springer Verlag: *Zeitschrift für Physik. The Sunday Times:* 'Doctors Call for a Total Ban on Tobacco Advertisements', and Review of *A portrait of Isaac Newton.* Taylor and Francis Ltd: *Philosophical Magazine.* Thames & Hudson Ltd, World Publishing Corporation and Harold Matson Company Inc.: *The doomsday book.* Sir George Thomson and Taylor and Francis Ltd: 'The Early History of Electron Diffraction'. *The Times:* 'Comet Airliner crashes in Mediterranean'; 'Comet Wreckage and Bodies found'; and 'Energy from the Atom'. *The Times Educational Supplement:* 'Examples of Feedback'. J. G. Trump and The Institute of Physics: *Reports on progress in physics.* United States Atomic Energy Commission: *Energy in the future.* University of California Press: *Lectures on gas theory.* University of Chicago Press: *Astrophysical Journal* and *The electron.* Verband Deutscher Physikalischer Gesellschaften: *Verhandlungen der deutschen physikalischen Gesellschaft.* Viking Press Inc. and Macmillan, London and Basingstoke: *A star called the sun.* Ward Lock Ltd: *The memoirs of Sherlock Holmes.* Xerox College Publishing: *The world of elementary particles.* Cover photograph by permission of Professor R. E. Orville of State University of New York.

Contents

Kobe includes chance ch 8 189–206

0 Patterns in physics

A note Why is this called chapter zero?

'"Rule Forty-two. All persons more than a mile high to leave the court."
Everybody looked at Alice.
"I'm not a mile high," said Alice.
"You are," said the King.
"Nearly two miles high," added the Queen.
"Well, I shan't go, at any rate," said Alice: "besides, that's not a regular rule: you invented it just now."
"It's the oldest rule in the book," said the King.
"Then it ought to be Number One," said Alice.'

L. Carroll (1865). *Alice's Adventures in Wonderland.*

This is not a 'regular' chapter but it is the 'oldest' chapter in physics.

Patterns in physics

'In 1796 a minor scandal occurred at the Greenwich Observatory: Maskelyne, the Astronomer Royal, dismissed one of his assistants because the latter's observations differed from his own by half a second to a whole second. Ten years later the German astronomer Bessel read about this incident in a history of the Greenwich Observatory. Bessel, who combined a highly original mind with meticulous precision in his observations, was puzzled by the frequent occurrence of similar timing mistakes by astronomers. It was a typical case of a "shift of attention" from the nuisance aspect of a trivial phenomenon to the investigation of its causes.

After ten years of comparing his own records with those of several other astronomers, Bessel was able to prove that there existed systematic and consistent differences between the speed with which each of them reacted to observed events; and he also succeeded in establishing the characteristic reaction-time—called "the personal equation"—of several of his colleagues.

These studies were continued by other astronomers over the next thirty years, in the course of which the development of more precise, automatic recording instruments made it possible to arrive at "absolute personal equations". Finally, fifty years after Bessel's discovery, von Helmholtz published a paper showing that the rate of conduction of impulses in nerves was of a definite, measurable order—and not, as had previously been assumed, practically instantaneous. Helmholtz was well acquainted with the work that astronomers had done on personal equations, and his experiments on the propagation of impulses in motor and sensory nerves followed their procedure and techniques. Helmholtz's discovery inaugurated the era of "mental Chronometry", and was a decisive step in the progress of neurophysiology and experimental psychology.'

A. Koestler (1964). *The act of creation.* Hutchinson/Macmillan, New York.

Facts which did not fit the established pattern; dismissal of the facts as experimental error; a relationship seen; the 'working' relationship given a theoretical background—this briefly summarizes the above extract and is a model that is often followed in science.

Many scientific discoveries can be called accidents; for example, Jansky's discovery of radio waves originating from outside the earth (see chapter 12). Jansky was investigating radio interference at the time and not looking for extra-terrestrial radio waves. Other discoveries have arisen from scientists feeling that something is not quite right, an experimental fact does not quite fit a theory or a theory seems to have some strange predictions—Leverrier and Adams predicted the existence of a new planet, Neptune, on the basis of slight irregularities of the orbit of the planet Uranus, irregularities which did not fit with the theoretical prediction based on Newton's theory of gravitation (see chapter 2). Scientific intuition might be the right phrase to describe many discoveries.

'"My friend Hastings will tell you that from the moment I received the first letter I was upset and disturbed. It seemed to me at once that there was something very wrong about the letter."

"You were quite right," said Franklin Clarke dryly.

"Yes. But there, at the very start, I made a grave error. I permitted my feeling—my very strong feeling about the letter to remain a mere impression. I treated it as though it had been an intuition. In a well-balanced, reasoning mind there is no such thing as an intuition—an inspired guess! You can guess, of course—and a guess is either right or wrong. If it is right you call it an intuition. If it is wrong you usually do not speak of it again. But what is often called an intuition is really an impression based on logical deduction or experience. When an expert feels that there is something wrong about a picture or a piece of furniture or the signature on a cheque he is really basing that feeling on a host of small signs and details. He has no need to go into them minutely—his experience obviates that—the net result is the definite impression that something is wrong. But it is not a guess, it is an impression based on experience.'

Agatha Christie (1936). *The ABC murders.* Collins/Dodd, Mead and Co. Inc.

To be a scientist you have to be a detective, a Hercule Poirot (the detective in the above extract) or Sherlock Holmes. Detective stories seem to follow a general sequence:

(a) A fact requiring explanation; perhaps a murder.

(b) Investigations to yield data; gathering clues.

(c) Forming an hypothesis, a pattern; putting the clues together to give a pattern and so a suspect for the crime.

(d) Looking for evidence to confirm the hypothesis, seeking the vital evidence to give a cast-iron case.

(e) The successful theory and the arrest of the murderer. There will almost inevitably be false data and false hypotheses which lead in the wrong direction, if only to make certain that the reader does not 'guess' the identity of the murderer too soon.

The above steps describe not only detective stories but also much of the way in which science progresses. The following is a physics sequence described in more detail in chapter 13.

(a) A fact; the spectrum lines of an element are characteristic of that element—they serve as a fingerprint.

(b) Determination of the wavelengths of all the lines in the hydrogen spectrum.

(c) Balmer derives an empirical relationship between the wavelengths—a pattern has emerged.

(d) Bohr produces a theory of the atom out of which Balmer's relationship emerges.

(e) Bohr's theory of the atom proves unsatisfactory.

(f) A wave theory replaces Bohr's theory, Balmer's relationship still comes out of this theory.

(g) The wave theory enables other predictions to be made—these check with experiments (e.g., the emission of alpha particles from nuclei).

(h) We have, for the present, a successful theory.

In one sense physics is a search for patterns among data. The data may have been in existence many years but no pattern seen, or regrouping of existing patterns of data may lead to a bigger simpler pattern.

Kepler used Tycho Brahé's data on planets to give a simple pattern for planetary behaviour. Newton combined the patterns of Kepler for the movements of the planets with the pattern of Galileo for the behaviour of objects falling to the surface of the earth to give a general simple pattern of universal gravitation.

In one sense physics is the search for patterns with which to describe the physical events of the world, and universe, around us. The patterns enable us to give order to events and make predictions. We expect that the sun will rise every morning in the east and every evening set in the west. We feel we can predict this because in our experience it always has done—what we mean is that we have made a large number of observations of the sun rising and setting and established a pattern and feel confident enough to make predictions based on this pattern. The immediate purpose of physics, indeed science, is to make predictions of events in nature. It is through this ability of man to make predictions that he has survived—he could not outrun or outfight other animals but he could outwit them.

Problems

1. How would you describe the patterns evident in the graphs shown in Fig. 0.1?

2. What pattern can you discern in the following data:

	Melting point, /K	*Molar latent heat of fusion* /kJ mole^{-1}
Lead	600	4·77
Copper	1360	13·05
Nickel	1730	17·61
Platinum	2040	19·66
Iridium	2730	26·36
Tungsten	3650	35·23

What molar latent heats would you forecast for the following:

Magnesium, melting point 923 K,
Vanadium, melting point 2190 K?

3. There is an error in the following data. Which is the incorrect boiling point? The information given in question 2 may help. Use your intuition to guess which point is in error and then try to analyse why you consider that point to be in error.

	Boiling point/K
Lead	2824
Copper	2855
Nickel	3100
Platinum	4100
Iridium	4400
Tungsten	5880

4. ' "Is there any other point to which you would wish to draw my attention?"

"To the curious incident of the dog in the night-time."

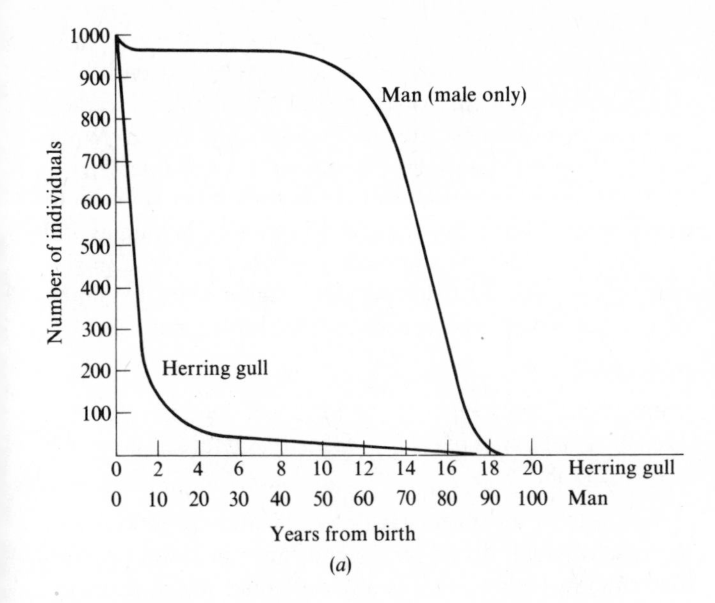

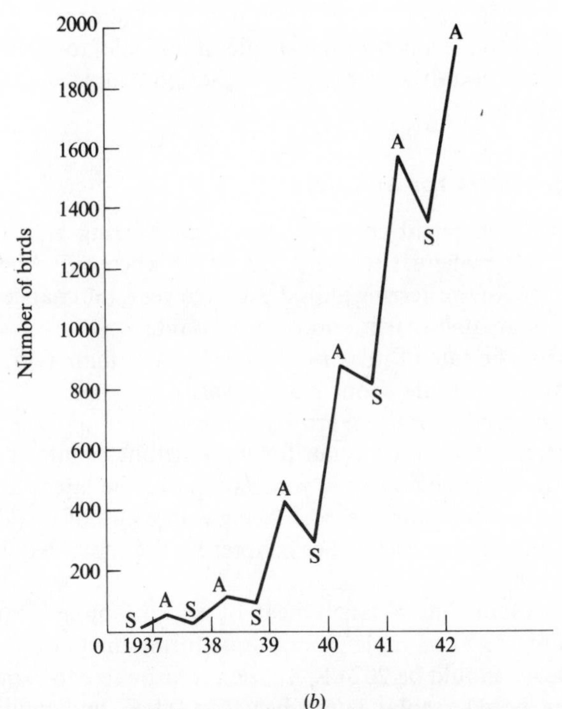

Fig. 0.1 (a) Survivorship records for man and herring gull. Both curves commence from 1000 newborn (Allison, A. (Ed.) Population control, *Article by C. Perrins, Fig. 7, © Penguin Books, 1970). (b) Number of pheasants on an island after the initial introduction of two cocks and six hens in 1937. A = Autumn count, S = Spring count (Allison, A. (Ed.)* Population control, *Article by C. Perrins, Fig. 4, © Penguin Books, 1970).*

"The dog did nothing in the night-time."

"That was the curious incident," remarked Sherlock Holmes,'

A. Conan Doyle (1892). 'The Silver Blaze', *The memoirs of Sherlock Holmes.*

A result of 'nothing happened' can be just as useful as a positive result. In Sherlock Holmes' case the non-barking dog clue led to the realization that the criminal must have been known to the dog.

Comment on the following facts:

Atoms are mainly empty space.

In most collisions between atoms the collision is perfectly elastic. The atoms bounce apart with no effects on the internal structure of either atom.

5. If you throw a six-sided die one hundred times how many times do you think it will land with a six uppermost? Explain how you arrived at your answer.

6. The following data give the years in which the summers in southern England were good and those in which they were bad. Can you discern any pattern which would suggest the years in which you should take a holiday in southern England?

'*Good*' (very warm, sunny and dry)	'*Bad*' (very cool, dull and wet)
1899	1888
1911	1890
1921	1903*
1933	1912
1947	1920
1949	1954
1959	1956

O. G. Sutton (1964). *Understanding weather.* Penguin.

* An explanation exists—a volcano exploded and put a lot of dust in the atmosphere.

7. Oxford and Cambridge Schools Examination Board. Nuffield Advanced Physics. Short answer paper. 1970.

'This question is about describing different kinds of scientific statements. Here are some possible descriptions of such statements:

"states an experimental fact"

"makes a hypothesis"

"quotes a scientific law"

"is a rough estimate"

"is a deduction from earlier statements".

You are asked to give a brief description of each of the numbered statements in the passage below. Your descriptions should use phrases like those given above; you may use, combine or adapt the phrases above or invent others of your own.

1. If we assume that in a gas the atoms have a radius of about 10^{-10} m and a mean separation of about ten atomic diameters . . .

2. . . . it is clear that an alpha particle in traversing several centimetres of the gas must encounter some thousands of atoms of gas.

3. Only a minute fraction of such encounters however produce any appreciable deflection of the alpha particle.

4. It is difficult to avoid the conclusion that the greater part of the atomic volume is effectively empty.'

8. The equation $(1 + x)^n$ can be expanded to form a series of terms x^0, x^1, x^2, x^3, x^4, etc.

For example:

$$(1 + x)^1 = 1x^0 + 1x^1 \qquad (x^0 = 1)$$
$$(1 + x)^2 = 1x^0 + 2x^1 + 1x^2$$

Guess the pattern taken by the series by filling in the blanks in the following table.

	x^0	x^1	x^2	x^3	x^4	x^5
$n = 1$	1	1	0	0	0	0
$n = 2$	1	2	1	0	0	0
$n = 3$	1	3	3	1	0	0
$n = 4$	...	...	...	...	...	...
$n = 5$	...	...	...	...	...	...

9. Guess the numbers that follow on the sequences given below.

(a) 24, 29, 34, 39, ...
(b) 6, 18, 54, 162, ...
(c) 4800, 1200, 300, ...
(d) 2, 4, 16, 256, ...

10. The electric force between two charged masses is given by

$$F_e = \frac{q_1 q_2}{4\pi\varepsilon_0 r^2}$$

The gravitational force between the two is

$$F_g = \frac{Gm_1 m_2}{r^2}$$

The ratio of the forces is

$$\frac{F_e}{F_g} = \frac{q_1 q_2}{4\pi\varepsilon_0 G m_1 m_2}$$

Suppose we consider our two objects to be two electrons; the ratio of the forces becomes $4{\cdot}1 \times 10^{41}$. If the two objects are protons the ratio becomes $1{\cdot}2 \times 10^{37}$. If we consider a proton and an electron the ratio becomes $2{\cdot}3 \times 10^{39}$.

The age of the universe is about 10^{10} years, i.e., about 3×10^{17} s. The time taken by light to traverse a proton is 3×10^{-24} s. The ratio of the two terms is 10^{41}.

Both the forces ratio and the times ratio are large numbers of a similar order of magnitude. Do you think there is any pattern here? If there was a pattern what could be the consequences?

Practical problems

1. Determine the densities of the particles present in dust and consider whether the results could be used to identify the dust and its origin. This could be of significance in crime detection.

A possible method for measuring densities would be to employ Archimedes' principle according to which a particle will float under a liquid surface if its weight is equal to the weight of fluid displaced. A variable density fluid can be obtained by mixing two fluids or by changing the temperature of a fluid.

2. Rubber as a material is characterized by the very large strains which can be obtained before breakage occurs. This is due to the structure of rubber in which the molecules, long-chain molecules, are not straight but coiled. When stress is applied the molecules uncoil. (See chapter 9.)

Determine the stress–strain graph for rubber right up to its breaking point—this may involve strains as high as 700 per cent. Consider the significance of your graph. Can you make a mechanical model, perhaps a chain of freely jointed rigid links, which would behave in a similar manner?

References

Schallamach. *Bull. Inst. Phy. & Phy. Soc.*, **18**, 215–21, 1967.
Treloar. *The physics of rubber elasticity*, Oxford: Clarendon Press, 1958.

3. The freezing of water is a crystallization process and in the case of small drops is of significance in cloud physics. Determine whether there is any difference in the temperature at which water freezes when it is in small drops; consider diameters from 0·1 (and smaller if possible) to 1·0 mm. The drops could be cooled on a thermo-couple under a microscope.

Suggestions for answers

1. (a) You might comment that most herring gulls die when young whereas with man most die when old. About 43 per cent of the herring gulls die in each year, the chance of dying in any one year is almost a constant.

(b) The rate of increase is slightly more than a doubling every two years, from 1939 onwards.

See chapter 5 for more growth patterns.

2. High latent heats occur for high melting points. You might have expected this if you thought of the latent heat and the melting point as both being a measure of the size of the interatomic forces. See chapter 9 for a more detailed discussion.

Magnesium has a latent heat of 8·95 kJ mole^{-1} and vanadium 17·57 kJ mole^{-1}, not quite fitting the pattern.

3. Lead should be 2024 K. It seems reasonable to expect melting points, molar latent heat capacities, and boiling points to follow similar patterns.

4. We might expect that atoms which are mainly empty space might pass straight through each other when they meet or certainly 'mix and interact'. The fact that they do not is very important. We can find an explanation in terms of the atomic electrons being considered to have a chance of being anywhere in an atom. See chapter 13 for a fuller discussion.

5. 100/6. If we do not expect the die to be biased then the chance of any one face being uppermost is 1 in 6. There is just one way in six of the die landing the way we want. The chance of this way happening is a measure of the frequency with which it occurs. See chapter 8 for a fuller discussion of chance.

6. There is a greater chance of a good summer in a year with an odd date than in one with an even date.

7. The following are my suggestions, you may disagree.
(1) Makes an hypothesis.
(2) A deduction from earlier statements.
(3) An experimental fact.
(4) A deduction.

8. 1, 4, 6, 4, 1, 0: 1, 5, 10, 10, 5, 1

$$(1 + x)^n = 1 + \frac{nx}{1} + \frac{n(n-1)}{1 \times 2}x^2 + \frac{n(n-1)(n-2)}{1 \times 2 \times 3}x^3 + \text{etc.}$$

This is called the Binomial theorem.

9. (a) 44
(b) 486
(c) 75
(d) 65 536

10. Dirac considers the approximate number 10^{39} to have significance—he believes there is a pattern. He feels there is a deep connection between cosmology and atomic theory. As the age of the universe increases Dirac believes that the ratio will increase and so therefore the universal gravitation constant G must decrease with time.

1 Motion

Teaching note Practical work appropriate to this chapter will be found in:
W. Bolton. *Physics Experiments and Projects*, Vol. 5, Pergamon, 1969
Nuffield O-level Physics, Guide to Experiments, 4, Longmans/Penguin, 1967.
Project Physics Course. Handbook. Holt, Reinhart and Winston, 1970.

1.1 Describing motion

What is meant by motion? Anybody answering this question would probably talk of position changing with time. But if we look at how the position of objects or parts of objects change with time it seems very complex. Look at the position–time traces for the planet Mars (Fig. 1.1(a)), for a point on the rim of a car wheel as the car moves steadily along a road (Fig. 1.1(b)), for a tennis ball during a game of tennis (Fig. 1.1(c)). To find laws governing motion and its causes we need ways of describing motion and of simplifying it, so that relationships can be readily established. Having arrived at relationships for a simplified case we can then try to apply them to complex motions.

For our simplified case we take the motion of a point in a straight line. We chose a point because we do not have to bother with the structure of the object and possible rotation. We chose a straight line because it seems to be the simplest. How can we describe the motion of a point? One way is to determine the position of the point at various instants of time and use the position–time data to specify the motion.

Let us take an example:

Position of point/m	0	0	2	4	6	10	15
Time/s	0	1	2	3	4	5	6

Position of point/m	18	20	22	23	23	23
Time/s	7	8	9	10	11	12

The table could be presented diagrammatically as a graph, Fig. 1.2. What is happening to the point? When we start our timing the point is at the zero on our distance scale.

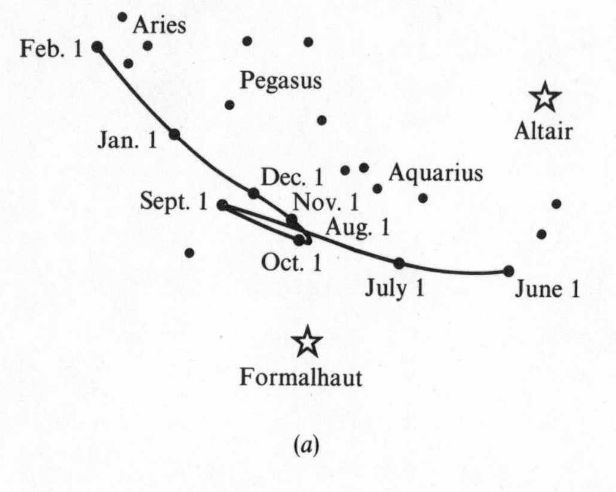

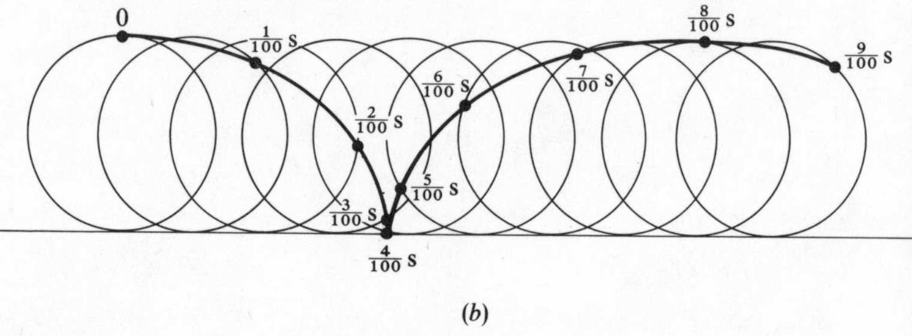

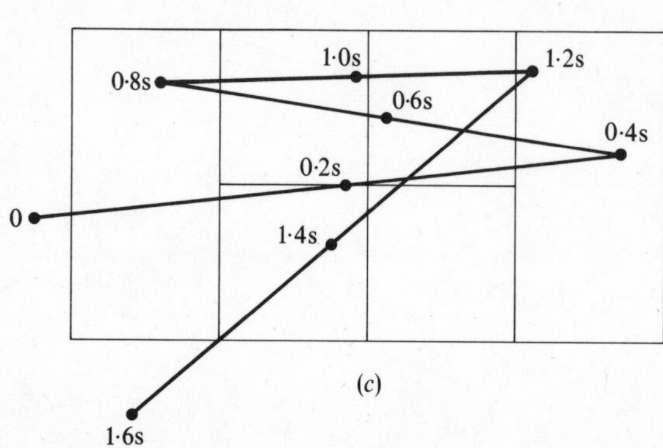

Fig. 1.1 (a) From R. H. Baker: Astronomy, *8th Ed. Published by Van Nostrand Reinhold Co., Copyright 1964 by Litton Educational Publishing. The position–time trace for the planet Mars with respect to the fixed stars. (b) The position–time trace for a point on the rim of a wheel. (c) The position–time trace for a tennis ball.*

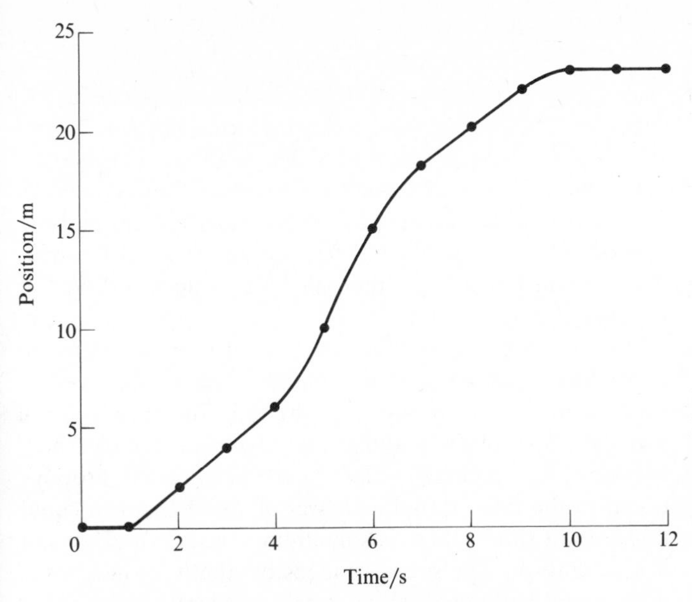

Fig. 1.2 Position–time graph.

After 1 s the point is still at the zero position. During the next second the point moves a distance of 2 m. We can describe this motion by saying that the point has an average speed of 2 m s^{-1} during the time interval 1 to 2 s.

$$\text{Average speed} = \frac{\text{distance covered in a time interval}}{\text{time interval}}$$

The word average is necessary because the point could have higher, and lower, speeds during the time interval. If our time interval had been from zero to 2 s then the average speed would have been 1 m s^{-1}: for part of this interval the speed was zero, for part 2 m s^{-1}.

Over the time intervals 2 to 3 s and 3 to 4 s the distance covered in a second remains constant. We say that the point is moving with uniform speed. The average speed is uniform when equal distances are covered in equal intervals of time. The speed is uniform if equal distances are covered in equal time intervals, however short the time intervals are.

We supposed our point to be moving in a straight line; would our definition of speed and that of uniform speed be changed if the point moved in any other path? Can we use the terms for say an object moving in a circular path? The term speed is used regardless of any change of direction of an object. We can have an object moving with constant speed round a circular path. A car could go round a race track with a constant speed of 100 km h^{-1} (at least theoretically if not in practice). Uniform speed can describe the motion of an object moving in a circular path—all that is required is that the object changes its position by equal distances in equal intervals of time, however short the time intervals.

The term velocity is used, instead of speed, to describe motion which is restricted to a straight line. Thus the average velocity is the distance covered in a straight line in a time interval divided by the time interval. An object can only have a constant velocity if it moves in a straight line and covers equal distances along that line in equal intervals of time. Thus an object may have a uniform speed but need not necessarily have a uniform velocity. The car moving round the race track with uniform speed is not moving with a uniform velocity.

Why do we have the two terms, speed and velocity? In many cases it is useful to know the speed at which an object can move without bothering about the direction. You can perhaps walk at a speed of 7 km h^{-1}. But if we need to know the position of an object after some time interval then we need the velocity. You could walk at 7 km h^{-1} in a circular path and at the end of an hour be back where you started; your speed would be 7 km h^{-1} and would give no information as to where you might be at the end of the hour, your average velocity over the hour would be zero and would indicate that at the end of an hour you would be found where you started.

When the velocity of an object changes we say it has accelerated.

$$\text{Average acceleration} = \frac{\text{change of velocity in a time interval}}{\text{time interval}}$$

If the velocity was 2 m s^{-1} at the beginning of a one-second time interval and 4 m s^{-1} at the end then the average acceleration was +2 m s^{-2}. The acceleration is positive because the velocity is increasing in the time interval. A negative acceleration would mean a decreasing velocity—the object was slowing down.

The acceleration is uniform when equal changes in velocity occur in equal intervals of time, however small the time intervals.

The reason for defining acceleration in terms of velocity rather than speed will become apparent in chapter 2.

We have in this section defined certain quantities, speed, velocity, and acceleration, and what we understand by constant speed, constant velocity, and constant acceleration. Why are the definitions in the form given? Why define speed in terms of the distance covered in a particular time interval? Why define acceleration in terms of the velocity change in a particular time interval? Why do we use these terms to describe motion?

The following quotation from Galileo may serve to answer the question. The work in which it appeared was in the form of a discourse between three characters.

> 'When, therefore, I observe a stone initially at rest falling from an elevated position and continually acquiring new increments of speed, why should I not believe that such increases take place in a manner which is exceedingly simple and rather obvious to everybody? If now we examine the matter carefully we find no addition or increment more simple than that which repeats itself always in the same manner. This we readily understand when we consider the intimate relationship between time and motion; for just as uniformity of motion is defined by and conceived through equal times and equal spaces (thus we call a motion uniform when equal distances are

traversed during equal time-intervals), so also we may, in a similar manner, through equal time-intervals, conceive additions of speed as taking place without complication; thus we may picture to our mind a motion as uniformly and continuously accelerated when, during any equal intervals of time whatever, equal increments of speed are given to it. ... A motion is said to be uniformly accelerated, when starting from rest, it acquires, during equal time intervals, equal increments of speed.

... Although I can offer no rational objection to this or indeed any other definition, devised by any author whomsoever, since all definitions are arbitrary, I may nevertheless without offence be allowed to doubt whether such a definition as the above, established in such an abstract manner, corresponds to and describes that kind of accelerated motion which we meet in nature in the case of freely falling bodies.... So far as I see at present, the definition might have been put a little more clearly perhaps without changing the fundamental idea, namely uniformly accelerated motion is such that its speed increases in proportion to the space traversed; so that, for example, the speed acquired by a body in falling four cubits would be double that acquired in falling two cubits...

... It is very comforting to me to have had such a companion in error; and moreover let me tell you that your proposition seems so highly probable that our author himself admitted, when I advanced this opinion to him, that he had for some time shared the same fallacy.'

G. Galileo (1638)
Dialogues concerning two new sciences.

Galileo's argument for defining acceleration in terms of change of velocity in an interval divided by the length of the time interval depends on a freely falling object being defined as having something natural about its motion and that natural something being called acceleration. A freely falling object is considered to have a constant acceleration.

Figure 1.3 shows a strobe photograph of a freely falling object. The photograph shows the positions occupied by the object at equally spaced successive time intervals. Figure 1.4 shows the distance–time graph for the motion. As can be seen, the object certainly does not cover equal distances in equal time intervals and so does not have a constant speed.

We can read off from the graph the distances covered in short time intervals and so obtain values of the average speed at different times during the fall. The speeds are in fact the slopes of the graph at the time instants concerned. Measure for yourself. The speed is directly proportional to the time. Equal increases of speed occur in equal intervals of time—there is something constant and we call it acceleration. The speed changes by about 10 m s^{-1} for each second change—we say the acceleration is about 10 m s^{-2}.

Not only is the acceleration constant for a freely falling object but all freely falling objects, in vacuo, have the same acceleration (symbol g). We need to specify a vacuum because an object falling in a medium has its motion affected by the medium. For dense objects falling in air over relatively short distances the effect of the medium is generally not noticeable. Galileo comments on a test of this acceleration being the same for all objects.

'But I, ..., who have made the test can assure you that a cannon ball weighing one or two hundred pounds, or even more, will not reach the ground by as much as a span ahead of a musket ball weighing only half a pound, provided both are dropped from a height of 200 cubits.'

G. Galileo (1638).
Dialogues concerning two new sciences.

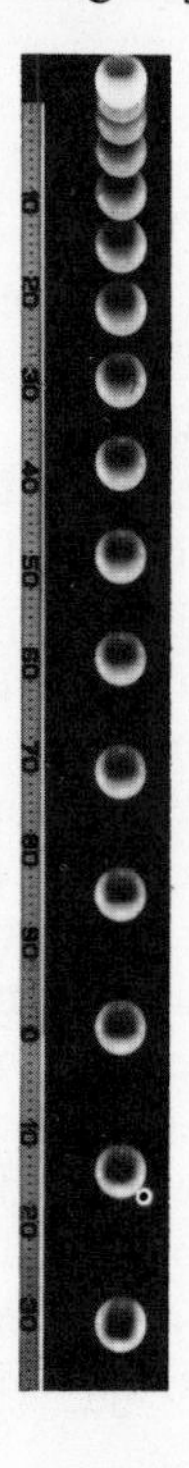

Fig. 1.3 Photograph from PSSC Physics, 2nd Ed., D. C. Heath and Co. Strobe photograph of a falling billiard ball. The position scale is in centimetres, and the time interval between the successive positions of the ball is 1/30 s.

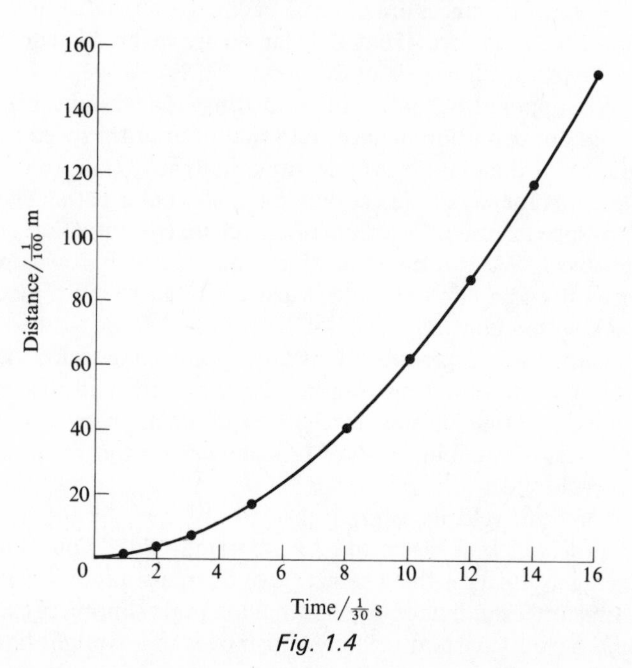

Fig. 1.4

If two objects take the same time to accelerate over the same distance then they must have the same acceleration. Galileo is reputed to have dropped objects from the Leaning Tower of Pisa in test of this constant acceleration idea. This, however, is considered by many to be a legend.

'..., the famous experiment of dropping cannon balls from the leaning Tower of Pisa was carried out not by Galileo but by his opponent, ... Coressio, and not in refutation, but in confirmation of the Aristotelian view that larger bodies must fall quicker than smaller ones.'

A. Koestler (1959) *The sleep walkers.* Hutchinson/Macmillan, New York.

'In studying the problem of falling bodies, Galileo, we know, made experiments in which he dropped objects from heights, and—notably in the Pisan days of his youth—from a tower. Whether the tower was the famous Leaning Tower of Pisa or some other tower we cannot say; the records that he kept merely tell us that it was from some tower or other. Later on his biographer Viviani, who knew Galileo during his last years, told a fascinating story which has since taken root on the Galileo legend. According to Viviani, Galileo, desiring to confute Aristotle, ascended the Leaning Tower of Pisa, "in the presence of all other teachers and philosophers and of all students", and "by repeated experiments" proved "that the velocity of moving bodies of the same composition, unequal in weight, moving through the same medium, do not attain the proportion of their weight, as Aristotle assigned to them, but rather that they move with equal velocity...." Since there is no record of this public demonstration in any other source, scholars have tended to doubt that it happened, especially since in its usual telling and retelling it becomes fancier each time. Whether Viviani made it up, or whether Galileo told it to him in his old age, not really remembering what had happened many decades earlier, we do not know....'

I. B. Cohen (1960). *The birth of a new physics.* Heinemann/Doubleday and Co. Inc.

A discussion of the way the term acceleration evolved is given in *The fabric of the heavens* by Toulmin and Goodfield (Penguin), chapter 8. They consider the first significant step in the definition of acceleration was made by scholars at Merton College, Oxford, between 1330 and 1350.

Motion in the absence of a force

Consider a simple pendulum swinging backwards and forwards. If there was no resistance to motion, how high would the pendulum bob rise if I let it fall from position X in Fig. 1.5(a)? Would it rise to a height above X, the same as X, or less than X?

Consider a ball rolling down a curved piece of track. How high up the other side in Fig. 1.5(b) will the ball rise?

What happens if as in Fig. 1.5(c) the ball rolls down the track and on to a level part of the track?

In (a) and (b) the bob or ball rises to the same level when frictional effects are small. In the case of (c) we should expect the ball to keep going until it reached the same height as it started from—this it never does, so it should keep on moving for ever. This means it cannot slow down, and there is thus no acceleration, when it runs along the horizontal part of the rail. Thus in the absence of a resultant

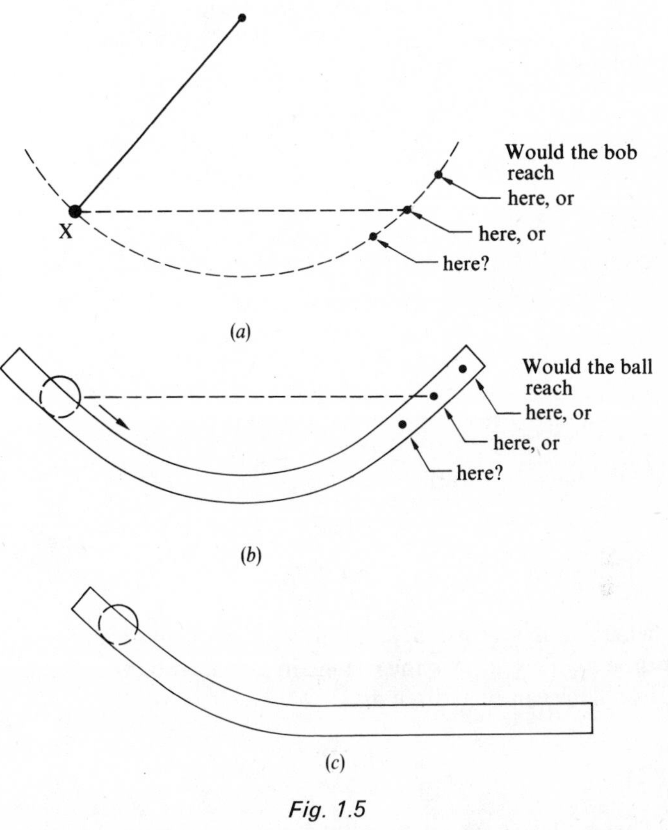

Fig. 1.5

force an object keeps on moving with a uniform velocity. This was first established by Galileo.

'... any velocity once imparted to a moving body will be rigidly maintained as long as the external causes of acceleration and retardation are removed, ... from this it follows that motion along a horizontal plane is perpetual; for if velocity be uniform, it cannot be diminished or slackened....'

G. Galileo (1638). *Dialogue concerning two new sciences.*

This is called the law of inertia—an object continues in motion with a uniform velocity unless acted on by a force.

If we give an object a push and send it sliding along a table or the floor it does not keep on moving—there is no uniform speed in a straight line. Why? We have a condition to the law of inertia—no resultant force acting on the object, for no acceleration. In the case of the object sliding along the table or floor it is rubbing against the table or floor. We say that there is a resultant force acting on the object, the force due to friction. Thus we would not expect the law of inertia to mean that the object would continue moving with uniform speed in a straight line—we have resultant forces acting.

1.2 Distance–time relationships

This section is only concerned with motion in a straight line.

Suppose we determine experimentally the positions of an object at different instants of time. The data can be used to plot a graph of distance, from some starting point, against time. What is the significance of the slope of such a graph?

Over a straight line portion of the graph, Fig. 1.6, the slope is

$$\frac{s_2 - s_1}{t_2 - t_1}$$

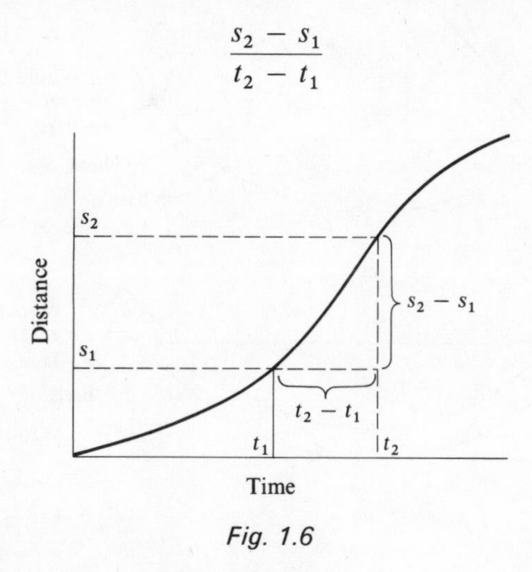

Fig. 1.6

where s_1 is the distance at time t_1, and s_2 the distance at time t_2. The slope is the change in s divided by the change in t. This can be written as

$$\frac{\Delta s}{\Delta t}$$

The Δ (delta) sign is used to denote a finite bit of some quantity.

The slope tells us the distance covered in the time taken. This quantity is known as velocity.

$$\text{Velocity} = \frac{\Delta s}{\Delta t}$$

The greater the velocity the greater the slope of the distance–time graph (Fig. 1.7).

If the velocity is not constant the graph will not be a straight line, i.e., the slope will not be constant. What about our velocity definition now? If we take large values of Δt in order to compute the slope then we are averaging out the velocity over the time interval Δt. The smaller the value of Δt the more accurate will be our value of the velocity for an instant of time. The velocity at some instant of time is the value of $\Delta s/\Delta t$ when Δt is very small, i.e., when Δt is nearly zero.

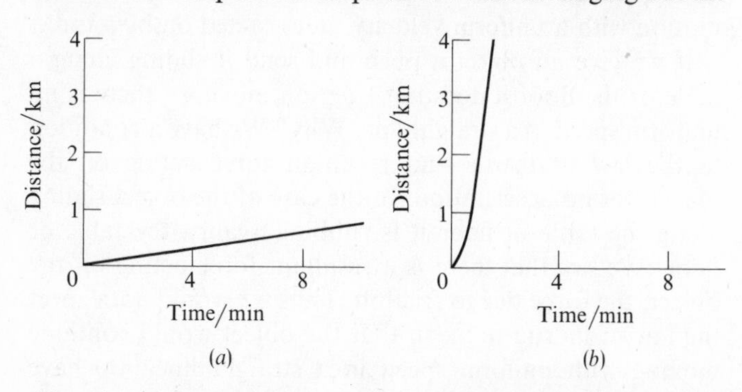

Fig. 1.7 Distance–time graphs for (a) a cyclist, (b) a car.

$$\text{Velocity} = \lim_{\Delta t \to 0} \frac{\Delta s}{\Delta t}$$

The above is the mathematical way of writing this, 'limit as Δt approaches zero' of $\Delta s/\Delta t$. A shorter way of writing this is $\mathrm{d}s/\mathrm{d}t$.

$$\text{Velocity} = \frac{\mathrm{d}s}{\mathrm{d}t}$$

The Δ's and the d's are part of the notation, like sine or cosine, and are not arithmetical symbols which can be cancelled. $\mathrm{d}s/\mathrm{d}t$ is called the derivative of s with respect to t.

Suppose we have a distance–time graph which can be represented by the equation (Fig. 1.8)

$$s = Kt^2$$

where K is some constant. How can we calculate the velocity at some instant of time? The velocity is the slope of the graph at the instant concerned. How then can we calculate the slope?

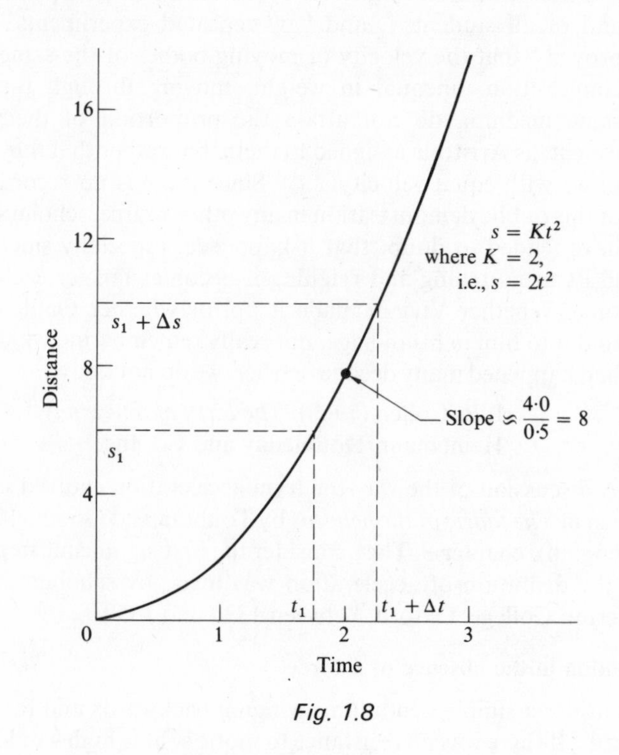

Fig. 1.8

Let s_1 be the distance at time t_1.

$$s_1 = Kt_1^2 \tag{1}$$

If we increase the time to $t_1 + \Delta t$ then the distance must also increase, to say $s_1 + \Delta s$. If both these values lie on the graph then

$$s_1 + \Delta s = K(t_1 + \Delta t)^2$$

Expanding the term in the brackets

$$s_1 + \Delta s = Kt_1^2 + 2Kt_1\,\Delta t + K(\Delta t)^2$$

If eq. (1) is subtracted from this

$$\Delta s = 2Kt_1\,\Delta t + K(\Delta t)^2$$

$$\frac{\Delta s}{\Delta t} = 2Kt_1 + K\,\Delta t$$

As Δt tends to zero so this value of $\Delta s/\Delta t$ will come nearer to being the value of the velocity at the time t_1. But the smaller we make Δt the smaller will the $K\,\Delta t$ term become. Hence in the limit when Δt is near to zero the velocity will be given by

$$\text{Velocity (at } t_1) = \frac{ds}{dt} = 2Kt_1$$

This process is known as differentiation.

With K = 2 this gives at $t_1 = 2$ a velocity or slope of distance–time graph of 8; this checks with a measurement of the graphical slope at $t = 2$.

The equation shows that the velocity depends on the time elapsed. The object whose motion is described by this equation must therefore be accelerating—a velocity depending on the time is not constant.

Velocity is the rate at which distance along a straight line is covered, with respect to time, and is the slope at any instant of the distance–time graph. If there is uniform velocity then equal distances along a straight line are covered in equal intervals of time, however short, and the slope of the graph is the same at all instants, during the constant velocity. When the velocity is not constant then the slope of the distance–time graph is not constant. Figure 1.9 shows several distance–time graphs and their corresponding velocity–time graphs. The velocity has been obtained from measurements of the slopes at various times of the distance–time graphs.

When the slope of the velocity–time graph is zero then the velocity is constant, there is no acceleration. If the slope of the velocity–time graph is positive, i.e., the velocity is increasing as the time increases, then there is a positive acceleration. If the slope is negative, i.e., the speed is decreasing as the time increases, then there is a negative acceleration. Acceleration is defined as

$$\text{Acceleration} = \frac{\text{change in velocity in a time interval}}{\text{time interval}}$$

$$= \frac{\Delta v}{\Delta t}$$

Acceleration is the slope of a velocity–time graph. In Fig. 1.9(a) the slope of the velocity–time graph is zero and thus there is no acceleration. In Fig. 1.9(b) the slope increases initially and then decreases to zero, initial acceleration which decreases to zero. In Fig. 1.9(c) the slope decreases to zero. The slope is negative and thus there is a negative acceleration.

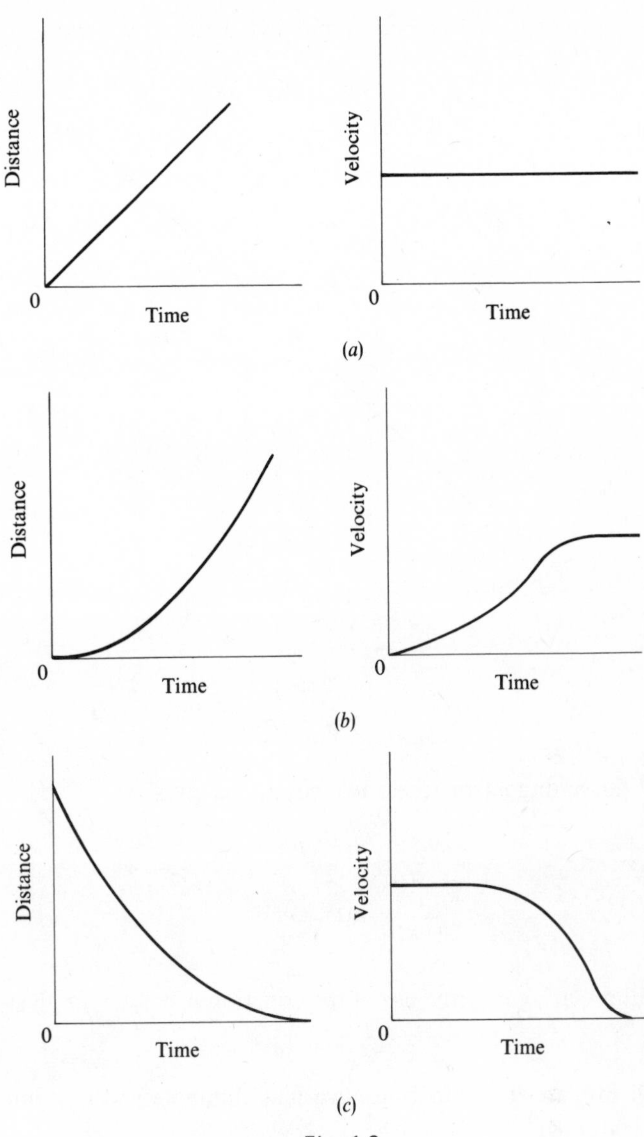

Fig. 1.9

Uniformly accelerated motion

The slope at any instant of the velocity–time graph gives the acceleration at that instant. For a uniform or constant acceleration we have a constant slope (Fig. 1.10).

Suppose we have a velocity of v_0 at time $t = 0$ and v at time t. Then

$$\text{Slope} = \frac{v - v_0}{t}$$

and thus acceleration, a, is given by

$$a = \frac{v - v_0}{t}$$

or

$$v = v_0 + at$$

The velocity is increasing uniformly with time, hence the average velocity $\bar{v}$ is given by

$$\bar{v} = \frac{v + v_0}{2}$$

But during time t the object moves a distance s. Hence

$$\bar{v} = \frac{s}{t}$$

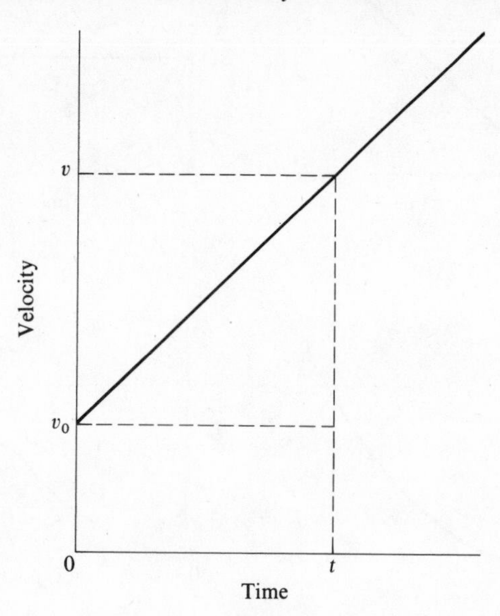

Fig. 1.10

Eliminating $\bar{v}$ from these two equations gives

$$\frac{s}{t} = \frac{v + v_0}{2}$$

$$s = \frac{v_0 t}{2} + \frac{vt}{2}$$

Eliminating v by the use of the equation $v = v_0 + at$ gives

$$s = v_0 t + \tfrac{1}{2}at^2$$

If instead of eliminating v we had eliminated t the result would be

$$v^2 = v_0^2 + 2as$$

As acceleration is the slope of the velocity–time graph we should be able, if we know the equation relating velocity and time, to use differentiation to obtain the acceleration.

$$\text{Acceleration} = \frac{\Delta v}{\Delta t}$$

At time t the velocity is v. Suppose we have the following relationship:

$$v = Kt, \quad K \text{ is a constant.}$$

After a time interval Δt the velocity has changed to $v + \Delta v$.

$$v + \Delta v = K(t + \Delta t)$$

Eliminating v and t from the two equations gives

$$\Delta v = K\,\Delta t$$

or

$$\text{Acceleration}\ \frac{\Delta v}{\Delta t} = K \quad \text{(a constant acceleration)}$$

Now we will use the same procedure for a non-uniform acceleration. We will start with the velocity–time equation,

$$v = Ct^2,$$

where C is some constant.

At time t the velocity is v, after a time interval Δt the velocity becomes $v + \Delta v$.

$$v + \Delta v = C(t + \Delta t)^2$$
$$= Ct^2 + 2Ct\,\Delta t + C(\Delta t)^2$$

Hence eliminating v gives

$$\Delta v = 2Ct\,\Delta t + C(\Delta t)^2$$

$$\frac{\Delta v}{\Delta t} = 2Ct + C\,\Delta t$$

In the limit as Δt tends to zero

$$\frac{dv}{dt} = 2Ct$$

Hence the acceleration $a = 2Ct$.

Road traffic

What is the stopping distance for a car? Does it depend on the speed of the car?

Two factors contribute to the stopping distance for a car: (a) the distance travelled during the reaction time of the driver, (b) the distance travelled during the deceleration of the car. If the reaction time of the driver is T and the car speed v then during this time the car travels a distance vT. If a is the mean deceleration then the distance travelled during this deceleration from speed v is $v^2/2a$ (from $v^2 = v_0^2 + 2as$). Thus the stopping distance is the sum of these two terms:

$$vT + \frac{v^2}{2a}$$

A typical reaction time is 0·7 s and deceleration 7·5 m s^{-2}. These yield the following stopping distances:

Speed/km h^{-1}	*Stopping distance*/m
30	10
60	30
90	60
120	100

If cars maintained the appropriate stopping distance apart, what is the speed with which cars should travel along a single carriageway to give the maximum number of cars flowing along the way per unit time? A point to remember is that the faster cars move, the further apart they will be. The time interval occurring between the rear bumper of one car passing a point and the rear bumper of the next car to pass that point is the sum of the time taken for the 'stopping distance' to be covered plus the time taken for the length

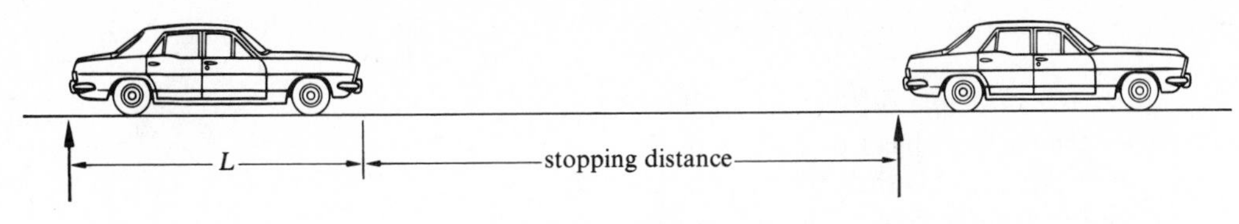

Fig. 1.11

of the car to pass (Fig. 1.11). This is therefore

$$\frac{S}{v} + \frac{L}{v}$$

where S is the stopping distance, L the length of a car, and v the speed with which the cars are travelling. Using our expression for the stopping distance gives

$$\text{Time taken} = \frac{1}{v}\left(vT + \frac{v^2}{2a}\right) + \frac{L}{v}$$

$$= T + \frac{v}{2a} + \frac{L}{v}$$

What does a graph of this look like? When $v = 0$ then the time is infinity. When v is infinity then the time is infinity. But what happens between these two values? We can try some of the values we already have; $T = 0{\cdot}7$ s, $a = 7{\cdot}5$ m s^{-2}. A typical value for L is 5 m. When we put these values in the formula we obtain the graph shown in Fig. 1.12.

The time taken appears to be a minimum between 30 and 40 km h^{-1}. This minimum value could have been estimated by the use of calculus. Differentiating the equation for the time t gives:

$$\frac{dt}{dv} = \frac{1}{2a} - \frac{L}{v^2}$$

When this is equated to zero, the slope of the graph is zero, the coordinates of the minimum are obtained.

$$0 = \frac{1}{2a} - \frac{L}{v^2}$$

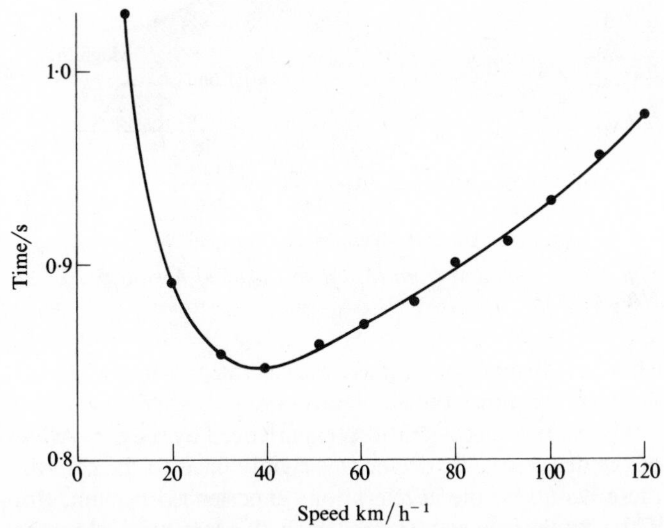

Fig. 1.12

Hence

$$v = \sqrt{(2aL)}$$

This is the value of the speed to give the minimum time. For the values previously used this gives v as 8·66 m s^{-1} or 31 km h^{-1}. The maximum number of cars passing a point will thus occur when they are moving at a speed of 31 km h^{-1}. Quite a number of assumptions have been made in this argument, for example the spacing between cars. Observations of cars have, however, shown maximum flow occurring with speeds in the range 30 to 50 km h^{-1}.

Further reading

A. J. Miller. 'Road Traffic Flow; some theories and their applications', *Endeavour* **24**, 143, 1965.

The accurate measurement of the acceleration due to gravity

Why do we bother about an accurate value for the acceleration due to gravity (see p. 8)? The acceleration can be measured in terms of standard lengths and times and if the acceleration is accurately known it is possible, with a known standard mass, to specify a force standard (see chapter 2).

$$\text{Force} = \text{weight} = mg$$

Such a standard is required in calibrating machines used to measure forces such as those occurring when rocket motors are fired. The current balance used to specify the ampere requires an accurate value of g in that the force between two current-carrying conductors is measured in terms of weights put in a balance pan.

Recent measurements of g have involved direct measurements on a falling body, whereas earlier measurements had been concerned with measurements on a pendulum. The time taken for a pendulum to complete one oscillation is related to the acceleration due to gravity. The measurements on a falling body were of the time taken for a freely falling body to cover a measured distance. A graduated scale has been used as the falling object and a photograph taken using a very sharp pulse of light occurring at known time intervals. Such free fall methods have given g results with uncertainties of about 1 or 2 parts in a million. Thulin at the International Bureau of Weights and Measures (1961) and Faller at the Palmer Laboratory, Princeton (1963) have used this method. Another method which has been used, by Cook at the National Physical Laboratory (1967), involves throwing an object upwards and measuring the time intervals at two points between the object passing them on the way up and on the way down. The uncertainty in this method for the g value is about a few parts in ten million.

Figure 1.13 illustrates the principle of this last method. T_A is the time interval between the object passing A on its way up and on its way down. $T_A/2$ is thus the time taken for the object to go from A to the top of its path. It is also the time taken for the object to freely fall from the top of its path to A. The velocity at A, in either the upwards or the downwards direction, is thus

$$v_A = g\frac{T_A}{2}$$

Similarly from point B

$$v_B = g\frac{T_B}{2}$$

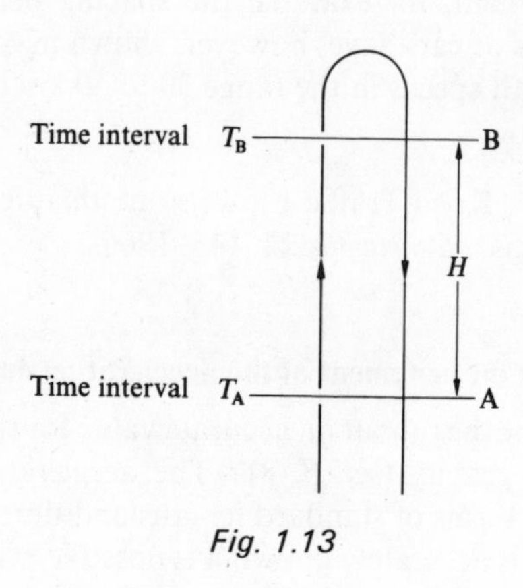

Fig. 1.13

But

$$v_B^2 = v_A^2 - 2gH$$

hence

$$g = \frac{8H}{T_A^2 - T_B^2}$$

The height H is sufficiently small for the variations with height of g to be insignificant.

Acceleration and man

What values of acceleration or deceleration are we likely to experience and what will be their effects?

Consider a car braking, not too severely, and coming to rest from 60 km h^{-1} in a distance of 15 m (about three car lengths). If we assume that the deceleration is uniform (unlikely, but the calculation is simpler than the non-uniform case and gives some idea of likely acceleration), then

$$v_0^2 + 2as = 0$$

hence

$$a = -9{\cdot}3 \text{ m s}^{-2}$$

or about the same as the acceleration due to gravity.

What would the acceleration have been if there had been a crash and the car had been very abruptly brought to a halt? Assume that the car hits an object which brings it to a halt in a distance of 1·5 m. The acceleration is now 93 m s^{-2} or about ten times the acceleration due to gravity.

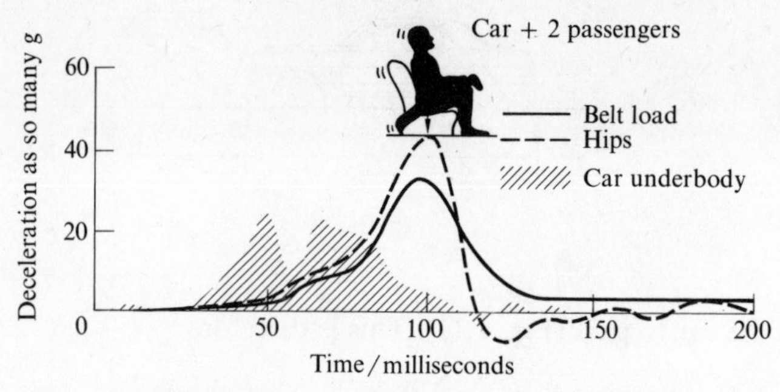

Fig. 1.14 Relative impact speed about 10 m s^{-1} for two cars in collision. The cars collapsed under impact by a distance of about 0·6 m. (Adapted from D. M. Severy, J. H. Matthewson, and A. W. Seigel. Proceedings of the Society of Automotive Engineers National Passenger Car Body and materials meeting, *Detroit, March 1958.)*

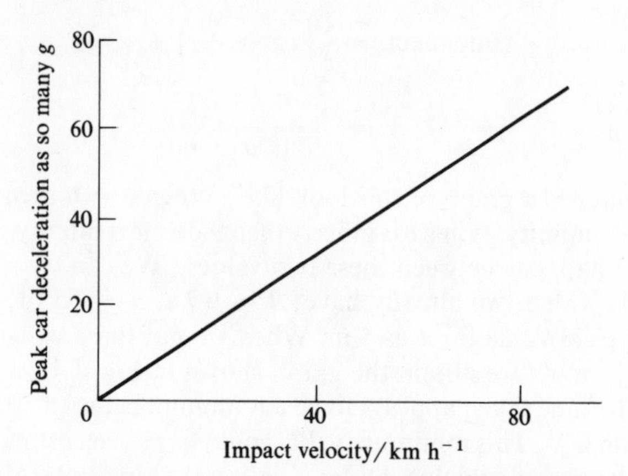

Fig. 1.15 Head-on car collision. (Adapted from D. M. Severy, J. H. Matthewson, and A. W. Seigel. Proceedings of the Society of Automotive Engineers National Passenger Car Body and materials meeting, *Detroit, March 1958.)*

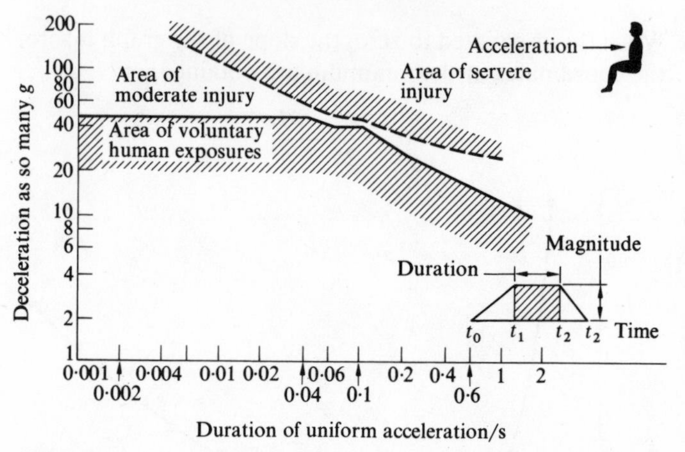

Fig. 1.16 Adapted from M. Eiband (1959) Memo 5-10-59E NASA.

These are the accelerations experienced by the car—what of the driver? If the driver was rigidly fixed to the car seat these would be the accelerations experienced by him. But if the driver was not restrained to the seat then when the car decelerated the driver would continue in motion until he hit the windscreen or some other part of the car. Only then would the deceleration occur for him. Thus if the car stops in 1·5 m the man may continue moving for some 0·5 m of this distance, roughly the distance from seat to windscreen, and thus he will be decelerated in a distance of 1·0 m and experience not a deceleration of 93 m s^{-2}, as the car does, but 140 m s^{-2}. The moral of this would seem to be that car safety belts should be worn.

The graphs shown in Figs. 1.14, 1.15, and 1.16 show some of the results of tests with car crashes.

1.3 Projectiles

The motion so far considered has only been in a straight line—can the results and concepts developed for the straight-line motion be applied to motion in two or three dimensions? Galileo considered this problem for projectiles in his work *Dialogues concerning two new sciences.*

> 'In the preceding pages we have discussed the properties of uniform motion and of motion naturally accelerated.... I now propose to set forth those properties which belong to a body whose motion is compounded of two other motions, namely, one uniform and one naturally accelerated.... This is the kind of motion seen in a moving projectile; its origin I conceive to be as follows: Imagine any particle projected along a horizontal plane without friction.... This particle will move along this same plane with a motion which is uniform and perpetual, provided the plane has no limits. But if the plane is limited and elevated, then the moving particle, which we imagine to be a heavy one, will on passing over the edge of the plane, acquire, in addition to its previous uniform and perpetual motion, a downward propensity due to its own weight; so that the resulting motion ... is compounded of one which is uniform and horizontal and of another which is vertical and naturally accelerated.'

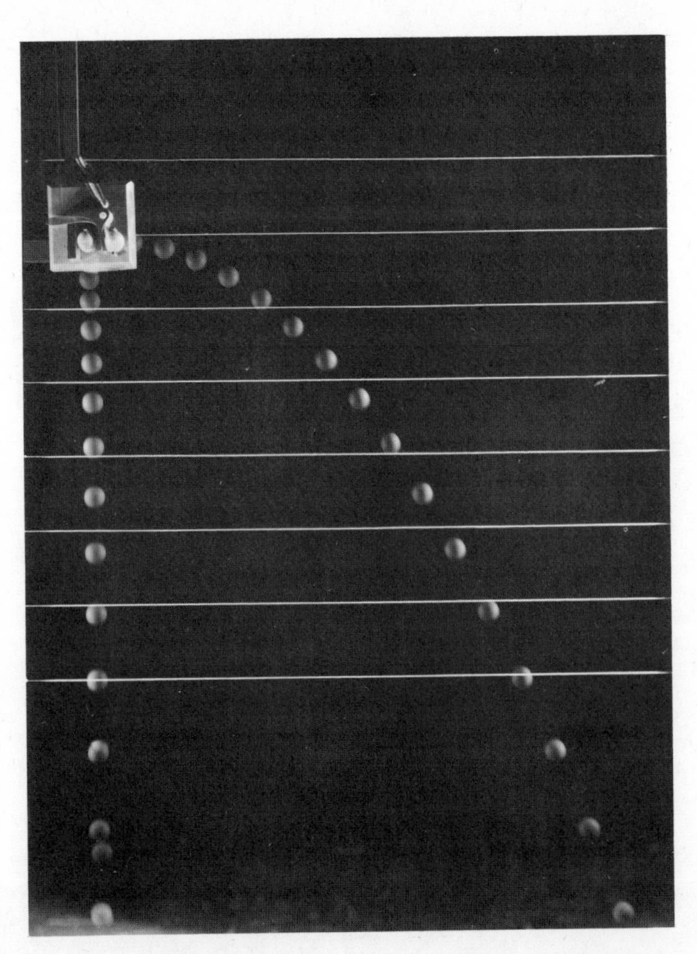

Fig. 1.17 PSSC Physics, *2nd Ed., D. C. Heath and Co. A flash photograph of two golf balls, one projected horizontally at the same time that the other was dropped. The strings are 15 cm apart and the interval between flashes was 1/30 s.*

The photograph, Fig. 1.17, shows an object falling freely from rest and another object which although freely falling has also received an initial horizontal motion. An interesting point emerges from an examination of the stroboscopic photograph—the two objects are at the same horizontal level at the same time. They are both falling with the same acceleration, the accleration due to gravity. The conclusion from this is that motions at right angles to each other are independent. The motion in one direction does not influence the motion in a direction at right angles to it.

The motion of a projectile is considerably simplified by considering it as due to two motions at right angles to each other.

Let us consider a specific example of a projectile—an object projected horizontally with a velocity of 15 m s^{-1}. We can determine the position of the object at any instant of time by working out how far in the horizontal direction it moves in the time and also how far in the vertical direction it has fallen due to gravity.

In the first second the object falls through a distance of $\frac{1}{2} \times 9{\cdot}8 \times 1^2$.

$$s = \tfrac{1}{2}at^2$$

Thus the vertical distance moved in one second is 4·9 m downwards. The horizontal distance moved is 15 × 1 m.

$$s = ut$$

Thus we can plot on a graph the position of the object after one second (Fig. 1.18).

After two seconds the object has fallen through a distance of $\frac{1}{2} \times 9{\cdot}8 \times 2^2$ or 19·6 m. The horizontal distance moved is 15 × 2 or 30 m.

After three seconds the object has fallen through a distance of $\frac{1}{2} \times 9{\cdot}8 \times 3^2$ or 44·1 m. The horizontal distance moved is 15 × 3 or 45 m. In this way we can construct a picture resembling the stroboscopic photograph.

The stroboscopic photograph gives a graph of vertical distance against horizontal distance—what is the equation of the line joining the images?

For the vertical motion $s = \frac{1}{2}gt^2$. Let us put this in the form where vertical distances are represented by y. Then

$$y = \tfrac{1}{2}gt^2$$

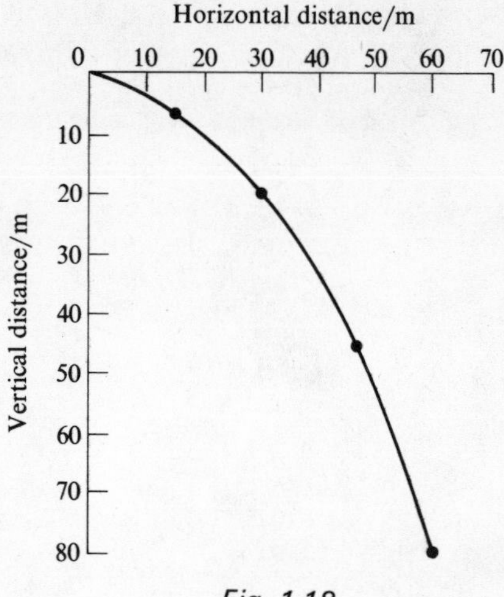

Fig. 1.18

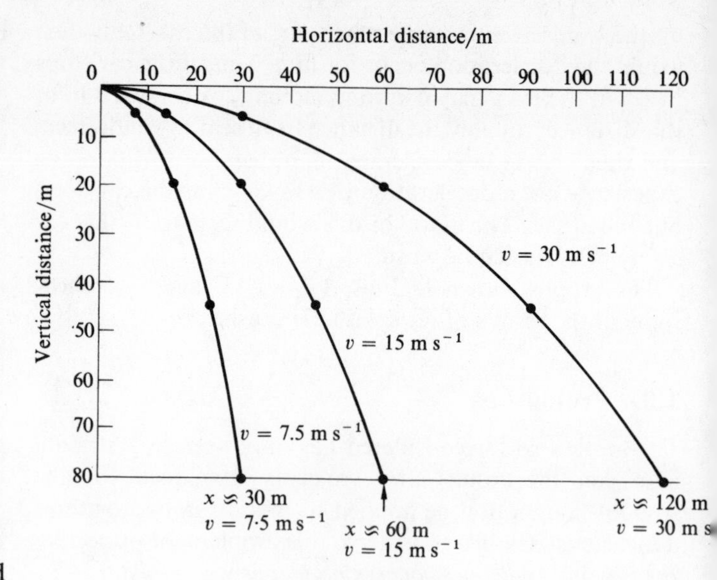

Fig. 1.19

This equation gives the y ordinates of the object, considered as a point, at different times.

For the horizontal motion $s = vt$, where v is the horizontal velocity. If we use x to represent distances in the horizontal direction, then

$$x = vt$$

Eliminating t between these two equations gives

$$y = \tfrac{1}{2}g\left(\frac{x}{v}\right)^2$$

$$y = \frac{g}{2v^2}x^2$$

This is the equation of the graph. It is in fact the equation of a parabola.

Satellites

The distance travelled along a horizontal plane by a projectile, given an initial horizontal velocity of v, is vt. The value of the time t is determined by the height from which the object falls. If we keep the height constant how does the distance depend on the horizontal velocity?

$$\text{Distance} \propto v \qquad \text{(Fig. 1.19)}$$

Now consider a projectile being given a horizontal velocity from the top of a high mountain on the earth. If the earth is flat then the distance from the mountain at which the projectile will hit the earth is directly proportional to the velocity. But the earth is not flat—what effect does this have? The projectile will move further along the surface of the earth than would occur with a flat earth. If the velocity is high enough it will not come down to earth—it will go into orbit (Fig. 1.20). During the entire motion of the projectile it is falling with the acceleration due to gravity.

How long would it take for the projectile to orbit the earth? We will assume that the satellite is close to the surface of the earth and that the acceleration due to gravity is therefore the same for the satellite as at the earth's surface.

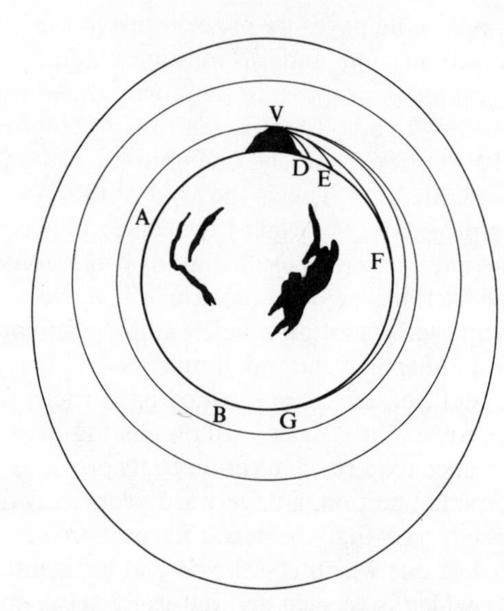

Fig. 1.20 I. Newton (1686). Principia, *Vol. 2.*

Consider a satellite 160 km above the earth's surface. The radius of the earth is approximately 6400 km. If the satellite remains the same distance from the earth, i.e., a circular orbit, then we know the path of our projectile. Figure 1.21 shows a scale diagram. But we can calculate how far the projectile should fall in, say, 200 s.

$$\begin{aligned} s &= \tfrac{1}{2}at^2 \\ &= \tfrac{1}{2} \times 9{\cdot}8 \times 200 \times 200 \\ &= 1{\cdot}96 \times 10^5 \text{ m} \\ &= 196 \text{ km} \end{aligned}$$

This is the 'fall distance' at right angles to the path of the satellite. The force of gravity is always at right angles to the circular path, on our diagram this can only be the case when the satellite has reached the position P. During this time the satellite must have travelled a distance of about 1600 km. This distance was measured from the diagram. This means

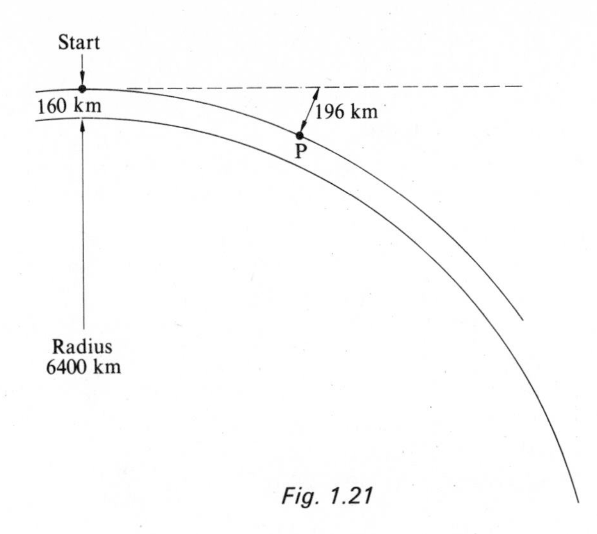

Fig. 1.21

that the speed of the satellite must be

$$\text{speed} = \frac{1600}{200} = 8 \text{ km s}^{-1}$$

How long will it take to go round the earth? The orbit has a radius of 6400 + 160 and hence the total distance covered in one orbit is $2\pi \times 6560$ or 41 200 km. At 8 km s^{-1} the time taken to cover this distance is therefore

$$\frac{41\,200}{8} = 5150 \text{ s}$$

This is about 86 minutes. This is a rough estimate because of the limited accuracy with which the diagram was drawn.

The intersecting chords theorem given in appendix 2B could have been used, instead of the scale diagram. It gives 5140 s. This is still a rough estimate—we have assumed that the acceleration due to gravity is the same at a height of 160 km as at the surface of the earth.

The first artificial satellite was launched on 4 October 1957 by the USSR. The time to cover one orbit was 96·2 minutes; the orbit was, however, not circular but an ellipse. The height above the earth's surface varied from 215 to 939 km. The first manned space flight, by Yuri Gagarin in 1961, was in an orbit whose height varied between 169 and 315 km above the earth's surface and had an orbital time of 89·3 minutes.

Further reading

G. Gamow. *Gravity*, Heinemann, Science Study Series No. 17, 1962.

The motion of the planets

The sun rises in the morning in the east and in the evening sets in the west. It seems the most natural thing to consider the earth stationary and the sun orbiting the earth. After all if the earth was moving surely we would somehow feel the motion. These views held the scene until the seventeenth century. Though superficially they seem to be the simplest answer they did introduce many complications. A simpler answer turned out to be that the earth is moving round the sun. In the following we briefly examine the arguments advanced both for and against the idea of the moving earth.

Ptolemy, A.D. 85–165, considered the earth to be the central item in the universe and all the other planets and the sun to revolve round the earth. The motion of the planets could not, however, be as simple as a circle because this conflicted with observations (see Fig. 1.1(a)). The planets had to move in epicycles round the earth. A simple epicycle is traced out by a point on the rim of a wheel which is rolling round the circumference of a circle. Figure 1.22(a) shows such an epicycle. To describe the entire solar system necessitated many epicycles. The model was, however, adequate for the precision of the data then known. With its aid tables were constructed giving the positions of the planets in the sky—a vital navigational aid.

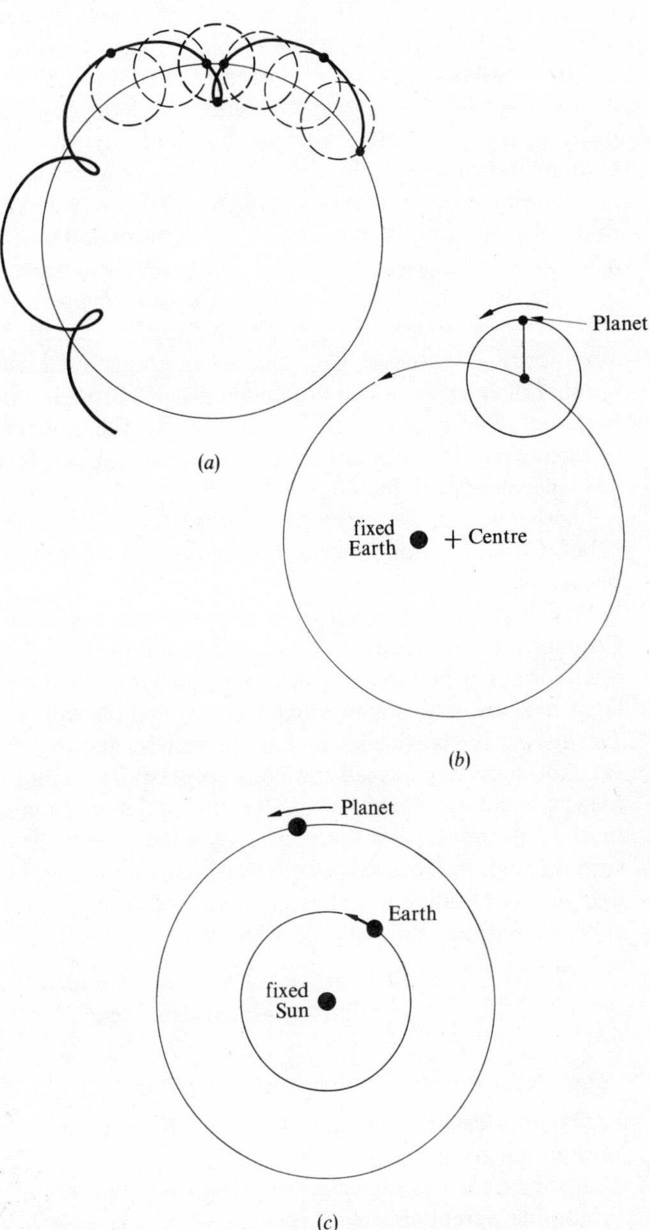

Fig. 1.22 (a) An epicycle. (b) The Ptolemy model (simplest form). (c) The Copernican model (simplest form).

Copernicus, 1473–1543, proposed a new model for the universe—the planets, including the earth, orbiting the sun (Fig. 1.22(c)). Copernicus could not, however, do without epicycles for his model to fit the data then known. His model turned out to be a stimulus to science and the start of new advances in astronomy, even if as a model it proved no better than that of Ptolemy in forecasting planetary positions. In concept it was simpler than Ptolemy's model.

'*The book that nobody read*

The Book of the Revolutions of the Heavenly Spheres (by Copernicus) was and is an all-time worst-seller.... The main reason for this neglect is the book's supreme unreadability. It is amusing to note that even the most conscientious modern scholars, when writing about Copernicus, unwittingly betray that they have not read him. The give-away is the number of epicycles in the Copernican system. At the end of his Commentariolus, Copernicus had announced "altogether, therefore, thirty-four circles suffice to explain the entire structure of the universe and the entire ballet of the planets". But the Commentariolus had merely been an optimistic preliminary announcement; when Copernicus got down to detail in the Revolutions, he was forced to add more and more wheels to his machinery, and their number grew to nearly fifty. But since he does not add them up anywhere, and there is no summary to his book, this fact has escaped attention.... Moreover, Copernicus had exaggerated the number of epicycles in the Ptolemaic system. Brought up to date by Peurbach in the fifteenth century, the number of circles required in the Ptolemaic system was not 80, as Copernicus said, but 40.

In other words, contrary to popular, and even academic belief, Copernicus did not reduce the number of circles, but increased them....

How was it possible that the faulty, self-contradictory Copernican theory, contained in an unreadable and unread book, rejected in its time, was to give rise, a century later, to a new philosophy which transformed the world? The answer is that the details did not matter, and that it was not necessary to read the book to grasp its essence. Ideas which have the power to alter the habits of human thought do not act on the conscious mind alone; they seep through to those deeper strata which are indifferent to logical contradictions. They influence not some specific concept, but the total outlook of the mind.'

A. Koestler (1959). *The sleepwalkers*. Hutchinson/Macmillan, New York.

'*Reasons for asserting the earth is motionless:*

1. David in Psalm 89: God has founded the earth and it shall not be moved.
2. Joshua bade the sun stand still—which would not be notable were it already at rest.
3. The earth is the heaviest element, therefore it more probably is at rest.
4. Everything loose on the earth seeks its rest on the earth, why should not the whole earth itself be at rest?
5. We always see half of the heavens and the fixed stars also in a great half circle, which we could not see if the earth moved, and especially if it declined to the north and south....
6. A stone or an arrow shot straight up falls straight down. But if the earth turned under it, from west to east, it must fall west of its starting point.
7. In such revolutions houses and towers would fall in heaps.
8. High and low tide could not exist; the flying of birds and the swimming of fish would be hindered and all would be in a state of dizziness.

And similarly on both sides.

Reasons for the belief that the earth is moved:

1. The sun, the most excellent, the greatest and the midmost star, rightly stands still like a king while all the other stars with the earth swing round it.
2. That you believe that the heavens revolve is due to ocular deception similar to that of a man on a ship leaving shore.
3. That Joshua bade the sun stand still Moses wrote for the people in accordance with the popular misconception.
4. As the planets are each a special created thing in the heavens, so the earth is a similar creation and similarly revolves.
5. The sun fitly rests at the center as the heart does in the middle of the human body.
6. Since the earth has in itself its especial *centrum*, a stone or an arrow falls freely out of the air again to its own *centrum* as do all earthly things.
7. The earth can move five miles in a second more readily than the sun can go forty miles in the same time.'

Voight (1667). *Der Kurstgunstigen Einfalt Mathematischer Raritäten Erstes Hundert*.

An argument against the Copernican view that the earth rotated round the sun was that freely falling objects would always fall to the west of the vertical. An object dropped from the mast of a moving ship should fall not at the base of the mast but behind the mast as during the time it was falling the ship would have moved. In 1640 Gassendi tried the experiment with a stone being dropped from the mast of a ship—the result was that the stone always hit the deck of the ship at the same position, regardless of whether the ship was moving or not. The explanation was that the stone before being released has the same horizontal velocity as the ship. When it falls it has a vertical acceleration, that due to gravity, and the horizontal velocity. In the horizontal direction it thus keeps pace with the ship. The path of the stone from the point of view of an observer not moving with the ship is that of a projectile, i.e., a parabolic path. From the point of view of a man on the ship who has the same horizontal velocity as the stone, the stone appears to be falling vertically. We on earth see falling objects in the same way that the man on the ship sees the stone falling from the top of the mast.

In 1609 Johann Kepler published the first two of what are known as Kepler's three planetary laws. The third law appeared in 1619. These laws can be written as:

1. All planets follow elliptical orbits with the sun located at one of the foci. (See appendix 1B.)
2. An imaginary line connecting the sun and a planet sweeps over equal areas of the planetary orbit in equal intervals of time.
3. The square of the periods of revolution of any two planets are as the cubes of their mean distances from the sun.

$$\frac{T_1^2}{R_1^3} = \frac{T_2^2}{R_2^3}$$

where T_1 is the period of revolution for planet 1, T_2 is the period for planet 2, R_1 is the mean distance from the sun of planet 1, and R_2 is the mean distance from the sun of planet 2.

An alternative way of expressing it is:

$$(\text{planetary year})^2 \propto (\text{mean distance from sun})^3$$

These were laws arrived at from an examination of experimental data—a search for simple models to fit the data. They were not arrived at as a result of some grand theoretical line of argument.

Kepler had an advantage over earlier astronomers: he had much better data on planetary positions. This was the result of work by Tycho Brahe (1546–1601).

'. . . Kepler's work, like Copernicus' sixty-five years earlier, was accessible only to trained astronomers, and, in spite of the great accuracy that Kepler was known to have achieved, many astronomers found his noncircular orbits and his new techniques for determining planetary velocities too strange and uncongenial for immediate acceptance. Until after the middle of the century a number of eminent European astronomers can be found trying to show that Kepler's accuracy can be duplicated with mathematically less radical systems. . . . not until the last decades of the seventeenth century did Kepler's Laws become the universally accepted basis for planetary computations even among the best practising European astronomers . . .'

T. S. Kuhn (1957). *The Copernican revolution.* Harvard Univ. Press.

The publication of Newton's *Principia* in 1687, in which he states the universal law of gravitation (see chapter 2), ended the debate, on whether the earth or sun moved, for most people. There were exceptions, however, in the next hundred years: still some people who doubted.

'Many common christians to this day firmly believe that the earth really stands still and that the sun moves all round the earth once a day: neither can they be easily persuaded out of this opinion, because they look upon themselves bound to believe what the Scripture asserts.'

S. Pike (1753). *The principles of natural philosophy extracted from the divine revelation.*

Further reading

S. Toulmin and J. Goodfield. *The fabric of the heavens*, Penguin, 1961.

A. Koestler. *The sleepwalkers*, Penguin, 1964.

Nuffield O-level physics background book. *Astronomy*, Penguin.

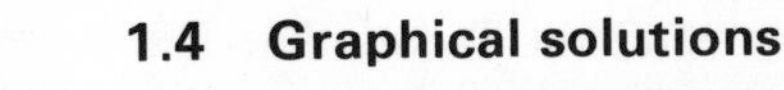

1.4 Graphical solutions

Consider an object moving with uniform velocity. Figure 1.24(a) shows the velocity–time graph for the motion. The distance travelled, s, during a time, t, is given by

$$s = vt$$

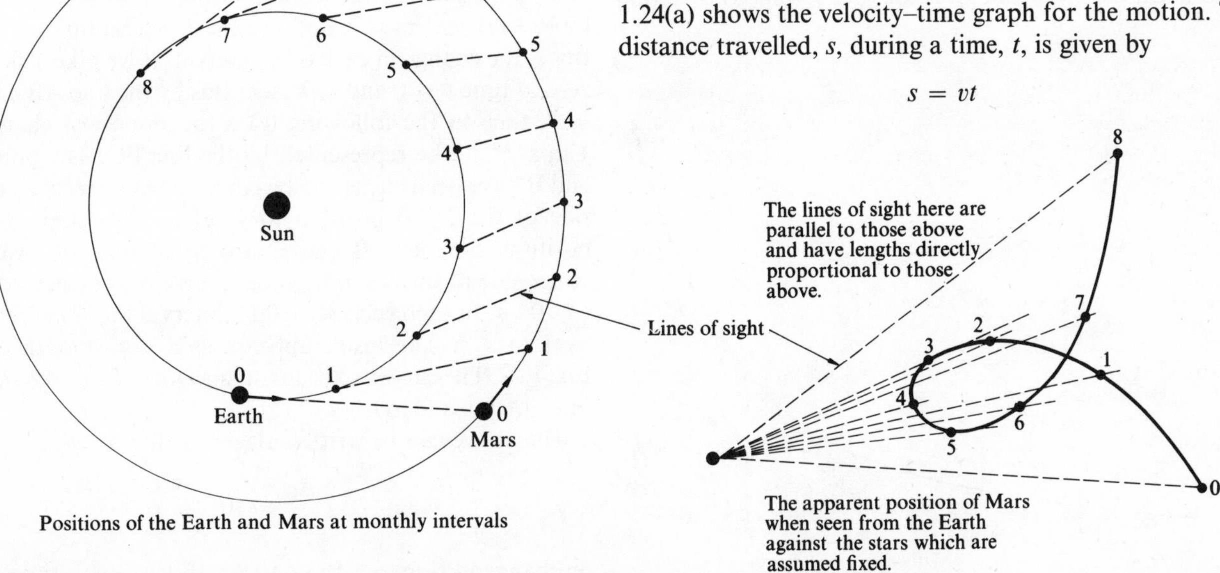

Fig. 1.23 Copernicus model used to explain the motion of Mars.

if $s = 0$ at time $t = 0$. But vt is the area under the graph between the times $t = 0$ and t. Is this a general rule—is the area under the velocity–time graph the distance travelled?

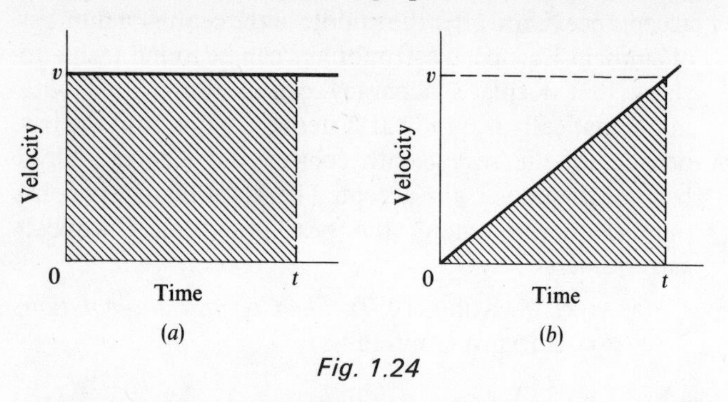

Fig. 1.24

What about a uniformly accelerated object? Figure 1.24(b) shows the velocity–time graph. The area under the graph between $t = 0$ and t is $\frac{1}{2}vt$. How does this compare with the calculated distance?

$$s = \tfrac{1}{2}at^2, \qquad v = at$$

If we eliminate a from the two equations we have

$$s = \tfrac{1}{2}vt$$

The area is equal to the distance covered.

Suppose we have a non-uniformly accelerated motion. Figure 1.25 shows a typical velocity–time graph for the first stage of a rocket. Can we determine the distance–time graph? We can find the distance covered in the first 10 s by estimating the area under the graph between time zero and time 10 s, in this case approximately 125 m. The distance covered in the first 20 s is the area between the time zero and time 20 s lines, 1000 m. In a similar manner we find the distance covered at various times and so plot the distance–time graph. Figure 1.26 shows the result. After 100 s the rocket has covered a distance of about 38 km. The first stage of the rocket, Saturn, which put the first men on the moon burns for 159 s and gives the rocket a speed of about

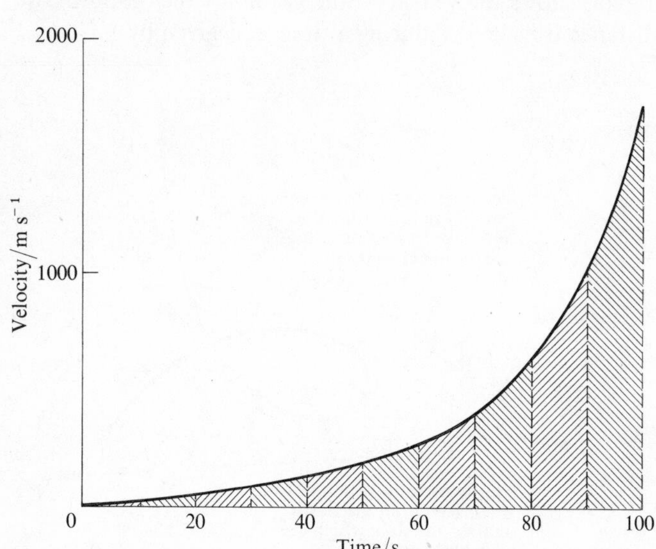

Fig. 1.25 Velocity–time graph for the first stage of a rocket.

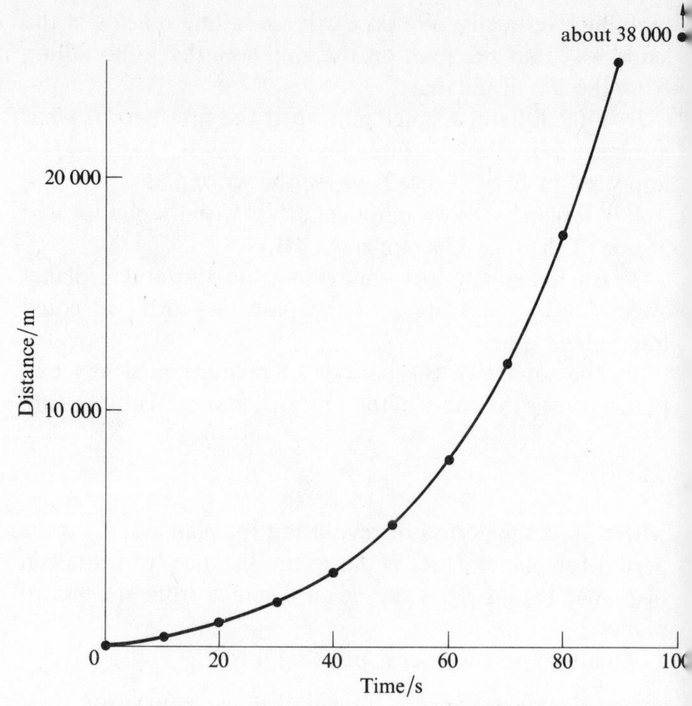

Fig. 1.26 Distance–time graph for the first stage of a rocket.

140 000 m s^{-1}. At the end of the time the rocket has covered a distance of about 61 km.

We could have obtained the distance–time graph in a different manner provided we knew the velocity–time relationship, or the acceleration–time relationship. For example, for a uniformly accelerated object, say a freely falling object,

$$\text{acceleration } a = \text{a constant} = 10 \text{ m s}^{-2}$$

Acceleration is the rate at which the velocity changes. But velocity is the slope of the distance–time graph at the instant concerned. Thus the rate at which the slope of the distance–time graph is changing is the acceleration, 10 m s^{-2}. Thus in 0·1 s the slope of the distance–time graph should change by $10 \times 0{\cdot}1 = 1$ m s^{-1}. Because the acceleration is constant this is the change in each 0·1 s interval. If we take a slope of zero at time $t = 0$ and represent this by the line AB on Fig. 1.27, then in the following 0·1 s the slope will change to 1 m s^{-1} and be represented by the line BC. The points A and B have been chosen to be equal time intervals on either side of the $t = 0$ position to give the zero slope at that position. Similarly B and C are equal time intervals on either side of the $t = 0{\cdot}1$ s time to give the correct slope at $t = 0{\cdot}1$ s. In each successive 0·1 s interval the slope changes by 1 m s^{-1}. The result appears as a non-smooth curve, Fig. 1.27. The curve can be made smoother if we take smaller time intervals than 0·1 s.

The above can be written algebraically as

$$\frac{\Delta v}{\Delta t} = 10$$

or change in slope $= \Delta v = 10\,\Delta t$. If very small time intervals are chosen, so small that we can say that Δt is approach-

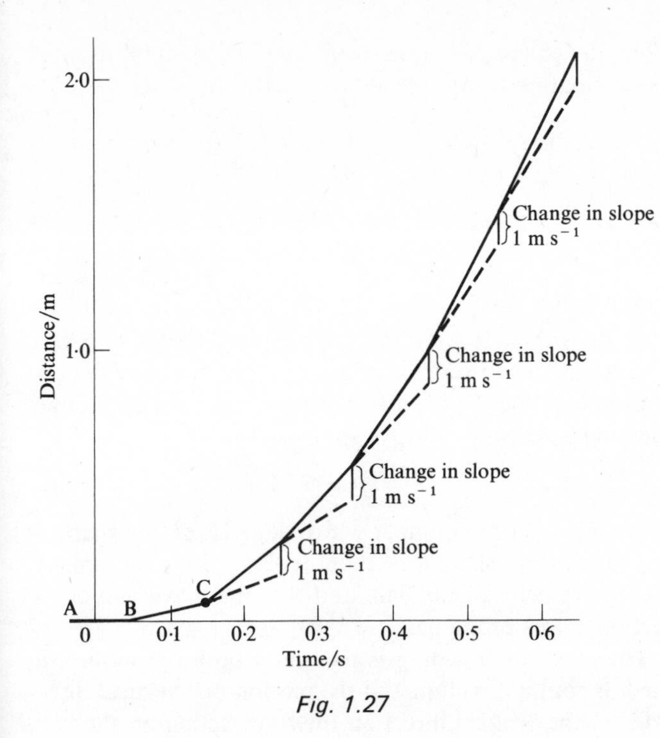

Fig. 1.27

ing zero, we write the equation as

$$\frac{dv}{dt} = 10$$

Those who have studied calculus could probably solve this differential equation by integration.

$$\int_0^v dv = \int_0^t 10\, dt$$

$$v = 10t$$

The limits used are the same as those chosen for the numerical form of integration used earlier. To obtain the equation for the distance–time graph we utilize the fact that

$$v = \frac{ds}{dt}$$

and so

$$\frac{ds}{dt} = 10t$$

$$\int_0^s ds = \int_0^t 10t\, dt$$

$$s = 10\frac{t^2}{2} = 5t^2$$

This is the equation corresponding to Fig. 1.27 with very small time intervals.

Projectiles in a vacuum

Consider an object thrown at some angle to the horizontal. To simplify the process we will consider the projectile to be moving in a vacuum, i.e., we neglect air resistance. How can we obtain the graph of height variation with horizontal distance? Section 1.3 gave one method. Here we will look at a method which we can use, later, when air resistance is taken into account.

The force acting on the projectile, when it has been released, is mg in the vertical direction, where m is the mass of the projectile and g the acceleration due to gravity. In the horizontal direction there are no forces acting on the projectile.

$$\text{Force} = -ma = -m\frac{\Delta(\text{slope of distance–time graph})}{\Delta t}$$

Thus for a graph of vertical height against time we have

$$\Delta(\text{slope}) = -g\,\Delta t,$$

m cancels. The minus sign arises because the force is opposing the initial motion. Taking $g = 10$ m s^{-2}, we have

$$\Delta(\text{slope}) = -10\,\Delta t$$

If we take time intervals of 0·2 s then the change in slope will be 2 m s^{-1} in each successive interval of 0·2 s. Figure 1.28 shows the resulting graph for a projectile with an initial velocity of 30 m s^{-1} at an angle of 45° to the horizontal. This is an initial vertical velocity component of 21·4 m s^{-1}.

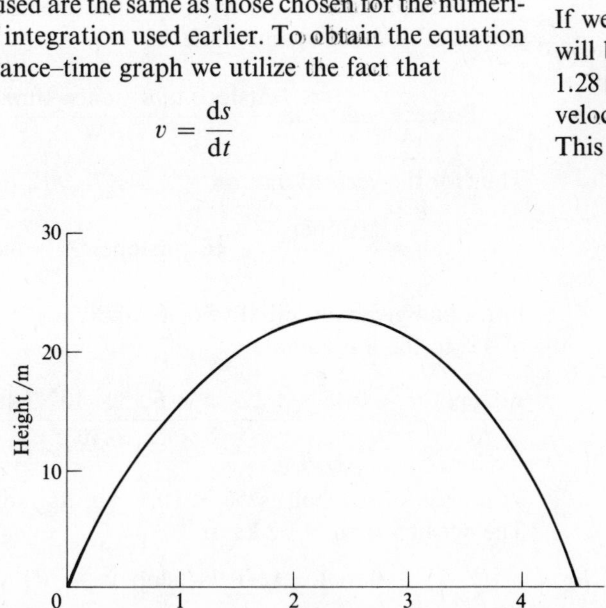

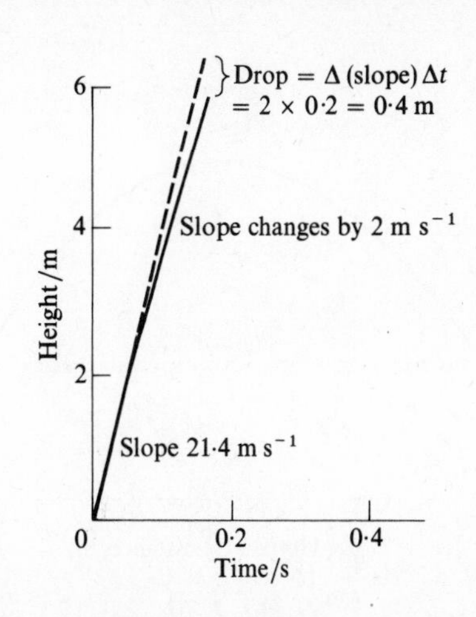

Fig. 1.28

Thus the graph starts off with this initial slope, decreases its slope by 2 m s^{-1} in the first 0·2 s interval (this is a drop of 0·4 m in 0·2 s), further decreases its slope by 2 m s^{-1} in the next 0·2 s interval, etc.

For the horizontal component there is no force. Thus the graph of distance against time for the horizontal direction is a straight line giving a slope of 21·4 m s^{-1}, the horizontal velocity component (Fig. 1.29).

Combining the graphs for the horizontal and vertical motions we can obtain a graph showing how the vertical distance of the projectile varies with horizontal distance (Fig. 1.30).

We could have used our equations for straight-line motion to obtain Figs. 1.28 and 1.29. The vertical motion is described by the equation

$$s = ut + \tfrac{1}{2}at^2$$

thus height

$$h = 21{\cdot}4t - 5t^2$$

and the horizontal motion by

$$s = 21{\cdot}4t$$

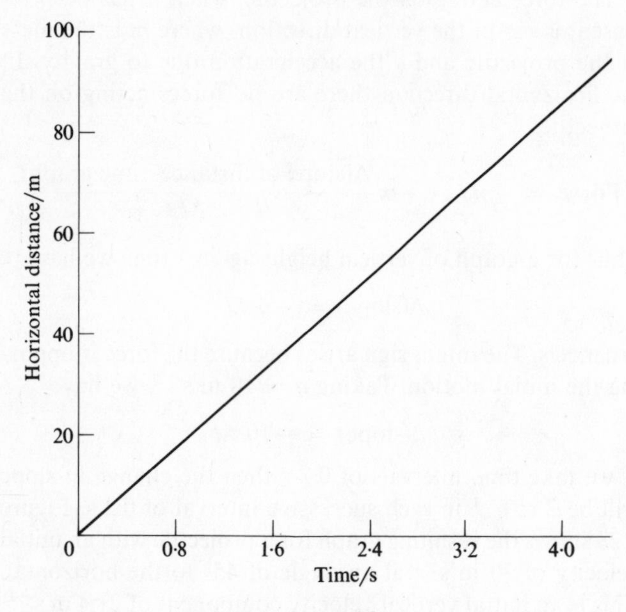

Fig. 1.29

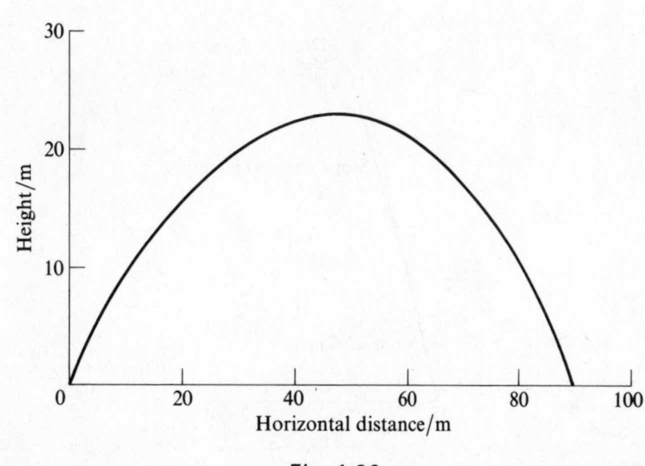

Fig. 1.30

Thus the equation linking height with horizontal distance is, eliminating t from the above equations,

$$h = 21{\cdot}4\left(\frac{s}{21{\cdot}4}\right) - 5\left(\frac{s}{21{\cdot}4}\right)^2$$

This is the equation of Fig. 1·30.

Projectiles in air

How does the path of a projectile in air differ from that in a vacuum? The presence of air will give rise to a drag or resistance force. For most of the balls used in ball games the drag force, assuming no spin, is given by

$$F = \tfrac{1}{2}C_D \rho v^2 A$$

where C_D is a drag number which depends on the shape of the ball (for a sphere it is about 0·45), ρ is the air density, v is the velocity of the ball, and A is the maximum cross-sectional area of the ball.

The effect of this drag is to reduce both the motion in the horizontal direction and the motion in a vertical direction. In the vertical direction the force acting on the ball, when in flight, is

$$-\tfrac{1}{2}C_D \rho v_V^2 A - mg$$

The minus signs are because both forces are opposing the motion. In the horizontal direction the force is just

$$-\tfrac{1}{2}C_D \rho v_H^2 A$$

v_V and v_H are the vertical and horizontal components of the velocity. Taking both these horizontal and vertical forces into account we need to produce a graph showing how the height of the projectile varies with the distance travelled in the horizontal direction (i.e., the range). Probably the simplest procedure is to treat each of the forces separately and produce, for both the vertical and horizontal directions, graphs of distance against time.

$$\text{Force} = ma = m\frac{\Delta(\text{slope of distance–time graph})}{\Delta t}$$

Thus for the vertical motion

$$m\frac{\Delta(\text{slope})}{\Delta t} = -\tfrac{1}{2}C_D \rho(\text{slope})^2 A - mg$$

For a ball given an initial velocity of 30 m s^{-1} at an angle of 45° to the horizontal,

$$\frac{\Delta(\text{slope})}{\Delta t} = \frac{-0{\cdot}45 \times 1{\cdot}2 \times \pi \times 6{\cdot}5^2 \times 10^{-4}(\text{slope})^2 - 10}{2 \times 56 \times 10^{-3}}$$

The mass of the ball is 56×10^{-3} kg, its radius 6·5 cm. The density of air is 1·2 kg m^{-3}.

$$\Delta(\text{slope}) = [-0{\cdot}064(\text{slope})^2 - 10]\,\Delta t$$

The initial slope is the vertical component of 30 m s^{-1}, i.e., 21·4 m s^{-1}.

After 0·2 s,

$$\Delta(\text{slope}) = [-0{\cdot}064 \times (21{\cdot}4)^2 - 10]0{\cdot}2$$
$$= -7{\cdot}6 \text{ m s}^{-1}$$

After a further 0·2 s,

$$\Delta(\text{slope}) = [-0{\cdot}064 \times (13{\cdot}8)^2 - 10]0{\cdot}2$$
$$= -4{\cdot}4 \text{ m s}^{-1}$$

After a further 0·2 s,

$$\Delta(\text{slope}) = [-0{\cdot}064 \times (9{\cdot}4)^2 - 10]0{\cdot}2$$
$$= -3{\cdot}1 \text{ m s}^{-1}$$

Figure 1.31 shows the result of continuing this process.

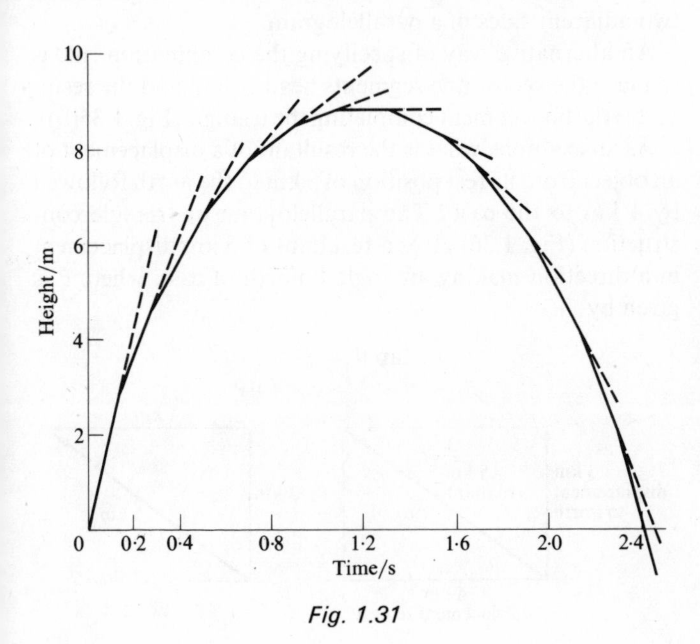

Fig. 1.31

For the horizontal motion we have

$$\Delta(\text{slope}) = -0{\cdot}064(\text{slope})^2\, \Delta t$$

The initial slope is the horizontal component of 30 m s^{-1}, i.e., 21·4 m s^{-1}.

After 0·2 s,

$$\Delta(\text{slope}) = -0{\cdot}064 \times (21{\cdot}4)^2 \times 0{\cdot}2$$
$$= -3{\cdot}8 \text{ m s}^{-1}$$

After a further 0·2 s,

$$\Delta(\text{slope}) = -0{\cdot}064 \times (17{\cdot}6)^2 \times 0{\cdot}2$$
$$= -3{\cdot}0 \text{ m s}^{-1}$$

After a further 0·2 s,

$$\Delta(\text{slope}) = -0{\cdot}064 \times (14{\cdot}6)^2 \times 0{\cdot}2$$
$$= -2{\cdot}7 \text{ m s}^{-1}$$

Figure 1.32 shows the result of continuing this process.

The vertical height graph, Fig. 1.31, shows that the ball will be in the air for about 2·4 s. The horizontal distance graph shows that in this time the horizontal distance covered will be 20·5 m. Combining the results from both graphs gives the vertical distance with horizontal distance graph (Fig. 1.33). The graph shows a number of features: the ball descends more steeply than it rises, the horizontal range is quite a lot smaller than it would be in a vacuum (see Fig. 1.30), the vertical height reached is also smaller.

The mass of a ball divided by its diameter squared has been taken as a measure of the ballistic properties of balls.

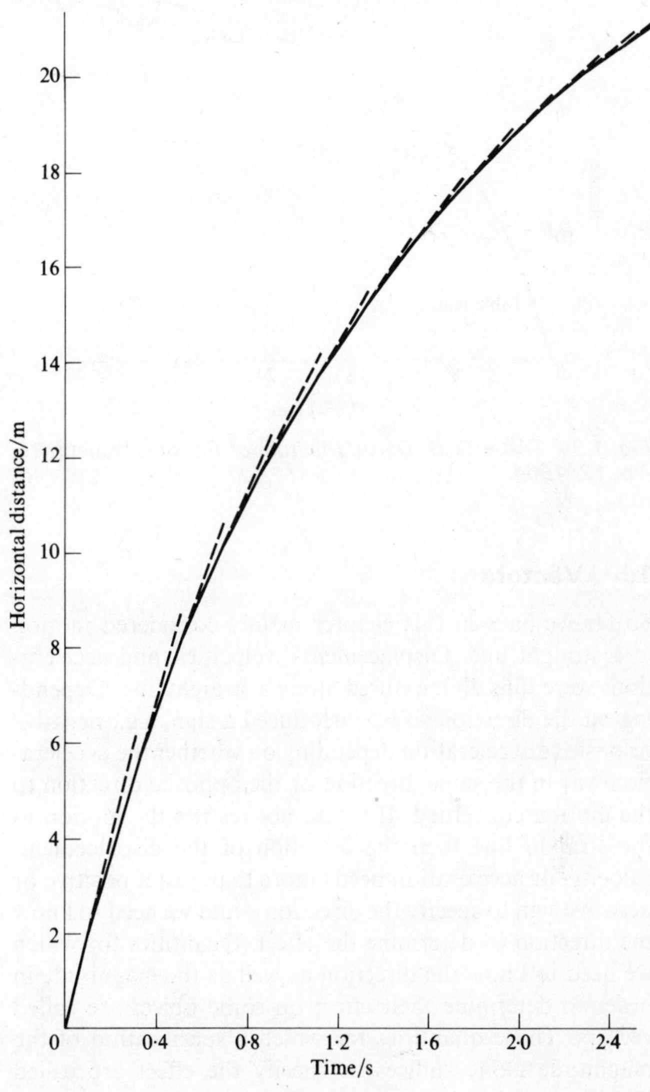

Fig. 1.32

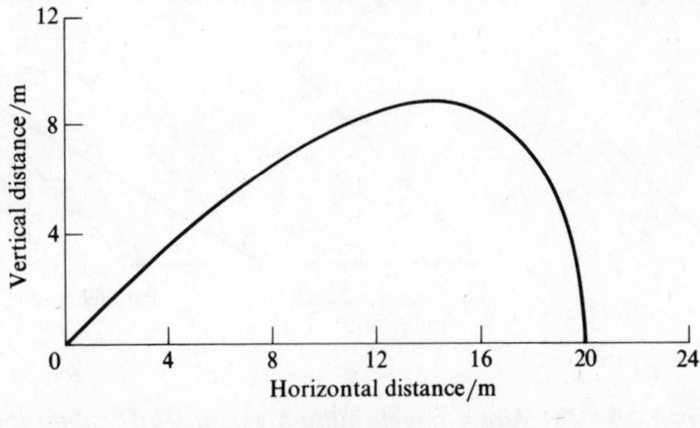

Fig. 1.33

The diameter squared term comes from the area term in the drag force and the mass term from the fact that the retardation produced is the drag force divided by mass. Figure 1.34 shows how the range varies with the value of m/d^2, for the same initial velocity. For the ball to be projected a long distance, a large mass and small cross-sectional area are required.

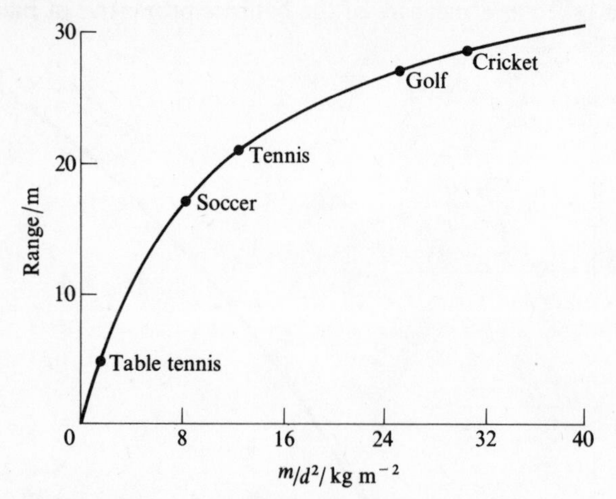

Fig. 1.34 After C. B. Daish, Institute of Physics. Bulletin **15**, *No. 12, 1964.*

1.5 Vectors

So far we have in this chapter mainly considered motion in a straight line. Displacements, velocities, and accelerations were thus all measured along a straight line. Depending on the direction so we introduced a sign, e.g., negative or positive acceleration depending on whether the acceleration was in the same direction or the opposite direction to the motion concerned. If we do not restrict the motion to the straight line then the direction of the displacement, velocity, or acceleration needs more than just a positive or negative sign to specify the direction—and we need to know the direction to determine the effect. Quantities for which we need to know the direction as well as the magnitude in order to determine their effect on some object are called vectors. Those quantities for which a specification of the magnitude alone suffices to specify the effect are called scalar quantities.

Displacements, velocities, and accelerations are examples of vectors; mass, volume, and temperature are examples of scalar quantities.

A convenient representation of a vector quantity is an arrow-headed line segment, the length of which is representative of the magnitude of the vector quantity and the direction of which specifies the direction of the vector quantity (Fig. 1.35(a)).

All vector quantities are found to combine in the same way—by what is called the parallelogram law. The vectors to be combined are represented by the arrow-headed line segments placed tail-to-tail. For a pair of vectors the resultant effect is equivalent to a vector represented by the line segment completing the diagonal of the parallelogram (Fig. 1.35(b)) resulting from using the two line segments as two adjacent sides in a parallelogram.

An alternative way of specifying the combination rule is to place the vector line segments head-to-tail and the resultant is the line segment completing the triangle (Fig. 1.35(b)).

As an example, what is the resultant of a displacement of an object from its rest position of 3 km to the north followed by 4 km to the east? The parallelogram or triangle construction (Fig. 1.36) gives a resultant of 5 km displacement in a direction making an angle θ north of east, where θ is given by

$$\tan \theta = \tfrac{3}{4}$$

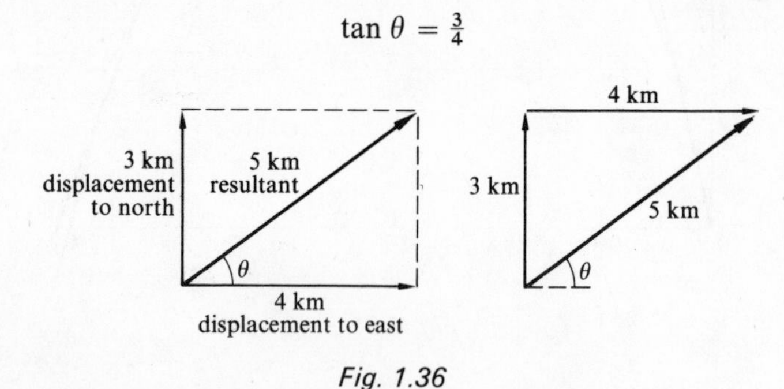

Fig. 1.36

The total journey is, however, 7 km; journeys are not vector quantities.

Velocities are vector quantities. Suppose you were to swim across a river, always keeping your body at right angles to the river bank (Fig. 1.37), at a velocity of 3 km h^{-1}. There is a current, parallel to the bank, of 4 km h^{-1}. What is your effective velocity? 5 km h^{-1} at an angle θ to the bank, where

$$\tan \theta = \tfrac{3}{4}$$

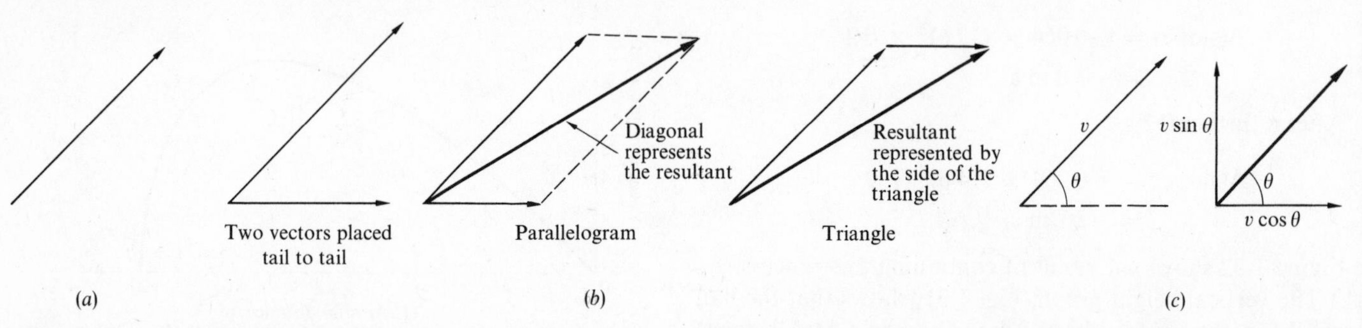

Fig. 1.35 (a) Arrow representing a vector. (b) Finding the resultant of two vectors. (c) Resolving a vector into two other vectors.

A single vector can be resolved into two other vectors by doing the above procedure in reverse (Fig. 1.35(c)). Thus a velocity v can be resolved into two velocities $v \cos \theta$ and $v \sin \theta$ at right angles to each other. We can obtain the same effect as the velocity v by considering the effects of the two independent velocities $v \cos \theta$ and $v \sin \theta$. This in fact was the procedure adopted when considering projectiles.

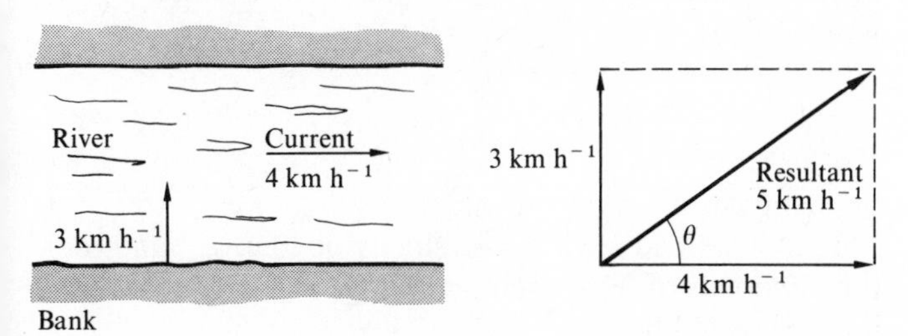

Fig. 1.37

If we have more than two vectors to be combined we can use the parallelogram for combining pairs of vectors and continue combining pairs until we are left with just one resultant (Fig. 1.38 shows an example of this). A probably simpler way of combining the vectors is to use the 'triangle' construction but putting all the vectors tail-to-head and taking the resultant as the vector needed to join the last head to the first tail (Fig. 1.39). This is equivalent to combining **A** and **B**, then combining the resultant with **C**, then this resultant with **D**.

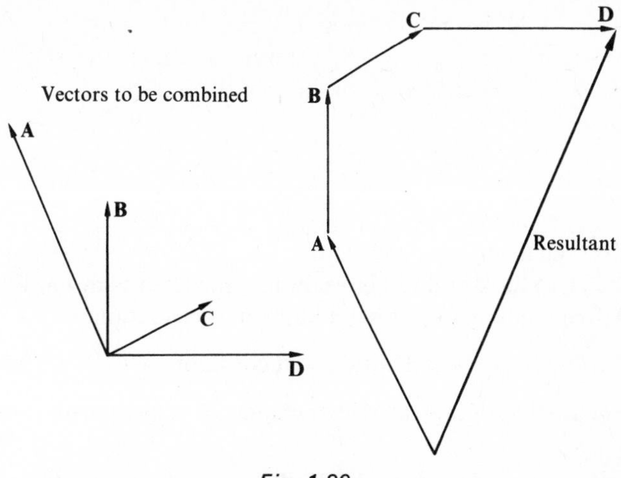

Fig. 1.39

Appendix 1A Differential equations

If we say that the velocity of an object is constant then we can write this as an equation of the form

$$v = C$$

where C represents the constant. But velocity is the rate of change of position with time. If the object covers a distance Δx in time Δt then the velocity is given by

$$v = \frac{\Delta x}{\Delta t}$$

This v is the average velocity over a distance Δx and time Δt. We can make this approximate to the instantaneous velocity by making Δt small. If we make Δt vanishingly small ($\mathrm{d}t$) we have what we can call the velocity at an instant.

Instantaneous velocity $v = \mathrm{d}x/\mathrm{d}t$.

For the case under consideration the velocity is a constant, thus

$$\frac{\mathrm{d}x}{\mathrm{d}t} = C$$

This is known as a differential equation and describes the motion of the object. Differential equations can be used to describe the rate of change of many quantities. We can describe the rate at which a radioactive isotope decays by the differential equation

$$\frac{\mathrm{d}N}{\mathrm{d}t} = -\lambda N$$

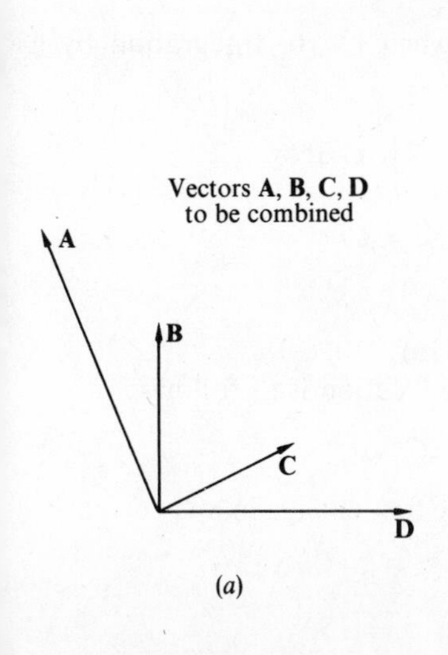

(a)

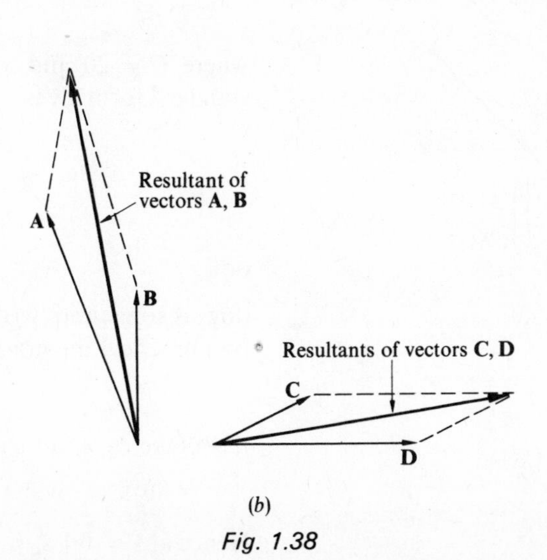

(b)

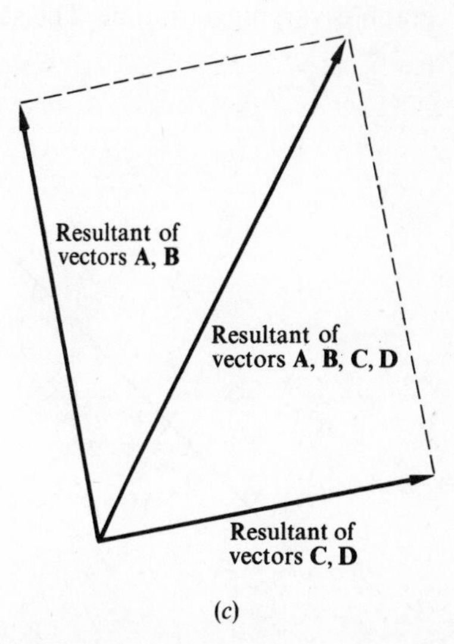

(c)

Fig. 1.38

where N is the number of atoms at some instant and λ a constant. We can describe the rate at which the electric potential V varies with distance r from a charge by the differential equation

$$\frac{dV}{dr} = -\frac{q}{4\pi\varepsilon_0 r^2}$$

q is the charge, ε_0 a constant.

Let us look at another example connected with motion. A freely falling object has a uniform acceleration.

$$\text{Acceleration} = \text{a constant } C$$

But acceleration is the rate of change of velocity with time.

$$\text{Acceleration} = \frac{dv}{dt}$$

The same arguments about instantaneous acceleration apply as occurred for velocity. Velocity is the rate of change of position with time, thus

$$\text{Acceleration} = \frac{d(dx/dt)}{dt}$$

We can rewrite this as

$$\text{Acceleration} = \frac{d^2x}{dt^2}$$

Thus the differential equation describing the constant acceleration of a freely falling object can be written as

$$\frac{d^2x}{dt^2} = C$$

This is known as a second-order differential equation, the ones previously considered being first-order equations.

If we take a graph of distance x against time t what can we identify as both the first- and second-order differential equations? The first-order equation involves dx/dt, the velocity. dx/dt is the rate at which x changes with time and thus is represented by the slope of the graph. The first-order equation thus tells us how the slope of the distance–time graph is varying with time. The second-order equation tells us how the velocity varies with time and thus how the rate of change of the slope with time depends on the time.

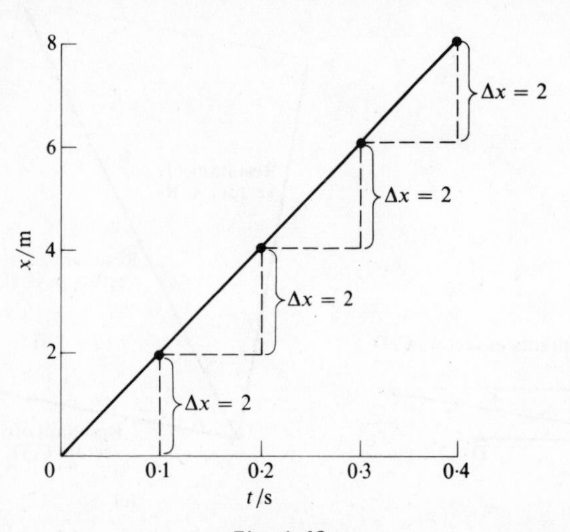

Fig. 1.40

Another way of looking at this is to consider finite differences in x and t instead of the infinitesimally small. A $\Delta x/\Delta t$ equation tells us how much we should change x for each small incremental change in t of Δt. Thus if the differential equation states that dx/dt is a constant C then in each interval of time Δt the value of x increases by Δx, where

$$\frac{\Delta x}{\Delta t} = C$$

or change in $x = \Delta x = C\,\Delta t$.

Thus if we had a constant velocity of 20 m s^{-1} and took time intervals of 0·1 s, then the change in x in the first time interval would be

$$\Delta x = C\,\Delta t = 20 \times 0{\cdot}1 = 2\text{ m}$$

If at time $t = 0$ the x value is 0 then we can use the change in x value to arrive at the position after 0·1 s (Fig. 1.40). Because C is a constant the change in x in the next 0·1 s time interval will be 2 m, and 2 m in the next 0·1 s interval, and so on. The graph gives the solution of the differential equation.

In this case we could have arrived at the solution by using a standard integration form.

$$\frac{dx}{dt} = C = 20$$

$$\int_0^x dx = \int_0^t 20\,dt$$

$$x = 20t$$

This is the equation of the graph in Fig. 1.40. Standard integral forms are not always possible, the previous method, known as numerical integration, can be used in quite complicated cases.

Let us consider another example:

$$\frac{dx}{dt} = Cx$$

where $C = 20$ and $x = 5$ when $t = 0$. Integration by a standard form gives

$$\int_5^x \frac{dx}{x} = \int_0^t C\,dt$$

$$\log_e (x/5) = Ct$$

or

$$x = 5\,e^{Ct}$$

($\log_e$ is sometimes written as ln)

By numerical integration the solution is as follows:

$$\Delta x = Cx\,\Delta t$$

If we take $\Delta t = 0{\cdot}1$ s then the first change in x is

$$\Delta x = 20 \times 5 \times 0{\cdot}1 = 10$$

Hence at t $= 0{\cdot}1$ s, $x = 10 + 5 = 15$.

In the next 0·1 s time interval the change in x is

$$\Delta x = 20 \times 15 \times 0{\cdot}1 = 30$$

In the next time interval

$$\Delta x = 20 \times 45 \times 0{\cdot}1 = 90$$

This can be continued until sufficient of the graph has been plotted (Fig. 1.41).

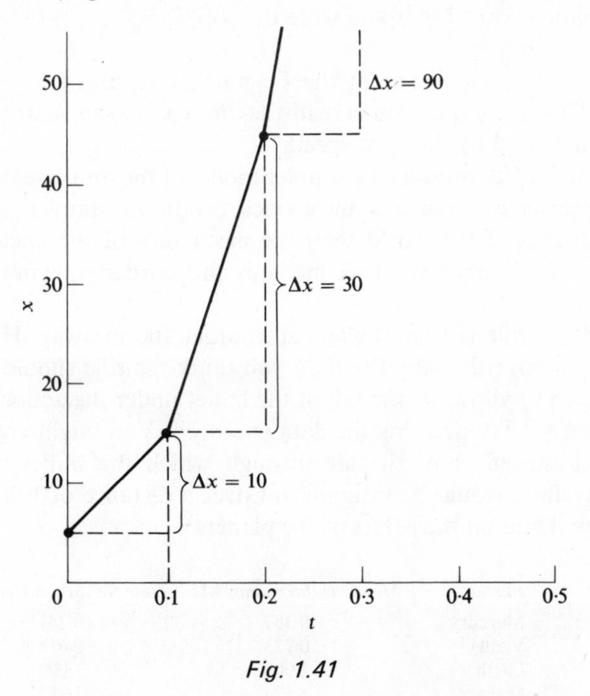

Fig. 1.41

An important use of numerical integration is the solution of second-order differential equations. For an object freely falling from rest we can under laboratory conditions assume a constant value for the acceleration due to gravity, say 10 m s^{-2}.

$$\text{Acceleration} = \frac{d^2x}{dt^2} = C = 10$$

We can rewrite this as

$$\frac{d}{dt}\left(\frac{dx}{dt}\right) = 10$$

dx/dt is the slope of the x against t graph

$$\frac{d}{dt}(\text{slope}) = 10$$

The rate of change of the slope is 10. The slope is the velocity.
For finite small changes we can write

$$\frac{\Delta}{\Delta t}(\text{slope}) = 10$$

or the change in slope is

$$\Delta\,(\text{slope}) = 10\,\Delta t$$

Figure 1.27 shows a graph produced by taking Δt intervals of 0·1 s. In each successive time interval the slope changes by $10 \times 0{\cdot}1 = 1 \text{ m s}^{-1}$.

In chapter 4 we meet another case, the solution of the equation

$$\frac{d^2x}{dt^2} = -Kx$$

This is the equation for an object oscillating with simple harmonic motion.

Appendix 1B The ellipse

The elliptical orbits of the planets are very close to being circular orbits and thus for simplicity in rough calculations we assume the orbits to be circular. An ellipse is the projection of a circle on one plane onto another plane inclined at some angle to the first plane (Fig. 1.42). The smaller the angle between the planes the more like a circle the ellipse is. For an angle θ between the planes the ellipse has axes of length $2r$ and $2r\cos\theta$, where r is the radius of the circle.

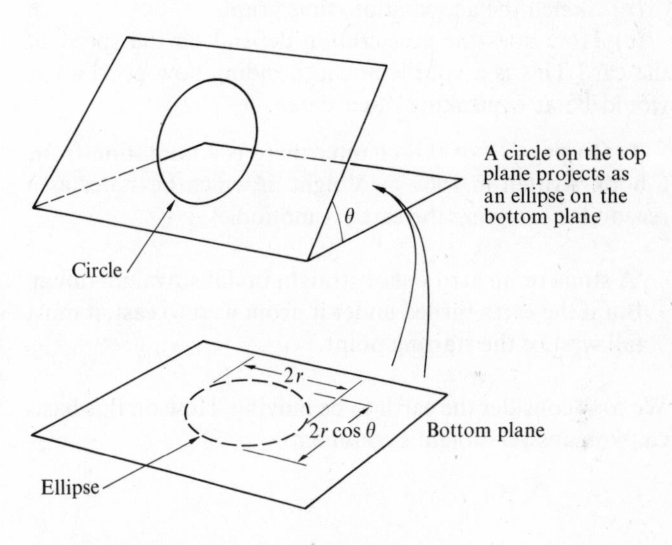

Fig. 1.42

For the earth

$$\frac{2r}{2r\cos\theta} = \frac{1{\cdot}000\,14}{1}$$

For Mars

$$\frac{2r}{2r\cos\theta} = \frac{1{\cdot}0043}{1}$$

One way of drawing an ellipse is to attach the ends of a piece of thin string to two pins which are then pressed into a sheet of paper. The length of the string should be equal to $2r$, the distance between the pins should be less than $2r$. A pencil pressed into the loop of the string and moved round the pins will trace out an ellipse (Fig. 1.43). The pin positions are known as the foci of the ellipse.

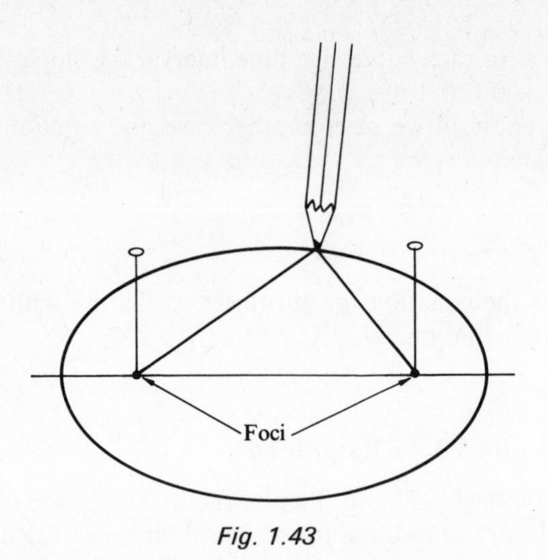

Fig. 1.43

Problems

1. Figure 1.44 shows the distance–time graph for a car with automatic transmission accelerating from rest.

(a) Sketch the velocity–time graph.
(b) Sketch the acceleration–time graph.
(c) How does the acceleration depend on the speed of the car? This is a vital factor in deciding how good a car would be at overtaking other cars.

2. On page 18 of this chapter there is a quotation from a book written in 1667 by Voight in which he states as a reason for asserting the earth is motionless:

> 'A stone or an arrow shot straight up falls straight down. But if the earth turned under it, from west to east, it must fall west of the starting point.'

We now consider the earth to be moving. How on this basis can you answer Voight's comment?

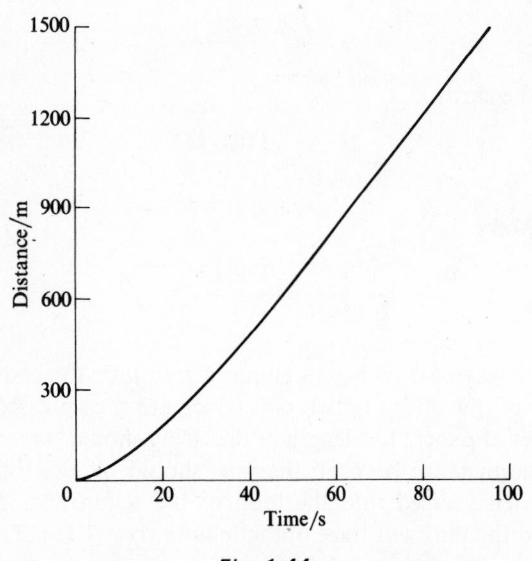

Fig. 1.44

3. A query is posed in *The Feynman Lectures on Physics* volume 1, page 8-3 by Feynman, Leighton and Sands:

> 'A lady in a car was stopped by a cop. The cop goes up to her and says, "Lady, you were going 60 miles an hour!" She says, "That's impossible, sir, I was travelling for only seven minutes. It is ridiculous—how can I go 60 miles an hour when I wasn't going an hour?' 'How would you answer her if you were the cop?'

(For an answer look at the Feynman reference.)

The above question is really asking you to say what you understand by the term speed.

4. Kepler produced a simpler model of the universe than Copernicus. Galileo's mechanics produced simpler descriptions of the world than the mechanics of the ancient Greeks. Explain what we mean by the word simple in this context.

5. A rifle is firing bullets at a target 100 m away. How high above the target bull do you think the rifle should be aimed to allow for the fall of the bullet under the action of gravity? Try guessing the data you need. You might argue that the effect of the air through which the bullet was travelling would be insignificant over a distance of 100 m.

6. Data on the orbits of the planets.

Planet	*Mean orbital radius*/AU	*Time for an orbit*/*years*
Mercury	0·387	0·241
Venus	0·723	0·615
Earth	1·000	1·000
Mars	1·524	1·881
Jupiter	5·203	11·862
Saturn	9·539	29·46
Uranus	19·182	84·01
Neptune	30·058	164·79
Pluto	39·439	248·43

1 AU = $149{\cdot}6 \times 10^9$ m

(a) Do the above data confirm Kepler's third law?

(b) There is a law called the Bode law, or sometimes the Titius–Bode law, which states that

$$r_n = 0{\cdot}4 + 0{\cdot}3\,(2)^{n-1}$$

where r_n is the radius of the orbit of the nth planet, the planets being numbered in sequence from the earth as one outwards. How well is this law obeyed? A group of asteroids occurs between the orbit of Mars and Jupiter.

(c) The asteroid belt extends from about an orbital radius of 2·5 to 3·0 AU. Calculate the range of times taken for the asteroids to orbit the sun.

7. Figure 1.45 shows how the height of a seedling varies with time. Plot a graph showing how the rate of growth varies with time.

8. Sketch the paths you would expect to see followed by the following items:

(a) the stars shooting out of an exploding firework rocket,
(b) the water coming from a garden hose when it is at different angles to the horizontal,
(c) apples falling from a tree when the tree is shaken.

(d) an artificial satellite orbiting the earth—the path need not be circular.

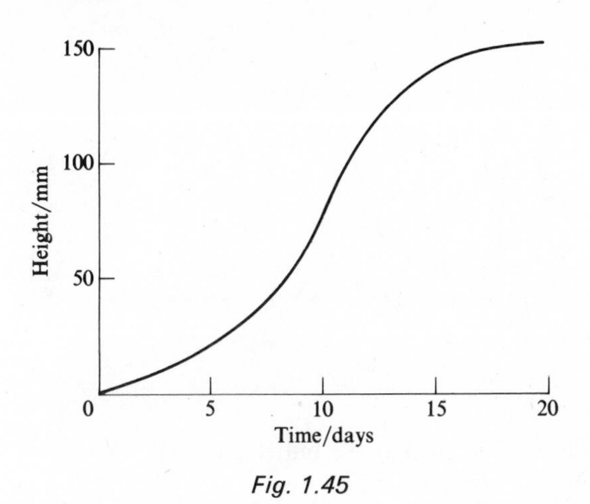

Fig. 1.45

9. The term 'scientific method' has been used by some to describe the sequence:

experimental observations;
postulation of a model, to fit the observations;
checking of the predictions of the model against further observations;
adjustment or replacement of the model as required by new observations.

How far do you think this pattern of behaviour has been followed by the evolution of the model of the universe?

10. Oxford and Cambridge Schools Examination Board. Nuffield Advanced Physics Special paper. 1971.

'A particle is projected at 100 m s^{-1} up a sloping plane, as shown in the diagram [Fig. 1.46], so that it is moving along a line inclined at 30° to the horizontal. By considering the velocity changes that happen during successive short time intervals begin to construct a distance–time graph for the particle. Your construction need not be accurate, but you should give enough notes to explain how an accurate plot would be made. Ignore friction forces between the plane and the particle, and assume that gravity exerts a force of 9·8 N kg^{-1} which does not vary with height.'

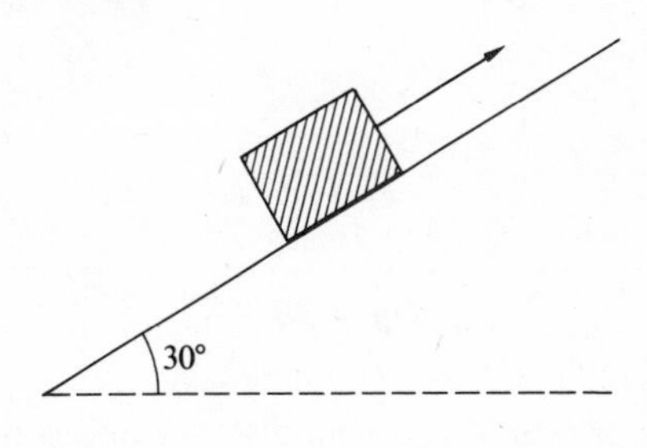

Fig. 1.46

Teaching note Further problems appropriate to this chapter will be found in:
E. M. Rogers. *Physics for the inquiring mind*, chapters 1, 2, 12–23, OUP, 1960.
Project physics course, Text 1 and 2, Holt, Reinhart, and Winston, 1970.
Physical science study committee. *Physics*, 2nd ed., chapters 5 and 6. Heath, 1965.

Practical problems

1. Study the motion of parachutes. Consider the following factors:

(a) Does the parachute continually accelerate during its motion or does it reach a constant velocity?
(b) How is the motion of the parachute affected by the load carried?
(c) How is the motion of the parachute affected by the size of the parachute?

The experiments can be performed with model parachutes.

2. Study the flow patterns associated with the movement of air over cliffs, hills, etc. The movement of air under such conditions is used by glider pilots to assist them in climbing to high altitudes.

One method of studying such movements is to observe the movement of smoke, and can be done with a model. The flow directions can also be indicated by tufts of wool or cotton attached to the model.

3. Study the behaviour of drifting snow and the resulting flow patterns. Consider the effect on the flow of the terrain, e.g., hills, trees, houses, etc. The problem is really one of aerodynamics in the consideration of the flow of snow as a fluid.

Snow can be simulated by flaked mica, borax, or sawdust.

Suggestions for answers

1. (a) See Fig. 1.47.
(b) See Fig. 1.48.
(c) Up to about 20 m s^{-1} there is an increase in acceleration with increasing speed, above that the acceleration decreases.

2. Both the earth and the stone or arrow have the same horizontal velocity and so keep pace with each other.

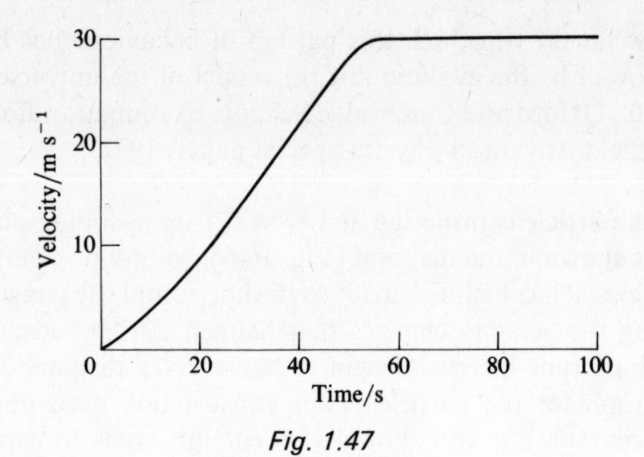

Fig. 1.47

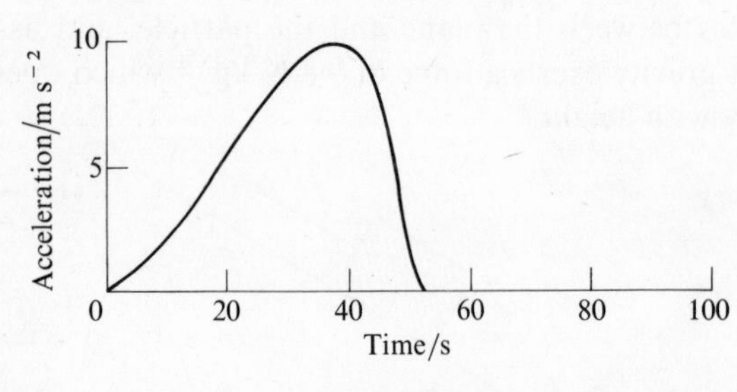

Fig. 1.48

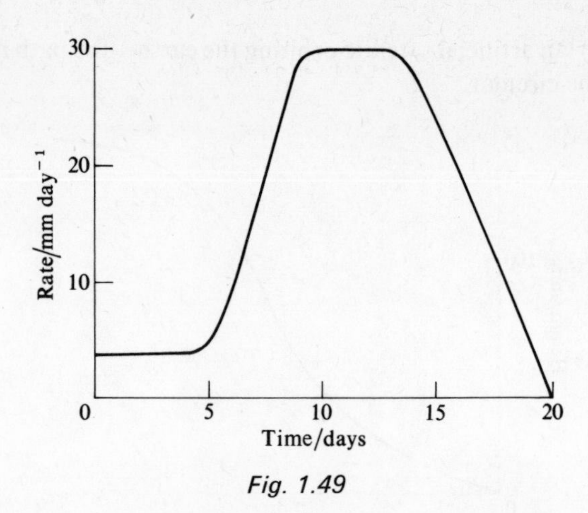

Fig. 1.49

3. Speed is a rate of covering distance and is not dependent on the total distance covered.

4. William of Ockham lived at the beginning of the fourteenth century and suggested that scientists should prefer the theory which uses in its explanation the smallest number of unknown quantities, the smallest number of variables. We regard the theory which has the smallest number of devices or arbitrary conditions as the simplest.

5. You might guess that a bullet has a velocity of a few thousand metres per second. Thus in travelling 100 m it will be in the air, and falling, for about 0·05 s. In this time it falls a vertical distance of about 0·012 m, that is 12 mm.

We have assumed that the bullet does not slow down over 100 m.

6. (a) You should find that R^3 is proportional to T^2.

(b) Bode's law works very well. Mercury has $n = -\infty$, Venus $n = 0$, Earth $n = 1$, Mars $n = 2$, the asteroids $n = 3$, Jupiter $n = 4$, and so on.

(c) 3·86 to 5·20 years.

7. See Fig. 1.49.

8. (a), (b), (c) parabolic paths like projectiles.

(d) Ellipses with the earth at a focus.

9. To this sequence we might add—introduction of a new model not because of further data but because a simpler model can be shown to fit the data.

10. This is the motion of a particle described by the equation

$$\begin{aligned}\text{Acceleration} &= -g \sin \theta \\ &= -9{\cdot}8 \sin 30^\circ \\ &= -4{\cdot}9 \text{ m s}^{-2}\end{aligned}$$

or rate of change of distance–time graph $= -4{\cdot}9 \text{ m s}^{-2}$.

2 Forces

Teaching note Practical work appropriate to this chapter will be found in:
W. Bolton. *Physics experiments and projects*, Vol. 5, Pergamon, 1969.
Nuffield O-level physics. Guide to experiments, 4 and 5, Longmans/Penguin, 1967.
Project physics course. Handbook, Holt, Reinhart and Winston, 1970.

2.1 'Let Newton Be!'

In 1686 Isaac Newton wrote in the preface to his book *Philosophia Naturalis Principia Mathematica* (*Mathematical Principles of Natural Philosophy*) the following. It could be read as the preface to this chapter of this book.

'We offer this work as mathematical principles of philosophy ... by the propositions mathematically demonstrated in the first book, we then derive from the celestial phenomena the forces of gravity with which bodies tend to the sun and the several planets. Then, from these forces, by other propositions which are also mathematical, we deduce the motions of the planets, the comets, the moon, and the sea. I wish we could derive the rest of the phenomena of nature by the same kind of reasoning from mechanical principles; for I am induced by many reasons to suspect that they may all depend upon forces by which the particles of bodies, by some causes hitherto unknown, are either mutually impelled towards each other, and cohere in regular figures, or are repelled and recede from each other; which forces being unknown, philosophers have hitherto attempted the search of nature in vain; but I hope the principles here laid down will afford some light either to that or some truer method of philosophy.'

I. Newton (1686), *Mathematical Principles of Natural Philosophy*.

In the early part of this chapter we concern ourselves with what are known as Newton's laws of motion. In these he laid down the concepts of force, mass, and momentum. The latter part of the chapter is concerned with an examination of the different types of forces.

Newton's laws of motion can be written as:

First Law

In the absence of a resultant force an object keeps moving with a uniform velocity or it remains at rest;

Second Law

The rate of change of momentum of a body is proportional to the resultant force acting on it, and occurs in the direction in which the force is acting,

this can be worded in a different manner as

the product of the mass and acceleration of a body is proportional to the force acting on it, the acceleration occurring in the direction in which the force is acting;

Third Law

If one body exerts a force on a second, the second exerts an equal and opposite force on the first,

this is often worded as

to every action there is an opposite and equal reaction.

These laws were of considerable importance in the development of physics.

'The discovery of the laws of dynamics, or the laws of motion, was a dramatic moment in the history of science. Before Newton's time, the motions of things like the planets were a mystery, but after Newton there was complete understanding....'

R. B. Feynman, R. P. Leighton, M. Sands.
The Feynman lectures on physics, Vol. 1,
1st ed., Addison Wesley, Reading, Mass., 1963.

'Nature and Nature's laws lay hid in night.
God said "Let Newton be," and all was light.'

A. Pope (1688–1744).

What was Newton, the man, like? The following views paint a far from pleasant picture.

'... in vulgar modern terms, Newton was profoundly neurotic of a not unfamiliar type, but—I should say from the records—a most extreme example. His deepest instincts were occult, esoteric, semantic—with profound shrinking from the world, a paralyzing fear of exposing his thoughts, his beliefs, his discoveries in all nakedness to the inspection and criticism of the world. . . .'

J. M. Keynes, 'Newton, the Man'.
The Royal Society (*1947*).
Newton Tercentenary Celebrations, *1946*. C.U.P.

'. . . He was a man of fearful rages and implacable in offence. He broke promises and withheld knowledge in order to advance his sole glory. His quarrels with Hooke, himself a man of genius, and with Leibnitz, who was his true peer, are among the ugliest in intellectual history. . . .'

Review of *A portrait of Isaac Newton*,
by F. E. Manuel (Oxford), in *The Sunday Times*
of 16 February 1969 by G. Steiner.

2.2 The cause of motion

What causes motion? What stops motion? Is any cause necessary for things to move? Why when you push an object along the bench does it soon stop moving? Why do the planets keep moving round the sun?

'A moving body comes to a standstill when the force which pushes it along can no longer so act as to push it.' This was the view of Aristotle. This seems to be reasonable—when you cease to push an object along a bench it comes to rest soon after you cease pushing. A force seems to be necessary to keep an object in motion. However, let us look at the situation more critically. We will start with the object, say a block of wood, at rest on the bench (Fig. 2.1). We apply

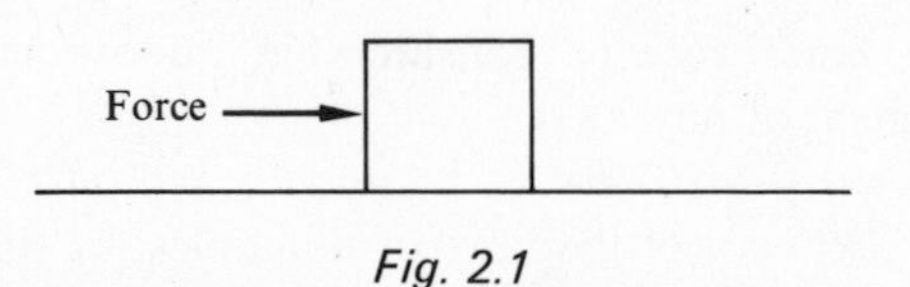

Fig. 2.1

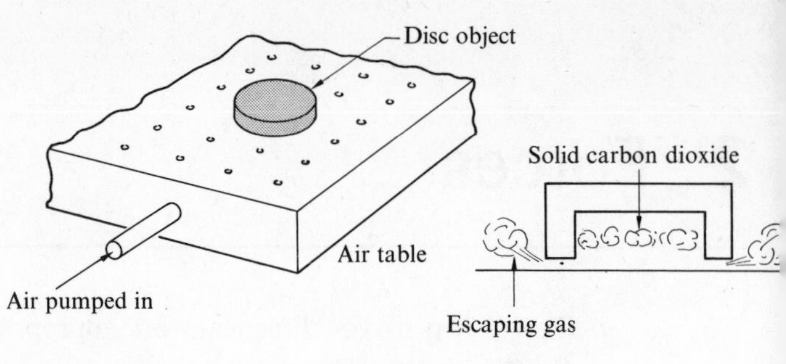

Fig. 2.2

a small force to the block—nothing happens. We increase the force and then at some particular value the block moves. If we increase the force, by pushing or pulling more, the object moves faster. When we stop pushing the block rapidly comes to rest. The result is, however, different if we put wheels on the block of wood: a very small force causes the block to move. Why the difference? Wheels—but why? Friction? Perhaps friction complicates the results and we have some force other than our push acting on the block.

It is possible to achieve nearly friction-free conditions by floating the block on a cushion of gas (like the hovercraft), Fig. 2.2. One way of doing this is by blowing air upwards through holes in the base on which the object floats. In an air table a flat base containing the holes is used. An alternative is for the object to blow gas downwards onto the base. An upturned tin lid containing solid carbon dioxide is one form. The carbon dioxide is very cold and on warming up vaporizes, the tin lid shape forces the escaping gas downwards. What happens when we push an object in the absence of friction? It keeps on moving with a uniform velocity, when no resultant force is acting on it. (We are only concerned with forces and motion in the horizontal plane—gravity still acts on the object but no motion in the vertical plane is possible.) In the absence of a resultant force an

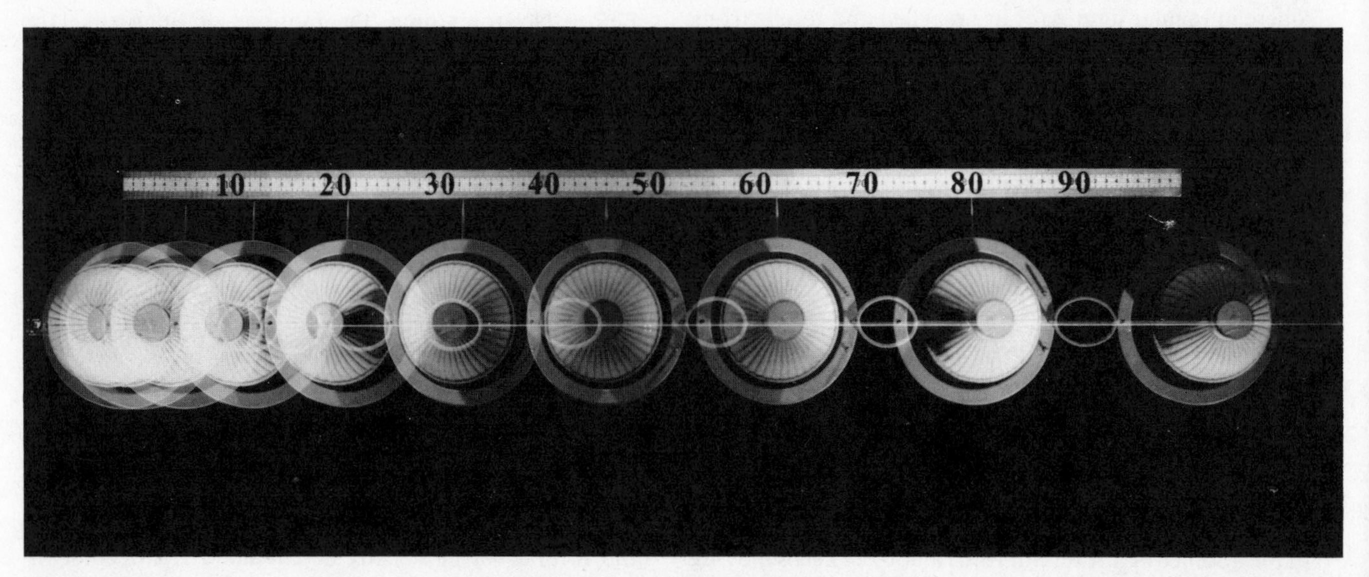

Fig. 2.3 PSSC Physics, *2nd Ed., D. C. Heath and Co. The flash photographs show the puck being pulled to the right.*

object keeps moving with a uniform velocity or it remains at rest. This is known as Newton's first law of motion.

This idea has already been encountered in the previous chapter. There a ball rolling down a curved piece of track was considered. When the rail is in the form of a U the ball rolls down one side of the U and up the other side to almost the same height. This occurs regardless of the shape of the U and regardless of whether the slopes of each arm of the U are the same. If it rolls down one side and the other side of the U is horizontal then the ball should keep rolling until it reaches the same height again—in fact in the absence of friction, never. It continues in motion with a uniform velocity.

What does a force do? For our frictionless object (Fig. 2.3), it causes a change in velocity. So for no resultant force we have rest or uniform velocity and when a force is applied the velocity is changed. What now do we think about the situation where an object was pushed along the bench. When the force was small no motion occurred. But a force should cause motion—the conclusion we are forced to is that there must be another force acting on the object which just cancels out the effect of the force we have applied (forces are vector quantities). As we increase our force so this opposing force increases, until at some particular value the opposing force ceases to increase and motion occurs because there is a resultant force acting on the object. This opposing force is known as the force of friction.

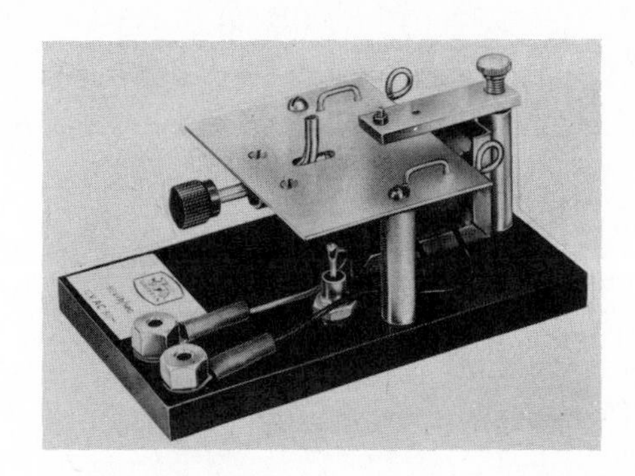

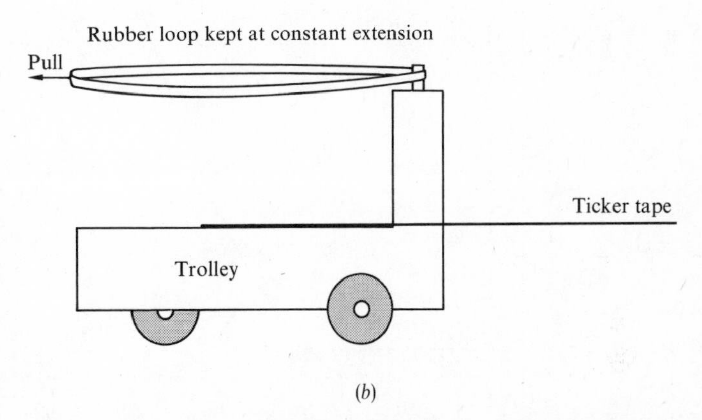

Fig. 2.4(a) Ticker tape vibrator. The tape passes through the two slots and the upper arm vibrates against the tape. (Griffin and George Ltd.) (b) Trolley being pulled with a constant force.

Change of velocity

Is there any simple relationship between force and the change in motion produced by the force? Let us simplify the matter—what happens when a constant force acts on an object? The first problem is how to produce a constant force. If we pull a piece of rubber between our hands and keep it stretched by the same amount—is the stretching force constant? A constant extension of a piece of rubber would seem to be a convenient way of defining a constant force. Now let us apply our constant force to an object and study its motion. A trolley is a convenient object and its motion can be studied if ticker tape is attached to it (Fig. 2.4) and passed through a vibrator which produces marks on the tape at regular time intervals. Figure 2.5 shows the ticker tape trace produced for a constant force. When this is plotted as a distance–time graph, a curve is found. A velocity–time graph of the motion, however, is a straight line. The motion is one of constant acceleration. A constant force produces a constant acceleration. Try this analysis with Fig. 2.3.

We have met constant acceleration before—a freely falling object. The conclusion would seem to be that a constant force is acting on the object during free fall.

What happens if, for the same object, we double the force? Double the force can be produced by using two identical lengths of rubber (Fig. 2.6) and giving both the same extension. Double the force gives double the acceleration. If we use three times the force we get three times the acceleration. The acceleration, a, is directly proportional to the force F.

$$a \propto F$$

We can write this as

$$F = ma$$

where m is the constant of proportionality. We call the constant m mass. The bigger this constant for a given force the smaller the acceleration. The greater the constant the more difficult it is to accelerate an object. What factors change this constant?

Suppose instead of using one trolley being pulled by the piece of rubber we use two trolleys stacked together, one on top of the other. The acceleration with the two trolleys is half that with the single trolley. With three trolleys the acceleration is one-third of that with the single trolley. F is kept constant.

Number of trolleys	*Acceleration*	*Value of m*
1	a	m
2	$a/2$	$2m$
3	$a/3$	$3m$

The product of a number representing the number of trolleys and the acceleration is a constant with a constant force. The value of m, the constant of proportionality in $F = ma$,

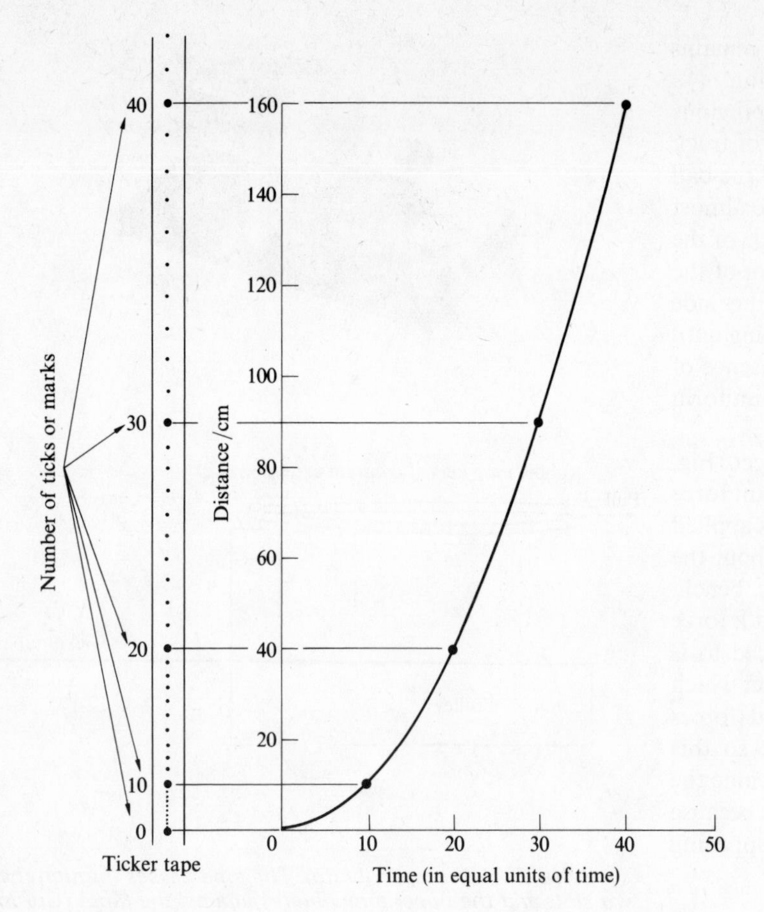

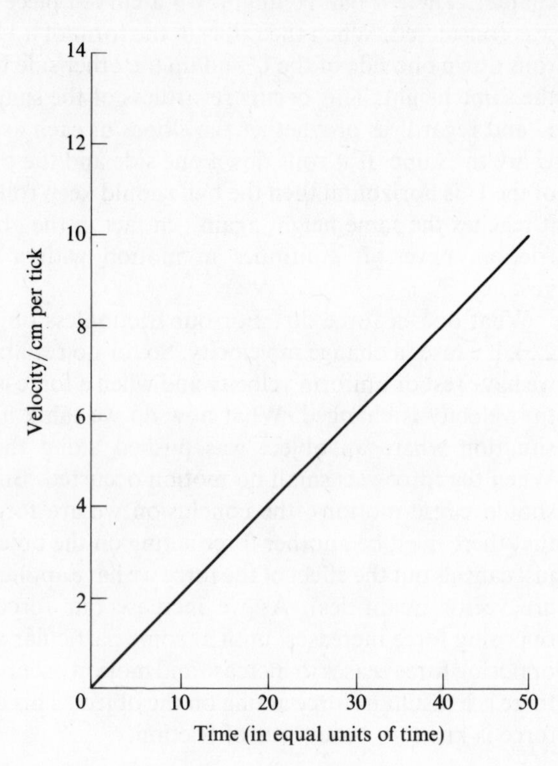

Fig. 2.5

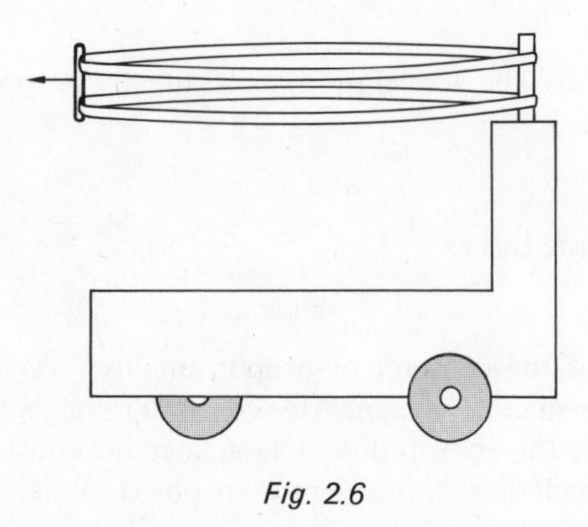

Fig. 2.6

is determined by the number of trolleys. Doubling the number of trolleys doubles the value of m.

What do we mean by saying that force is a vector quantity? If we apply two lengths of rubber to our trolley the resulting acceleration will depend on the directions in which the rubber strips pull on the trolley (Fig. 2.7). Our equation is concerned with the resultant force. The equation is one form of what is known as Newton's second law of motion.

The units of acceleration are m s^{-2}; what are the units of mass and force? As the two quantities are related by the equation we need only specify a standard for one; the other can be defined in terms of it. In fact we adopt a piece of material as a standard mass, called 1 kg. The unit of force, called the newton (N), is then defined as that force which produces an acceleration of 1 m s^{-2} for a mass of 1 kg.

The international standard of mass, the kilogramme, is kept in Paris. Many countries keep copies of this standard and compare their copies with the standard in Paris.

For a particular object Newton's law tells us that the ratio force/acceleration is a constant. We call the constant the mass. There is nothing in the law restricting it to any initial state for our object. It could be at rest, relative to the earth, or moving at 100 km h^{-1} in a fast car. It turns out that, experimentally determined, at very high speeds the ratio is no longer a constant—our law is not correct at these high speeds. The speeds have to be close to the speed of light before the ratio departs from its constant value. Einstein's mechanics have to be used at such speeds. Figure 2.8 shows how the force/acceleration ratio changes with speed. The force/acceleration ratio we call mass, thus mass varies with speed. The following equation describes this variation.

$$m_v = \frac{m_0}{(1 - v^2/c^2)^{1/2}}$$

m_v = mass at speed v, m_0 = rest mass, c = speed of light. The above discussion is very vague, we have to be very careful to specify what we mean by acceleration and force when high speeds are involved. See chapter 12 for more details or the book *Special Relativity* by A. P. French, (Nelson & Sons Ltd, 1968).

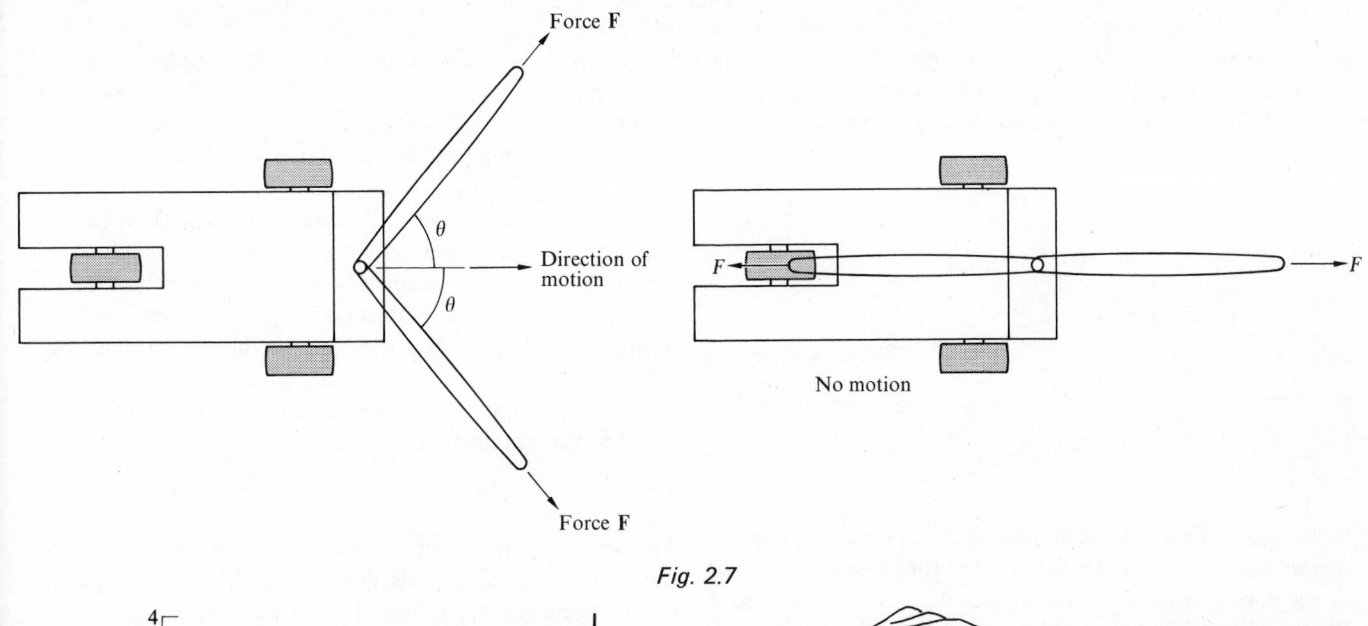

Fig. 2.7

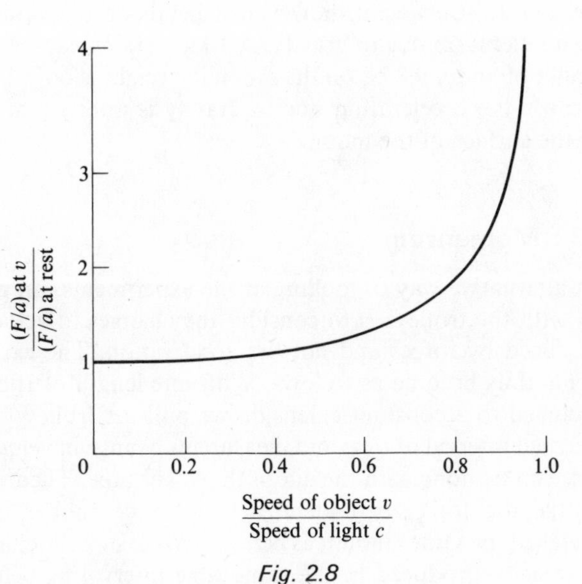

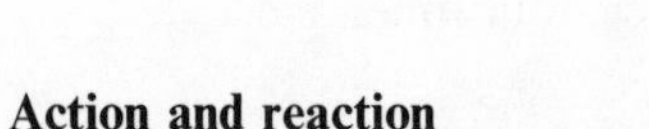

Fig. 2.8

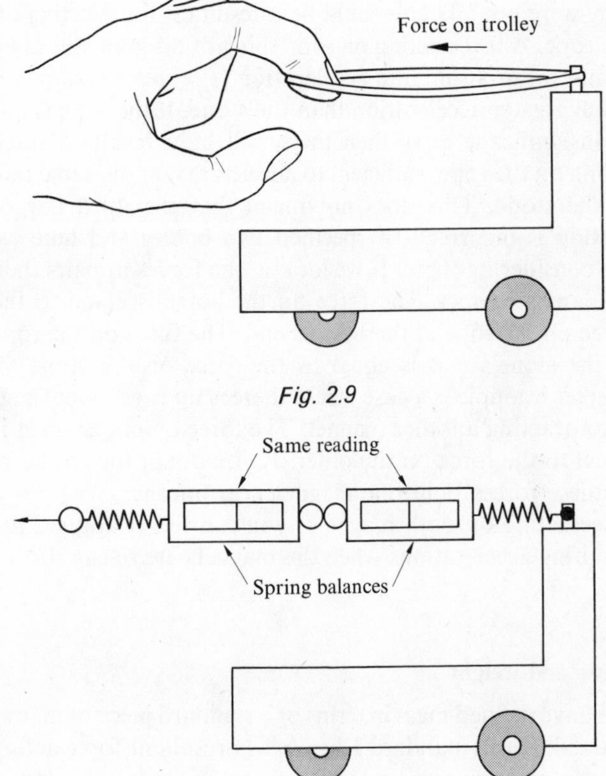

Fig. 2.9

Fig. 2.10

Action and reaction

Let us take a closer look at the forces acting when we pulled the trolley with a length of rubber (Fig. 2.9). The rubber pulled the trolley—after all the trolley did accelerate. Are there any other forces? If we pull on one end of a length of rubber—does it extend? Forces in opposite directions are needed at each end of the piece of rubber. When we pull the trolley a force is applied to the trolley and a force in the opposite direction is applied to our hand. What are the relative magnitudes of the two forces? If we replace the length of rubber by two spring balances hooked together and then pull we find that the readings on the two balances are the same. (Fig. 2.10.)

Newton dealt with this in what he called the third law of motion.

If one body exerts a force on a second, the second exerts an equal and opposite force on the first.

The force exerted by the first body on the second is known as the action, the opposite and equal force acting on the first object is known as the reaction. Briefly the law can be described as—action and reaction are always equal.

> 'If you press a stone with your finger, the finger is also pressed by the stone. If a horse draws a stone tied by a rope, the horse (if I may say so) will be equally drawn back towards the stone; for the distended rope, by the same endeavour to relax or unbend itself, will draw the horse as much towards the stone as it does the stone towards the horse. . . .'
>
> I. Newton (1687) *Principia.*

Let us take a closer look at the horse drawing a stone tied by a rope (Fig. 2.11). There is a force acting on the rope at the horse's end. Because the rope resists being extended there will be an oppositely directed force at the stone's end

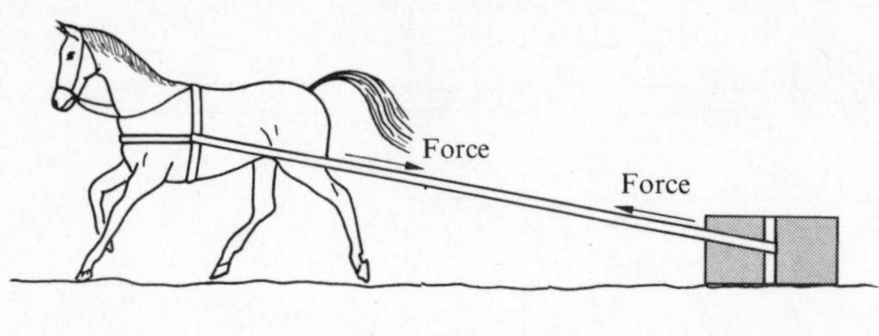

Fig. 2.11

of the rope. If the rope is of insignificant mass, by comparison with the stone and the horse, then the forces at each end of the rope are opposite and equal. What would happen if they were not? There would be a resultant force acting on the rope. A force acting on something of insignificant mass would mean an infinite acceleration ($F = ma$), certainly a much greater acceleration than the stone. If the rope is not of insignificant mass then there will be a resultant force acting on the rope, sufficient to accelerate it at the same rate as the stone. This does not mean that the third law of motion is incorrect—it specified two bodies and here we are considering three. If we look at the forces in pairs then the law is correct. The force on the horse is equal to the force on the rope at the horse's end. The force on the rope at the stone's end is equal to the force on the stone. A simpler example is a case where there is no rope—one magnet attracting another magnet. The force on one magnet is equal to the force on the other. If you doubt the existence of these forces hold one magnet near another—there is a force acting on both magnets: you can feel it and see the resulting accelerations when the magnets are released.

Mass and weight

We have defined mass in terms of a standard piece of matter and called this standard 1 kg. A 1 N resultant force acting on this mass gives it an acceleration of 1 m s^{-2}. If we have a 2 kg mass then the 1 N force only produces an acceleration of 0·5 m s^{-2}. Our scale of mass is determined by the acceleration produced by a force. Mass is the property of a body which represents its inertia. Our definition of mass has been defined independently of its situation: on the moon or on the earth the same force is needed to accelerate a 1 kg mass at 1 m s^{-2}.

If we allow a mass to fall freely it will accelerate with the acceleration due to gravity, about 9·8 m s^{-2} near the surface of the earth. An accelerating mass means a force—the force of gravity.

$$\text{Gravitational force} = mg$$

Because the acceleration due to gravity is the same for all masses then the gravitational force is directly proportional to the mass.

What is weight? If we put a mass on the pan of a spring balance the spring becomes extended and the amount of the extension can be used to tell us the force necessary to keep the mass at rest relative to the earth's surface. This force is called the weight of the object. An object has weight when we prevent it freely accelerating under the action of the gravitational force. The object is weightless when it is freely falling. Astronauts when they are weightless are freely falling under the action of the local gravitational force.

If we prevent an object accelerating we must apply a force equal to the gravitational force. Thus

$$\text{weight} = \text{gravitational force} = mg$$

The weight of an object is thus proportional to the mass of the object. The weight, however, depends on the value of the acceleration due to gravity. A 1 kg mass on earth has a weight of about 9·8 N, on the moon its weight is only 1·6 N because the acceleration due to gravity is only 1·6 m s^{-2} on the surface of the moon.

2.3 Momentum

An alternative way of looking at the experiments, on page 33, with the trolleys is to consider the changes in velocity produced by forces and not the acceleration. The experiments thus become as follows. With one length of rubber stretched to a constant extension we pull the trolley for a specified interval of time and measure its change in velocity. This can be done with the aid of the ticker tape. Then with double the force, two identical lengths of rubber both stretched the same amount as before, we measure the change in velocity produced in the same time interval as before. Then with three times the force, four times, etc.

Force	*Change in velocity in time t*
F	v
$2F$	$2v$
$3F$	$3v$
$4F$	$4v$

The change in velocity is proportional to the force acting on the trolley.

What happens if we keep the force constant and for a particular trolley increase the time over which we determine the change in velocity?

Time	*Change in velocity*
t	v
$2t$	$2v$
$3t$	$3v$
$4t$	$4v$

The change in velocity is directly proportional to the time.

Suppose we now keep the force and time constant and

use first one trolley, then two trolleys stacked together, then three, etc.

Number of trolleys	*Change in velocity*
1	v
2	$v/2$
3	$v/3$

The change in velocity is inversely proportional to the number of trolleys used. Let us say that this factor associated with number of trolleys is mass. The bigger the mass the less the change in velocity in a given time.

These various relationships can be summarized as

Change in velocity $\Delta v \propto F$

with mass m, time interval Δt, constant.

Change in velocity $\Delta v \propto \Delta t$

with mass m, force F, constant.

Change in velocity $\Delta v \propto \dfrac{1}{m}$

with force F, time interval Δt, constant.

These three relationships can be brought together in one equation

$$\Delta v \propto \frac{F(\Delta t)}{m}$$

or

$$F \propto \frac{m(\Delta v)}{\Delta t}$$

The proportionality sign can be replaced by an equality sign by introducing a constant of proportionality. However, we can choose to make this constant of proportionality one. Thus

$$F = \frac{m(\Delta v)}{\Delta t}$$

We can thus define a force unit, the newton, as that resultant force which acting on a standard mass, called 1 kg, causes it to change its velocity by 1 m s^{-1} in a time interval of 1 s. This is the same definition of force as we obtained using the equation $F = ma$. Either can be used as a basis for the definition.

The product of mass and velocity, mv, is given a name—momentum. $m(\Delta v)$ is the change in momentum produced by the force in time Δt, assuming m to remain constant.

$$F = \text{rate of change of momentum}$$

Thus momentum, like velocity and force, is a vector quantity.

Newton's second law of motion can therefore appear in the form—the rate of change of momentum of a body is proportional to the force acting on it, and occurs in the direction in which that force is acting. This is essentially the form in which Newton first wrote the law.

We can have conditions where both velocity and mass are changing or even just the mass changing by mass being ejected from a system (see the section in this chapter on rockets). Our force equation works in all these cases if we take the change in momentum as the change in the product mv.

$$\text{Force} = \frac{\Delta(mv)}{\Delta t}$$

The momentum changes in a time Δt; the time over which the force is considered to act. The product $F\,\Delta t$ is known as the impulse.

$$\text{Impulse} = F\,\Delta t = \Delta(mv)$$

A big impulse means a big change in momentum.

If you jump off a wall and land on the ground 'stiffly' you may break a bone or in some way suffer damage. If, however, you land with your knees flexing as you hit the ground there is less chance of damage. In both cases there was the same change in momentum and so the same impulse. In the 'stiff' interaction between you and the earth Δt was smaller than when you flexed your knees. Thus the force acting on you was greater in the 'stiff' landing.

Conservation of momentum

What happens when two bodies collide—what can we say about the forces? According to Newton's third law the force exerted by the first body on the second body must at any instant of time be equal and opposite to the force exerted by the second body on the first body. This must apply however the force varies during the approach and collision of the two bodies. Thus at every instant the rate of change of momentum of the first body must be opposite and equal to that of the second body.

$$F_1 = -F_2$$

$$\frac{\Delta(mv)_1}{\Delta t} = -\frac{\Delta(mv)_2}{\Delta t}$$

If we take the same time interval then the change of momentum of the first body must be opposite and equal to the change in momentum of the second body. If the velocity of the first body before any interaction between the two bodies was v_1 and after the collision the velocity is v_1' when the two bodies no longer interact, similarly v_2 before the collision and v_2' after the collision for the second body, then

$$m_1(v_1' - v_1) = -m_2(v_2' - v_2)$$

$$m_1v_1 + m_2v_2 = m_1v_1' + m_2v_2'$$

The vector sum of the momentum before a collision is equal to the sum after the collision. This is known as the conservation of momentum.

Measurements of momentum before and after collisions (Fig. 2.12) all show that momentum is conserved. Newton's third law has led to the conservation of momentum law. We could have started with the experimental results showing the conservation of momentum and deduced Newton's third law. It is the experimental fact that momentum is conserved that establishes the third law.

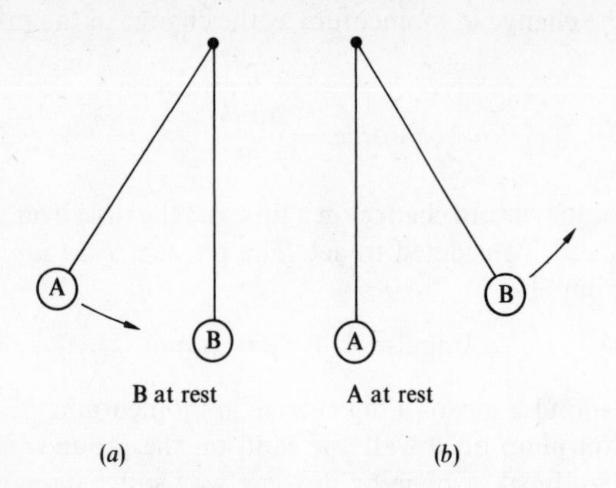

Fig. 2.12 Collision between two identical pendulum bobs. A hits B, initially at rest, and transfers all its momentum to B. The result is that A comes to rest and B moves off.

Conservation of momentum seems to be applicable to all linear events. Collisions between billiard balls or between balloons or electrons or atoms, these all obey the conservation of momentum law. Anywhere a change in momentum occurs for one object there must be a corresponding change in momentum for another body. You drop a book—the book accelerates and so its momentum is increasing. The book accelerates because of the gravitational attraction of the earth. The earth, however, is also attracted by the book, the forces are opposite and equal, and thus the earth accelerates towards the book and gains momentum at the same rate as the book; momentum is conserved. A speeding car brakes and comes to a halt—its momentum decreases and becomes zero (relative to the earth). The loss in momentum of the car is exactly balanced by the change in momentum of the earth. One way of visualizing the fact that there must be an interaction between the car and the earth for the car to stop is to consider the car to be speeding along on a sheet of ice—an almost no-interaction situation.

Collisions

We have argued that in any collision momentum is conserved. We can therefore write down a relationship involving the velocities of the colliding bodies both before and after the collision. These are the velocities before the bodies collide, or interact, and after. What happens during the collision time? During the collision time object 1 experiences a force because of the proximity of object 2, and vice versa. Because they both experience forces they will accelerate. This acceleration will continue until the separation of the two objects is such that the forces become insignificant. Let us look at a specific case—two trolleys held together with a spring in compression between them (Fig. 2.13). When the trolleys are released they spring apart. They accelerate under the action of the force they impart to each other.

$$F_1 = m_1a_1 \quad \text{and} \quad F_2 = m_2a_2$$

But at any instant these forces must be opposite and equal

$$F_1 = -F_2$$

$$m_1a_1 = -m_2a_2$$

The forces decrease as the trolleys fly apart; after some particular separation the forces will be zero when the trolleys are no longer coupled by the spring. Because of this the accelerations of the two trolleys will change with time. When the trolleys are no longer coupled by the spring the forces will be zero and thus the accelerations will be zero. The above equation thus indicates the relationship between the accelerations at any instant of time.

Figure 2.14(a) shows a typical distance–time graph for two such trolleys. When 'contact' ceases the distance–time graph shows a constant velocity for each trolley. Figure 2.14(b) shows the corresponding velocity–time graph and Fig. 2.14(c) the acceleration–time graph.

What would happen if we changed the spring between the trolleys and so changed the 'contact' time? If the 'contact' time is shortened the trolleys reach their constant velocities quicker (Fig. 2.15). The values of the velocities depends only on the masses of the trolleys and not on the 'contact' time.

Does the conservation of momentum apply at any instant during the collision time? The momentum before the collision, in this case, is zero. Therefore if momentum is conserved at any instant we must have

$$m_1v_1 = m_2v_2$$

where v_1 and v_2 are the velocities at some particular instant of time. Rearranging the equation gives

$$\frac{v_1}{v_2} = \frac{m_2}{m_1}$$

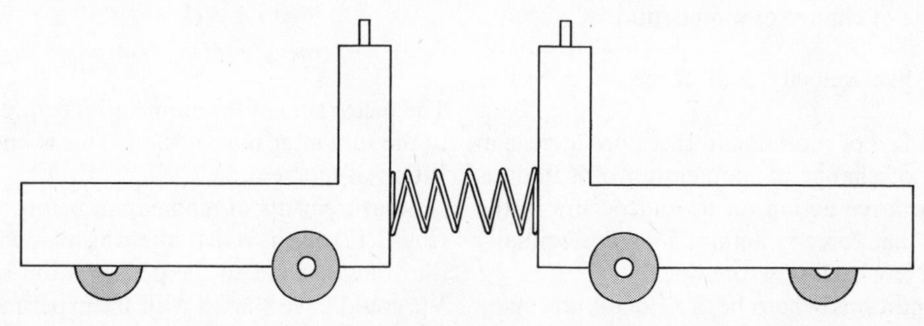

Fig. 2.13

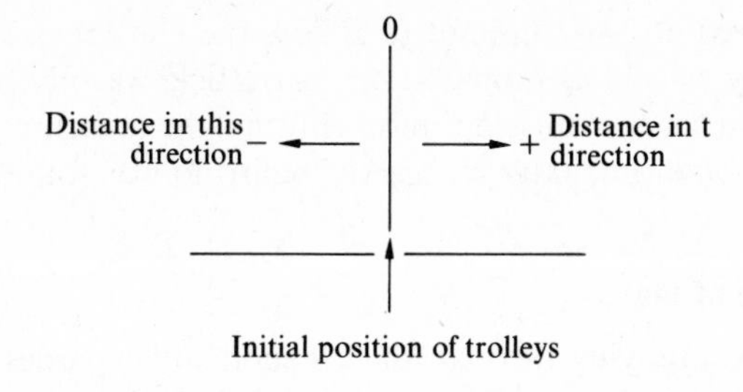

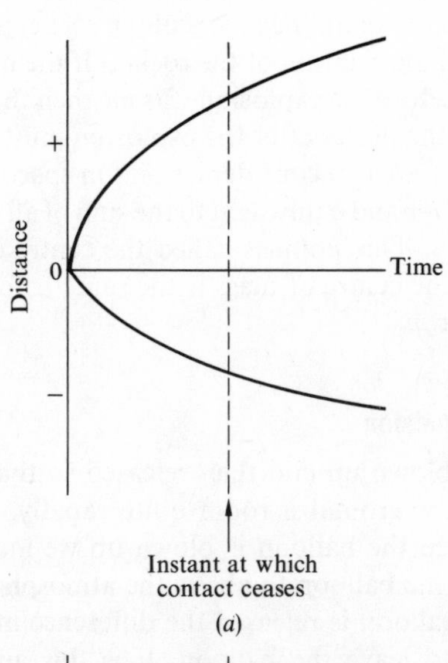

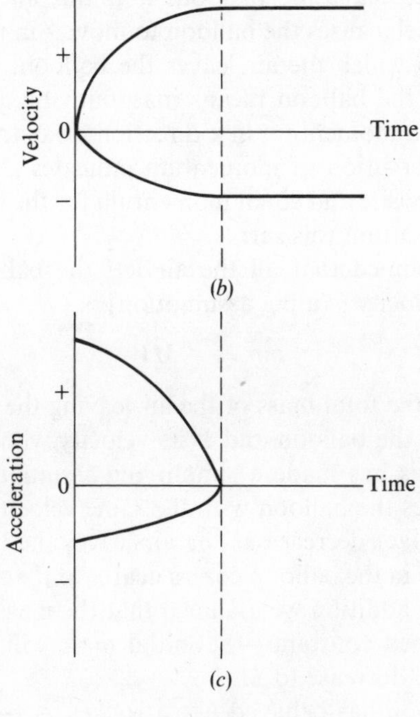

Fig. 2.14

The ratio of the velocities at any instant is the inverse ratio of the masses. This we can check on the velocity–time graph. At any instant the ratio of the velocities is a constant (Fig. 2.16)—we can apply the conservation of momentum to any instant in a collision.

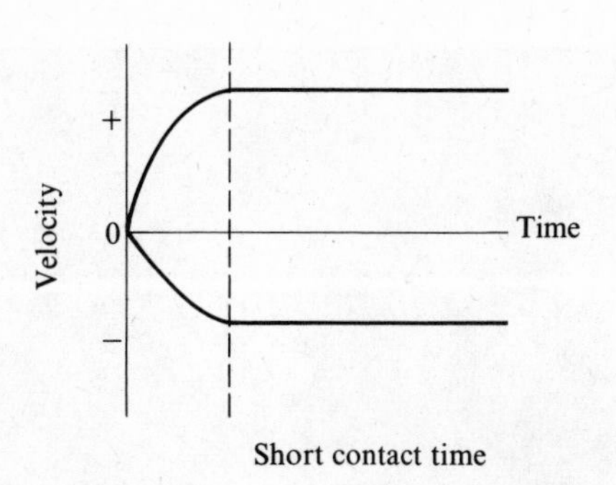

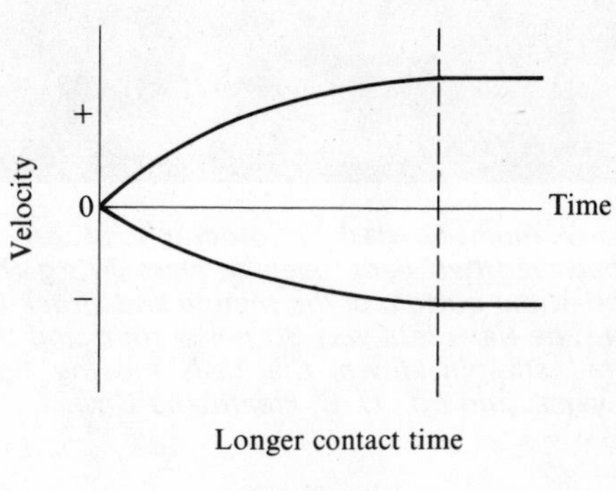

Fig. 2.15

Here in fact is a method of comparing masses which does not rely on a measurement of force.

So far only collisions involving motions along a single straight line have been considered. What happens if the collisions are not restricted to a single line but can involve any directions in a plane, i.e., in two dimensions? Figure 2.17 shows the results of a collision between two balls, the dotted ball entering the picture at the bottom and striking the initially stationary striped ball. The momentum of the striped ball after the collision was 0·481 kg m s^{-1}, the dotted ball had a momentum of 0·559 kg m s^{-1} before the collision and 0·247 kg m s^{-1} afterwards. If we sum the momenta before the collision, 0·559 kg m s^{-1}, and those afterwards, 0·728 kg m s^{-1}, without regard to direction they are not

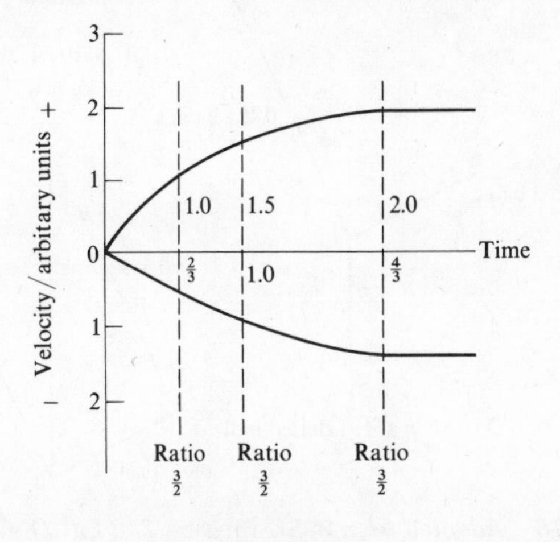

Fig. 2.16

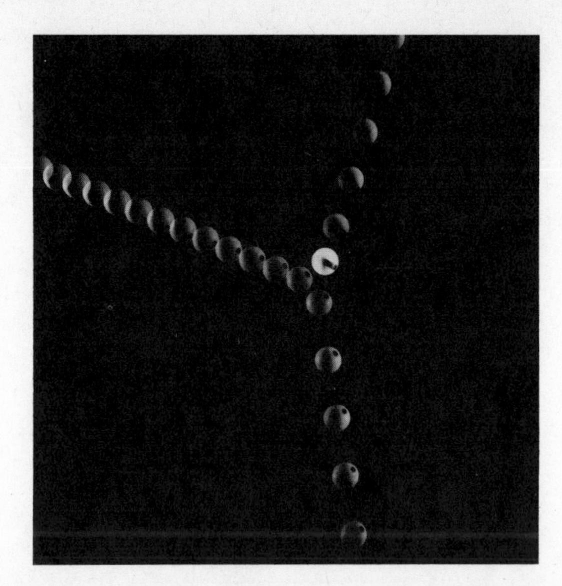

*Fig. 2.17 A multiple flash photograph of an off-centre collision between two balls, each of mass 173 g. The dotted ball entered at the bottom of the picture and struck the striped ball at rest. The flash rate was 30 per second and the camera was pointed straight down, the balls moving horizontally. (*PSSC Physics*, 2nd Ed., D. C. Heath and Co.)*

equal. Taking directions into account, i.e., treating momentum as a vector quantity, does show (Fig. 2.18) that momentum is conserved.

An alternative to the drawing of the vector diagram (see chapter 1) is to resolve the momenta into two perpendicular directions. Momentum must be conserved for the components in both the directions.

$$0{\cdot}247 \sin \theta = 0{\cdot}481 \sin \phi$$

$$0{\cdot}559 = 0{\cdot}247 \cos \theta + 0{\cdot}481 \cos \phi$$

The conservation of momentum law or principle is considered to be valid in all circumstances. If the results of an experiment appear to conflict with the conservation of momentum law then we suspect the results, not the law. The emission of beta particles from radioactive substances appeared at one time not to follow the conservation laws (energy as well as momentum). A particle was invented to account for the 'missing' momentum—the neutrino. It was quite some time later before the neutrino was found.

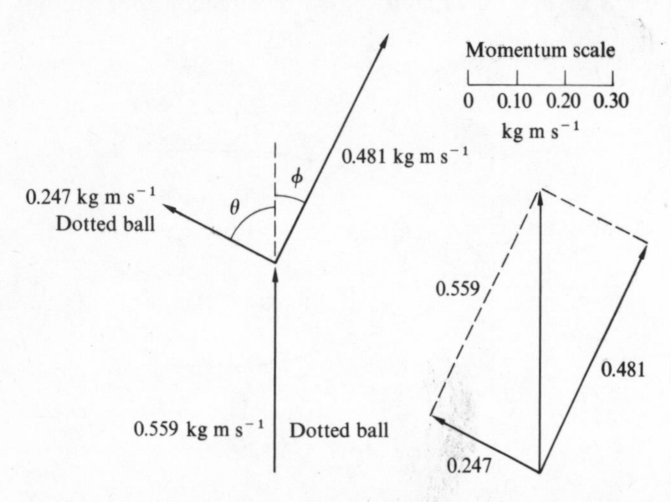

Fig. 2.18 Adapted from PSSC Physics, *2nd Ed., D. C. Heath and Co.*

Centre of mass

A rocket moving through the air suddenly explodes into a shower of fragments. The rocket initially had momentum—what has happened to it after the explosion? If momentum is conserved then the sum, taking into account direction, of the momenta of all the pieces of the rocket must be equal to the initial momentum of the rocket. If the momentum of the rocket before the explosion was mv then the momentum sum for all the pieces after the explosion must be mv. After the explosion we can consider a point in space moving with momentum mv and equivalent to the sum of all the momenta of the pieces. This point is called the centre of mass. The velocity of the centre of mass is the same before the explosion as after, v.

Rocket propulsion

A balloon blown up and then released so that the air can escape moves around a room quite rapidly. Why does it move? When the balloon is blown up we increase the air pressure in the balloon to above the atmospheric pressure. When the balloon is released the difference in air pressure causes air to leave the balloon. It is this air leaving the balloon which causes the balloon to move—in the opposite direction to which the air leaves the balloon. Air moving away from the balloon means mass moving away with a velocity, i.e., momentum in a direction away from the balloon. Conservation of momentum indicates that we must have an opposite and equal momentum for the balloon; the initial momentum was zero.

If we assumed that all the air left the balloon with a constant velocity v (a big assumption)

$$mv = -MV$$

where m is the total mass of the air leaving the balloon, M the mass of the balloon and V its velocity when all the air has left. This has made a simplifying assumption that all the air leaves the balloon with the same velocity; the velocity will in fact decrease as the air leaves the balloon and the pressure in the balloon comes nearer to the atmospheric pressure. In addition we assumed that the mass of the balloon remained constant—the initial mass will be $M + m$ and this will decrease to M.

The balloon has behaved like a rocket. A rocket moves because fuel is burnt inside the rocket and the products of combustion emerge from the rear of the rocket with a high velocity. The term thrust is used for the rate of change of momentum.

Consider a small interval of time Δt during which mass Δm of fuel is ejected from a rocket. Let m be the mass of the rocket plus fuel during this instant. Δm is considered to be so small that the mass of rocket plus fuel can be considered constant during this small time interval. The ejection of this fuel causes an increase in the velocity of the rocket of ΔV. The velocity of the ejected fuel is taken as being at constant velocity v.

By the conservation of momentum law—

$$m\,\Delta V = -(\Delta m)v$$

The change in momentum of the rocket is equal and opposite to the momentum change produced by the movement of the fuel. In many rocket motors v is a constant, during the time the fuel is being consumed.

Rearranging the equation gives—

$$\frac{\Delta m}{m} = -\frac{\Delta V}{v}$$

As m decreases, due to the fuel being used, the change in velocity of the rocket will increase.

How does the velocity of the rocket vary with time? In each successive interval of time Δt a mass Δm is ejected and the rocket's mass decreases by Δm. Let us work out a rough solution for this change in mass to be $\Delta m = M/10$. This assumes that the rate at which the rocket uses fuel is constant, i.e., Δm is the same for each interval of time. This is generally the case. Also we will take a particular value for v of $1{\cdot}5 \times 10^3$ m s^{-1}.

Starting from rest, the mass of rocket plus fuel is M. Hence the change in velocity of the rocket is given by

$$\frac{M}{10}\cdot\frac{1}{M} = \frac{-\Delta V}{1{\cdot}5 \times 10^3}$$

$$\Delta V = -150 \text{ m s}^{-1}.$$

The velocity of the rocket after a time Δt is therefore 150 m s^{-1}. The minus sign is because the rocket's velocity is in the opposite direction to that of the fuel.

In the next interval of time the mass of the rocket plus fuel is

$$M - \frac{M}{10} = \frac{9M}{10}.$$

Hence the change in velocity of the rocket is given by

$$\frac{M}{10}\cdot\frac{10}{9M} = \frac{-\Delta V}{1{\cdot}5 \times 10^3}$$

$$\Delta V = -167 \text{ m s}^{-1}$$

The velocity of the rocket after a time $2\,\Delta t$ is therefore 337 m s^{-1}.

In the next interval of time the mass of the rocket plus fuel is

$$\frac{9M}{10} - \frac{M}{10} = \frac{8M}{10}.$$

Hence the change in velocity of the rocket is given by

$$\frac{M}{10}\cdot\frac{10}{8M} = \frac{-\Delta V}{1{\cdot}5 \times 10^3}$$

$$\Delta V = -185 \text{ m s}^{-1}.$$

The velocity of the rocket after a time $3\,\Delta t$ is therefore 522 m s^{-1}.

In a similar manner we find that:

after $4\,\Delta t$, $\Delta V = -214$ m s^{-1} and the rocket velocity is 736 m s^{-1};
after $5\,\Delta t$, $\Delta V = -250$ m s^{-1} and the rocket velocity is 986 m s^{-1};
after $6\,\Delta t$, $\Delta V = -300$ m s^{-1} and the rocket velocity is 1286 m s^{-1};
after $7\,\Delta t$, $\Delta V = -375$ m s^{-1} and the rocket velocity is 1661 m s^{-1}.

This would be the final velocity of the rocket if the fuel initially constituted 70 per cent of the total mass of the rocket plus fuel. The graph (Fig. 2.19) of velocity against time would be much smoother if smaller time intervals, and hence smaller values of Δm, had been considered.

All rockets, whether they be fireworks or those used to

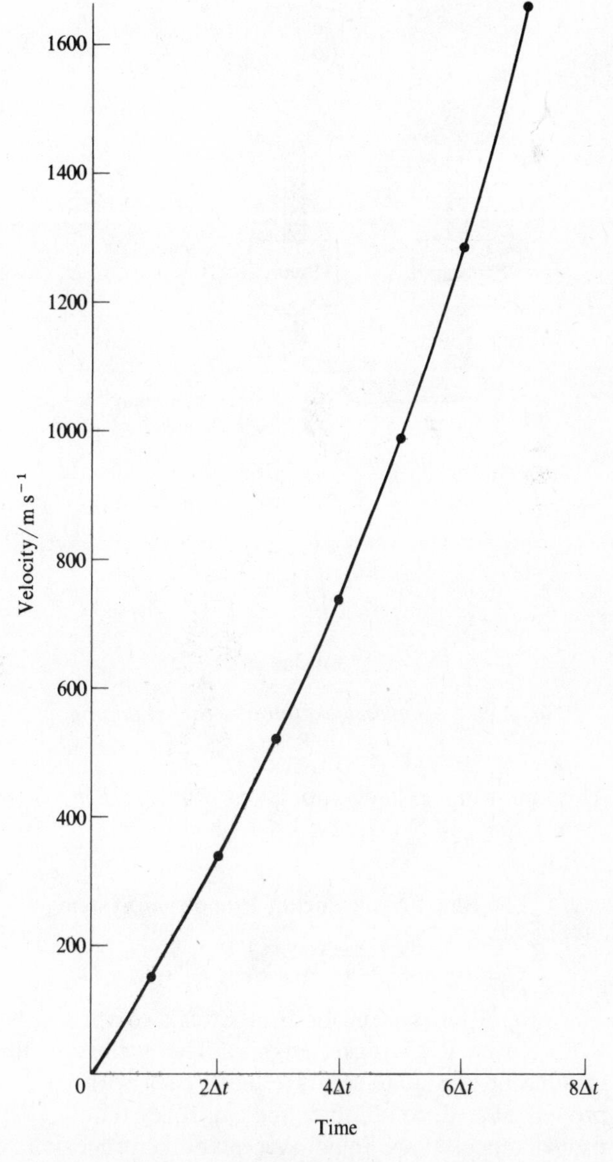

Fig. 2.19

put men on the moon, work on the same principle: the expulsion of mass at a high velocity from the rear of the rocket. In chemical propellant rockets the fuel and the oxidant when mixed together at the right temperature react and the resulting product emerges from the rocket exhaust with a high velocity. A typical fuel would be a hydrocarbon, such as kerosene, and as oxidant liquid oxygen (LOX). Liquid oxygen boils at 89 K, i.e., −184°C, and it is this which is responsible for the ice which forms, and the clouds of white condensed water vapour, before launch. The fuel and oxidant are stored in separate tanks within the rocket and only mix during firing, when they are forced into the combustion chamber by either gas stored under high pressures or by pumps driven by a turbine (Fig. 2.20). Such an arrangement would constitute a rocket engine. A large rocket would have more than one such engine.

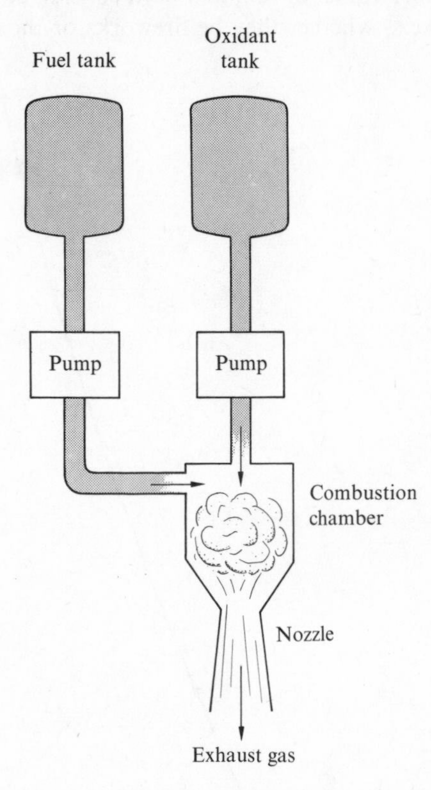

Fig. 2.20 Simplified sketch of a rocket engine.

The following extracts are taken from *J. Brit. Interplanetary Soc.*, **18** No. 7, 259, 1962.

'The Blue Streak Rocket Propulsion System

By A. V. Cleaver, F.R.Ae.S
Chief Engineer (Rocket Propulsion), Rolls-Royce Ltd.

The propulsion system for Blue Streak comprises two Rolls-Royce RZ.2 rocket engines. The engines in this installation are quite separate units, each with its own propellant feed, combustion and control system. ... The propellants used are liquid oxygen and kerosene. ... As will be seen from Fig. 2.21, each engine has a turbo pump set which feeds the propellants to the thrust chamber. ... The propellant pumps are driven by a turbine via a simple spur reduction gear; the turbine is supplied from the separate gas generator which burns a very fuel-rich mixture of the propellants. ... Propellants for starting the engine are supplied from pressurized ground-mounted tanks, and ignition in the gas generator and thrust chamber is by pyrotechnic igniters. ... All valves are opened and closed by pneumatic (nitrogen) servo-pressures. The starting sequence is broadly as follows, ...

(1) The engine start signal causes the ground start tanks and lubricating oil tank to be pressurized.
(2) Firing of the thrust chamber igniter is initiated.

Fig. 2.21 (a) RZ.2 Rocket engine. (Rolls Royce (1971) Ltd, Industrial and Marine Division). (b, page 43) Blue Streak propulsion bay (Rolls Royce (1971) Ltd, Industrial and Marine Division). (c, page 44) RZ.2 starting sequence (Rolls Royce (1971) Ltd, Industrial and Marine Division).

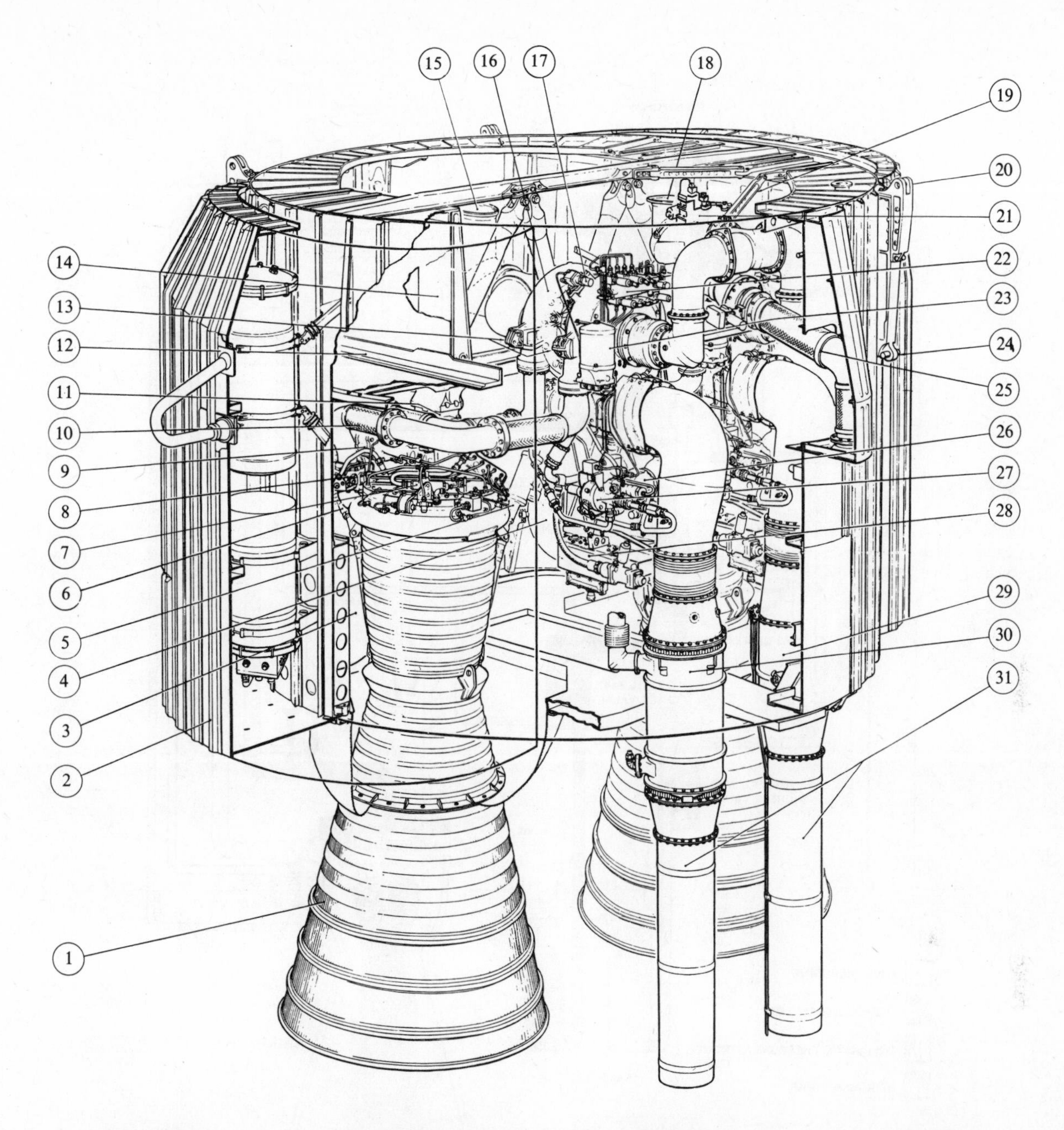

1. Tubular wall thrust chamber
2. Equipment fairing
3. Engine oil tank
4. Liquid nitrogen bottle
5. Pitch control ram
6. Yaw control ram
7. Igniter fuel valve
8. Main fuel valve
9. Main LOX valve
10. Propellant flexible
11. Gimbal mounting
12. Main motor beam
13. Pump mounting
14. Thrust bracket
15. LOX inlet to pumps
16. LOX pump
17. Reference pressure loader
18. Fuel inlet to pumps
19. Turbopump vee frame
20. Attachment to tank bay
21. Fuel tank valve
22. Pneumatic manifold
23. Engine relay boxes
24. Launcher bracket
25. Main fuel probe
26. Gas generator
27. LOX regulator
28. Instrumentation box
29. Heat exchanger (nitrogen)
30. Heat exchanger (LOX)
31. Turbine exhaust

Fig. 2.21(b)

(3) When an electrical link in the thrust chamber igniter burns through, opening of the main liquid oxygen valve... and fuel igniter valve is initiated. Thus liquid oxygen under tank head, and a pilot flow of fuel under start tank pressure enter the thrust chamber and are ignited.

(4) A LOX-rich flame is established which breaks an ignition detector wire stretched across the thrust chamber exit.

(5) Breaking of this wire initiates "main stage" by firing the gas generator igniters. Links in these igniters burn through, and signal the main fuel valves to open.

(6) Fuel valve opening initiates opening of the gas generator blade valve—thus supplying hot gas to the turbine, which accelerates the pumps and begins to feed propellants at high pressure to the thrust chamber and gas generator.

(7) When pump outlet pressures exceed start tank

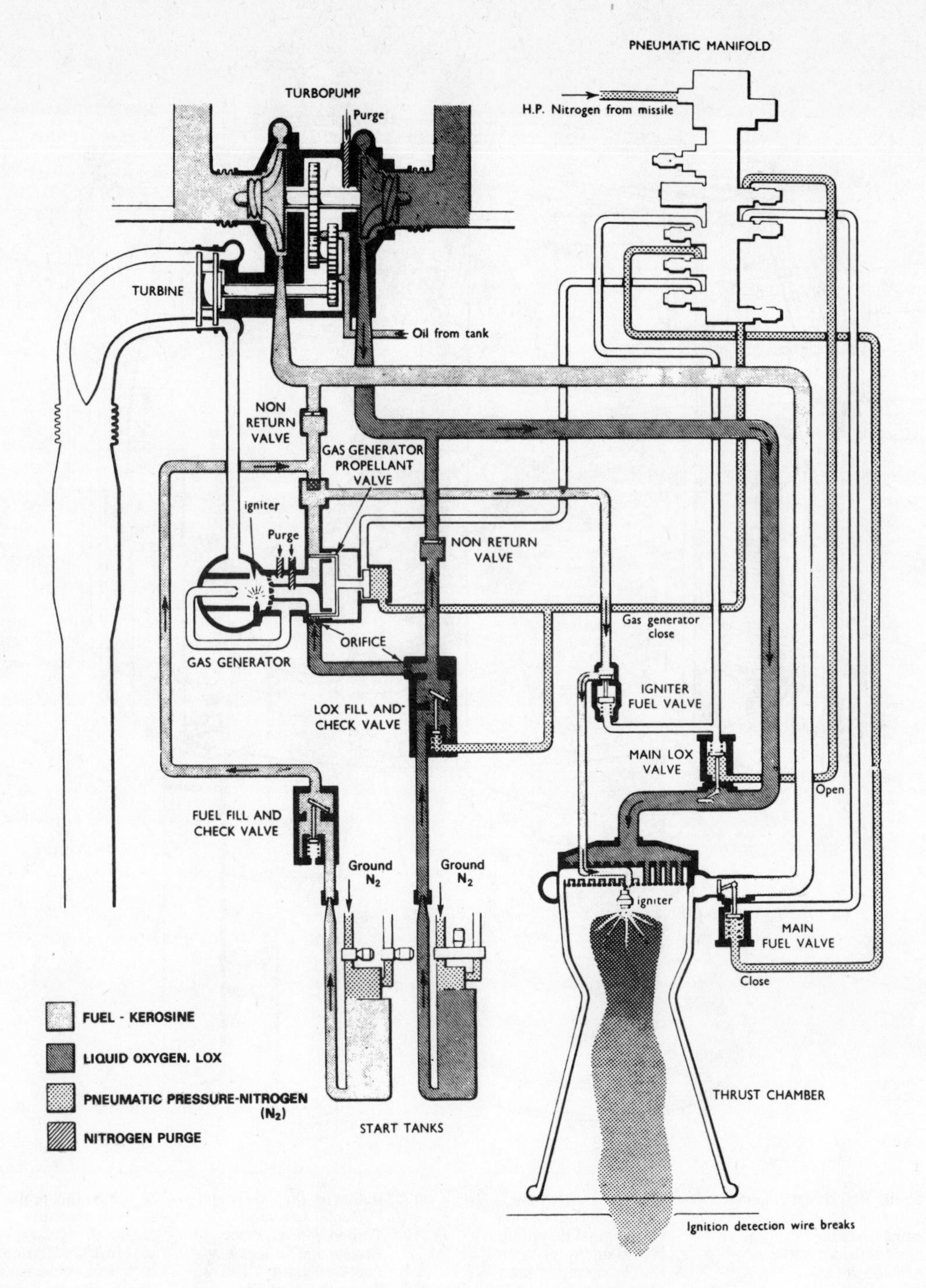

Fig. 2.21(c)

pressures the fill-and-check valves at the exit from the start tanks are vented. The engine is now fully self-sustaining and the start sequence is complete.'

The recoil of rifles

When a rifle is fired the butt jerks back against the shoulder. This recoil is a consequence of the bullet being fired; momentum must be conserved. Two factors contribute to the recoil—the momentum given to the bullet and the momentum given to the gases produced by the explosion. This last item can amount to some 40 per cent of the total forward momentum.

If the momentum of the bullet, after being fired, is mv; the momentum of the gases leaving the barrel after the explosion $m'v'$, then the momentum of the gun is given by

$$MV = mv + m'v'$$

For a typical rifle, $M = 3{\cdot}6$ kg, $V = 3{\cdot}8$ m s^{-1}, $m = 10$ g,

$v = 900\ \mathrm{m\ s^{-1}}$. This gives as the momentum due to the gases leaving the barrel

$$m'v' = 3{\cdot}6 \times 3{\cdot}8 - 0{\cdot}01 \times 900$$
$$= 4{\cdot}7\ \mathrm{kg\ m\ s^{-1}}$$

This is about 34 per cent of the total momentum given to bullet and gas.

If the momentum of the gas can be reduced then the recoil momentum will be reduced. The momentum of the gas is due to the rapid build-up of gas pressure behind the bullet while it is in the barrel. The gas is produced by the explosion and the high temperature produced contributes to the pressure build up. Pressures above atmospheric pressure of the order of $10^5\ \mathrm{N\ m^{-2}}$ are produced. The pressure is needed to get the bullet going but need not give all its momentum to the recoil if it does not all stream from the end of the barrel after the bullet has left. In recoil control devices the gas is diverted out of vents to the side or upwards instead of straight forward.

Collisions with atoms

Suppose there was a small metal object located somewhere in a haystack. One way of locating the object and estimating its size would be to fire bullets at the haystack. Those bullets that did not encounter the object would pass straight through the stack and would not be significantly deviated from their straight line paths. Those bullets that did encounter the object would be deviated from their straight line paths (Fig. 2.22). By raining bullets on the haystack

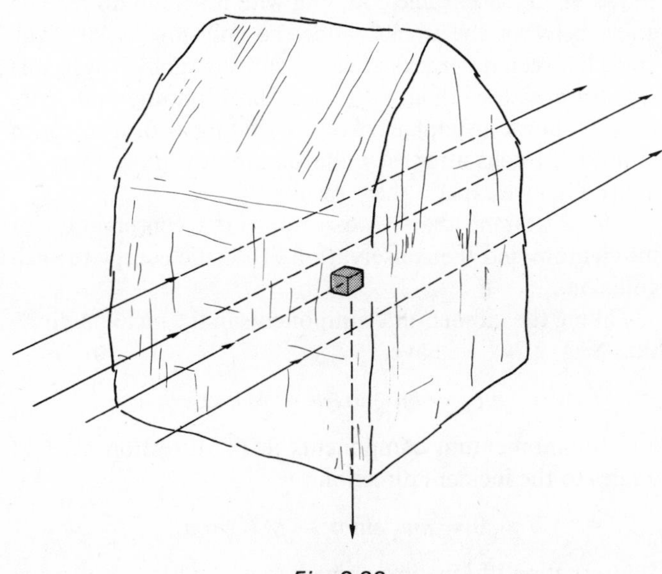

Fig. 2.22

information about the object inside the haystack can be obtained. This is essentially the technique used by scientists to obtain information about atoms. For projectiles they use, instead of bullets, atomic particles such as alpha particles or protons (see chapter 13).

In this section we take a look at some samples of this type of work. A very significant set of experiments were those that established the existence of nuclei within atoms. The projectiles were alpha particles emitted at high speed from a radioactive source, the target was a thin gold foil.

> 'Now I myself was very interested in the next stage, . . . and I would like to use this example to show how you often stumble upon facts by accident. In the early days I had observed the scattering of alpha particles, and Dr. Geiger in my laboratory had examined it in detail. He found, in thin pieces of heavy metal, that the scattering was usually small, of the order of one degree. One day Geiger came to me and said, "Don't you think young Marsden, whom I am training in radioactive methods, ought to begin a small research?" Now I had thought that too, so I said, "Why not let him see if any alpha particles can be scattered through a large angle?" I may tell you in confidence that I did not believe that they would be, since we knew that the alpha particle was a very fast massive particle, with a great deal of energy, and you could show that if the scattering was due to the accumulated effect of a number of small scatterings the chance of alpha particles being scattered backwards was very small. Then I remember two or three days later Geiger coming to me in great excitement and saying, "We have been able to get some of the alpha particles coming backwards . . ." It was the most incredible event that has happened to me in my life. It was almost as incredible as if you fired a 15-inch shell at a piece of tissue paper and it came back and hit you. On consideration I realized that the scattering backwards must be the result of a single collision, and when I made calculations I saw that it was impossible to get anything of that order of magnitude unless you took a system in which the greater part of the mass of the atom was concentrated in a minute nucleus. It was then that I had the idea of an atom with a minute massive centre carrying a charge.'
>
> E. Rutherford (1937). 'The development of the theory of atomic structure', *Background to Modern Science* (1938), Ed. J. Needham, Page 1. C.U.P.

The calculations Rutherford made were based on the use of the conservation of momentum, the conservation of energy, the inverse square law force between electric charges, and geometry. The experimental results agreed with the idea of a nuclear atom.

Cloud chambers enable us to see the paths followed by atomic particles. The particles leave tracks in the vapour used in the chamber, like the vapour trails left by aircraft in the sky. Figure 2.23 shows photographs taken for alpha particles in different gases. The photographs show, in addition to the normal alpha tracks, collisions between alpha particles and gas atoms. Figure 2.23(a) shows an alpha particle colliding with a hydrogen atom. The short dense track after the collision is the alpha particle and the long

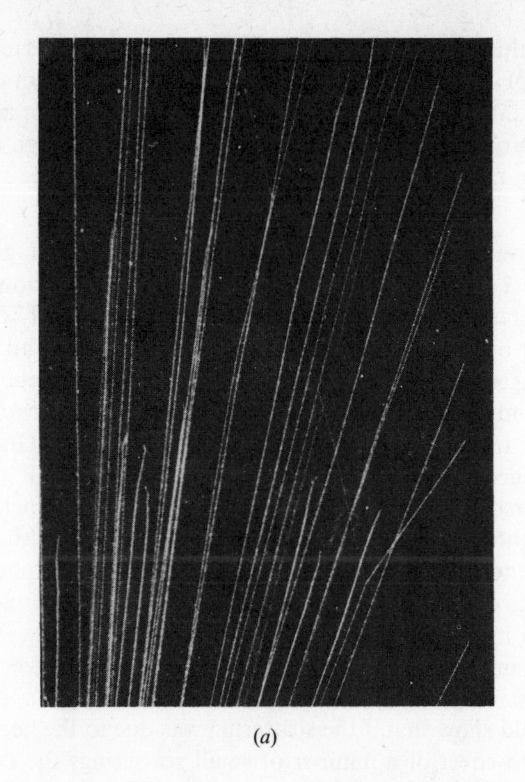
(a)

(c)

(b)

Fig. 2.23 (a) Scattering in hydrogen. (b) Scattering in helium. (c) Scattering in nitrogen. (Photo. Science Museum, London) (Photographs reproduced by permission of Lord Blackett.)

thin track is that due to the hydrogen atom. Figure 2.23(b) shows an alpha particle colliding with a helium atom. The angle between the tracks, after the collision, is 90°. The angle between the tracks after the collision in hydrogen was significantly less than 90°. The third photograph, Fig. 2.23(c) shows an alpha particle colliding with a nitrogen atom, the tracks after the collision are now more than 90° apart. Can we explain these results?

Let us assume that we can apply the conservation of momentum and the conservation of kinetic energy to such collisions.

Taking the momentum components in the incident direction (Fig. 2.24), we have

$$mv_1 = mv_2 \cos\phi + MV\cos\theta$$

and for momentum components in the direction at right angles to the incident direction

$$0 = mv_2 \sin\phi - MV\sin\theta$$

Conservation of kinetic energy gives

$$\tfrac{1}{2}mv_1^2 = \tfrac{1}{2}mv_2^2 + \tfrac{1}{2}MV^2$$

The three velocity terms v_1, v_2, and V can be eliminated to give

$$\frac{m}{M} = \frac{\sin(2\theta + \phi)}{\sin\phi}$$

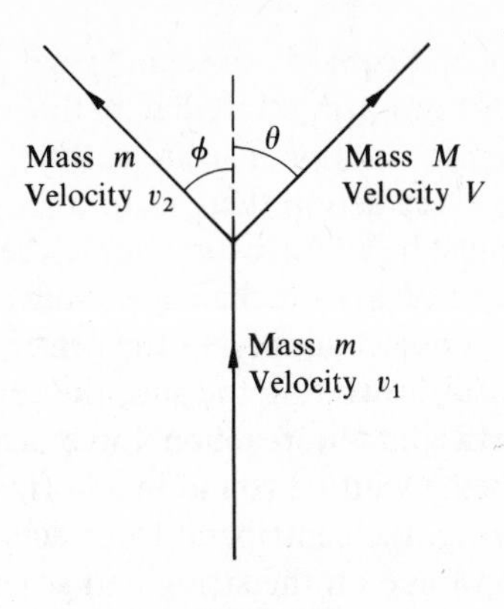

Fig. 2.24

When $m = M$ the equation simplifies to $\phi + \theta = 90°$. The alpha particle collision in helium gave 90° between the tracks—evidence that alpha particles are in fact helium nuclei. In hydrogen we can measure the angles and we get the result that the alpha particle mass is 3·96 times that of a hydrogen atom. In nitrogen the angles give us that the alpha particle mass is about 4/14 that of a nitrogen atom.

In the case of the collision between an alpha particle and a helium atom when the masses of the particles are all the same we can use the triangle or parallelogram of vectors to show that the angle between the tracks must be 90° after the collision. With equal masses the conservation of kinetic energy equation becomes (see chapter 3)

$$v_1^2 = v_2^2 + V^2$$

The only way we can draw the vector diagram for v_1, v_2, and V so that this equation applies is for v_1 to be the hypotenuse of a right-angled triangle (Fig. 2.25).

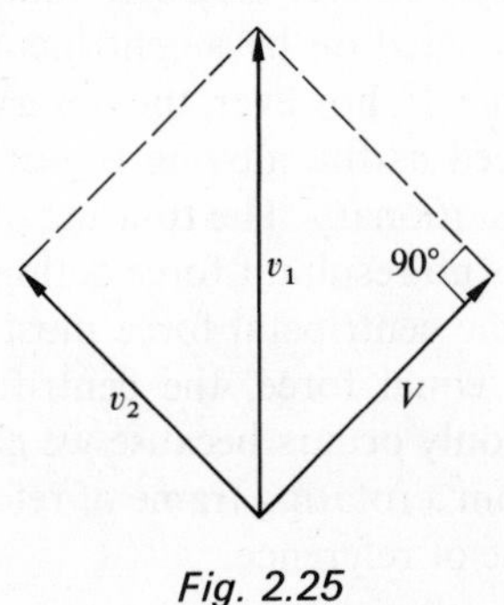

Fig. 2.25

The mass of the neutron was first obtained from applying just these same conservation equations (see chapter 13).

'Particular interest has recently been shown in the disintegration of elements beryllium and boron. It was found by Bothe and Becker that these elements emitted a penetrating radiation, apparently of the gamma type, when bombarded by polonium alpha particles. A few months ago, Mme. Curie-Joliot and M. Joliot made the striking observation that these radiations have the property of ejecting protons with high speeds from matter containing hydrogen. . . . It was found that the radiation ejects particles not only from hydrogen but from helium, lithium, beryllium, etc. . . . In each case the particles appear to be recoil atoms of the elements. . . .

A satisfactory explanation of the experimental results was obtained by supposing that the radiation consists not of quanta but of particles of mass 1 and charge 0, or neutrons.

In the case of two elements, hydrogen and nitrogen, the ranges of the recoil atoms have been measured with fair accuracy, and from these their maximum velocities were deduced. They are $3{\cdot}3 \times 10^9$ cm s^{-1} and $4{\cdot}7 \times 10^8$ cm s^{-1} respectively. Let M, V be the mass and maximum velocity of the particles of which the radiation consists. Then the maximum velocity which can be given to a hydrogen nucleus is

$$u_H = \frac{2M}{M+1} V$$

and to a nitrogen nucleus

$$u_N = \frac{2M}{M+14} V$$

Hence

$$\frac{M+14}{M+1} = \frac{u_H}{u_N} = \frac{3{\cdot}3 \times 10^9}{4{\cdot}7 \times 10^8}$$

and

$$M = 1{\cdot}15$$

Within the error of experiment M may be taken as 1, . . .

Note by J. Chadwick at the end of paper by E. Rutherford, *Proc. Roy. Soc* **A136**, 735, 1932.

The mass equation used by Chadwick can be arrived at by eliminating the velocity of the neutron after the collision in the conservation of momentum and kinetic energy equations. The maximum velocities are realized when the atoms are knocked on in the same direction as the incident neutron, i.e., $\theta = 0°$.

2.4 Motion in a circle

An object continues in uniform motion in a straight line, i.e., constant velocity, or remains at rest if no force is acting on it. Thus when an object moves in a circular path there must be a force acting on it, moving it from the straight line path. This force can be provided by the tension in a string, for an object being whirled round in a horizontal circle on the end of a string, or gravity for the earth orbiting the sun. The force, whatever its origin, is known as a centripetal force. Objects moving in circular paths are not in equilibrium, there must be a resultant force otherwise there would be only straight line motion.

Consider an object of mass m, moving in a circular path of radius R with a speed v. At some instant it is at point P,

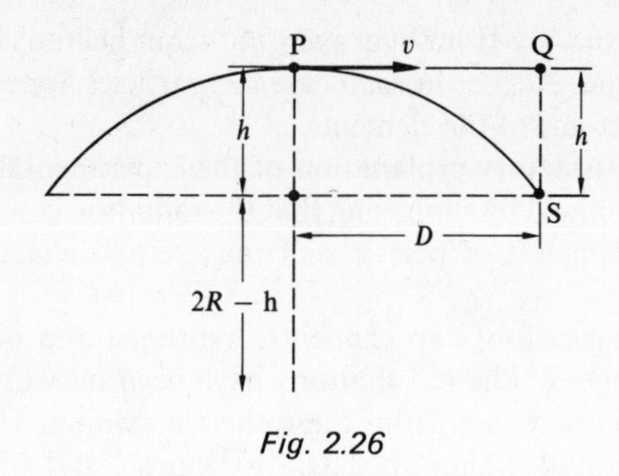

Fig. 2.26

Fig. 2.26. In the absence of a centripetal force it would move to point Q in, say, time t. However, it moves to S. The centripetal force has caused the mass to fall through the distance QS or h.

$$h(2R - h) = D^2$$

(Intersecting chords theorem see Appendix 2B)

$$2Rh - h^2 = D^2$$

If we only consider a small interval of time, i.e., t is small, then $2Rh \gg h^2$ and we can write the expression as

$$2Rh = D^2$$

But D is the horizontal distance travelled in time t. As the horizontal speed is v at P and there is no force acting in this direction then the distance D must be covered at a speed v.

$$v = \frac{D}{t}$$

Hence

$$h = \frac{v^2t^2}{2R}$$

At P there is no velocity in the vertical direction, i.e., QS direction. h is, however, the distance fallen in time t. Using $s = \frac{1}{2}at^2$

$$\frac{v^2t^2}{2R} = \frac{at^2}{2}$$

Hence the object accelerates with an acceleration given by

$$a = \frac{v^2}{R}$$

The direction of the acceleration is towards the centre of the circle.

The centripetal force produces this acceleration and hence this force is

$$F = \frac{mv^2}{R}$$

If the centripetal force is made zero the object will fly off along a tangent to the circle.

The term centrifugal force is often used incorrectly, to denote a force acting outwards, along a radius, to balance the centripetal force. The forces on a rotating object are not balanced—if there was a centrifugal force, acting on the object, opposite and equal to the centripetal force, then the object would not move in a circular path. An unbalanced force is necessary for motion in a circle. But it can be argued that because to every action there is an opposite and equal reaction there must be another force somewhere. The action and reaction do not act on the same body. The reaction force acts on the object supplying the centripetal force. In the case of the earth orbiting the sun, the centripetal force acts on the earth and the reaction force acts on the sun. For an object being whirled round in a horizontal circle on the end of a string, the centripetal force acts on the object and the reaction force on the string and so on the hand of the person holding the string (Fig. 2.27).

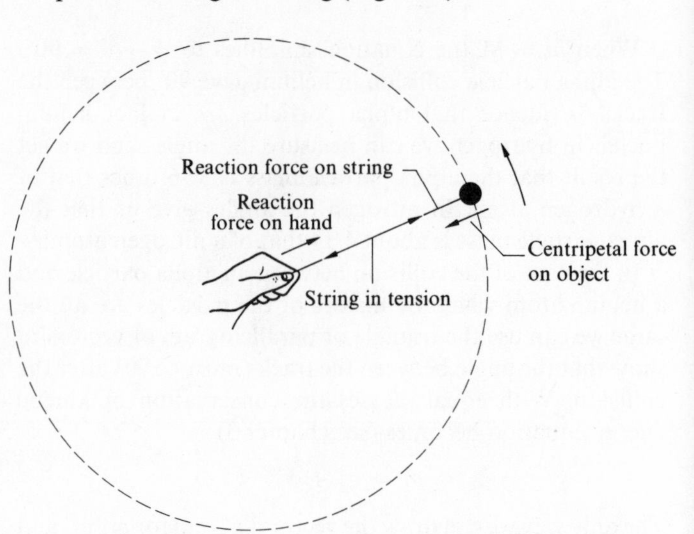

Fig. 2.27 Rotating object viewed by a stationary observer.

For a stationary observer an object moves in a circular path because it is acted on by a centripetal force; there is no centrifugal force. If, however, the observer rotates at the same angular speed as the moving object then the object will appear to be stationary. The rotating observer can then argue that there is no resultant force acting on the rotating object and thus the centripetal force must be balanced by an opposite and equal force, the centrifugal force. This centrifugal force only occurs because we chose to view the moving object from a rotating frame of reference instead of a stationary frame of reference.

Angular motion and forces

A particle moving in a circular path or a point on a spinning rigid body can have its motion described in terms of the angle swept out in unit time. We defined (chapter 1) linear velocity as the distance covered, in a straight line, in unit time—in a similar manner we can define angular velocity as the angle of rotation in unit time.

$$\text{Linear velocity} = \frac{\text{distance change}}{\text{time taken}}, \qquad v = \frac{\Delta s}{\Delta t}$$

$$\text{Angular velocity} = \frac{\text{angle rotated}}{\text{time taken}}, \qquad \omega = \frac{\Delta\theta}{\Delta t}$$

The angular velocity tells us how far a point has gone round in a unit time. The angle is usually measured in radians and thus the angular velocity could have units of radians per second. 2π radians are equal to 360°.

If a point moves along the arc of a circle, Fig. 2.28, then the angle of rotation, in radians, is equal to the distance moved along the arc divided by the radius.

$$\theta = \frac{\text{distance moved along arc}}{\text{radius}}$$

Thus for one complete rotation the distance covered is the circumference, $2\pi r$, and thus

$$\theta = \frac{2\pi r}{r} = 2\pi.$$

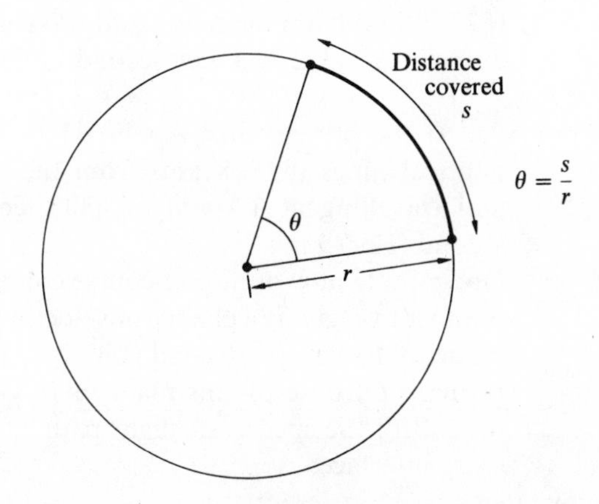

Fig. 2.28

One complete rotation is a movement through 360° and thus 2π radians are equal to 360°.

How is the angular velocity related to speed? If a point moves a distance s along the arc of a circle in time t then the speed, if assumed uniform, is

$$v = \frac{s}{t}$$

But in moving this distance an angle θ has been swept out.

$$\theta = \frac{s}{r}$$

Thus we have

$$v = \frac{\theta r}{t}$$

θ is the angle swept out in time t, thus

$$\omega = \frac{\theta}{t}$$

and therefore

$$v = r\omega.$$

Angular acceleration is the rate of change of angular velocity with time.

A force is needed to produce a linear acceleration—what is needed to produce an angular acceleration? We say that a torque—a rotary or twisting force—is needed to produce an angular acceleration. What is torque and how is it related to force? If you want to push open a door, i.e., cause rotation, it is easiest if the push is applied as far from the hinges as possible. Door handles are usually situated as far from the hinges as possible because least effort is then needed. (Try opening a door by pushing near the hinges.) Thus to produce angular acceleration of a door requires a force but also the result depends on the distance from the centre of rotation at which the force is applied. Torque thus depends on the force and the distance from the pivot at which the force is applied.

We can do an experiment with the door, or better still a turntable, by applying a force at different distances from the axis of rotation and determining the change in angular velocity produced. Known forces can be applied tangentially to the turntable or door by using a calibrated spring balance. The results show that the rate of change of angular velocity with time is, for a constant force, directly proportional to the distance from the hinge axis at which the force is applied. If this distance is kept constant then the rate of change of angular velocity is directly proportional to the force. Thus:

$$\text{Angular acceleration} \propto \text{distance from axis}$$

and

$$\text{Angular acceleration} \propto \text{force}$$

The product of the tangential force F and the distance x of the point of application of the force from the axis of rotation is called the torque.

$$\text{Torque } \tau = Fx$$

Thus:

$$\text{Angular acceleration} \propto \tau$$

With linear motion we have $F = ma$, the constant of proportionality relating force and acceleration is mass. What is the constant of proportionality relating torque and angular acceleration? Suppose we try different masses on the turntable—how does this affect the angular acceleration produced by a given torque? Changing the mass certainly has an effect, but changing the position of a mass on the turntable also changes the angular acceleration. The angular acceleration, for a given torque, depends on both the mass and its position. When we were discussing mass and weight, page 36, we used the words 'Mass is a property of a body which represents its inertia'—we should qualify that to refer to linear acceleration. Mass tells us how easy it is to accelerate (linearly) an object. For rotation we need a different quantity, involving mass and distance, to tell us how easy it is to produce an angular acceleration for an object. This quantity we call the moment of inertia, symbol I.

$$\text{Torque} = (\text{moment of inertia}) \times (\text{angular acceleration})$$

If you sit in one of those swivelling typist's chairs with your arms outstretched and large masses in your hands, whirl round and then pull your hands, and the masses, close in to your body a remarkable thing happens—your angular velocity increases (you go round faster). A skater slowly spinning with arms outstretched spins much faster when he pulls his arms in. In both these cases the moment of inertia has been changed by the movement of the arms. This is one of the important differences between mass and moment of inertia—the mass of a body cannot be changed, the moment of inertia can be. Why does changing the moment of inertia change the speed of spinning? For this we will introduce a term called angular momentum.

$$\text{Torque } \tau = Fx$$

But

$$F = \text{rate of change of linear momentum}$$

therefore

$$\tau = x\frac{\mathrm{d}(mv)}{\mathrm{d}t}$$

We can make torque the rate of change of some quantity, as force is the rate of change of momentum, by calling xmv the angular momentum.

$$\tau = \text{rate of change of angular momentum}$$

Angular momentum of a particle is the magnitude of the linear momentum times the perpendicular distance to the pivot axis. Thus in the case of somebody sitting in a typist's chair with outstretched hands holding masses—the angular momentum of each mass is the product of the mass, its velocity, and the distance of the mass to the pivot axis. Moving the masses closer to the pivot axis decreases the angular momentum of each mass if the velocity is unchanged. But the velocity increases and if measurements are made we find that the angular momentum remains unchanged. If the distance of the masses from the pivot axis is halved the velocity is doubled. Angular momentum is conserved, if no external torques act on the system considered. Putting the torque equal to zero in our equation means that the rate of change of angular momentum must be zero, i.e., the angular momentum must remain constant.

2.5 Types of forces

Rather surprisingly when one considers all the apparently different types of forces, e.g., gravitational, electric, atomic, nuclear, frictional, magnetic, molecular, etc., there are only four different types. These are gravitational, electric, strong nuclear, and weak nuclear. All the other apparently different types of forces can be explained in terms of these four basic forms. Thus magnetic forces can be completely described in terms of electric forces when charged particles are moving. The term magnetic force is not theoretically necessary. Atomic forces can be explained by the electric and nuclear forces.

Gravitational forces

The following are extracts from the announcements made by the Apollo 11 flight controllers during the flight of that spacecraft back from the first manned landing on the moon (17–19 July 1969). All times are in British Summer Time.

'5.57 am (Apollo 11 leaves moon orbit.)

12.36 pm This is Apollo control. Apollo 11 is 18 243 nautical miles (37 750 km) from the moon. Velocity is 4426 feet per second (1327·8 m s^{-1})....

3.32 pm Apollo 11 is 25 857 nautical miles (47 835 km) from the moon heading towards home at 4338 feet per second (1301·9 m s^{-1}).

4.32 pm At this time Apollo 11 is 28 421 nautical miles (52 579 km) from the moon and travelling at a speed of 4322 feet per second (1296·6 m s^{-1})....

6.02 pm ... At the present time Apollo 11 is 32 253 nautical miles (59 668 km) from the moon and travelling at a speed of 4303 feet per second (1290·9 m s^{-1}).

8.36 pm One minute now until mid-course correction number five, giving a change of velocity retrograde of 4·8 feet per second (1·4 m s^{-1}). The primary purpose of this manoeuvre will be to control the spacecraft flight path angle at entry interface.

One day later

7.50 pm ... at the present time Apollo 11 is 106 482 nautical miles (197 002 km) from the earth. The velocity is 5607 feet per second (1682·4 m s^{-1}).

One day later

5.36 pm At blackout we were showing velocity 36 237 feet per second (10 871·1 m s^{-1})....

5.50 pm (Splashdown) Our condition is excellent....'

P. Ryan (1969). *The invasion of the moon 1957–70*, Penguin.

Apart from the one rocket burn for the mid-course correction, only a small velocity change, the spacecraft was freely travelling from the moon to earth. Yet the velocity first decreases as the craft moves away from the moon and then builds up a very high value on re-entry to the earth's atmosphere (Fig. 2.29). What force produces these velocity changes?

The only forces acting on the spacecraft are the gravitational attractions between the craft and both the earth and the moon. If you, standing on the earth's surface, release a stone it falls to the earth—there is a force of attraction on it due to the presence of the earth. The spacecraft slows down on leaving the moon because the force acting on it due to the moon is greater than that due to the earth. After a distance of about 60 000 km from the moon the force on the craft due to the earth must be about equal to the force due

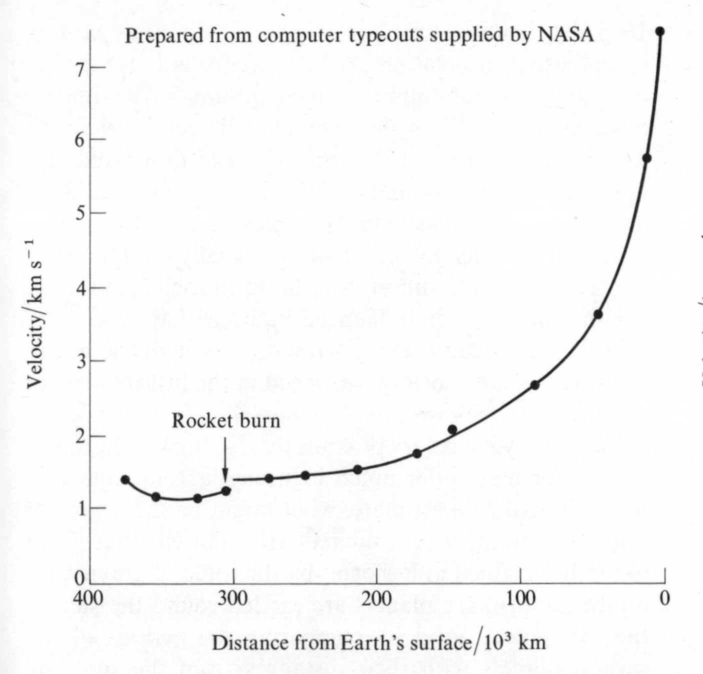

Fig. 2.29

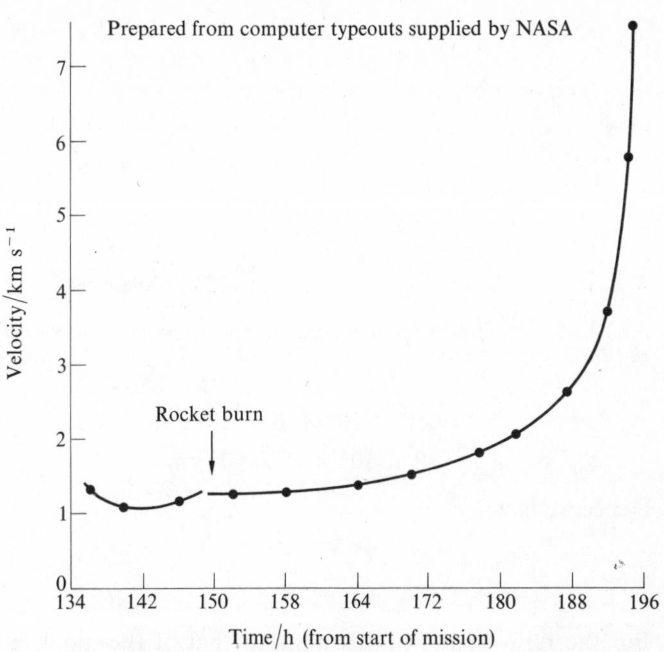

Fig. 2.30

to the moon because the velocity is almost constant. From that point on the craft accelerates towards the earth due to the force of attraction between it and the earth becoming greater than that between it and the moon. This explanation of the velocity–distance graph seems feasible—we have, however, to reckon with a gravitational force of attraction which varies with the separation between two bodies.

We can use our data to calculate the resultant force acting on the craft at different distances from the earth, or moon.

$$\text{Force} = ma$$

Thus

$$\text{Force} = m \text{ (slope of velocity–time graph)}$$

m is the mass of the spacecraft.

Taking values of the slope from the velocity–time graph (Fig. 2.30) results in Fig. 2.31 showing how the acceleration, and thus the force, varies with distance from the earth. An examination of this graph shows that, roughly, the acceleration decreases by a factor of four when the distance decreases by a factor of two for distance values out to about 200×10^3 km. This means that the force acting on the spacecraft varies as the reciprocal of the distance squared. Closer examination of the data shows that this is the case when the distance is measured from the centre of the earth and not the surface (see chapter 11).

$$\text{Force} \propto \frac{1}{r^2}$$

The moon also exerts a force on the craft and thus the net force is the difference between the force due to the earth and that due to the moon. This force due to the moon only becomes apparent at distances greater than about 200×10^3 km. At about 350×10^3 km the two forces are approxi-

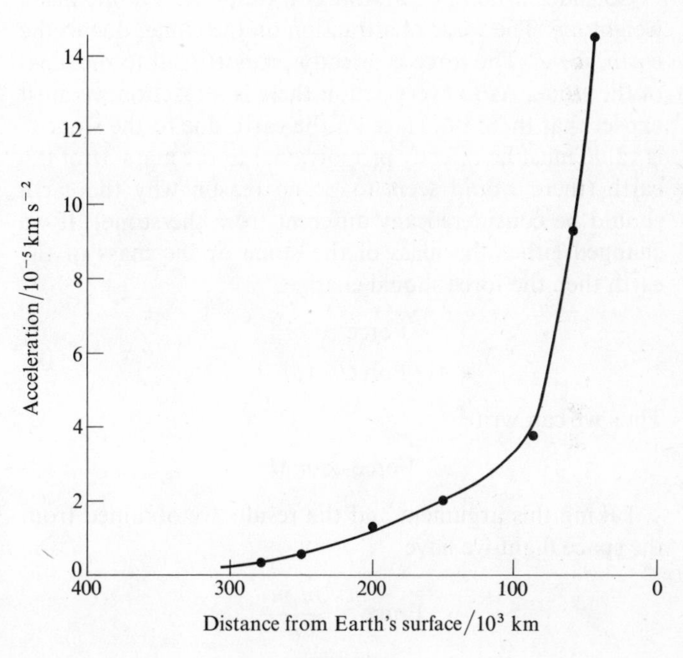

Fig. 2.31

mately opposite and equal and there is virtually no acceleration. Beyond that distance the force due to the moon is greater than that due to the earth (Fig. 2.32). At the equal force point we must have

$$\frac{C_e}{r_e^2} = \frac{C_m}{r_m^2}$$

where C_e and C_m are the constants of proportionality for the forces due to the earth and the moon, r_e and r_m are the distances to the equal forces point from the centres of the earth and the moon.

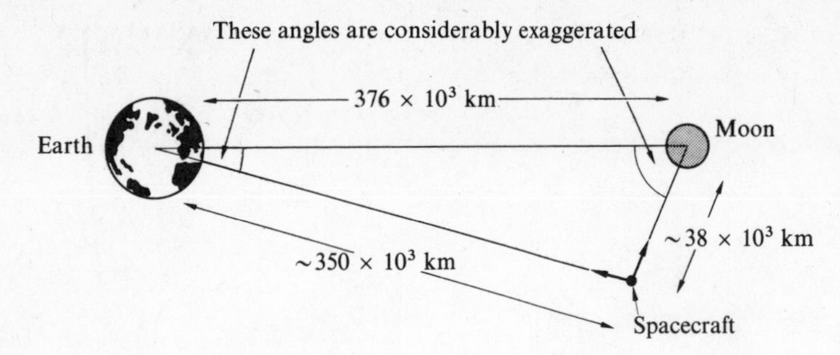

Fig. 2.32 Position of a spacecraft when there is virtually no net force on the craft.

$$r_e = 350 \times 10^3 + 6 \times 10^3 \text{ km}$$
$$r_m = 38 \times 10^3 + 2 \times 10^3 \text{ km}$$

Hence we have

$$\frac{C_e}{C_m} = 79$$

But the ratio of the earth's mass to that of the moon is about 81—a coincidence?

Consider a more down-to-earth example, a stone has a weight *mg*. The force of attraction on the stone, due to the earth, is *mg*. The force is directly proportional to the mass of the stone. As to every action there is a reaction we must expect that there is a force on the earth due to the stone—is value must be directly proportional to the mass M of the earth (there would seem to be no reason why the earth should be considered any different from the stone). If we changed either the mass of the stone or the mass of the earth then the force should change.

$$\text{Force} \propto m$$
$$\text{Force} \propto M$$

Thus we can write

$$\text{Force} \propto mM$$

Taking this argument and the results we obtained from the space flight we have

$$\text{Force} \propto \frac{m_1 m_2}{r^2}$$

where m_1 and m_2 are the two masses, a distance r apart, which produce a force on each other. We can write the equation as

$$\text{Force} = G\frac{m_1 m_2}{r^2}$$

where G is a constant, called the Universal constant of gravitation. Its value is $6{\cdot}67 \times 10^{-11}$ N m² kg⁻².

This relation was first arrived at by Newton. The following extract from *The background to Newton's Principia*, by J. Herival, gives the background to this deduction.

'*Pemberton:* The first thoughts, which gave rise to his Principia, he had, when he retired from Cambridge in 1666 on account of the plague. As he sat alone in a garden, he fell into a speculation on the power of gravity: that as this power is not found sensibly diminished at the remotest distance from the center of the earth, to which we can rise, neither at the tops of the loftiest buildings, nor even on the summits of the highest mountains; it appeared to him reasonable to conclude, that this power must extend much farther than was usually thought; why not as high as the moon, said he to himself? and if so, her motion must be influenced by it; perhaps she is retained in her orbit thereby. However, though the power of gravity is not sensibly weakened in the little change of distance, at which we can place ourselves from the center of the earth, yet it is very possible that, so high as the moon this power may differ much in strength from what it is here. To make an estimate, what might be the degree of this diminution, he considered with himself, that if the moon be retained in her orbit by the force of gravity, no doubt the primary planets are carried round the sun by the like power. And by comparing the periods of the several planets with their distances from the sun,[1] he found, that if any power like gravity held them in their courses, its strength must decrease in the duplicate proportion of the increase of distance. This he concluded by supposing them to move in perfect circles concentrical to the sun, from which the orbits of the greatest part of them do not much differ. Supposing therefore the power of gravity, when extended to the moon, to decrease in the same manner, he computed whether that force would be sufficient to keep the moon in her orbit. In this computation, being absent from books, he took the common estimate in use among geographers and our seamen, before Norwood had measured the earth, that 60 English miles were contained in one degree of latitude on the surface of the earth. But as this is a very faulty supposition, each degree containing about 69½ of our miles,[2] his computation did not answer expectation; whence he concluded, that some other cause must at least join with the action of the power of gravity on the moon. On this account he laid aside for that time any farther thoughts upon this matter.

"After dinner on 15th April 1726 the weather being warm we went into the garden and drank tea under the shade of some apple trees, only he and myself. Amidst other discourse, he told me, he was just in the same situation, as when formerly, the notion of gravitation came into his mind. It was occasioned by the fall of an apple, as he sat in a contemplative mood."

[1] By means of Kepler's Third Law of Planetary Motion.
[2] Cajori points out that this implies 1 mile = 5280 ft whereas it seems more probable that Newton would have taken 1 mile = 5000 ft.'

The sequence of Newton's argument can be summarized as follows:

Kepler had already established, experimentally, that the time, T, taken for one complete orbit of the sun is related to the radius, R, of the orbit by

$$T^2 \propto R^3$$

The force, F, necessary to cause a planet to move in a circular orbit about the sun is given by

$$F = m\frac{v^2}{R}$$

where m is the mass of the planet, v its speed in orbit.

But

$$v = \frac{2\pi R}{T}$$

Thus

$$F = \frac{4\pi^2 mR}{T^2}$$

To satisfy the relationship found by Kepler we must have

$$F \propto \frac{1}{R^2}$$

In this way the inverse square law for gravitation can be arrived at. The arguments for the mass terms proceed in the way the arguments for the stone were conducted.

No distinction has been made in this text between inertial and gravitational mass. Inertial mass is the mass defined in terms of the acceleration produced by a force, perhaps that produced by a stretched spring. Gravitational mass is the mass defined by the force of attraction between objects. Both can be defined independently. But it turns out that the gravitational mass is always proportional to the inertial mass. This equivalence is the basis of Einstein's general theory of gravitation.

There have been many experiments to determine G, the universal constant of gravitation. Some of the earlier experiments were concerned with the measurement of the deviation of a pendulum bob line from the vertical due to the presence of a large mass such as a mountain (Fig. 2.33). The force on the bob due to the mountain is

$$F' = \frac{GM'm}{r^2}$$

where M' is the mass of the mountain and r is the distance of its centre of mass from the pendulum bob. m is the mass of the bob. The force on the bob due to the earth is mg.

$$F = mg$$

g is the acceleration due to gravity.

$$\frac{F'}{F} = \frac{GM'}{r^2 g}$$

But

$$\frac{F'}{F} = \tan\theta$$

Hence

$$G = \frac{r^2 g \tan\theta}{M'}$$

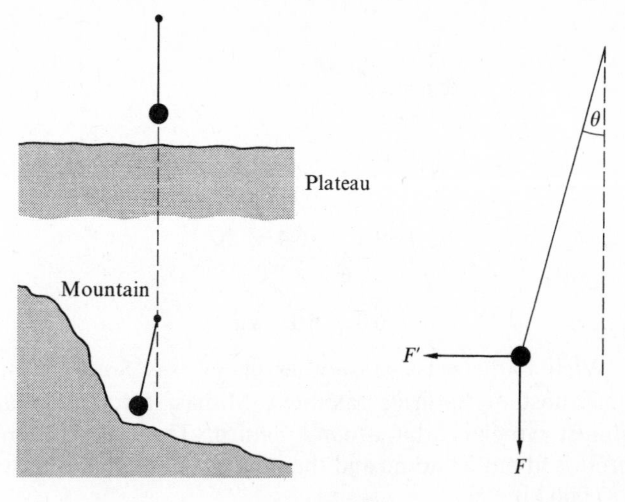

Fig. 2.33

A difficulty of this experiment is the determination of the mass of the mountain and the distance of its centre of mass from the bob.

Later experiments generally involved masses in a laboratory. In essence these consisted of a light rod carrying a small metal sphere at each end, the rod being suspended by a fine fibre. When two other spheres are brought close to the suspended ones (Fig. 2.34) the gravitational forces cause the rod to rotate slightly. The forces necessary to twist the fibre and bring the rod back to its original position can be measured and thus the force of attraction between the spheres determined. The force is, however, very small and this thus means that errors become difficult to eliminate.

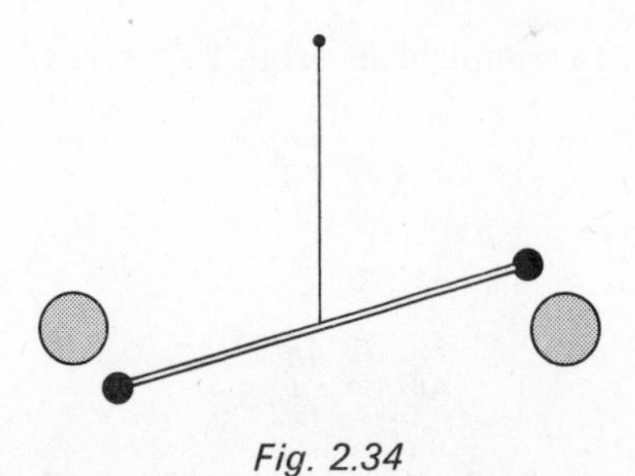

Fig. 2.34

What is the mass of the earth? We can estimate this by the use of our gravitation equation.

Consider a mass m, perhaps a stone, near the earth's surface. The force on the mass will be

$$F = mg$$

where g is the acceleration due to gravity. But the force must also be

$$F = G\frac{mM}{R^2}$$

where M is the mass of the earth, R the radius of the earth.

Thus

$$mg = G\frac{mM}{R^2}$$

$$M = \frac{gR^2}{G}$$

$$= \frac{9{\cdot}8 \times (6{\cdot}4 \times 10^6)^2}{6{\cdot}7 \times 10^{-11}}$$

$$= 6{\cdot}0 \times 10^{24} \text{ kg}$$

What is the mass of another planet, say Saturn? The innermost of Saturn's satellites, Mimas, revolves in an almost circular orbit around Saturn. The time for one orbit is about 23 hours and the radius of the orbit is about 187 000 km.

The force acting on the satellite to keep it moving in a circular path must be

$$F = \frac{mv^2}{r}$$

where r is the radius of the orbit, m the mass of the satellite, and v its speed in orbit. But

$$F = G\frac{mM}{r^2}$$

where M is the mass of Saturn. Thus

$$\frac{mv^2}{r} = G\frac{mM}{r^2}$$

$$M = \frac{v^2 r}{G}$$

But the time to complete an orbit, T, is related to v.

$$v = \frac{2\pi r}{T}$$

Hence

$$M = \frac{4\pi^2 r^3}{GT^2}$$

Using our data gives for the mass M of Saturn $5{\cdot}6 \times 10^{26}$ kg.

This method can be used to determine the mass of any planet, or star, around which another mass rotates. We could have used the rotation of the moon around the earth to give us the mass of the earth.

Our galaxy is spinning, that is the stars in that galaxy are rotating around some central point. Our sun is about 30 000 light-years from the centre and rotating at about one orbit every 2×10^8 years. If M is the mass of the galaxy contained within our sun's orbit, then

$$M = \frac{4\pi^2(3 \times 10^4 \times 9{\cdot}5 \times 10^{15})^3}{6{\cdot}7 \times 10^{-11}(2 \times 10^8 \times 3{\cdot}1 \times 10^7)^2}$$

1 light-year $= 9{\cdot}5 \times 10^{15}$ m
1 year $= 3{\cdot}1 \times 10^7$ s

This gives M, the mass of the galaxy contained within the sun's orbit, as about 10^{41} kg.

What is the origin of gravitational forces? Is there any mechanism we can suggest? At present the answer is—we do not know.

'So far I have accounted for the phenomena presented to us by the heavens and the sea by means of the force of gravity.... But hitherto I have not been able to discover the cause of those properties from phenomena, and I frame no hypotheses.... To us it is enough that gravity does really exist, and act according to the laws which we have explained, and abundantly serves to account for all the motions of the celestial bodies and of our sea.'

Newton, *Principia*.

'Many mechanisms for gravitation have been suggested. It is interesting to consider one of these, which many people have thought of from time to time. At first, one is quite excited and happy when he "discovers" it, but he soon finds that it is not correct. It was first discovered about 1750. Suppose there were many particles moving in space at a very high speed in all directions and being only slightly absorbed in going through matter. When they are absorbed, they give an impulse to the earth. However, since there are as many going one way as another, the impulses all balance. But when the sun is nearby, the particles coming toward the earth through the sun are partially absorbed, so fewer of them are coming from the sun than are coming from the other side. Therefore, the earth feels a net impulse towards the sun and it does not take one long to see that it is inversely as the square of the distance—because of the variation of the solid angle that the sun subtends as we vary the distance. What is wrong with the machinery? It involves some new consequences which are not true. This particular idea has the following trouble: the earth, in moving around the sun, would impinge on more particles which are coming from its forward side than from its hind side (when you run in the rain, the rain in your face is stronger than that on the back of your head!). Therefore there would be more impulse given the earth from the front, and the earth would feel a resistance to motion and would be slowing up in its orbit. One can calculate how long it would take for the earth to stop as a result of this resistance, and it would not take long enough for the earth to still be in its orbit, so this mechanism does not work. No machinery has ever been invented that "explains" gravity without also predicting some other phenomenon that does not exist.'

R. B. Feynman, R. P. Leighton, M. Sands.
The Feynman lectures on Physics, vol. 1., (1963). Addison Wesley, Reading, Mass.

Further reading

G. Gamow. *Gravity*, Heinemann Science Study Series, No. 17, 1962.

The discovery of Neptune

The earth, or indeed any planet, in orbit round the sun is acted on by forces due not only to the gravitational force of the sun but those of all the other planets. These other forces cause a planet to deviate from the elliptical path it would follow if it alone was orbiting the sun. These effects are called perturbations.

When Copernicus produced his model of the universe only six planets were known: Mercury, Venus, Earth, Mars, Jupiter, and Saturn. The invention of the telescope (see chapter 6) led to the discovery of Jupiter's moons by Galileo in 1610 and in the same year Saturn's rings. In 1781 telescope observations revealed the existence of a new planet—Uranus. It was shown to be a planet because the size of its image in the telescope changed when different magnifying power telescopes were used; the image size does not change when stars are observed (they are so far away that they behave as points). The planet was discovered by W. Herschel. His observations, coupled with those on old star maps where it was recorded as a star, enabled its orbit to be plotted.

The orbit, however, did not precisely fit the ellipse shape, even when perturbations due to the presence of Jupiter and Saturn were allowed for. Figure 2.35(a) shows the unexplained deviations of Uranus's orbit from that expected. The solution to the problem was proposed, independently, by J. C. Adams and U. J. Le Verrier—there was another planet farther out than Uranus and this produced the unexplained deviation. They both theoretically worked out the position that the new planet should occupy—it was found close to the predicted position and called Neptune.

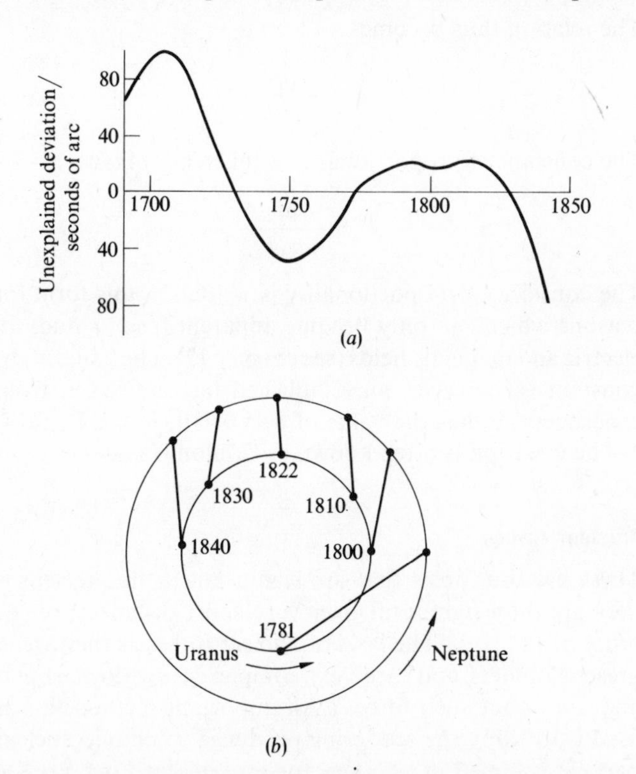

Fig. 2.35

Figure 2.35(b) shows how its position changes relative to that of Uranus—compare this figure with the maximum and minimum deviations of Uranus in Fig. 2.35(a). In the period immediately prior to 1822 Neptune's pull speeded up Uranus, after that its pull slowed Uranus down.

Irregularities in the orbit of Neptune led in 1930 to the discovery of the planet Pluto by C. Tombaugh. Pluto is much smaller than the other outer planets and has an orbit which is not in the same plane as all the other planets. It is thought that it might be an escaped moon of Neptune.

The planet Mercury also shows unexplained perturbations. Le Verrier estimated from these deviations the position of a new planet, which he called Vulcan, between Mercury and the sun. No such planet has been found. An explanation of the deviation was supplied by Einstein replacing Newton's gravitational theory by a new theory—the general theory of relativity.

Electrical forces

If you rub a plastic comb (or pen or other similar object) against a wool material it becomes charged. This shows by the comb attracting small pieces of paper or other light objects. We call this attractive force an electrical force. Such forces have been known for many centuries. It was not, however, until about 1730 that it was realized that electrical forces were not always attractive forces but could be repulsive.

> '... there are two distinct electricities, very different from each other: one of these I call vitreous electricity; the other, resinous electricity. The first is that of (rubbed) glass, rock crystal, precious stones, hair of animals, wool, and many other bodies. The second is that of (rubbed) amber, copal, gum lac, silk, thread, paper, and a vast number of other substances.
>
> The characteristic of these electricities is that a body of, say, the vitreous electricity repels all such as are of the same electricity; and on the contrary, attracts all those of the resinous electricity. ...'
>
> C. F. de C. Dufay (1733–4). *Phil. Trans*, **38**, 258.

The rubbed plastic comb has to be put fairly close to objects to pick them up; the repulsion, or attraction, between two charged objects is much more noticeable when they are close together—the conclusion from these observations is that the electric force depends on the distance away from the charged object. What is the relationship linking force and distance?

Though the answer to this question had, by analogy with gravitation, been earlier guessed by J. Priestley in 1767, it was Coulomb in 1785 who first experimentally obtained the answer.

> '... It follows from these trials that the repulsive force which two balls exert on each other when they are electrified with the same kind of electricity is inversely propor-

tional to the square of the distance (between the centres of the balls).'

C. A. Coulomb (1785). *Memoires de l'Academie Royale des Sciences.*

Coulomb obtained his experimental results by the use of a torsion balance. In this a charged ball was situated near to a second similarly charged ball. The second ball was attached to the end of an insulator rod which itself was horizontally suspended by a wire attached to its midpoint (Fig. 2.36). The other end of the wire was fixed. The force of repulsion between the spheres caused the wire to twist.

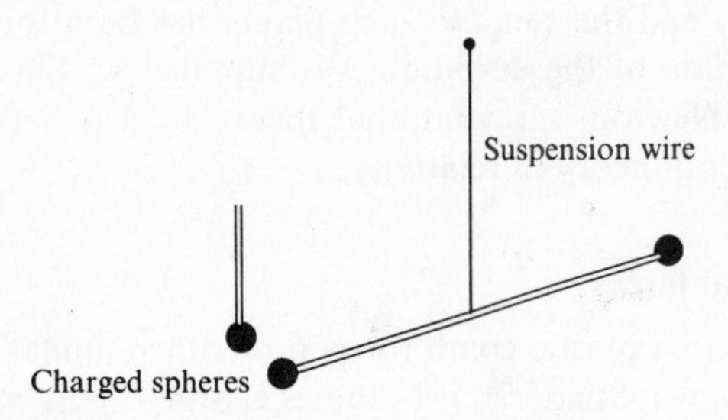

Fig. 2.36 Schematic diagram of Coulomb's torsion balance.

In an earlier experiment Coulomb had found the torque necessary to twist the wire through different angles and so was enabled to determine the force of repulsion.

We can repeat Coulomb's experiment in the laboratory with simpler apparatus. Two graphite-coated pith or expanded polystyrene balls, one on the end of an insulator rod and the other suspended by two insulator threads, can be used (Fig. 2.37). Coulomb used pith balls, because they

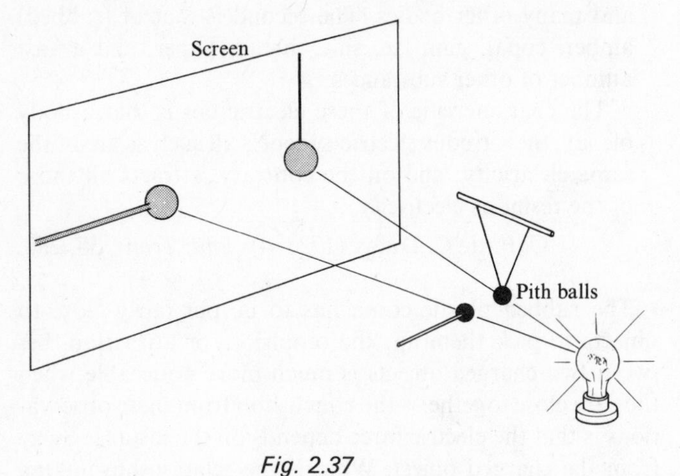

Fig. 2.37

are very light. When the pith ball on the rod is brought up to the similarly charged suspended ball deflection of the suspended ball occurs. This deflection can be easily measured by casting shadows of the balls onto a distant screen. The deflection is proportional to the force (for small deflections). You can readily check this by 'deflecting' a suspended weight with a spring balance (Fig. 2.38). The deflection is proportional to the force reading given by the spring

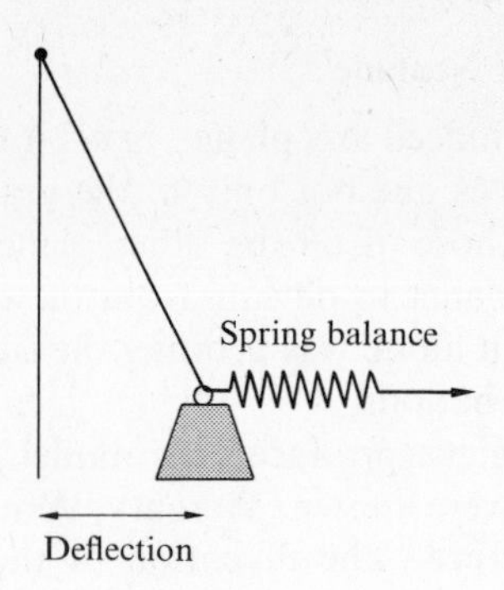

Fig. 2.38

balance. The result—the force is inversely proportional to the square of the distance between the centres of the balls

$$F \propto \frac{1}{r^2}$$

The force also depends on the amount of charge carried by the balls. If at a constant distance between the two balls we reduce the charge on one of the balls by a factor of two we find that the force is reduced by a factor of two. The reduction of the charge by a factor of two can be produced by putting in contact with the charged ball an equal size but uncharged ball. The charge then becomes equally shared between the two. If the charge on both balls is reduced by a factor of two then the force is reduced by a factor of four. More results all show that the force is directly proportional to the product of the charges on both balls.

$$F \propto q_1 q_2$$

The relation thus becomes

$$F \propto \frac{q_1 q_2}{r^2}$$

The constant of proportionality is taken as $1/4\pi\varepsilon_0$.

$$F = \frac{q_1 q_2}{4\pi\varepsilon_0 r_2}$$

The constant of proportionality is written in this form for reasons which can only become apparent from a study of electric and magnetic fields (see chapter 12). The value of the constant is, however, an established fact arrived at from experiment. ε_0 has the value of $8{\cdot}85 \times 10^{-12}\ \mathrm{C^2N^{-1}m^{-2}}$.

The equation is often known as Coulomb's law.

Nuclear forces

These are the forces that are significant in the nucleus—they are only significant over very short distances, of the order of 10^{-15} m. The laws of nuclear forces as they are at present 'understood' are very complex. How do scientists find out about such forces? An important method that is used is to study the scattering produced when one nuclear particle is aimed at another, for example protons fired at neutrons. The stronger the force between the particles the

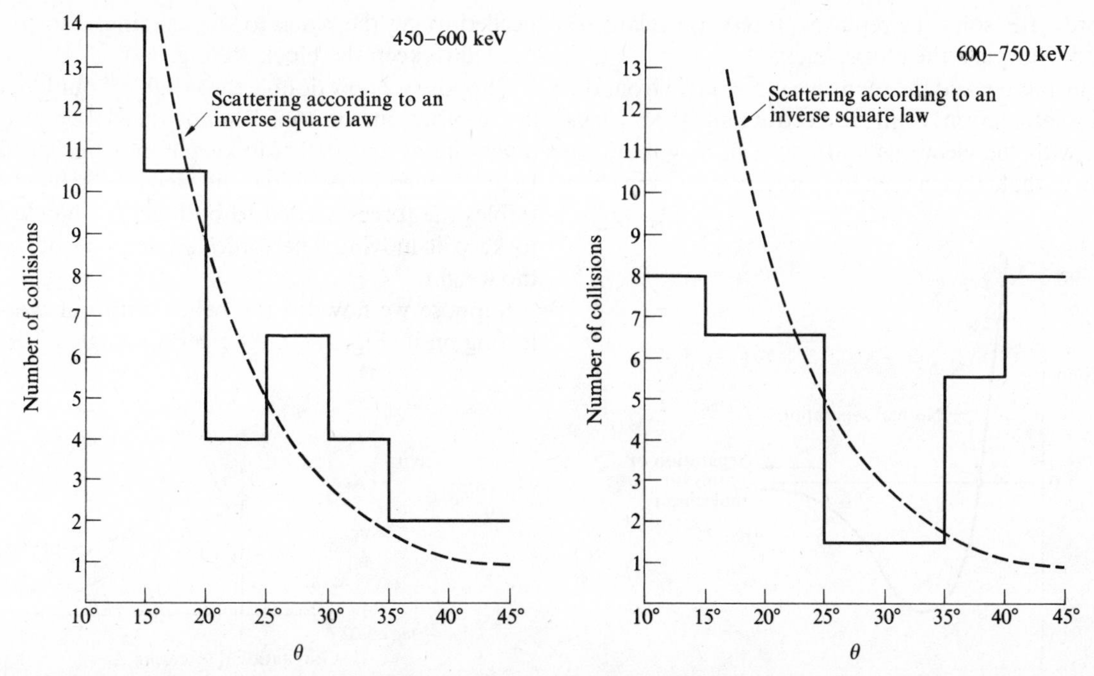

Fig. 2.39 7340 photographs were taken of high energy proton tracks in hydrogen. When the energy of the protons was in the 450–600 keV region the scattering was according to an inverse square law. With energies between 600 and 750 keV the scattering departs from the inverse square law. (Taken from M. G. White Physical Review, **49**, *309, 1936.)*

greater will be the deflection of the projectile from its straight line path.

As an example of such methods consider the scattering of protons by protons. We would probably expect that the scattering could be explained by the use of an inverse square law of repulsion (our electric force). With some slight modification to take account of the identity of the particles this was thought to be the case—until 1935. In that year White working at Berkeley found evidence that the scattering was not completely described by the modified inverse square law (Fig. 2.39). In 1936 Tuve, Heydenberg, and Hafstad, using a Van de Graaff machine to produce their projectiles (see chapter 13), were able to confirm that the modified inverse square law was insufficient. Breit, Condon, and Present in 1936 were able to show that the deviation from the modified inverse square law could be explained if there was a short range attractive force between protons, as well as the inverse square law force.

Other work has shown that there are nuclear forces (short range) between protons and neutrons, between neutrons and neutrons, and between many other particles. The forces can, however, be grouped into two basic forms, the weak and the strong interactions. The force that holds the neutrons and protons together in a nucleus is the result of a strong interaction. In such an interaction pions are exchanged between the two particles. Pions, sometimes called π-mesons, have masses of about 270 times that of an electron. When a neutron and proton are close together, the proton can emit a pion which is rapidly absorbed by the neutron. This has the result of changing the proton into a neutron and the neutron into a proton. This process continues—the proton and neutron continually changing from one to the other. This results in an attractive force between the two particles (Fig. 2.40). The process involved with a strong nuclear force takes place very rapidly—lasting only about 10^{-23} s. Some interactions are, however, found to take place much more slowly—a mere 10^{-10} s or longer. These slower interactions are known as the weak interactions.

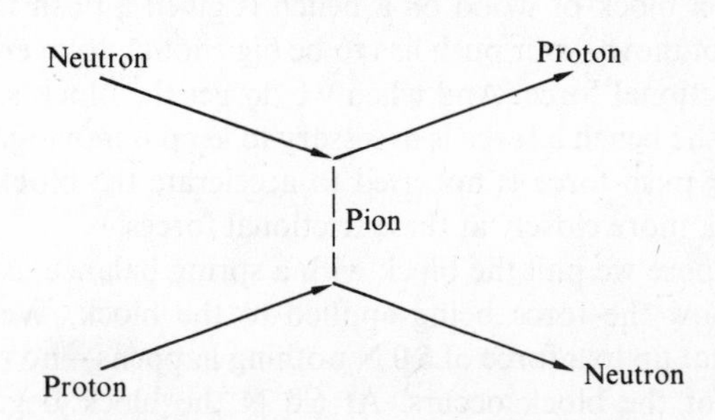

Fig. 2.40 Neutron–proton interaction.

Molecular or atomic forces

The atoms in a solid stick together to form a solid. There must therefore be some force of attraction between atoms. If we squeeze a solid we push the atoms closer together. This requires a force and when the force is removed the solid springs back (if the force was not too large). The atoms are thus held apart by repulsive forces and held together by attractive forces. At the normal atomic spacing these attractive forces must be equal in magnitude to the repulsive forces—hence stability. If we extend the solid the attractive forces are relatively greater and try to pull the atoms back, if we compress the solid the repulsive forces are relatively greater and try to push the atoms back.

Both the attractive and the repulsive forces arise from the very complex interactions of all the electrons and the nucleus in an atom with the electrons and nuclei in neighbouring atoms. A curve that is often used to represent the summation

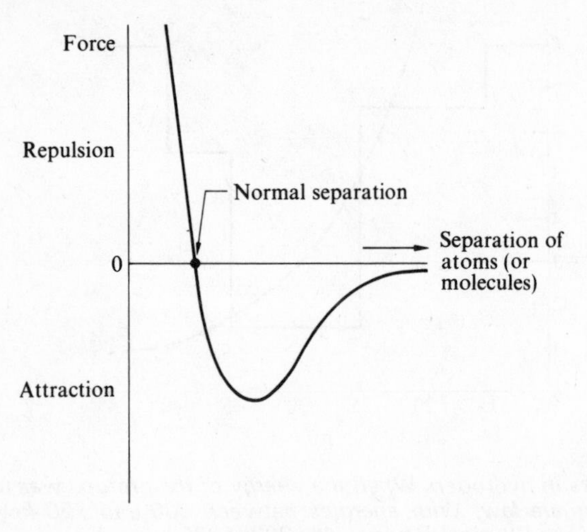

Fig. 2.41

of these effects is shown in Fig. 2.41. This curve will be deduced in chapter 3 for ionic crystals. Similar forms, however, can be used to describe the forces between atoms or molecules.

Frictional forces

When a block of wood on a bench is given a push it may well not move—our push has to be big enough to overcome the frictional force. And when we do get the block sliding along the bench a force is necessary to keep it moving. Also all our push force is not used to accelerate the block. Let us look more closely at these frictional forces.

Suppose we pull the block with a spring balance, so that we know the force being applied to the block. We may find that up to a force of 6·0 N nothing happens—no movement of the block occurs. At 6·0 N the block begins to move. If we repeat the experiment a number of times we may get results of 5·5 N, 6·5 N, 5·7 N, 6·1 N, etc. The results seem to be quite variable—but roughly 6·0 N. For forces above roughly 6·0 N the block accelerates. Suppose we measure the acceleration, say with a vibrator and ticker tape, then we calculate the resultant force acting on the block from $F = ma$. When our spring balance reads 8·0 N the acceleration measurements indicate a force of about 3·0 N. Thus 5·0 N has been used to overcome the frictional force acting on the moving block. When our spring balance reads 10·0 N the acceleration measurements indicate a force of about 5·0 N, a frictional force of 5·0 N. If we just apply 6·0 N to get the block moving we find that once it has started moving it is possible to reduce the force to about 5·0 N and still keep it moving. To summarize: a force of 6·0 N is needed to get the block to start sliding, a force of 5·0 N is needed to keep the block sliding.

Suppose we now double the weight of our block by stacking another one on top. We now need 12·0 N to start the block sliding and 10·0 N to keep it moving. The forces have been doubled by doubling the weight. Trebling the weight trebles the forces needed to both get the block sliding and to keep it moving The forces seem to be proportional to the weight.

Suppose we now tilt the bench with our wooden block resting on it (Fig. 2.42). At a certain angle of tilt the block

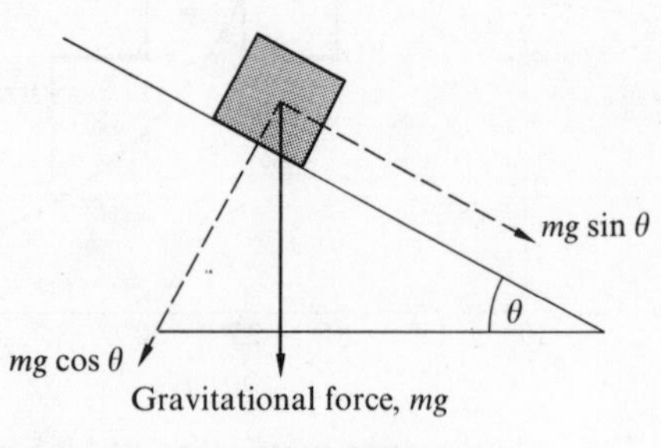

Fig. 2.42

begins to slide. For our block the angle is about 17°. If we stack two blocks together the angle at which they slip turns out to still be about 17°, the same angle occurs with three blocks. How does this compare with our earlier result that the frictional force seemed to be proportional to the weight? Changing the weight makes no difference to the angle at which slipping starts. The force pulling the block along the bench in this case is the component of the gravitational force (or weight) acting along the bench, i.e., $mg \sin \theta$. The other component is at right angles to the bench, i.e., $mg \cos \theta$. This is the force holding the block on the surface (sometimes called the normal force). Perhaps the frictional force is proportional to the normal force. If this is so then if we double the weight (mg to $2mg$) we double both the force pulling the block and the normal force.

$$\frac{mg \sin \theta}{mg \cos \theta} = \text{a constant}$$

In the case of our block the constant $\tan \theta$ is about 0·3. Thus we have

$$\frac{\text{Force to start motion}}{\text{normal force}} = 0{\cdot}3$$

or

$$\frac{F}{N} = \mu \quad \text{(a constant)}$$

The constant μ is called the coefficient of static friction. If we had taken the results for the moving block

$$\frac{\text{Force to continue motion}}{\text{normal force}} = \mu$$

we would have obtained a value for the constant of 0·2. This is called the coefficient of kinetic or dynamic friction.

Typical values of these coefficients are

	Static coefficient	*Dynamic coefficient*
Steel sliding on steel:	0·74	0·57
Glass sliding on glass:	0·94	0·40
Oak on oak (parallel to grain):	0·62	0·48

A surprising fact is that the force to cause the sliding, or the force to continue sliding, is independent of the surface area of the block (Fig. 2.43).

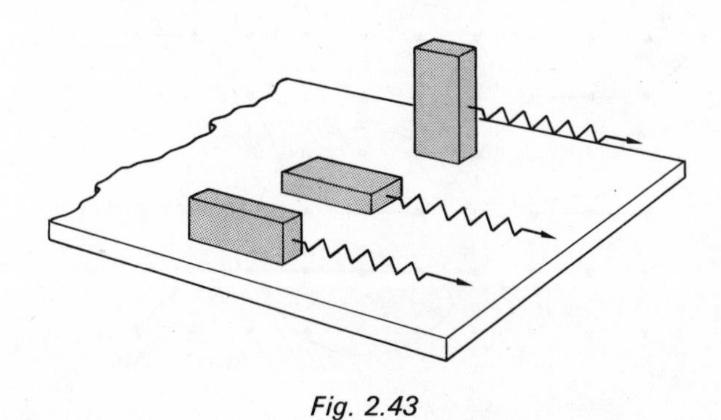

Fig. 2.43

The experimental results almost invariably show considerable scatter. This might be expected when we slide a wooden block on a wooden bench. Wood is not uniform throughout and it is possible that we may not use precisely the same part of the bench each time. But if we do the experiment with a metal block sliding on a smooth metal surface we still get a scatter of results. Perhaps the surfaces are not clean. Most metals are coated with oxides and other impurities. What is the coefficient of friction if we remove all these impurities and have the clean metal sliding on a clean metal surface? Such an experiment has to be carried out in a vacuum, to avoid oxide layers forming. The result—clean metal surfaces do not slide over one another, they adhere together. For clean metals the coefficient of friction is very high. A slight trace of oxygen produces a thin layer of oxides which can drop the coefficient down to about 2. With prolonged exposure to oxygen the coefficient drops to about 0·6.

How can we explain the experimental results and the origin of frictional forces? If we put two metal surfaces very close together we would probably expect that the atoms in one surface would strongly attract the atoms in the other surface. If the two sets of atoms came as close as the normal separation between the atoms in the metal then there would effectively be just one block of metal—seizure and a very high coefficient of friction (Fig. 2.44(a)). This is, however, an unrealistic picture—metal surfaces are not perfectly smooth. At the best we would have a number of small regions in contact (Fig. 2.44(b)) and small localized regions where seizure occurs. These may, however, still be sufficient to give a very high coefficient of friction. Now, what about metals in contact in air? The real area of contact between two metals is very small.

Area of contact between flat steel surfaces

Load /N	*True area of contact*/cm^2	*Fraction of macroscopic area in contact*
500 × 9·8	0·05	1/400
100 × 9·8	0·01	1/2000
5 × 9·8	0·0005	1/40 000
2 × 9·8	0·0002	1/100 000

F. P. Bowden and D. Tabor (1956), *Friction and Lubrication*, Methuen.

The results in the table were obtained by measurements of the electrical resistance between the two surfaces in contact. The real area of contact is much smaller than the area of the sliding surface. Also it is proportional to the load. The pressures at these small contact areas are very large—100 × 9·8 N on an area of 0·01 cm^2. These high pressures are sufficient to break through some of the oxide and dirt layers on the metal surfaces and give very close contact on an atomic scale. Local points of seizure will occur. The area over which seizure occurs increases as the force holding the surfaces together increases. Thus to produce sliding we have to overcome these interatomic forces at the seizure points. The frictional force is proportional to the normal force because the true area of contact is proportional to the normal force. The frictional force is proportional to the real area of contact.

The area results given in the table were for two steel flats 21 cm^2 in area. Essentially the same results were obtained for steel flats only 1 cm^2 in area—the real area of contact is independent of the surface area. Hence we have an explanation of the experimental result that the frictional force is independent of the surface area.

When you walk you push against the floor and rely on frictional forces to stop you falling and to enable you to move forwards. The force applied to the floor has two components—a horizontal component and a vertical component (Fig. 2.45). The magnitude of these forces, and the direction of the horizontal component, changes during the walking process. Figure 2.46 shows a typical set of records of the forces for a subject walking in a straight line. Slipping will occur if the ratio of the horizontal component to the vertical component is equal to or greater than the coefficient of friction. Coefficients of friction between shoe and floor thus should be at least 0·4.

Appendix 2A Elliptical orbits

Given Newton's inverse square law of gravitation, can we forecast the orbit of a planet orbiting the sun or a satellite around some planet if the orbit is not circular? Sputnik 1, the first artificial satellite (launched 4 October 1957), had an orbit which took it from a nearest distance to the earth's surface of 213 km out as far as 965 km. In the case of the planets orbiting the sun, their motion is almost circular.

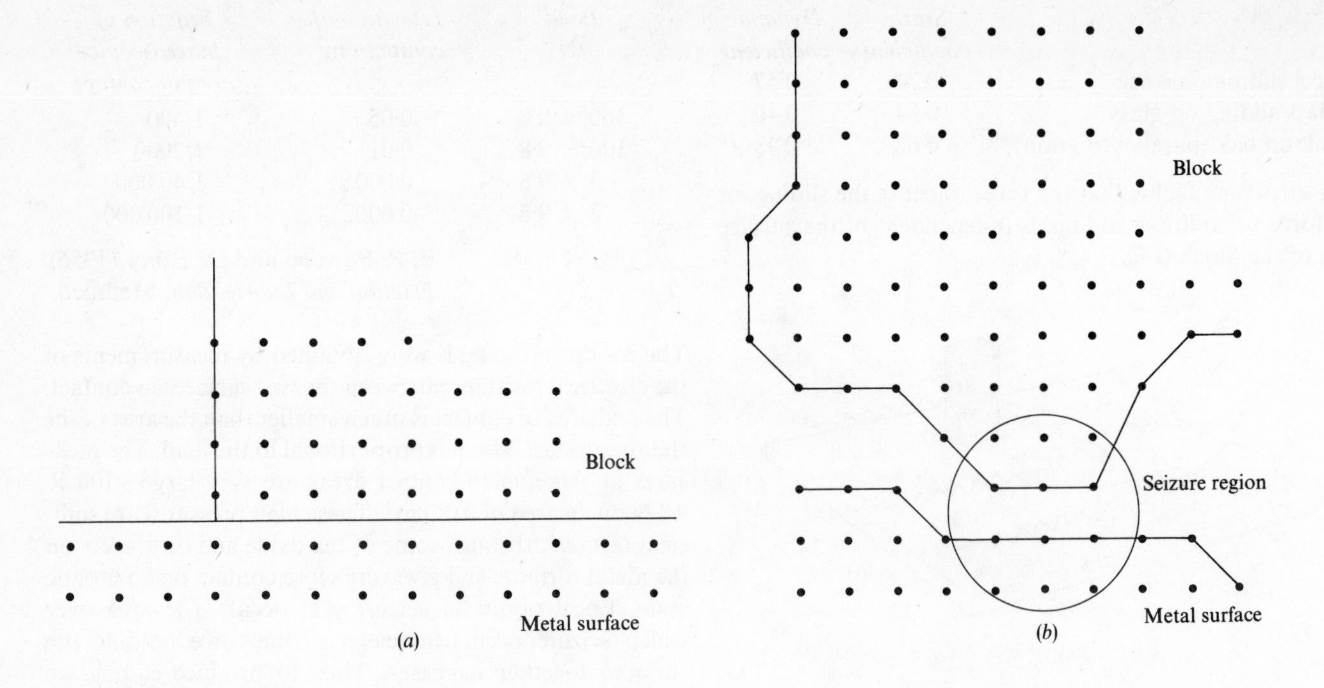

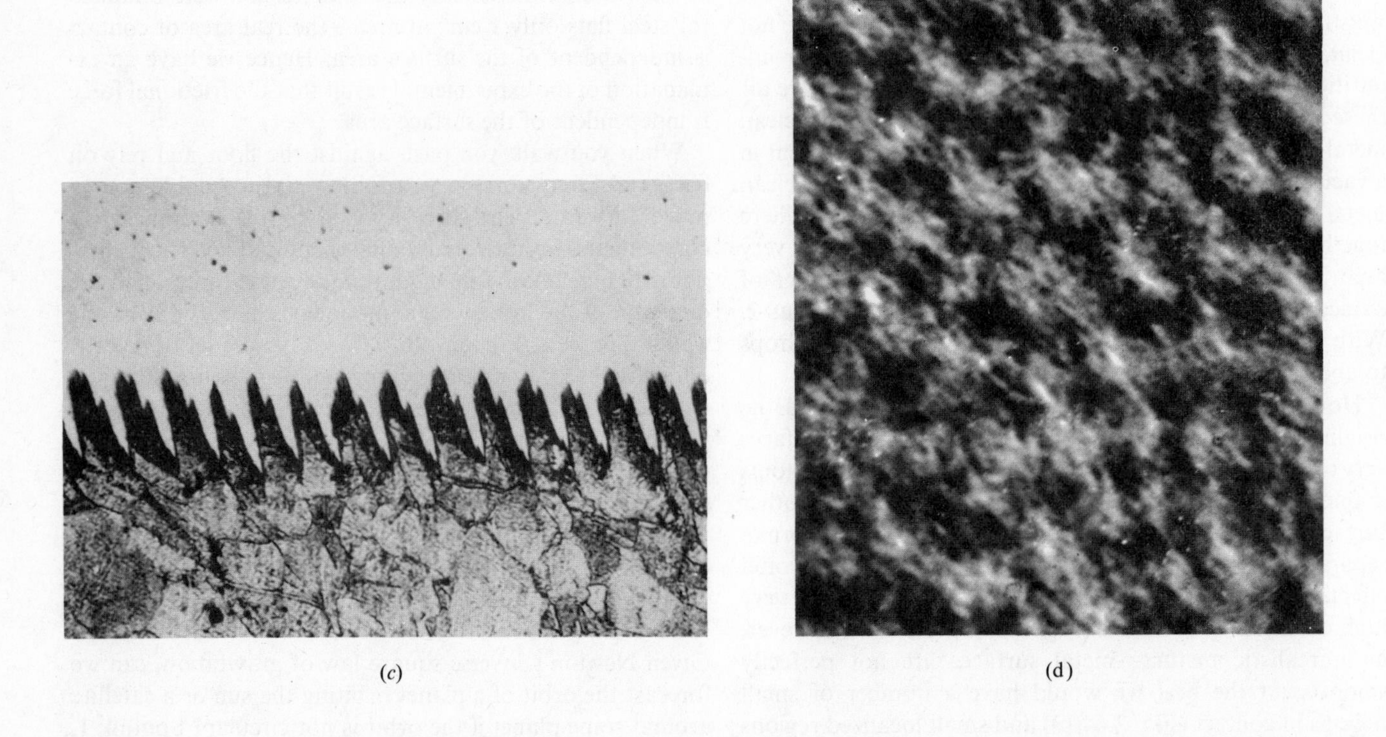

Fig. 2.44 (a) Seizure condition with perfectly smooth metals. (b) Seizure under realistic conditions. (c) Oblique section of finely turned copper surface. The irregularities are about 5×10^{-6} m. (d) Electron micrograph of electrolytically polished aluminium surface. The surface irregularities are between 10^{-8} and 10^{-7} m high. (F. P. Bowden and D. Tabor (1956). Friction and lubrication, *Methuen.)*

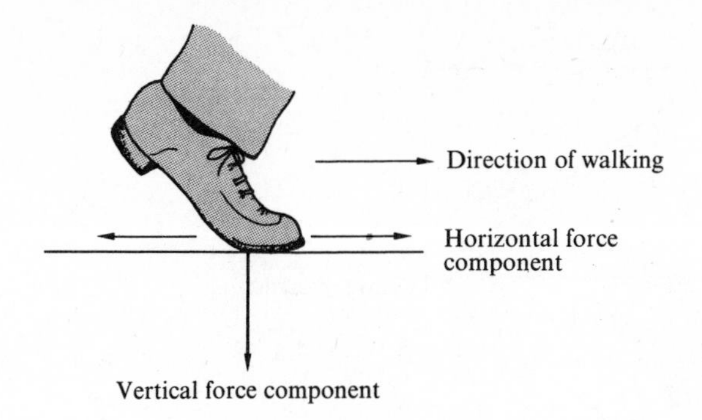

Fig. 2.45

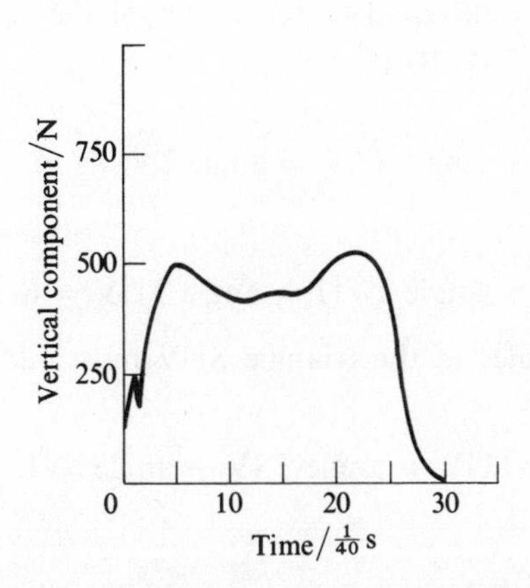

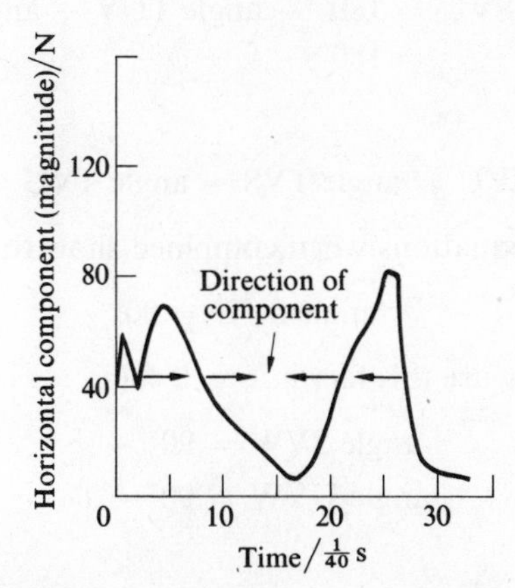

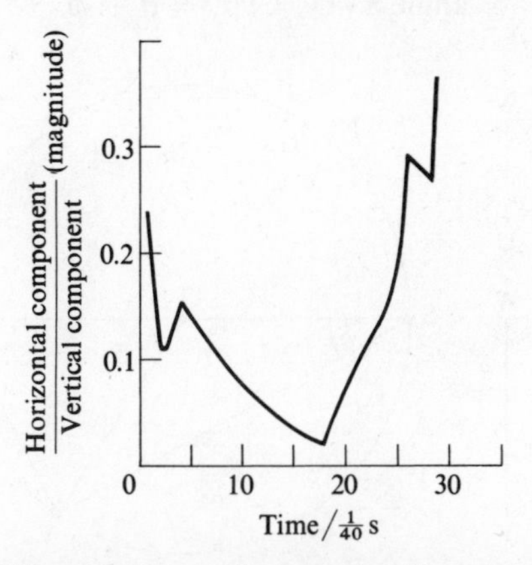

Fig. 2.46 'The forces applied to the floor by the foot in walking' F. C. Harper (1962). From NBS Research Paper 32. Reprinted by permission of the Controller HMSO.

The ratio of the maximum diameter of the earth's orbit to its minimum value is only 1·000 14. Mars, the planet whose investigation by Kepler led to the idea of elliptical orbits, has a ratio of maximum to minimum diameters of 1·0043.

A method of determining the path of an object moving in a non-circular orbit is to use essentially the same method as we used in chapter 1 to determine the path of a projectile. Indeed the argument is the same as that given by Newton, also in chapter 1, for projectiles fired from the top of a mountain. Those with low velocity will fall to the earth, those with a high enough velocity will go into orbit around the earth, those with too high a velocity will escape. To find the path of a projectile we have to determine two quantities —its distance–time graph in the x direction and its distance–time graph in the y direction. In the case of a projectile on a flat earth, there is a constant velocity in the x direction and a constant acceleration in the y direction. For a projectile orbiting a spherical earth there are accelerations in both the x and y directions—non-constant accelerations, because the acceleration depends on the inverse square of the distance of the projectile from the centre of the earth.

Force in x direction (F_x)
$= -m$ (acceleration in x direction)
Force in y direction (F_y)
$= -m$ (acceleration in y direction)

The minus sign is because as x and y increase the force becomes less. The resultant force on the projectile is given by $F = GM_m/r^2$, where r is the radial distance of the projectile from the centre of the earth. The force in the x direction and the force in the y direction are the resultant force resolved in these two directions (Fig. 2.47).

$$F_x = F\cos\theta$$
$$F_y = F\sin\theta$$

But

$$\cos\theta = \frac{x}{r} \quad \text{and} \quad \sin\theta = \frac{y}{r},$$

hence

$$F_x = \frac{-Fx}{r} = \frac{-GMmx}{r^3}$$

$$F_y = \frac{-Fy}{r} = \frac{-GMmy}{r^3}$$

Thus we have for the accelerations:

in the x direction,

$$a_x = \frac{-GMx}{r^3}$$

in the y direction

$$a_y = \frac{-GMy}{r^3}$$

The radial distance (r) is related to x and y by:

$$r^2 = x^2 + y^2$$

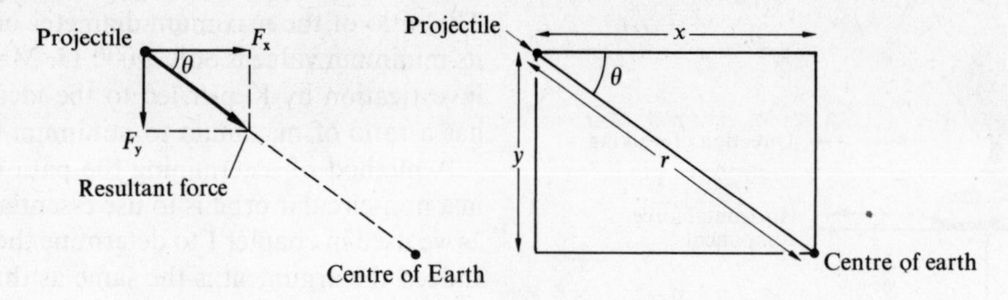

Fig. 2.47

Acceleration is the rate of change of slope with time of a distance–time graph, hence the equations needed to plot the graphs for both the x and the y directions are:

$$\frac{\Delta\,(\text{slope})}{\Delta\,(\text{time})} = -GM\frac{x}{(x^2 + y^2)^{3/2}}$$

for x against time graph, and

$$\frac{\Delta\,(\text{slope})}{\Delta\,(\text{time})} = -GM\frac{y}{(y^2 + x^2)^{3/2}}$$

for y against time graph.

A plot of such equations is given in *The Feynman Lectures on Physics*, Feynman, Leighton and Sands, Vol. 1, pp. 9–6.

For a satellite close to the earth's surface, an orbital speed of $7{\cdot}9 \times 10^3$ m s^{-1} gives a circular orbit, a speed of between 7·9 and $11{\cdot}3 \times 10^3$ m s^{-1} gives an elliptical orbit, speeds greater than this result in the 'escape' of the satellite. The value of the circular orbit speed is obtained by equating $F = GMm/R^2$ and $F = mv^2/R^2$, taking R as the radius of the earth.

Appendix 2B Intersecting chords theorem

A useful theorem that is used in both this and other chapters is that known as the theorem of intersecting chords. A chord is any straight line passing from one point on the circumference of a circle to another point. Figure 2.48(a) shows two such chords with an intersection point. The chords have been chosen to cut each other at right angles.

In Fig. 2.48(b)

$$ST = TU = TV$$

as these are all equal to the radius of the circle. T is the centre of the circle.

Thus

$$\text{angle TSV} = \text{angle TVS} = \phi$$

and

$$\text{angle TVU} = \text{angle TUV} = \theta$$

All the angles in the triangle SUV must add up to 180°. Thus

$$\text{angle TUV} + \text{angle TSV} + \text{angle SVU} = 180°$$

or

$$\begin{aligned}\text{angle SVU} &= 180° - \text{angle TUV} - \text{angle TSV}\\ &= 180° - \theta - \phi\end{aligned}$$

but

$$\text{angle SVU} = \text{angle TVS} + \text{angle TVU} = \theta + \phi$$

These two equations when combined show that

$$\text{angle SVU} = 90°$$

We can now use this fact in Fig. 2.48(a).

$$\text{angle SVW} = 90° - \phi$$
$$\text{angle UVW} = 90° - \theta$$

But

$$\text{angle SVU} = 90° = \theta + \phi$$

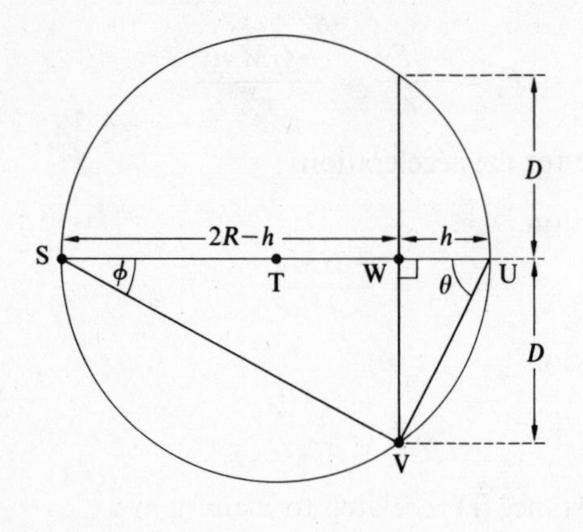

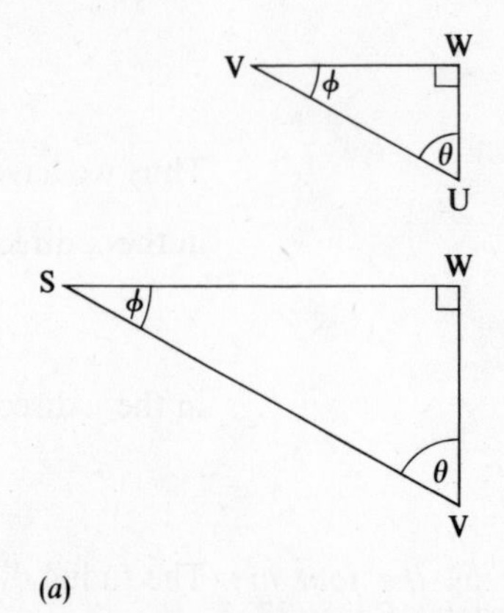

(a)

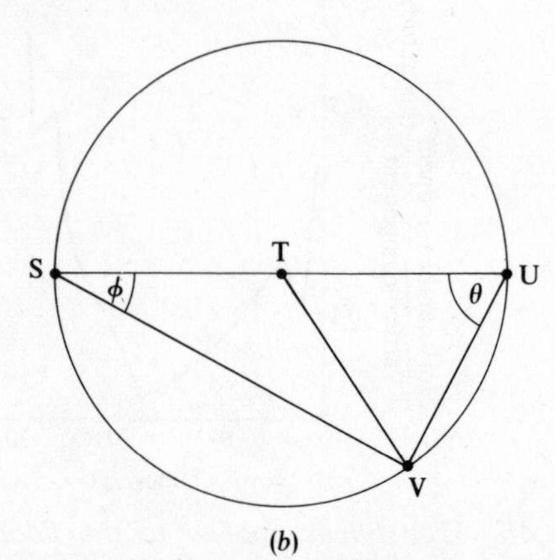

(b)

Fig. 2.48

therefore

$$\text{angle SVW} = \theta$$

and

$$\text{angle UVW} = \phi$$

Hence the triangles VWU and SWV are similar, i.e., they contain all the same angles.

So for the sides we can write

$$\frac{WV}{WS} = \frac{WU}{WV}$$

or

$$\frac{D}{2R - h} = \frac{h}{D}$$

$$D^2 = h(2R - h)$$

Problems

1. Oxford and Cambridge Schools Examination Board. Nuffield Advanced Physics, 1970. Q8 short answer paper.

'An alpha-particle moves near a massive nucleus N. The diagram [Fig. 2.49] represents part of the motion of the alpha-particle: it has a velocity v_a at A and a velocity v_b at B, a short time later. The arrows on the diagram show the directions but not the sizes of v_a and v_b.

(*a*) Why is the direction of v_b different from the direction of v_a?

(*b*) Give an argument to decide whether the size of the velocity v_b is greater than or is less than the size of v_a.

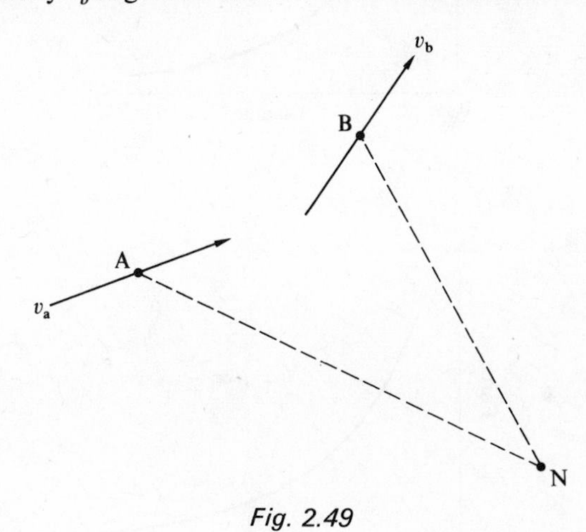

Fig. 2.49

2. Draw graphs showing how the force between a pair of particles varies with the distance between the particles for (a) two tennis balls, (b) an alpha particle and a nucleus, (c) two planets.

3. B. Russell in his book *A history of western philosophy* (1945) states:

'... the conception of "force" ... has been found to be superfluous.'

What do you think he means? (See Feynman, Leighton, Sands. *The Feynman lectures on physics*, Vol. 1, p. 12–1.)

4. Predict how the weight of an astronaut will vary from the time of the rocket taking off from the earth to his orbiting the earth in the spacecraft.

5. How could an astronaut measure (a) mass and (b) weight on the surface of the moon?

6. Predict the course of events after the following collisions:

(a) a 0·1 kg mass moving at 0·2 m s^{-1} collides with and sticks to a stationary 0·2 kg mass,

(b) a proton moving at 10^5 m s^{-1} collides with a stationary proton,

7. If you jump from a wall onto the ground and land stiffly you experience a greater force (and a greater chance of broken limbs) than if you relax and bend your knees. Why?

8. (a) What is the orbital time of a satellite which appears to be stationary in the sky?

(b) What is the distance from the earth's surface of such a satellite?

Mass of the earth = 6×10^{24} kg

Radius of the earth = $6{\cdot}4 \times 10^6$ m

9. You drop an object and it accelerates downwards towards the surface of the earth. Because it is accelerating its velocity is increasing and so its momentum is increasing. How can this be reconciled with the principle of the conservation of momentum?

10. '... We can imagine that this complicated array of moving things which constitutes "the world" is something like a great chess game being played by the gods, and we are observers of the game. We do not know what the rules of the game are; all we are allowed to do is to watch the playing. Of course, if we watch long enough, we may eventually catch on to a few of the rules. The rules of the game are what we mean by fundamental physics. Even if we knew every rule, however, we might not be able to understand why a particular move is made in the game, merely because it is too complicated and our minds are limited. If you play chess you must know that it is easy to learn all the rules, and yet it is often very hard to select the best move or to understand why a player moves as he does. So it is in nature, only much more so; but we may be able at least to find all the rules. Actually, we do not have all the rules now.'

R. B. Feynman, R. P. Leighton, M. Sands. *The Feynman Lectures on Physics*. Vol. 1 (1963) Addison Wesley, Reading, Mass.

(a) What does Feynman understand by the term 'fundamental physics'?

(b) Give an example drawn from this chapter of a 'fundamental physics' rule.

(c) Use your example to either support or oppose the argument given by Feynman.

Teaching note Further problems appropriate to this chapter will be found in:

E. M. Rogers. *Physics for the inquiring mind*, chapters 7 and 8, OUP, 1960.
Project physics course, Texts 1 and 3, Holt, Reinhart and Winston, 1970.
Physical science study committee. *Physics*, 2nd ed., chapters 19, 20, 22, Heath, 1965.
Nuffield O-level physics. Questions book 4, Longmans/Penguin, 1966.

Practical problems

1. Investigate either the frictional behaviour of skates on ice or skis on snow. Consider the following factors.

(a) Temperature

(b) The effect of speed

(c) In the case of skis consider the effect of commercial ski lacquers.

Before embarking on this project establish whether the laws of friction between surfaces apply to these conditions.

2. One problem that presents itself when aircraft or ships are used in cold weather is the formation of ice on surfaces. De-icing of such surfaces is a major concern. Investigate this problem. The following indicates a possible breakdown of the problem:

(a) For a clean metal surface how does the specific adhesion, i.e., the force per unit area, depend on the temperature?

(b) Do the above results depend on the solids considered?

(c) What happens if impurities are present in the water forming the ice (e.g., salt)?

(d) What happens if the surfaces are coated with plastics such as polystyrene, PTFE, etc.?

On the basis of your results consider whether it is possible to reduce or eliminate icing.

3. Investigate the bouncing of ball bearings on glass sheets.

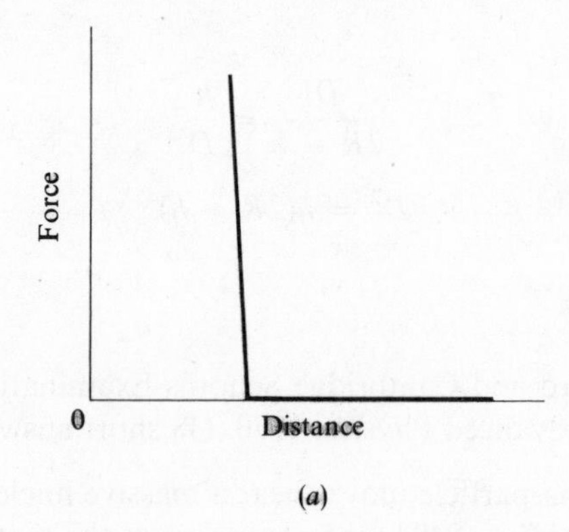

(a)

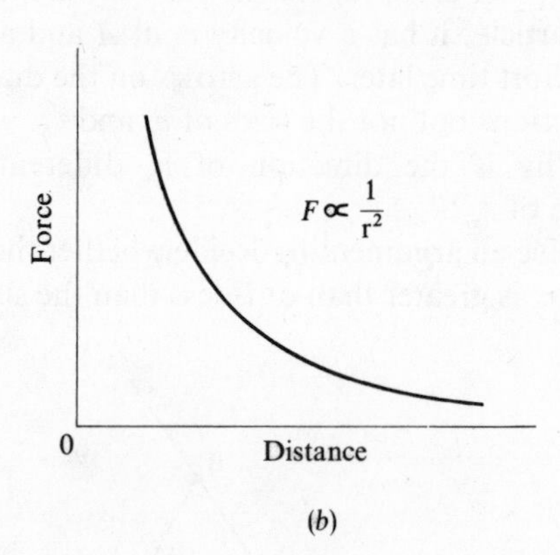

(b)

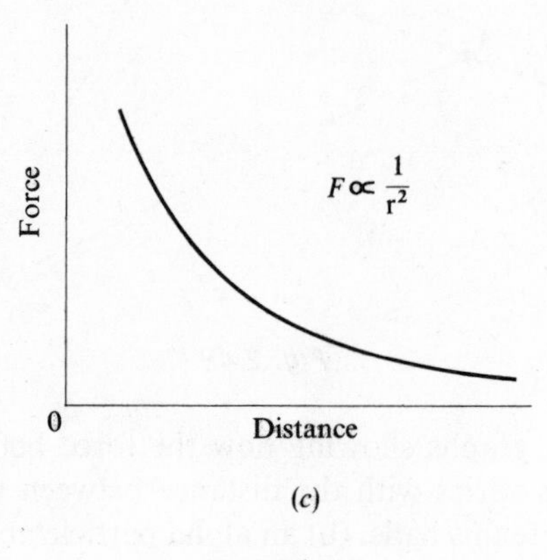

(c)

Fig. 2.50

Suggestions for answers

1. (a) A repulsive force is acting on the particle and the force is not along the line of the particle's path. In the distance AB the particle has been turned away from its original path by this force. The force is the Coulomb law force.

(b) The distance AN is greater than the distance BN. This would indicate that the magnitude of the velocity at B is less than that at A. The velocity of the alpha particle will be a minimum at the nearest point in its path to N and be bigger the greater the distance from N.

2. See Fig. 2.50.

3. We only see the effects of force—acceleration or a change of momentum. According to Russell it would be possible to describe all happenings without introducing the term force. Think of describing all events by momentum changes. On this argument we would describe everything by observable changes in position. The idea has, however, drawbacks—it is impossible to discuss on this idea the concept of a source of force. The sun can be considered a source of gravitational force. See the Feynmann reference given in the question.

4. Before take off weight normal, during take off weight increasing as the rocket accelerates, weight above normal until the rocket motors stop firing, weight then becomes zero when the only forces acting on both the spacecraft and the astronauts are gravitational forces.

5. (a) Measure a force and an acceleration or use a beam balance which compares the forces on two masses or use the conservation of momentum for a collision between the mass and a standard mass.

(b) A spring balance.

6. (a) They both move off in the same direction with a velocity of $0{\cdot}02/0{\cdot}3 \text{ m s}^{-1}$.

(b) The angle between the two proton tracks after the collision will be 90°, the ratio of the two velocities being $\tan\theta$, where θ is the angle between one of the proton tracks and the initial proton direction.

7. The impulse Ft is equal to the change of momentum. The bigger t is the smaller will be the force experienced. The longer the time over which the momentum changes the smaller is the force.

8. (a) 24 hours

(b) $v = \dfrac{2\pi R}{T}$, thus $R^3 = \dfrac{GMT^2}{4\pi^2}$

$R = 4{\cdot}23 \times 10^7$ m, hence the distance above the earth is $35{\cdot}9 \times 10^6$ m.

9. The object experiences a force towards the earth, the earth experiences a force towards the object. Both will accelerate and both will have momentum. The momentum of the object at any instant will be equal and opposite to the momentum of the earth.

10. (a) The rules of the game—the laws of physics.

(b) Conservation of momentum.

(c) The laws are obtained from watching nature, doing experiments.

Our rules are invariably a simplification of natural events. A tennis ball being hit by a tennis racket is a complex situation—one of our rules we can apply to this event is the conservation of momentum, but there is far more to the event than just this.

3 Energy

Teaching note Practical work appropriate to this chapter will be found in:
W. Bolton. *Physics experiments and projects*, Vol. 5, Pergamon, 1969.
Nuffield O-level physics. Guide to experiments, 4, Longmans/Penguin, 1967.
Project physics course. Handbook, Holt, Reinhart and Winston, 1970.

3.1 What is energy?

It is hard to give a general definition of energy. The following quotations are the ways a number of authors have defined energy.

> 'We shall now start with a very earthy, and rather vague, description of energy as something we pay for, the product of fuel.'
>
> E. M. Rogers (1960).
> *Physics for the Inquiring Mind*, OUP.

> 'For the moment we shall say that energy is the essential thing involved in jobs—not the creation of energy but its transfer from one form to another.'
>
> *PSSC Physics*, Heath.

> '"Energy" is not a substance, fluid, paint, or fuel which is smeared on bodies and rubbed off from one to another. We use this term to denote a construct—numbers, calculated in a certain prescribed way, that are found by theory and experiment to preserve a remarkably simple relationship in very diverse physical phenomena.'
>
> A. B. Arons (1965). *Development of concepts of physics*, Addison Wesley, Reading, Mass.

> 'It is important to realize that in physics today, we have no knowledge of what energy is. We do not have a picture that energy comes in little blobs of a definite amount. It is not that way. However, there are formulas for calculating some numerical quantity, and when we add it all together it gives "28"—always the same number. It is an abstract thing in that it does not tell us the mechanism or the reasons for the various formulas.
>
> R. B. Feynman, R. P. Leighton, M. Sands (1963).
> *The Feynman Lectures in Physics, Vol. 1*,
> Addison Wesley, Reading, Mass.

Energy—the product of fuels, the essential 'thing' involved in jobs, a number calculated in a certain prescribed way. This chapter is concerned with the search for these numbers and the ways of calculating them. They will all describe the result of a fuel being expended. In addition they will always add up to the same number.

Transferring energy

If fuel is burnt, i.e., used up, cars can be made to move, aeroplanes fly, machines 'run' and do jobs. Fuel enables jobs to be done. The fuel provides something we call energy.

When a fuel provides energy—what are the results? If the fuel is the petrol in a car the result is motion. The petrol could be supplied to a motor operating a hoist lifting bricks up the side of a building under construction. Here the fuel results in a gain in height for bricks. The fuel could be used in a furnace to provide a change in temperature. A change in velocity, a change in height, and a change in temperature are three possible outcomes of energy being released by a fuel. The cornflakes you ate for breakfast are a fuel supplying energy to enable you to run, to climb steps, to keep warm, and just to exist.

The motor could be used to lift bricks up to the top of a building, a gain in height, and then the bricks could fall down to the ground again, a gain in velocity. It appears as if the fuel supplies energy which results in a height change and that the bricks somehow at the new height have energy which can be released and result in a change in velocity. There is energy associated with height of an object. This is called potential energy. There is energy associated with velocity of an object, called kinetic energy. The process of transferring energy from a fuel to an object and producing a change in potential or kinetic energy is known as work. The process of producing a change in temperature by means of an energy transfer is known as heat.

We say we are working, in the physical sense, when we move objects against the force of gravity (lifting objects) or friction (pushing objects) or against some force. To do work we require energy. This we obtain by eating, food is the fuel which enables us to work. When we work we are transferring the 'food energy' from us to some object—we lose energy and the object gains energy. A heavy manual worker needs more food fuel than an office worker.

In physics the term work is used to denote the energy transferred either to or from an object. After the movement both the objects involved are in some way in a different state—a different level above the ground, a different velocity. Work measures the energy transferred from object X to object Y which results in a force moving an object through some distance.

Work is defined as the product of the force and the distance moved, along the line of action of the force. Thus if a weight of 5 N is lifted vertically through 10 m the energy transferred to the object by work is $5 \times 10 = 50$ N m. A newton metre (N m) is known as a joule (J). If we pull a trolley with a constant horizontal force of 5 N along a horizontal table a distance of 10 m then the work is 50 J.

Why is work defined in this way? If we did experiments with a motor doing work, perhaps lifting objects, we would find that, approximately, for the same quantity of fuel used the same value of force–distance product was obtained. Thus, for the same fuel quantity, doubling the force halves the distance through which it moves the object.

Why do we have the term work in physics? Why do we want to calculate it? If we know the work done then we know the amount of energy being transferred to or from some object. This does not tell us what the result will be, i.e., whether the object will move faster or gain height or both, but it does tell us the total energy being transferred by mechanical means. (Similar arguments can be presented for the term heat.) As will be seen in the sections in this chapter on potential energy and kinetic energy the work definition enables us to calculate what the changes in velocity and height can be if we know which one, or how both, are changing.

The definition of work as the force–distance product gives us a quantity that is found by theory and experiment to preserve a remarkably simple relationship in very diverse phenomena: a relationship called the conservation of energy.

If we lift a weight a force is being moved through a distance and energy is needed to do the work. This energy is supplied by the food fuel we 'burn' inside us. This seems an obvious case of fuel energy being used and a reason for eating. But what about the situation when we do no work but still get tired? Getting tired would seem to be a reasonable indication that we are using energy. If we hold a weight and do not move it then no force is moved through a distance and thus no work done—however, we still get tired. If the weight were on a table, still at rest, no work is done—why should there be any difference between the table supporting the weight and us supporting the weight? The answer is that 'physiological' work is done for us to support a weight. The muscles in our arms need to be continuously supplied with nerve impulses for a muscle to remain in the 'tense' position for supporting weights. The muscle is not able to remain in the 'tense' position without nerve impulses; when they cease the muscle relaxes. Some animals have muscles which will remain in the 'tense' or 'set' position without being fed with nerve impulses. The muscles in the clam are of this type, the shell can be kept open at any position without energy being expended to keep it at that position.

Power is the rate at which energy is transferred from one object to another. The unit of power is the watt and is 1 J s^{-1}. A typical car engine has a possible power of about 75 kW whereas a man only averages out at about 25 W. The first stage of the Saturn rocket used to put the first men on the moon had a power of $1{\cdot}2 \times 10^{11}$ W at lift-off.

3.2 Potential energy

If we push a mass against a spring and compress the spring we need to expend energy. If we release the spring and mass, the spring pushes the mass through some distance. In compressing the spring we transferred energy to the spring from us, when the spring is released it transfers energy to the mass and causes it to move. We can talk about the compressed spring having a store of energy, elastic energy, or strain energy. A general term used to describe the energy which a body possesses by virtue of being in some particular position is called potential energy.

For masses near the earth's surface there is a force due to the earth's gravitational pull on the mass. As this force is product of the mass and acceleration, mg, lifting a weight through a vertical height h means an energy transfer to the mass of mgh. The energy transferred is known as gravitational potential energy.

$$\text{Potential energy} = mgh$$

This assumes that the force remains constant over the distance h. If this is not the case the total potential energy in moving through h can be obtained by considering the mass to be moved through a number of small steps in h and summing the result of all the steps.

$$\text{Potential energy} = F_1\,\Delta h_1 + F_2\,\Delta h_2 + F_3\,\Delta h_3 + \cdots$$

where F_1 is the force acting over the small distance Δh_1, F_2 the force over distance Δh_2, etc. In calculus notation—

$$\text{Potential energy} = \int_0^h F\,\mathrm{d}h$$

If we pull a trolley up a smooth hill, how much energy is needed? This can be measured by using a spring balance to measure the force pulling the trolley and a rule to measure the distance the trolley is pulled (Fig. 3.1). If the force is kept constant (F) and the distance is s then the energy used is Fs. The gain in potential energy is mgh, where m is the mass of the trolley and h the vertical height through which the trolley is moved. If all the energy is used to produce a

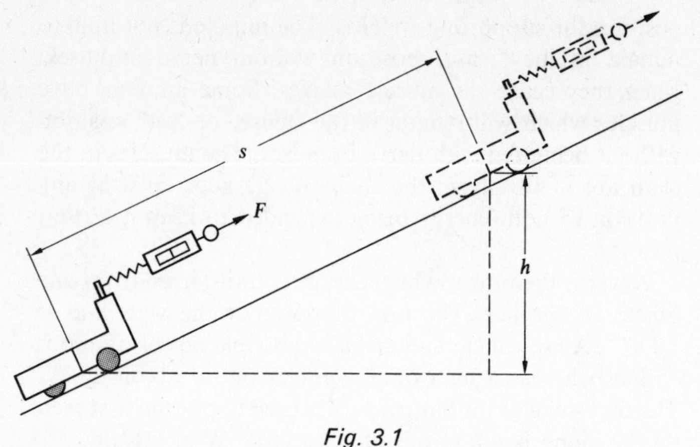

Fig. 3.1

gain in height, e.g., the trolley gains no kinetic energy and keeps the same velocity all the way up the slope, then

$$Fs = mgh$$

Gravitational potential energy is the energy resulting from a mass having a particular position in a gravitational field. Other forms of potential energy exist. For example a charged object in an electric field has electric potential energy.

The force between two charged particles is given by

$$F = \frac{q_1 q_2}{4\pi\varepsilon_0 r^2}$$

where q_1 and q_2 are the charges carried by the particles, ε_0 is a constant, and r is the distance between the centres of the two particles. If we move one of the particles a small distance Δr farther away from the other particle then the change in potential energy is $F\,\Delta r$ or

$$\text{Change in potential energy} = \frac{q_1 q_2}{4\pi\varepsilon_0 r^2}\cdot\Delta r$$

We have chosen a small distance Δr because we are assuming that the force remains constant over that distance. If we want to find the change in potential energy for moving a large distance then we must break the movement down into a small number of steps and find the changes for each small change in Δr.

It is often convenient to measure potential energy changes relative to some zero or origin point. In the case of the two charged particles the zero point is taken to be when one charged particle is an infinite distance from the other. Infinity is the distance at which the force between the particles becomes vanishingly small. Thus if we say that the potential energy of a charged particle at some point is X then this is the energy needed to fetch the particle from infinity to that point. To find this energy X we have to fetch the particle in a large number of small steps, Δr, and sum the energies needed for each step (Fig. 3.2).

If we squash a piece of rubber, energy is needed and the squashed rubber has potential energy. The rubber is just like the spring. If we squash a gas we can consider the

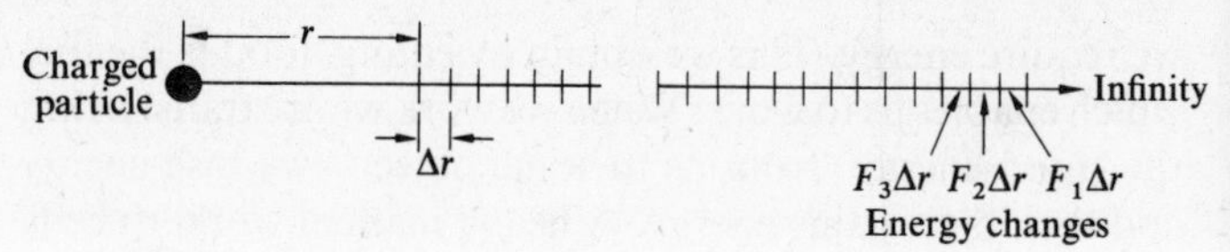

Total energy change = $F_1\Delta r + F_2\Delta r + \cdots + F_n\Delta r = X$

Fig. 3.2

squashed gas to have potential energy—energy is used to squash the gas. Suppose we think of the gas enclosed by a piston in a cylinder. When we push the piston in by an amount Δh we need to supply an energy of $F\,\Delta h$, where F is the force opposing the movement of the piston. The force is provided by the gas pressure P.

$$P = \frac{F}{A}$$

where A is the surface area of the piston. Hence the energy supplied is $PA\,\Delta h$. But $A\,\Delta h$ is the change in volume of the gas, ΔV, thus we can write

$$\text{Energy needed} = P\,\Delta V$$

If there is no temperature rise of the gas this will be the potential energy stored in the gas due to the movement of the piston.

Force–distance graphs

If an object acted on by a constant force moves through a distance then the energy transfer, the work, is the product of the force and the distance.

$$\text{Energy transfer} = Fs$$

On a graph of the force against distance (Fig. 3.3a) Fs is the area under the line between the distance equal to zero and equal to s lines. The area gives the energy transferred.

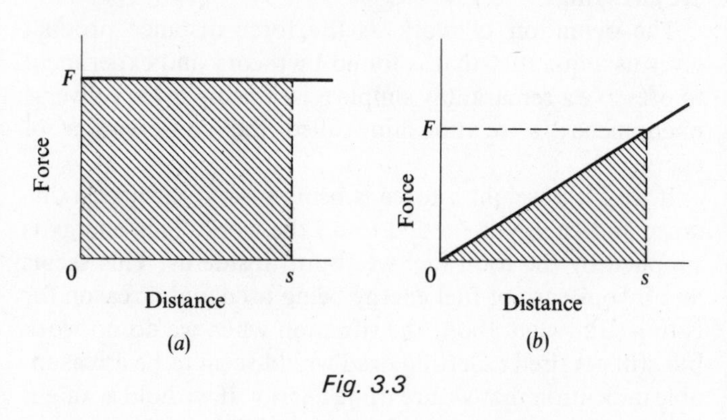

Fig. 3.3

If the force is not constant over the distance the energy transfer can be obtained from the area under the graph. For example, when a spring is extended the force is proportional to the extension. Figure 3.3(b) shows the force–distance graph. The energy transferred when the extension is changed from zero to s is therefore in this case $\frac{1}{2}Fs$, where F is the force at extension s.

We can write a general equation for the area under a force–distance graph:

$$\text{Energy transfer} = \int_{s_1}^{s_2} F\,\mathrm{d}s$$

or in words, the energy transferred in moving from distances s_1 to s_2 is the area under the graph of force against distance between the distance equal to s_2 and the distance equal to s_1 lines. The energy transferred can be obtained from determining the area under the graph or, in the case where the equation relating F and s is known, by integration.

As an example we will look at the energy transfer for a spring being extended. If the spring obeys the relationship $F = ks$, i.e., the extension is proportional to the force, then the energy transfer in extending it from $s = 0$ to s is

$$\text{Energy transfer} = \int_0^s F\,\mathrm{d}s = \int_0^s ks\,\mathrm{d}s$$

$$= \frac{ks^2}{2}$$

This can be written as $\frac{1}{2}Fs$, when the equation $F = ks$ is used to eliminate the k. This is the same result as we obtained from looking at the area under the graph.

For another example we look at the energy change involved in fetching a charged particle from infinity up to a distance r from another charged particle. The force between the two charged particles can be written as (section 2.5)

$$F = \frac{K}{r^2}$$

where r is the distance between the particles and K some constant. Figure 3.4 shows a graph of F against r.

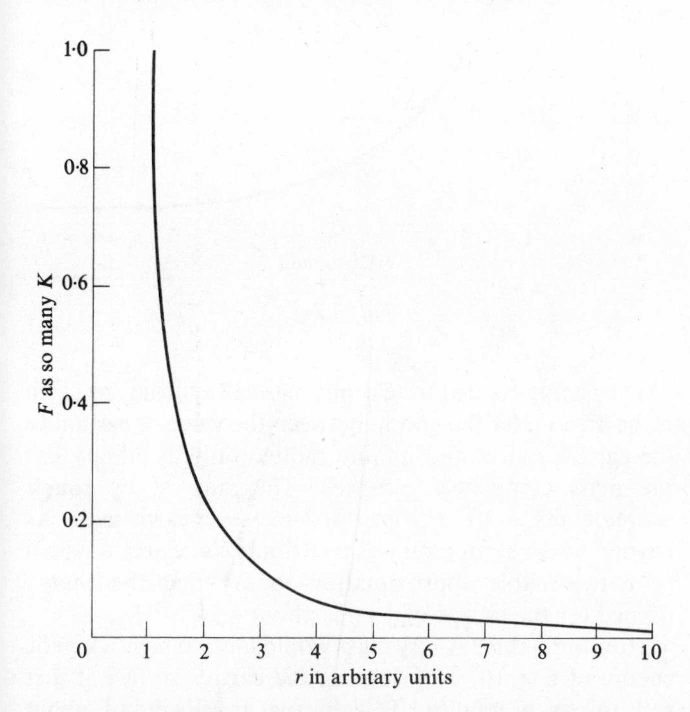

Fig. 3.4

Area between		(very approx.)
$r = \infty$ and $r =$	10	0·040 K
10	9	0·020 K
9	8	0·024 K
8	7	0·028 K
7	6	0·036 K
6	5	0·040 K
5	4	0·052 K
4	3	0·088 K
3	2	0·176 K
2	1	0·480 K

Thus the total areas up to particular r values are

r	∞	9	8	7	6	5	4	3	2	1
Area × K	0	0·060	0·084	0·112	0·148	0·188	0·240	0·338	0·504	0·984

Thus the energy needed to fetch a charge from infinity up to $r = 1$ is about 0·984K or 1K. The energy in fact varies as $1/r$ as a graph of the above energies against $1/r$ shows (Fig. 3.5). The results fit an equation

$$\text{Potential energy} = \frac{K}{r}$$

According to this equation the potential energy falls as r increases. For an attractive force this is not correct. Decreasing r decreases the potential energy. Think of the potential energy of a mass above the earth's surface—the higher it is above the surface the greater the potential energy. We can make our equation fit this behaviour by writing it as

$$\text{Potential energy} = -\frac{K}{r}$$

Now the smaller r becomes the more negative the potential energy becomes.

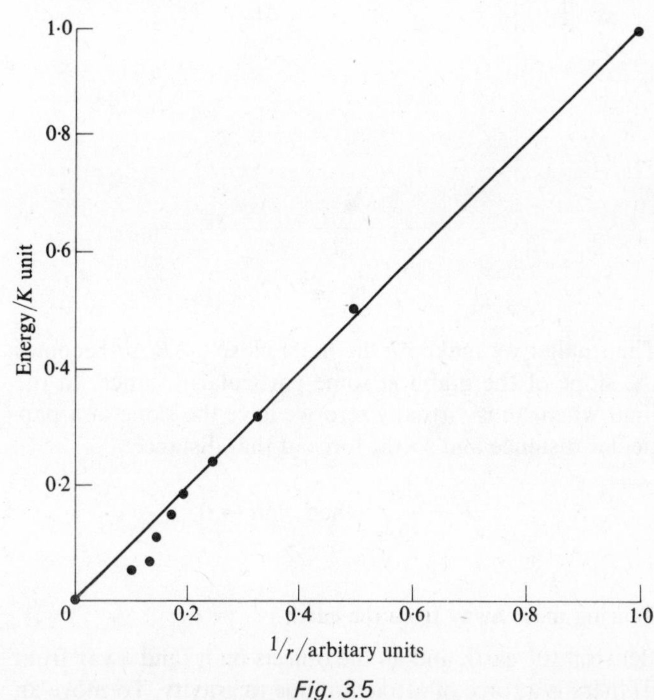

Fig. 3.5

This result could have been obtained by integration.

$$\text{Potential energy} = \int_{\infty}^{r} F\,\mathrm{d}r = \int_{\infty}^{r} \frac{K}{r^2}\,\mathrm{d}r = -\frac{K}{r}$$

Thus at points in space around a charged object we can associate certain potential energies as being the energies needed to put a charge at those points. If we double the charge we put at a point we double the energy needed, the force between two charges is directly proportional to the sizes of the charges. It is convenient to talk of the potential energy per unit charge for a point—this is called the potential.

$$\text{Potential} = \frac{\text{potential energy}}{\text{charge}}$$

Potential energy–distance graphs

If a weight mg moves through a small vertical distance Δh then the energy transferred is $mg\ \Delta h$, the product of the force (F) acting on the mass and the distance.
Change in energy $= F\,\Delta h = \Delta E$

Thus we can write

$$F = \frac{\Delta E}{\Delta h}$$

For a graph of E against h (Fig. 3.6), $\Delta E/\Delta h$ is the slope or steepness of the graph. Thus the slope of an E against h graph gives the force acting at the distance h considered.

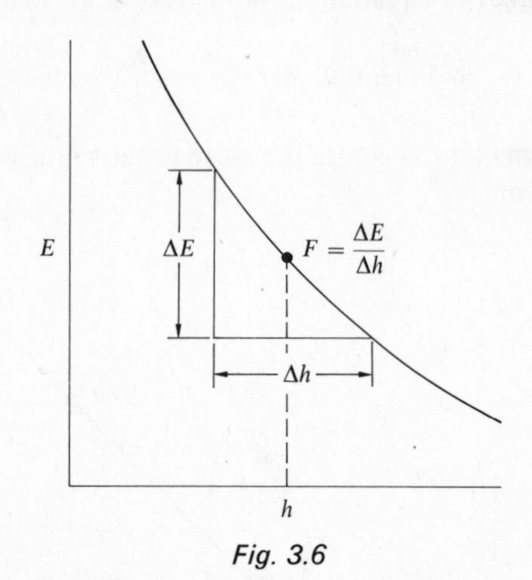

Fig. 3.6

The smaller we make Δh the more closely $\Delta E/\Delta h$ becomes the slope of the graph at some particular distance. In the limit when Δh is virtually zero we have the slope at a particular distance and so the force at that distance.

$$F = \frac{\mathrm{d}E}{\mathrm{d}h} \quad \text{when} \quad \Delta h \to 0$$

Tearing mass away from the earth

Between the earth and all the objects on it (and away from it) there is a force of attraction due to gravity. To move an object away from the earth requires energy—a force is required to move through a distance.

The force of attraction, due to gravity, between two masses m and M a distance r apart is (section 2.5)

$$F = G\frac{mM}{r^2}$$

where G is a constant, $6{\cdot}67 \times 10^{-11}\ \mathrm{m^3\ kg^{-1}\ s^{-2}}$.

Let us consider moving a 1 kg mass away from the surface of the earth. The mass of the earth is 6×10^{24} kg. Thus we have

$$F = \frac{3{\cdot}8 \times 10^{14}}{r^2}$$

and can plot a graph showing how the force varies with distance. r is measured from the centre of the earth to the centre of the 1 kg mass. The radius of the earth is 6371 km and thus we could use the formula to calculate the force on a 1 kg mass at the earth's surface—but this we know, it is the weight of the 1 kg mass, i.e., 9·8 N. At twice the earth's radius the force will be 9·8/4; the force varies as $1/r^2$. Figure 3.7 shows the resulting graph.

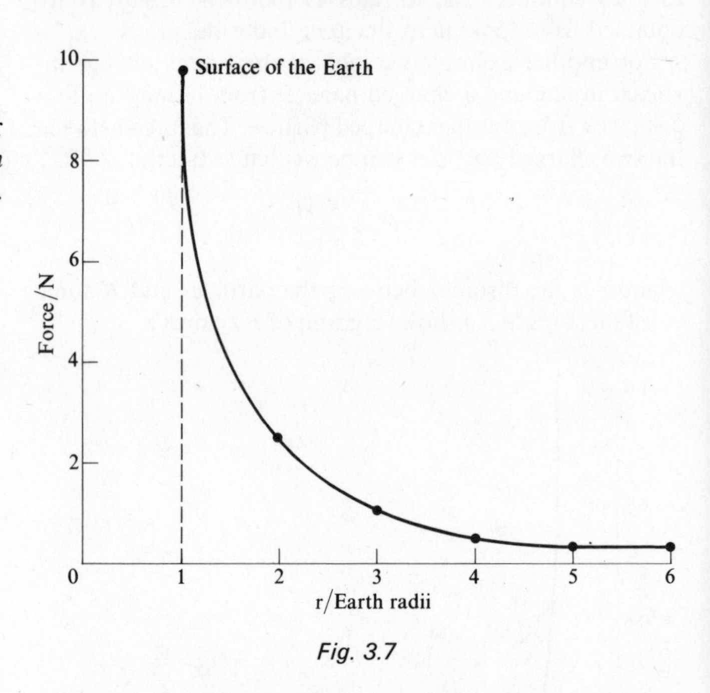

Fig. 3.7

The energy needed to tear our mass away from the earth is the area under the graph, between the value of r equal to the earth's radius and infinite radius (only at infinity has the mass completely escaped). The area is, by rough estimate, $5{\cdot}2 \times 10^7$ J from $r = 1$ to $r = 6$ earth radii. As beyond $r = 6$ earth radii will contribute some area it would be a reasonable approximation to say that the energy needed for the escape of 1 kg is about 6×10^7 J.

How big is this? A 1 kg mass would have to have a kinetic energy of 6×10^7 J on leaving the earth's surface if it is just to reach infinity. This means a velocity of about $10^4\ \mathrm{m\ s^{-1}}$. This is known as the escape velocity.

The area could have been obtained by integration.

$$\text{Energy} = \int_{r=6371\times10^3}^{r=\infty} \frac{3{\cdot}8 \times 10^{14}}{r^2}\,dr$$

$$= 5{\cdot}97 \times 10^7\ \text{J}$$

Using symbols we have (see page 69)

$$\text{Energy} = \frac{GMm}{R}$$

where R is the earth's radius. The escape velocity is obtained by equating this energy to $\frac{1}{2}mv^2$. Hence

$$v = \sqrt{\left(\frac{2GM}{R}\right)}$$

Notice that the escape velocity is the same for all masses leaving the earth's surface. The equation can be simplified as we know the force at the earth's surface in terms of the acceleration due to gravity g.

$$mg = \frac{GMm}{R^2}$$

Thus escape velocity v is given by

$$v = \sqrt{(2gR)}$$

This is $11{\cdot}2 \times 10^3$ m s^{-1} at the earth's surface.

All space ships going to other planets or the moon must reach a velocity of this order. In the Apollo 11 shot which landed the first men on the moon, the sequence of events which got the crew on their way to the moon was: first stage rocket brings velocity up to $2{\cdot}4 \times 10^3$ m s^{-1}, second stage brings velocity up to $6{\cdot}7 \times 10^3$ m s^{-1}, third stage puts the spacecraft in an earth orbit before bringing the velocity up to $10{\cdot}7 \times 10^3$ m s^{-1} (the velocity not to escape the earth but to get as far as the moon).

Tearing ions apart

To separate two oppositely charged objects requires energy. To separate them their force of attraction has to be overcome and the charged objects moved through a distance.

We can make a rough calculation of the energy needed to pull a pair of ions apart, e.g., the sodium and chlorine ions in a sodium chloride crystal. The force of attraction between the ions is

$$F = \frac{-q^2}{4\pi\varepsilon_0 r^2}$$

where q is the charge on the ions, $1{\cdot}6 \times 10^{-19}$ C, r the distance apart of the two ions, and ε_0 is a constant, $8{\cdot}85 \times 10^{-12}$ C^2 N^{-1} m^{-2}. The minus sign is because the charges are of opposite sign. We will take the distance apart of the ions to be the lattice spacing in the sodium chloride crystal, $2{\cdot}8 \times 10^{-10}$ m. The equation shows how the force varies with distance. We want to change the distance from $2{\cdot}8 \times 10^{-10}$ m to infinity, i.e., pull the ions completely apart. We could use the equation to plot a graph of force F against distance r and then obtain a value for the energy by measuring the area under the graph (Fig. 3.8) between

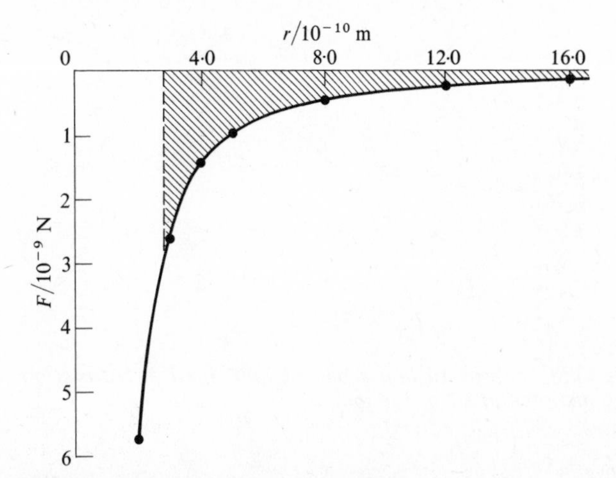

Fig. 3.8 The force of attraction between ions as a function of their distance apart.

the lines $r = 2{\cdot}8 \times 10^{-10}$ m and $r = \infty$ or use the integration

$$\int_{2\cdot8\times10^{-10}}^{\infty} F\,\text{d}r = \text{Potential energy.}$$

A rough estimate of the area under the graph suggests that the energy will be of the order of 8×10^{-19} J. Integration to obtain a more accurate result is as follows

$$\text{Potential energy} = \int_{2\cdot8\times10^{-10}}^{\infty} F\,\text{d}r = \int_{2\cdot8\times10^{-10}}^{\infty} \frac{-q^2}{4\pi\varepsilon_0 r^2}\,\text{d}r$$

$$= \frac{-q^2}{4\pi\varepsilon_0}\left[-\frac{1}{r}\right]_{2\cdot8\times10^{-10}}^{\infty}$$

$$= \frac{-q^2}{4\pi\varepsilon_0 2{\cdot}8 \times 10^{-10}}$$

$$= -8{\cdot}2 \times 10^{-19}\ \text{J.}$$

If we wanted to find how the energy needed to tear apart ions depended on the separation of the ions, r, we could use the force–distance r graph and measure the area between different values of r and infinity or use integration. The result in both cases is shown in Fig. 3.9 and is described by the equation

$$\text{Potential energy} = \frac{-q^2}{4\pi\varepsilon_0 r}$$

Experimental results for the tear-apart energy for a pair of ions in a sodium chloride crystal give $12{\cdot}7 \times 10^{-19}$ J. The theoretical result $8{\cdot}2 \times 10^{-19}$ J is thus of the right order.

What factors could explain the discrepancy between the theoretical and experimental results? We considered a pair of ions in isolation from the rest of the ions in the crystal—perhaps the other ions will affect the result? Around any

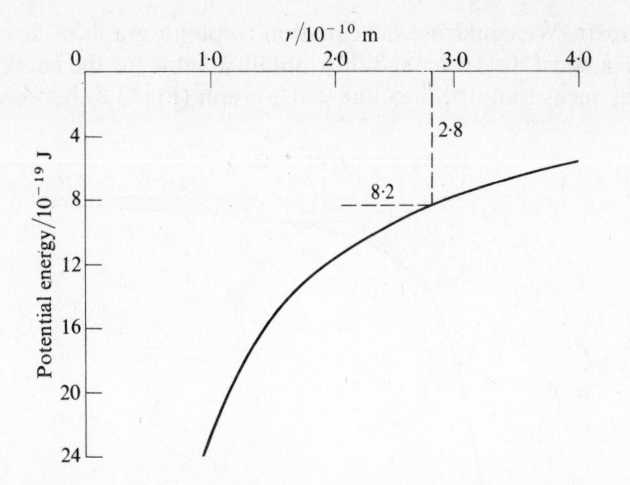

Fig. 3.9 Energy to tear a pair of ions apart as a function of the distance apart of the ions.

one ion there are six neighbours of opposite sign at distance $2{\cdot}8 \times 10^{-10}$ m, twelve neighbours of the same sign at distance $\sqrt{2} \times 2{\cdot}8 \times 10^{-10}$ m, eight neighbours of opposite sign at distance $\sqrt{3} \times 2{\cdot}8 \times 10^{-10}$ m, etc. (Fig. 3.10). We can find the energy to tear away this one ion by calculating the energy needed to take away each of these other groups of ions from our one ion. Thus the energy needed to take away the six near neighbours with opposite sign is $-8{\cdot}2 \times 10^{-19} \times 6$ J, the energy needed for the twelve neighbours of the same sign is $+8{\cdot}2 \times 10^{-19} \times 12/\sqrt{2}$ J, the energy needed to take away the eight neighbours of the opposite sign is $-8{\cdot}2 \times 10^{-19} \times 8/\sqrt{3}$ J, and so on with all the other ions in the crystal. Summing the terms so far gives a result of $-17{\cdot}2 \times 10^{-19}$ J. If all the terms are considered the result becomes $-14{\cdot}3 \times 10^{-19}$ J.

This result is higher than the experimental result, what else have we forgotten? If there were only attractive forces between oppositely charged ions as Coulomb's law predicts the crystal would collapse. Repulsive forces must exist. Perhaps these account for the difference between the theoretical and experimental results. If we make this assumption then repulsive forces must account for an energy difference of $1{\cdot}6 \times 10^{-19}$ J. The existence of the repulsive forces meant that $1{\cdot}6 \times 10^{-19}$ J less of energy was needed to remove an ion. What was the repulsive force acting on the ion at $2{\cdot}8 \times 10^{-10}$ m which gave this amount of energy?

If an ion moved a small distance Δr, energy $F\,\Delta r$ would be transformed. F is the force acting on the ion. Thus the change in energy $\Delta(\text{energy}) = -F\,\Delta r$

$$F = -\frac{\Delta(\text{energy})}{\Delta r}$$

The force is therefore the slope of the graph of energy against distance (Fig. 3.11), in our case the slope at $r = 2{\cdot}8 \times 10^{-10}$ m. But we do not know how the energy varies with distance for the repulsive force. We do, however, know how the energy varies with distance for the attractive force and at $2{\cdot}8 \times 10^{-10}$ m the attractive force must be equal to the repulsive force as the ions are in equilibrium at this distance. The slope of the graph gives a value of $-5{\cdot}1 \times 10^{-9}$ N for the force. The force of repulsion at $2{\cdot}8 \times 10^{-10}$ m from an ion is thus $+5{\cdot}1 \times 10^{-9}$ N.

We could have obtained this result by differentiating the equation showing how the energy transformed varies with distance, modified to include the effects of other ions.

$$\text{Potential energy} = -\frac{1{\cdot}747\, q^2}{4\pi\varepsilon_0 r}$$

$$F = -\frac{\text{d(energy)}}{\text{d}r} = -\frac{1{\cdot}747\, q^2}{4\pi\varepsilon_0 r^2}$$

At $r = 2{\cdot}8 \times 10^{-10}$ m this gives $F = -5{\cdot}1 \times 10^{-9}$ N. 1·747 is the factor the energy must be multiplied by to convert the $-8{\cdot}2 \times 10^{-19}$ J result, for a pair of ions, to that of $-14{\cdot}3 \times 10^{-19}$ J, the result for an ion in the midst of a crystal. 1·747 is called the Madelung constant. Its value depends on the crystal structure.

Knowing the repulsion energy and the force at $2{\cdot}8 \times 10^{-10}$ m it is possible to arrive at equations showing how the repulsive energy and the force vary with distance.

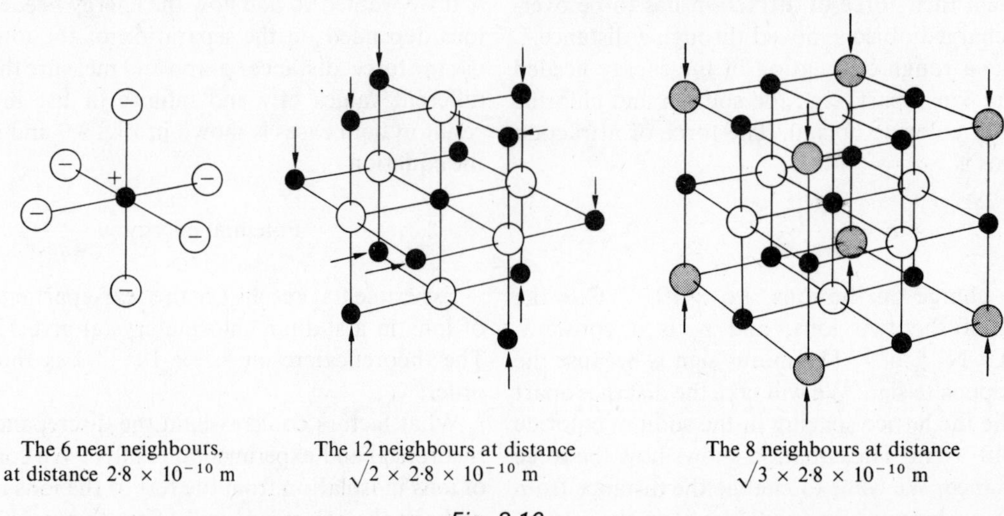

Fig. 3.10

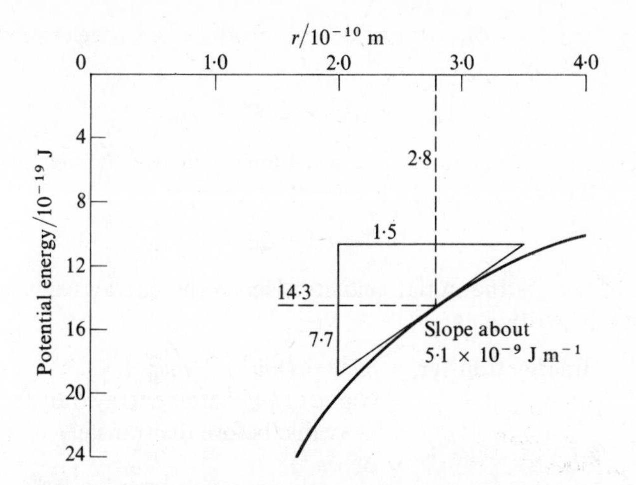

Fig. 3.11 Energy to tear an ion from a crystal as a function of lattice spacing.

For the attractive force

$$\text{Energy} \propto \frac{1}{r} \text{ and } F \propto \frac{1}{r^2}.$$

Let us make a guess and suggest that for the repulsive force

$$\text{Energy} \propto \frac{1}{r^n}$$

and thus

$$\text{energy} = \frac{K}{r^n}$$

$$F = -\frac{\text{d(energy)}}{\text{d}r} = \frac{nK}{r^{n+1}}$$

Thus

$$\frac{F}{\text{(energy)}} = \frac{n}{r}$$

Check: for the attractive force $n = 1$. At $2{\cdot}8 \times 10^{-10}$ m $F = 5{\cdot}1 \times 10^{-9}$ N and energy $= 14{\cdot}3 \times 10^{-19}$ J,

$$\frac{F}{\text{(energy)}} = \frac{5{\cdot}1 \times 10^{-9}}{14{\cdot}3 \times 10^{-19}} = \frac{1}{2{\cdot}8 \times 10^{-10}}$$

Putting in the values for the repulsive force and energy we have

$$\frac{F}{\text{(energy)}} = \frac{5{\cdot}1 \times 10^{-9}}{1{\cdot}6 \times 10^{-19}} = \frac{n}{2{\cdot}8 \times 10^{-10}}$$

$$n = 8{\cdot}9 \text{ or approximately } 9$$

Thus the repulsive energy and force relationships could be of the form

$$\text{energy} = \frac{K}{r^9}, \qquad F = \frac{nK}{r^{10}}$$

Figure 3.12 shows how the repulsive force, the attractive force, and the net force on an ion varies with distance. The attractive force is calculated from

$$F = -\frac{1{\cdot}747\, q^2}{4\pi\varepsilon_0 r^2}$$

and the repulsive force from

$$F = \frac{9 \times 1{\cdot}747\, q^2}{4\pi\varepsilon_0 r^{10}}$$

Figure 3.13 shows how the energy varies with distance for the attractive electrical forces, the repulsive forces, and the net result when the energies are summed. The net result could have been directly plotted from the equation

$$\text{Energy} = \frac{1{\cdot}747\, q^2}{4\pi\varepsilon_0 r^9} - \frac{1{\cdot}747\, q^2}{4\pi\varepsilon_0 r}$$

At $r = 2{\cdot}8 \times 10^{-10}$ m the net force on an ion is zero and the net potential energy is at a minimum.

What deductions can be made from these graphs? Compressing the crystal moves the ions closer together. When this happens the net force changes from zero to become repulsive (Fig. 3.14). The repulsive forces are now greater than the attractive forces. Extending the crystal moves the ions further apart. When this happens the net force changes from zero to become attractive, the attractive forces are now greater than the repulsive forces. In both cases, compression or attraction, the result is a force opposing the action. This could be the reason why when a solid is picked up it keeps together, e.g., why when you pick up a ruler by one end the other end follows.

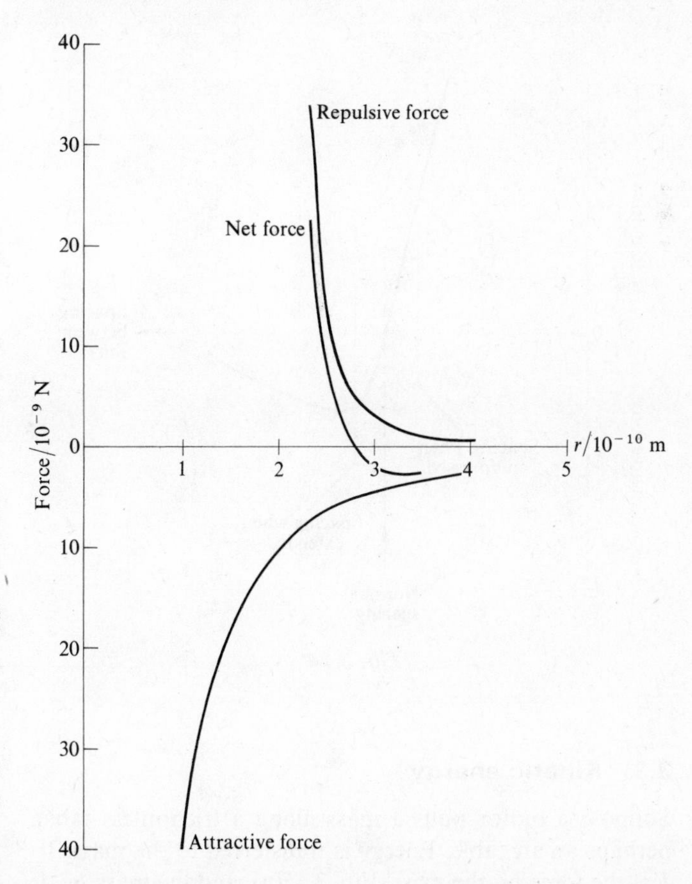

Fig. 3.12

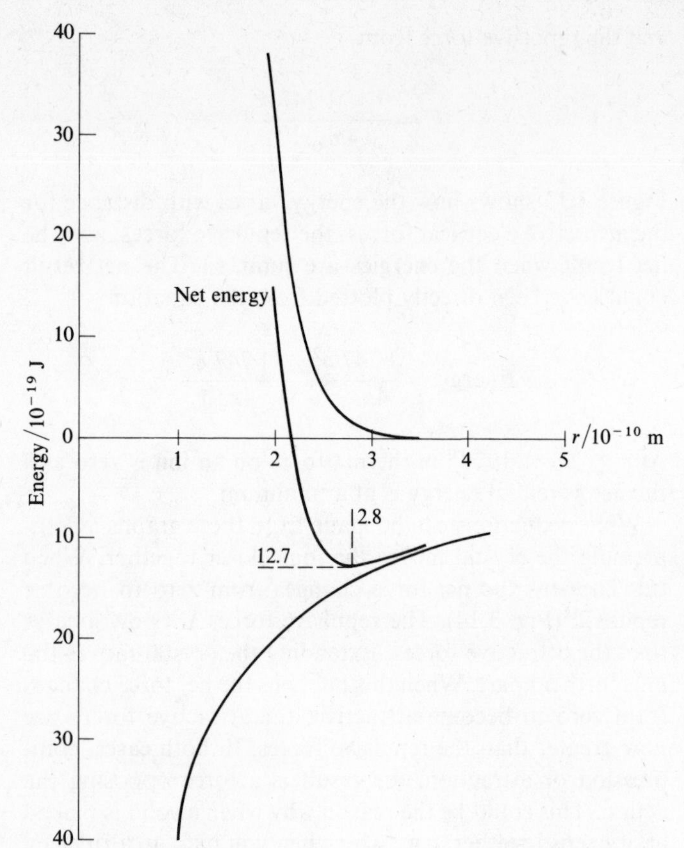

Fig. 3.13

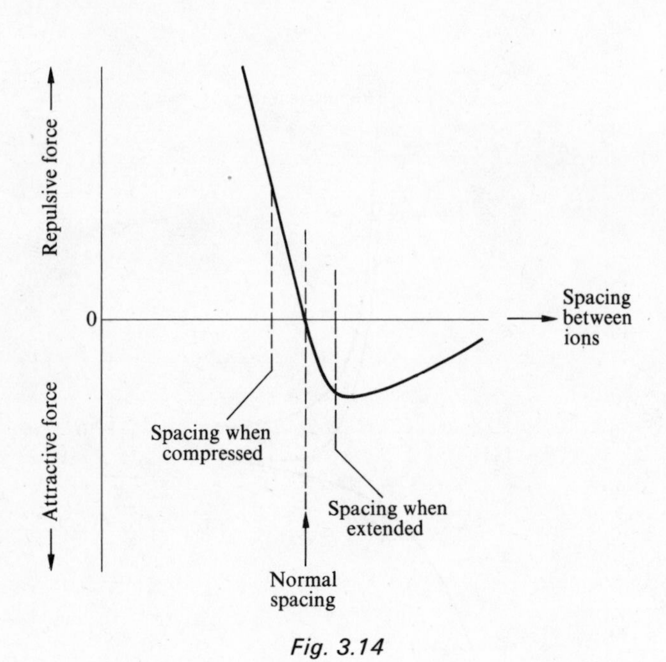

Fig. 3.14

3.3 Kinetic energy

Suppose a motor pulls a mass along a frictionless table, perhaps an air table. Energy is transferred to the mass. If F is the force on the mass, Fig. 3.15(a), and the mass, m, is pulled a distance s then the energy transferred is Fs. But the force acting on the mass will produce an acceleration a, given by $F = ma$.

$$\text{Energy transfer} = Fs = mas$$

The mass accelerates over a distance s, hence the velocity v at the end of that distance will be given by

$$v^2 = v_0^2 + 2as$$

where v_0 is the initial velocity. Hence the energy transfer can be written as

$$\begin{aligned}\text{Energy transfer} &= mas = \tfrac{1}{2}mv^2 - \tfrac{1}{2}mv_0^2 \\ &= (\text{value of } \tfrac{1}{2}mv^2 \text{ after energy transfer}) \\ &\quad - (\text{value before the transfer}) \\ &= \Delta(\tfrac{1}{2}mv^2)\end{aligned}$$

$\Delta(\frac{1}{2}mv^2)$ is the gain in energy of the mass as a result of the energy transferred to it from the motor. This is the energy used to change the state of motion of the mass. It is called kinetic energy.

The above argument still applies when the mass is not moving along a frictionless table. The start of such an argument for conditions with friction is as follows. The motor pulls the mass with a force F_1 and transfers energy F_1s (Fig. 3.15(b)). The frictional force acting on the mass produces an opposing force F_2 and thus pulls the motor with this force. The energy transferred to the motor is thus F_2s.

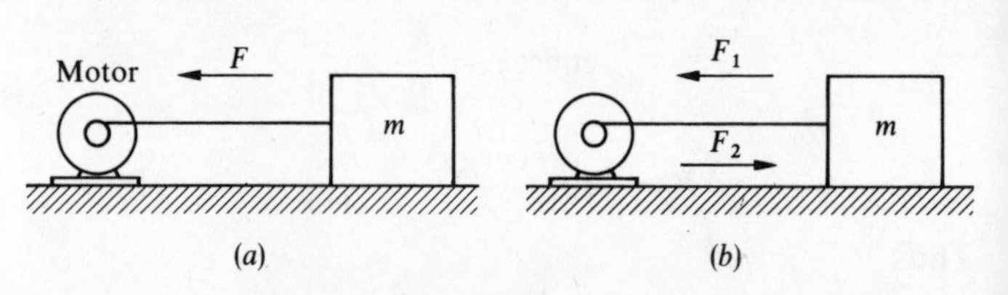

Fig. 3.15

As far as the mass is concerned it is pulling the motor and moving it towards itself. The net transfer of energy to the mass is $F_1s - F_2s = (F_1 - F_2)s$. This resultant force $(F_1 - F_2)$ produces the acceleration. The argument then continues as previously.

What happens when an object slows down? The object must lose kinetic energy—where does it go? Let us take a simple case—a ball being thrown vertically upwards. As the ball moves upwards it slows down, its kinetic energy decreases. It is, however, gaining height against the force of gravity and thus gaining potential energy. Is the gain in potential energy related to the loss in kinetic energy? If the initial velocity was v_0 and the velocity at a vertical distance h above the point of release v, then

$$\text{Change in kinetic energy} = \tfrac{1}{2}mv^2 - \tfrac{1}{2}mv_0^2 = \tfrac{1}{2}m(v^2 - v_0^2)$$

Using the equation $v^2 = v_0^2 + 2as$, we can write

$$v^2 = v_0^2 - 2gh$$

Hence the change in kinetic energy $= \frac{1}{2}m(2gh) = mgh$. The change in kinetic energy is equal to the gain in potential energy.

What happens when a ball drops? There is a gain in kinetic energy at the expense of the potential energy. What happens when the ball hits the ground and stops? The potential energy has been changed into kinetic energy, but what happens to this kinetic energy when the ball is stopped by the ground? If you hit an object with a hammer both the object and the hammer become warm. The same thing occurs when the ball hits the ground—there is a rise in temperature. The kinetic energy has been transferred through heat to increased atomic motion in the ball and the ground.

Work, kinetic, and potential energies

Why define work as force × distance? The reason given earlier in this chapter was that this definition gives a quantity from which, if we assumed that energy was conserved, we could calculate changes in height and velocity.

By utilizing Newton's laws we have, in this chapter, argued what changes in height or velocity should occur when a resultant force acts on a body. These results can be checked experimentally and are found to be in agreement with the theory.

$$Fs = \Delta(mgh) + \Delta(\tfrac{1}{2}mv^2)$$

'... there are formulas for calculating some numerical quantity, and when we add it all together it gives "28" —always the same number.'

R. B. Feynman, R. P. Leighton, M. Sands. (1963).
The Feynman Lectures in Physics, Vol. 1,
Addison Wesley, Reading, Mass.

Our formula is not quite complete—it doesn't always give the same number, only sometimes. The exceptions seem to be apparent when our force produces a change in temperature in addition to either a change in height or velocity.

Equivalence of mass and energy

According to Einstein's mechanics the mass of a body appears to change with speed (see chapter 2).

$$m_v = \frac{m_0}{\sqrt{(1 - v^2/c^2)}}$$

where m_v is the mass at speed v, m_0 is the rest mass, and c is the speed of light.

Rearranging this equation,

$$m_v = m_0(1 - v^2/c^2)^{-1/2}$$

Expanding this by the Binomial theorem (see chapter 0, question 8) gives

$$m_v \approx m_0(1 + \tfrac{1}{2}v^2/c^2 + \tfrac{3}{8}v^4/c^4 + \cdots)$$

If we have $v \ll c$, then we can neglect all but the first two terms.

$$m_v \approx m_0(1 + \tfrac{1}{2}v^2/c^2)$$
$$m_v \approx m_0 + \tfrac{1}{2}m_0v^2/c^2$$

The second term is the increase in mass due to the speed of the body.

Rearranging the equation gives

$$m_vc^2 \approx m_0c^2 + \tfrac{1}{2}m_0v^2$$

The term $\frac{1}{2}m_0v^2$ is the kinetic energy of the body according to Newtonian mechanics. Thus by giving the body a speed v we have produced a change in kinetic energy which is given by

Change in kinetic energy $\approx (\Delta m)c^2$

We can extend this by assuming, as Einstein did, that the energy of a body always equals mc^2.

$$\text{Energy} = mc^2$$

Thus when an object slows down there is a change in mass and hence a release of energy. In the large-scale everyday world this effect is very small and generally not noticeable. Figure 2.9 shows how mass varies with speed—only at speeds near to the velocity of light does the mass differ significantly from the rest mass.

Temperature changes

Sliding an object along a rough floor produces a rise in temperature and we say that some of the mechanical energy has been transferred to the surroundings to give the temperature rise.

'Without the recognition of a causal connection between motion and heat, it is just as difficult to explain the production of heat as it is to give any account of the motion that disappears.... We prefer the assumption that heat proceeds from motion, to the assumption of a cause without effect and of an effect without a cause....'

J. R. Mayer (1842).
Annalen der Chemie und Pharmacie.

The word heat is used to describe the transfer of energy which results in a temperature change (or phase change). Do we get the same amount of energy transferred (heat) every time we 'use up' mechanical energy?

'The apparatus ... consisted of a brass paddle wheel working horizontally in a can of water. Motion could be communicated to this paddle by means of weights, pulleys, &c....

The paddle moved with great resistance in the can of water, so that the weights (each of four pounds [1·8 kg]) descended at the slow rate of about one foot [30 cm] per second. The height of the pulleys from the ground was twelve yards [8·4 m], and consequently, when the weights had descended through that distance, they had to be wound up again in order to renew the motion of the paddle. After this operation had been repeated sixteen times, the increase of the temperature of the water was ascertained by means of a very sensible [sensitive] and accurate thermometer.

A series of nine experiments was performed in the above manner, and nine experiments were made in order to eliminate the cooling or heating effects of the atmosphere ... it appeared that for each degree [this degree is a unit equivalent to about 1000 J] of heat evolved by the friction of the water, a mechanical power equal to that which can raise a weight of 890 lbs [403 kg] to the height of one foot [30 cm], had been expended.'

J. P. Joule (1845) *Phil. Mag.*, **27**, 205

After further experiments and taking into account the work of other experimenters Joule was able to conclude:

'1st. That the quantity of heat produced by the friction of bodies, whether solid or liquid, is always proportional to the quantity of force [we would use the word energy] expended. And,

2nd. That the quantity of heat capable of increasing the temperature of a pound [0·45 kg] of water ... by 1° Fahr [0·56 deg C], requires for its evolution the expenditure of a mechanical force [energy] represented by the fall of 772 lbs [350 kg] through the space of one foot [30 cm].'

J. P. Joule (1850), *Phil. Trans. Roy. Soc.*, **140**, 61.

Thus a change in temperature can be produced when 'motion disappears'. We say that mechanical energy has been transformed to molecular kinetic energy—a temperature change.

The size of this temperature change is directly proportional to the mechanical energy transformed. The amount of energy needed to produce a specific temperature change in a given body is a constant. Thus 1 kg of water always needs 4180 J to give a 1 deg C temperature change. (The word 'always' should be taken to mean always working at the same temperatures, e.g., 18 to 19°C. Every time the experiment is done, between these two temperatures, the same energy is needed.)

'... You will therefore be surprised to hear that until very recently the universal opinion has been that living force [kinetic energy] could be absolutely and irrevocably destroyed at any one's option. Thus, when a weight falls to the ground, it has been generally supposed that its living force is absolutely annihilated, and that the labour which may have been expended in raising it to the elevation from which it fell has been entirely thrown away and wasted, without the production of any permanent effect whatever. We might reason, à priori, that such absolute destruction of living force cannot possibly take place, because it is manifestly absurd to suppose that the powers with which God has endowed matter can be destroyed any more than that they can be created by man's agency; but we are not left with this argument alone, decisive as it must be to every unprejudiced mind. The common experience of every one teaches him that living force is not destroyed by the friction or collision of bodies.... Experiment ... has shown that, wherever living force is apparently destroyed, an equivalent is produced which in process of time may be reconverted into living force. This equivalent is heat. Experiment has shown that wherever living force is apparently destroyed or absorbed, heat is produced.... In these conversions nothing is ever lost. The same quantity of heat will always be converted into the same quantity of living force....'

J. Joule (1847). 'On matter, living force, and heat', *The scientific papers of James Prescott Joule*, The Physical Society, 1884, p. 265.

Fuels

Coal is a fuel, mainly carbon. When carbon burns in oxygen carbon dioxide is produced.

$$C + O_2 \rightarrow CO_2$$

The reaction when once started produces a rise in temperature, showing that energy is being transferred from the carbon to the surroundings. One mole, i.e., 12 g, of carbon produces 393 000 J of energy. Thus by breaking the bond between two oxygen atoms and combining them with a carbon atom we release energy.

The human body uses the energy stored in food in a similar manner to an engine burning coal. The reaction in the body yields carbon dioxide, water, and energy. Typical values for the energy released per kilogramme of food are:

Butter	27 000 J
Cheese	15 600 J
Sugar	14 000 J
Beef	12 000 J
Bread (white)	9 500 J
Potato	2 700 J

The energy we use to perform jobs can be estimated by the amount of carbon dioxide we breathe out. Typical energy values are:

Resting in bed	66 J s^{-1}
Walking	340 J s^{-1}
Cycling	490 J s^{-1}

3.4 Electrical energy

We can make a motor work by using a fuel such as petrol or by electricity. Electricity would thus seem to be a sort of 'fuel' or form of energy.

If we supply a current to an electric motor we can measure the energy transferred per second by allowing the motor to pull a weight (Fig. 3.16) and measuring the distance covered by the weight in one second. The experiment neglects the electrical energy used to produce a temperature change of the motor. Roughly the results of such an experiment indicate that the energy transformed per second is proportional to the square of the current. An alternative experiment is to

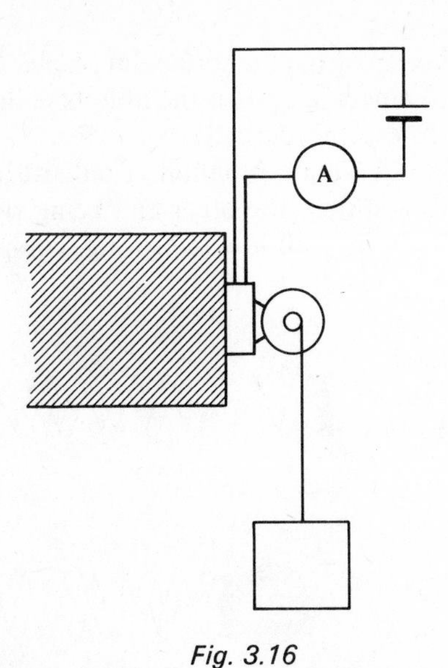

Fig. 3.16

use all the electrical energy to produce a temperature change, e.g., an immersion heater in a block of some substance, and then find what mechanical energy has to be transformed to give the same temperature change. Such an experiment shows that the energy transformed is proportional to the square of the current (see section 10.1 for further details).

$$\text{Energy transferred per second} \propto I^2$$

The constant of proportionality is the resistance of the coil (see section 10.1).

$$\text{Energy transferred per second} = RI^2$$

Hence as $V = IR$

$$\text{Energy transferred per second} = IV$$

The units of energy are joules, current amps, voltage volts, resistance ohms.

3.5 Forms of energy and conservation

Petrol in a car engine enables the car to gain kinetic and perhaps potential energy. The energy in the fuel is transformed into mechanical energy. A fuel could, however, have been used just to provide a temperature change, e.g., coal on a fire. In that case the energy is transferred by means of heat to give a temperature change. The heating or the movement of the car could have been produced by using electricity. Electricity is another way by which energy can be transferred. Nuclear reactors can be used to provide energy—thus we have another form of energy.

Energy can exist in many forms and can be changed from one form to another. There is a firm belief that whatever the form and whatever the changes we make energy is always conserved. That is, if we add up all the energy in some system before an event and then add up all the bits after the event we will have the same amount of energy. This is known as the law of conservation of energy. The belief in this law is so strong that physicists will believe something to be wrong with their experiments or that they have missed something if calculations in an experiment seem to show that energy is not conserved.

One example of this was the early experiments with radioactivity.

'We have discovered that the salts of radium constantly release heat.

An iron-constantan thermo-electric couple, whose junctions are surrounded, one by radium-bearing barium chloride and the other by pure barium chloride, in fact shows a difference in temperature between the two substances. . . .

We have attempted a quantitative evaluation of the heat developed by radium in a given time.

For that purpose, we first compared this heat with that developed by an electric current of known intensity in a wire of known resistance. . . .

One g. of radium develops a quantity of heat of the order of 100 small calories [400 J] per hour. . . .

The hypothesis of a continuous modification of the atom is not the only one compatible with the development of heat by radium. This development of heat may still be explained by supposing that the radium makes use of an external energy of unknown nature.'

P. Curie and A. Laborde (1903). *Comptes rendus de l'Academie des Sciences*, Paris, **136**, 673.

'. . . Kelvin had an idea. . . . Perhaps, he suggested, there was some kind of energy that one could not detect . . . floating around in space, and perhaps radium had the property of absorbing it, like a fountain, and shooting it out, and that was what was observed. Even in those days people were perfectly willing to balance the books on conservation of energy in such a manner.'

E. Condon (Oct. 1962). *Physics Today*, p. 44.

The following quotations give the views of various scientists on the conservation of energy.

'A remark of Poincaré is often quoted to the effect that if we ever found the conservation law for energy appearing to fail we would recover it by inventing a new form of energy. This it seems to me is a misleadingly partial characterization of the situation. If in any specific situation the law apparently failed, we would doubtless first try to maintain the law by inventing a new form of energy, but when we had invented it we would demand that it be a function (of numbers, or parameters, that describe the state of the system) and that the law would continue to hold for all the infinite variety of combinations into which the new parameters might be made to enter. Whether conservation would continue to hold under such extended conditions, could be determined only by experiment. The energy concept is very far from being merely a convention.

P. W. Bridgman (1941). *The nature of thermodynamics*, Harvard Univ. Press.

'Most people seem to believe this [the conservation of energy] firmly; mathematicians because they believe it is a fact of observation; observers because they believe it is a theorem of mathematics; philosophers because they believe it is aesthetically satisfying, or because they believe no inference based upon it has ever been proven false, or because they believe new forms of energy can always be invented to make it true. A few neither believe nor disbelieve it; these people maintain that the First Law is a procedure for bookkeeping energy changes, and about bookkeeping procedures it should be asked, not are they true or false, but are they useful.'

H. A. Bent (1965). *The second Law*, Oxford Univ. Press, New York.

'There is a fact, or if you wish, a law, governing all natural phenomena that are known to date. There is no known exception to this law—it is exact so far as we know. The law is called the conservation of energy. It states that there is a certain quantity, which we call energy, that does not change in the manifold changes which nature undergoes. That is a most abstract idea, because it is a mathematical principle; it says that there is a numerical quantity which does not change when something happens. It is not a description of a mechanism, or anything concrete; it is just a strange fact that we can calculate some number and when we finish watching nature go through her tricks and calculate the number again, it is the same. (Something like the bishop on a red square, and after a number of moves—details unknown—it is still on some red square. It is a law of this nature.)'

R. B. Feynman, R. P. Leighton, M. Sands (1963). *The Feynman Lectures on Physics*, Vol. 1, Addison-Wesley, Reading, Mass.

'The transactions of the material universe appear to be conducted, as it were, on a system of credit. Each transaction consists of the transfer of so much credit or energy from one body to another. This act of transfer or payment is called work. The energy so transferred does not retain any character by which it can be identified when it passes from one form to another.'

J. C. Maxwell (1877). *Matter and motion.*

Conservation of kinetic energy

When two objects collide momentum is conserved. Is kinetic energy conserved in collisions? If no kinetic energy is converted into other forms of energy, kinetic energy must be conserved. This means no change in potential energy, the collision takes place on a horizontal plane. This means no change in temperature when the objects collide. Under what conditions can this occur?

Let us look at a simple situation, a billiard ball hitting the free end of a spring, the other end being rigidly fixed (Fig. 3.17). When the ball hits the spring, compression of

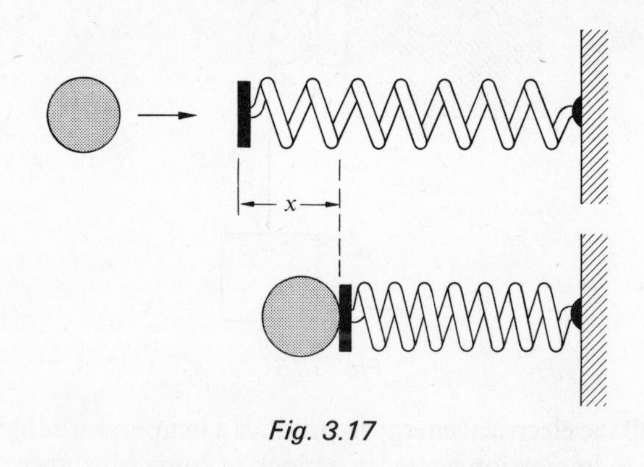

Fig. 3.17

the spring occurs. The kinetic energy of the ball is transferred to the elastic energy of the spring. If the ball is to regain all its initial kinetic energy then when the spring re-extends all the elastic energy produced by the collision must be converted back to kinetic energy. The energy transferred by the ball doing work on the spring must be equal to the energy transferred by the spring doing work on the ball. In other words,

$$\underbrace{\int_0^s F\,\mathrm{d}s}_{\text{during compression of the spring}} = \underbrace{-\int_s^0 F\,\mathrm{d}s}_{\text{during extension of the spring}}$$

This equation only states that the area under the force–distance graph, between the distance equal to zero and s lines (Fig. 3.18), must be the same during the compression

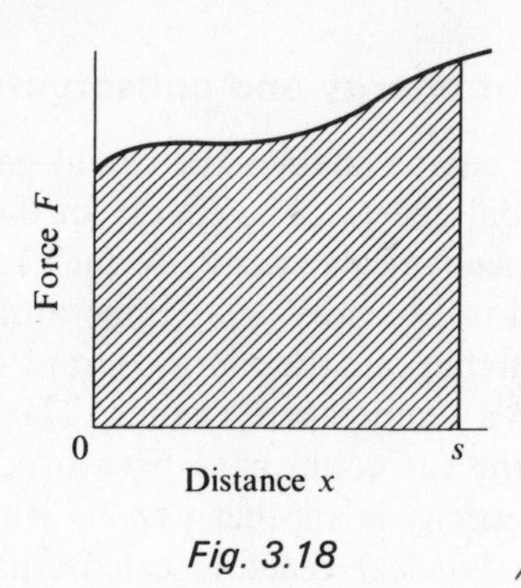

Fig. 3.18

and the extension. It does not matter how the force changes with compression distance x as long as the force–distance curve is the same during the compression as the extension. Such a collision is known as an elastic collision. In the case of the collision between the ball and the spring a perfectly elastic collision does not occur. It is possible for it to be very nearly elastic, to such an extent that any discrepancy is difficult to detect.

When two billiard balls collide the explanation is the same—the balls become elastically strained during the collision. The collision is perfectly elastic if the force–distance graph for billiard balls under compression is the same as when they become uncompressed.

Energy as an additive property

A very important property of energy is that it is additive. If we take 5 ml of water and add a further 3 ml then the result is 8 ml—volume is an additive property. Mass is another additive property. Temperature is, however, not an additive property—the temperature of an object is not equal to the sum of the temperatures of the constituent parts. If we take two beakers of water, each at a temperature of 20°C, and pour them into just one of the beakers the result is still a temperature of 20°C—not 40°C. If a certain mass of fuel will release 50 J of energy and another mass 30 J then putting the two masses together will give 80 J of energy—energies are additive.

3.6 Heat and temperature

When a hot and a cold object are put in contact does energy 'flow' between the two until they both have the same amount of energy or both are at the same temperature?

> 'Even without the help of thermometers, we can perceive a tendency of heat to diffuse itself from any hotter body to the cooler ones around it, until the heat is distributed among them in such a manner that none of them is disposed to take any more from the rest. The heat is thus brought into a state of equilibrium.
>
> This equilibrium is somewhat curious. We find that, when all mutual action is ended, a thermometer applied to any one of the bodies undergoes the same degree of expansion. Therefore the temperature of them all is the same. No previous acquaintance with the peculiar relation of each body to heat could have assured us of this, and we owe the discovery entirely to the thermometer. We must therefore adopt, as one of the most general laws of heat, the principle that all bodies communicating freely with one another, and exposed to no inequality of external action, acquire the same temperature, as indicated by a thermometer. . . .'
>
> J. Black, 1759–1762, published 1803.

Do Black's observations seem obvious? Would you expect, on the basis of your scientific knowledge, that two objects put in contact will come to the same temperature and not to the same heat energy content? (At this point in the book we will not try to define temperature as anything more than—what a thermometer reads. Chapter 8 will discuss the meaning of temperature.) How do we know that the same temperature does not mean the same energy content?

> 'It was formerly a common supposition that the quantities of heat required to increase the temperatures of different bodies by the same number of degrees were directly proportional to the quantities of matter in them (and thus to their weights); and, therefore, when the bodies were of equal volumes, that the quantities of heat were proportional to their densities (and thus to their weights per unit volume). But very soon after I began work on this subject (anno 1760), I perceived that this opinion was a mistake, and that the quantities of heat which different kinds of matter must receive to raise their temperatures by an equal number of degrees are not in proportion to the quantity of matter in each, but in proportions widely different from this, and for which no general principle or reason could be assigned. . . .
>
> . . . let us suppose the water to be at 100°F [the Fahrenheit scale of temperature] and that an equal volume of warm quicksilver at 150°F is suddenly mixed and agitated with it. We know that the temperature midway between 100° and 150°F is 125°F, and we know that this middle temperature would be produced by mixing cold water at 100°F with an equal volume of warm water at 150°F, the temperature of the warm water being lowered by 25 degrees while that of the cold is raised by just as much. But when warm quicksilver is used in place of warm water, the temperature of the mixture turns out to be only 120°F, instead of 125°F. The quicksilver, therefore, has cooled through 30 degrees, while the water has become warmer by 20 degrees only; and yet the quantity of heat which the water has gained is the very same as that which the quicksilver has lost. This shows that the same quantity of heat has more effect in warming quicksilver than in warming an equal volume of water, and therefore that a smaller quantity of it is sufficient for increasing the temperature of quicksilver by the same number of degrees. . . .
>
> J. Black, 1759–1762, published 1803.

Different materials, same mass or same volume, at the same temperature produce different temperature changes when placed in thermal contact with say cold water. Materials at the same temperature do not pass onto another different temperature object the same amount of energy. The energy associated with a body at a definite temperature depends on the nature of the body.

When a substance is heated, energy is supplied to it and causes a rise in temperature of the substance. What factors determine the temperature change? One way of experimentally investigating this is to supply the energy by an immersion heater. Electrical energy can be easily measured with a voltmeter and an ammeter (Fig. 3.19). (Energy per second = IV.) The resulting temperature change can be measured with a thermometer. The following are typical results for a block of aluminium, mass 1 kg.

Electrical energy supplied	*Temperature change*
500 J	0·5 deg C
1000 J	1·1 deg C
1500 J	1·6 deg C
2000 J	2·2 deg C

(An immersion heater was used which for a 12 V supply gave 50 J per second. Readings were taken after successive intervals of 10 s.)

To a reasonable approximation the results show that the temperature change is directly proportional to the energy supplied.

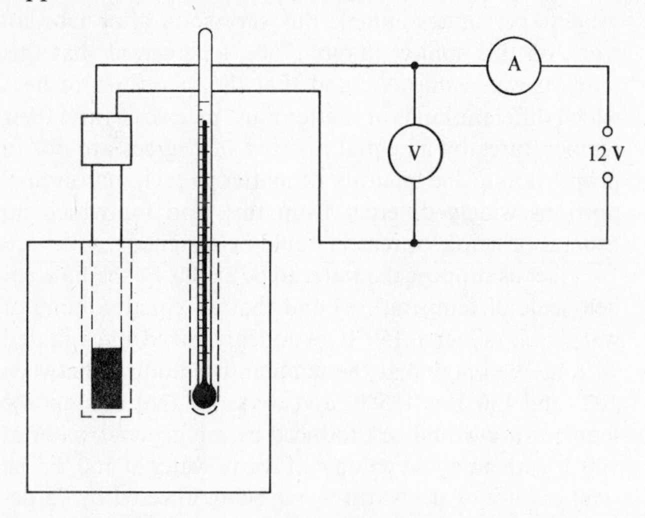

Fig. 3.19

If an aluminium block of twice the mass is used 1000 J is needed to give a temperature change of about 0·5 deg C. Doubling the mass halves the temperature change. For a given energy supply the product of mass m and temperature change Δt is a constant. Doubling the energy doubles the product of mass and temperature change.

$$m\,\Delta t \propto \text{energy supplied}$$

or by introducing some constant, C,

$$Cm\,\Delta t = \text{energy supplied}$$

If a different material is used the same type of relationship is found; there is, however, a different constant of proportionality. The constant, C, is called the specific heat capacity of the substance. Thus the specific heat capacity is the energy needed to raise the temperature of 1 kg of a substance by 1 deg C. The unit of specific heat capacity is therefore J kg^{-1} deg C^{-1}.

The concept of specific heat capacity was introduced by J. Black about the middle of the eighteenth century.

For the aluminium block the results give a specific heat capacity of about 900 J kg^{-1} deg C^{-1}. More precise measurements show that specific heat capacities are not constant but vary, slightly, with temperature (Fig. 3.20). The results also depend on the physical circumstances of the object when the heat energy is supplied. The most noticeable differences occur with gases—the specific heat capacity measured for a gas at constant pressure being significantly different from the result obtained at constant volume. The differences are much less marked with solids and liquids—the results normally quoted being at constant pressure (atmospheric pressure). Typical specific heat capacities:

	Temperature /deg C	*Specific heat capacity* *at constant pressure* /J kg^{-1} deg C^{-1}	*at constant volume* / J kg^{-1} deg C^{-1}
Air (dry)	20	1006	718
Carbon dioxide	20	897	690
Aluminium	20	890	867
Copper	20	380	375

See chapters 8 and 9 for further discussion of specific heat capacities.

The following is an example of an industrial specific heat capacity measurement.

'INTRODUCTION

For the calculation of thermal transients occurring in a reactor, due, for example, to changes in power level, a knowledge of the specific heat [capacity] of the materials involved is required. An adiabatic calorimeter similar to

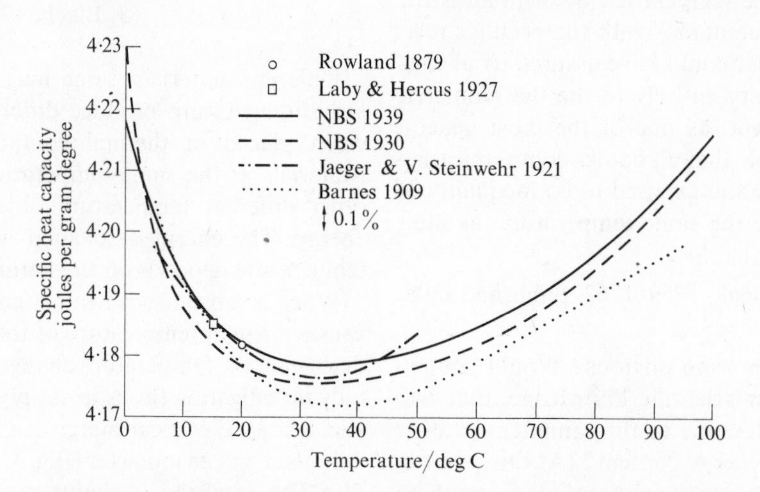

Fig. 3.20 Specific heat capacity of water. (Taken from Osborne, Stimson, Ginnings, J. Res. Nat. Bureau Standards *23, 246, 1939.)*

that described by Armstrong, has been constructed for measurements on some fuel and canning materials.

In this method the temperature of the specimen is increased by supplying a known amount of electrical energy, while loss of heat is reduced by a guard system surrounding the specimen and maintained at the same temperature. The specimen is in two similar halves, providing a symmetrical arrangement around the internal heater and also ease of heater assembly. The specimen size is 12 cm^3, so the method allows specific heat [capacity] determinations to be made on materials which are only available in small amounts.

THEORY

The amount of heat q required to increase the temperature of a mass m by $\Delta\theta$ is given by

$$q = ms\,\Delta\theta \qquad (1)$$

where s is specific heat [capacity].

If this heat is provided electrically in time t,

$$q = EIt = ms\,\Delta\theta \qquad (2)$$

where E = voltage
I = current

Some of the heat supplied may be (a) absorbed by material close to the specimen, or (b) lost by conduction along leads attached to the specimen.

Equation (2) may be re-written

$$q = EIt = ms\,\Delta\theta + \sum(m's'\,\Delta\theta') + \sum\left(\frac{KA\theta}{l}\cdot t\right) \qquad (3)$$

where $\sum(m's'\,\Delta\theta)$ represents the summation of effect (a) and $\sum\left(\frac{KA\theta}{l}\cdot t\right)$ the summation of conduction losses (b) by various paths of conductivity K, cross-sectional area A and thermal gradient θ/l.

It will be observed that the magnitude of (b) may be found by measuring the rate of supply of energy necessary to maintain the system at constant temperature. Loss (a) may then be determined by varying the mass m' and extrapolating to $m' = 0$. Alternatively, if a measurement is carried out with a material of known specific heat, the overall value of both error terms for particular temperature conditions may be obtained.'

W. H. Martin and J. Rixon (1959). *J. Sci. Instrum.*, **36**, 179.

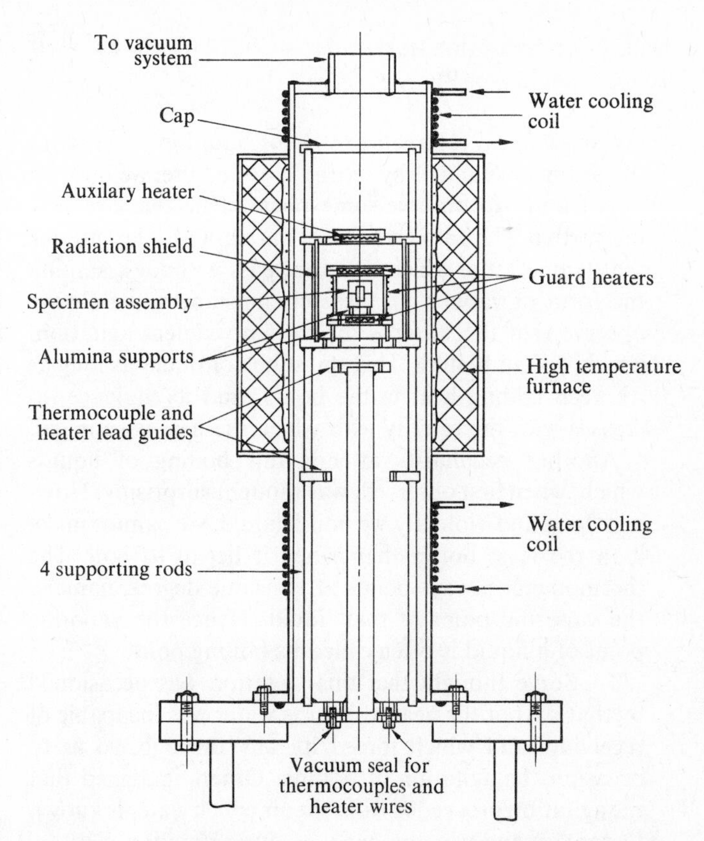

Fig. 3.21 General assembly of specific heat capacity apparatus.

Latent heat

Suppose that instead of trying to measure the specific heat capacity of aluminium we used a block of ice. If the ice was say at an initial temperature of $-10°C$ and we supplied enough energy to raise the temperature by a few degrees, then the result would be a specific heat capacity of about 2000 J kg^{-1} deg C^{-1}. This is quite different from the specific heat capacity of liquid water, about 4200 J kg^{-1} deg C^{-1} at 10°C. What happens to the values of specific heat capacity if measurements are taken within the interval -10 to $+10°C$? The specific heat capacity of ice increases slightly as the temperature is taken from $-10°C$ to 0°C. At 0°C the specific heat capacity would seem to be infinite in that energy is absorbed by the ice without any change in temperature. The ice changes to liquid water. From 0°C up to 10°C the specific heat capacity of the liquid falls slightly (Fig. 3.22).

The amount of energy absorbed in making the transition from ice at 0°C to liquid water at 0°C is proportional to the mass of water. The energy needed for 1 kg is known as the specific latent heat. In the case of water the specific latent heat is 334 000 J kg^{-1}

A similar situation occurs when the liquid changes to a vapour; energy is needed for the change. The specific latent

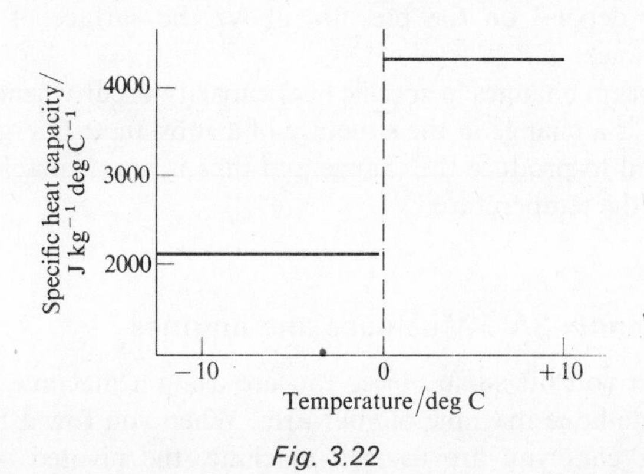

Fig. 3.22

heat of vaporization for water is 2 270 000 J kg^{-1} at 100°C (see chapter 9).

'When we heat a large quantity of liquid in a vessel, in the ordinary manner, by setting it on the fire, we have an opportunity to observe some phenomena that are very instructive. The liquid gradually warms, and at last attains that temperature which it cannot pass without assuming the form of vapour. In these circumstances, we always observe that the water is thrown into violent agitation, which we call boiling. This agitation continues as long as we keep adding heat to the liquid, and its violence increases with the celerity with which the heat is supplied.

Another peculiarity attends this boiling of liquids which, when first observed, was thought surprising. However long and violently we boil a liquid, we cannot make it in the least hotter than when it began to boil. The thermometer always points at the same degree, namely, the vaporific point of that liquid. Hence the vaporific point of a liquid is often called its boiling point.

... Some thought that this agitation was occasioned by that part of the heat which the water was incapable of receiving, and which forced its way through, so as to occasion the agitation of boiling. Others imagined that the agitation proceeded from the air which water is known to contain and which is expelled during boiling....

A more just explanation will occur to any person who will take the trouble to consider this subject with patience and attention. In the ordinary manner of heating water, the source of heat is applied to the lower parts of the liquid. If the pressure on the surface of the water (such as that due to the atmosphere) be not increased, the water will soon acquire the highest temperature that it can attain without assuming the form of vapour. Any subsequent addition of heat must therefore convert into vapour that part of the water which it enters. As this added heat all enters at the bottom of the liquid, elastic vapour [steam] is constantly produced there, which, because it weighs almost nothing, must rise through the surrounding water and appear to be thrown up to the surface with violence, ...'

J. Black, 1759–1762, published 1803.

The temperatures at which fusion and vaporization take place depend on the pressure above the surface of the substance.

Abrupt changes in specific heat capacity occur whenever there is a change in the structure of a substance. Energy is needed to produce the change and thus is not available to raise the temperature.

Appendix 3A Machines and engines

When you lift some object you are using a machine, the muscle-bone machine of your arm. When you row a boat using oars you are using a machine, the pivoted oars. Pulleys, screws, water wheels, gears are all examples of machines—machines that have been in use for many centuries. A machine can be defined as 'a contrivance for redirecting effort to better advantage' (A. R. Ubbelohde, *Man and energy*, Hutchinson 1954). Thus a large force applied over a short distance may be redirected to give a small force applied over a long distance, a small force over a long distance may be redirected to give a large force over a short distance. The contrivance for making such changes is called a machine.

One of the simplest machines is the lever. Figure 3.23 shows some forms of lever. In all the examples we have the product of the force applied and the distance of the point of application to the pivot equal to the product of the load force and its point of application distance from the pivot. The product of a force and the distance, along the line of action of the force, through which its point of application

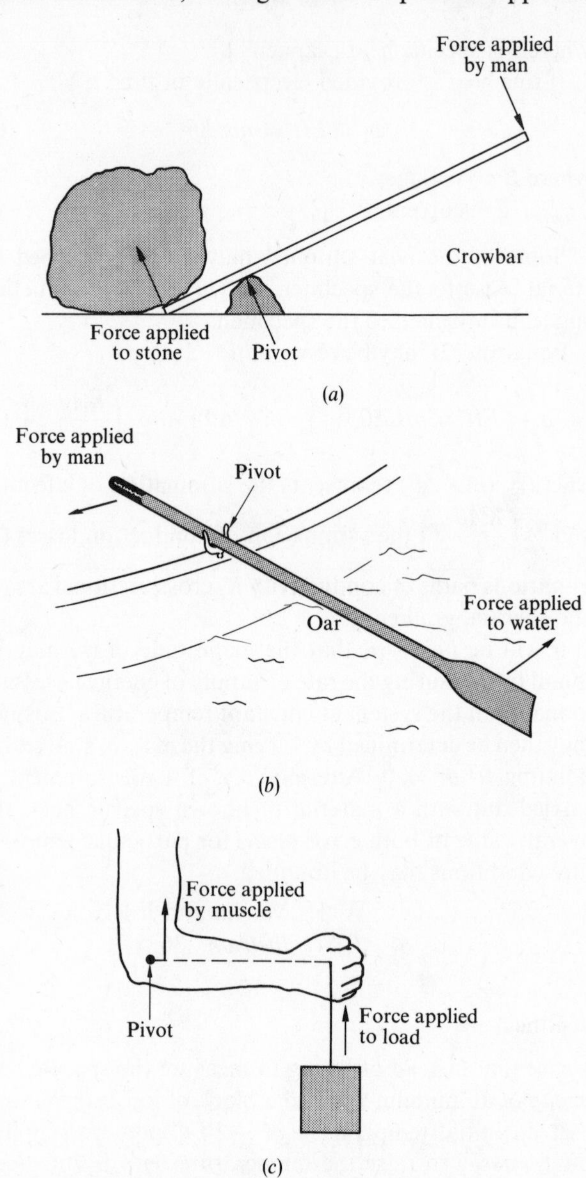

Fig. 3.23 (a) Using a crowbar to move a large stone. (b) Using an oar to propel a boat through water. (c) Lifting a load.

moves is called work and is a measure of the energy transferred. The energy taken from the 'source', e.g., the man in (a) and (b), is transferred to the load and as energy is conserved, then that energy transferred at one end of the system is equal to that transferred at the other end, hence the identity of the force–distance product. The lever is a means of transferring energy.

In Fig. 3.23(a) the force applied by the man is less than that applied to the stone, the distance of the point of application of the man's force from the pivot is greater than that from the point of application of the force to the stone from the pivot. In Fig. 3.23(b) the force applied by the man is greater than that applied to the water by the oar, the distance of the point of application of the man's force from the pivot is less than that from the point of application of the force to the water from the pivot. In Fig. 3.23(c) the force applied by the muscle is greater than that applied to the load. Figures 3.23(c) and (b) are examples of what might be called 'distance magnifier' machines, Fig. 3.23(a) being an example of a 'force magnifier' machine.

A pulley wheel acts like a lever where the distances from the force applied by the man and that applied to the object to the pivot are equal, the pulley merely giving a change in direction of the force (Fig. 3.24). The distance through

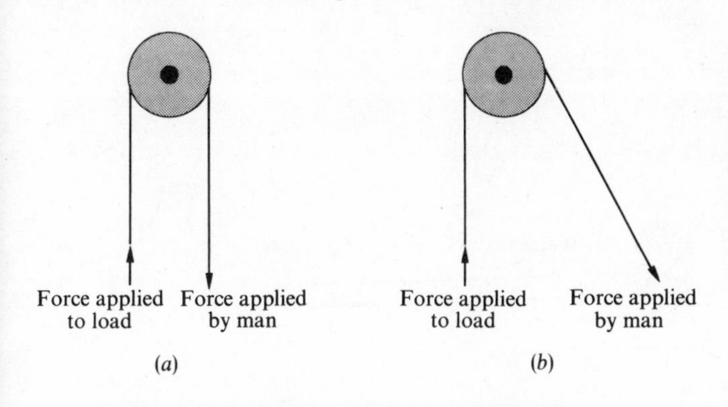

Fig. 3.24 Single pulley wheel diagrams.

which the force applied by the man moves must be equal to the distance moved by the load, they are at the opposite ends of the same rope. If we take energy to be conserved then this means that the forces must be equal. This applies regardless of whether we have the situation represented by Fig. 3.24(a) or (b).

Figure 3.25 shows two more complex pulley systems. For the system represented by (a) the distance through which the man's force moves is twice that which the load moves through. The force applied to the load is thus twice that exerted by the man. In (b) the distance through which the man's force moves is four times that which the load moves through, hence the force applied to the load is four times that applied by the man. One way of working out these distances is to consider how each of the ropes must be shortened if the load is to be lifted through say 1 m. The total amount of shortening must be equal to the distance over which the man must exert his pull if the rope remains taut.

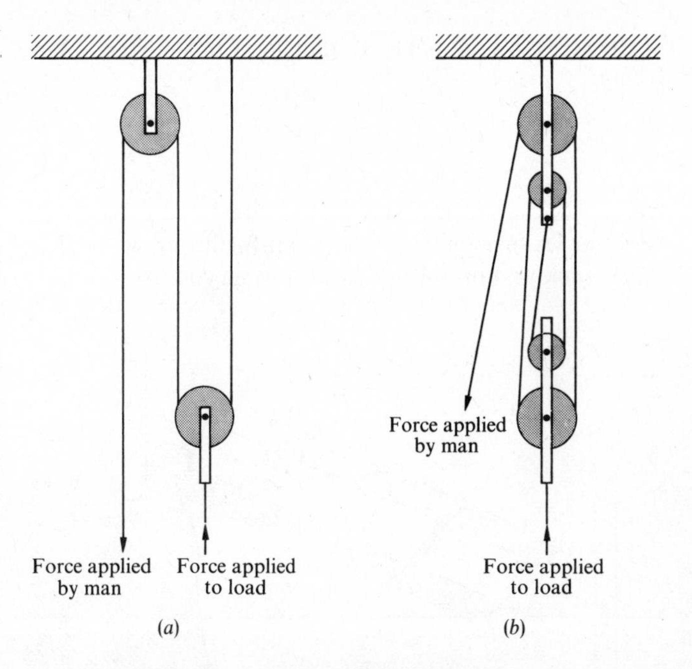

Fig. 3.25 Multiple pulley wheel diagrams.

The use of pulleys and levers as machines dates back to well before Christ. Archimedes (third century B.C.) devised some pulley systems. In erecting the pyramids (3000 to 2000 B.C.) the Ancient Egyptians used a very simple machine, the inclined plane. The Great Pyramid at Giza is about 160 m high and is constructed of some two and a half million stones, the mass of each being about $2\frac{1}{2}$ tonnes (1 tonne is 1000 kg). A force of $2{\cdot}5 \times 10^6$ N has thus to be applied to lift one of these stones vertically. To push it along an inclined plane the force needed is that to overcome the component of the gravitational attraction down the plane, $mg \sin\theta$ for a plane making an angle θ with the horizontal, and the frictional force, $\mu mg \cos\theta$, where μ is the coefficient of friction. The frictional term can be made quite low by using lubricants; the Egyptians poured water under the runners on which they slid the stone blocks. With a plane inclined at 10° and taking μ as 0·1 the force needed to slide a $2\frac{1}{2}$ tonne block is about $0{\cdot}75 \times 10^6$ N, about one-third of that needed for direct lifting.

A screw can be considered as an inclined plane wrapped round a cylinder (Fig. 3.26). With a helical thread of constant pitch, one revolution of the screw changes the height

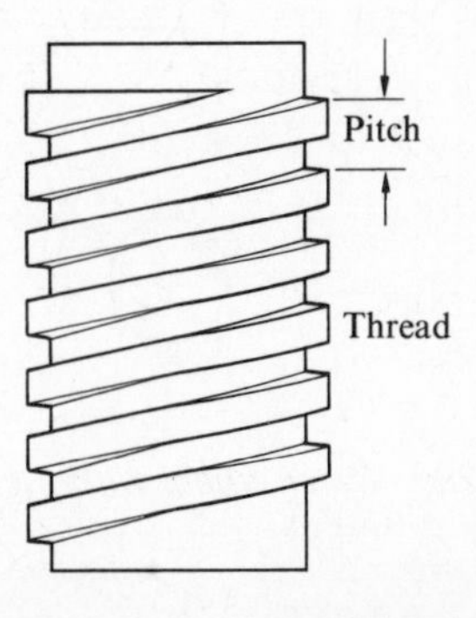

Fig. 3.26 Threaded rod.

of the screw by the pitch. The angle of this inclined plane is thus

$$\tan \theta = \frac{p}{2\pi r}$$

where r is the mean radius of the screw and p the pitch. Archimedes invented a screw pump for lifting water (Fig. 3.27); such pumps are still used in some countries.

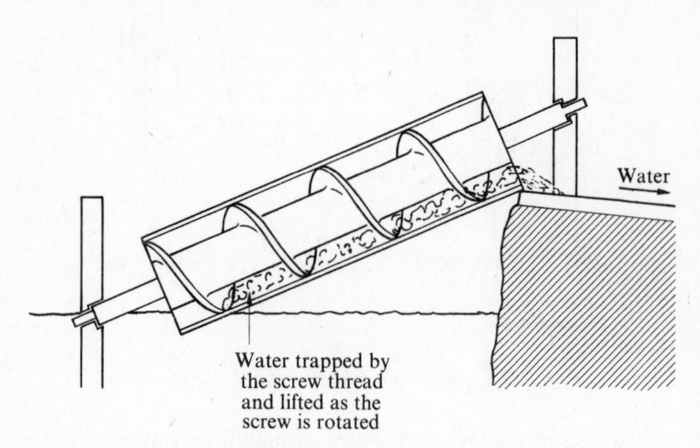

Fig. 3.27 Screw thread lifting water.

Many Greek and Roman inventions were concerned with the raising of water. During the first century B.C. water wheels for this purpose were in use in many parts of the Roman Empire (Fig. 3.28). The rotating wheel, rotated by hand or horse, trapped water in its buckets and carried it up out of the river before spilling it into a channel at the top of the wheel. Such wheels, generally driven by oxen or donkeys, are still in use in many countries. The animals walk in a horizontal circular path and turn a capstan. This converts the rotation about a vertical axis into a rotation about a horizontal axis (Fig. 3.29).

The Romans were responsible for the spread of water mills throughout their empire—these utilized water power to produce rotation of a wheel. In one form the rotation was produced by blades on the wheel dipping into a flowing river, in another form where there was a drop in the water level the water from the upper level was passed along a chute to the top of the wheel and caused the rotation by falling onto the blades (Fig. 3.30).

An essential feature of water mills was the use of gearing. By this means the water wheel was able to rotate at a different speed to that of the working wheel, perhaps for grinding corn. When two different diameter wheels, in contact at a point on their circumferences, rotate without slipping then

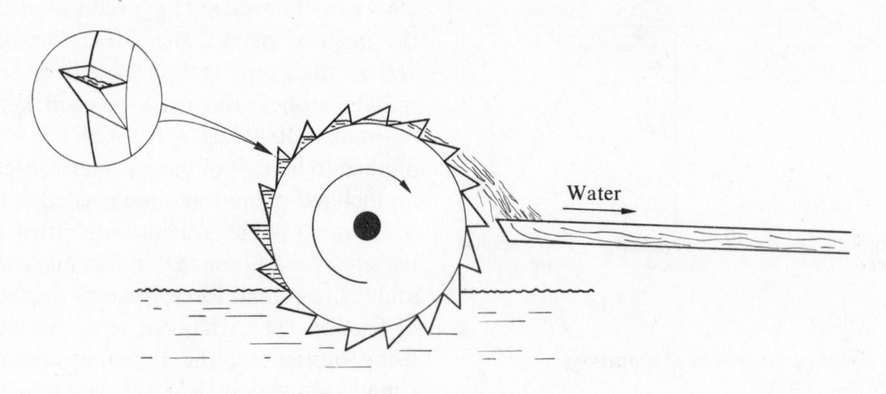

Fig. 3.28 Water-raising wheel.

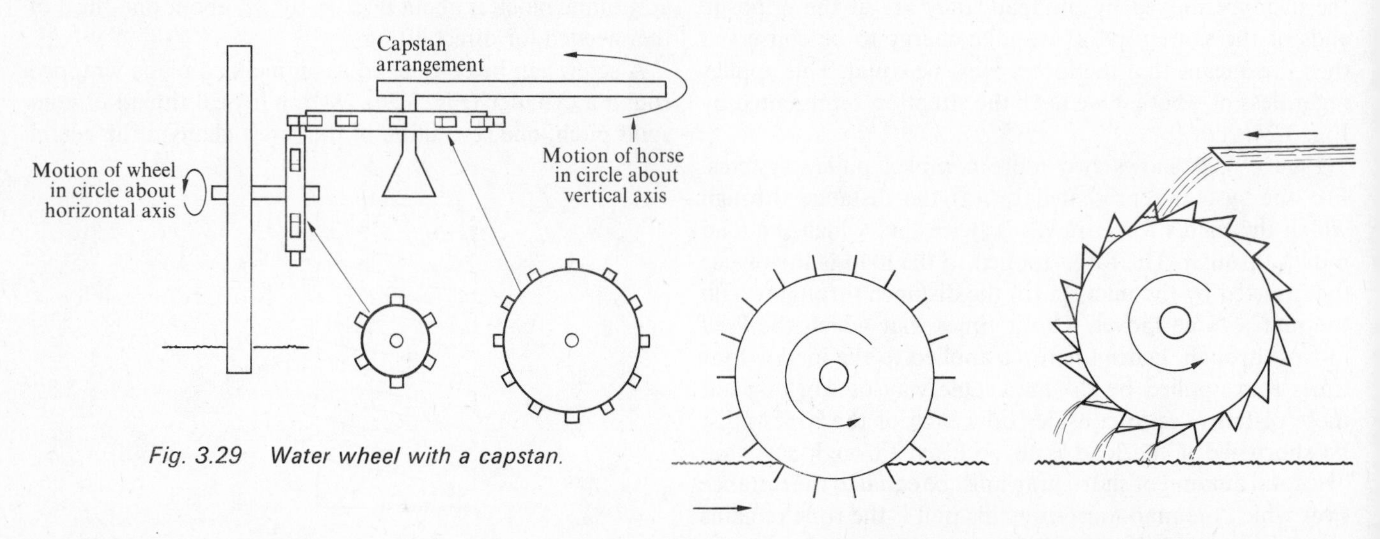

Fig. 3.29 Water wheel with a capstan.

Fig. 3.30 Two forms of water mill.

the ratio of their angular speeds is in the inverse ratio of their diameters. Contact without slipping, or very little, is achieved by putting teeth on the wheels (Fig. 3.31).

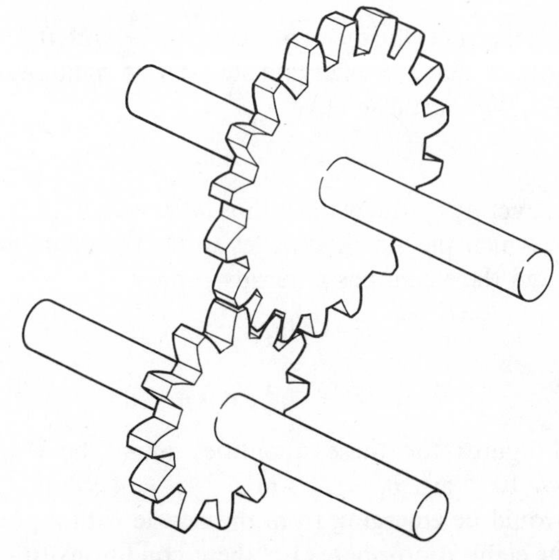

Fig. 3.31 Gear wheels.

Water mills probably originated in the first century B.C. They and the later windmill, probably developed about the seventh century A.D., allowed men to harness energy sources other than those of their fellow humans or beasts. These early mills were only capable of producing power of some 500 W. The eighteenth century saw the emergence of another energy source which man was able to utilize, steam. The early steam engines functioned on the principle—steam used to expel a piston from a chamber, the steam condenses and so produces a partial vacuum, this drop in pressure to below atmospheric pressure on one side of the piston causes it to be pulled into the chamber, the steam is then admitted to the chamber and expels the piston, and so the cycle repeats itself. Figure 3.32 shows the engine invented by T. Newcomen which worked on this principle. The steam was made to condense by water being injected into the chamber.

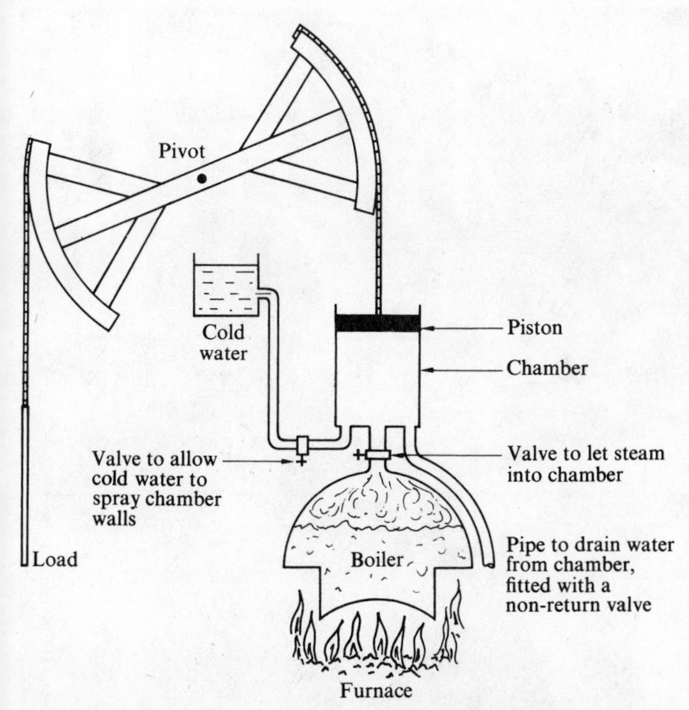

Fig. 3.32 Simplified diagram of a Newcomen engine.

With the Newcomen engine the steam pressure remains reasonably constant as the piston is pushed outwards, the steam from the boiler is flowing into the chamber during the entire operation. When the steam is condensed there is an abrupt drop in pressure followed by the piston moving back into the chamber. The pressure–volume changes can be roughly represented by Fig. 3.33. The work done at each

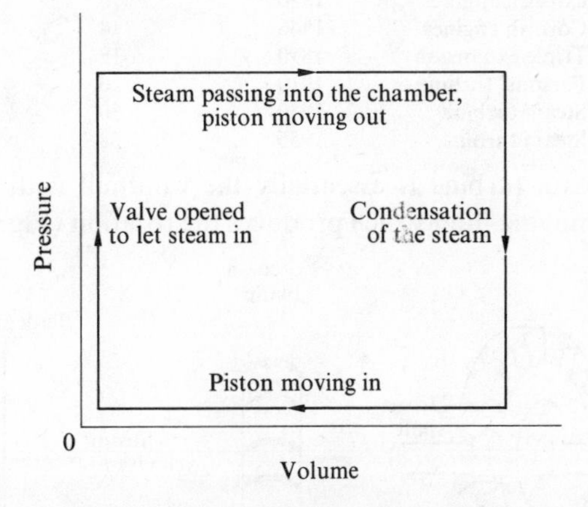

Fig. 3.33 Pressure–volume relationship for a Newcomen engine.

stroke is the area under the graph (see earlier in this chapter). In this case this will be the product of the steam pressure and the chamber volume. The Newcomen engine of 1712 had a piston 0·54 m in diameter and a chamber 2·4 m long, hence the chamber volume was about 0·55 m^3. The steam was at atmospheric pressure, 10^5 N m^{-2}, and thus the theoretical energy per stroke was $0{\cdot}55 \times 10^5$ J. The engine made a stroke every 5 seconds so the power would be $0{\cdot}11 \times 10^5$ W or 11 kW. The engine in fact was able to lift 46 kg of water through 46 m on each stroke; this means about $0{\cdot}21 \times 10^5$ J at each stroke, a power of about 4·2 kW. Newcomen was able to increase the power of his engines by increasing the volume of the chamber—this meant a much larger boiler to produce the steam. Later workers made significant power improvements by increasing the steam pressure.

The higher pressure is produced by higher boiler temperatures. Steam at one atmosphere pressure means the water boiling at 100°C, twice the atmospheric pressure means a temperature of about 121 deg C, six atmospheres means 159 deg C (see chapter 10 and the section in which vapour pressure is discussed). J. Watt made a significant improvement when he made the engine double acting—the steam was applied first to one side and then to the other of the piston. The early engines, even with these improvements, still had very low efficiencies—less than 5 per cent. Efficiency is the percentage of the energy input which is available at the output, i.e., as mechanical energy. Much of the energy liberated by the fuel in the boiler was used in increasing the temperature of the surroundings or when the steam was produced not all the steam was used to produce the expansion. Only about 12 per cent of the steam in the Newcomen engine was available for the expansion, the rest condensed on the colder walls of the chamber.

The efficiencies of steam engines show a significant increase over the years, as the following table shows.

	Year	*Efficiency (approx.) /per cent*
Newcomen	1712	2
Watt	1770	3
Watt	1796	4
Cornish engines	1830	10
Cornish engines	1846	14
Triple expansion	1890	18
Parsons' turbine	1910	20
Steam turbine	1950	30
Steam turbine	1955	38

A steam turbine is essentially the windmill with steam as the moving fluid which produces the rotation (Fig. 3.34).

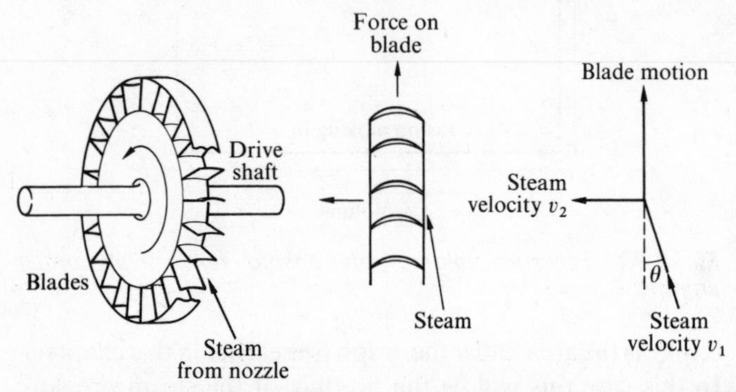

Fig. 3.34 Impulse turbine.

For steam with a mass per second m impinging on a blade we can write for the change in momentum of the blade per second

$$mv_1 \cos\theta$$

where θ is the angle the incident steam makes with the blade. This assumes that the emerging steam is at right angles to the blade. We must also have

$$mv_1 \sin\theta = mv_2$$

The force acting on the blade is thus $mv_1 \cos\theta$. If the nozzle through which the steam emerges to hit the blade has an area A and the steam has a density ρ then

$$m = Av_1\rho$$

Hence

$$\text{force} = Av_1^2\rho\cos\theta$$

Typical figures for these quantities might be $\theta = 20°$, $A = 5 \times 10^{-4}\ \text{m}^2$, $v_1 = 500\ \text{m s}^{-1}$, $\rho = 4\ \text{kg m}^{-3}$. The steam would be emerging from the nozzle with a pressure of about eight atmospheres for these conditions to occur. This results in a force of about 500 N on a single blade.

The above form of turbine is known as an impulse turbine in that the steam strikes the blades and gives them an impulse. Another form of turbine is the reaction turbine (Fig. 3.35). In the impulse turbine the steam must have a high velocity when it impinges on the blades, in the reaction turbine the steam supply has to be at high pressure (this does not involve steam molecules having a high velocity in some particular direction). The steam can escape from the high to a low pressure region by passing through the blades of the turbine. The steam emerges at high velocity. The situation is comparable to a garden hosepipe or sprinkler,

Fig. 3.36 (a) Thorpe Marsh Power Station. The turbine hall (Central Generating Board).

the water at high pressure passes through the nozzle to emerge as high speed water. As is certainly evident with the water sprinklers which revolve the nozzle experiences a force. This is the so called reaction force that is experienced by the blades of the turbine. The force causing rotation will be equal to the rate of change of momentum of the steam in a direction at right angles to the drive shaft.

Many of the large turbines use both the impulse and the reaction methods of operation, having a number of stages each with a number of turbine rotors with blades. Turbines are used with steam or water or hot gases. In the gas turbine hot gas at high pressure is generated by internal combustion and then passed to the turbine section. Turbines with powers of the order of many thousands of kilowatts are now quite common (Fig. 3.36).

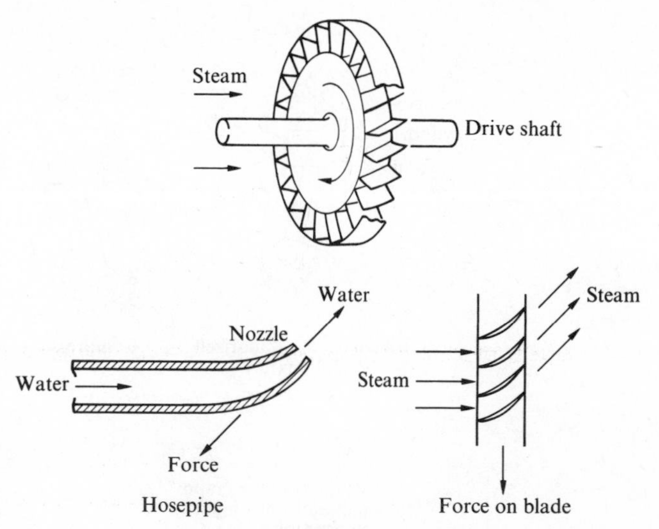

Fig. 3.35 Reaction turbine.

One engine that must not be omitted in this brief survey is the internal combustion engine. Figure 3.37 shows the basic sequence of operations in a four-stroke engine. The sequence is briefly—intake of a mixture of air and vaporized petrol; compression of this mixture, this raises the temperature of the mixture; ignition by a spark, the mixture burns and the hot gases push the piston down; the gases remaining in the chamber are ejected. Figure 3.38 shows the pressure–volume graph for the gases in the cylinder. Compression gives an increase in pressure, a decrease in volume, and an increase in temperature. At ignition there is a sudden increase in pressure and then the pressure drops as the piston moves down. The area enclosed by the graph is a measure of the energy made available. This can be made larger by increasing the compression ratio, i.e., reducing the volume by a greater fraction during the compression part of the sequence. A limit on this ratio is set by the fact that the temperature rise resulting from the compression must not become so high that the gas ignites.

Further reading

J. F. Sandfort. *Heat engines*, Science Study Series, No. 22, Heinemann, 1964.

H. Hodges. *Technology in the Ancient World*, Penguin, 1970. This is useful for details of ancient machines.

Fig. 3.36 Thorpe Marsh Power Station (Central Electricity Generating Board). (b) View of the turbines (Reyrolle Parsons).

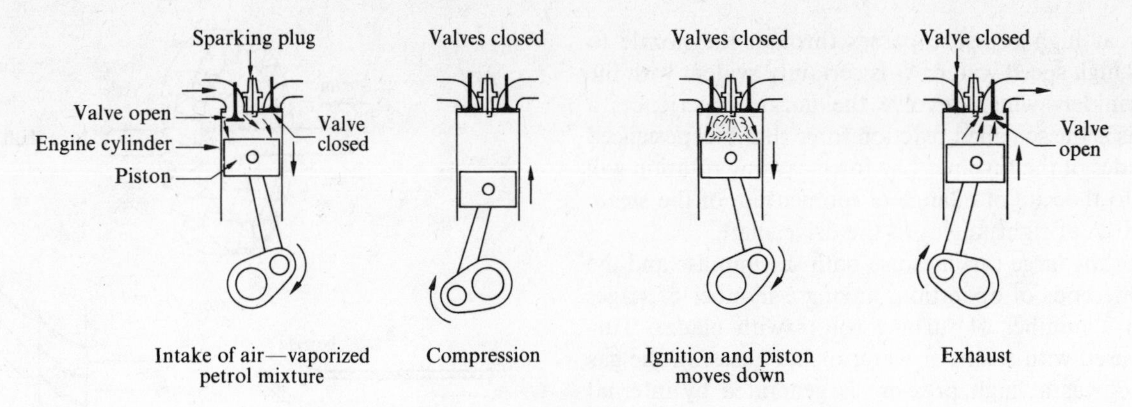

Fig. 3.37

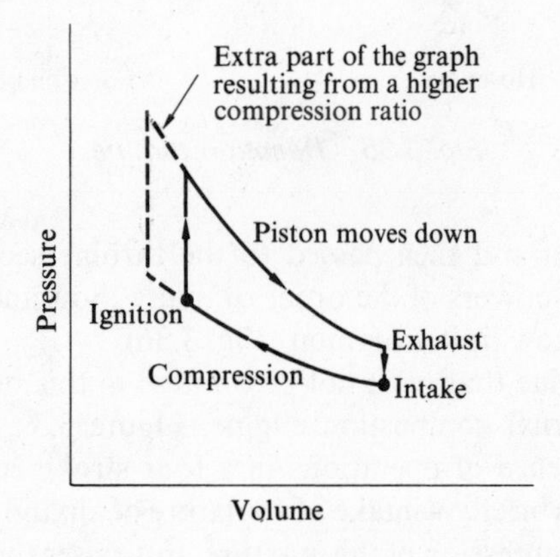

Fig. 3.38

Appendix 3B World energy supplies

'Energy universally and increasingly has become a major subject of public interest; its indispensability to the orderly functioning and steady growth of a nation's economy has come to be regarded as axiomatic. In practically every country, short- and long-term expectations of energy requirements are the object of continuing statistical study and reappraisal at the highest government and industry levels.'

F. Felix (July 1964). *IEEE Spectrum*, 81.

Energy is more than just a physicist's concept; it is a vital factor in a country's economy. Low cost energy is vital to a country's growth. In this section we will take a look at the energy supplies of the world and look for answers to the questions—How much energy do we use a year?—What are the world's energy supplies?—Have we enough energy stocks to continue as at present?

To start with—How much energy are we using?

	Energy supplied in 1947 /Q	/*per cent*
Coal	0·048	52
Fluid fuels	0·024	26
Wood and wastes	0·020	21
Water power	0·001	1
Total	0·093	100

(Taken from P. Putnam (1953), *Energy in the future*, Macmillan/van Nostrand Reinhold. One Q unit is 10^{21} J.)

By 1964 the energy used per year by the world had increased to 0·14 Q—this is about 40 per cent higher than the 1949 total. In 1850 the energy used was only 0·01 Q. The rate of growth of the world energy demand has averaged about 2·2 per cent per year since 1860. Recently the rate has increased to about 3 per cent. Are there enough energy sources left in the world for this rate of use of energy to continue?

Period of time	*Total of energy used/Q*
Up to 1850	9
Up to 1900	10
Up to 1950	13
Up to 2000	26 at 3 per cent growth, 35 at 5 per cent growth
Up to 2050	85 at 3 per cent growth, 500 at 5 per cent growth.

(Figures taken from P. Putnam (1953), *Energy in the future*, Macmillan/van Nostrand Reinhold.)

'... the rate of extraction of coal and oil gives rise to anxiety as to the future sources of energy for power production....'

Warne (1950).

	World energy reserves recoverable at present (Estimated 1950) /Q
Coal	21·0
Oil gas	5·0
Oil shale	1·0
Tar sands	0·2
Total	27·2

This would only take us to about the year 2000 at the 5 per cent rate of growth of demand. Additional energy can be obtained, indirectly in many instances, from the sun.

	Minimum plausible energy in 100 years/Q
Solar heat collectors	5
Fuel woods	1·5
Farm wastes	1·0
Water power	0·5
Wind power	0·1
Various other methods	0·126
Total	8·226

This coupled with our fossil fuels is not going to take us very far in time. Nuclear fuels would appear to be the means by which the energy demand for the future can be met. In 1950 the reserves of thorium and uranium then found was sufficient to supply about 575 Q—enough to take us at least half-way through the next century. Present estimates put the figure much higher.

'We see no real limitation on the availability of nuclear fuel. Therefore we may look forward to the widespread application of such sources of power to the future economy and technology of the world.'

Oppenheimer (1947).

Unfortunately an increase in the number of power stations using nuclear fuels means an increase in the amount of highly dangerous radioactive waste that has to be disposed of. The environment may be seriously affected by our demand for more energy.

Radioactivity is not the only pollution problem resulting from an increased demand for energy. Thermal pollution is a problem now and will become more serious in the future. All power stations need cooling water. In some cases large cooling towers are used, in other cases river water is used. The result is an increase in the temperature of the environment. This can have serious effects on the local scale, e.g., changing the balance of nature in rivers used for cooling, and on the larger scale the climate.

'In many parts of the US the capacity of the environment to absorb heat is going to be the limiting condition for the use of nuclear power.... The projected 1980 power output will need half the total runoff of US rivers for cooling purposes, except in the four months of the year when rivers are in flood....

...sharks are now entering Southampton water, attracted by the warmer temperature. Some US rivers may start to boil by 1980, and by 2010 may evaporate completely, according to two Rutgers University experts. The disposal of heat may be the limiting factor in power supply....'

G. R. Taylor (1970). *The Doomsday Book*, Thames and Hudson, 276.

'...As power production increases the inland sites will one by one become saturated from the standpoint of heat rejection capability, or thermal pollution. Only the very largest rivers, lakes and sea coast sites will be able to add new generating capacity....

...I see the typical energy centre (in the future) ... as a floating complex some 15 km offshore. It would be located, where possible, in water at least 100 metres deep in order to reach the colder bottom waters below the mixed zone of the sea.... As the bottom waters are rich in marine nutrients, the 40 million cubic metres of warm water discharged from the condensers would support a large fish population....'

R. P. Hammond (Jan. 1959). *Science Journal*, 34.

Further reading

A. R. Ubbelohde. *Man and energy*, Penguin, 1963.

Appendix 3C Horse power and man power

How much work can a horse do in a day? How much work can a man do in a day? Comparisons have given one horse equivalent to $2\frac{1}{2}$ men (G. Agricola 1556), one horse equivalent to 14 men (Schulze 1783), one horse equivalent to 30 men (Fuel Conservation, British Productivity Council, 1953). In 1782 J. Watt adopted a unit of power of 'a horsepower' so that comparisons could be made between the 'newly' invented steam engines and horses. He considered that a horse could pull 150 pounds at a velocity of 2·5 miles per hour, a power of 33 000 foot pounds per minute (0·745 kW). The early steam engines had powers of about 10 horse power.

Various experimental determinations have been made of the work a man can do in a day. Coulomb in 1798 arrived at a figure of about 2×10^6 J from the fact that some marines had climbed to a height of 2923 m in $7\frac{3}{4}$ hours (considered to be a reasonable length day). He assumed a marine had a mass of 70 kg, thus the potential energy gained in the 'day' was $70 \times 10 \times 2923$ J. This value of Coulomb's is an average of about 80 W of power over the $7\frac{3}{4}$ hours, i.e., about 1/9 horse power.

Problems

1. 'A conservation law is a statement of constancy in nature....

...The strong hint emerging from recent studies of elementary particles is that the only inhibition imposed upon the chaotic flux of events in the world of the very small is that imposed by the conservation laws. Everything that can happen without violating a conservation law does happen.

This new view of democracy in nature—freedom under law—represents a revolutionary change in man's view of natural law. The older view of a fundamental law of nature was that it must be a law of permission. It defined what can (and must) happen in natural phenomena. According to the new view, the more fundamental law is a law of prohibition. It defines what cannot happen. A conservation law is, in effect, a law of prohibition. It prohibits any phenomenon that would change the conserved quantity....

K. W. Ford (1963) *The world of elementary particles*, Blaisdell, p. 81.

(a) How would you define a conservation law, without using the word 'conserved'?

(b) Give examples of laws of permission and of prohibition that occur in everyday life.

(c) Are Newton's laws of motion laws of permission or prohibition?

2. In the following extract from the book *Heat Engines* by J. F. Sandfort (1962 Science Study Series No. 22, Heinemann/Doubleday) the word 'energy' is used—what is its meaning in this context?

'Man is an earth-bound animal, of only average size, and physically weak as animals go, yet he dominates the earth. As a matter of fact, he has loosened the shackles of the earth's great gravitational field and moved in the new frontiers of outer space. How has this come about? It happened because man discovered the vast reservoirs of energy available to him on earth and learned how to control them. Unlocking this secret has made possible the remarkable growth of our modern technological civilization. It is the key that has turned certain nations into great military powers and has made possible the highest standard of living of all time.'

3. Oxford and Cambridge Schools Examination Board. Nuffield Advanced Physics, 1970.

'The graph [Fig. 3.39] shows how the force of attraction on a 1 kg mass towards the earth varies with its distance from the centre of the earth. It shows that the force at a distance equal to the radius R of the earth is 10 newtons.

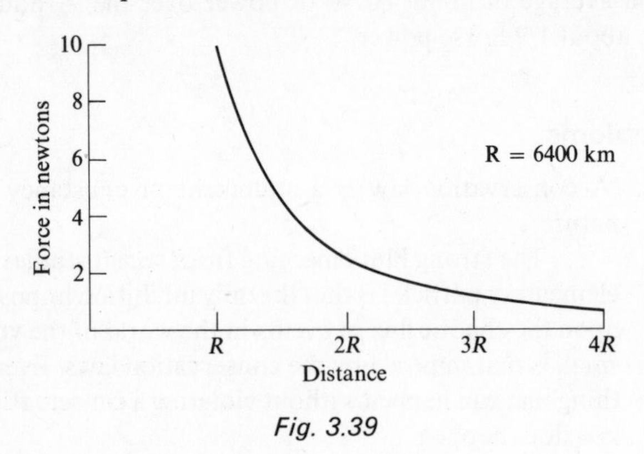

Fig. 3.39

(*a*) Calculate the force on the mass at distances of (i) $9R$ (ii) $10R$ from the earth's centre.

(*b*) Shade in on the graph an area which gives the energy change in moving the mass from $3R$ to $2R$.

(*c*) Make a rough estimate of the increase in kinetic energy when the mass falls from $10R$ to $9R$.'

4. A catapult is a means of converting the potential energy of a stretched piece of rubber into kinetic energy for some stone.

(a) What is the change in potential energy of a strip of rubber when it is stretched by ΔL? Make an assumption about the behaviour of rubber when stretched (see chapter 9).

(b) How does the velocity of the stone immediately after release from the catapult depend on the amount by which the catapult rubber is stretched? You will need to make some assumptions.

5. It has been reported that it took 100 000 men 20 years to build the Great Pyramid of Gizeh. The pyramid was originally about 150 m high and had a square base of side 230 m. The pyramid is almost completely solid stone, density about 2500 kg m^{-3}.

(a) Estimate the total gravitational potential energy of the pyramid relative to the ground level.

(b) How much energy had to be supplied by each of the men involved in the building of the pyramid? (See appendix 11C for a discussion of the energy produced by man.)

6. Make estimates for the following:

(a) your kinetic energy when running,

(b) the change in potential energy of an athelete doing the high jump,

(c) the electrical energy 'used' by an electric light left on for an entire day,

(d) the energy transferred to the ground when a 'super-ball' bounces on it,

(e) the energy needed to melt an iceberg,

7. How does the escape velocity for an object from the moon compare with that from the earth?

	Radius /km	*Mass* /10^{23} kg
Earth	6378	59·76
Moon	1738	0·7350

The earth has an atmosphere—the moon has none, or virtually none. Could this be anticipated from the sizes of the two objects?

8. Figure 3.40 shows the force–distance graphs for three springs.

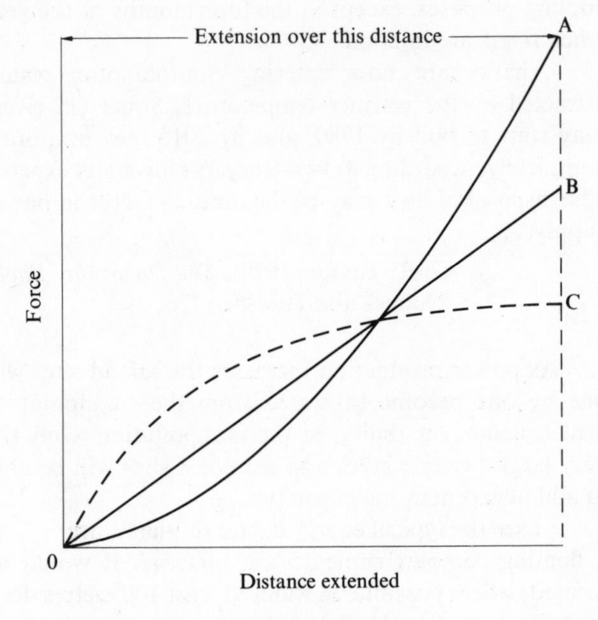

Fig. 3.40

(a) Which of the three springs needs the most energy to be stretched through the marked distance?

(b) Sketch the potential energy–distance graph for the spring A.

9. A mid-course correction was made by the Apollo 11 spacecraft on its passage back to the earth from the first moon landing (see chapter 2). The rocket motors were fired for 3 s and changed the velocity of the craft from 1531·56 to 1527·16 m s^{-1}. The spacecraft has a mass of $34{\cdot}5 \times 10^3$ kg. What energy must be released by the fuel?

10. Oxford and Cambridge Schools Examination Board. Nuffield Advanced Physics, 1971.

'Use the data given below to make *quantitative* predictions, estimates or comparisons concerning man's use of the fuel resources (coal, oil etc.) stored in the Earth.

In particular, you should consider the following problems:

(*a*) The possibility of using up these fuel resources, and any need there may be to find alternative sources of energy.

(*b*) Any differences between groups of people in the amount of energy they use.

Data

Energy reaching Earth from sun: 5×10^{24} joules per year.

Maximum energy likely to be available from hydroelectric power if all such sources were used: 5×10^{19} joules per year.

Total energy stored in all known reserves of fuels (coal, oil etc.) 10^{23} joules.

Total energy stored in all known reserves of fuels (coal, oil etc.) that can be extracted economically at present: 10^{22} joules.

Total energy stored in known oil deposits that can be extracted economically at present: 10^{21} joules.

Time taken to produce the Earth's store of coal, oil and other fuels: 10^8 years.

Energy used by one car in a year: 3×10^{10} joules.

Work energy one labourer can produce in one year: 10^{10} joules.

Total energy used in 1964		*Population*
WORLD	10^{20} joules	$3\,000 \times 10^6$
North America	4×10^{19} joules	200×10^6
Western Europe	2×10^{19} joules	300×10^6
Middle East and Africa	$0{\cdot}3 \times 10^{19}$ joules	300×10^6
Asia	2×10^{19} joules	1700×10^6

Teaching note Further problems appropriate to this chapter will be found in:
E. M. Rogers. *Physics for the inquiring mind*, chapters 26, 27, 28, 29, OUP, 1960.
Physical science study committee. *Physics*, 2nd ed., chapters 23, 24, 25, Heath, 1965.
Project physics course. Text 3, Holt, Reinhart and Winston, 1970.
Nuffield O-level physics. Question book 4, Longmans/Penguin, 1966.
Nuffield Advanced physics. Unit 3 Student's book, Penguin, 1972.

Practical problems

1. In the conventional laboratory condenser energy is transferred from a hot to a cooler fluid. Consider the design of the device and its efficiency. Produce a new or modified design of a condenser which would be more efficient.

In order that the cooling process be most rapid as large an area of the hot liquid as is possible should be in contact with the cold surface. This is often obtained by the use of metal vanes protruding from the hot surface. Another possibility is to pass the fluid through a large number of small-bore pipes instead of a single large-bore tube.

2. A desert is not a completely lifeless and waterless waste. Life in the form of insects such as ants, locusts, grasshoppers, and spiders does exist. Some water does exist below the surface of the sand. It is mainly below the surface that the insects live during the day. The reason for the water being below the surface is the lower temperatures existing there by virtue of an insulating layer of sand above. A typical example of temperatures for wind-blown sand are: air temperature 43°C, surface 72°C, and a considerable drop in temperature to about 50°C in a matter of centimetres below the surface.

Investigate the temperature distribution with depth for different size sand particles. On the basis of your results state the type of surface conditions under which most water and insect life could be expected.

3. When copper sulphate is heated to a sufficiently high temperature dehydration occurs. Whenever phase changes occur the temperature remains constant until the phase change is complete. Thus if heat is supplied to a copper sulphate sample the temperature will increase steadily—with the exception of the temperatures at which phase changes occur. These temperatures can be most easily detected by comparing the temperatures of samples of copper sulphate and a material which shows no phase changes in that region—sodium chloride can be used—while heat is supplied at the same rate to both samples.

Devise a suitable form of apparatus so that the temperatures at which phase changes occur in a sample can be detected. Try crystalline copper sulphate as a sample.

Reference

Borchard. *J. Chem. Ed.*, **33**, 103, 1956.

Suggestions for answers

1. (a) Whenever a quantity is counted it gives the same number. This is only one possible answer, you can no doubt think of many other forms.

(b) Prohibition: keep off the grass. Permission: keep left. A law of permission tells you what you can do, a law of prohibition tells you what you cannot do.

(c) The laws can be read in both prohibition and permission forms. A force enables you to have an acceleration: no acceleration is permitted without a force.

2. I leave this to you—read the beginning of this chapter again if in doubt.

3. (a) 10/81 N, 10/100 N.

(b) The area between the vertical lines at $2R$ and $3R$ and the curve.

(c) $0{\cdot}35R$ J.

4. (a) Assume that the extension is directly proportional to the force. Then change in potential energy is equal to $\frac{1}{2}F\,\Delta L$.

(b) Assume that all the potential energy of the rubber is given to the stone. This involves assuming that the mass of the rubber is insignificant so that we can ignore the kinetic energy of the rubber and also that the rubber converts the potential energy only to kinetic energy.

$$\tfrac{1}{2}F\,\Delta L = \tfrac{1}{2}mv^2$$

where m is the mass of the stone and v its velocity.

5. One way of doing this calculation is to consider the pyramid as a number of layers and find the potential energy for each layer in turn. Figure 3.41 illustrates this. If we slice the pyramid into horizontal 10 m thick slices then the potential energy of the bottom slice is

$$10 \times \left(\frac{115 + 107}{2}\right)^2 \times 2500 \times 9{\cdot}8 \times \frac{10}{2}$$

thickness × (mean side length)^2
× density × g × mean height

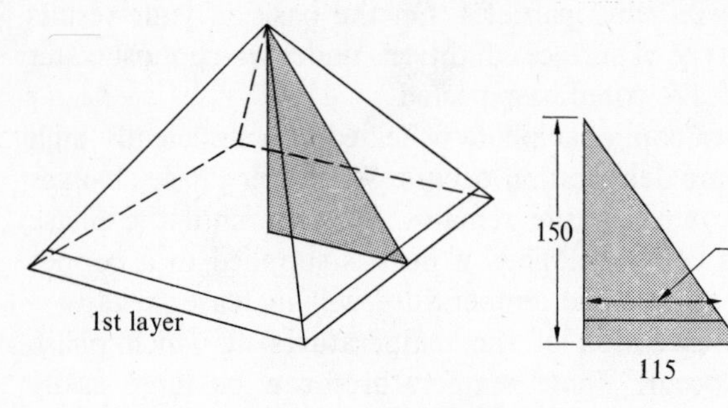

Fig. 3.41

The next slice will have a potential energy of

$$10 \times \left(\frac{107 + 99}{2}\right)^2 \times 2500 \times 9{\cdot}8 \times \left(\frac{10}{2} + 10\right)$$

The next slice will have a potential energy of

$$10 \times \left(\frac{99 + 91}{2}\right)^2 \times 2500 \times 9{\cdot}8 \times \left(\frac{10}{2} + 20\right)$$

Counting the energy for the fifteen layers gives a total of about $2{\cdot}4 \times 10^{12}$ J. This is the energy that had to be supplied by the men building the pyramid. Thus each man had to supply, assuming equality of labour, $2{\cdot}4 \times 10^7$ J over the years. Compare this with Appendix 3C and consider the efficiency.

6. (a) $\frac{1}{2} \times 60 \times 250 = 7500$ J

(b) $60 \times 9{\cdot}8 \times 2 = 1180$ J

(c) For a 100 W lamp, $100 \times 24 \times 3600 = 8\,640\,000$ J

(d) If the ball bounces up to 9/10 of its initial height of say 1 m the change in potential energy is

$$mg\,(1 - 0{\cdot}9)$$

If the ball has a mass of say 50 g then the change is about 50 J. This is the energy transferred to the surroundings, mainly the ground.

(e) Suppose the iceberg has a volume of 20 m^3. This would be a mass of about 2×10^4 kg. The energy needed to melt this ice, change from ice at 0°C to water at 0°C, will be $2 \times 10^4 \times 3{\cdot}3 \times 10^5$ J, i.e., $6{\cdot}6 \times 10^9$ J.

7. The escape velocity from the earth is 4·7 times greater than that from the moon. The lower the escape velocity the more difficult it is for a planet to retain an atmosphere as the molecular velocities of the atmospheric gases can be such that they escape.

8. (a) The energies are approximately the same.

(b) See Fig. 3.42.

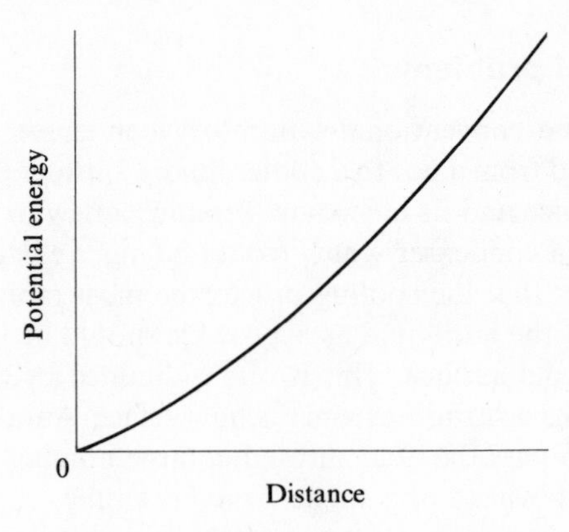

Fig. 3.42

9. $\frac{1}{2} \times 34{\cdot}5 \times 10^3 \times (1531{\cdot}56^2 - 1527{\cdot}16^2)$ J

10. This is left to you. See appendices 3A and 11A.

4 Oscillations

Teaching note Practical work appropriate to this chapter will be found in:
W. Bolton. *Physics experiments and projects*, Vol. 5, Pergamon, 1969.
Nuffield Advanced Physics. Unit 4 Teacher's Guide, Penguin, 1972.

4.1 What are oscillations?

What is an oscillation? Something that goes wig-wag! Can we produce a scientist's description? If we study the motion of a number of objects we can perhaps answer the question. If a mass is suspended from a spring, pulled down and then released, oscillations occur (Fig. 4.1). A marble rolling in a curved track, the marble oscillates backwards and forwards about its rest position. A mass suspended from the end of a piece of string (a simple pendulum), when the mass is displaced from its rest position and released oscillations occur. A trolley tethered between two supports on a horizontal bench by springs will oscillate when the trolley is displaced from its rest position and then released. A ruler clamped at one end to a bench oscillates when the free end is depressed and then released. What do we do, for these and other examples, to get oscillations? The masses are pulled away from their rest position and then released. A restoring force then pulls them back and they seem to overshoot the rest position. This restoring force must exist otherwise they would not move when released. Because there is a force then we must have an acceleration. The restoring force is always directed towards the central rest position—thus the acceleration is always directed towards the central rest position.

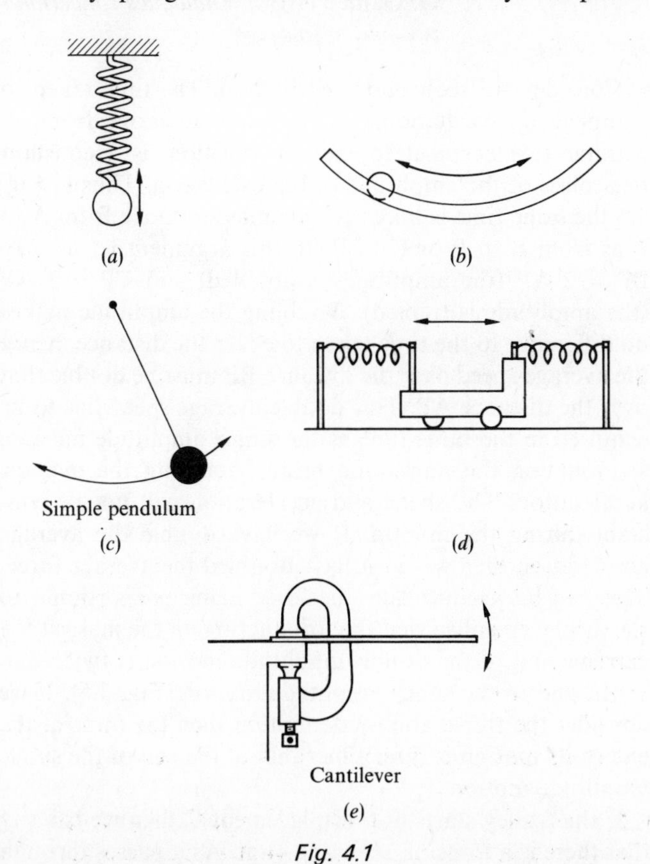

Fig. 4.1

We can determine the distance–time graph for an oscillating object by taking a strobe photograph—Fig. 4.2 shows one for a pendulum—or using ticker tape and a vibrator. A vibrating metal strip puts marks on a strip of paper at regular time intervals and if the tape is attached at one end to say the trolley a trace similar to Fig. 4.3 is obtained as the trolley moves from its maximum displacement on one side of the rest position to its maximum displacement on the other.

Figure 4.4 shows some distance–time graphs. Some oscillations seem to have the property that the same time is taken for each complete oscillation. Such oscillators are known as isochronous, and they maintain this constant time property regardless of amplitude changes due to damping.

How do we know that these masses take the same time for each oscillation? Because we measured it with a clock. But what is a clock? Something that keeps regular time due to the regular oscillation of some device such as a spring or a pendulum. Are our oscillations constant time oscillations because we use this property to measure time and in fact have a closed circle argument? What was used to measure time before people knew of isochronous oscillators? The movement of the sun across the sky and the passage of day and night—we could define a unit of time as the time occurring between two successive appearances of the sun at maximum elevation (a day). We could define our time unit in terms of our heart or pulse beat. Both these definitions give reasonable agreement regarding the regularity of time,

Fig. 4.2 Strobe photograph of a simple pendulum. (Educational Development Center Inc.).

Fig. 4.3 Ticker tape trace.

the day is a constant number of heart beats. We thus have natural time units which we can use to determine whether an oscillation is isochronous.

Galileo is reputed to have watched a swinging lamp during a service in the Cathedral of Pisa in 1538 and timed the oscillations by the use of his pulse. The period of time taken for one complete oscillation was found constant regardless of the amplitude (not, however, the very large amplitude).

> 'Thousands of times I have observed vibrations especially in churches where lamps, suspended by long cords, had been inadvertently set in motion; but the most which I could infer from these observations was that the view of those who think that such vibrations are maintained by the medium is highly improbable: for, in that case, the air must needs have considerable judgment and little else to do but kill time by pushing to and fro a pendent weight with perfect regularity. But I never dreamed of learning that one and the same body, when suspended from a string a hundred cubits long and pulled aside through an arc of 90° or even 1° or $\frac{1}{2}$°, would employ the same time in passing through the least as through the largest of these arcs; and, indeed, it still strikes me as somewhat unlikely.'
>
> G. Galileo (1638). *Dialogues concerning two new sciences.*

Consider an isochronous oscillation. The time taken to complete an oscillation, i.e., the time to move from an extreme displacement to the rest position, is a constant regardless of the amplitude of the oscillation. Thus in Fig. 4.5 the same time is taken for the mass to move from A to P as from B to P or C to P. In this argument let us take BP = 2 AP (the amplitude is doubled) and CP = 3 AP (the amplitude is trebled). Doubling the amplitude makes no difference to the time taken to cover the distance, hence the average speed over the distance BP must be double that over the distance AP. This double average speed has to be acquired in the same time as the single amplitude motion, so doubling the amplitude means doubling the average acceleration. The speed and acceleration will not be constant during the motion. If we have double the average acceleration then we must have doubled the average force. This can be produced by the force being proportional to the displacement so that the force acting on the mass at the extreme end of the double amplitude motion is twice that at the end of the single amplitude motion (Fig. 4.6). If we consider the treble amplitude motion then the force at the end of its motion is three times that at the end of the single amplitude motion.

If the trolley starts at a displacement C then we can say that there is a force of $3F$ acting on it. As it passes through

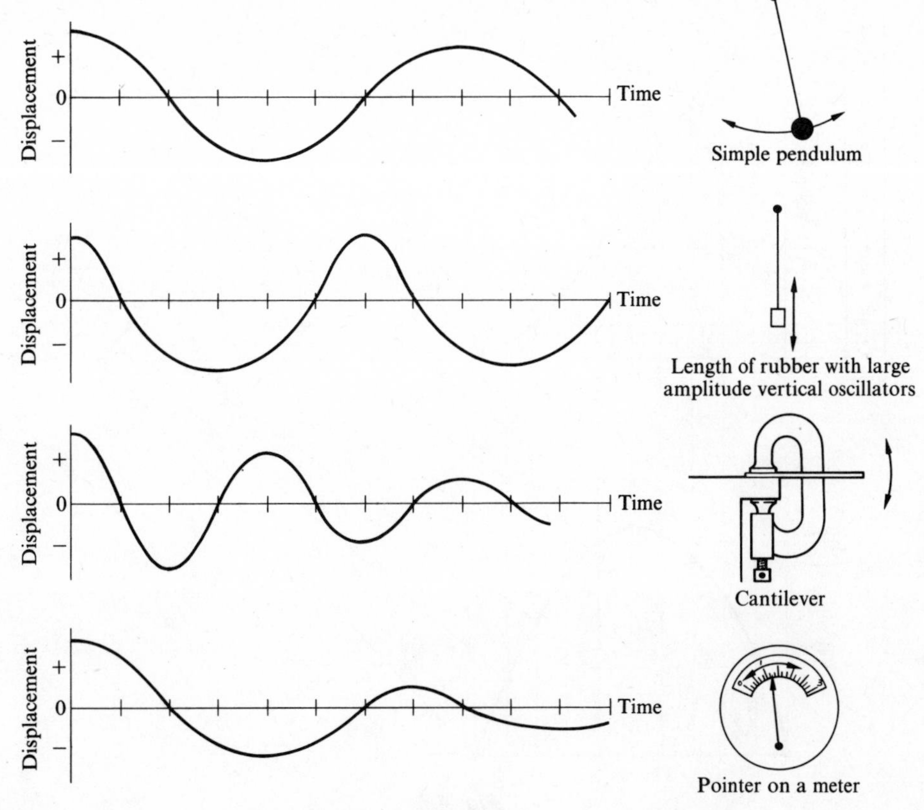

Fig. 4.4

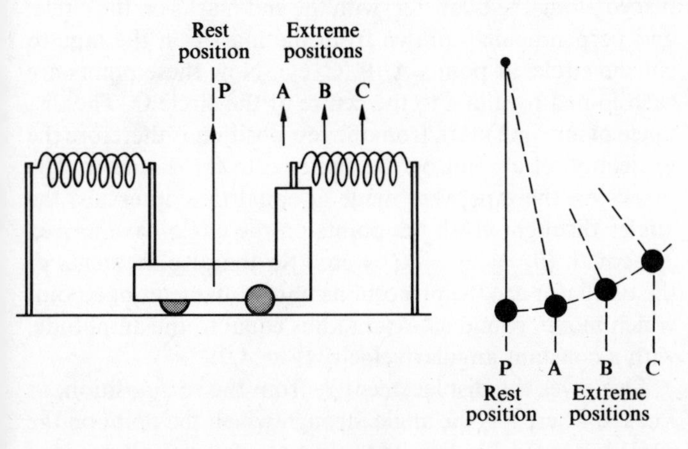

Fig. 4.5

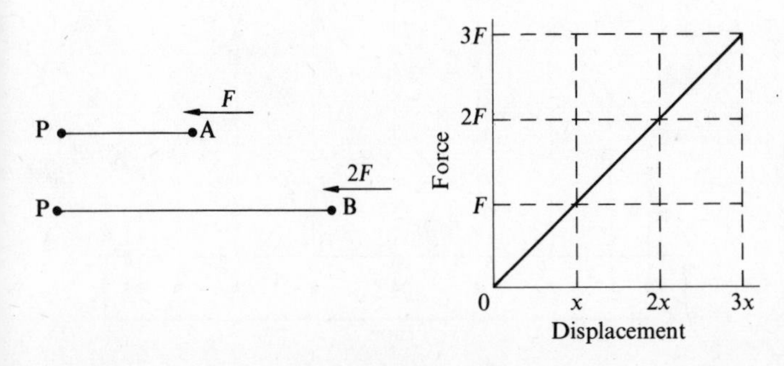

Fig. 4.6

point B the force has dropped to $2F$ and when it reaches A the force is only F. As the acceleration at any instant is proportional to the force acting on the mass then the acceleration must be proportional to the displacement, from the rest position.

This can readily be checked from the ticker tape or stroboscopic results obtained with isochronous oscillations. These are distance–time records. As velocity is the rate of variation of distance with time, the slope of the graph of distance against time then a velocity–time graph can be produced. Acceleration is the rate of change of velocity with time, the slope of the graph of velocity with time. Hence we can obtain the accelerations at different displacements (Fig. 4.7). The acceleration really is proportional to the displacement from the rest position.

$$\text{Acceleration} \propto -\text{displacement}$$

The minus sign is because the acceleration is always opposing an increase in displacement.

Section 4.2 in this chapter shows how we can arrive at the displacement–time graph for motions following the above equation.

What can we say about the energy of an oscillating mass? At an extreme position the mass is stationary and hence has no kinetic energy. When it passes through the rest position the velocity of the mass is a maximum (check this against Fig. 4.7). The mass thus has maximum kinetic energy when it passes through the rest position. Where does

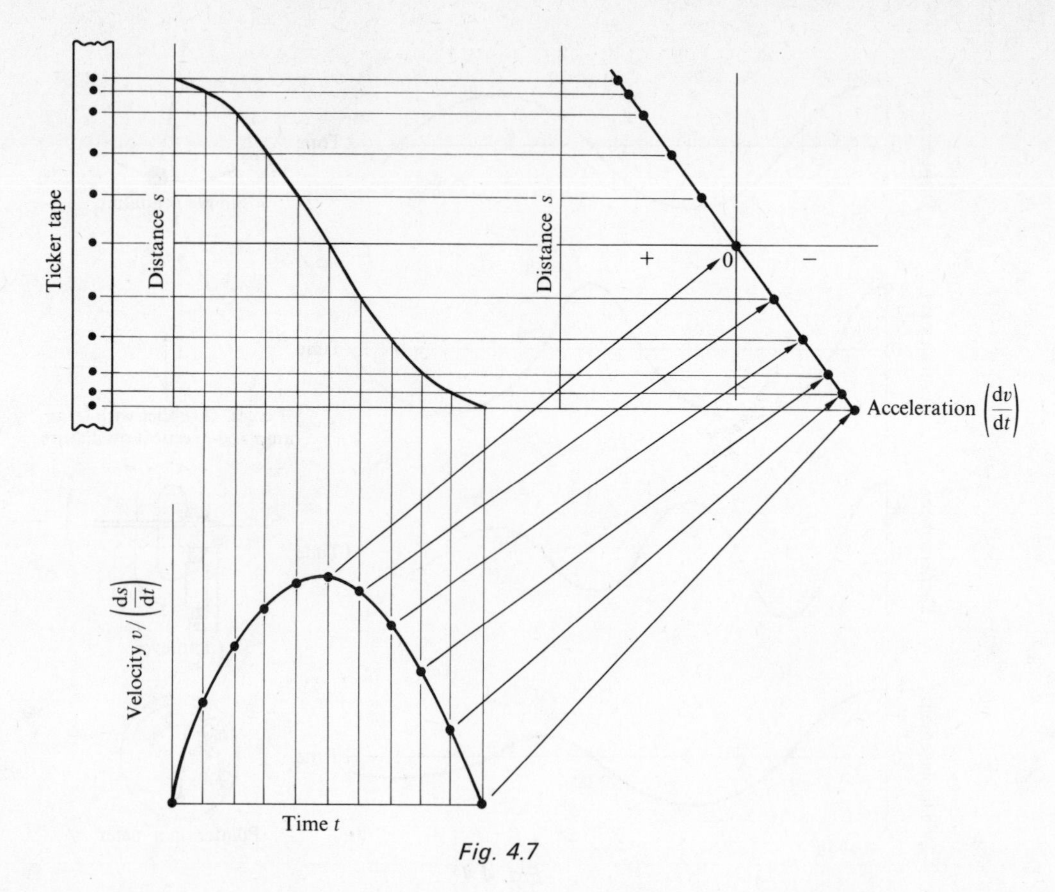

Fig. 4.7

this kinetic energy come from? When the mass is at an extreme displacement a spring is compressed or extended or a mass has been raised above the rest position or in some way the mass has some potential energy. When the mass moves from an extreme position potential energy is converted into kinetic energy. As the mass moves away from the rest position kinetic energy is converted into potential energy. The sum of the potential and kinetic energies at any instant is a constant, if no energy is transferred to the surroundings.

How can we tell if energy is transferred to the surroundings? If energy is transferred out of the oscillating system then the potential energy after one oscillation will be less than the initial potential energy. The potential energy after the second oscillation will be less than after the first oscillation. If there is less energy it means that the spring will be extended or compressed less or the mass not raised so high above the rest position. The maximum displacement (known as the amplitude) from the rest position becomes less. We say the oscillation is damped.

Mapping the motion onto a circle

If we know the equation relating the displacement with time we can arrive at the way the acceleration varies with distance by differentiation. The variation of distance with time, is certainly not a straight line relationship. What is it? Because I know the answer I am going to suggest that we try the following method to arrive at the equation. We take the ticker tape trace and draw a semicircle equal to the amplitude of the oscillation (Fig. 4.8). The tape is then placed along the diameter, with the end marks on the circle, and perpendiculars drawn from the marks on the tape to cut the circle, at points A, B, C, etc. Now these points are each joined by a line to the centre of the circle O. The distance of any one mark from the rest position is therefore the projection of a point on the circle on to the diameter. The marks on the tape were made at equal time intervals, the angles through which the points on the circle have moved are equal. ($\theta_1 = \theta_2 = \theta_3 =$ etc.) So the displacements of the oscillator are the projections onto a diameter of a point which moves round a circle, radius equal to the amplitude, with a constant angular velocity (Fig. 4.9).

This gives the displacement x, from the rest position, as $r \cos \theta$, where θ is the angle through which the point on the circle has moved in time t. At time $t = 0$ the oscillator is at

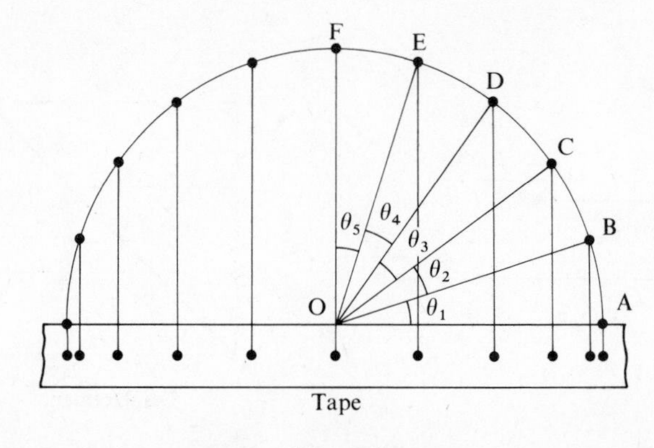

Fig. 4.8

an extreme position. But the radius is equal to the amplitude of the motion, A. Hence

$$x = A\cos\theta$$

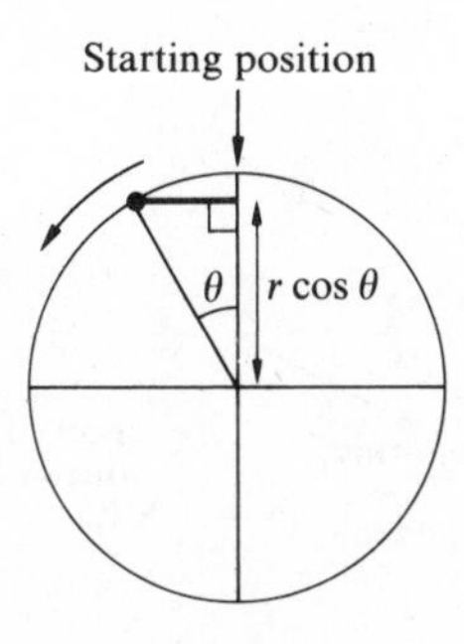

Fig. 4.9

For one complete oscillation the point moves completely round the circle. If T is the time for one complete oscillation (known as the periodic time) then the angle covered in this time is 2π radians (2π radians equal 360°). In time t the point has moved through angle θ. If ω is the angular velocity then

$$\omega = \frac{2\pi}{T} = \frac{\theta}{t}$$

Hence we can write θ as

$$\theta = \omega t = \frac{2\pi t}{T}$$

and the displacement as

$$x = A\cos\omega t = A\cos\frac{2\pi t}{T}$$

Figure 4.10 shows the graph of this equation.

Differentiation can be used to obtain the velocity and acceleration variations with time. The velocity is the slope at an instant of the distance–time graph, i.e., $\mathrm{d}x/\mathrm{d}t$. Hence

$$v = \frac{\mathrm{d}x}{\mathrm{d}t} = -A\omega\sin\omega t$$

This agrees with the graph obtained in Fig. 4.7. Acceleration is the slope at any instant of the velocity–time graph, i.e., $\mathrm{d}v/\mathrm{d}t$. Hence

$$a = \frac{\mathrm{d}v}{\mathrm{d}t} = -A\omega^2\cos\omega t$$

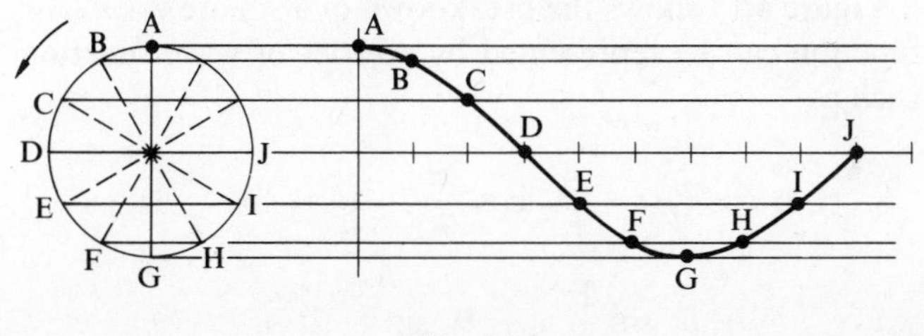

Fig. 4.10

But $x = A\cos\omega t$, hence

$$a = -\omega^2 x$$

The acceleration is directly proportional to the displacement.

What is the significance of the minus sign in the velocity and acceleration equations? As the displacement from the rest position increases the velocity decreases. As the displacement increases so the acceleration becomes more negative, i.e., the retardation increases. As the displacement becomes smaller, the mass is approaching the rest position, the acceleration becomes smaller and at the rest position is zero (look again at Fig. 4.7).

The force trying to restore the mass to its original rest position will therefore be

$$F = ma = -m\omega^2 x$$

The restoring force is proportional to the displacement from the rest position and always directed towards that position (the mass always accelerates towards the rest position). If the mass was on the end of a spring then provided the spring is only extended or compressed within the elastic region

$$F \propto x$$

The force is proportional to the extension,

or $$F = -kx$$

where k is the force constant. Hence

$$k = m\omega^2$$

But

$$\omega = \frac{2\pi}{T}$$

Therefore

$$T = 2\pi\sqrt{\frac{m}{k}}$$

This is the periodic time of the oscillator. This equation tells us how the time taken to complete one oscillation depends on the mass and the force constant of the spring. Often the term frequency is used instead of periodic time. Frequency, f, is the number of oscillations per second and hence

$$f = \frac{1}{T}$$

as T is the time for one oscillation. The frequency unit is the hertz (Hz). One oscillation or cycle per second is one hertz.

$$f = \frac{1}{2\pi}\sqrt{\frac{k}{m}}$$

If we do an experiment with a mass on the end of a vertically suspended spring then doubling the mass on the spring increases the time for one oscillation by a factor of $\sqrt{2}$ and decreases the frequency by $1/\sqrt{2}$. If the mass is quadrupled then the frequency is halved. Suppose we double the force constant of the spring (sometimes loosely referred to as the stiffness of the spring) by using two identical springs, side by side, to support the mass then the frequency increases by a factor of $\sqrt{2}$.

The simple pendulum

> 'As to the times of vibration of bodies suspended by threads of different lengths, they bear to each other the same proportion as the square roots of the lengths of the thread; or one might say the lengths are to each other as the squares of the times; so that if one wishes to make the vibration-time of one pendulum twice that of another, he must make its suspension four times as long.'
>
> Galileo (1638).

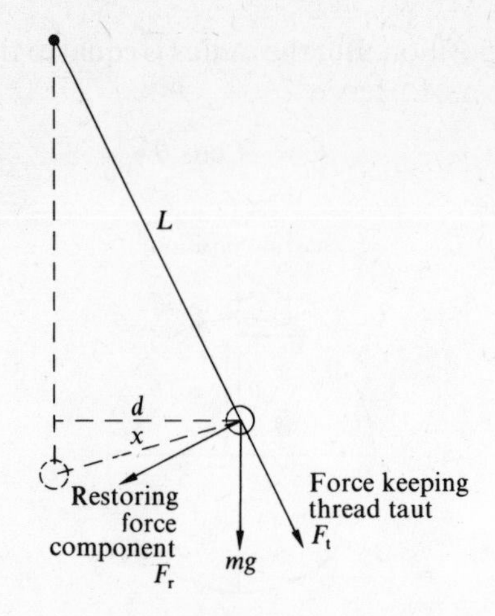

Fig. 4.11

With the simple pendulum the force trying to restore the bob (the mass at the end of the thread) to its rest position when it is pulled to one side is gravity (Fig. 4.11). The weight of the bob can be broken down into two components, one along the thread keeping it taut and the other perpendicular to the thread moving it towards the rest position, i.e., the restoring force. From the similar triangles in Fig. 4.12,

$$\frac{F_r}{mg} = \frac{d}{L}$$

Therefore

$$F_r = \frac{mg}{L} d$$

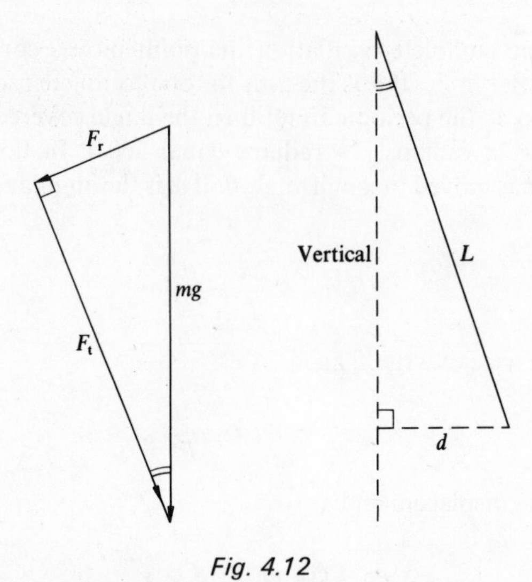

Fig. 4.12

As long as the arc through which the pendulum swings is small the displacement along the arc x is a reasonable approximation to d. Hence we can write

$$F_r = \frac{mg}{L} x$$

The restoring force is proportional to the displacement and thus the motion is isochronous.

By comparison with our equation for the restoring force for the spring, i.e., $F = kx$, our force constant with a simple pendulum is $k = mg/L$. Hence the frequency of oscillation is given by

$$f = \frac{1}{2\pi}\sqrt{\frac{g}{L}}$$

This in fact confirms the results quoted by Galileo at the beginning of this section. Galileo, however, in our earlier quotation in this chapter said

> 'But I never dreamed of learning that one and the same body, when suspended from a string a hundred cubits long and pulled aside through an arc of 90° or even 1° or $\frac{1}{2}$°, would employ the same time in passing through the least as through the largest of these arcs. . . .'

A pendulum swinging through 90° would not give the same frequency as one swinging through a small arc—our approximation would not hold. Was Galileo's timing too inaccurate or did his results only apply to small arcs and he generalized them?

Simple harmonic motion

The isochronous motion of the spring-supported mass, the tethered trolley, the pendulum, and many others all have a displacement–time graph in the form of a cosine curve if damping is insignificant. Because of this they all have an acceleration proportional to the displacement from the rest position and directed towards it. This type of motion is known as simple harmonic motion. It is important because it is a very common form of motion. Also any repetitive motion can be broken down into simple harmonic components (this is known as Fourier's theorem).

Figure 4.13 shows the breakdown of a square wave. Any function can be represented by a series of wave functions such as

$$A_0 + A_1 \cos\frac{2\pi}{T}t + A_2 \cos\frac{2\pi}{T}2t + A_3 \cos\frac{2\pi}{T}3t + \cdots$$

$$+ B_1 \sin\frac{2\pi}{T}t + B_2 \sin\frac{2\pi}{T}2t + \cdots$$

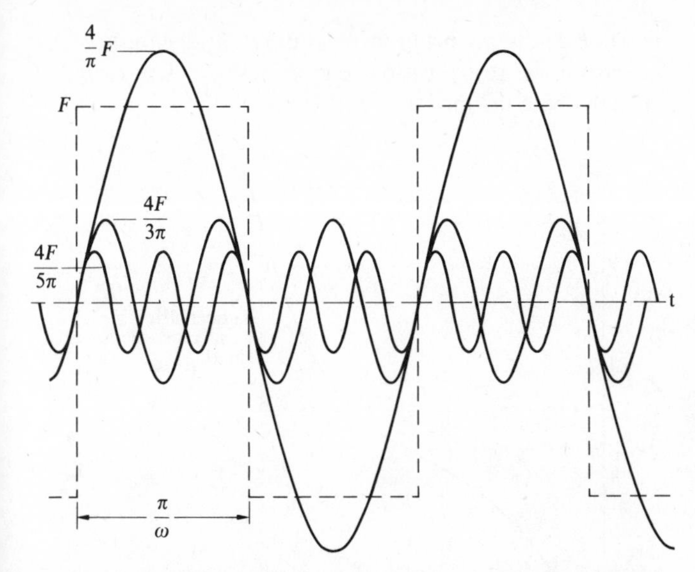

Fig. 4.13 The first, second and third trigonometric components required by a Fourier representation of a square wave. (L. S. Jacobsen, and Ayre (1958). Engineering vibrations, *McGraw-Hill Book Company.)*

4.2 Solving the acceleration equation

For an object oscillating freely with simple harmonic motion we have the relation

$$\text{Acceleration } a \propto -\text{displacement } x$$

$$(\text{or Restoring force} \propto -\text{displacement } x)$$

or using the constants of proportionality previously obtained

$$a = -\omega^2 x$$

(or $F = ma = -kx$, thus $a = -kx/m$ and $\omega^2 = k/m$)
But acceleration is the change of velocity with time;

$$a = \frac{\Delta v}{\Delta t}$$

where Δv is the velocity change occurring in time Δt. Velocity is the rate at which distance is covered with time;

$$v = \frac{\Delta x}{\Delta t}$$

where Δx is the distance covered in time Δt. Thus we can write for acceleration

$$a = \frac{\Delta(\Delta x/\Delta t)}{\Delta t};$$

acceleration is the rate of change with time of $(\Delta x/\Delta t)$. Thus for our simple harmonic motion equation we can write

$$\frac{\Delta(\Delta x/\Delta t)}{\Delta t} = -\omega^2 x$$

On a graph of displacement x against time t, $(\Delta x/\Delta t)$ is the average slope over an interval of time Δt.

(For this solution it is convenient to use finite intervals, hence the Δ sign, instead of infinitesimally small intervals. Hence all the velocities and accelerations are average values over some finite time interval.)

Thus our equation can be written:

$$\frac{\Delta(\text{slope})}{\Delta t} = -\omega^2 x$$

or

$$\text{Change in slope} = -\omega^2 x\,\Delta t$$

To plot a graph of x against t we start at some arbitrary value of the slope, calculate the change in slope during the next time interval, plot the new x position given by this new slope, use the new x value to find the change in slope that will occur in the next time interval, and so on. Let us take a specific case; at time $t = 0$ the displacement to be a maximum of 10 cm, thus velocity (initial slope) to be zero, and $\omega^2 = 20\ \text{s}^{-2}$. We will take time intervals of 0·1 s. (See Fig. 4.14 for the step by step plotting.)

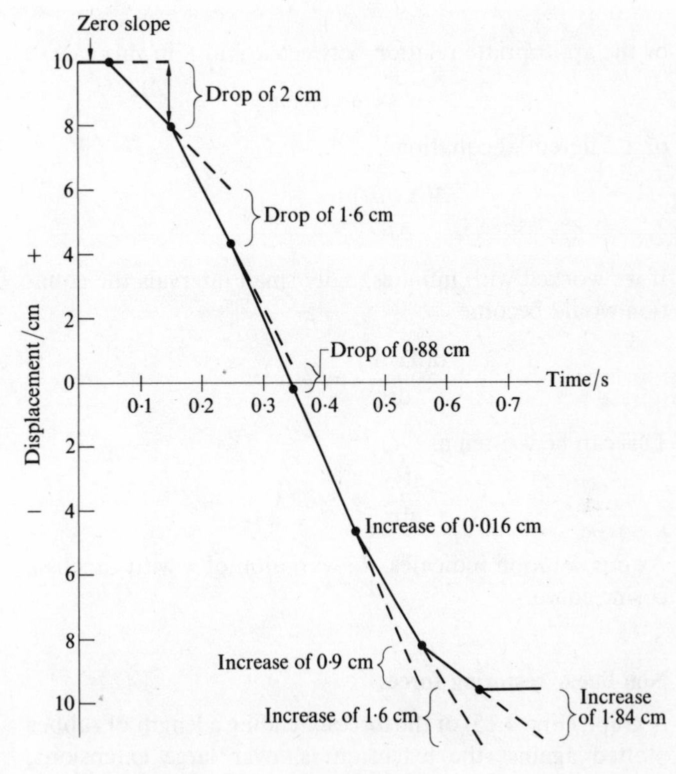

Fig. 4.14

Time 0·0 s, slope 0.

Time 0·1 s, slope change $= -20 \times 10 \times 0{\cdot}1 = -20$ cm s^{-1}. Over a time interval of 0·1 s this change in slope means a drop of 2 cm.

Time 0·2 s, slope change $= -20 \times 8 \times 0{\cdot}1 = -16$ cm s^{-1}. Over a time interval of 0·1 s this is a drop from the previous slope of 1·6 cm.

Time 0·3 s, slope change $= -20 \times 4{\cdot}4 \times 0{\cdot}1 = -8{\cdot}8$ cm s^{-1}. This is a drop of 0·88 cm, in the next 0·1 s, from the previous slope.

Time 0·4 s, slope change $= -20 \times (-0{\cdot}08) \times 0{\cdot}1 = +0{\cdot}16$ cm s^{-1}. This is an increase in slope and hence, in the next 0·1 s, an increase of 0·016 cm from the previous slope.

Time 0·5 s, slope change $= -20 \times (-4{\cdot}5) \times 0{\cdot}1 = +9\ \text{cm s}^{-1}$. This is an increase of 0·9 cm, in the next 0·1 s, from the previous slope.

Time 0·6 s, slope change $= -20 \times (-8) \times 0{\cdot}1 = +16\ \text{cm s}^{-1}$. An increase of 1·6 cm, in the next 0·1 s, from the previous slope.

Time 0·7 s, slope change $= -20 \times (-9{\cdot}2) \times 0{\cdot}1 = +18{\cdot}4\ \text{cm s}^{-1}$. An increase of 1·84 cm, in the next 0·1 s, from the previous slope.

This last result starts the graph moving back towards the axis again. The result of our graph plotting is the displacement–time graph. Our oscillating object takes about 2 × 0·7 s to complete one oscillation.

The graph shows that, within the accuracy of the plotting, the object moves from 10 cm on one side of the zero mark to 10 cm on the other. The motion is undamped. We can describe the motion by

$$a = -\omega^2 x$$

or the appropriate relation between x and t, in this case

$$x = A \cos \omega t$$

or a differential equation

$$\frac{\Delta(\Delta x/\Delta t)}{\Delta t} = -\omega^2 x$$

If we worked with infinitesimally small intervals the equation would become

$$\frac{\mathrm{d}(\mathrm{d}x/\mathrm{d}t)}{\mathrm{d}t} = -\omega^2 x$$

This can be written as

$$\frac{\mathrm{d}^2 x}{\mathrm{d}t^2} = -\omega^2 x$$

As our solution indicates, the variation of x with time is a cosine curve.

Non-linear restoring forces

A graph (Fig. 4.15) of the force extending a length of rubber plotted against the extension is, over large extensions, non-linear. Consider a length of rubber hung vertically with a mass on the lower end. Let us take the rubber giving the graph shown in Fig. 4.15. If the mass used is 180 g then the force will be about 1·8 N. This will produce an extension of 0·4 m. To extend the rubber by a further 0·1 m an extra force of 0·3 N is required, to extend it 0·2 m an extra force of 0·8 N is required, for an extra 0·3 m extension a force of 1·8 N is required. The further extension is not directly proportional to the extra force needed to produce it. If we pull our suspended mass down by say 0·3 m and then release it there will be a restoring force. This restoring force will not, however, be directly proportional to the distance of the mass from the rest position (i.e., the extra extension). We can read off the force–extension graph the value of the restoring force at different distances of the mass from the rest position, e.g., 0·3 N at 0·1 m displacement, 0·8 N at 0·2 m, 1·8 N at 0·3 m.

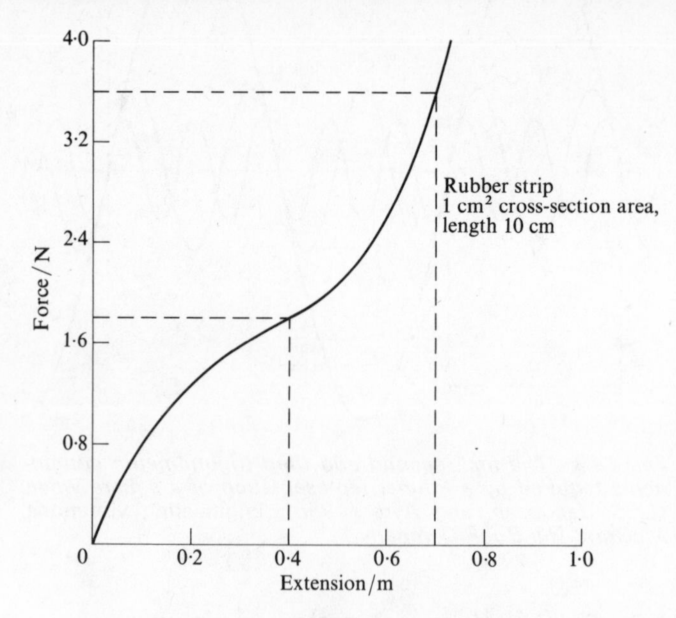

Fig. 4.15

The restoring force F will cause the mass m to accelerate towards the rest position:

$$\text{Acceleration} = -\frac{F}{m}$$

The minus sign is because the object decelerates as the force increases, i.e., as F increases the acceleration becomes more negative.

But

$$\text{Acceleration} = \frac{\Delta(\text{velocity})}{\Delta t} = \frac{\Delta(\text{slope of distance–time graph})}{\Delta t}$$

Thus we have

$$\Delta(\text{slope}) = -\frac{F}{m}\,\Delta t$$

Knowing how F changes with distance of the mass from the rest position, we can build up the distance–time graph by estimating the slope changes in successive intervals of time.

We will take a mass of 180 g attached to the rubber strip used to give the graph given in Fig. 4.15 and pull the strip down a further 0·3 m. The restoring force for this extra extension is 1·8 N.

Thus with time intervals of 0·1 s,

$$\Delta(\text{slope}) = -\frac{1{\cdot}8}{0{\cdot}18} \times 0{\cdot}1 = -1\ \text{m s}^{-1}$$

This is a decrease in distance from the rest position of 0·1 m in 0·1 s (Fig. 4.16).

In the next 0·1 s:

$$\Delta(\text{slope}) = -\frac{0{\cdot}8}{0{\cdot}18} \times 0{\cdot}1 = -0{\cdot}44 \text{ m s}^{-1}$$

This is a drop in distance of 0·044 m in 0·1 s.

In the next 0·1 s:

$$\Delta(\text{slope}) = -\frac{(-0{\cdot}12)}{0{\cdot}18} \times 0{\cdot}1 = +0{\cdot}06 \text{ m s}^{-1}$$

This is an increase of 0·006 m.

Figure 4.16 shows this continued for 1 s. A point to notice is that the graph is not symmetrical about the time axis.

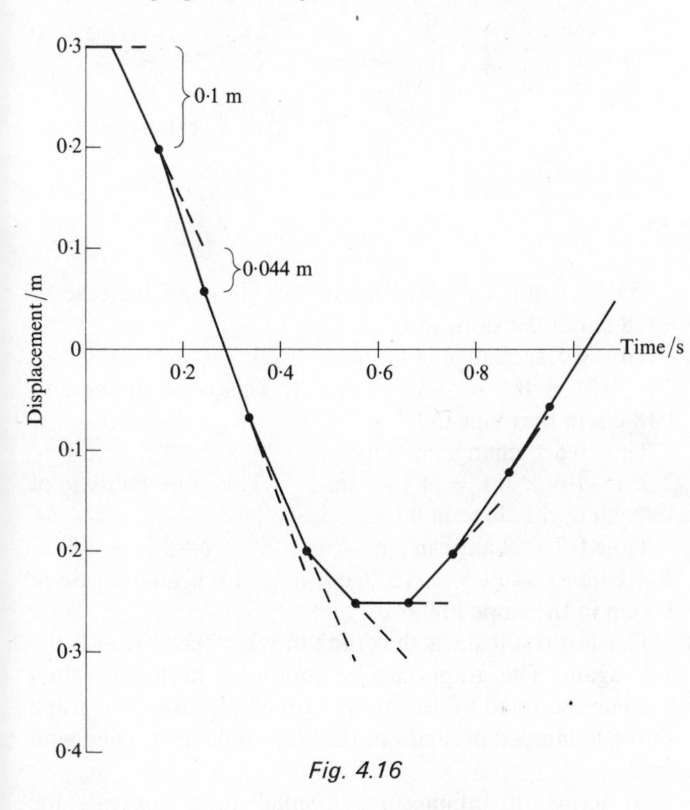

Fig. 4.16

Damping

Oscillations do not continue indefinitely, the motion dies away with time. We say the motion is damped. Figure 4.17 shows how the displacement varies with time for an oscillator over a number of oscillations. The oscillator could be a vertically spring-mounted oscillating bob dipping into oil. The amplitude varies with time in a very significant manner —in this case the amplitude goes from +6 to −3 to $+1\frac{1}{2}$ to $-\frac{3}{4}$ to $+\frac{3}{8}$, etc. How can we explain this form of damping where the amplitude decreases by a constant fraction in equal intervals of time?

The damping is caused by friction. Let us suppose we have a frictional force that is directly proportional to the velocity of the oscillating mass. The greater the velocity the greater the frictional force opposing the motion. If we double the velocity we double the frictional force. The maximum velocity of an oscillating mass is proportional to the amplitude. Thus the maximum frictional force is

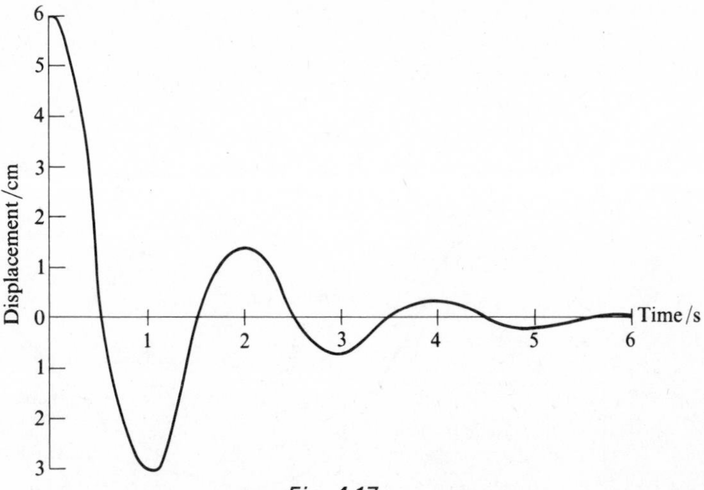

Fig. 4.17

directly proportional to the amplitude. Indeed the frictional force at any instant will be proportional to the displacement at that instant. Thus the restoring force at every instant will be made smaller by some constant factor. Suppose the frictional force reduces the restoring force by 10 per cent. After one oscillation the amplitude will have been reduced by 10 per cent. This will be true regardless of the amplitude considered. Thus if the initial amplitude was 20 cm, then after one oscillation it would be reduced by 10 per cent to 18 cm, after another oscillation the amplitude would be reduced by 10 per cent to 16·2 cm, in the next oscillation a further 10 per cent reduction would occur to 14·58 cm, and so on. The amplitude decreases by a constant factor after each successive oscillation, in this case 10 per cent.

$$\left(\begin{array}{c}\text{Change in amplitude} \\ \text{per oscillation}\end{array}\right) = -\frac{10}{100}(\text{amplitude})$$

or in symbols

$$\frac{\Delta A}{\Delta t} = -CA$$

where C is a constant. See chapter 5 for the solution of this equation—the amplitude decays exponentially.

The displacement–time graph with damping

How can we modify the acceleration equation for an oscillator in order to take account of damping? A general equation is not feasible because damping is so varied. There are, however, many instances where the damping is due to a frictional force which is directly proportional to the velocity of the moving object. An object moving in a liquid has such a frictional force. When the object is stationary there is no frictional force, when it moves the frictional force comes into effect and the faster it moves the greater the frictional force.

$$\text{Frictional force} \propto \text{velocity}$$

or using a constant of proportionality c,

$$\text{Frictional force} = cv$$

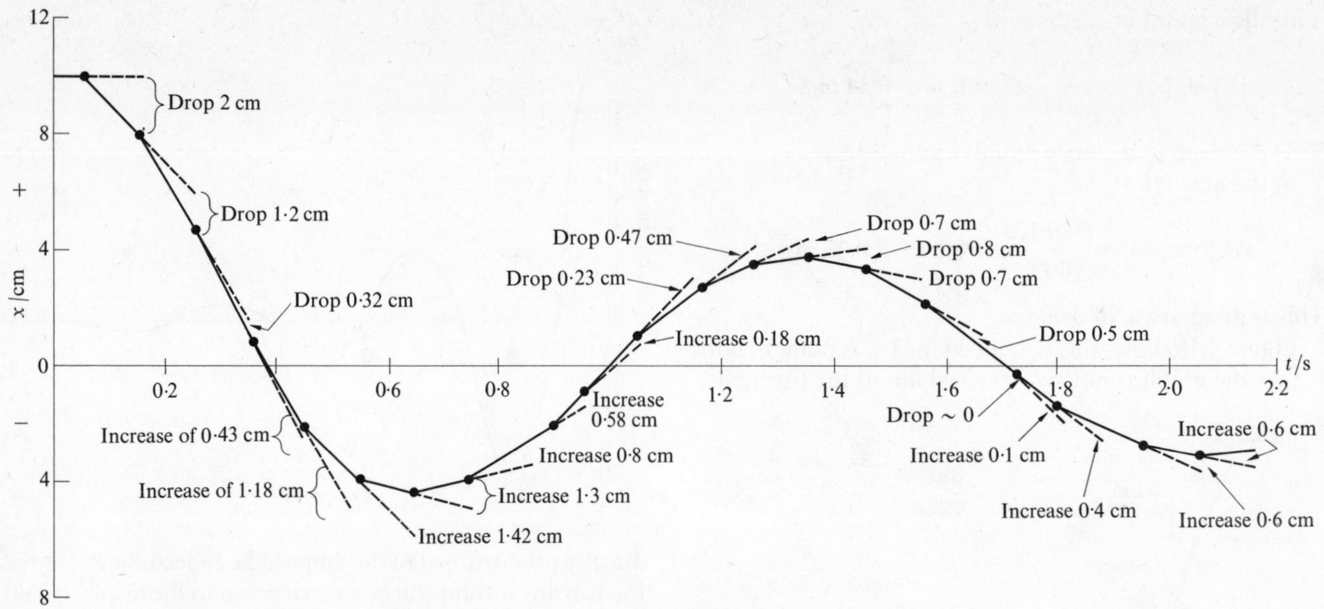

Fig. 4.18

But velocity is the rate at which distance is covered, thus

$$\text{Frictional force} = c\frac{\Delta x}{\Delta t}$$

The force now acting on an oscillating mass is

$$\text{Restoring force} - \text{frictional force}$$

The resultant force acting on the mass is ma, where a is the acceleration.

Thus

$$ma = \text{restoring force} - \text{frictional force}$$

$$ma = -\omega^2 mx - c\frac{\Delta x}{\Delta t}$$

or

$$m\frac{\Delta(\Delta x/\Delta t)}{\Delta t} = -\omega^2 mx - c\frac{\Delta x}{\Delta t}$$

or

$$\text{Change in slope} = -\omega^2 x\,\Delta t - \frac{c(\text{slope})}{m}\Delta t$$

If we take the same initial conditions as before, i.e., at $t = 0$ the displacement is a maximum of 10 cm, the velocity is zero and $\omega^2 = 20\ \text{s}^{-2}$. For the frictional term we will take c/m to be $2\ \text{s}^{-1}$. Figure 4·18 shows the resulting graph.

Time 0·0 s, slope 0.

Time 0·1 s, change in slope $= -20 \times 10 \times 0{\cdot}1 + 0 = -20\ \text{cm s}^{-1}$

This is a drop of 2 cm in the slope in 0·1 s.

Time 0·2 s, change in slope $= -20 \times 8 \times 0{\cdot}1 - 2 \times (-20) \times 0{\cdot}1 = -12\ \text{cm s}^{-1}$

This is a drop of 1·2 cm in the slope in 0·1 s.

Time 0·3 s, change in slope $= -20 \times 4{\cdot}8 \times 0{\cdot}1 - 2 \times (-32) \times 0{\cdot}1 = -3{\cdot}2\ \text{cm s}^{-1}$

This is a drop of 0·32 cm in the slope in 0·1 s.

Time 0·4 s, change in slope $= -20 \times 1{\cdot}28 \times 0{\cdot}1 - 2 \times (-35{\cdot}2) \times 0{\cdot}1 = +4{\cdot}28\ \text{cm s}^{-1}$. This is an increase of 0·428 cm in the slope in 0·1 s.

Time 0·5 s, change in slope $= -20 \times (-2{\cdot}8) \times 0{\cdot}1 - 2(-30{\cdot}9) \times 0{\cdot}1 = +11{\cdot}8\ \text{cm s}^{-1}$. This is an increase of 1·18 cm in the slope in 0·1 s.

Time 0·6 s, change in slope $= -20 \times (-5{\cdot}2) \times 0{\cdot}1 - 2 \times (-19) \times 0{\cdot}1 = +14{\cdot}2\ \text{cm s}^{-1}$. This is an increase of 1·42 cm in the slope in 0·1 s.

Time 0·7 s, change in slope $= -20 \times (-6{\cdot}3) \times 0{\cdot}1 - 2 \times (-5{\cdot}8) \times 0{\cdot}1 = +12{\cdot}76\ \text{cm s}^{-1}$. This is an increase of 1·3 cm in the slope in 0·1 s.

This last result starts the graph moving back towards the axis again. The graph can be continued until the values become too small to permit even rough plotting. The graph shows a damped oscillation, the amplitude decreasing with time.

In terms of infinitesimally small time intervals the differential equations for undamped motion can be written as

$$m\frac{d^2x}{dt^2} + m\omega x = 0$$

and for damped motion

$$m\frac{d^2x}{dt^2} + c\frac{dx}{dt} + m\omega x = 0$$

In addition to changing the amplitude, damping increases the time taken per oscillation. Figure 4.19 shows some typical results. When the damping is sufficiently large, oscillations cease and the displaced mass just slowly returns to the rest position. This is known as critical damping.

The pointers of many instruments, such as ammeters, are damped. This is to prevent the oscillation continuing for so long that it is difficult to obtain the instrument reading. Critical damping would seem to be what is needed—a current passes through the ammeter and it swings out to the current reading with no oscillation. Unfortunately this often takes too long and it is more common to use damping near the critical damping rather than at it. This means that some oscillation occurs but is rapidly damped out.

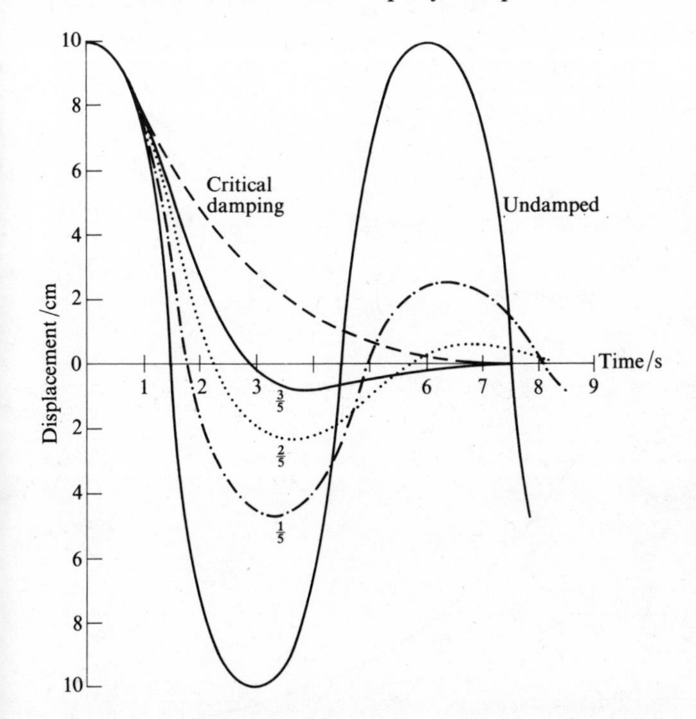

Fig. 4.19 Damped oscillations. The numbers alongside the curves refer to the damping constant, c, as a fraction of its value at critical damping. (Adapted from L. S. Jacobsen and Ayre (1958). Engineering vibrations, *McGraw-Hill Book Company.)*

4.3 Energy

What about the energy of an oscillating mass? Well, the form of the energy is continually changing—kinetic to potential energy and then back again and so on. The potential energy is a maximum when the mass is at one of the extreme points of the oscillation, the kinetic energy is a maximum when the mass is passing through the rest position. When the mass is passing through the rest position all the energy is kinetic energy.

$$\text{Kinetic energy at this point} = T = \tfrac{1}{2}mv^2$$

where m is the mass and v the velocity of the mass at that point.

But

$$v = -A\omega \sin \omega t$$

Hence

$$T = \tfrac{1}{2}m(A\omega \sin \omega t)^2$$

At the rest position $\sin \omega t = 1$. Therefore

$$T = \tfrac{1}{2}m\omega^2 A^2$$

If we ignore any effects of damping then the total energy at any instant must be equal to $\frac{1}{2}m\omega^2A^2$. At an extreme displacement all the energy is potential energy, hence the potential energy at these points U must be equal to $\frac{1}{2}m\omega^2A^2$. At any instant the energy is given by

$$T + U = \tfrac{1}{2}m\omega^2 A^2$$

We can write the equation in terms of frequency, f, as $\omega = 2\pi f$.

$$T + U = \tfrac{1}{2}m4\pi^2 f^2 A^2 = 2\pi^2 m f^2 A^2$$

An important point here is that the energy is proportional to the square of the amplitude.

Potential energy–distance graphs

For an oscillator, how does the potential energy vary with distance from the rest position? The potential energy could be the energy stored in the springs for the tethered trolley or the energy due to the change in height of a pendulum bob above some datum line. Let us consider the simple pendulum (Fig. 4.20). The potential energy (mgh) is directly proportional to the height of the pendulum bob above some horizontal datum line. In this case we will take the datum line such that the potential energy is zero when the pendulum is at its rest position. Thus a graph can be plotted of potential energy against the distance the bob is displaced from its rest position (s) (Fig. 4.21).

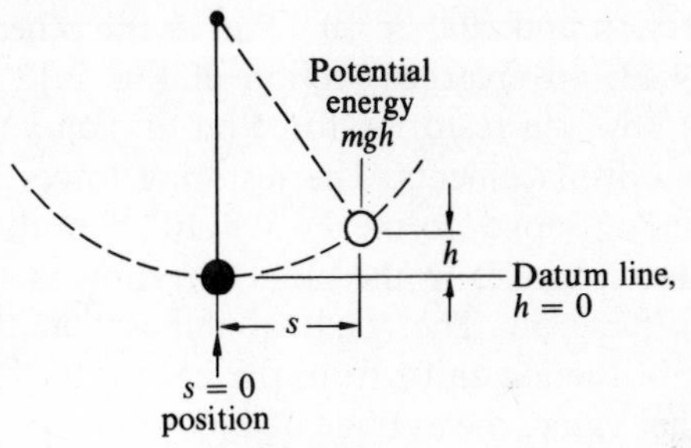

Fig. 4.20

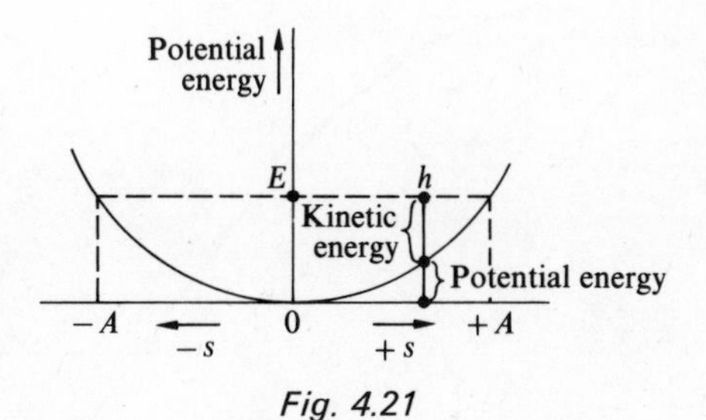

Fig. 4.21

If the maximum amplitude of the pendulum is $+A$ and $-A$ then E is the maximum energy of the bob. When the pendulum bob is at either of the extreme displacements then all the energy E will be potential energy (at $+A$ and $-A$). When $s = 0$ and the bob is at the rest position there is no potential energy and the kinetic energy is E. At any displacement the potential energy plus the kinetic energy is E, thus the potential energy at a particular displacement can be read from the curve and the kinetic energy is the energy between the point on the curve and the potential energy $= E$ line.

Though this argument has been given for a simple pendulum it applies to any oscillator.

Total energy at any instant = PE + KE at that instant.

The relation between potential energy and distance need not be symmetrical about the zero distance line. An important oscillator with a non-symmetrical potential energy–distance graph is that of an ion oscillating in a crystal. In chapter 3 of this book the potential energy–distance graph (Fig. 3.13) for an ion in a sodium chloride crystal was derived. An enlarged section of that figure is shown in Fig. 4.22. The potential energy, for this figure, is not zero at the rest position of an ion—it just means that we have put our zero potential energy datum line in a different position. The rest position of an ion we will take as the lowest point on the potential energy curve—this is in fact the position at which the net force on the ion is zero. Instead of measuring displacements from this rest position we have measured them from some other position—the rest position of the nearest ion. In the case of the simple pendulum, oscillation is produced when we pull the bob to one side and then release it—when we give it some potential energy and then allow this to be converted into kinetic energy. If we supply energy of $0{\cdot}2 \times 10^{-19}$ J to the sodium chloride ion, by perhaps heating the crystal and raising its temperature, then the ion can be displaced to $2{\cdot}65 \times 10^{-10}$ m in one direction and $3{\cdot}00 \times 10^{-10}$ m in the other direction. Figure 4.23, an enlarged section of Fig. 3.12 in chapter 3 shows how the restoring force on the ion varies within these two displacements. The restoring forces are greater when the ion moves to the $2{\cdot}65 \times 10^{-10}$ m displacement than when it moves in the other direction to the $3{\cdot}00 \times 10^{-10}$ m position. The result of this is that the average position of the ion shifts from the $2{\cdot}8 \times 10^{-10}$m position to a higher value, the average of 2·65 and $3{\cdot}00 \times 10^{-10}$ m, i.e., about $2{\cdot}85 \times 10^{-10}$ m. If all the ions in the crystal receive this energy then the average distance between ions will change from 2·80 to $2{\cdot}825 \times 10^{-10}$ m—the change in temperature has resulted in expansion. We have thus an

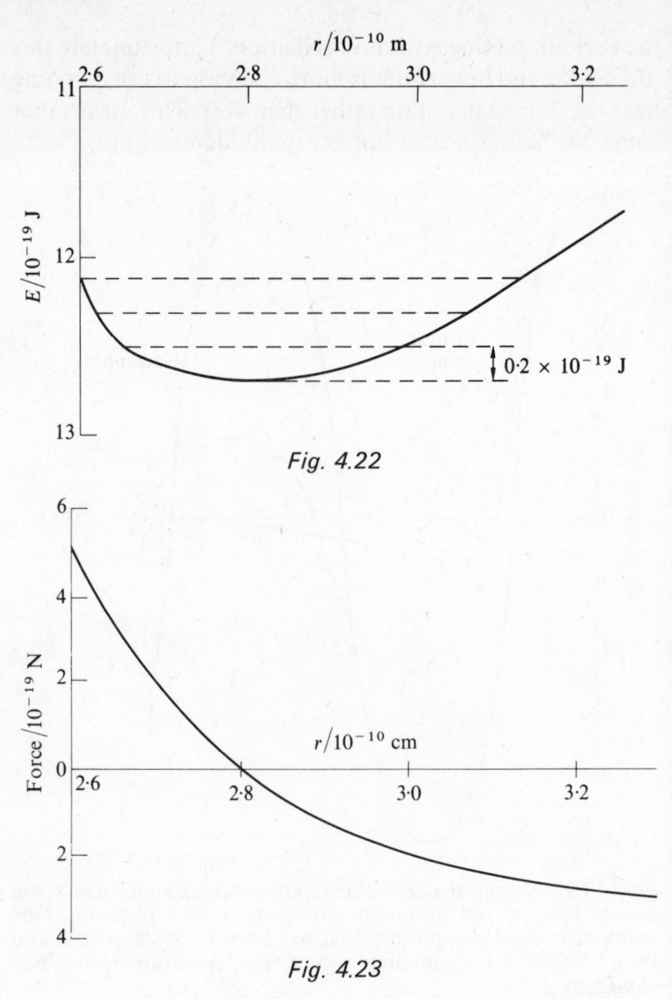

Fig. 4.22

Fig. 4.23

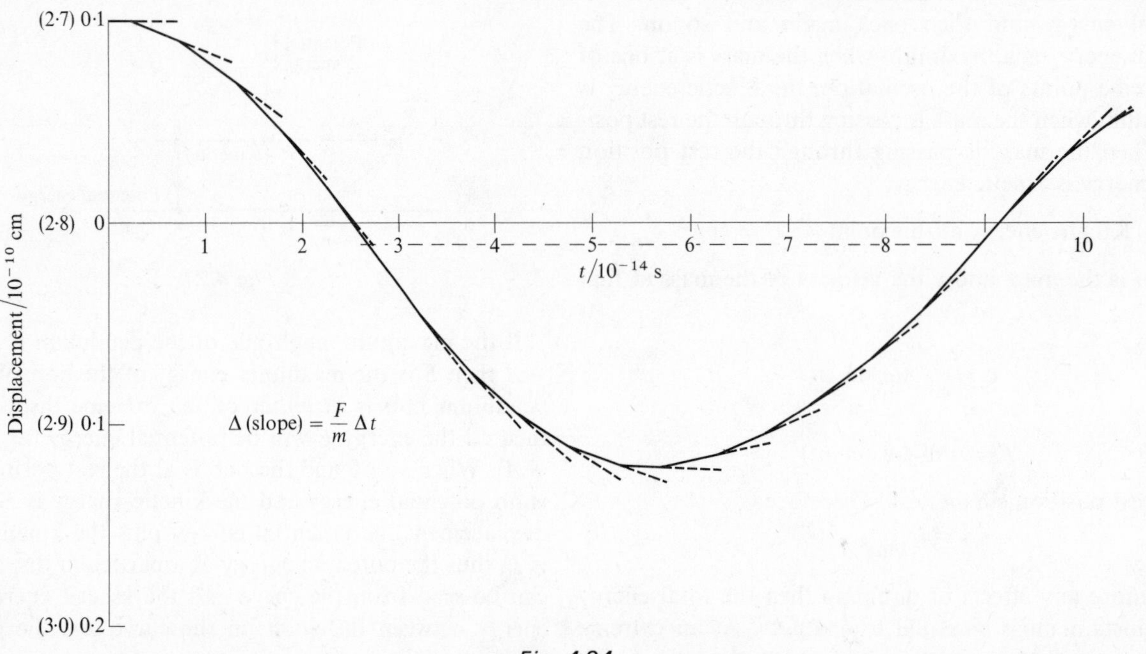

Fig. 4.24

explanation of why sodium chloride expands when its temperature rises. (Figure 4.24 shows a plot of the distance–time graph derived using the force–distance information given in Fig. 4.23. The curve is not symmetrical about the zero displacement line but about a line displaced about $0{\cdot}01 \times 10^{-10}$ m from the zero position. This is given by a supply of about $0{\cdot}1 \times 10^{-19}$ J.)

We can take our considerations of thermal expansion further. If instead of adding $0{\cdot}2 \times 10^{-19}$ J energy we added $0{\cdot}4 \times 10^{-19}$ J we find that the mean separation of the ions becomes halfway between 2·62 and $3{\cdot}06 \times 10^{-10}$ m, i.e., $2{\cdot}84 \times 10^{-10}$ m, an expansion of $0{\cdot}04 \times 10^{-10}$ m. Adding $0{\cdot}6 \times 10^{-19}$ J gives an expansion of about $0{\cdot}06 \times 10^{-10}$ m. The expansion is roughly proportional to the energy added. If we assume that the temperature of the sodium chloride is proportional to the energy added we get the result that the expansion should be proportional to the temperature. This is in fact the case, the expansion is nearly proportional to the temperature.

The coefficient of linear expansion is the change in length per unit per degree.

$$\text{Coefficient} = \frac{\Delta L}{L\,\Delta T}$$

For the sodium chloride crystal the change in length, for an initial length of $2{\cdot}8 \times 10^{-10}$ m, is about $0{\cdot}02 \times 10^{-10}$ m when energy $0{\cdot}2 \times 10^{-19}$ J is supplied. The temperature change produced by this energy will be of the order of

$$\frac{0{\cdot}2 \times 10^{-19}}{1{\cdot}4 \times 10^{-23}} = 1400 \text{ K}$$

The equation used to determine the temperature was $E = kT$ where k is a constant called Boltzmann's constant and has the value $1{\cdot}4 \times 10^{-23}$ J K^{-1}. This temperature change results in a coefficient of expansion of about

$$\frac{0{\cdot}02 \times 10^{-10}}{2{\cdot}8 \times 10^{-10} \times 1400} \approx 10^{-5} \text{ K}^{-1}$$

This is of the order of magnitude obtained by experiment.

Sodium chloride melts at 1070 K—in our analysis we have assumed that the ions could keep on oscillating, however big we made the amplitude, without any changes in the structure of the crystal occurring. Using $E = kT$, this temperature will be realized with energy

$$1{\cdot}4 \times 10^{-23} \times 1070 = 0{\cdot}15 \times 10^{-19} \text{ J.}$$

This energy addition to the crystal gives ions oscillating between about 2·67 and $2{\cdot}98 \times 10^{-10}$ m. This is an amplitude of oscillation of

$$\frac{(2{\cdot}98 - 2{\cdot}67)10^{-10}}{2}$$

or about $0{\cdot}21 \times 10^{-10}$ m. This is about 7·5 per cent of the spacing between the ions.

4.4 Resonance

When the pendulum or the mass on the end of the spring was considered a force was applied to displace the mass from its equilibrium position and then removed. The only force then acting on the mass, which caused motion, was the restoring force. The restoring force varies with time and is proportional to the displacement of the mass from the rest position. What happens if an external force, continually causing displacements, is continually applied to the mass?

When a simple pendulum is oscillating the only force acting on the bob is the restoring force, proportional to the displacement of the bob from its rest position. The frequency of the oscillation of the pendulum is a constant. Now what happens if every time the bob reaches a maximum displacement it is given a push? A force is being applied to the bob at the rate of once every complete oscillation, i.e., at the same frequency as that of the pendulum. When this happens large amplitudes build up. If the frequency of the applied force is different to that of the pendulum the amplitude produced is not so great. One method of continually applying a force to a pendulum is to suspend it from a support through which the force can be applied. A string stretched between two supports and the pendulum hung from it enables this to be done (Fig. 4.25). The force is supplied by another pendulum with a larger mass bob attached to the same string. As this oscillates a varying force is applied to the pendulum. The frequency of this applied force can be varied by changing the length of the driver pendulum (the frequency is inversely proportional to the square root of the length). Figure 4.26 shows a typical set of results. When the applied frequency is the same as the natural frequency a large amplitude oscillation is produced. This effect is known as resonance.

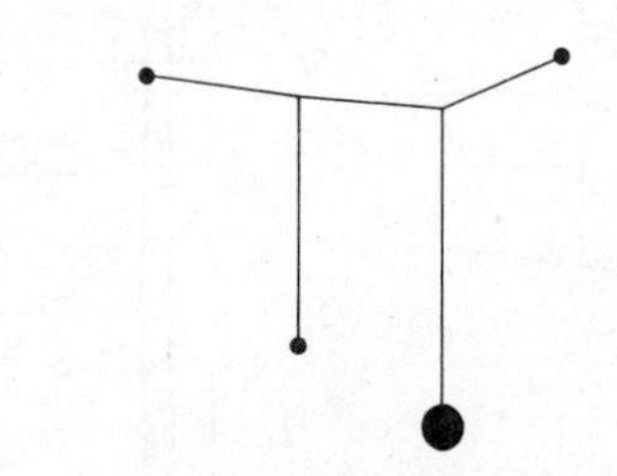

Fig. 4.25

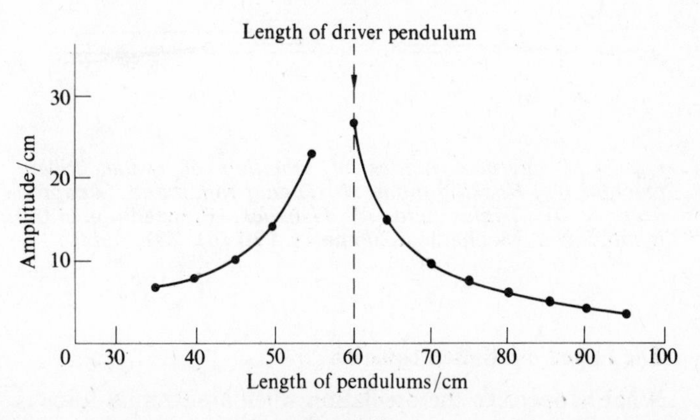

Fig. 4.26

Look what happens to a drilling machine when the applied frequency becomes equal to one of the natural frequencies of the machine (Fig. 4.27)—large amplitudes build up and accurate work is not feasible. Similarly when you are vibrated at one of your natural frequencies, for example 3 Hz for your thorax-abdomen system, large amplitude oscillations can build up (Fig. 4.28), and it can feel uncomfortable or even painful. Big amplitudes can build up in a ship when the forcing frequency is the same as a resonant frequency (Fig. 4.29). Resonant frequencies and the build up of large amplitudes occurs with structures both large and small. Ions in a sodium chloride crystal can be made to oscillate with large amplitudes when the applied frequency (infrared radiation about 61 μm) is equal to the natural frequency of the ions (see later in this chapter).

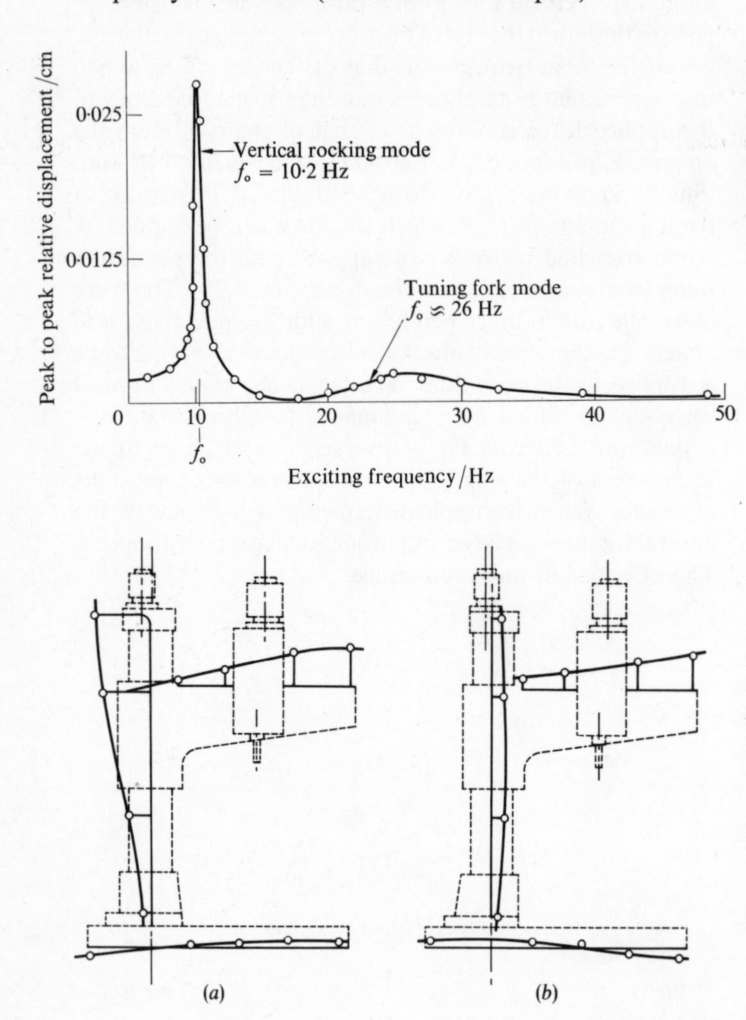

Fig. 4.27 Vertical modes of vibration of radial drilling machine. (a) Rocking mode. (b) Tuning-fork mode. (Adapted from S. A. Tobias and W. Fishwick. Proceedings of the Institution of Mechanical Engineers, **170** *(6), 232, 1956.)*

The forced oscillation equation

What happens to the oscillation when an external force is applied to a freely oscillating object? If this force is F then

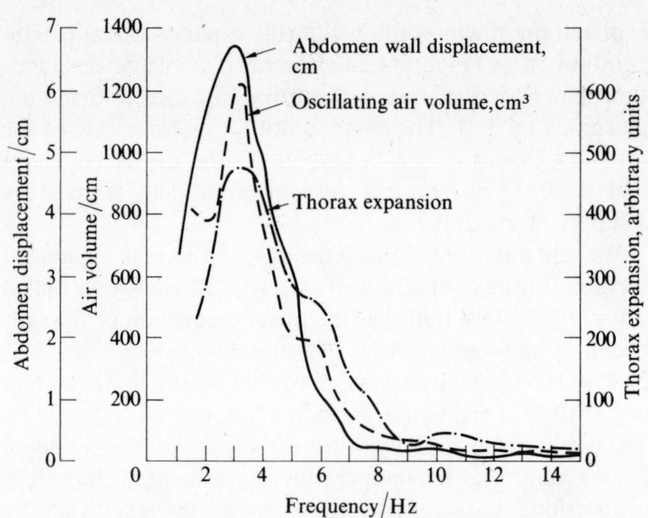

Fig. 4.28 Typical response curves of the thorax-abdomen system of a human subject in the supine position exposed to longitudinal vibrations. The displacement of the abdominal wall (5 cm below umbilicus), the air volume oscillating through the mouth, and the variations in thorax circumference are shown per g *longitudinal acceleration. (R. R. Coermann,* et al. Aerospace Medicine, **31**, *443, 1960. Aerospace Medical Association.)*

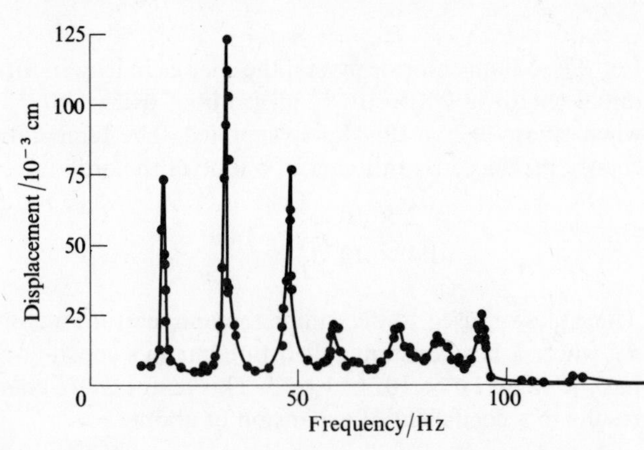

Fig. 4.29 Resonance curve for vertical vibration of excitation of an attack transport ship. (N. H. Jasper (Dec. 1948). David Taylor Model Basin Report 699. *Courtesy of the US Navy Department.)*

the force acting on the object is

$$F + \text{restoring force}$$

These forces cause the object to accelerate, hence we have

$$ma = F + \text{restoring force}$$

If the restoring force is proportional to the displacement the motion would, in the absence of the applied force, be simple harmonic.

$$\text{Restoring force} = -kx$$

or

$$\text{Restoring force} = -\omega^2 mx$$

Hence we can write

$$m\frac{\Delta(\Delta x/\Delta t)}{\Delta t} = F - \omega^2 mx$$

or

$$\Delta(\text{slope}) = \frac{F}{m}\Delta t - \omega^2 x \,\Delta t$$

If we know how the applied force varies with time we can solve this equation in a manner similar to the solution of the simple harmonic equation.

Suppose F varies with time according to

$$F = F_a \cos \omega_a t,$$

ω_a is not necessarily the same as ω.

$$\omega = 2\pi f$$

Thus the frequency of the applied force is not necessarily the same as that at which the object freely oscillates. In our simple harmonic motion solution we took the following values; $\omega^2 = 20\ \text{s}^{-2}$, $x = 10$ cm at $t = 0$, $v = 0$ at $t = 0$, and time intervals of $\Delta t = 0{\cdot}1$ s. We will use the same values for the oscillator and take for the applied force the values; $F_a/m = 100\ \text{cm s}^{-2}$, and $\omega_a^2 = 19\ \text{s}^{-2}$. Figure 4.30 shows a graph of $(F_a/m)\cos\omega_a t$ variation with time. The force has thus a frequency of 0·69 Hz and the natural frequency of the oscillator is 0·71 Hz.

Figure 4.31 shows the result of plotting the displacement–time graph step by step. The amplitude builds up with each oscillation. This build-up only occurs when the forcing frequency is close to the natural frequency of the object. Figure 4.32 shows the results that are obtained when other forcing frequencies are used. When the forcing frequency is equal to the natural frequency we say that resonance is occurring.

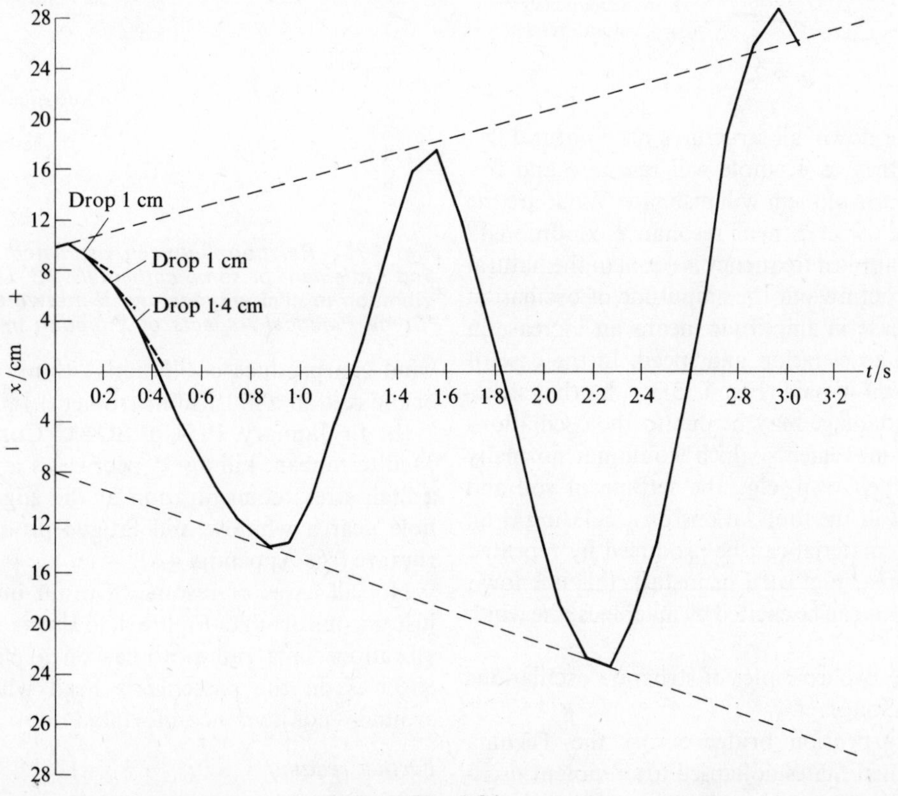

Fig. 4.31

Vibrations of structures

The resonant frequency of the Empire State Building in New York is 1/8 Hz (one complete oscillation every 8 s), the new Severn bridge has a (vertical) frequency of about 0·14 Hz (one complete oscillation about every 7 s), your body as a whole has a resonant frequency of about 5 Hz

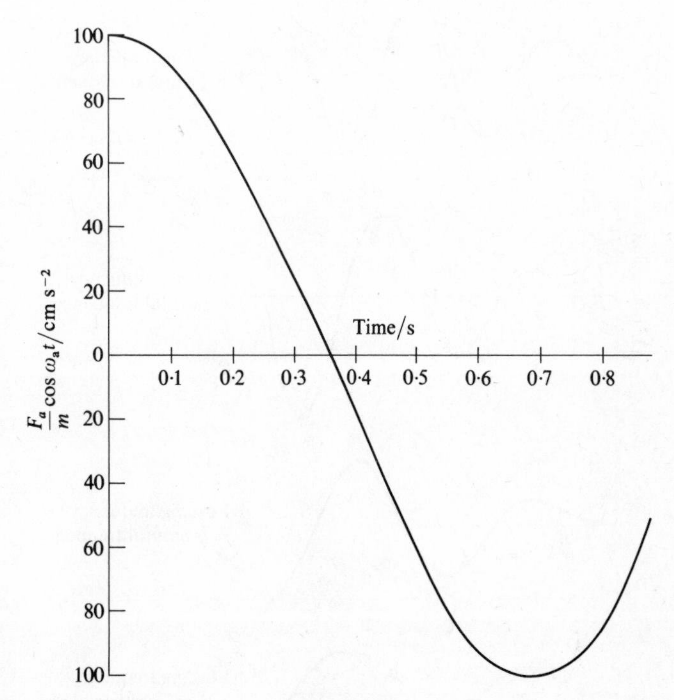

Fig. 4.30

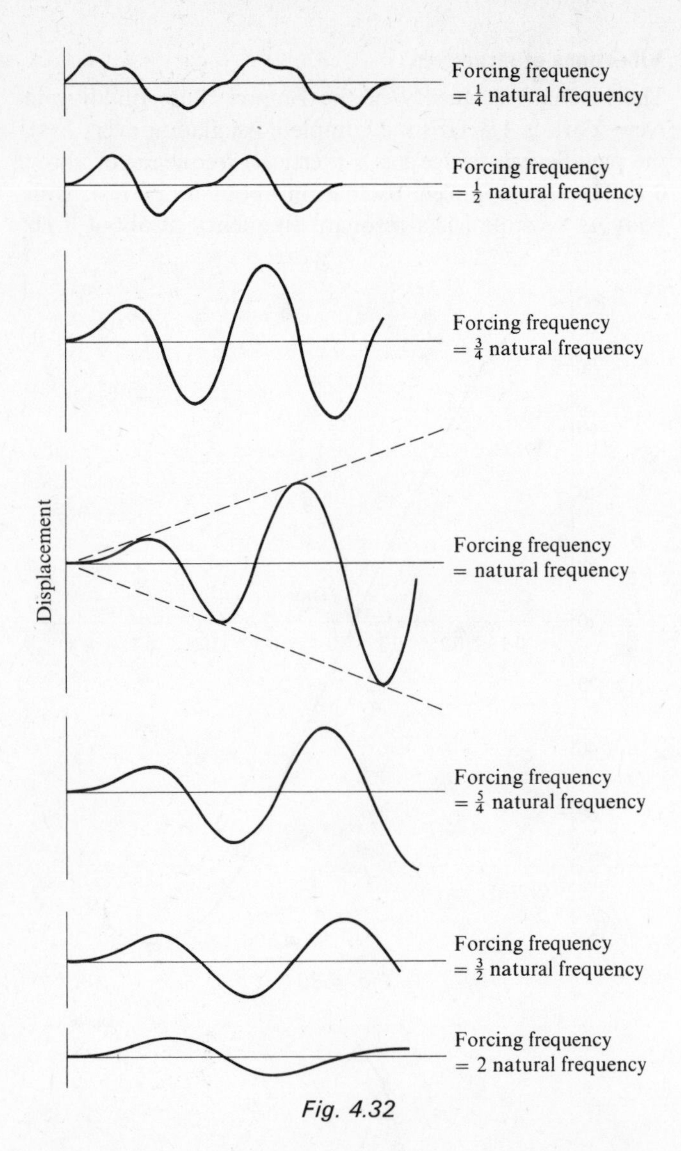

Fig. 4.32

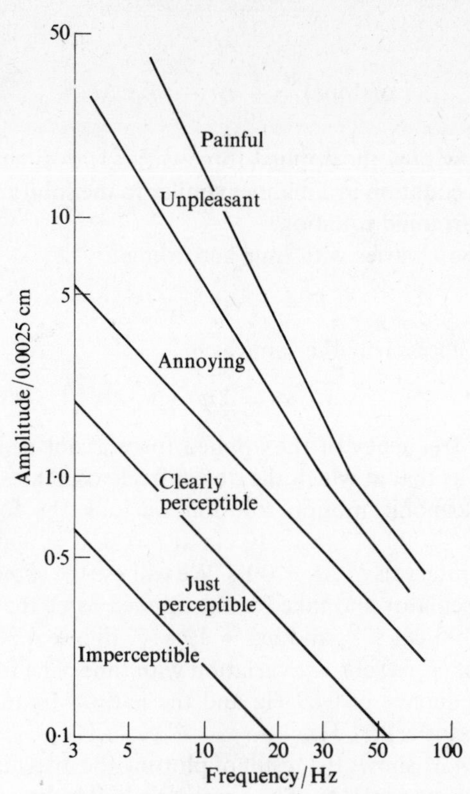

Fig. 4.33 Human sensitivity to vibration.

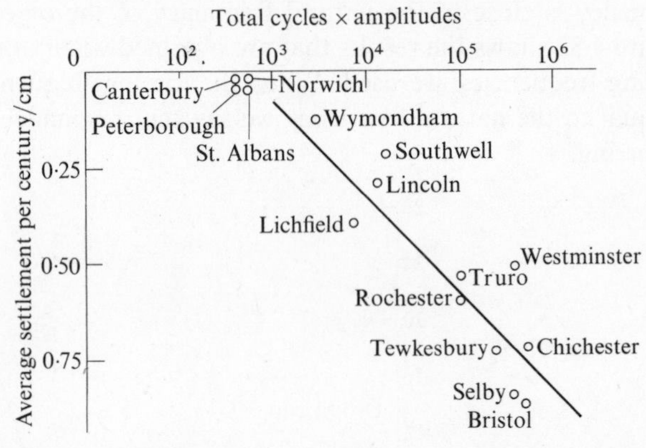

Fig. 4.34 Relation between estimated history of vibration and settlement of some cathedrals. (B. O. Skipp, Ed. (1966). Vibration in civil engineering, *Butterworth. J. H. A. Crockett 'Some Practical Aspects of Vibration in Civil Engineering'.)*

when you are sitting down, all structures have natural frequencies at which they as a whole will resonate and frequencies at which parts of them will resonate. What are the effects of resonance (or even near resonance conditions)?

At resonance the applied frequency is equal to the natural frequency of the structure and the amplitude of oscillation builds up. An increase in amplitude means an increase in both the maximum acceleration and speed. In the case of humans this can result in pain (Fig. 4.33), with other structures damage. The damage may be due to the oscillations triggering off stress movements which would not normally have occurred on their own, e.g., the settling of soil and resulting movement in the foundations of a building (Fig. 4.34). Fracture of a material can be produced by repeated flexing (try it with a strip of card or metal); this is known as fatigue. Oscillations can be excited by machines, the wind, earthquakes, etc.

The following are two examples of structure oscillations which resulted in damage:

(a) In 1940 a suspension bridge across the Tacoma Narrows in the United States collapsed after violent oscillations produced by a 19 m s^{-1} wind (Fig. 4.35) (a steady wind can produce oscillations—think of a fluttering flag, or the reed in a musical instrument).

(b) In January 1954 a BOAC Comet crashed in the Mediterranean, killing 35 people, as a result diagnosed as a high stress concentration at the edge of a countersunk hole near a window and fatigue producing the resulting rupture (see Appendix 4A).

Not all cases of resonance result in damage, some are just uncomfortable. Figure 4.36 shows the effect of vertical vibrations in a rail-motorcar on a passenger. The main effect is on the passenger's head which keeps bobbing around—not a very comfortable ride.

Further reading

R. E. D. Bishop. *Vibration*, CUP.

Fig. 4.35 Torsional oscillations of the Tacoma Narrows Bridge. (Bulletin 116, University of Washington.)

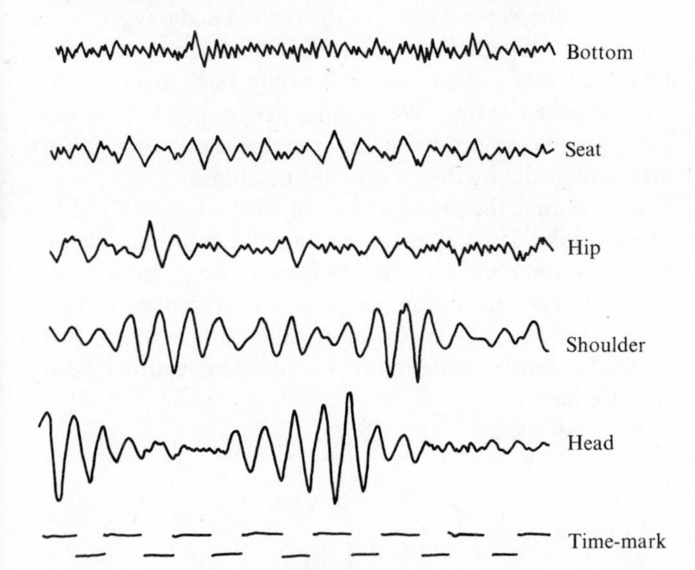

Fig. 4.36 Vertical vibrations of seat and man in a rail-motorcar. The top two records show car vibration, the lower three show vibration amplitudes of the subject. Time marker, 1 s. (D. Dieckmann. Ergonomics, **1** *353, 1958.)*

Molecular vibration spectra

If we have a mass oscillating at the end of a spring we can calculate the force constant or stiffness of the spring from measurements of the frequency of oscillation and a knowledge of the mass. Suppose the spring and the mass were so small that we could not see when oscillations were occurring and certainly not directly measure the frequency, how could we obtain a value for the frequency of oscillation? If we applied a wide range of different frequency forces then the one at the right frequency would set our mass and spring system into oscillation, i.e., resonance would occur. When this occurs energy must be absorbed by the mass-spring system. If we can detect the frequency at which energy absorption occurs then we have determined the oscillation frequency.

When white light is incident on a gas, or a liquid, or a solid, such as HCl, we find that at a certain frequency in the infrared, $8{\cdot}658 \times 10^{13}$ Hz, absorption occurs (Fig. 4.37).

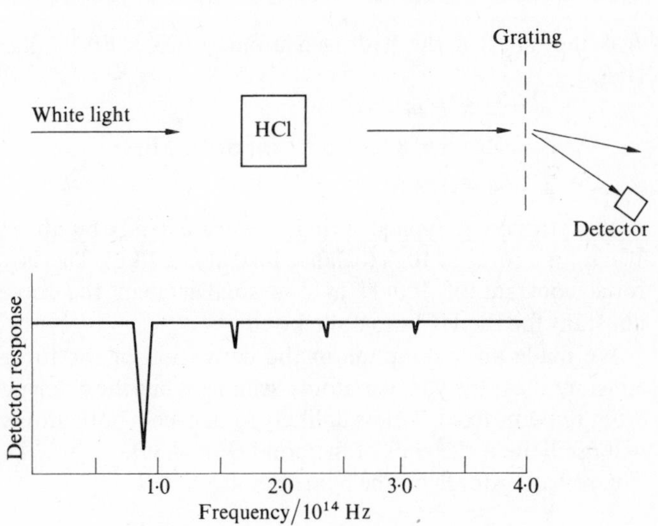

Fig. 4.37 Absorption spectrum of HCl.

The absorbed energy is re-radiated in all directions. Absorption also occurs at twice this frequency, three times this frequency, four times, etc., though each succeeding absorption is weaker. Similar absorption results are produced, at different frequencies, for other diatomic molecules.

Molecule	*Strong absorption frequency, /10^{13} Hz*
HF	8·721
HCl	8·658
HBr	7·677
HI	6·690
CO	6·429
NO	5·628

Is this absorption of light similar to the energy absorption considered for the mass and spring system? With a diatomic molecule we can consider the two atoms to be at the ends of a bond, rather like two masses attached to the ends of a spring. If light is some kind of oscillating electrical disturbance then it might drive the atoms into oscillation (Fig. 4.38). White light contains a wide range of frequencies and thus absorption could be expected at the frequencies corresponding to resonance for the molecule.

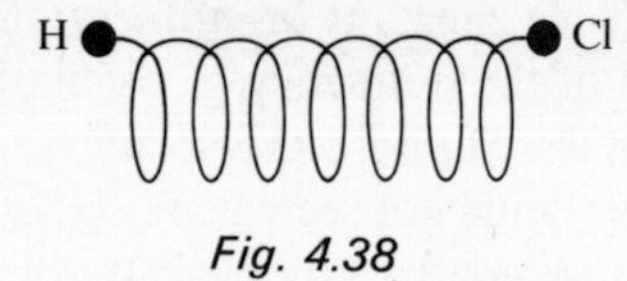

Fig. 4.38

Let us try to calculate the strength of the bond in the HCl molecule. We will make a simplifying assumption that the mass of the chlorine atom is so great compared with that of the hydrogen atom that it remains at rest and only the hydrogen atom oscillates at the end of the bond. If the restoring force acting on the hydrogen atom is proportional to its displacement from its rest position then the oscillation should be simple harmonic. For such a motion—

$$\text{Frequency } f = \frac{1}{2\pi}\sqrt{\frac{k}{m}}$$

m is the mass of the hydrogen atom, $1{\cdot}673 \times 10^{-27}$ kg. Hence

$$\begin{aligned} k &= 2^2\pi^2 f^2 m \\ &= 4\pi^2\, 8{\cdot}658^2 \times 10^{26} \times 1{\cdot}673 \times 10^{-27} \\ &= 495{\cdot}0 \text{ N m}^{-1} \end{aligned}$$

Is this strong? A typical spring balance extends by about 10 cm for a force of 10 N (balance load about 1 kg). This is a force constant of 100 N m^{-1}— smaller than the force constant for the HCl molecule bond.

We made an assumption in the derivation of the force constant that the hydrogen atom oscillates while the chlorine atom remains fixed. This is unlikely to be true—both atoms will oscillate at the ends of the bond (Fig. 4.39).

The restoring force on the hydrogen atom $= k\,(x_2 - x_1)$

$$= m_1 a_1$$

where m_1 is the mass of the hydrogen atom, x_1 its displacement, and a_1 its acceleration.

But

$$a_1 = -\omega^2 x_1$$

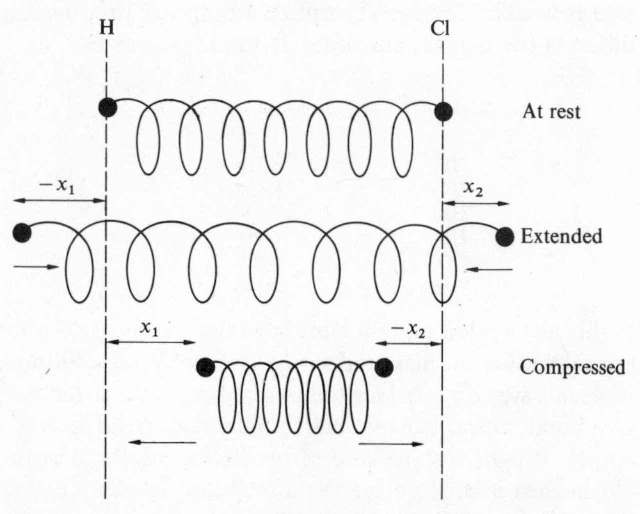

Fig. 4.39

Thus

$$k(x_2 - x_1) = -m_1\omega^2 x_1$$

The restoring force on the chlorine atom $= -k(x_2 - x_1)$

$$= m_2 a_2$$

where m_2 is the mass of the chlorine atom, x_2 its displacement and a_2 its acceleration.

But

$$a_2 = -\omega^2 x_2$$

Thus

$$-k(x_2 - x_1) = -m_2\omega^2 x_2$$

Eliminating x_2 and x_1 gives—

$$\omega^2 = \frac{k(m_1 + m_2)}{m_1 m_2}$$

Hence

$$f = \frac{1}{2\pi}\sqrt{\left[\frac{k}{m_1 m_2/(m_1 + m_2)}\right]}$$

$m_1 m_2/(m_1 + m_2)$ is known as the reduced mass.

For the HCl molecule the effect of this factor on the calculated value of the bond force constant is to decrease the value of 495·0 N m^{-1} by a factor of $(35 \times 1)/(35 + 1)$ to 481·1 N m^{-1}. The chlorine atom is assumed to have a mass of thirty-five times that of the hydrogen atom.

If the calculations are repeated for the other diatomic molecules given near the beginning of this section then the following results are obtained.

Molecule	*Force constant*/N m^{-1}
HF	970
HCl	480
HBr	410
HI	320
CO	1840
NO	1530

The CO molecule is obviously much more rigid than the HI molecule. We can think of the carbon and oxygen atoms being held together by a tightly coiled spring while the hydrogen and iodine atoms are only held together by a loosely coiled spring. We could perhaps expect that more energy would be needed to separate the carbon and oxygen atoms than the hydrogen and iodine atoms.

How would the absorption frequency change if the isotopes involved in a molecule changed? Let us take the HCl molecule and consider what happens if the hydrogen atom is replaced by deuterium, an isotope of hydrogen having a mass twice that of the conventional hydrogen atom. If the force constant is assumed to be not significantly affected then we have—

For hydrogen

$$f_H = \frac{1}{2\pi}\sqrt{\frac{k}{m}}$$

For deuterium

$$f_D = \frac{1}{2\pi}\sqrt{\frac{k}{2m}}$$

Hence

$$f_H = f_D\sqrt{2}$$

The effect of the chlorine atom oscillations has been ignored. Thus we would expect DCl to have an absorption frequency lower than HCl by a factor of about $\sqrt{2}$, i.e., about $6{\cdot}18 \times 10^{13}$ Hz. The observed frequency is $6{\cdot}30 \times 10^{13}$ Hz.

Vibration spectra for ions in solids

The force constants for the diatomic molecules considered, gases, are of the order of 1000 N m^{-1}. What are the values for solids? Sodium chloride crystals have a strong absorption at a wavelength of about 6×10^{-5} m (Fig. 4.40). This is an absorption frequency of about 5×10^{12} Hz. Sodium has an atomic mass of 23 and chlorine 35·5. If we take the mass of the oscillating ion as $30 \times 1{\cdot}66 \times 10^{-27}$ kg a crude approximation is being made but the order of the force constant should be feasible. The ions in a sodium chloride crystal are acted on by forces due to a large number of ions surrounding them and not, as in our considerations of gases, just one other ion. Using these figures in our oscillation equation

$$f = \frac{1}{2\pi}\sqrt{\frac{k}{m}}$$

gives k as about 50 N m^{-1}. This must be regarded as a very approximate value.

Figure 4.23, in this chapter, derived from considerations of the energy needed to tear the ions apart in a sodium chloride crystal, shows how the force on an ion varies with the separation of the ion from its neighbours. At the normal separation of the ions, $2{\cdot}80 \times 10^{-10}$ m, a force of about $0{\cdot}2 \times 10^{-9}$ N is needed to move the ions farther apart to a separation of $2{\cdot}81 \times 10^{-10}$ m or to push them closer together to a separation of $2{\cdot}79 \times 10^{-10}$ m. The force needed per metre is thus about 200 N m^{-1}. Considering the approximations made this is reasonable agreement.

The following are the strong absorption frequencies of various alkali halide crystals.

	Frequency $/10^{-12}$ Hz
LiCl	6·12
LiBr	5·13
LiI	4·32
NaCl	4·92
NaBr	4·05
NaI	3·51
KCl	4·32
KBr	3·48
KI	3·03

(Randall, Fuller, Montgomery. *Solid State Commun*, **2**, 273, 1964)

There is a pattern. The chloride has a higher frequency than the bromide, which in turn has a higher frequency than the iodide for each alkali. This is what we might expect if the values of k were much the same for all the halides of a particular alkali and the variations were due to mainly the mass variation. The same thing seems to be the case for the various alkalis. The bigger the mass of either the alkali or the halide the lower the frequency (Fig. 4.41).

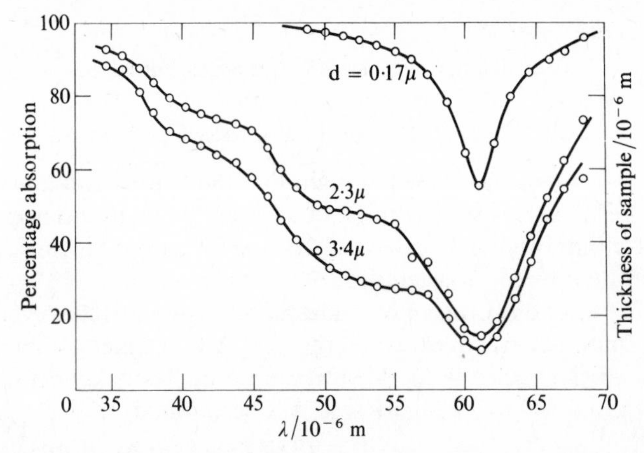

Fig. 4.40 Barnes, Zeitschrift für Physik, **75**, *47, 1932, Part 11–12.*

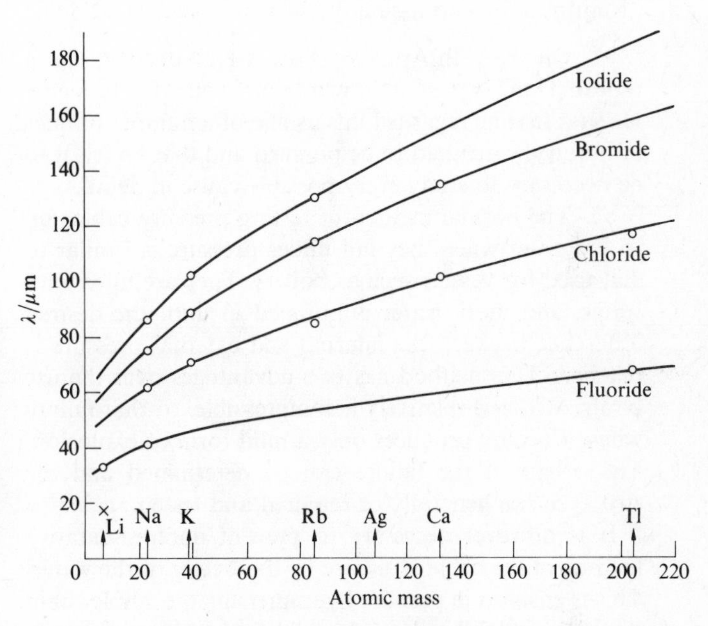

Fig. 4.41 Adapted from Barnes, Zeitschrift für Physik, **75**, *731, 1932. Part 11–12.*

Appendix 4A The Comet disaster

The first crash

'Comet Airliner Crashes in Mediterranean

10 Children Among 35 on Board Feared Lost

A B.O.A.C. Comet jet airliner flying from Singapore to London crashed in the Mediterranean yesterday morning, about 20 minutes after leaving Rome on the last stage of its flight.

On board were 29 passengers and a crew of six. Last night 15 bodies had been recovered from the sea. Among the passengers was Mr. Chester Wilmot, the former war correspondent. Ten children are reported to have been on board.'

(*The Times*, 11 January, 1954.)

The second crash

'Comet Wreckage and Bodies Found

WITHDRAWAL OF AIRWORTHINESS CERTIFICATE

It was announced last night that the United Kingdom certificate of airworthiness of all Comet aircraft has been withdrawn pending detailed investigations into the causes of the recent disasters.

After one of the most widespread air-sea searches ever undertaken, wreckage of the B.O.A.C. Comet airliner which crashed in the Mediterranean on Thursday night, about 200 miles from Rome, has been found.'

(*The Times*, 10 April, 1954.)

The official report (H.M.S.O Cap 127 1955)
Reprinted by permission of the Controller, H.M.S.O.

'61. On the 18th April Sir Arnold Hall decided that a repeated loading test of the whole cabin ought to be made. He said that he regarded this as one of a number of lines of inquiry which had to be pursued and that he felt it to be necessary to study every possible cause in detail.

62. The normal method of testing pressure cabins up to the point when they fail under pressure is similar to that used for vessels such as boilers. They are filled with water, and more water is pumped in until the desired difference between the internal and external pressure is reached. This method has two advantages over the use of air. Water is relatively incompressible, so that failure when it occurs produces only a mild form of explosion. The origin of the failure can be determined and the structure can generally be repaired and tested again.

It is however necessary to prevent unrepresentative loading of the cabin structure by the weight of the water. This is ensured in practice by immersing the whole cabin in a tank, and filling the tank and the cabin simultaneously with water. Pressure in the cabin is then raised by pumping in water from the space outside it.

Cycles of loading, to the same or different levels of pressure as desired are applied by a suitable routine of pumping.

63. By a remarkable effort, to which de Havillands and the firms who built the tank contributed to the full and by the use of all the resources of R.A.E., repeated loading tests began early in June on aircraft G–ALYU (Yoke Uncle). The object of the tests was to simulate the conditions of a series of pressurised flights. To this end the cabin and wings were repeatedly subjected to a cycle of loading as far as possible equivalent to that to which they would be subjected in the period between take-off and landing. In addition to one application of cabin pressure, fluctuating loads were applied to the wings in bending to reproduce the effect of such gusts as might be expected in normal conditions, although the contribution of gust loads to the stresses in the cabin structure, compared with that made by the internal pressure, was in general small.

64. Yoke Uncle had made 1,230 pressurised flights before the test and after the equivalent of a further 1,830 such flights, making a total of 3,060, the cabin structure failed, the starting point of the failure being the corner of one of the cabin windows. The fact that the failure occurred during one of the proving tests to 11 lb/sq. in. is not thought significant since the crack would have spread in very much the same way after a few more applications of the working pressure. Examination of the failure provided evidence of fatigue at the point where the crack would be most likely to start, namely near the edge of the skin at the corner of the window.

71. At the time of the Elba accident Yoke Peter had made 1,290 pressurised flights and at the time of the Naples accident Yoke Yoke had made 900 pressurised flights. Sir Arnold Hall said in evidence that in the light of the experiment on Yoke Uncle, and of the measurements and calculation of stress referred to above, he considered that the cabin of Yoke Peter had reached a point in its life when it could be said to be in danger of failure from fatigue, and that the cabin of Yoke Yoke would similarly be in danger.

It was therefore decided to make a search of an area some miles long in the sea below the path of the aircraft working towards Rome from the area where the main items were recovered. As the depth of the sea increased rapidly in this direction, the only practicable method was trawling.

73. As a result of the new search R.A.E. received a piece of cabin skin, which had been found by an Italian fishing boat. It was identified as coming from the centre of the top of the cabin approximately over the front spar of the wing. . . . It contained the two windows in which lie the aerials which are part of the A.D.F. (Automatic Direction Finding) equipment.

75. By examination of the piece containing the A.D.F. windows and the adjacent pieces it was established that it was here that the first fracture of the cabin structure of Yoke Peter occurred. In general terms, it took the form of a split along the top centre of the cabin along a line approximately fore and aft passing through corners of the windows as shown in Figure 17. [Fig. 4.42.] The direction in which the fracture spread was determined by examination of the lines of separation of the material.

79. It is my opinion that the fundamental cause of the failure of the cabin structure was that there existed around the corners of the windows and other cut-outs a level of stress higher than is consistent with a long life of the cabin, bearing in mind the unavoidable existence of points, within the areas of generally high stress, at which it will be still further raised by relatively local influences, such as the countersunk hole near the starboard rear corner, and the small crack with its 'locating' hole near the port forward corner. I find it impossible to say, definitely, on any evidence before me, which of these operated first. But, since the existence of fatigue near the bolt hole is established, I think it the more probable.'

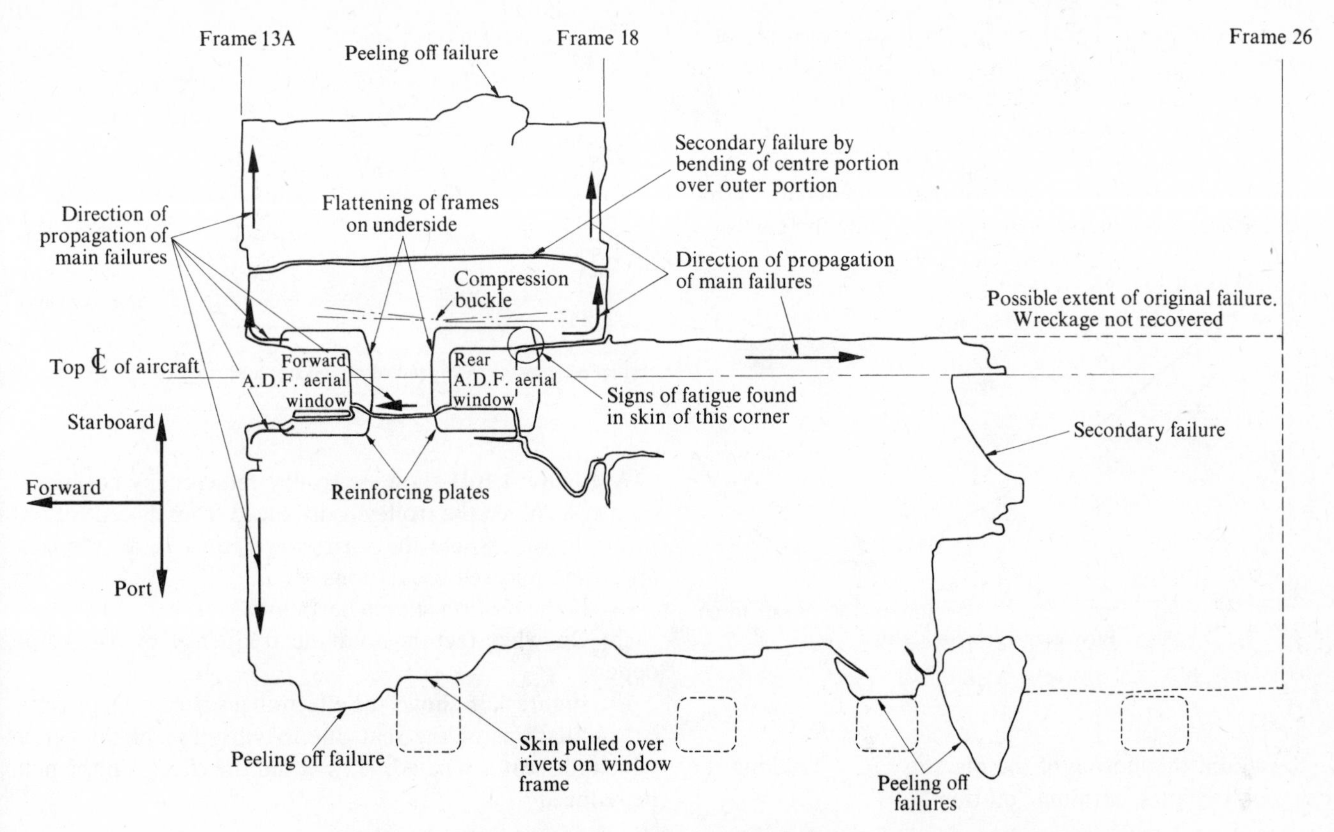

Fig. 4.42 Salient features of disruption of pressure cabin—G—ALYP.

Problems

1. Oxford and Cambridge Schools Examination Board. Nuffield Advanced Physics. 1970. Special paper.

'The diagram [Fig. 4.43] shows a spring carrying a mass A suspended from a support which may either be held stationary or which may be moved up and down to perform a simple harmonic motion of small constant amplitude. The frequency of motion of the support may be varied. With the support stationary the period of vertical oscillations of mass A is 0·5 s.

(*a*) With the support stationary, mass A is replaced by mass, B, which has the same volume as mass A but twice the density. Describe the small vertical oscillations of mass B.

(*b*) Mass B is now replaced by mass A and the frequency of the support is increased step by step up to 5 Hz. Sketch a graph to illustrate the variation of the maximum amplitude of mass A's vertical oscillations with frequency of the support.

(*c*) Describe the motion of mass A when the frequency of the support is 2·5 Hz.

(*d*) Mass A is replaced by mass C (of equal mass but 1/5 density of A) and the frequency again increased in steps up to 5 Hz. Compare the results with those in (*b*).

(*e*) One can think of the above system in considering the behaviour of a car when driven over a road with fairly regular bumps or ridges. What makes some cars more comfortable than others to ride in over such a road? How would you expect a 'comfortable' car to behave on sharp corners?'

2. At what time during the release of an 'undamped' pendulum, from its maximum displacement, are the potential and kinetic energies equal?

3. For an object oscillating with simple harmonic motion state when (a) the acceleration, (b) the velocity, (c) the potential energy, (d) the kinetic energy, and (e) the restoring force are a maximum.

4. The force constant for the bond between two oxygen atoms in the oxygen molecule is about 1170 N m^{-1}. At what frequency would you expect strong absorption?

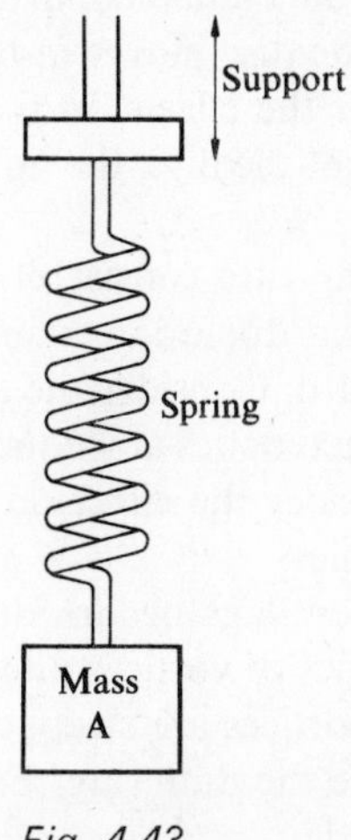

Fig. 4.43

5. How would changing the suspension springs of a car, i.e., changing the force constant, change the behaviour of the car when it goes over a bump in the road?

6. What is 'simple' about simple harmonic motion?

7. Figure 4.44 shows how the threshold of audibility of sound varies with frequency for the human ear. What would seem to be the resonant frequency for the ear?

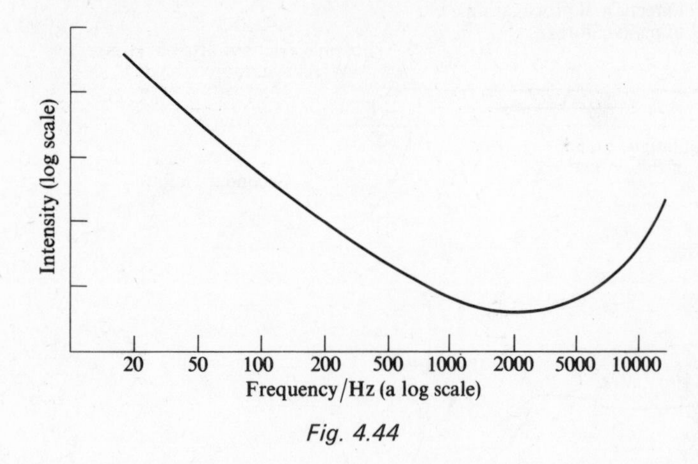

Fig. 4.44

8. Would the motion of the mass in Fig. 4.45 along the channel be simple harmonic motion?

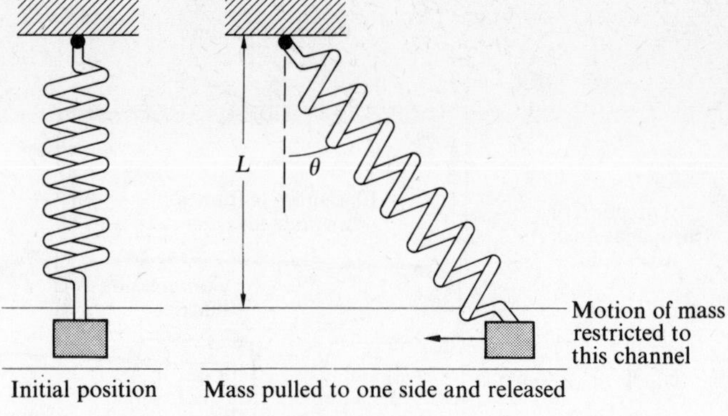

Fig. 4.45

9. Figure 4.1(d) shows a trolley tethered by springs at each end. When the trolley is displaced from its central rest position, we assume the springs are equal in all respects, and when released oscillations occur.

(a) Is the motion simple harmonic?

(b) On what factors does the frequency of oscillation depend?

10. Figure 4.36 shows the effect on a seated man of vertical oscillations of the seat due to vibration of the car in which the seat is located. How could the effects on the man be reduced?

Teaching note Further problems appropriate to this chapter will be found in: *Nuffield Advanced Physics. Unit 4 Student's book*, Penguin, 1972.

Practical problems

1. Investigate the fluid dash pot method of damping the motion of instrument parts. This consists of a flat disc, attached to the instrument pointer in some cases, and which is immersed in oil contained in a small pot. When the pointer moves, the disc moves in the oil and so damps the motion. Consider the effects on the damping of such variables as the area of the disc, the liquid used, and the size of the container.

2. Many ceiling tiles consist of a perforated layer of a material such as hardboard over an absorbent layer of glass wool or fibre board. Consider the effect of the holes on the absorption characteristics at frequencies within the audible range. Also consider the effect on the results of changing the size of the holes.

3. Fluttering of flags occurs due to turbulent flow of air forming a series of vortices alternately on either side of the flag. These vortices are swept along the flag by the air flow and produce the fluttering. Examine this fluttering at different air speeds.

Suggestions for answers

1. (a) New period $= \sqrt{2} \times 0{\cdot}5$ s

(b) Similar to Fig. 4.26. Resonance occurs at 2 Hz.

(c) See Fig. 4.32 where the forcing frequency θ is equal to 5/4 of the natural frequency. This shows the amplitude building up. Every 2 s the forcing and natural oscillations will be in step. Thus every 2 s the mass will give a large amplitude oscillation. Halfway between these maximum amplitude oscillations the forced and natural oscillations will not be in step and thus a minimum amplitude oscillation will occur. The amplitude of the oscillation will fluctuate with a period of 2 s. This phenomenon is called beats. The beat frequency is equal to the difference between the natural and forcing frequencies, 0·5 Hz.

(d) The damping will change and the maximum amplitudes reached will be larger than those in (c). Otherwise the motion will be the same.

(e) When the frequency of the forcing vibration produced by the road is the same as the natural vibration frequency of the car springs resonance will occur. The more comfortable car has a natural frequency far removed from the forcing frequencies that may arise. When a car goes round a sharp corner the springs are given a sudden deflection, oscillations will occur unless the motion is heavily damped.

2. After 1/8 of an oscillation which starts at maximum amplitude.

3. (a) At the extreme displacement positions.
(b) At the mid oscillation position.
(c) At the extreme displacement positions.
(d) At the mid oscillation position.
(e) At the extreme displacement positions.

4. $4{\cdot}7 \times 10^{13}$ Hz.

5. The natural frequency is proportional to $\sqrt{k}$, where k is the force constant. Changing the springs will change frequency of the oscillations produced by the sudden deflection of the springs produced by the bump. See the answer to question 1(e).

6. Probably no motion is simple harmonic—it would have to be undamped. However, simple harmonic motion is the basic form of oscillation which is almost followed by many oscillating objects. It is a common denominator.

7. About 2000 Hz.

8. The motion is not simple harmonic motion because the restoring force is not proportional to the displacement.

$$\text{Force causing the spring extension} = k\left(\frac{L}{\cos\theta} - L\right)$$

$$\text{The restoring force is } F\sin\theta = k\left(\frac{L}{\cos\theta} - L\right)\sin\theta$$

9. When the trolley is displaced to one side by a distance x the extension of one spring is increased to $X + x$ and the extension of the other spring decreased to $X - x$. The resultant force on the trolley is thus $k(X + x) - k(X - x)$ or $2kx$. The restoring force is thus proportional to the displacement and thus the motion is simple harmonic.

$$f = \frac{1}{2\pi}\sqrt{\left(\frac{2k}{\text{mass}}\right)}$$

10. The effects can be reduced by increasing the damping of the oscillation—perhaps changing the cushion on the seat. Another way would be to change the frequency applied to the man. The source of the forcing frequency could be natural frequencies in the car being excited by external shocks.

5 Growth and decay

Teaching note Practical work appropriate for the radioactive decay examples discussed in this chapter will be found in:

W. Bolton. *Physics experiments and projects*, Vol. 3, Pergamon, 1968.
Nuffield Advanced Physics. Unit 5 Teacher's Guide, Penguin, 1972.

5.1 The population explosion

The world's population in the Stone Age has been estimated at about a million. By 1850 it had grown to about one thousand million, by about 1930 it was two thousand million, by 1960 three thousand million. When will it reach four and then five thousand million? When will the earth be unable to support its population? What does the future hold?

Figure 5.1 shows how the population has changed over the years, Fig. 5.2 shows how the rate at which the population grows has changed with the years. In the seventh century the growth rate has been estimated at about zero. This is the case where the death rate equals the birth rate. In 1961 the growth rate was 2·55 per cent per year. This represents, in a world population of about three thousand million, an increase of 76·5 million people in one year.

$$(\text{Change in population per year}) = \frac{2{\cdot}55}{100} \times \text{population}$$

If this growth rate was maintained then the following population would be produced:

Year	*Change in population /millions*	*Total population /millions*
1961		3095
1962	78·9	3174
1963	80·9	3255
1964	83·0	3338
1965	85·1	3423
1966	87·2	3510
1967	89·5	3600
1968	91·8	3692
1969	94·1	3786
1970	96·5	3883
1971	99·0	3982
1972	101·5	4084

The four thousand million should, according to this estimate, have been reached by 1972. This is about 11 years

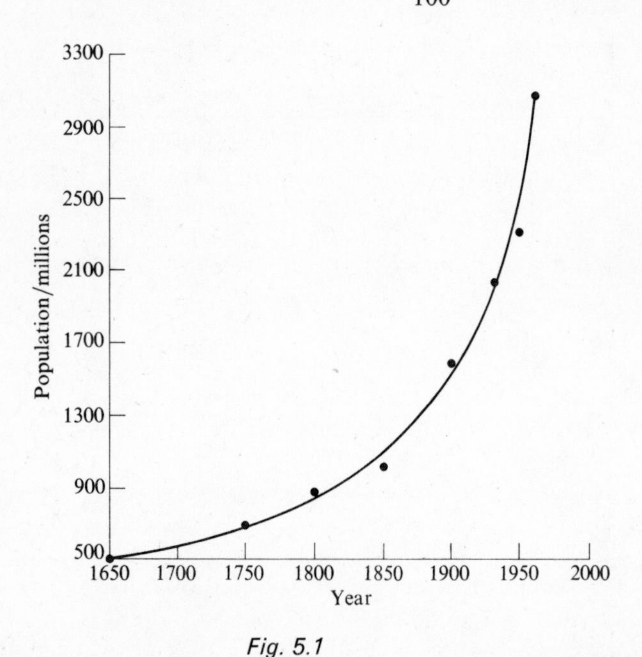

Fig. 5.1

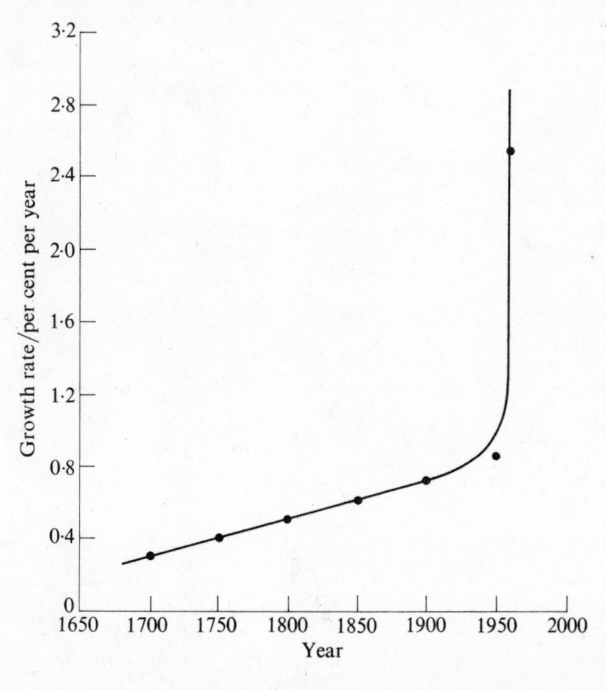

Fig. 5.2

for a change of a thousand million; the previous thousand million took 30 years.

The equation used to obtain the population data was of the form—

(Change in population per year) = constant × population

or in symbols—

$$\frac{\Delta N}{\Delta t} = kN$$

where ΔN is the change in population, initially N, occurring in a time interval Δt, k is the constant. This equation is in fact telling us how the slope of the population–time graph changes with time (the slope is kN).

If we keep on using the 1961 growth rate, 2·55 per cent per year, the five thousand million mark will be reached in the 'eighties, the six thousand million mark in the 'nineties and by the end of this century the seven thousand million will have been reached. A vital question is whether this rate can be maintained. In the USA the growth rate is now decreasing. The 1961 rate was 1·8 per cent, the 1962 rate 1·75 per cent. The following table has been taken from a paper by F. Felix and gives a forecast of the growth rates and resulting population for the USA. His estimates are (1964) in reasonable agreement with other estimates.

Year	*Rate of increase /per cent*	*Population /millions*
1961	1·8	183·7
1962	1·75	189·9
1980	1·44	247·8
2000	1·22	322·4
2050	0·75	525·0
2100	0·45	706·3
2150	0·28	845·9
2200	0·17	943·1
2250	0·11	1009·2
2300	0·06	1054·9

(F. Felix. *IEEE Spectrum*, July 1964, 99.)

Perhaps the world growth rate will stop rising and decline.

Why do decreases in the growth rate occur? Affluent parents have less children? Contraceptives? A change in the number of fertile women in the community? A reduction in the area of land available per person? The high stress due to living in an industrialized society? These are but a few causes. The following research on the changes in the population of deer on an island may be of significance in forecasting the trend for the human population.

James Island has an area of about 2·5 km^2 and lies about 1 km off the shore of Chesapeake Bay. In 1916 four or five Sika deer were released on the uninhabited island. By 1955 there were about 280 to 300 deer. This was about one deer for every 8000 m^2 of land (this is about 90 m separation between each deer). In the first three months of 1958 over half the deer died. The following year more deer died. The deaths reduced the population to about 80 deer. Why did the deer die? The answer was not starvation, all the deer that were examined were in excellent condition. The examination did, however, show one difference between the deer dying in 1958 and samples taken in 1955: the adrenals were much larger. Adrenal glands become larger in response to continued stress. The deer population collapsed because of the high stress conditions brought about by high density living.

The land area of the earth is about 5×10^{14} m^2. Much of this area cannot be lived in because it is too mountainous or too arid. The 1960 world population of about 3×10^9 people thus had an average of about 10^5 m^2 each, or if spread uniformly over the earth nearest neighbours would be about 300 m away. By the end of this century this distance should be down to about 130 m. The collapse of the deer population occurred at a spacing of about 90 m. The stresses in deer may be different from humans and 90 m may be a large separation for humans.

Another possible cause of a collapse of the world population was forecast by Malthus in 1798.

'According to a table of Euler, calculated on a mortality of one to thirty-six, if the births be to the deaths in the proportion of three to one, the period of doubling (of the population) will be only twelve years and four-fifths.... Sir William Petty supposes a doubling possible in so short a time as ten years.... It may safely be pronounced, therefore, that population, when unchecked, goes on doubling itself every twenty-five years, or increases in a geometrical ratio. (i.e., 1, 2, 4, 8, 16, 32, etc.)....

It may be fairly pronounced...that considering the present average state of the earth, the means of subsistence, under circumstances the most favourable to human industry, could not possibly be made to increase faster than in an arithmetical ratio. (i.e., 1, 2, 3, 4, 5, 6, etc.)

The necessary effects of these two different rates of increase, when brought together, will be very striking. Let us call the population of this island eleven millions; and suppose the present produce equal to the easy support of such a number. In the first twenty-five years the population would be twenty-two millions, and the food being also doubled, the means of subsistence would be equal to this increase. In the next twenty-five years the population would be forty-four millions, and the means of subsistence only equal to the support of thirty-three millions. In the next period the population would be eighty-eight millions and the means of subsistence just equal to the support of half that number....

Taking the whole earth instead of this island,...the human species would increase as the numbers, 1, 2, 4, 8, 16, 32, 64, 128, 256, and subsistence as 1, 2, 3, 4, 5, 6, 7, 8, 9...

The ultimate check to population appears then to be a want of food arising necessarily from the different ratios according to which population and food increases. But this ultimate check is never the immediate check, except in cases of actual famine.

The immediate check may be stated to consist in all

those customs, and all those diseases, which seem to be generated by a scarcity of the means of subsistence; and all those causes, independent of this scarcity, whether of a moral or physical nature, which tend prematurely to weaken and destroy the human frame. . . .'

T. R. Malthus (1798)
An essay on the principle of population.

5.2 The growth equation

With a constant growth rate the changes in population ΔN in a time interval Δt are given by

$$\frac{\Delta N}{\Delta t} = kN$$

where k is the growth rate and N the total population at the instant considered. In a given time interval the change in population is directly proportional to the growth rate and the number of people existing at that time. Thus with a growth rate of 2·55 per cent per year and a population of 3095 millions the population will increase by 78·9 millions in one year. If we had twice the population then the increase per year would be doubled. Because the change in population per year depends on the population each successive change will be greater than the previous one and so the curve will grow, i.e., N will become bigger.

$\Delta N/\Delta t$ is the slope of the graph at N, t. The slope is directly proportional to N and thus, as t increases and more and more ΔN increments are added to N, the slope becomes steeper as time progresses. The equation describes how the number N changes with time, considered in intervals of Δt, when the change (ΔN) in any interval of time is some fixed fraction (or percentage) k of the number present at that time. Let us consider another example, the way in which a sum of money earning interest changes with time. Suppose we have £100 in an account which gives 10 per cent interest per year.

Initially £100

After 1 year the interest paid will be $\frac{1}{10}$ of £100 per year or

$$\frac{\Delta N}{1} = \frac{1}{10} \times 100$$

(In fact this is our growth equation)

$$\Delta N = £10$$

After 1 year the money will have grown to £110.

After 2 years the interest paid will be

$$\frac{\Delta N}{1} = \frac{1}{10} \times 110$$

$$\Delta N = £11$$

After 2 years the money will have grown to £121.

After 3 years the interest paid will be

$$\frac{\Delta N}{1} = \frac{1}{10} \times 121$$

$$\Delta N = £12{\cdot}1$$

After 3 years the money will have grown to £133·1.

And so the money will keep on growing. The change in money, i.e., the interest paid, will increase each year because the total money is increasing each year. We can plot a graph showing how N, the total money, varies with time (Fig. 5.3).

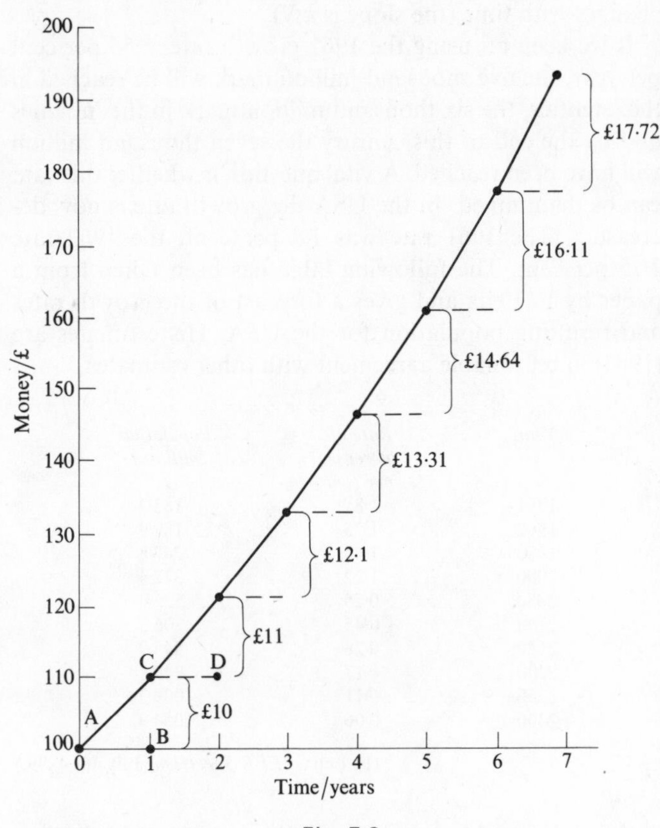

Fig. 5.3

Rather than just plot the points given by the above calculations we will build up the graph step by step. The reason for doing this is that the technique will be useful later in this chapter and gives a greater insight into the significance of the operation specified by the growth equation.

At time $t = 0$ the total money is £100. If there was no growth, i.e., no interest, then after 1 year there would still be £100. The rate of growth at 1 year would be zero. This would mean no change in money from the value at time $t = 0$. The value at time $t = 0$ is zero and represented on the graph by the point A. The unchanged value at $t = 1$ year is given by continuing the line to B. But interest is earned and the slope of the graph becomes

$$\text{Slope} = kN = £10 \text{ per year.}$$

If we just consider 1 year then the change in the amount of money is

$$\text{Change in money} = \Delta N = kN\,\Delta t$$
$$= £10$$

This change is represented on the graph by the line going from A to C instead of A to B. If no further interest was earned in the next year the graph would progress to point D. But because interest is earned we must add to point D an amount ΔN given by

$$\text{Change in money} = kN\,\Delta t = £11$$

In the next interval of time we must add £12·1. In this way we can build up the graph, the procedure being:

read off from the graph the last value of N,
calculate the change in N during the next time interval,
use this value to plot the new value of N.

Our growth equation tells us, at each value of N, the change in N that will occur in the next interval of time.

What is the equation relating money (N) and time (t)? This would be the equation represented by the graph, Fig. 5.3.

After 0 years, money $= N_0$
After 1 year, money $= N_0 + kN_0 = N_0(1 + k) = N_1$
After 2 years, money $= N_1 + kN_1 = N_1(1 + k)$
$= N_0(1 + k)^2 = N_2$
After 3 years, money $= N_2 + kN_2 = N_0(1 + k)^3$
After t years, money $= N_0(1 + k)^t$

Thus we can describe the graphical curve by the equation

$$N = N_0(1 + k)^t$$

After 10 years our £100 would, at 10 per cent per year interest, have grown to

$$N = 100\left(1 + \frac{1}{10}\right)^{10}$$
$$= £259{\cdot}4$$

An alternative way of writing the equation is

$$N = N_0 a^t$$

where a is a constant.

$$a = 1 + k$$

Thus in the case of the £100 at 10 per cent interest the equation can be written as

$$N = 100(1{\cdot}1)^t$$

If the interest rate had been 20 per cent then the equation would be

$$N = 100(1{\cdot}2)^t$$

There is another way of writing this equation where the interest rate is in the power term.

$$N = N_0\,e^{kt}$$

k is the interest rate and 'e' a number. For this equation to be identical with our previous equation we must have

$$e^{kt} = a^t = (1 + k)^t$$

For the 10 per cent interest rate this means that

$$e^{0{\cdot}1} = 1{\cdot}1$$

For the 20 per cent interest rate

$$e^{0{\cdot}2} = 1{\cdot}2$$

Are these two values of 'e' identical? With the 10 per cent interest 'e' has the value 2·59, with 20 per cent the value is 2·49. The following table shows the values of 'e' for different interest rates.

Interest rate /per cent	e
100	2
20	2·49
10	2·59
1	2·70
0·1	2·717
0·01	2·718

As the interest rate becomes low so the value of 'e' seems to be tending to a value of 2·718. What is the significance of this?

Look at Fig. 5.3 giving the money variation with time for a 10 per cent interest rate and compare it with Fig. 5.4 giving the comparable graph (different scale) for 100 per cent interest rate. The bigger the interest rate the more discontinuous the graph, i.e., the changes in N per time interval become larger. In both these cases the interest was worked out at the end of each year.

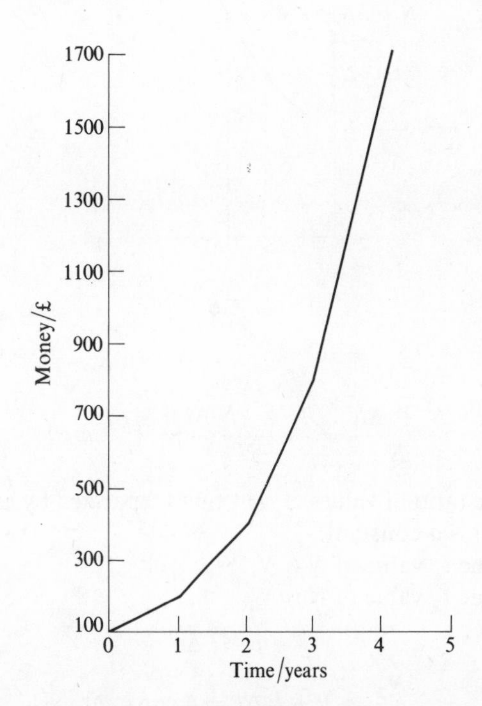

Fig. 5.4

The value of 'e' becomes constant when we have small interest rates. This is when we have a smooth graph relating money and time. We could have achieved a smooth curve by taking smaller intervals of time, e.g., calculating the interest every month or every day. Our equation

$$N = N_0 \, e^{kt}$$

where 'e' is a constant having the value 2·718, is thus for the case where our N or t intervals between plotted points are very small, i.e., in the limiting case when $\Delta t \to 0$.

This equation can also be used to describe the way in which the population increases with time, if there is a constant growth rate.

As $\Delta N/\Delta t = kN$ we must be able to write, for our limiting case when $\Delta t \to 0$,

$$\frac{dN}{dt} = kN_0 \, e^{kt}$$

The rate at which N varies with time t is also proportional to e^{kt} and thus a graph of dN/dt against t should be a similar shape to the graph of N against t. The graph is of the same form, that given by e^{kt}, but multiplied by a different constant, kN_0 instead of just N_0.

The constant ratio curve

In successive intervals of time the value of N changes from N to $N + \Delta N$ (Fig. 5.5).

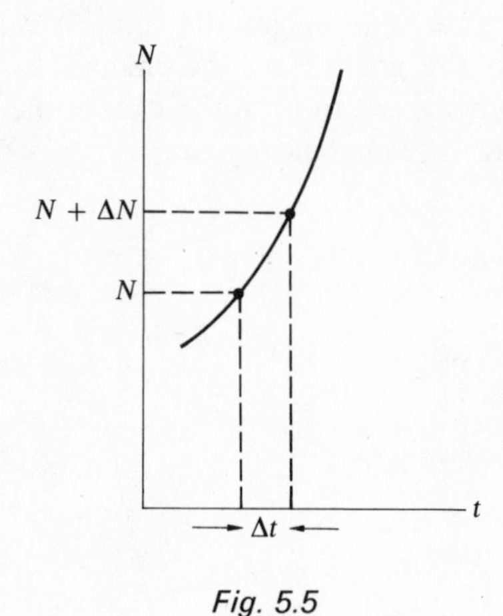

Fig. 5.5

$$\frac{N + \Delta N}{N} = \frac{N + kN\,\Delta t}{N} = 1 + k\,\Delta t$$

Thus the ratio of values of N at times separated by an interval of Δt is a constant.

At time t_1 value of N is N_1.
At time t_2 value of N is N_2,

$$t_2 - t_1 = \Delta t,$$

$$\frac{N_2}{N_1} = 1 + k\,\Delta t = \text{a constant}$$

At time t_3 value of N is N_3,

$$t_3 - t_2 = \Delta t,$$

$$\frac{N_3}{N_2} = 1 + k\,\Delta t = \text{the same constant}$$

Similarly for values of N spaced at successive intervals of Δt,

$$\frac{N_2}{N_1} = \frac{N_3}{N_2} = \frac{N_4}{N_3} = \frac{N_5}{N_4} = \text{a constant}$$

The curve grows by constant ratios.

Series

The numbers 1, 2, 3, 4, 5, 6, 7, 8, . . . form a series in which successive terms are obtained by adding 1. In the series 3, 6, 9, 12, 15, 18, 21, . . . successive terms are formed by adding 3. Such series are known as arithmetic series. The numbers 2, 4, 8, 16, 32, 64, 128, . . . form a series in which successive terms are formed by multiplying by 2. Another series is 3, 9, 27, 81, 243, . . . where successive terms are obtained by multiplying by 3. Series in which the successive terms are produced by multiplying the previous term by a constant factor are known as geometric series. If the variation of some quantity with time can be represented by the terms in a geometric series then we have a constant ratio curve. Each successive time interval gives a quantity which is some constant factor times the quantity in the previous time interval. In algebraic symbols we have for the successive terms, $a, a \times a, a \times a \times a, a \times a \times a \times a, a \times a \times a \times a \times a, \ldots$ or $a, a^2, a^3, a^4, a^5 \ldots$

In the case of the interest rate of 10 per cent for the £100 we have

After 0 years, total $= 100 \times 1{\cdot}1^0 =$ £100. Any quantity raised to the power 0 is 1.
After 1 year, total $= 100 \times 1{\cdot}1^1 =$ £111.
After 2 years, total $= 100 \times 1{\cdot}1^2 =$ £121.
After 3 years, total $= 100 \times 1{\cdot}1^3 =$ £133·1.

Logarithmic graphs

$$N = N_0 \, e^{kt}$$

Taking logarithms to base 'e' gives

$$\log_e N = kt + \log_e N_0$$

This equation is of the form

$$y = mx + c,$$

the equation of a straight line. A graph of the logarithm of N plotted against time t will be a straight line with a slope of k. Logarithms to base 10 can be used without affecting the shape of the graph.

$$\log_{10} N = kt \log_{10} e + \log_{10} N_0$$

The slope is just multiplied by a constant of $\log_{10} e$ when logarithms to base 10 are used.

Exponential changes

Whenever some quantity (N) changes by an amount (ΔN) that is directly proportional to how much of that quantity was already present then the growth is called exponential. An exponential growth can be described by the equations

$$\frac{dN}{dt} = kN$$

$$N = N\,e^{kt}$$

A graph of the logarithm of the quantity N against time t is a straight line.

Because we can write the rate of change equation in the form

$$\frac{dN}{dt} = kN_0\,e^{kt}$$

then a graph of the rate of change of N with t has the same form as that of N against t.

5.3 Decay

'A thick layer of thorium oxide was enclosed in a narrow rectangular paper vessel..., made up of two thicknesses of foolscap paper. The paper cut off the regular radiation almost entirely, but allowed the emanation to pass through....A slow current of air from an aspirator or gasometer,...was passed through the apparatus. The current of air, in its passage by the thorium oxide, carried away the radioactive particles with it...[Fig. 5.6], shows the relation existing between the current through the gas and the time.... It will be observed that the current through the gas diminishes in a geometrical progression (i.e. 32, 16, 8, 4, 2, 1) with the time. It can easily be shown, ..., that the current through the gas is proportional to the intensity of the radiation emitted by the radio-active particles. We therefore see that the intensity of the radiation given out by the radio-active particles falls off in a geometrical progression with time....'

E. Rutherford (1900). *Phil. Mag.* (5) **49**, 1.

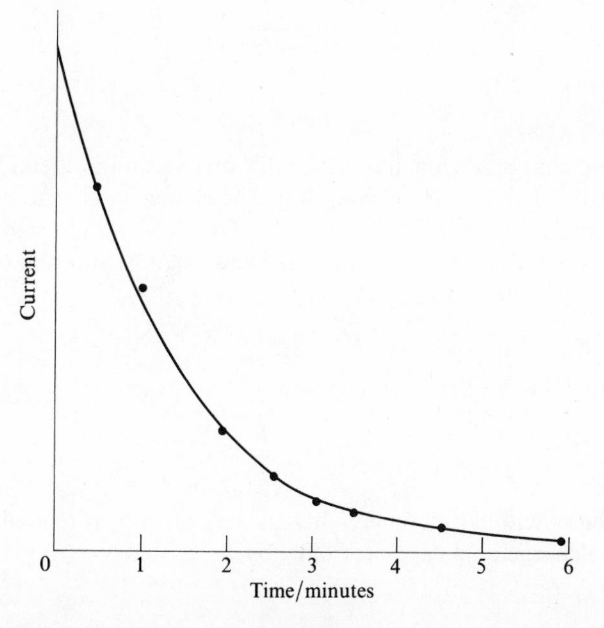

Fig. 5.6

The radioactivity of a radioactive isotope decreases with time and as Rutherford observed the decrease is in a geometrical progression with time. This means that, in the case of the isotope used by Rutherford, in one minute the activity decreases by a factor of two. In the next minute it decreases by a factor of two from what it was at the beginning of that minute. This does not mean that at the end of two minutes the activity is zero. If the activity initially was 32 units, then after one minute the activity would be 16 units, after the second minute it would be 8 units, after a third minute 4 units. The term 'half-life' is often used to describe the time taken for a quantity to decrease by a factor of two. In this case the half-life is 1 minute.

A geometric series is characteristic of an exponential change. In such a change the rate at which a quantity changes is proportional to the amount present. The decay equation is of the form

$$\frac{\Delta N}{\Delta t} = -kN$$

ΔN is the change in the number of atoms in a time interval Δt, N is the number of atoms present and k the disintegration rate. The minus sign is because in each succeeding time interval the number N is getting less. The value of ΔN we calculate will be negative, indicating that it has to be subtracted from the preceding value of N.

Consider a numerical example. Suppose we have 1000 radioactive atoms with a decay constant of 0·1 per minute (i.e., a 10 per cent decrease per minute).

At $t = 0$ number of atoms = 1000.

At $t = 1$ minute

$$\text{Change in number of atoms } \Delta N = -\frac{1000}{10}$$

and thus the number of atoms will be 1000 − 100 = 900.

At $t = 2$ minutes

$$\text{Change in number of atoms } \Delta N = -\frac{900}{10}$$

and thus the number of atoms will be 900 − 90 = 810.

At $t = 3$ minutes

$$\text{Change in number of atoms } \Delta N = -\frac{810}{10}$$

and thus the number of atoms will be 810 − 81 = 729.

At $t = 4$ minutes

$$\text{Change in number of atoms } \Delta N = -\frac{729}{10}$$

and thus the number of atoms will be 729 − 73 = 646.

The numbers 1000, 900, 810, 729, 646 form a geometric series. The series is formed by multiplying 1000 by the factor $0{\cdot}9^t$.

Number of atoms after time $t = 1000 \times 0{\cdot}9^t$ where t is the time in minutes. In general terms the equation is

$$N = N_0(1 - k)^t$$

Following the lines used with the growth of interest example on page 120 we write

$$N = N_0\,\mathrm{e}^{-kt}$$

The half-life is the time taken for N to change from N_0 to $N_0/2$. Thus

$$\text{Half-life} = \frac{\log_e 2}{k} = \frac{0{\cdot}693}{k}$$

We can generate such a decay by using the ideas of chance. Suppose we have six-sided die. The chances of it, when thrown, landing with a five uppermost are one in six. There are six possible ways the die can land, one way giving the required result. If we throw 12 dice then we would expect to have 2 showing a five. We might get none or we might get 12 showing a five, but the most likely result is just 2 showing a five. If we repeated the throwing of the 12 dice a large number of times then the predominant result would be 2 fives per 12 dice.

Kerrich in his book *An experimental introduction to the theory of probability* (Witwatersand Univ. Press) gives the results of a large number of coin-tossing experiments. A coin has just two ways of landing, heads or tails. The chance of obtaining heads as the result is one in two ($\frac{1}{2}$). The number of heads recorded in 10 consecutive sets of 1000 tosses were

502, 511, 497, 519, 504, 476, 507, 518, 504, 529

The chance of obtaining heads is very close to $\frac{1}{2}$.

How is die-throwing or coin-tossing related to radioactivity? Suppose we have 1000 dice and the chance of one landing with a marked face uppermost is 1/10. (The die is a theoretical one with ten sides!) On throwing the 1000 dice we would expect to obtain 1/10 of them with the marked face uppermost, i.e., 100. Suppose we say that these 'dice' have decayed and remove them from our 1000. We are then left with 900. We throw these and 1/10 will be expected to land with the marked face uppermost. These 90 dice are then removed, leaving 810. Again we throw the dice and this time 81 decay, leaving 729. The decay is identical with that forecast by the use of our decay equation.

$$\frac{\Delta N}{\Delta t} = -(\text{chance of decay})\,N$$

There is strong evidence that radioactivity is ruled by the laws of chance. Because chance only tells us the likely result we would expect that when our 1000 atoms decay we might sometimes have 100 decaying when 'thrown' (i.e., in unit time) or we might have 90 or 110 or indeed any number from 0 to 1000. The most likely number is 100. Such fluctuations are found when counts of rate of decay of radioactive substances are made. The exponential gives the mean behaviour.

Radioactivity is discussed in more detail in chapter 13.

The discharge of a capacitor

A capacitor is charged and then allowed to discharge through a resistor. How does the current through the resistor change with time during the discharge? How does the charge on a capacitor plate change with time during the discharge? Suppose we charge a 1000 μF (i.e., 10^{-3} F) capacitor by putting 10 V across it. The charge on a capacitor plate will be

$$Q = CV = 10^{-3} \times 10 = 10^{-2} = 0{\cdot}01 \text{ coulomb}$$

When we discharge it the current (Fig. 5.7) is initially 1 mA (i.e., 10^{-3} A). So in one second the charge lost from the capacitor plate is given by

$$I = \frac{\Delta Q}{\Delta t}$$

and thus charge lost $\Delta Q = 10^{-3} \times 1 = 10^{-3}$ coulomb.

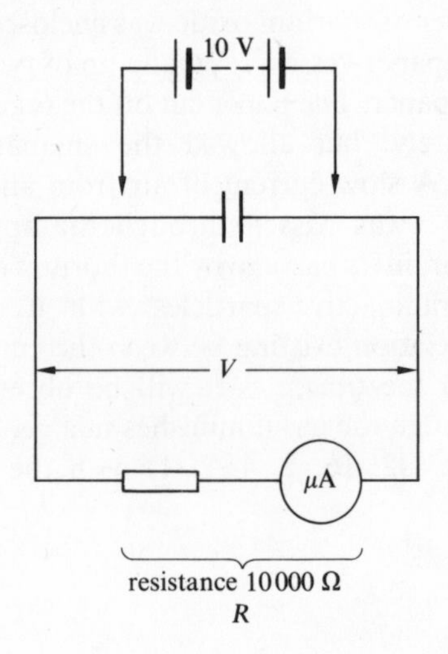

Fig. 5.7

The charge left on the plate after one second is therefore $0{\cdot}01 - 0{\cdot}001 = 0{\cdot}009$ coulomb. The change in charge is in any time interval Δt given by $-I\,\Delta t$. The minus sign is because there is a drop in charge and we must subtract our change from the total charge.

$$\Delta Q = -I\,\Delta t$$

But $V = IR$, hence we can write

$$\Delta Q = -\frac{V}{R}\Delta t$$

The potential difference V across the capacitor is related to the charge on a capacitor plate by

$$V = \frac{Q}{C}$$

hence

$$\Delta Q = -\frac{Q}{CR}\,\Delta t$$

The rate of change of charge on a capacitor plate is directly proportional to the charge on the plate. In our example with a resistance of 10 000 Ω and a capacitance of 10^{-3} F the change in charge per second is

$$\Delta Q \text{ per second} = -0{\cdot}1\,Q$$

At time $t = 0$, $Q = 0{\cdot}01$ C.
At time $t = 1$ s,

$$\Delta Q = -0{\cdot}1 \times 0{\cdot}01 = -0{\cdot}001 \text{ C}$$

Thus after 1 s the charge will be 0·01 − 0·001 = 0·009 C.
At time $t = 2$ s,

$$\Delta Q = -0{\cdot}1 \times 0{\cdot}009 = -0{\cdot}0009 \text{ C.}$$

Thus after 2 s the charge will be 0·009 − 0·0009 = 0·0081 C.
At time $t = 3$ s,

$$\Delta Q = -0{\cdot}1 \times 0{\cdot}0081 = -0{\cdot}00081 \text{ C.}$$

Thus after 3 s the charge will be 0·0081 − 0·00081 = 0·00729 C.

After 4 s the charge will be 0·006561 C, after 5 s 0·0059049 C, after 6 s 0·00531441 C, etc. Figure 5.8 shows a graph of the charge variation with time.

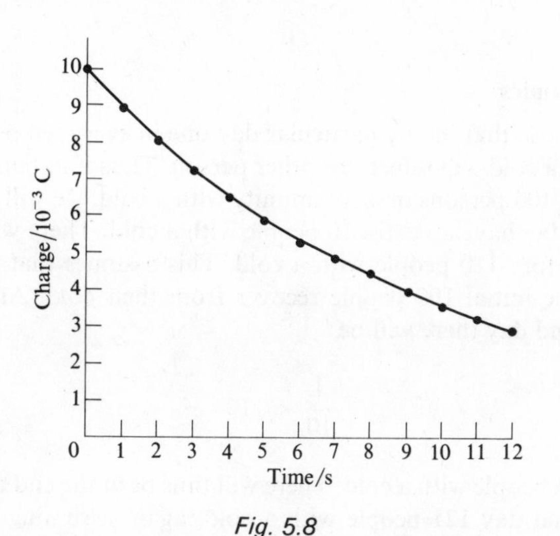

Fig. 5.8

How does the current vary with time?

$$I = \frac{\Delta Q}{\Delta t} = -\frac{Q}{CR}$$

We have already found how Q varies with time so we can state that the current variation with time will be of the same form as the charge variation with time, the current at any instant being the charge on the plate divided by CR.

This is discussed in more detail in chapters 10 and 11.

5.4 Growth and decay

Many radioactive isotopes do not decay into a stable isotope but into another radioactive isotope which itself decays. There may be many radioactive isotopes in the sequence before a stable isotope is reached. For example the thorium isotope (mass 232) decays by alpha emission to give a radium isotope (mass 228), the half-life being 1·90 years. The radium, however, decays to give an actinium isotope (mass 228) by beta emission, half-life 6·1 hours. In turn the actinium decays. The final stable end product is an isotope of lead. How does the amount of radium, or actinium, present vary with time?

Consider isotope A decaying into isotope B which in turn decays into isotope C. The amount of B being produced per unit time is equal to the rate at which B is produced by the decay of A minus the rate at which B decays into C. The rate at which A decays into B is given by

$$\frac{\Delta A}{\Delta t} = -k_A A$$

Thus the rate at which B is produced from A is given by

$$\frac{\Delta B'}{\Delta t} = k_A A$$

k_A is the rate constant for the decay of A into B. The rate at which B decays into C is given by

$$\frac{\Delta B''}{\Delta t} = -k_B B$$

Thus the rate at which the amount of B present increases with time is given by

$$\frac{\Delta B}{\Delta t} = k_A A - k_B B$$

Consider the decay of parent A to daughter B to be similar to that of the thorium decay to radium, i.e., one with a long half-life (very small value of k), and the decay of B to C to be similar to that of radium to actinium, i.e., a short half-life (large value of k). If we consider time intervals of an hour then over a period of say ten hours there will have been very little change in the amount of A (ΔA will be very small because of the smallness of k.) The result of this will be that the production of B from the decay of A will be virtually constant over the period we are considering. Let us suppose that this is constant at 10 000 atoms of B being produced per hour.

At $t = 0$, $B = 0$.

We start with only thorium present.

At $t = 1$ hour, 10 000 atoms of B have been produced. If $k_B = 1/10$ per hour, then about 1000 of these atoms will decay. The total of B atoms after 1 hour is thus 9000.

At $t = 2$ hours, a further 10 000 atoms of B will have been produced. During this second hour 1/10 of the 19 000 B atoms will decay. The total after 2 hours is thus 17 100.

At $t = 3$ hours, a further 10 000 atoms of B will have been produced. Of the 27 100 atoms 2710 will decay leaving 24 390.

At $t = 4$ hours, a further 10 000 atoms of B will have been produced. Of the 34 390 atoms 3439 will decay to leave 30 951.

Figure 5.9 shows the resulting graph. The number of atoms of B rises until the rate of production of B is equal to the rate at which it decays. That happens, when $k_B B = 10\,000$. When this happens the amount of B present will remain constant.

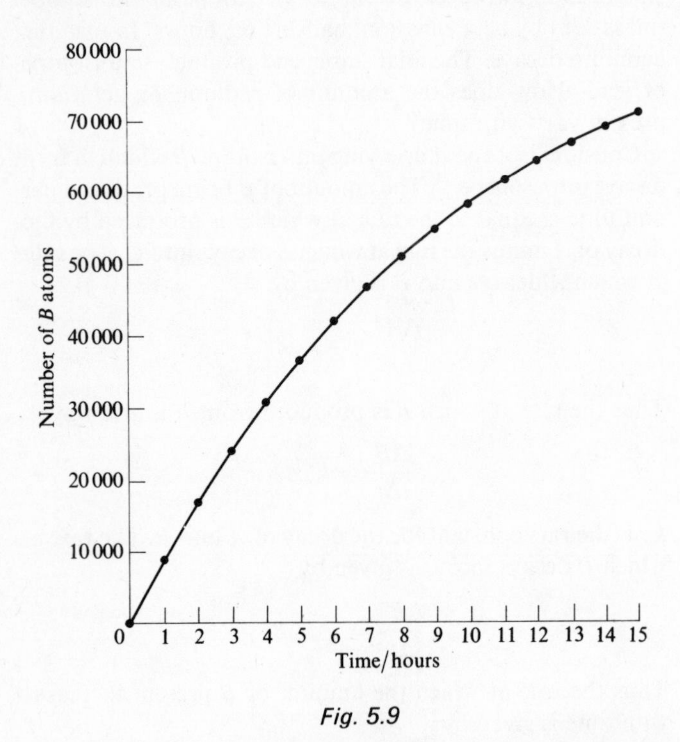

Fig. 5.9

In our example we considered the parent A to decay at a steady rate—not changing the number of disintegrations per hour during the period considered. This results in a graph, Fig. 5.10, where the number of B atoms rises to a maximum and remains at that maximum. This is only

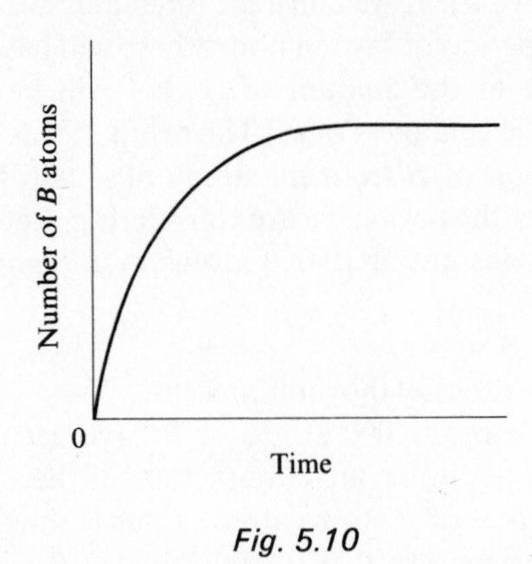

Fig. 5.10

theoretically possible—the parent A will decay with time. The result of taking this into account is to give a graph similar to that in Fig. 5.11. Here the parent A decays slowly with time and the daughter B rises to a maximum and then

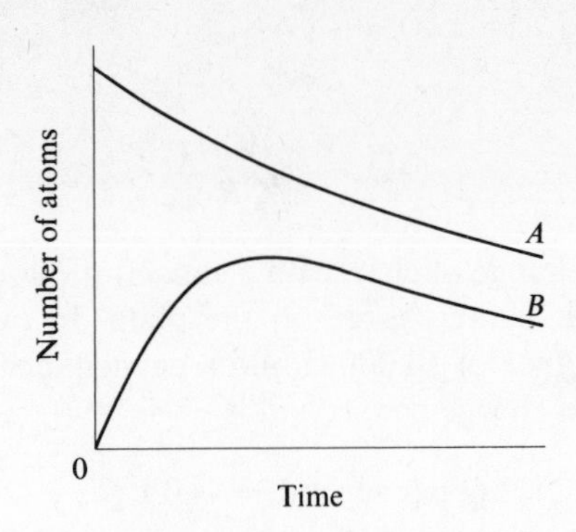

Fig. 5.11 Long life parent (A) with shorter life daughter (B).

decays at the same rate as the parent. If the parent has a shorter life than the daughter a graph like Fig. 5.12 occurs. Here the daughter B rises to a maximum and then decays in the same manner as it would do in the absence of A.

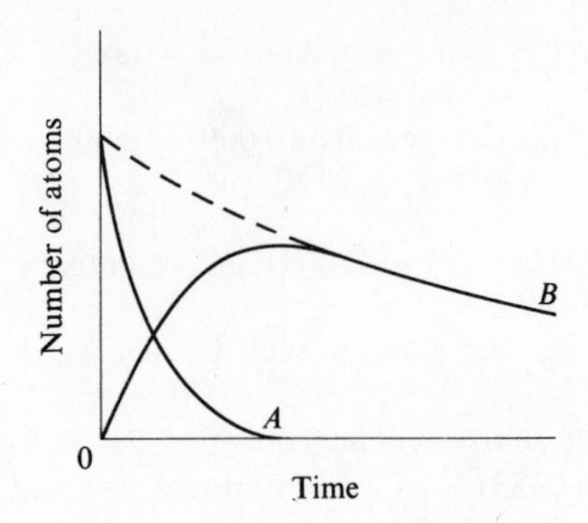

Fig. 5.12 Short lived parent (A) with longer life daughter (B).

Epidemics

Suppose that in any particular day one in every ten people with a cold will infect one other person. Thus if initially we have 100 persons in a community with a cold, we will after one day have an extra 10 people with a cold. There will be therefore 110 people with a cold. This assumes that none of the initial 100 people recover from their cold. After a second day there will be

$$\frac{1}{10} \times 110$$

extra people with a cold. There will thus be at the end of the second day 121 people with a cold, again assuming none recover. After three days there will be 133, after four days 146, after five days 161, after six days 177, after seven days 195. The number of people catching a cold in any one day is

$$\frac{1}{10}\text{(number of people with a cold)}$$

or

$$\frac{\Delta N}{\Delta t} = \frac{1}{10} N$$

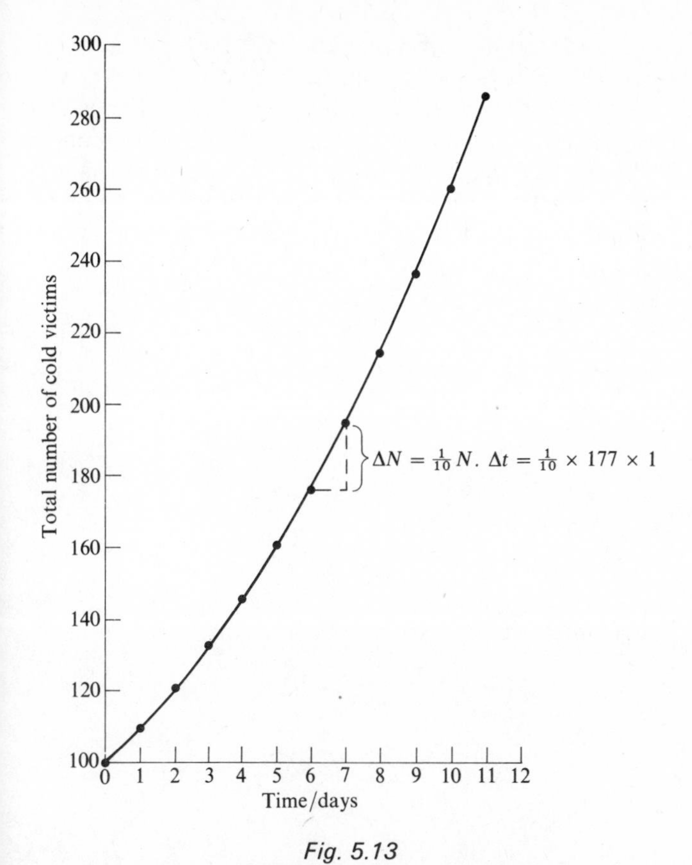

Fig. 5.13

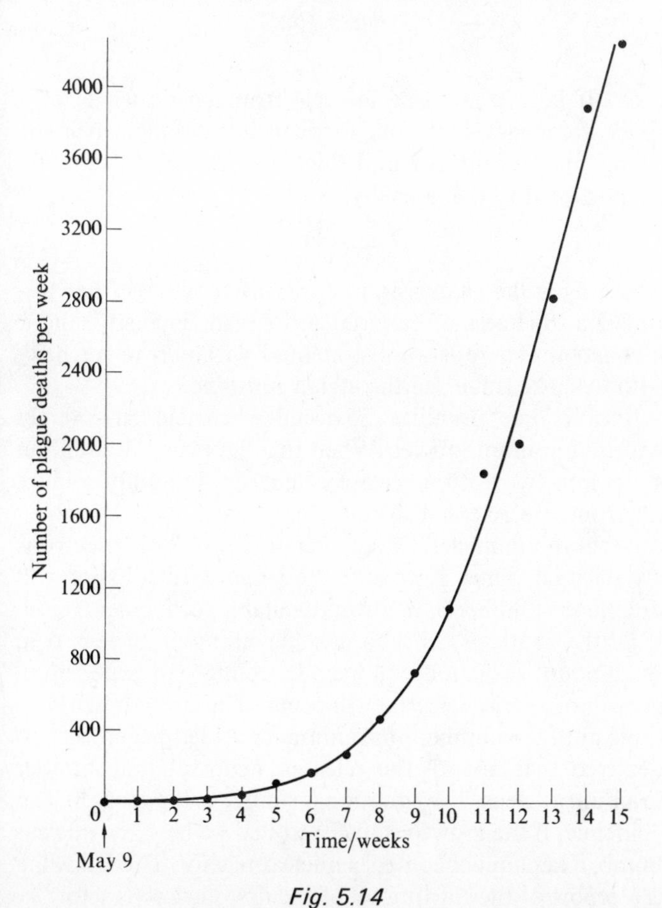

Fig. 5.14

where ΔN is the number of extra people catching a cold in a time interval Δt, N is the total number of people with a cold at any instant. The situation is just like the compound interest problem discussed earlier in this chapter.

Figure 5.13 is the graph produced using these data. Is this really representative of the spread of a cold in a community? One factor not taken into account is that people are only capable of infecting other people with a cold over a certain period of time—they tend to recover. Thus the total number of people capable of infecting others after say 7 days will not be as high as 195 because some of them will no longer be capable of infecting others. Another factor which may limit the growth is the size of the community in contact with the cold victims—after a while there may be very few people without a cold who could be infected. Some people may be immune to colds.

How realistic is this model for the spread of a cold, or indeed any ailment spread in a similar manner to a cold? The following is the record for the spread of an epidemic, the way in which the number of deaths changed from week to week during the Great Plague of 1665 in London.

Week ending		*Plague deaths per week*	*Total number of deaths*
April	25	0	0
May	2	0	0
	9	9	9
	16	3	12
	23	14	26
	30	17	43
June	6	43	86
	13	112	198
	20	168	366
	27	267	633
July	4	470	1103
	11	725	1828
	18	1089	2917
	25	1843	4760
August	1	2010	6770
	8	2817	9587
	15	3880	13 467
	22	4237	17 704
	29	6102	23 806
Sept.	5	6988	30 794
	12	6544	37 338
	19	7165	44 503
	26	5533	50 036
Oct.	3	4929	54 965
	10	4327	59 292

C. Creighton (1965). *A history of epidemics in Britain*, Vol. 1 Cass.

Figure 5.14 shows how the number of deaths per week changed with time. The early part of that curve can be obtained by assuming that the growth rate is 100 per cent per week.

$$\frac{\Delta N}{\Delta t} = 1{\cdot}0N$$

Thus if we start with three deaths in the week ending 2 May we have the following:

Week ending	*Plague deaths per week*	*Total number of deaths*
May 2		3
9	3	6
16	6	12
23	12	24
30	24	48
June 6	48	96
13	96	192
20	192	384
27	384	768
July 4	768	1536
11	1536	3072
18	3072	6144
25	6144	12 288

Up to about 4 July the epidemic has grown according to our equation. From then on the epidemic lags behind our estimate. After 5 September the number of deaths per week actually begins to drop. There is a very obvious reason for the epidemic departing from our equation—a shortage of possible victims occurs. Our equation assumes an infinite number of victims and no restrictions on the contact between those with the plague and those capable of being infected.

The equation often used to describe an epidemic, where there is an isolated community and all members of that community either have the disease or are able to catch it, is of the form

$$\frac{\Delta B}{\Delta t} = kAB$$

where A is the number of susceptibles and B the number of infectives. The number of people in the population is $A + B$. Where the number of infectives is small compared with the number of susceptibles the equation becomes of the form we have already used

$$\frac{\Delta B}{\Delta t} = (\text{constant})\ B$$

Where people are removed, i.e., become immune or die or just become better, from the population involved in either giving or receiving the disease the equation becomes modified to

$$\frac{\Delta B}{\Delta t} = kAB - CB$$

It is assumed that the rate of removal is proportional to the number of infectives.

5.5 Exponential equations

There are many situations which can be described by an exponential equation. The following are some examples of such cases.

(a) In each year of its life the chance of a robin dying remains the same. This is unlike humans and is thought to be typical of birds. Humans have greater chances of dying in old age and while very young than during the rest of their life. The chance of a robin dying in any one year is about $\frac{2}{3}$. Thus if we start off with 120 robins, in their first year of life $\frac{2}{3}$ of them will die, leaving 40. Of these 40, two-thirds will die in their second year, leaving about 27. At the end of their third year 9 will be left, at the end of four years about 3, at the end of five years 1. Figure 5.15 is based on data in a book by Lack on *The life of the robin* (Witherby, 1943) and reproduced in the book by Sawyer, *The search for pattern* (Penguin).

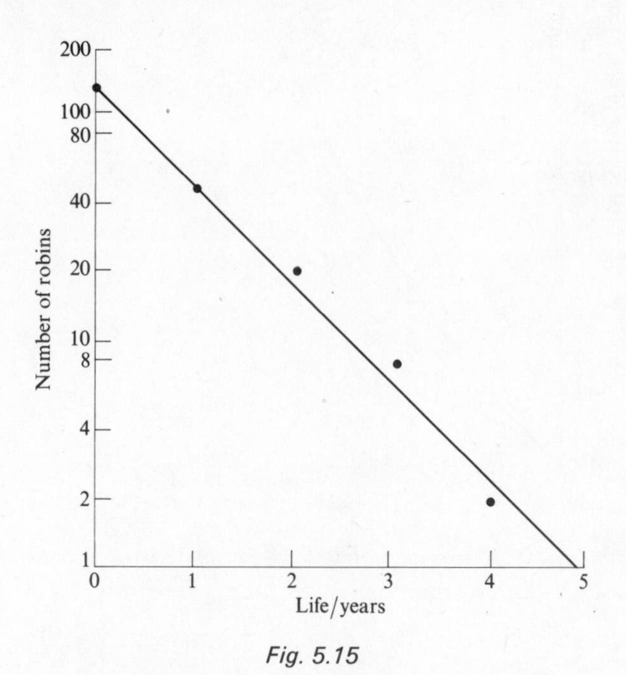

Fig. 5.15

(b) If light, or indeed any electromagnetic wave, of a single frequency is passed through a substance then the change in intensity per unit thickness traversed is directly proportional to the intensity.

$$\Delta I = -kI\,\Delta x$$

where ΔI is the change in intensity after the light has traversed a thickness of material Δx, I is the intensity, and k a constant. There is an exponential variation in intensity with distance traversed through a substance.

(c) Fission of uranium 235 occurs when neutrons interact with the uranium nuclei. When this happens the uranium splits into two with a release of energy. In addition more neutrons are released. These can cause further fission in other uranium nuclei. On an average about 2 or 3 neutrons are released. Thus, if we take the 2 figure, each fission will double the number of neutrons available for further fission. If all the neutrons can find another uranium nucleus then the amount of fission will keep doubling—an exponential growth of energy released will occur. This growth will continue until the number of uranium nuclei left begin to be so depleted that not all the released neutrons find another uranium nucleus. The situation is similar to the growth of an epidemic. If the growth is uncontrolled we have the nuclear bomb, if kept under control a nuclear reactor. The following is a report of the starting up of the first nuclear reactor.

...at 09.45 Fermi gave the order for the control bars to be withdrawn from the pile. The low whine of the small engine was heard. All eyes were turned on the indicator dials.

But the people on the balcony soon began watching the dials showing the production of neutrons, which had begun to move faster when the control bars were withdrawn, and the recording apparatus with its graph of the reaction....

At 10.37 Fermi said in a calm voice, without taking his eyes off the dials: 'Withdraw it (the control bar) another four yards, George.' The sound of the meters speeded up. The graph went higher. 'This graph will climb to here and then flatten out again,' said Fermi. And a few minutes later the needle reached the point indicated, but remained at that level.

Seven minutes later Fermi had the bar withdrawn another foot. Again, the meters indicated their activity and the graph rose. Then it became horizontal again....

At 15.20—another six inches. At 15.25—another foot. 'The point will be reached very shortly,' Fermi said to Compton. 'The graph will continue to rise, without any levelling out.'... He was working with his slide rule very quickly now.... But Fermi suddenly gave a wide smile. 'The chain reaction has begun. The graph is exponential,' he announced, closing his slide rule.

The first self-sustaining chain reaction continued for twenty-eight minutes....

C. Alarice and E. Trapnell. 'The greatest experiment of all time', Reprinted in *Bulletin of the International Agency for Atomic Energy*, December 1962.

Problems

1. In section 4.2 the damping of an object oscillating with simple harmonic motion was described, see Fig. 4.17. The amplitude decreases according to the equation

$$\frac{\Delta A}{\Delta t} = -CA$$

where A is the amplitude and C a constant.

(a) Describe how the amplitudes of successive oscillations change with time.

(b) If the value of C was doubled how would the oscillation change?

(c) Sketch a graph relating A and t.

2. A radioactive isotope has a half-life of 1 hour. If the amount present at some 'zero' time is M, state the amounts that will be present after 1 hour, 2 hours, 3 hours. How does the rate of decay at 2 hours compare with that at 'zero' time?

3. Figure 5.16 shows how the number of cars in Britain has changed over the years.

(a) Is the rate at which the number of cars per year increases a constant (approximately)?

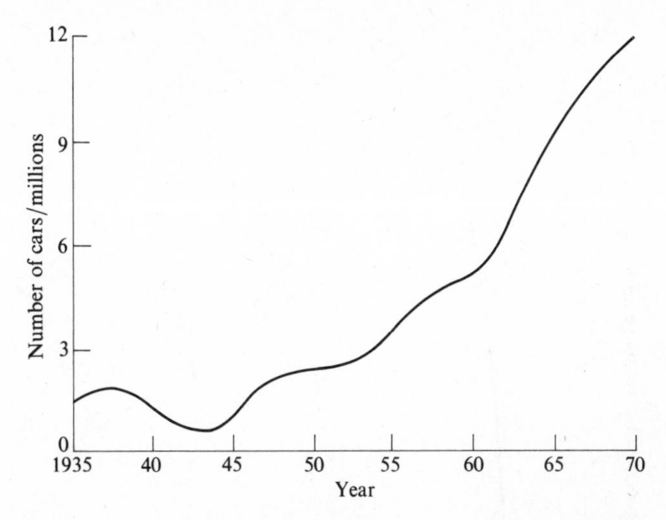

Fig. 5.16 From Britain in Statistics, *A. F. Sillitoe, Penguin, 1971.*

(b) Write down an approximate equation to describe the curve.

4. In 1950 the birth rate in Britain was 16 per thousand inhabitants and the death rate 12 per thousand.

(a) If nobody dies how would the population change with time?

(b) If nobody was born how would the population change with time?

(c) On the basis of the above birth and death rates, how does the population change with time?

Assume that there is no immigration and no emigration and that both the death and birth rates remain constant.

5. The rate at which the feminine sex change from one fashion, say the mini skirt, to a new fashion, the midi skirt, might be considered to be proportional to the number of people who have changed to the new fashion. If this were so how would you expect the number of those wearing the mini skirt, and those wearing the midi skirt, to change with time?

6. Figure 5.17 shows how the amount of the radioactive isotope carbon 14, in excess of the normal level, has changed over recent years. Large amounts of carbon 14 were produced in the atmosphere by nuclear bomb explosions in 1961 and 1962.

The authors state that the graph can be described by the following equation

$$\text{Carbon 14 excess} = 108\,(e^{-0\cdot 1t} - e^{-0\cdot 7t})$$

where t is the time in years measured from 1963. How could you check that this equation fitted the graph?

The dose rate resulting from a gram of carbon 14 in the body is about 14 disintegrations per second. The total dose received by the body due to this excess carbon 14 is thus proportional to

$$\int_0^t (\text{amount})\, dt$$

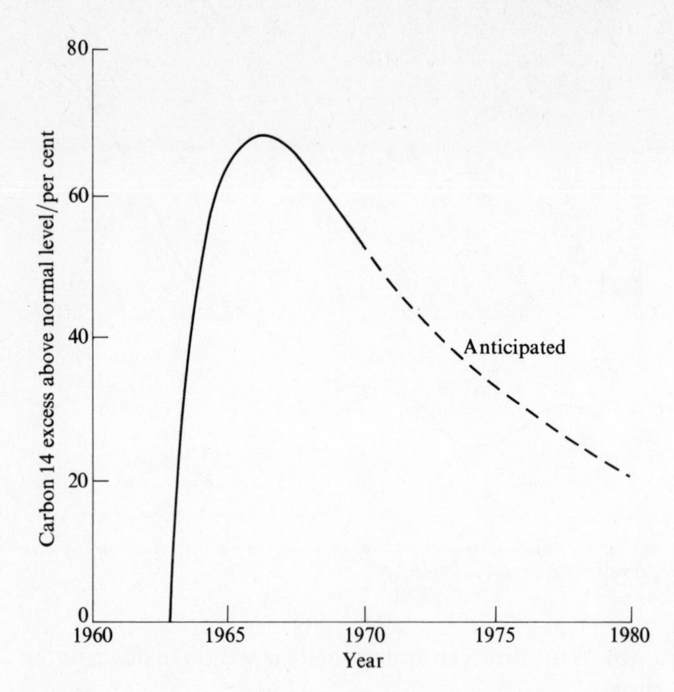

Fig. 5.17 From R. Nydal, K. Lövseth, and O. Syrstad, Nature, *Lond.*, **232**, *418, 1971.*

Either by the use of the graph or the equation determine how the dose received between 1963 and 1967 compares with the possible total dose that could be received by a child born in 1963.

7. An equation is just as much a model of a situation as a stack of spheres is a model of a crystal or a map a model of the countryside.

Discuss the above statement.

8. Which of the graphs in Fig. 5.18 are of exponential changes?

9. How would you describe what an exponential variation with time is, to say an arts student who has never met the term or studied physics?

10. What two quantities would you plot to obtain straight line graphs for

(a) $y = e^{-cx}$?

(b) $y = e^{-c/x}$?

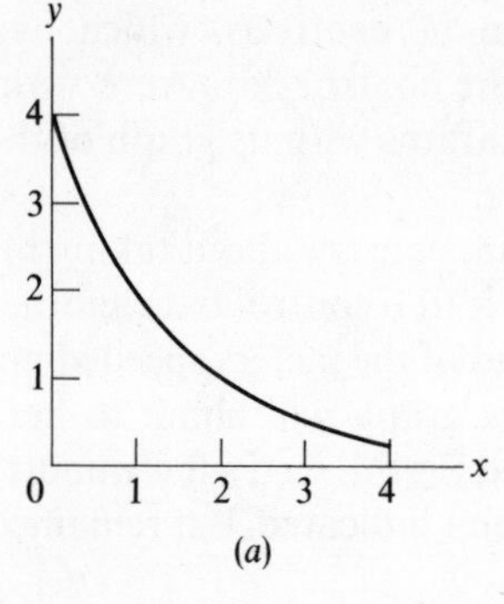

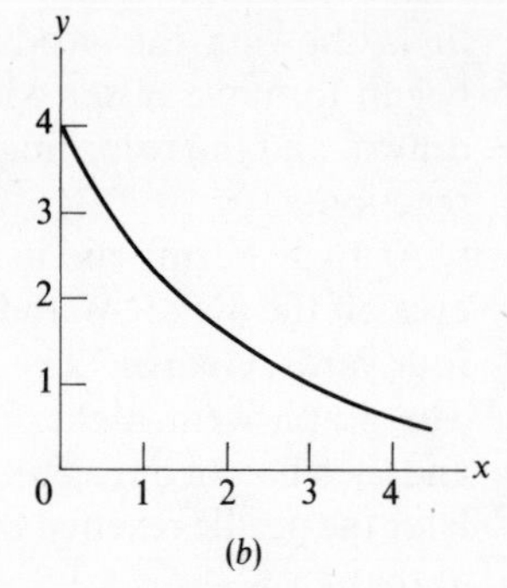

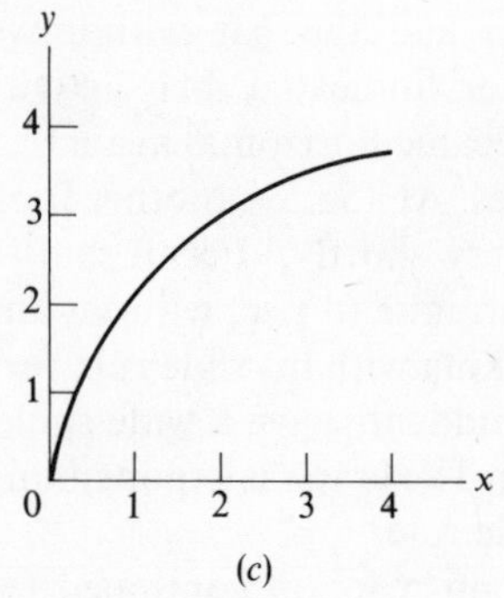

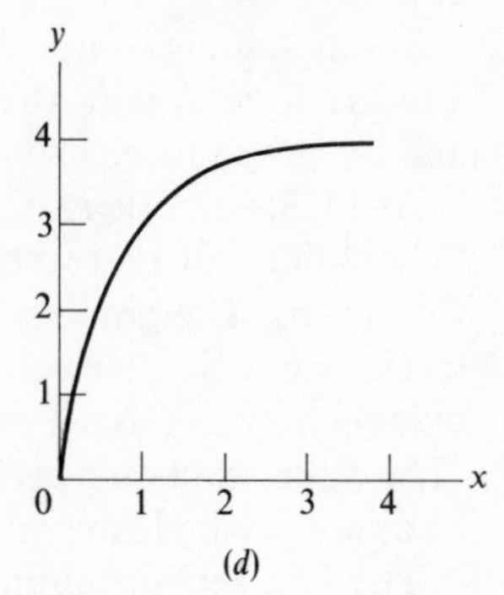

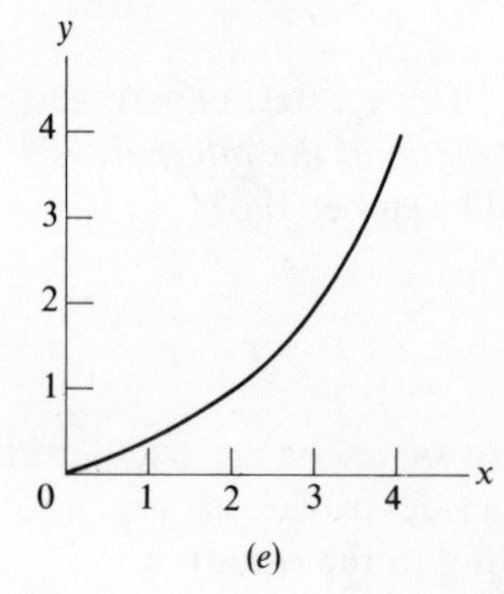

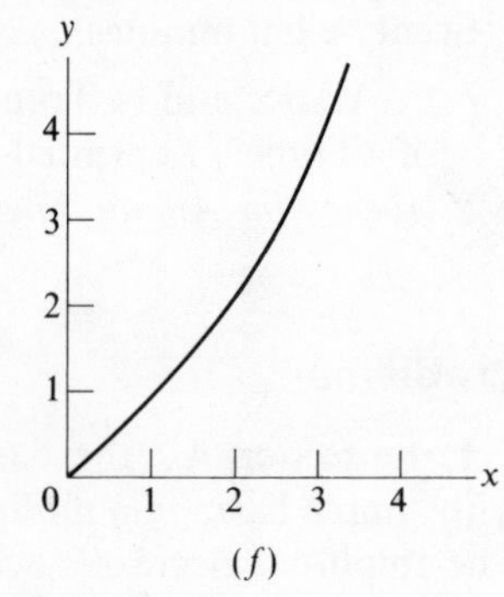

Fig. 5.18

Teaching note Further problems appropriate to this chapter will be found in: *Nuffield advanced physics. Unit 5 Student's book*, Penguin, 1972.

Practical problems

1. Investigate the factors determining the rate of cooling of a hot object in air.

Could you devise a method to determine wind speed based on your results?

2. In 1847 De Senarmont carried out a series of investigations into the flow of heat through crystals. A thin plate of the crystal was coated both sides with wax and heat was applied at one point on the plate by either focusing the sun's radiation to a point on the plate or by passing a current through a wire which passed through a hole in the plate. As the heat flowed out from this point so the wax melted and thus a series of isothermals could be plotted. Senarmont found that for isotropic substances such as glass and cubic crystals the isothermals were circles while for other crystals they were generally ellipses.

By either this or other suitable methods study the flow of heat through crystals. For a particular substance, plates should be cut at different directions from a crystal; try calcite.

Reference

De Senarmont. *Ann. de Chimie et de Physique* 3e, **21**, **22**, **23**, 1847–48.

3. Investigate the factors which determine the rate at which liquids rise in a vertical capillary tube. Possible factors are—surface tension, radius of tube, coefficient of viscosity, density, and angle of contact. The rate of movement will doubtless depend on the height of the liquid at some instant compared with the final height.

This investigation is of significance in considering the flow of sap in plants or the wetting of fabrics or paper chromatography.

Also consider the factors affecting the rate of movement of liquids through blotting paper.

Reference

Washburn. *Phys. Rev.*, **17**, 273, 1921.

Suggestions for answers

1. (a) The amplitudes if successive oscillations are in a constant ratio.

(b) The rate of change of amplitude with time would be doubled and thus the motion would die away more rapidly.

(c) An exponential decrease should be drawn.

2. M/2, M/4, M/8.
The rate of decay is one quarter of that at 'zero' time.

3. (a) A constant rate of increase would mean a straight line graph, the slope has to be constant. This does not appear to be even approximately true.

(b) The curve is approximately an exponential variation, the number of cars doubling about every 7 years.

$$\frac{\Delta N}{\Delta t} = +\frac{1}{7} N, \text{ approximately,}$$

or

$$N = N_0 \, e^{+t/7}, \text{ approximately.}$$

4. (a) Exponential growth.

$$\frac{\Delta N}{\Delta t} = \frac{16}{1000} N$$

(b) Exponential decay.

$$\frac{\Delta N}{\Delta t} = -\frac{12}{1000} N$$

(c) Growth and decay

$$\frac{\Delta N}{\Delta t} = \frac{16}{1000} N - \frac{12}{1000} N$$

5. For the change in the number of those wearing the midi

$$\frac{\Delta N}{\Delta t} = kN$$

For the change in the number of those wearing the mini

$$\frac{\Delta N}{\Delta t} = N_0 - kN$$

Like the epidemic the fashion change may run out of victims or some might recover. See the note on epidemics for more information on how to describe such trends.

6. You could plot graphs of $e^{-0 \cdot 1t}$ and $e^{-0 \cdot 7t}$ and subtract the two.

The dose comparison is a comparison of the areas under the graph. By estimating the relative areas I obtained a ratio of about 1/7.

7. A model is some representation of a situation. An equation, such as acceleration = a constant, enables us to picture the motion. It does not, like other models, give us all the information about the situation. Models tend to be appropriate to one particular aspect of the situation. Our moving object could be bright red and have a star shape—our equation does not tell us.

8. (a) and (b) are exponential decays, (e) and (f) are exponential growths. Curves (c) and (d) are $(4 - e^{-ct})$.

9. Phrases such as 'constant ratio change, i.e., ...' and 'rate of change proportional to the amount present' might be useful in such an explanation.

10. (a) $\log y$ against x
(b) $\log y$ against $1/x$

6 Waves

Teaching note Practical work appropriate to this chapter will be found in:
W. Bolton. *Physics experiments and projects*, Vol. 2, Pergamon, 1968.
Nuffield advanced physics. Unit 4 Teachers' Guide, Penguin, 1972.
Nuffield O-level physics. Guide to experiments 3, Longmans/Penguin, 1967.

6.1 Sound

How does sound travel through air? The fact that sound requires a material medium for its travel can readily be shown; a bell ringing in a vacuum cannot be heard. Air or some other medium is necessary for sound propagation. The experiment with a bell, actually a watch with an alarm, was first made by Boyle in 1660. The production of sound always involves some motion at the source; a bell set in vibration by a blow, the oscillation of a violin string, the thud produced by a falling object, the sound of an explosion. The need for a medium and the movement of part of a sound source led to the idea that sound travels through a medium by virtue of the medium being pushed and pulled, by compression and rarefaction. Ideas like this were held by the Ancient Greeks. That sound takes a finite time to cover a distance, i.e., has a finite speed, has also been known for quite some time. The first measurements of the speed were made at the beginning of the seventeenth century; the methods used being the measurement of the time interval between a sound being produced and its echo coming back by reflection at some distant reflector, and the measurement by an observer some distance from a gun of the time interval between his seeing the flash of the explosion at the gun and hearing the sound of the explosion. The speed of sound in air is about 340 m s^{-1}.

Let us consider more closely how the sound wave could travel through a medium. The atoms in a solid or the molecules in a liquid or gas can be displaced from their equilibrium positions. If, however, they move then neighbouring atoms or molecules must experience a force because an atom or molecule has either moved closer to them or further away and disrupted the equilibrium arrangement. Let us represent this by a model, a series of trolleys linked by springs (Fig. 6.1). When one trolley is displaced, along the direction of the line of trolleys, then in time the next trolley will become displaced, then the next one, and so on down the line. Figure 6.2 illustrates this movement of a compression pulse down the line of trolleys. If the end trolley is steadily pushed in with a constant speed u then in time t it will have moved a distance ut. If the compression travels along the line of trolleys with a speed v then in this time t it will have moved a distance vt. If x is the distance between trolley centres then the number of trolleys in this distance vt will be vt/x. The number of springs is equal to the number of trolleys, so the number of springs compressed in time t will be vt/x. As the end trolley was moved in a distance ut and this resulted in vt/x springs being compressed, the average compression of a spring is

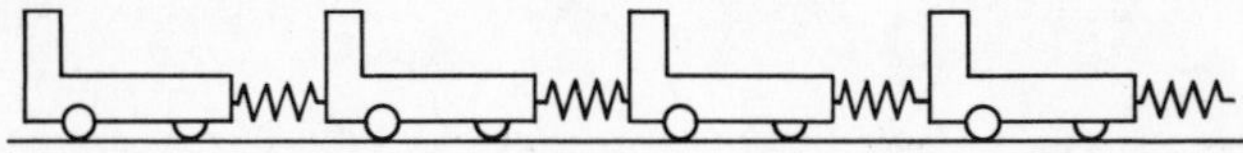

$$\frac{ut}{vt/x} = \frac{ux}{v}$$

As the end trolley is being pushed with a constant force at a constant speed there must be equal and opposite forces acting on it. Similarly the next trolley along the line will have equal and opposite forces acting on it when it is moving with constant speed; the springs each side of it are equally compressed. The force acting on a trolley due to a spring being compressed by ux/v will be kux/v if the compression is proportional to the force, k is the force constant (the springs are assumed to obey Hooke's law). A trolley which is just being moved from rest will have this force exerted on one side of it by a compressed spring, the other spring on the other side will not yet have been compressed. The unbalanced force acting on this trolley will be kux/v. During time t this force will have acted on vt/x trolleys and caused them to gain a momentum of mu, where m is the mass of a trolley. The rate of change of momentum is thus $(vt/x)mu$ in time t.

As force = rate of change of momentum

$$\frac{kux}{v} = \frac{vt}{x}\frac{mu}{t}$$

Thus

$$v = x\sqrt{\frac{k}{m}}$$

etc.

Fig. 6.1

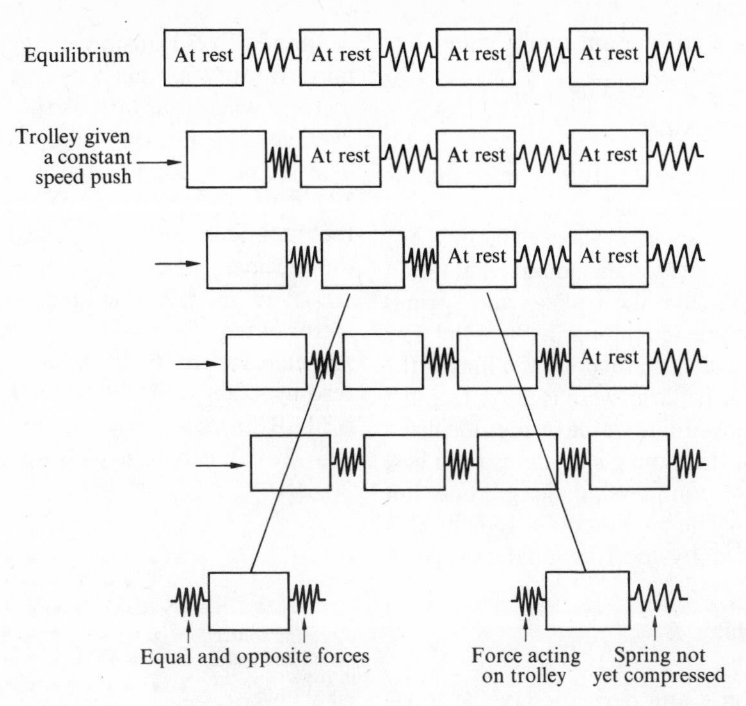

Fig. 6.2

The speed of the compression pulse thus depends on the spacing of the trolleys, the force constant of the springs and the mass of the trolleys.

Could this equation be applicable to compression waves in a medium? x could be the spacing between atoms, or molecules, m the mass of the atom, or molecule, and k the force constant for the bond joining atoms. A material would be three dimensional, Fig. 6.3, and thus our push on the end trolley would be a push on many atoms. However, for each

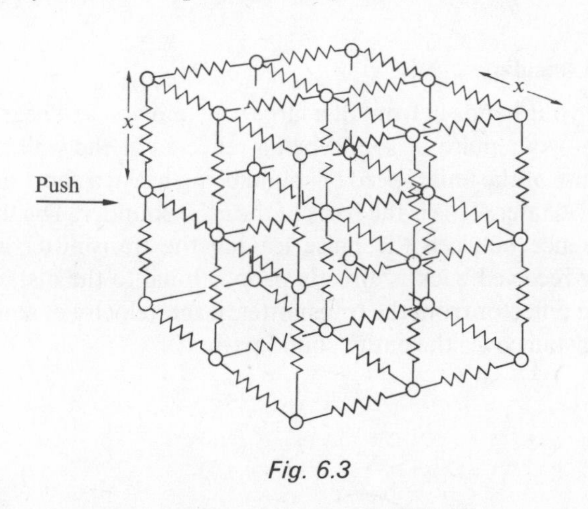

Fig. 6.3

atom there is one spring (bond), along the direction of the push, and the situation could be similar to that of the trolleys. There are, however, in the simple cubic array portrayed in Fig. 6.3, springs (bonds) in a direction at right angles to the push. These would have no effect if the compression acted over the entire face of the solid.

To use our equation to calculate the speed of a compression wave in a material we need to know the force constant of the bond. For a solid with a simple cubic array of atoms we can obtain the force constant in terms of the Young's modulus of the material.

$$\text{Young's modulus, } E = \frac{\text{stress}}{\text{strain}}$$

provided we have a linear stress–strain relationship.

$$\text{Stress} = \frac{\text{force}}{\text{area}}$$

For the piece of solid represented by Fig. 6.3 we have per pair of atoms

$$\text{Stress} = \frac{k\,\Delta x}{x^2}$$

where Δx is the amount of compression.

$$\text{Strain} = \frac{\text{compression}}{\text{length}}$$

$$= \frac{\Delta x}{x}$$

Thus

$$E = \frac{k\,\Delta x}{x^2}\frac{x}{\Delta x}$$

$$E = \frac{k}{x}$$

The speed equation then becomes—

$$v = x\sqrt{\frac{Ex}{m}}$$

$$= \sqrt{\frac{E}{m/x^3}}$$

But m/x^3 is the density, for a cubic atom array. Thus

$$v = \sqrt{\frac{\text{Young's modulus}}{\text{density}}}$$

For copper, Young's modulus is 13×10^{10} N m^{-1} and the density 8930 kg m^{-3}. Used in the equation they give a speed of 3800 m s^{-1}. The measured speed is about 3800 m s^{-1}.

This seems a reasonable model for a solid but what about a gas? A piece of solid can, like the trolleys and springs model, be pulled or compressed along its length. The motions of our atoms, or trolleys, are along the line of the force. Our masses, atoms or trolleys, were localized. In the case of a gas, or indeed any fluid, we have non-localized masses. We cannot refer to the Young's modulus for a gas, because this is stretching or compressing along a line, but must use the bulk modulus. The bulk modulus is defined as the pressure change divided by the fractional change in volume.

$$\text{Bulk modulus } K = \frac{\Delta p}{(\Delta V/V)}$$

We can by a re-examination of our derivation of the equation for the trolley model of the solid arrive at an equation for a gas. The result is

$$v = \sqrt{\frac{\text{Bulk modulus}}{\text{density}}}$$

If instead of just giving our row of trolleys a steady push we applied a force which pushed then pulled then pushed then pulled, etc., on the end trolley we would generate a sequence of compressions which would travel along the row of trolleys (Fig. 6.4). At some instant of time we would find trolleys, in one region, grouped closer together, in another region farther apart, than the initial equal distance apart state. We have a series of alternate compressions and rarefactions. In the case of a gas we would have alternate regions of pressure above the normal gas pressure and regions of pressure below the normal (Fig. 6.5(a)). If we plot the displacement of the trolleys with their positions along the row a wave pattern is found (Fig. 6.5(b)). The distance between successive compressions (or successive rarefactions) is the wavelength. The frequency is the number of waves produced per second, i.e., the number of compressions produced per second.

Sound waves transmit energy. The peak of a thunderclap may transmit as much as one watt through each square metre, a whisper as little as 10^{-9} watt per square metre. The average conversation is about 10^{-6} watt per square metre. The quietest sound that can be just heard is about 10^{-12} W m^{-2}. For convenience another unit is used for these sound intensities. The scale is logarithmic and is called the decibel scale. The zero of the scale is taken as 10^{-12} W m^{-2}. If the sound intensity is increased by a factor of ten, to 10^{-11} W m^{-2}, we say the intensity is 10 dB. An increase to 10^{-10} W m^{-2}, i.e., an increase from the zero by a factor of 100, is 20 dB ($100 = 10^2$). 10^{-9} W m^{-2} is 30 dB (an increase by a factor of 1000, i.e., 10^3). 10 W m^{-2}, i.e., 130 dB, is painful to the ears.

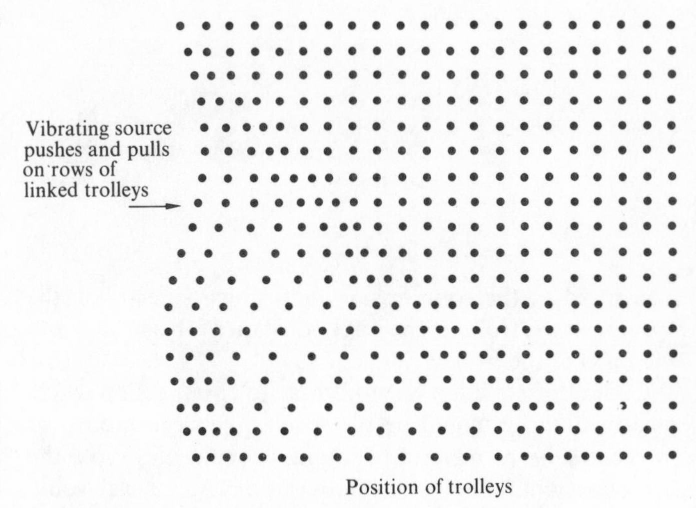

Fig. 6.4

Echo sounding

Clap your hands in front of a large wall and you can hear an echo—your pulse of sound being reflected at the wall and because of the finite speed of sound being heard a short time interval later. This is the basis of the echo sounder. The time difference between the pulse leaving the transmitter and being received back is directly proportional to the distance of the reflector from the transmitter, if the velocity of sound is constant over the entire sound path.

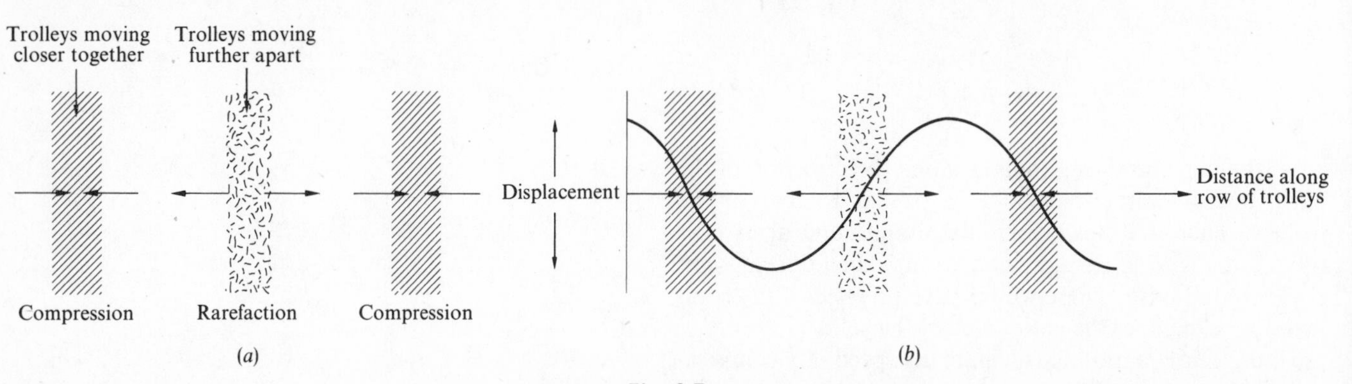

Fig. 6.5

Figure 6.6 shows the output from an echo sounder on board a ship. The output appears as a chart where horizontal distance along the chart represents horizontal distance traversed by the ship and vertical distances are proportional to the time taken for an echo to be received at the ship. The vertical distances are thus proportional to the vertical distances. The upper trace is of the sea surface, the other traces are echoes. More than one echo is detected at any one depth, due to the sound pulse travelling down to the sea bottom, being reflected to the sea surface, again reflected and going back down to the sea bed again. The number of echoes detected depends on the hardness of the sea bed. The amount of blur of any one reflection depends on the smoothness of the sea bed. The entire pulse may not be reflected at the sea bed but penetrate and be reflected at a lower rock stratum: some such strata are visible in Fig. 6.6.

Fig. 6.6 Echo-sounder chart. A recording of strata in the bed of the river Crouch at Burnham showing faults. (Multiple echoes from sea bed and strata are seen.) (Kelvin Hughes, a division of Smith Industries Ltd.)

One application of the ultrasonic echo principle is in medicine. The depth and thickness of organs in the human body can be determined; useful information when tumerous growths are suspected or when it is necessary to judge the maturation of the embryo during pregnancy.

6.2 Let there be light

'And God said, Let there be light: and there was light.'

Genesis, 1:1–5.

What is light? What is colour? Does light come from objects and enter the eye or does the eye send light out to detect the object? This was a question debated by the Ancient Greeks.

'...they (the gods) caused the pure fire within us, which is akin to that of day, to flow through the eyes in a smooth and dense stream;...whenever the stream of vision is surrounded by mid-day light, it flows out like unto like, and coalescing therewith it forms one kindred substance along the path of the eye's vision, wheresoever the fire which streams from within collides with an obstructing object without....'

Plato, about 400 B.C.

'...if vision were produced by means of a fire emitted by the eye, like the light emitted by a lantern, why then are we not able to see in the dark? To say that this light is extinguished as it spreads out in the darkness,..., is completely meaningless. How can light be extinguished?

...It is completely absurd to maintain that we can see because of "something" which is emitted by the eye and that this "something" may reach as far as the stars or until it meets another "something" coming towards it, as other philosophers have suggested.'

Aristotle, about 350 B.C.

Even as late as the end of the fifteenth century there were still doubts as to whether vision was due to 'something' sent out from the eye or 'something' sent out from the object.

'Some thinkers of good authority have said that vision takes place by means of an emission of a visual spirit which travels from the eye to the object seen.... There are others however, who think that the species or images of the objects seen and the spirit emitted by the eye meet in the interposed medium....'

Reisch (1486). *Margarita philosophica.*

It was only at the beginning of the seventeenth century that light came to be accepted as 'something' flowing out from luminous objects.

'Light has the property of flowing or of being emitted by its source towards a distant place.'

Kepler (1604). *Ad Vitellionem Paralipomena.*

In the seventeenth century two main theories of light were proposed—the corpuscular theory of Newton and the wave theory of Huygens. In the following pages these theories are considered in some detail.

Newton's theory

'Are not the Rays of Light very small Bodies emitted from shining substances?'

Newton (1704). *Opticks.*

One of the reasons Newton gave for not advocating a wave theory for light was the difficulty of explaining straight line motion of light by such a theory. With a particle theory he had no such difficulty.

'The waves on the surface of stagnating water, passing by the sides of a broad obstacle which stops part of them, bend afterwards and dilate themselves gradually into the quiet water behind the obstacle. The Waves, Pulses or Vibrations of the Air, wherein Sounds consist, bend manifestly, though not so much as the Waves of Water.... But Light is never known to follow crooked Passages nor to bend into the shadow....'

Newton (1704). *Opticks.*

Newton not only put forward a theory for light but did a considerable amount of experimental work, for which the theory was offered as explanation. A vital group of experiments were those concerning the refraction of light by prisms.

The point of view then current, in the mid seventeenth century, was that light was colourless. Colours were seen because light extracted them from objects and somehow communicated them to the eye. There was no thought that white light might be made up of coloured components, i.e., the colour was in the light. Newton's experiments with the refraction of light by prisms led to the eventual acceptance of 'the light being coloured'.

'In the year 1666,... I procured me a triangular glass prism to try therewith the celebrated phenomena of colours. And in order thereto, having darkened my chamber, and made a small hole in my window-shuts, to let in a convenient quantity of the sun's light, I placed my prism at its entrance, that it might be thereby refracted to the opposite wall. It was at first a very pleasing divertisement, to view the vivid and intense colours produced thereby; but after a while applying myself to consider them more circumspectly, I became surprised, to see them in an oblong form;... [Newton then describes how parts of this first spectrum were isolated and passed through a second prism]...the light, tending to that end of the image, towards which the refraction of the first prism was made, did in the second prism suffer a refraction considerably greater than the light tending to the other end. And so the true cause of the length of that image was detected to be no other, than the light is not similar or homogenial, but consists of different rays, some of which are more refrangible than others; so that without any difference in their incidence on the same medium, some shall be more refracted than others;.... The least refrangible rays are all disposed to exhibit a red colour,... the most refrangible rays are all disposed to exhibit a deep violet colour,.... When any one sort of rays hath been well parted from those of other kinds, it hath afterwards obstinately retained its colour, notwithstanding my utmost endeavours to change it.... I have often with admiration beheld that all the colours of the prism being made to converge and thereby to be again mixed, as they were in the light before it was incident upon the prism, reproduced light, entirely and perfectly white,....'

Newton. *Principia Mathematica.*

The basic points arising from this last extract are:

(a) white light is a mixture of all the colours of the spectrum,

(b) the different colours of light are refracted by a prism by differing amounts,

(c) red light is refracted less than violet light.

Figure 6.7 shows the basic arrangements of Newton's experiments with the prisms.

Newton's conclusions were not without their critics.

'Light is the simplest matter we have knowledge of, the least capable of analysis, the most homogeneous. It is not a compound body. Least of all is it compounded of coloured lights. Every coloured light is darker than colourless light. Brightness cannot be compounded of darkness.... The colours are excited in the light, not developed out of the light....'

Goethe (1810?)

How can refraction be explained? Why do light rays bend on passing from one medium to another? Newton's explanation is as follows.

'That bodies refract light by acting upon its rays in lines perpendicular to their surfaces.'

Newton (1704). *Opticks.*

Bodies are considered to refract light by virtue of an attractive force, perpendicular to the surface of a body, which acts on the light corpuscles.

'...that motion of the ray which is parallel to the refracting plane, will suffer no alteration by that force; and that motion which is perpendicular to it will be altered....'

Newton (1704). *Opticks.*

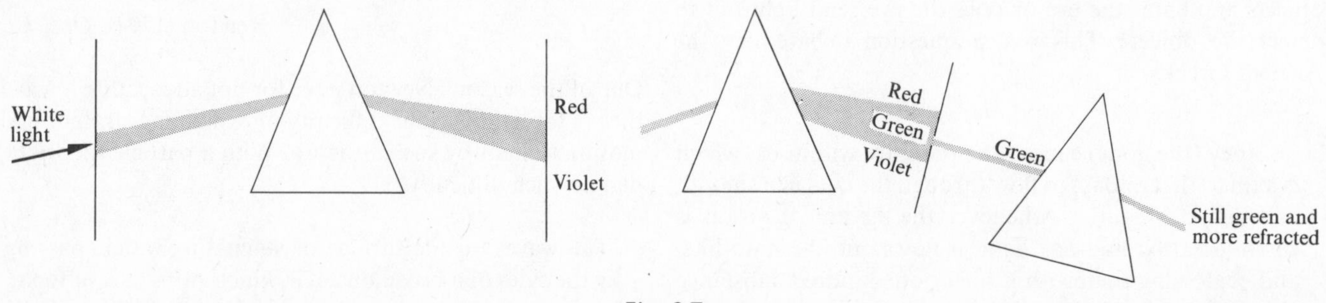

Fig. 6.7

This leads to the result that the horizontal components of the ray's velocity before refraction must be equal to that after refraction; both components will be acted on by the force in the same way (Fig. 6.8). Thus

$$v \sin i = v' \sin r$$

$$\frac{\sin i}{\sin r} = \frac{v'}{v}$$

The ratio of the sine of the angle of incidence to the sine of the angle of refraction is a constant, the constant being equal to the ratio of the velocity in the refracting medium to that in the incident medium. For this to be correct the velocity of light in the refracting medium must be greater than that in the incident medium.

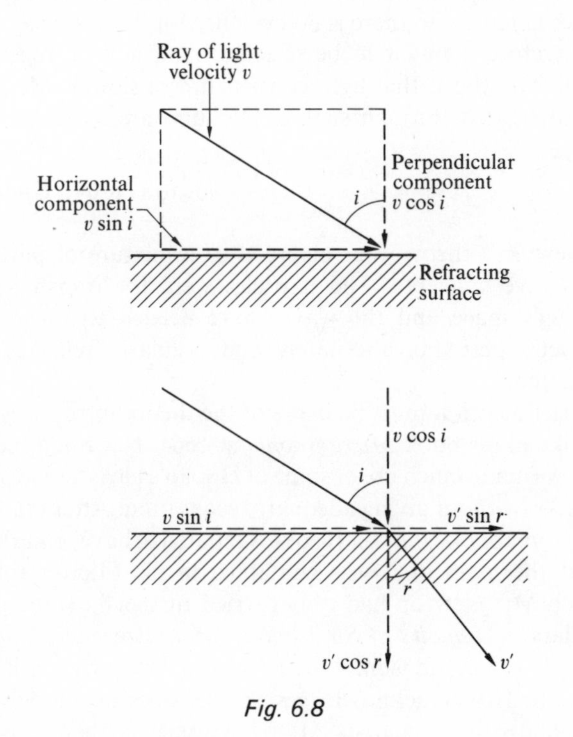

Fig. 6.8

> 'If light be swifter in bodies than in vacuo, in the proportion of the sines which measure the refraction of the bodies, . . .'
>
> Newton (1704). *Opticks.*

This was a distinguishing mark of Newton's theory. The wave theory led to the reverse result, a velocity higher in a vacuum than in an object.

Newton devoted a lot of time to what is now known as Newton's rings, an interference pattern formed by a thin air gap between a lens and a flat surface.

> 'It has been observed by others, that transparent Substances, as Glass, Water, Air, & etc., when made very thin by being blown into Bubbles, or otherwise formed into Plates, do exhibit various Colours according to their various thinness, altho' at a greater thickness they appear very clear and colourless. . . .
>
> I took two Object-glasses, the one a Plano-convex for a fourteen Foot Telescope, and the other a large double Convex for one of about fifty Foot; and upon this, laying the other with its plane side downwards, I pressed them slowly together, to make the colours successively emerge in the middle of the Circles, and then slowly lifted the upper Glass from the lower to make them successively vanish again in the same place. . . .'
>
> Newton (1704). *Opticks.*

The light pattern produced when a curved lens surface is put in contact with a flat surface is now known as Newton's rings. The pattern is a series of concentric rings. If monochromatic light, i.e., light of just one colour (one frequency), is used the rings are alternately bright and dark. If white light is used, as Newton did, the result is a series of concentric coloured rings, with inner edges violet shading through the spectrum to red outer edges. How were the rings produced? The following is Newton's answer to the problem.

> 'Every ray of light in its passage through any refracting surface is put into a certain transient constitution or state, which in the progress of the ray returns at equal intervals, and disposes the ray at every return to be easily transmitted through the next refracting surface, and between the returns to be easily reflected by it. . . .
>
> The returns of the disposition of any ray to be reflected I will call its Fits of easy Reflection, and those of its disposition to be transmitted its Fits of easy Transmission, and the space it passes between every return and the next return, the Interval of its Fits. . . .'
>
> Newton (1704). *Opticks.*

To put the above extract in more modern language: every ray of light in its passage through a medium has periodic variations which allow it to be either easily reflected or transmitted at a refracting surface. The dark rings are where the light is refracted and thus not reflected, the bright rings are where reflection occurs.

> 'If the rays which paint the colour in the confine of yellow and orange pass perpendicularly out of any medium into air, the intervals of their Fits of easy Reflection are the 1/89 000th of an inch. And of the same length are the intervals of their Fits of easy transmission. . . .'
>
> Newton (1704). *Opticks.*

Newton had found that a change in the thickness of the air film between his lenses of 1/89 000 inch (or 3×10^{-7} m) was needed to progress from one orange ring to another (Fig. 6.9). The method of calculating the thickness of the air film is given in Fig. 6.10. Thus every 1/89 000 inch along a light path in air there is, according to Newton, a change from a Fit of easy Reflection to a Fit of easy Transmission. Newton's corpuscles were having to possess a periodic nature.

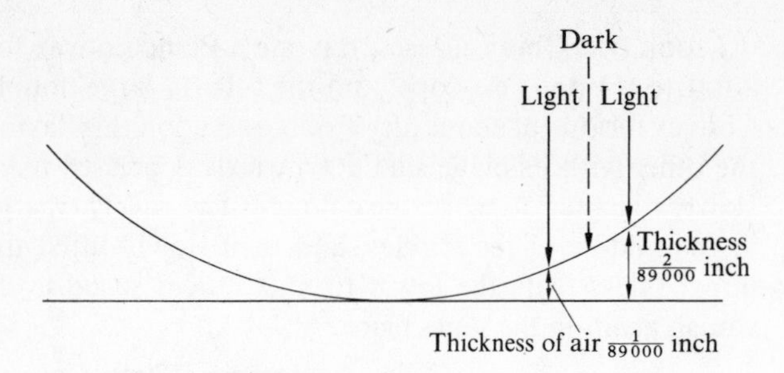

Fig. 6.9

Newton also made observations on the shadows cast by objects.

> 'I made in a piece of lead a small hole with a pin, whose breadth was the 42nd part of an inch. For 21 of those pins laid together took up the breadth of half an inch. Through this hole I let into my darken'd chamber a beam of the sun's light, and found that the shadows of hairs, thread, pins, straws, and such like slender substances placed in this beam of light, were considerably broader than they ought to be, if the rays of light passed on by these bodies in right lines. . . . The shadows of all bodies in this light were border'd with three parallel fringes or bands of colour'd light, . . .'
>
> Newton (1704). *Opticks.*

'Light does not travel in straight lines' is the interpretation that can be placed on these experiments of Newton's. Despite these results Newton, only a few pages later in his book, considers that light cannot be waves because waves bend into shadows. Newton's experiments indicated a broadening of the shadow and this he felt could be explained if surfaces exerted forces on the light corpuscles, similar to those he felt to be necessary to explain refraction.

> 'But light is never known to follow crooked passages nor to bend into the shadow. . . . The rays which pass very near to the edges of any body, are bent a little by the action of the body, as we shew'd above, but this bending is not towards but from the shadow. . . .'
>
> Newton (1704). *Opticks.*

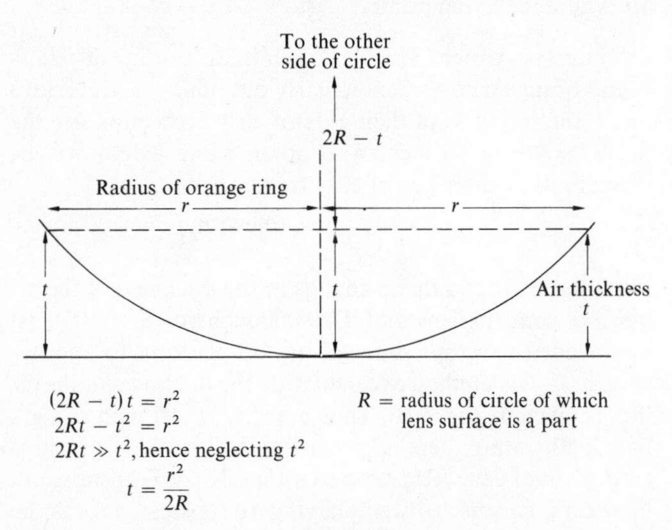

Fig. 6.10

An argument Newton advances against the wave theory is that waves, in his opinion, need a medium. The existence of a medium between the sun and the planets would, he considers, slow down the motion of the planets.

> 'Are not all hypotheses erroneous, in which light is supposed to consist in Pression or Motion, propagated through a fluid medium? . . . And as it (a medium) is of no use, and hinders the operations of Nature, and makes her languish, so there is no evidence for its existence, and therefore it ought to be rejected. And if it be rejected, the hypothesis that light consists in Pression or Motion, propagated through such a Medium, are rejected with it. . . .'
>
> Newton (1704). *Opticks.*

Newton's theory thus emerged as a mixture of particle and wave concepts. The particles were for transmission through space and the waves were needed to make the particles periodic, alternately and regularly reflected or refracted.

Brief mention must be made of the theory introduced by Hooke in his book *Micrographia* in 1664. Newton appears to have made much use of some of Hooke's ideas and experiments—without giving adequate recognition. After reading Newton's book *Opticks*, Hooke is reputed to have remarked, 'that the main of it was contained in his Micrographia, which Mr. Newton had only carried further in some particulars.' (*Memoirs of Sir Isaac Newton*, Brewster, Vol. 1, p. 138, 1855, Johnson, 1965.) Newton in a private, later, letter to Hooke acknowledges that he does owe a debt to him and other scientists. 'If I have seen farther, it is by standing on the shoulders of giants.'

Huygens' theory

Huygens wrote his book *Treatise on Light* in 1678 but it was not published until 1690. Huygens wanted to improve it but somehow never got round to it.

> 'It is inconceivable to doubt that light consists in the motion of some sort of matter. For whether one considers its production, one sees that here upon earth it is chiefly engendered by fire and flame which contain without doubt bodies that are in rapid motion, since they dissolve and melt many other bodies, even the most solid; or whether one considers its effects, one sees that when light is collected, as by concave mirrors, it has the property of burning as a fire does, that is to say it disunites the particles of bodies. . . .
>
> Further, when one considers the extreme speed with which light spreads on every side, and how, when it comes

from different regions, even from those directly opposite, the rays traverse one another without hindrance, one may well understand that when we see a luminous object, it cannot be by any transport of matter coming to us from this object, in the way in which a shot or an arrow traverses the air;.... It is then in some other way that light spreads; and that which can lead us to comprehend it is the knowledge which we have of the spreading of sound in the air.

We know that by means of the air, which is an invisible and impalpable body, sound spreads around the spot where it has been produced, by a movement which is passed on successively from one part of the air to another; and that the spreading of this movement, taking place equally rapidly on all sides, ought to form spherical surfaces ever enlarging and which strike our ears. Now there is no doubt at all that light also comes from the luminous body to our eyes by some movement impressed on the matter which is between the two; since, as we have already seen, it cannot be by the transport of a body which passes from one to the other. If, in addition, light takes time for its passage... it will follow that this movement, impressed on the intervening matter, is successive; and consequently it spreads, as sound does, by spherical surfaces and waves: for I will call them waves from their resemblance to those which are seen to be formed in water when a stone is thrown into it, and which present a successive spreading as circles, ...'

C. Huygens (1690). *Treatise on light.*

One of the key arguments advanced by Huygens concerned the passage of two rays of light through each other. He could not imagine rays of particles passing through each other without some effect being noticeable; he could, however, imagine waves passing through each other. He draws a comparison between light and sound. Sound travels through a medium, a medium being essential, by means of compressions and rarefactions—does light do the same? Huygens considers that light could travel in the same way as sound and introduces the idea of a medium existing between the earth and the sun (he calls it 'etheral matter'). He realizes that the medium is not air because when air is removed sound will not pass but light still will. The idea of light and sound being similar to water waves is a fruitful idea which he explores in some depth, using it to explain both reflection and refraction.

'... each little region of a luminous body, such as the sun, a candle, or a burning coal, generates its own waves of which that region is the centre. Thus in the flame of a candle, having distinguished the points A, B, C, concentric circles described about each of these points represent the waves which come from them. And one must imagine the same about every point of the surface and of the part within the flame [Fig. 6.11]....

... each particle of matter in which a wave spreads, ought not to communicate its motion only to the next particle which is in the straight line drawn from the luminous point, but that it also imparts some of it necessarily to all the others which touch it and which oppose themselves to its motion. So it arises that around each particle there is made a wave of which that particle is the centre....'

Huygens (1690). *Treatise on light.*

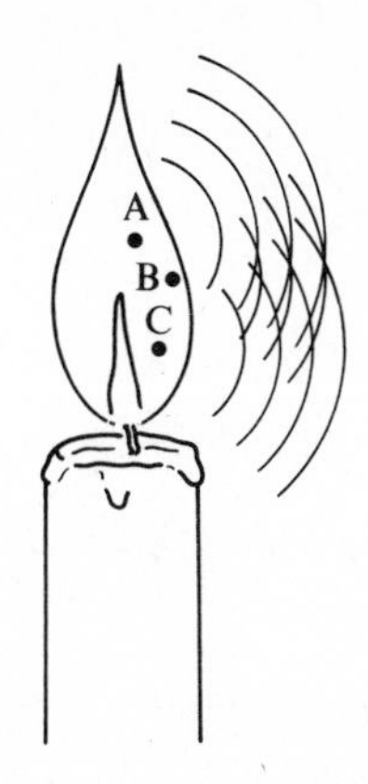

Fig. 6.11 Diagram after Huygens, Treatise on light.

Huygens thinks in term of a medium composed of particles which pass on the wave disturbance in a similar way that air particles pass on the compressions and rarefactions of sound. The vital point, however, in this argument is the idea of each part of a source being a centre of waves and also each part of a wavefront being a source of secondary waves (see Fig. 6.12 for details of this construction). This idea is used to explain both reflection and refraction.

'Let there be a surface AB, plane and polished [Fig. 6.13], ... and let a line AC... represent a portion of a wave of light, the centre of which is so distant that this portion AC may be considered as a straight line;.... The piece C of the wave AC, will in a certain space of time advance as far as the plane AB at B, following the straight line CB, which may be supposed to come from the luminous centre, and which in consequence is perpendicular to AC. Now in the same space of time the portion A of the same wave, which has been hindered from communicating its movement beyond the plane AB, or at least partly so, ought to have continued its movement in the matter which

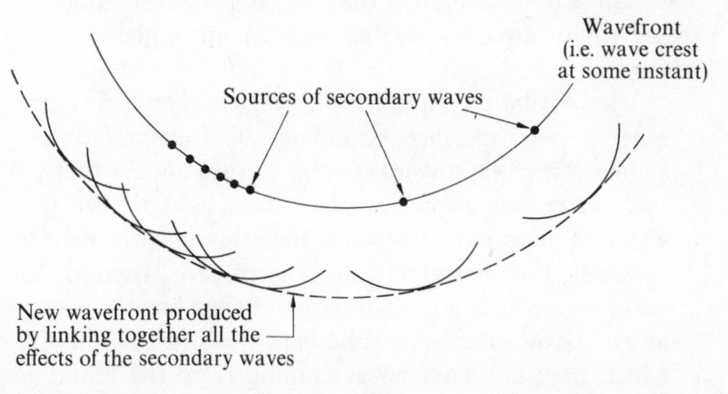

Fig. 6.12

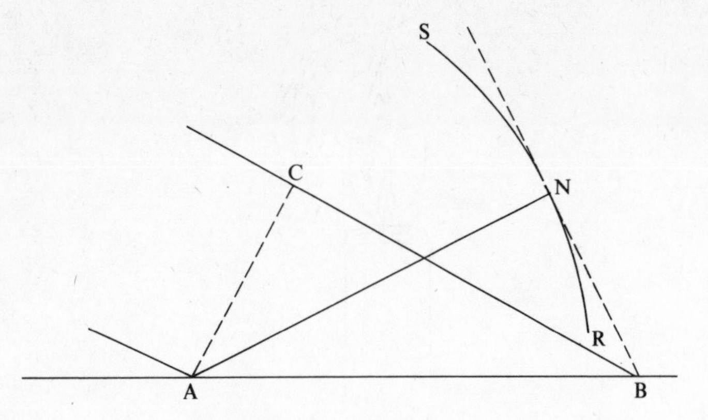

Fig. 6.13 Diagram after Huygens, Treatise on light.

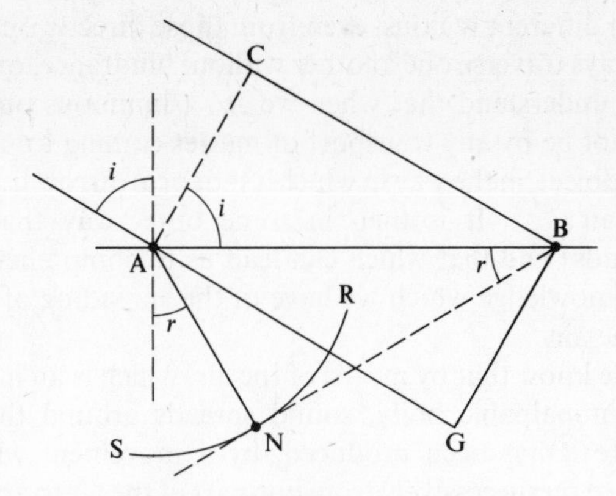

Fig. 6.14 Adapted from Huygens, Treatise on light.

is above this plane, and this along a distance equal to CB, making its own partial spherical wave.... Which wave is here represented by the circumference SNR, the centre of which is A, and its semi-diameter AN equal to CB.... Now it is apparent here that the angle of reflexion is made equal to the angle of incidence. For the triangles ACB, BNA being rectangular and having the side AB common, and the side CB equal to NA, it follows that the angles opposite to these sides will be equal, and therefore also the angles CBA, NAB. But as CB, perpendicular to CA, marks the direction of the incident ray, so AN, perpendicular to the wave BN, marks the direction of the reflected ray; hence these rays are equally inclined to the plane AB.'

Huygens (1690). *Treatise on light.*

We can summarize Huygens' method of dealing with reflection as:
AC is the wavefront at some instant, for a source so distant that the wavefront is straight. The secondary wave starting at C travels to B in the same time that the secondary wave at A travels to N. As both waves travel in the same medium CB = AN. As the wavefronts, AC and BN, are perpendicular to the rays, this makes the two triangles ACB and ANB identical. Thus the angle of incidence is equal to the angle of reflection.

Huygens explained refraction by a similar method to that he adopted for reflection—a difference being that he had to assume that the speed of light in a medium such as glass was less than that in air in order that the ray of light should bend in the 'right' direction on passing from air to glass.

'... let AB be the straight line [Fig. 6.14] which represents a plane surface bounding the transparent substances which lie towards C and towards N.... Let the line AC represent a portion of a wave of light, the centre of which is supposed so distant that this portion may be considered as a straight line. The piece C, then, of the wave AC, will in a certain space of time have advanced as far as the plane AB following the straight line CB, which may be imagined as coming from the luminous centre, and which consequently will cut AC at right angles. Now in the same time the piece A would have come to G along the line AG, equal and parallel to CB; and all the portion of wave AC would be at GB if the matter of the transparent body transmitted the movement of the wave as quickly as the matter of the ether. But let us suppose that it transmits this movement less quickly, by one-third, for instance. Movement will then be spread from the point A, in the matter of the transparent body through a distance equal to two-thirds of CB, making its own particular spherical wave.... This wave is then represented by the circumference SNR, the centre of which is A, and its semi-diameter equal to two-thirds of CB.... It is then BN, ..., which terminates the movement that the wave AC has communicated within the transparent body, ... And for that reason this line, ..., is the propagation of the wave AC at the moment when its piece C has reached B....'

C. Huygens (1690). *Treatise on light.*

To summarize: the wavefront AC would in the absence of the surface AB move to GB, the portion at C does move to B but the portion at A is slowed down by passing into the medium below the line AB. If the speed of light in this medium is 2/3 of that in air then the arc SNR represents the wave arising from the wave-centre at A, SN = 2CB/3. The line from B tangent to this arc gives the refracted wavefront NB.

$$\frac{CB}{AB} = \sin i$$

$$\frac{AN}{AB} = \sin r$$

Thus

$$\frac{CB}{AN} = \frac{\sin i}{\sin r}$$

But CB is the distance travelled in air in time t, and AN is the distance travelled in the medium in the same time t. Thus

$$\frac{CB}{AN} = \frac{\text{speed of light in air}}{\text{speed of light in medium}} = \frac{\sin i}{\sin r}$$

As the refractive index, i.e., the ratio sin i/sin r, is greater than one when the light goes from air to a medium such as glass then according to Huygens the velocity of light in air is greater than that in glass. This is the reverse of what Newton obtained from his theory. At the time of these theories nobody had measured the velocity of light in glass, only a crude value of the velocity of light in a vacuum had been obtained by Huygens from the data of Römer (see chapter 12).

Which theory?

In Newton's time there seemed to be but one answer—Newton's theory, not that of Huygens. Why was this so? A vital factor was the prestige of Newton—he did not believe in the wave theory though he did write 'Tis true, that from my theory I argue the corporeity of light; but I do it without any absolute positiveness'. Also Newton's theory did offer an explanation for all the facts then known about light. Huygens could not explain the straight line propagation of light; his one hope seemed to be that two beams of light can cross each other without any noticeable effect but particles could not be envisaged as doing just this. This was not enough to overthrow Newton's theory.

There was one noticeable difference between the two theories—according to Newton's theory the explanation of refraction requires the speed of light to be greater in say glass than in a vacuum (or air), according to Huygens the speed should be greater in a vacuum than in glass. This would seem to be a vital point which could be tested and used to decide between the two theories. Unfortunately experimental techniques were not sufficiently advanced for the measurements to be made. It was only some time after other evidence had decided in favour of the wave theory that the experiments could be performed (Foucault 1850—almost two centuries after Newton).

The factor which later decided in favour of the wave theory was a series of experiments by Young, at the beginning of the nineteenth century, in which he found that under certain conditions light plus light gave darkness, i.e., destructive interference. This could not be easily explained by Newton's corpuscles of light but was capable of a simple explanation by Huygens' waves.

> 'I made a small hole in a window shutter, and covered it with a piece of thick paper, which I perforated with a fine needle.... I brought into the sunbeam a slip of card, about one thirtieth of an inch [about 0·8 mm] in breadth, and observed its shadow, either on the wall, or on other cards held at different distances. Besides the fringes of colours on each side of the shadow, the shadow itself was divided by similar parallel fringes, of smaller dimensions, differing in number, according to the distance at which the shadow was observed, but leaving the middle of the shadow always white. Now these fringes were the joint effects of the portions of light passing on each side of the slip of card, and inflected, or rather diffracted, into the shadow. For, a little screen being placed a few inches from the card, so as to receive either edge of the shadow on its margin, all the fringes which had been observed in the shadow on the wall immediately disappeared, . . .'
>
> T. Young (1804). *Phil. Trans. Roy. Soc. Lond.*, **94**.

In this experiment light was bending, or to use the physics term, diffracting, round the edges of a strip of card. The diffracted light interfered to give regions of darkness and light, interference fringes. In the middle of the shadow was a light fringe.

> 'If we now proceed to examine the dimensions of the fringes, under different circumstances, we may calculate the differences of the lengths of the paths described by the portions of the light, which have thus been proved to be concerned in producing those fringes; and we shall find, that where the lengths are equal, the light always remains white; but that, where either the brightest light, or the light of any given colour, disappears and reappears, a first, a second, or a third time, the differences of the lengths of the paths of the two portions are in arithmetical progression, . . .'
>
> Young (1804). *Phil. Trans. Roy. Soc. Lond.*, **94**.

When there is no difference in path length between two portions of light there is a white fringe, constructive interference. Hence the white fringe in the middle of the shadow of the card strip. Whenever the path difference is 0 or 1 or 2 or 3 or 4 times some particular path difference there is constructive interference for some particular colour.

> 'Supposing the light of any given colour to consist of undulations, of a given breadth, or of a given frequency, it follows that these undulations must be liable to those effects which we have already examined in the case of the waves of water, and the pulses of sound. It has been shown that two equal series of waves, proceeding from centres near each other, may be seen to destroy each other's effects at certain points, and at other points to redouble them; and the beating of two sounds has been explained from a similar interference. We are now to apply the same principle to the alternate union and extinction of colours.
>
> In order that the effects of two portions of light may thus be combined, it is necessary that they be derived from the same origin, and that they arrive at the same point by different paths, in directions not much deviating from each other. This deviation may be produced in one or both of the portions by diffraction, by reflection, by refraction, or by any of these effects combined; but the simplest case appears to be, when a beam of homogenous light falls on a screen in which there are two very small holes or slits, which may be considered as centres of divergence, from whence the light is diffracted in every direction. In this case, when the two newly formed beams

are received on a surface placed so as to intercept them, their light is divided by dark stripes into portions nearly equal, but becoming wider as the surface is more remote from the apertures, so as to subtend very nearly equal angles from the apertures at all distances, and wider also in the same proportion as the apertures are closer together. The middle of the two portions is always light, and the bright stripes on each side are at such distances, that the light coming to them from one of the apertures, must have passed through a longer space than that which comes from the other, by an interval which is equal to the breadth of one, two, three, or more of the supposed undulations, while the intervening dark spaces correspond to a difference of half of a supposed undulation of one and a half, of two and a half, or more.

From a comparison of various experiments, it appears that the breadth of the undulations constituting the extreme red light must be supposed to be, in air, about one 36 thousandth of an inch [about 7×10^{-7} m], and those of the extreme violet about one 60 thousandth [about 4×10^{-7} m]; . . . .'

T. Young (1807). *Course of lectures on natural philosophy and the mechanical arts*, Lecture 39.

Figure 6.15 gives the basic idea of Young's experiment. The quantity referred to as the breadth of the undulations is now called the wavelength. The results obtained by Young are of the right order of magnitude. Young's work was, however, not without its critics.

'. . . the absurdity of this [Young] writer's "law of interference" as it pleases him to call one of the most incomprehensible suppositions that we remember to have met with. . . .'

Proc. Am. Academy Science. **13**, 37, 1888.

This was then the scene at the beginning of the nineteenth century—Newton's theory still held by many but experimental evidence emerging which needed a wave theory. As the century progressed so the wave theory came to be the accepted theory.

6.3 The properties of waves

To a water wave, or indeed any wave, we can ascribe a frequency, a wavelength, and a velocity. The frequency is the number of waves per second which pass a point in the wave path or we can think of it as the number of waves produced per second. One wave is taken as being from one crest to the next crest; this distance is known as the wavelength. Waves move and thus we can refer to their velocity. We must distinguish between the wave moving and the water moving. A wave does not carry the water with it. The wave breaking on a beach may have started a long way out at sea; the water has not, however, been brought by the wave this long distance. The water velocity is not the same as the wave velocity. A cork or a piece of wood bobbing up and down on the water surface does not move along with the wave, the water behaves just like the cork.

If the frequency is f then f waves of length λ, the wavelength, must pass a point per second. The distance travelled by a wave is thus $f\lambda$ in one second. The velocity v is given by the distance travelled per second and is thus

$$v = f\lambda$$

The unit of frequency is s^{-1}. This is given the name Hertz (Hz). 1 Hz is one cycle per second. If the wavelength is in metres then the velocity is m s^{-1}.

To get some ideas of the properties of water waves take a look at the waves in your bath water or the sink. A point to notice here is that the direction of the wave motion is at right angles to the line of crests or troughs. Later we will find it convenient to refer to the wavefront; this is the line of a crest, or trough, at some instant. With plane waves the wave fronts would be straight, with circular waves circular.

With plane waves reflected from a straight barrier the angle of incidence is equal to the angle of reflection—just like in optics. Figure 6.16 shows circular waves being reflected from a straight barrier. The angle of incidence is still equal to the angle of reflection and, in addition, we can begin to think in terms of an image source of waves. The image would have to be as far behind the barrier as the object source is in front in order to give the results shown in the photograph. Again this is just like reflection from a plane mirror in light. The same rules apply to both the reflection of light and the reflection of water waves.

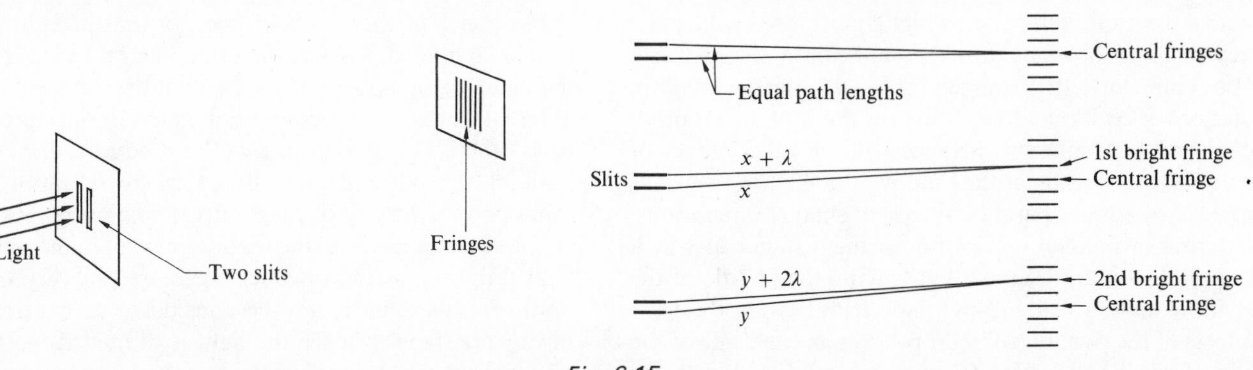

Fig. 6.15

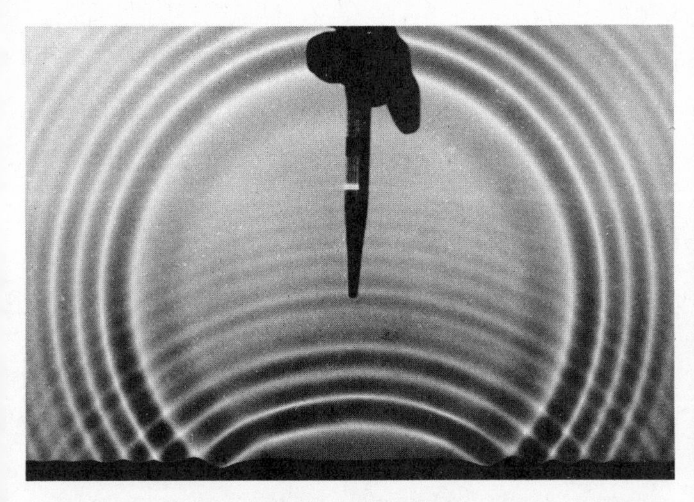

Fig. 6.16 PSSC Physics, *2nd Ed., D. C. Heath and Co.*

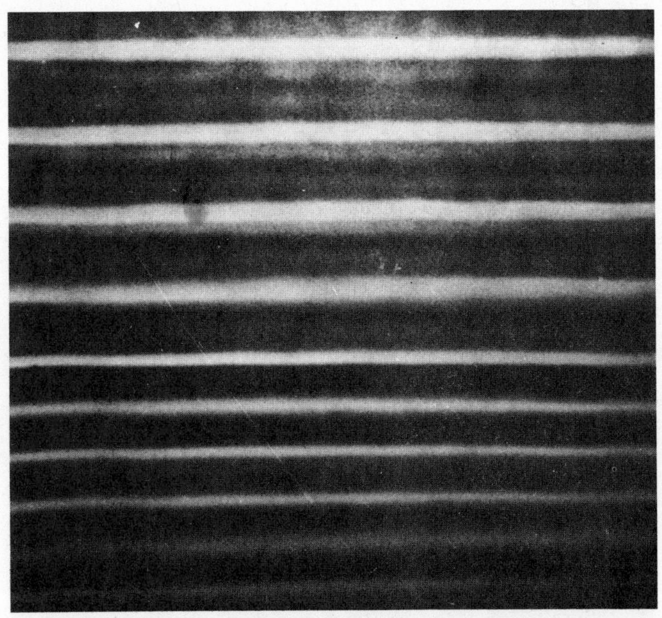

Fig. 6.17 PSSC Physics, *2nd Ed., D. C. Heath and Co. Passage of water waves from deep to shallow water.*

Figure 6.17 shows what happens when water waves pass from deep water to shallow water. There is a change in wavelength. The wavelength in the shallow water is less than the wavelength in the deep water. The frequency of the waves in both depths of water is the same—it was checked by using a stroboscope to 'stop' the motion; both sections were 'stopped' by the same stroboscope frequency. As the frequency has not changed then there must be a change in velocity of the wave when the water depth changes, the velocity being lower in the shallower water. In the figure the water waves were straight and parallel to the change in water depth step. What happens when the step is at an angle to the wave front? The waves are refracted (see Fig. 6.19). Is there any relationship between the angles of incidence and refraction?

$$\frac{\lambda_1}{PQ} = \sin i; \qquad \frac{\lambda_2}{PQ} = \sin r \qquad \text{(Fig. 6.18)}$$

Thus

$$\frac{\sin i}{\sin r} = \frac{\lambda_1}{\lambda_2}$$

But as the wavelengths are directly proportional to the wave velocities,

$$\frac{\sin i}{\sin r} = \frac{v_1}{v_2}$$

The ratio of the sine of the angle of incidence to the sine of the angle of refraction is a constant, the ratio of the velocities in the two depths. This ratio depends on the frequency of the water waves, as Fig. 6.19 shows. A change in frequency

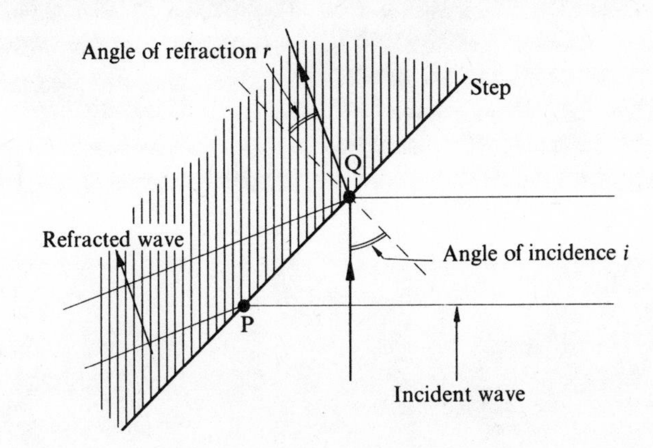

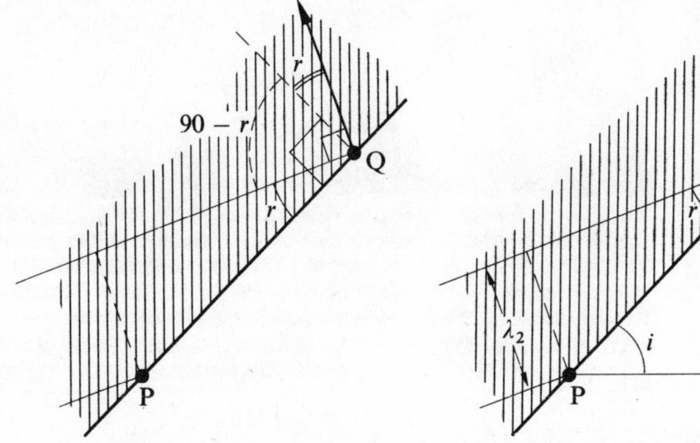

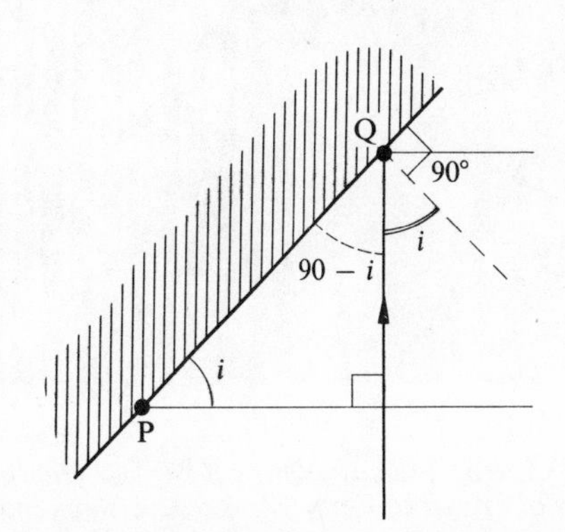

Fig. 6.18

means a change in the angle of refraction, the angle of incidence being kept constant. The different colours of light are refracted by differing amounts—an effect similar to that shown by the water waves.

Figure 6.20 shows what happens when water waves pass through apertures of different widths. When the wavelength is about the same as the size of the aperture considerable bending of the wave occurs when it passes through the aperture. This bending is called diffraction. The smaller the wavelength is the less the amount of diffraction.

Figure 6.22 shows what happens when the water waves produced by two point sources meet. The waves interfere and produce regions of calm water. This effect is called interference. As will be noticed from the two photographs, the effect changes if the wavelength is changed (or if the separation of the two sources is changed).

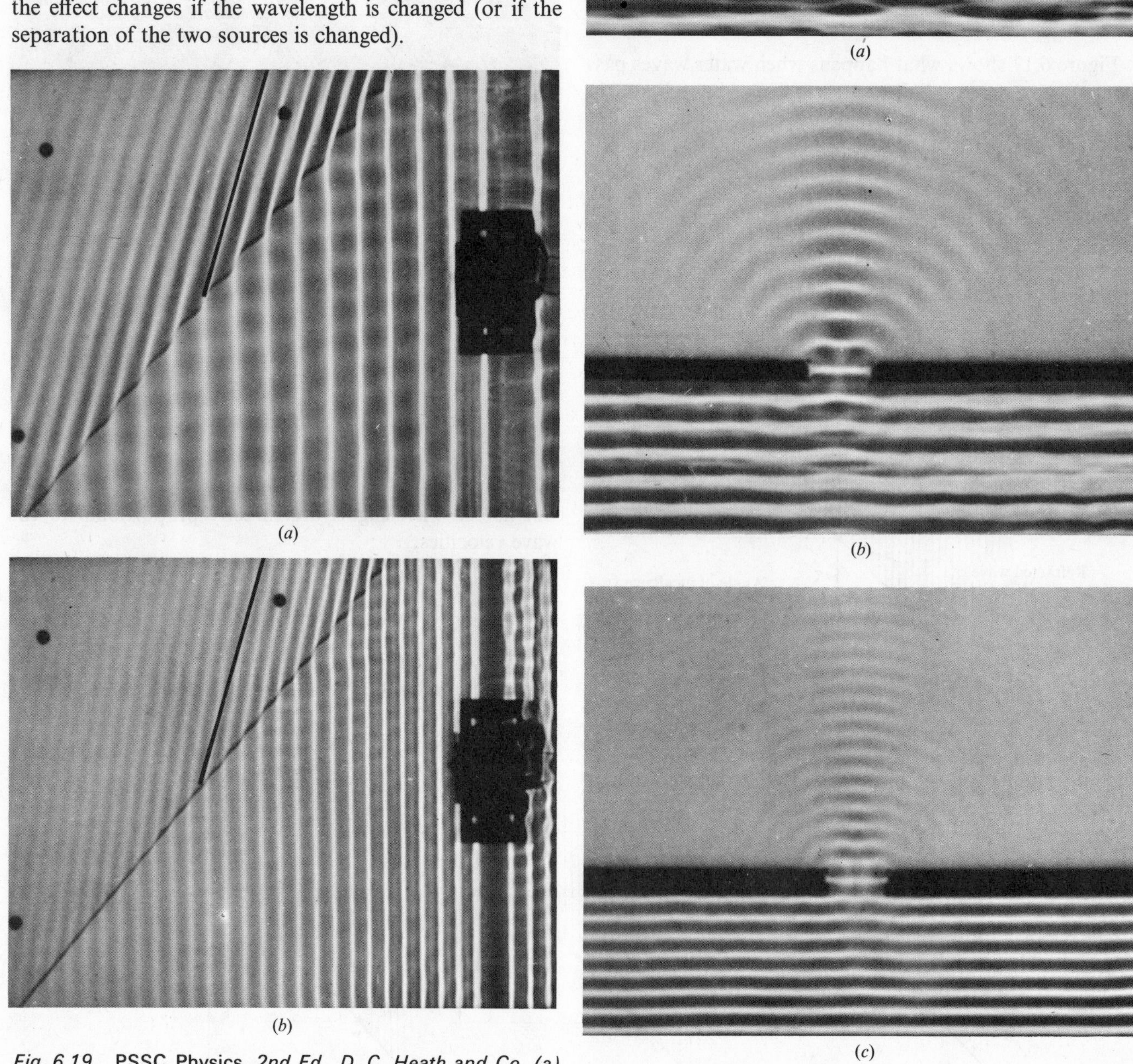

(a) (b)

Fig. 6.19 PSSC Physics, *2nd Ed.,* D. C. Heath and Co. *(a) Refraction of low-frequency waves. The black marker is placed parallel to the refracted waves. (b) Refraction of high-frequency waves.*

(a) (b) (c)

*Fig. 6.20 Three views of water waves of different wavelength passing through the same aperture (*PSSC Physics, *2nd Ed., D. C. Heath and Co.)*

Fig. 6.21 Moro Bay, California. (General Library, University of California, Berkeley, California.)

To summarize: water waves show reflection, refraction, diffraction, and interference. The aerial photograph, Fig. 6.21, of Moro Bay, California, if examined closely shows all these effects. Particularly noticeable is the diffraction of the waves on passing through the gap in the breakwater.

Interference of water waves

What happens when two waves meet? Do the waves pass right through each other? Do the waves bounce off each other? Do they cancel out each other and leave no disturbance?

Figure 6.22 shows what happens when two water waves meet. At certain positions the waves from the two sources have cancelled each other and give undisturbed water. It would seem natural to expect that where a wave trough met a wave crest they would cancel. Let us use this as a basis for an attempt at an explanation of the interference pattern. Figure 6.23(a) shows the waves moving out from a single point source, Fig. 6.23(b) shows waves from two point sources when we consider no interaction between the waves. If we take into account interaction between the two waves in (b) we arrive at the result shown in (c), regions of undisturbed water are produced (and regions of larger troughs and larger crests). Figure 6.23(d) shows diagram (b) extended to cover many more waves. Compare this with the photograph of water waves in Fig. 6.22(b), the wavelength and the source separation have been made the same. The diagram agrees with the photograph. The regions of undisturbed water occur when a crest meets a trough—when there is a path difference between the two waves reaching the point of half a wavelength (half a wavelength

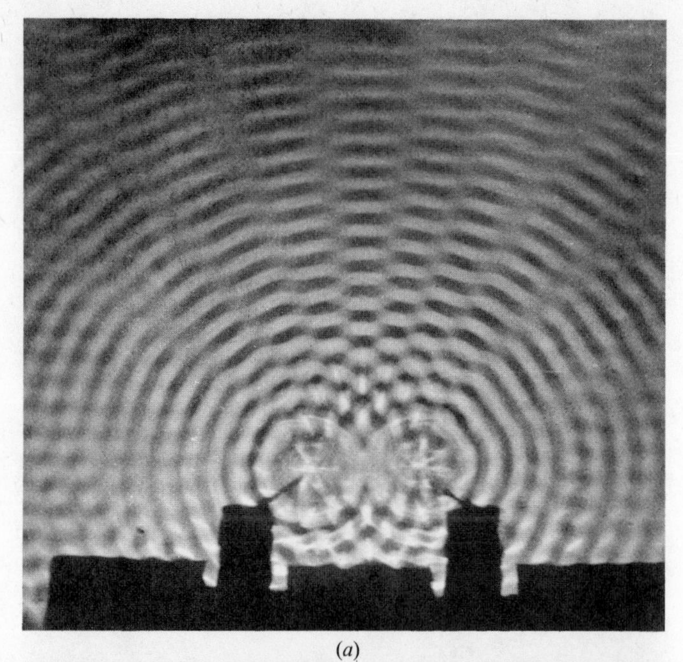

(a)

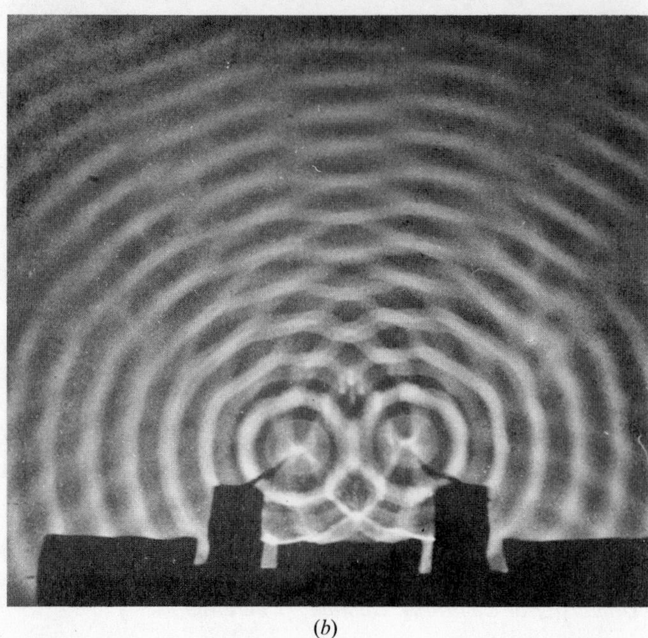

(b)

Fig. 6.22 PSSC Physics, *2nd Ed., D. C. Heath and Co. Interference pattern from two point sources: (a) separation of two sources about ten times the wavelength; (b) separation about five times the wavelength. (In fact the separation was kept constant and the wavelength changed.)*

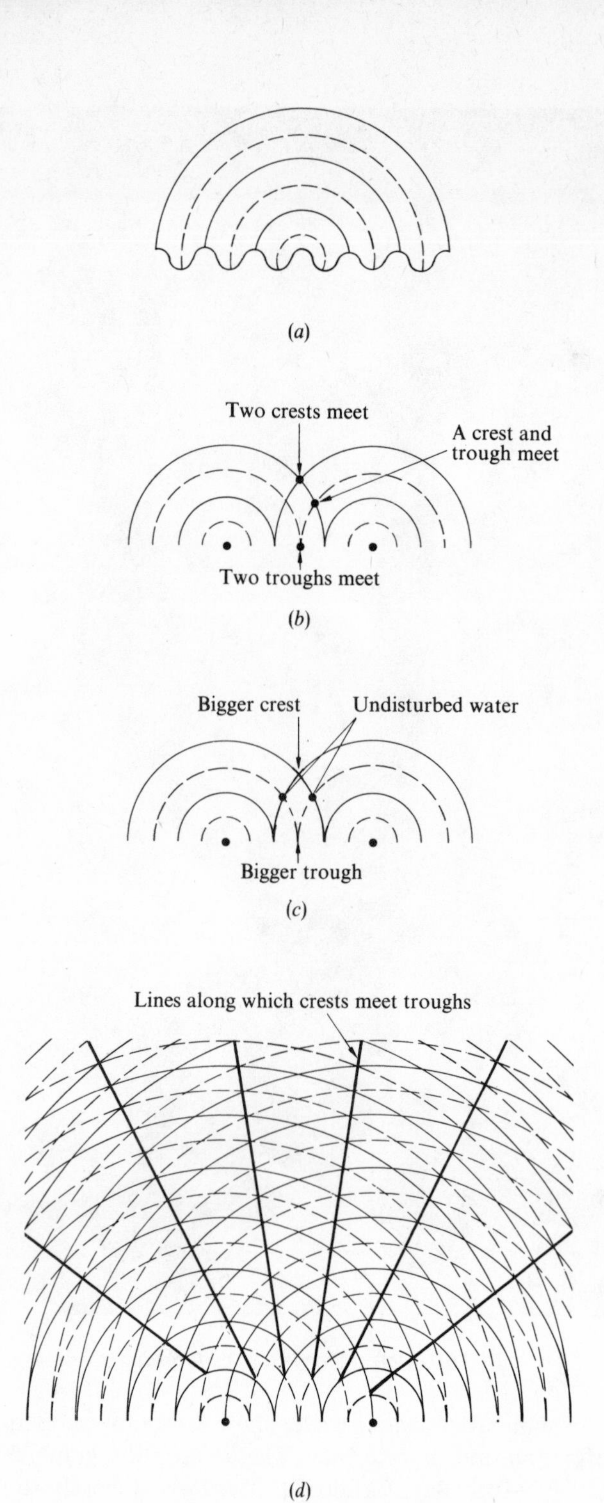

Fig. 6.23 (a) Waves spreading out from one point source. (b) Waves from two point sources without interference being considered. (c) The result of the two waves in (b) meeting when we consider interference. (d) Many waves from two point sources, without interference being considered.

is the distance between a wave trough and the next crest), or three half wavelengths, or five half wavelengths, or seven half wavelengths (Fig. 6.24), i.e., $(n + \frac{1}{2})\lambda$, where $n = 0, 1, 2, 3, 4$, etc. The regions of maximum disturbance, crest meeting with crest or trough meeting with trough, occur where the path difference is zero, one wavelength, two wavelengths, three wavelengths, i.e., $n\lambda$.

The photograph of the waves from two sources interfering was a picture of the waves at a single instant of time—the wave pattern produced by the water waves travelling out across the surface was frozen. Would the positions of undisturbed water change if we took a photograph at some later time? The answer is no—the undisturbed water positions do not change, provided we do not alter the vibrators producing the waves. The reason for this is that

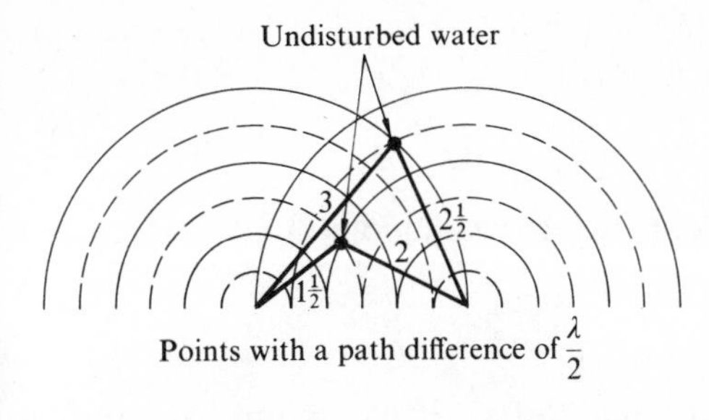

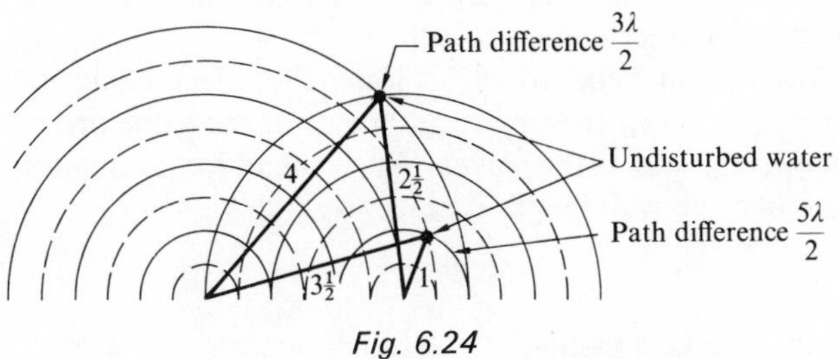

Fig. 6.24

the path differences to any point remain constant—for cancellation of the waves we do not necessarily need a trough to meet a crest, we can have any height of disturbance above the water surface cancelled by a corresponding depression below the water surface (Fig. 6.25). Similarly the maximum displacement positions do not change with time. In such a position the waves from the two sources are always giving displacements in the same direction, with the undisturbed water position the two waves always give displacements which are in opposite directions—whenever one gives a displacement above the surface the other wave gives a corresponding displacement below the surface. The pattern of maximum and minimum displacements remains stationary although it is formed by travelling waves. The interference pattern is sometimes called stationary or standing waves.

Interference patterns are produced whenever waves meet. A simple case is where a wave interferes with its own reflection from some barrier. The incident wave moving up to the barrier meets the reflected wave coming back from the barrier.

With plane waves incident on a plane reflector and the wavefronts parallel to the reflector the positions of undisturbed water are a series of lines parallel to the reflector. How far apart are these undisturbed water positions, these destructive interference positions? Consider A and B to be two successive destructive interference positions (Fig. 6.26). For the two waves arriving at A to cancel we must have

$$(W + Y) - (Z + X) = (n + \tfrac{1}{2})\lambda$$

For B to be the next position of cancellation we must have

$$(W + Y + X) - Z = (n + 1 + 1\tfrac{1}{2})\lambda$$

Subtracting the first equation from this one gives

$$2X = \lambda$$

X, the distance apart of successive destructive interference positions, is equal to half the wavelength.

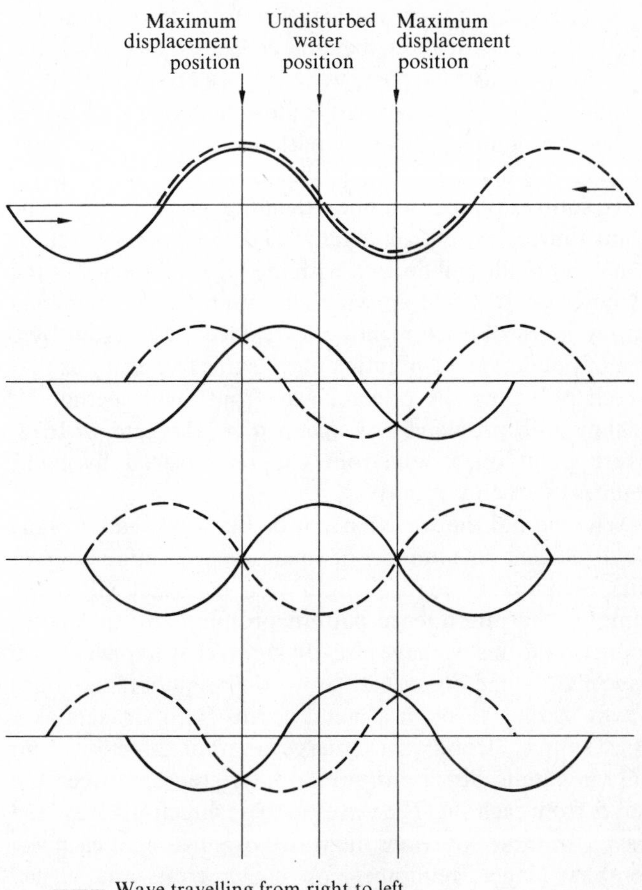

Fig. 6.25

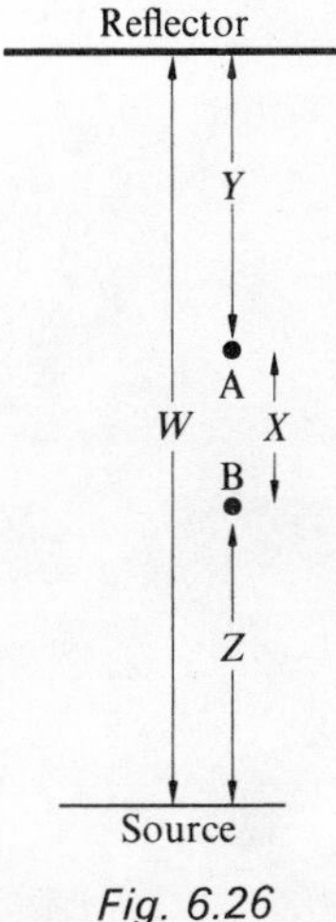

Fig. 6.26

Diffraction of water waves

Water waves bend round corners—they show diffraction. As Fig. 6.20 shows, the bending is most noticeable with slits when the slit aperture has a width equal to the wavelength of the waves. The bending is not, however, restricted to slits but occurs round any barrier. The diffraction of plane waves through slits does, however, reveal a significant fact—

when the slit is about one wavelength wide the resulting wave pattern is just about the same as that which is produced by a point source: the plane wave has been converted into a circular wave. What if the slit is more than one wavelength wide? The same diffraction pattern as of a plane wave passing through a slit can be produced by a line of point wave sources placed in a line extending across the slit. The point sources are all run together, i.e., when one produces a maximum they all do, and with the same frequency as the plane wave. It would appear that a useful way of tackling diffraction problems might be to replace a wavefront by a line of point sources vibrating at the same frequency as that which produced the original wavefront. This method of dealing with problems was invented by Huygens in 1678. Every point on a wavefront can be replaced by point sources of secondary waves.

What would the wave pattern be like if we had a plane wave incident on a number of equally spaced slits? For two slits, each about one wavelength wide, the result will be the same as the interference pattern produced by two point sources and this we have already met. What happens if we have many slits? Figure 6.27 shows the result of plane water waves passing through a metal comb. Each slit acts as a point source. Strong waves emerge in certain directions, and only in certain directions, due to interference between the waves from each slit. The waves in these directions are plane waves. In these directions the waves originating at each slit combine to give reinforcement, bigger crests and bigger troughs. The path difference between these waves must therefore be whole numbers of wavelengths. The path difference between waves from successive slits going in a direction at an angle θ to the straight through direction is CB (Fig. 6.28).

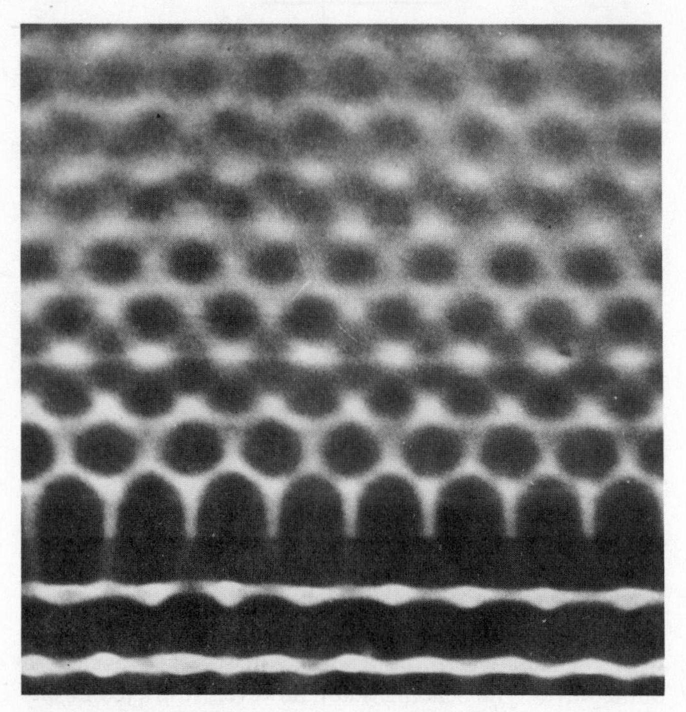

Fig. 6.27 Plane waves passing through a metal comb. (D. J. A. Dyson.)

But CB = AC sin θ and thus the condition for this direction to give reinforcement is

$$AC \sin \theta = n\lambda$$

AC is the distance between the midpoints of the slits, d, and thus

$$d \sin \theta = n\lambda$$

At angles other than those satisfying this relationship destructive interference occurs.

Waves can bend round objects, the effect being most noticeable when the object has a size of the same order as the wavelength of the waves. When the object is smaller in size than the wavelength it casts no shadow.

Light as a wave motion

Interference and diffraction are characteristics of wave motion—has light these characteristics? Two lamps side by side do not seem to give interference patterns, i.e., patches of darkness alternating with light patches; light does not seem to bend round corners but to travel in straight lines and give sharp shadows. With water waves the effects are, however, only noticeable when the two sources are only a matter of a few wavelengths apart or slits or objects are only a few wavelengths wide; perhaps the wavelength of light is very small, or perhaps light is not a wave motion.

With slits cut with a razor blade or a needle we can observe patterns that look very much like the interference and diffraction patterns produced by water waves. There is, however, one point which is different from water waves—we cannot use two separate sources of light and obtain interference, we must in some way divide the light from one source into two, or more, parts and look for the interference produced between the light from the separate parts. The two source interference obtained with water waves can be achieved with light if we put two slits in front of the single source and obtain interference between the light coming from the slits (Fig. 6.29). This experiment is known as Young's experiment as it was first performed by Young in 1803. The maxima (and minima), i.e., the fringes as they are termed, are farther apart for red light than blue-violet light. This would suggest that light does behave like a wave motion and that the wavelength depends on the colour of the light.

With the two slits there is just one line on the screen for which the light paths from the two slits are equal. This is the central bright fringe, constructive interference occurring. At all other points on the screen there is a path difference between the lengths of the light paths from the two slits. If the path difference is a whole number of wavelengths a bright fringe occurs, if the path difference is an odd number of half wavelengths there is a dark fringe. At any point, for example P in Fig. 6.30, the path difference is $S_2P - S_1P$. Because of the smallness of the wavelength of light the distance D has to be considerably greater than either d or x—we need path differences of the order of the wavelength and small multiples of this.

$$(S_1P)^2 = D^2 + \left(x - \frac{d}{2}\right)^2, \text{ by Pythagoras}$$

$$= D^2 + x^2 - xd + \frac{d^2}{4}$$

$$(S_2P)^2 = D^2 + \left(x + \frac{d}{2}\right)^2$$

$$= D^2 + x^2 + xd + \frac{d^2}{4}$$

$$(S_2P)^2 - (S_1P)^2 = 2xd$$

$$S_2P - S_1P = \frac{2xd}{S_2P + S_1P}$$

But because $D \gg x$ or d,

$$S_2P \approx S_1P \approx D$$

Hence path difference $= xd/D = \lambda$, for the first bright fringe. The fringes are equally spaced and xd/D is the spacing between the fringes.

If white light is incident on the two slits then the only position on the screen which will give a bright fringe for all the wavelengths is the centre fringe, no path difference. The central fringe will be white with on either side overlapping fringes for each wavelength in the light.

Another way of demonstrating interference with light is to use a thin film to produce a path difference between the light reflected from the top surface of the film and light from the lower surface (Fig. 6.31). The colours of a thin soap film are a good example of this type of interference. Where the thickness of the soap film is the right value to give constructive interference between the light reflected from the upper and lower film surfaces brightness occurs. Where the two reflections are out of step darkness occurs. With white light a series of fringes are produced because of the different

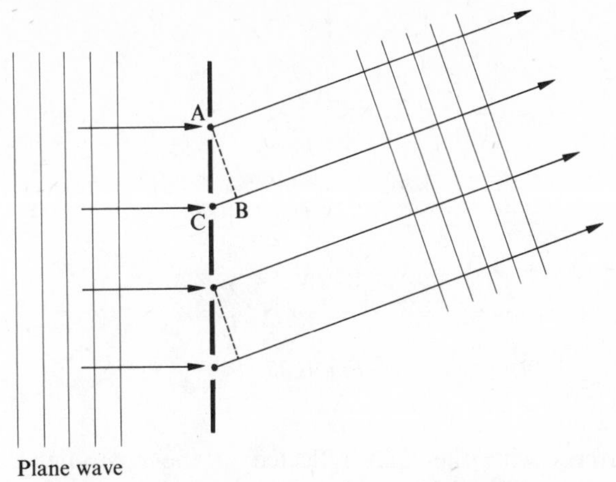

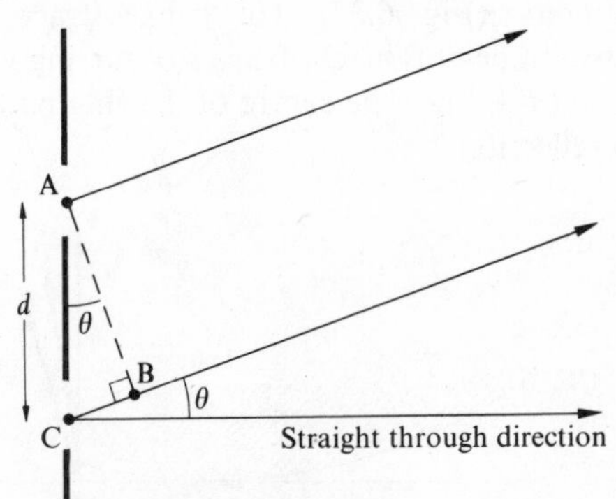

Fig. 6.28

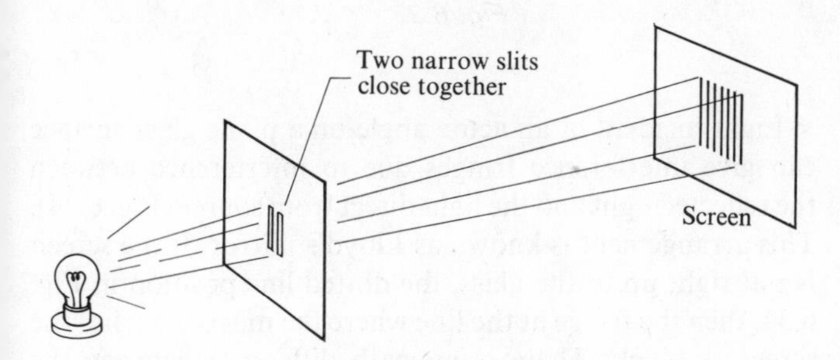

Fig. 6.29

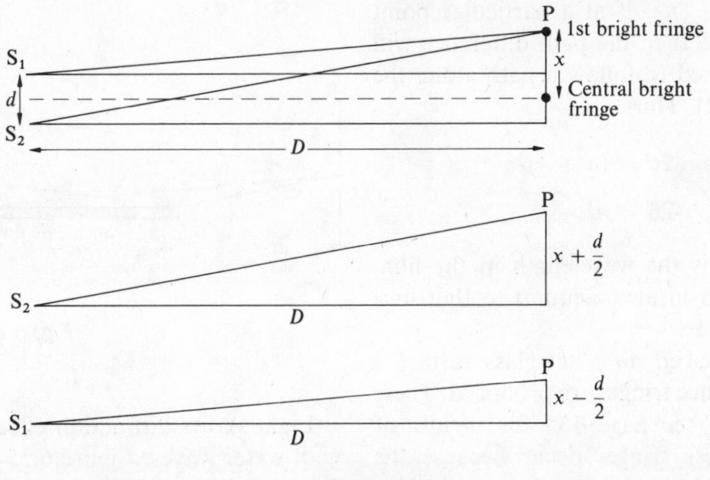

Fig. 6.30

Fig. 6.31 Interference in monochromatic light reflected from the two surfaces of a thin air film. (PSSC Physics, 2nd Ed., D. C. Heath and Co.)

wavelengths involved. The top of the soap film is thinner than the bottom of the film, because of the soap solution draining to the bottom. A significant point is that the thinnest part of the film is black—no reflection. With zero path difference we would expect brightness—instead as the thickness tends to zero we have darkness. The thickness of the soap film in the 'black' region is only that of a few soap molecules and is less than the wavelength of the light used. We can explain the 'blackness' by considering that one of the reflections introduces a phase change of $\lambda/2$, i.e., a crest becomes reflected as a trough. An experiment with a just one reflection (details given later on this page) at a glass surface show that this phase change occurs when reflection takes place at a less to more 'dense' medium, e.g., air to glass or air to soap film. No phase change occurs at a reflection from a more to less 'dense' medium, e.g., internal reflections at glass to air or soap solution to air. There is thus a bright fringe where the path difference between the light reflected from the two surfaces is $\lambda/2$, $3\lambda/2$, $5\lambda/2, \ldots (n + \frac{1}{2})\lambda$. Darkness occurs when the path difference is a whole number of wavelengths: λ, 2λ, $3\lambda, \ldots n\lambda$. If at a particular point the film has a thickness of d then the path difference will be $2d$, the light being assumed to follow a path along the normal to the film (Fig. 6.32). Thus

for brightness $\quad 2d = (n + \frac{1}{2})\lambda$

for darkness $\quad 2d = n\lambda$

The wavelength used here is the wavelength in the film. The ratio of the wavelength in air (vacuum) to that in a medium is the refractive index.

When a convex lens is placed on a flat glass surface a series of concentric interference fringes are produced. These are called Newton's rings (see page 135 for details of Newton's experiment). These fringes occur because the light reflected from the inside surface of the convex lens

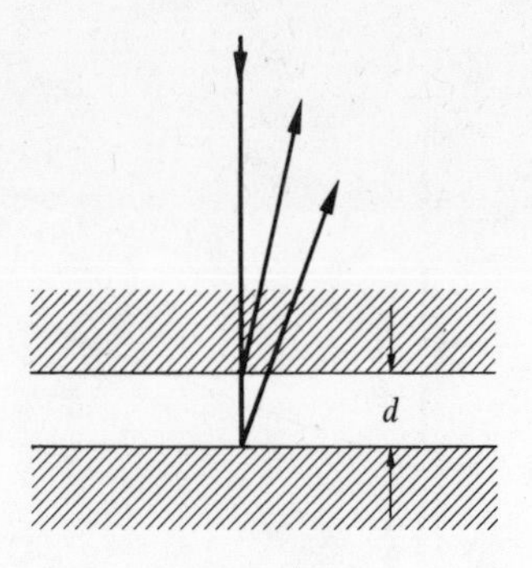

Fig. 6.32

interferes with the light reflected at the air-to-flat-glass boundary (Fig. 6.33). The fringes trace out the lines of constant height, bright fringes occurring wherever we have $2d = (n + \frac{1}{2})\lambda$. The centre of the ring pattern is black, i.e., no reflection.

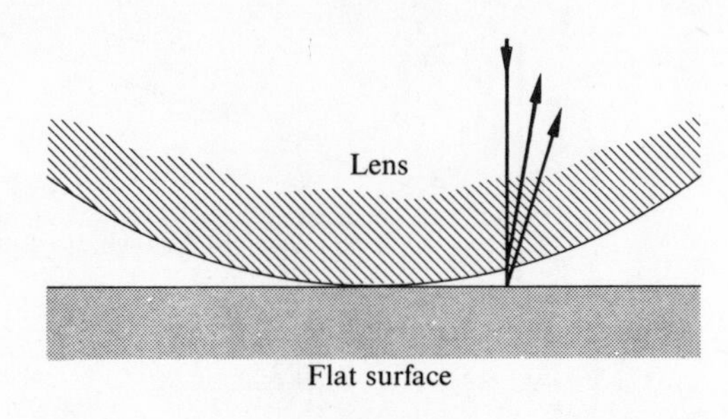

Fig. 6.33

Light incident at an acute angle on a plane glass surface can give interference fringes due to interference between the reflected light and the light direct from source (Fig. 6.34). This arrangement is known as Lloyd's mirror. If the screen is put right up to the glass, the dotted line position in Fig. 6.34, then the fringe at the line where the mirror touches the screen is black. There is no path difference between the reflected and direct light at this line—the destructive interference is because of a phase change of half a wavelength by the light on its reflection.

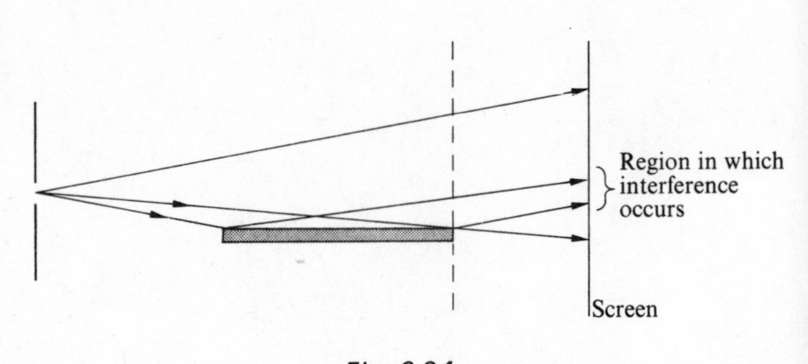

Fig. 6.34

Light shows diffraction effects similar to those obtained with water waves. Figure 6.35 shows the diffraction effects from (a) 1 slit, (b) 2 slits, (c) 3 slits, (d) 4 slits, (e) 5 slits,

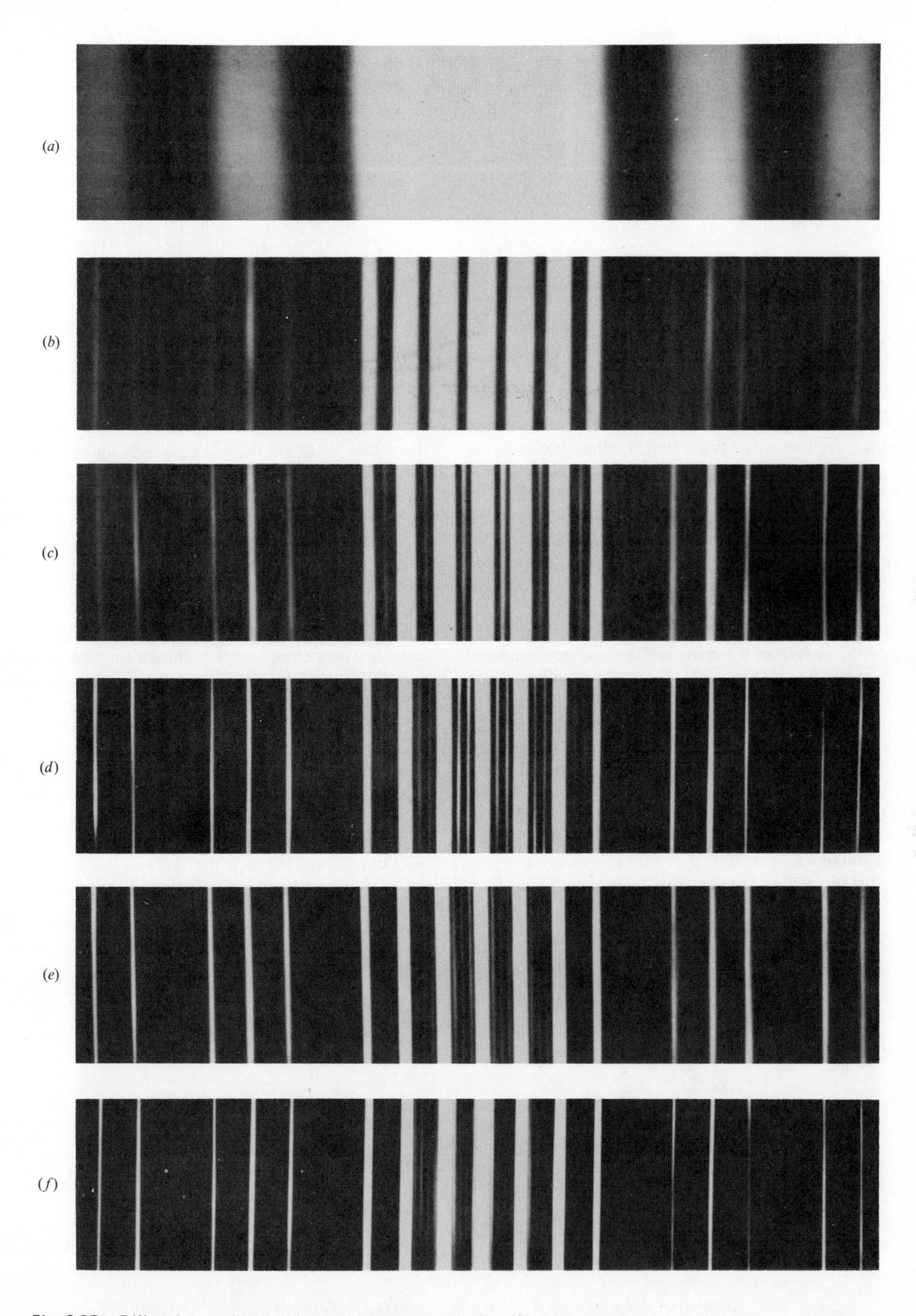

Fig. 6.35 Diffraction at slits: (a) 1 slit, (b) 2 slits, (c) 3 slits, (d) 4 slits, (e) 5 slits, (f) 6 slits. (R. S. Longhurst (1957). Geometrical and physical optics, *Longman.)*

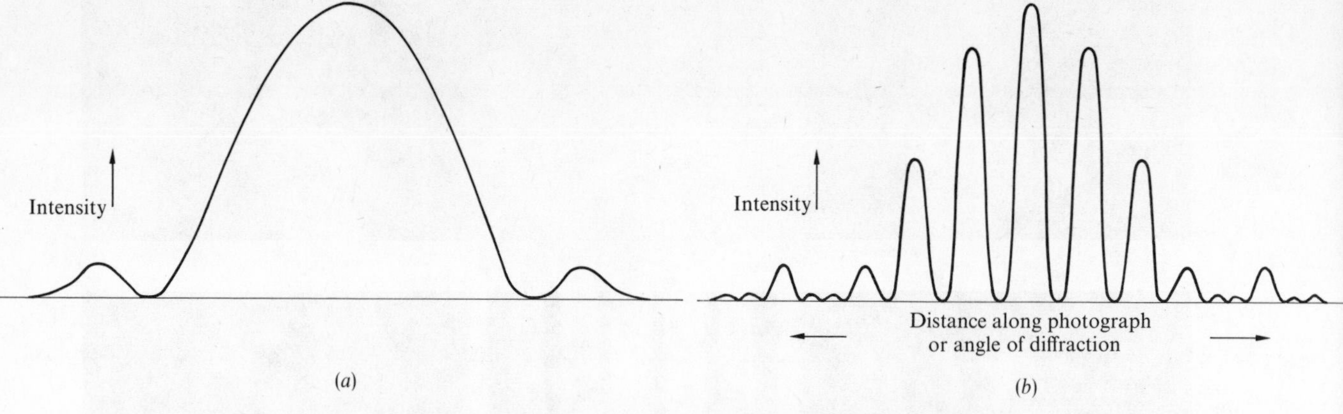

Fig. 6.36 Diffraction pattern (a) single slit (b) two slits.

(f) 6 slits when illuminated with plane waves of a single frequency. An immediate point that is noticeable from the photographs is that the positions of the maxima, where more than one slit is involved, do not change as the number of slits increases but the sharpness of the maxima increases. The intensities of all the maxima are governed by the diffraction pattern of the single slit (Fig. 6.36). The fringe maxima all lie within the maxima of the single slit diffraction pattern.

Consider first the single slit diffraction pattern. Each point on the wavefront in the slit, width b, can be considered as a source of waves (Fig. 6.37). These waves will interfere. Will there be a maximum, constructive interference, or a minimum, destructive interference, in a direction making an angle θ to the axis? For the light coming from the extreme edges of the slit, at this angle θ, there is a path difference of $b \sin \theta$, plane waves are incident on the slit. Light from a point half way along the slit will have a path difference of $(b/2) \sin \theta$ from the light coming from the two edges of the slit. What happens if $(b/2) \sin \theta = \lambda/2$? Light from half way along the slit will give destructive interference with light from the bottom of the slit. In fact for every point in the bottom half of the slit there is a point in the top half of the slit a distance $(b/2) \sin \theta$ from it. So when $(b/2) \sin \theta = \lambda/2$ there is destructive interference between points in the lower part of the slit and comparable points in the upper half of the slit. There will thus be a central maximum, because it is the same distance from comparable points in the lower and upper parts of the slit, and the first minimum will occur when $\sin \theta = \lambda/b$.

If instead of finding the condition for destructive interference between comparable points in the two halves of a slit we divide the slit width into quarters and find the condition for destructive interference between the light from comparable points in two quarters then we obtain the angle for the second minimum, i.e., $(b/4) \sin \theta = \lambda/2$ or $\sin \theta = 2\lambda/b$. The third destructive interference position can be found by dividing the slit width into sixths, $(b/6) \sin \theta = \lambda/2$ or $\sin \theta = 3\lambda/b$. In general, minima occur when $\sin \theta = n\lambda/b$.

With two slits, each of width b, a distance d apart (Fig. 6.38) the problem is much the same as the single slit diffraction. We have to consider the interference, at a particu-

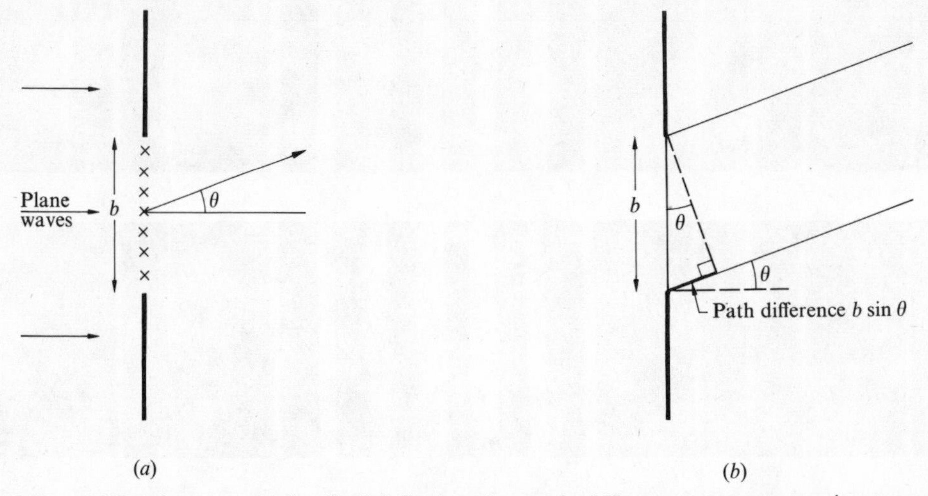

Fig. 6.37 Diffraction at a single slit. (a) Each point marked X represents a secondary source of waves. (b) For the light from secondary sources at each extremity of the slit the path difference is $b \sin \theta$.

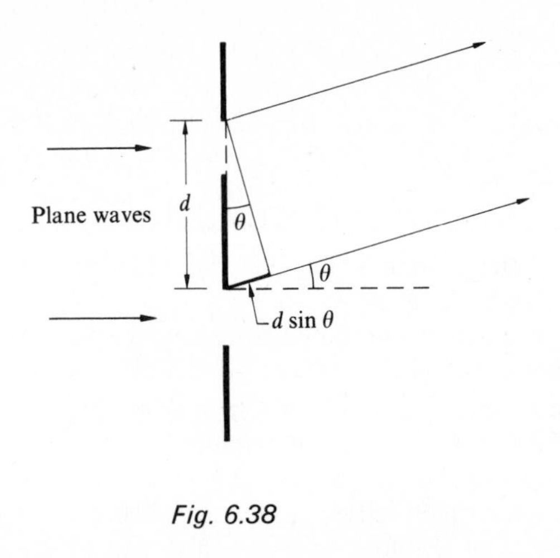

Fig. 6.38

lar diffraction angle, due to waves coming from two pieces of wavefront a distance d apart instead of immediately adjacent as in the single slit. The light coming from each slit will separately give the single slit diffraction pattern and so we can look for interference between the light emerging from the slits in the single slit maxima. For diffraction angles less than those given by $\sin\theta = \lambda/b$ we can get interference between the light coming from the two slits. At angles less than this, destructive interference will occur when $d\sin\theta = n\lambda/2$. Light from the top of the upper slit destructively interferes with that from the top of the lower slit, light from comparable points in the lower slit destructively interferes with light from comparable points in the upper slit. Constructive interference occurs for angles given by $d\sin\theta = n\lambda$; central maximum when $n = 0$, first maximum when $n = 1$, etc. This condition for maxima is equivalent to that obtained on page 147 for Young's slits, i.e., fringe spacing xd/D.

With three slits the same conditions will occur. When there is constructive interference between the light from the top two slits then at the same angle there will be constructive interference from the middle slit and the lower slit (Fig. 6.39). Thus the first maximum occurs when $d\sin\theta = \lambda$. With two slits there was only one point between this maximum and the central maximum when the light from the two slits cancelled. With three slits, if the path difference between the top two slits is $\lambda/2$ destructive interference will occur between the light from those slits; the third slit will, however, have a path difference of λ from the light coming through the upper slit and thus the intensity will be reduced but not zero for $d\sin\theta = \lambda/2$. Destructive interference in fact occurs when $d\sin\theta = \lambda/3$ and when $d\sin\theta = 2\lambda/3$. Between the zero maximum and the first maximum there appears a low intensity maximum, at $\sin\theta = \lambda/(2d)$, and two minima. The effect of this is to give sharper maxima.

Increasing the number of slits does not change the positions of the main maxima, they still occur for $d\sin\theta = \lambda$. Between this maximum and the zero maximum there are, however, many more angles for which destructive interference occurs—the result, very sharp maxima.

With very large numbers of slits, the arrangement is generally called a diffraction grating. Very sharp maxima

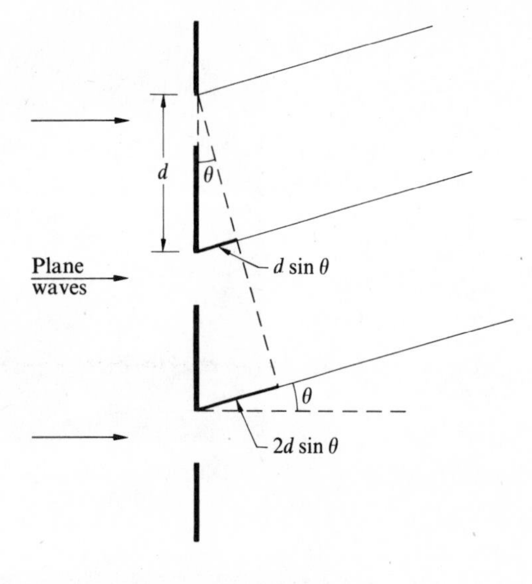

Fig. 6.39

are produced. The condition for maxima with slits, equally spaced with separation d between successive slits, on which plane waves are incident along the normal is

$$d\sin\theta = n\lambda$$

where n has the value 0, 1, 2, 3, 4, etc. The $n = 0$ condition is when $\theta = 0$ and the light from every slit follows the same path length to the maximum. This is called the zero order maximum. When $n = 1$ we have $d\sin\theta = \lambda$; there is a path difference of λ between the path lengths from successive slits. This is called the first order maximum. When $n = 2$ we have the second order maximum, the path difference being 2λ between successive slits. This is in fact the pattern shown by the water waves in Figure 6.27.

The angle at which the maxima are formed depends on the wavelength of the light. Thus if white light is incident on the grating, each maximum, with the exception of the zero order, appears as a spectrum. Diffraction gratings are much used in spectroscopy because of this ability to spread light out into a spectrum (see Fig. 7.6).

We do not need to have slits to show diffraction and interference effects; any shaped aperture will do. Figure 6.40(a) shows the pattern resulting from a random array of small circular holes—the pattern is a series of concentric circles. This is the pattern you see if you view a street lamp through a mist—a halo pattern. Figure 6.40(b) shows the diffraction pattern if the holes are in a regular array. The circular pattern has broken down into discrete maxima. (Similar patterns will be met in chapter 9.)

Blooming of lenses

Surfaces can be made non-reflecting, for one wavelength, by coating them with a thin layer of a transparent material. The thickness of the material must be such that the light reflected from the front surface of the film destructively

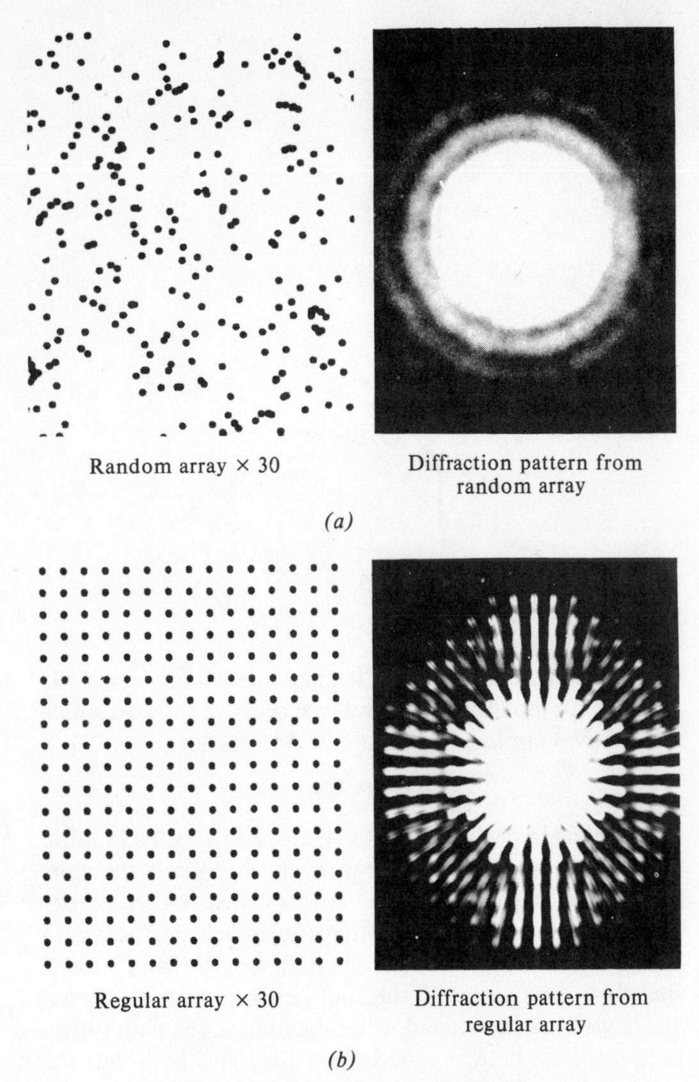

Fig. 6.40 (Ealing Scientific Ltd.)

interferes with that from the rear surface (Fig. 6.41). This will occur when the material coating the surface has a refractive index intermediate between that of the glass, if a glass surface is being coated, and air. It is assumed that the surface is in air. The first reflection is at an air-to-coating interface, i.e., less to more optically 'dense', and thus a phase change occurs. The second reflection occurs at the coating-to-glass interface, i.e., less to more optically

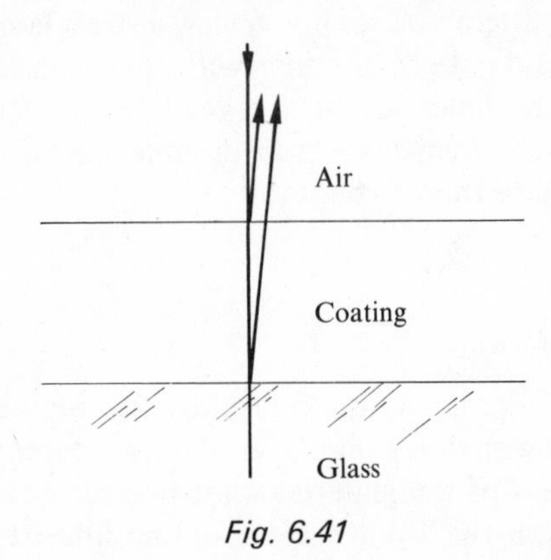

Fig. 6.41

'dense', and thus a phase change occurs. The two phase changes are the same and thus destructive interference occurs when the coating thickness is $\lambda/4$, a path difference of $\lambda/2$ between the two reflected rays. For the destruction to be complete the two reflected rays must be of equal intensity. This can be achieved by making the refractive index of the coating equal to the square root of the refractive index of the glass.

Camera lenses are coated to reduce the amount of light reflected at the lens surfaces and so give more light transmitted through to the film surface; interference conditions giving zero reflection give maximum transmission, and vice versa—energy is conserved. The coating is known as 'blooming'. Such lenses have a purplish hue by reflected light because the coating has a thickness which gives destructive interference only for one wavelength and this is chosen near the centre of the visible spectrum, the light being reflected is thus, in the main, the red and the blue.

Resolving power

When light passes through an aperture we obtain a diffraction pattern instead of sharply defined images. The aperture could be the objective lens of a telescope or a microscope. If we have two objects we will obtain two diffraction patterns. How close can these patterns be for us to say that there are two objects and not just one? The patterns will overlap and we will see the combined effect of both. There will not generally be interference between the two lots of diffracted light because they come from different objects or parts of an object. Figure 6.42 shows the diffraction patterns from three point sources produced by lenses of different diameters. As the diameter of the lens increases the resolution improves. Rayleigh suggested a resolution criterion—two sources are resolved if the central maximum from one diffraction pattern falls no nearer than the first minimum of the other pattern. Figure 6.43 shows this position. Looking at the photographs of the point sources makes this seem a reasonable criterion.

For slits the first minimum is at an angle θ given by $\sin\theta = \lambda/b$, where b is the width of the slit. If the light from two objects passes through a slit the minimum angle between their two central maxima at which resolution occurs is thus given by $\sin\theta = \lambda/b$. If circular holes are used instead of slits the angle is given by $\sin\theta = 1{\cdot}22\ \lambda/b$. The pupil of the eye has a diameter of about 3 mm and thus for visible light the minimum angle of resolution for the eye is

$$\sin\theta = \frac{1{\cdot}22 \times 5 \times 10^{-7}}{3 \times 10^{-3}} = 0{\cdot}00020$$

This is about 50 seconds. Optical defects in the eye do not enable this limit to be achieved.

The Jodrell Bank radiotelescope (Fig. 12.14) has a paraboloidal mirror about 80 m in diameter. At a radio wavelength of 0·2 m the minimum angle of resolution is

$$\sin\theta = \frac{1{\cdot}22 \times 0{\cdot}2}{80} = 0{\cdot}00305$$

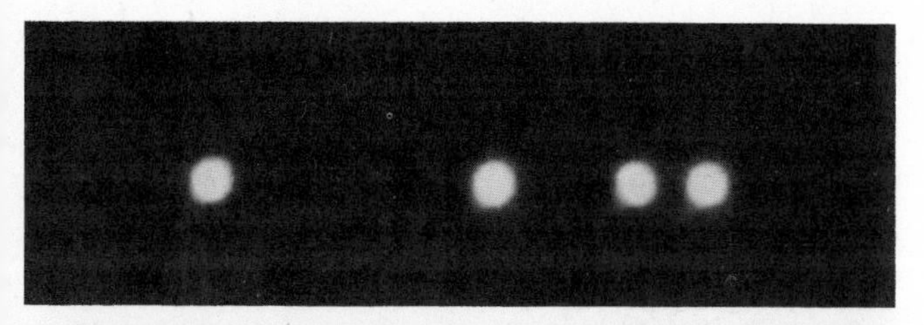

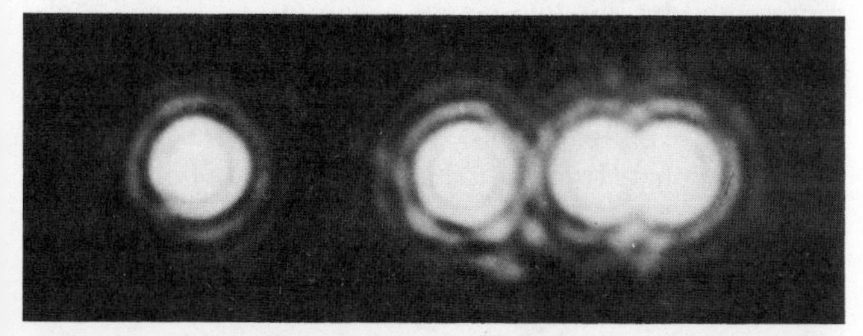

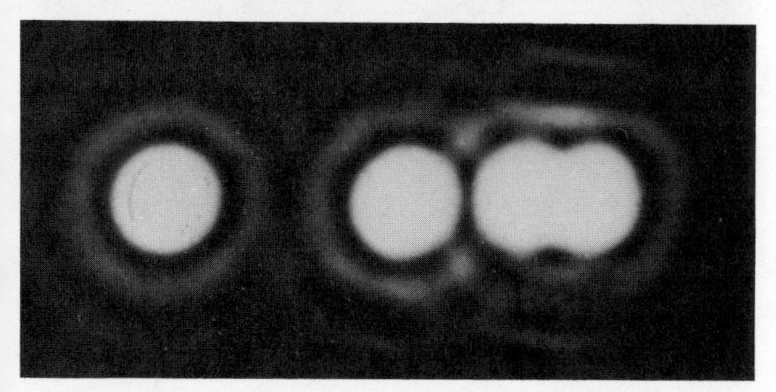

Fig. 6.42 Diffraction pattern from three point sources produced by lenses of different diameters. (F. W. Sears and M. W. Zemansky (1960). College physics, *3rd Ed., Addison Wesley.)*

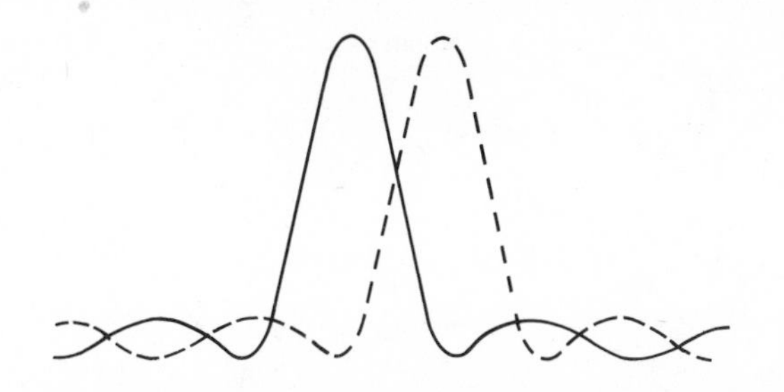

Fig. 6.43 Diffraction patterns due to two sources.

This large telescope is not as good for resolving radio sources as the eye is for resolving light sources.

Coherent and incoherent light

The simplest way of showing interference is to have two point sources separated by a distance comparable to a few wavelengths. The sources should emit single frequency waves and there should be a constant phase difference between the two sets of waves. The phase difference can be zero, the vital point is that it should not fluctuate. Such sources are said to be coherent.

In the case of light we have emission with a finite frequency bandwidth, i.e., a spread of frequency instead of precisely one frequency. The effect of this is to give changes of phase. If the bandwidth is Δf then for two equally monochromatic but independent sources the frequency components at each end of the frequency band will get out of phase by 180° in a time of $1/\Delta f$. This is called the coherence time. For time intervals less than the coherence time we can think of the phase difference between the two sources remaining essentially constant. For visible light from say a discharge tube the coherence time is of the order of 10^{-9} to 10^{-8} s, a bandwidth of 10^8 to 10^9 Hz for frequencies of the order of 10^{15} Hz (frequency monochromatic to about 1 part in a million). Thus interference with two separate light sources can produce fringes but the fringe positions change every 10^{-8} s. Because the eye cannot distinguish events less than 1/20 s apart no fringes are seen. These fringes can be detected, using photomultipliers which respond much more rapidly than the eye (Brown and Twiss, *Nature*, **178**, 1447, 1956).

Two coherent light sources can be obtained by dividing the light from one source into two parts and looking for interference between these two parts. Both sources will have the same fluctuating phase changes.

Lasers can have coherence times of the order of a hundredth of a second or even a second.

Transverse or longitudinal?

Sound is a longitudinal wave, the direction of the particle displacement being along the direction of the wave motion. There is another type of wave motion, transverse waves in which the direction of the particle displacement is at right angles to the direction the wave is travelling (Fig. 6.44). What type of wave motion are light waves? Huygens assumed in his wave theory that light was like sound waves and thus

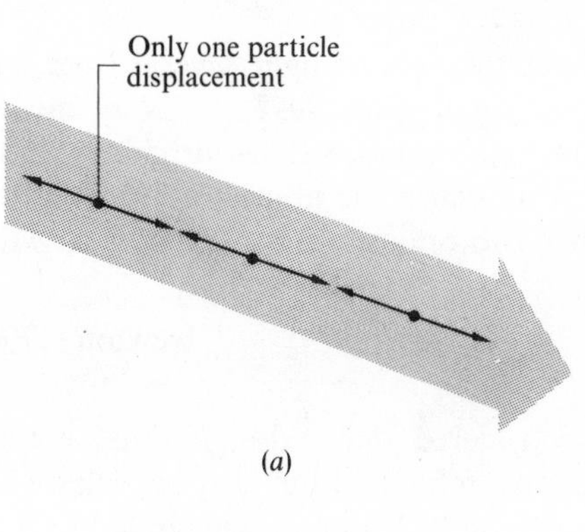

(*a*)

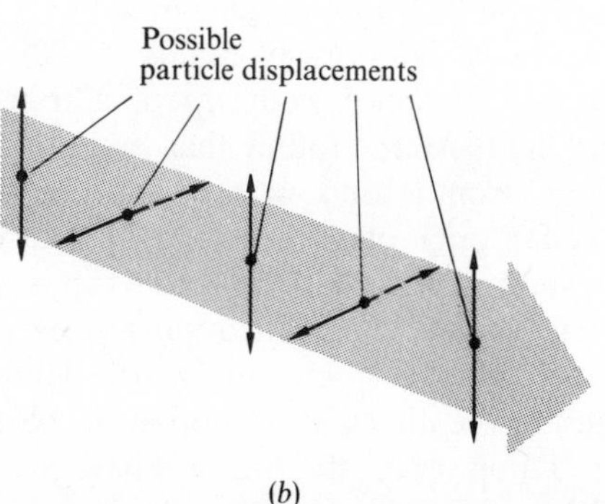

(*b*)

Fig. 6.44 (a) Longitudinal wave. (b) Transverse wave.

a longitudinal wave. Young also made the same assumption. The space between the sun and the earth was filled with an ether through which the compression wave of light could travel. An ether was considered necessary because it was realized that sound requires a medium in which to produce its compressions and thus light must have a medium.

The idea of the transverse wave nature of light arose from experiments with calcite crystals, known as Iceland Spa. Bartholinus reported in 1669 that this crystal exhibited double refraction, i.e., two refracted rays are produced from just one incident ray.

> 'As my investigation of this crystal proceeded there showed itself a wonderful and extraordinary phenomenon: objects which are looked at through the crystal do not show, as in the case of other transparent bodies, a single refracted image, but they appear double. . . .'
>
> E. Bartholinus (1669). *Experimenta Crystalli Islandici.*

One of the refracted rays was found to obey the 'normal' laws of refraction, i.e., the incident ray, the refracted ray, and the normal were all in the same plane and the ratio $\sin i/\sin r$ was a constant; the other ray did not and was thus called the 'extraordinary' ray.

Huygens, in his wave theory of light, was unable to offer any explanation for the two refracted rays—he still considered light to have a longitudinal nature. With a longitudinal wave there is only one mode of particle displacement, along the direction in which the wave is travelling, and thus there was difficulty in seeing any reason why the light should split into two refracted rays. Newton, with his corpuscular theory, offered the explanation:

> 'Have not the rays of light several sides, endued with several original properties? . . . Every ray of light has therefore two opposite sides, originally endued with a property on which the unusual refraction depends, and the other two opposite sides not endued with that property. . . .'
>
> Newton (1704). *Opticks.*

Newton considered that a longitudinal wave could not explain double refraction but that particles whose properties depended on their orientation could.

It was not until the work of Malus in 1808 that double refraction was explained in terms of a transverse wave nature for light. Malus found that when he viewed the light reflected from a window through a doubly refracting calcite crystal only one image was produced. Double refraction and this effect with reflection can be explained in terms of transverse waves. With a transverse wave there are a number of possible modes of particle vibration, all at right angles to the direction of motion of the wave. When double refraction occurs the two rays have planes of vibration at right angles to each other. Light reflected from a sheet of glass is, at one particular angle of incidence, only at one plane of vibration, the plane at right angles being refracted. The plane of vibration is generally referred to as the plane of polarization and light vibrating in just one plane is called plane polarized (Fig. 6.45(c)). Unpolarized light consists of all possible planes of vibration (Fig. 6.45(a)). Diagrammatically the refraction and reflection of light can be represented by Fig. 6.45(b). Unpolarized light can be represented by just two planes of vibration at right angles to each other—all the other components can be resolved into those two directions.

In 1852 Herapath discovered that crystals of iodoquinine sulphate transmitted just one plane of polarization,

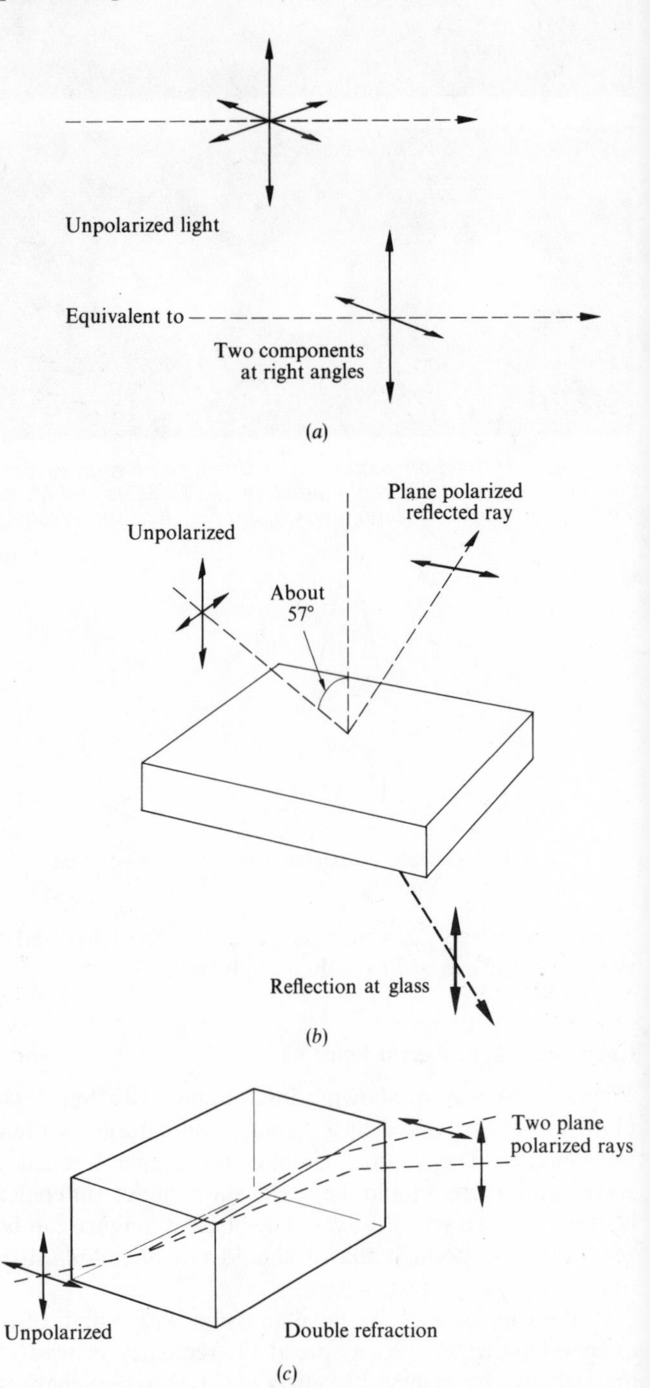

Fig. 6.45

absorbing other planes. In 1938 Land, while still an undergraduate at Harvard, used this discovery of Herapath to produce the material now known as 'Polaroid'. Land instead of trying to produce large crystals of the very fragile iodo-quinine sulphate, made a material with large numbers of small crystals all aligned in the same direction. The manufacturing process is briefly—

a sheet of plastic consisting of long hydrocarbon chains is stretched to line up the molecules;

the sheet is dipped in iodine, iodine atoms becoming attached to the hydrocarbon chains;

the result is chains of crystals absorbing all but one plane of polarization.

If unpolarized light is incident on a sheet of 'Polaroid' the transmitted light is plane polarized. If a second sheet of 'Polaroid' is placed in the path of the plane polarized light then no light is passed if the crystal axes of the second 'Polaroid' are at right angles to those of the first 'Polaroid'. If the crystal axes of the two 'Polaroids' are parallel light emerges from the second 'Polaroid'. Plane polarized light only passes through a 'Polaroid' if the crystal axes are in the correct direction for that plane (Fig. 6.46).

Another piece of evidence exists for the transverse nature of light—two rays of light polarized at right angles to each other will not interfere under conditions where they would interfere if polarized in the same plane. Experiments showing this null effect were first performed by Arago and Fresnel in 1819.

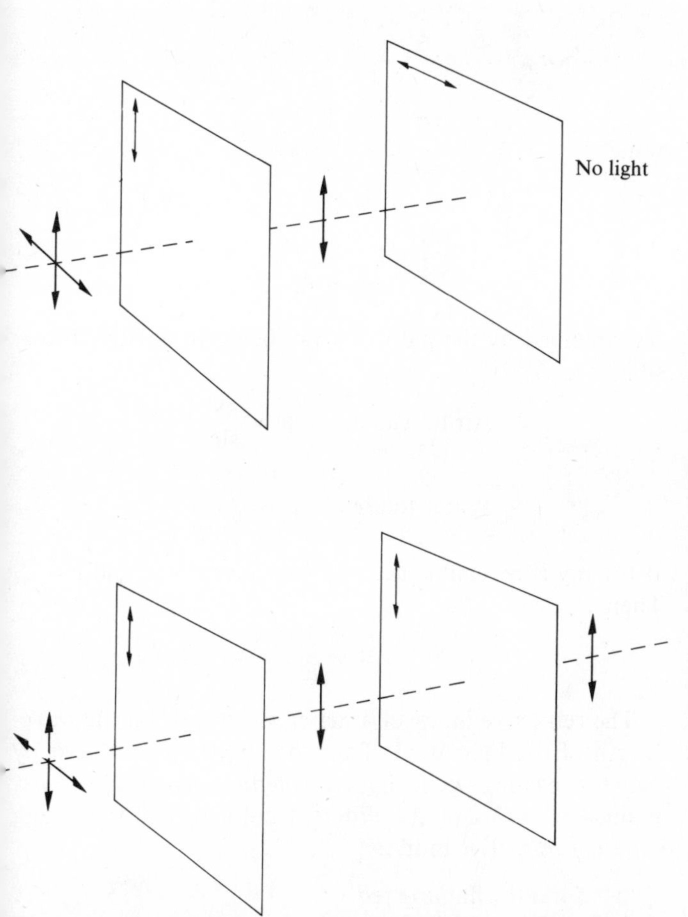

Fig. 6.46

'. . . It has been known for a long time that if one cuts two very narrow slits close together in a thin screen and illuminates them by a single luminous point, there will be produced behind the screen a series of bright bands resulting from the meeting of the rays passing through the right-hand slit with those passing through the left. [Description of how and why Arago used a pile of reflecting plates to produce plane polarized light. A set of plates was put in front of each slit.] . . . The light transmitted by these plates was almost completely polarized when the angle of incidence was about thirty degrees. And it was exactly at this angle that the plates were inclined when they were placed in front of the slits in the copper screen.

When the two planes of incidence were parallel, i.e., when the plates were inclined in the same direction, up and down, for instance, one could distinctly see the interference bands produced by the two polarized pencils. In fact, they behave exactly as two rays of ordinary light. But if one of the piles be rotated about the incident ray until the two planes of incidence are at right angles to each other, the first pile, say, remaining inclined up and down while the second is inclined from right to left, then the two emergent pencils will be polarized at right angles to each other and will not, on meeting, produce any interference bands.'

Arago and Fresnel (1819),
Annales de Chimie et de Physique, 288.

Light incident at a particular angle of incidence on a dielectric such as glass, or mica as Arago used, is reflected completely plane polarized. The transmitted, i.e., refracted, light consists of 100 per cent of the light with its plane at right angles to the reflected light and, in the case of glass, about 85 per cent with its plane in the same direction as the reflected light—only a small percentage is reflected. However, if the light passes through a number of plates the intensity of the reflected component is reduced at each boundary, because some is reflected, and the light with its plane at right angles to the reflected component remains undiminished. Arago found that 15 plates of mica were necessary to give transmitted light with virtually only one plane of polarization. The reflected light has in fact its electric vector (see chapter 12) at right angles to the plane of incidence, the transmitted light has its electric vector in the plane of incidence. Rotating a plate, or pile of plates, about the incident ray while still maintaining the correct angle of incidence rotates the plane of polarization of the transmitted (and reflected) ray.

6.4 Reflection and refraction

The law of reflection, the angle of incidence is equal to the angle of reflection (Fig. 6.47), was probably first discovered

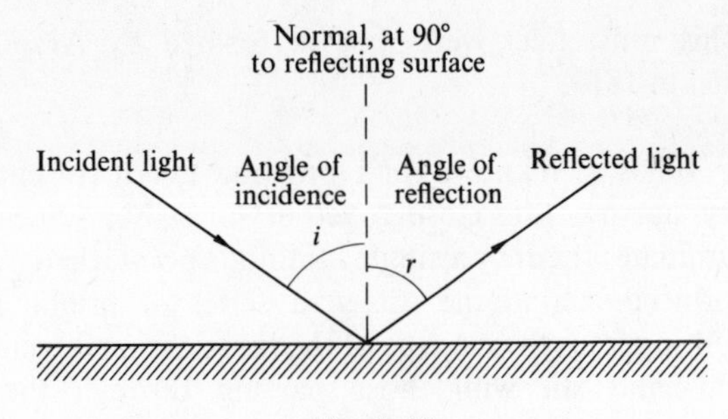

Fig. 6.47

as far back as the Roman empire. The law was certainly known in the second century A.D. The statement that the incident ray, the normal to the surface at the point where light meets the surface, and the reflected ray all lie in the same plane does not appear to have been given until about A.D. 1100.

The law of refraction was not discovered until much later. Ptolemy in the second century A.D. gave a table of angles of incidence and angles of refraction. The following is an extract:

Angle of incidence	*Angle of refraction*	
	Air to water	*Air to glass*
10°	8°	7°
20°	15½°	13½°
30°	22½°	20½°
40°	28°	25°
50°	35°	30°
60°	40½°	34½°
70°	45°	38½°
80°	50°	42°

Ptolemy incorrectly concluded, even for his own data, that the ratio of the angle of incidence to the angle of refraction was a constant for the same pair of media. Alhazen in A.D. 1100 realized that the incident ray, the normal to the surface at the point where the light meets the surface, and the refracted ray all lie in the same plane. Kepler, early seventeenth century, tried to find a law relating the incident angle and the refracted angle, but without success. He did, however, discover that for light passing from a medium such as glass into air there is a certain angle beyond which no refracted ray occurs but internal reflection only occurs. The angle at which this occurs is known as the critical angle (Fig. 6.48). The law that the ratio of the sine of the angle of

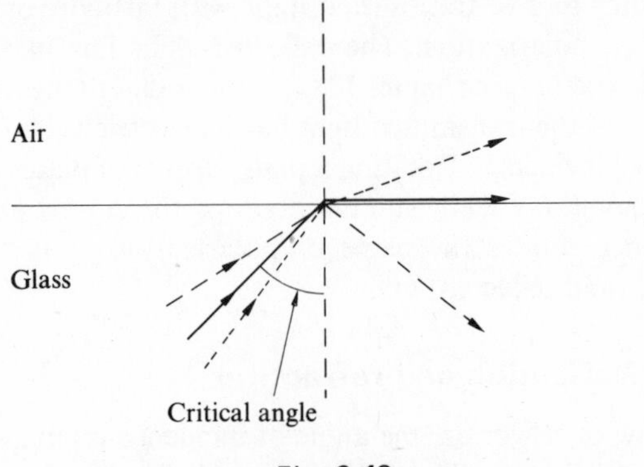

Fig. 6.48

incidence to the sine of the angle of refraction is a constant for a particular pair of media was first discovered by Snell in 1626, though not expressed in that form. The first expression of the law in terms of sines was by Descartes. He saw Snell's work prior to publication and then published the law as his own. The law offered no explanation as to why refraction occurred but merely was a way of writing the experimental results.

Using Ptolemy's results for air to water:

Angle of incidence i	*Angle of refraction* r	i/r	$\sin i/\sin r$
10°	8°	1·25	1·25
20°	15½°	1·29	1·28
30°	22½°	1·33	1·31
40°	28°	1·43	1·37
50°	35°	1·43	1·33
60°	40½°	1·48	1·33
70°	45°	1·56	1·33
80°	50°	1·60	1·29

The ratio of the sines is known as the refractive index (μ). The refractive index in going from air to water is the reciprocal of the refractive index for light going from water to air. For example if the refractive index in going from air to water is 4/3 then the refractive index in going from water to air is 3/4. This is borne out by experimental results and can

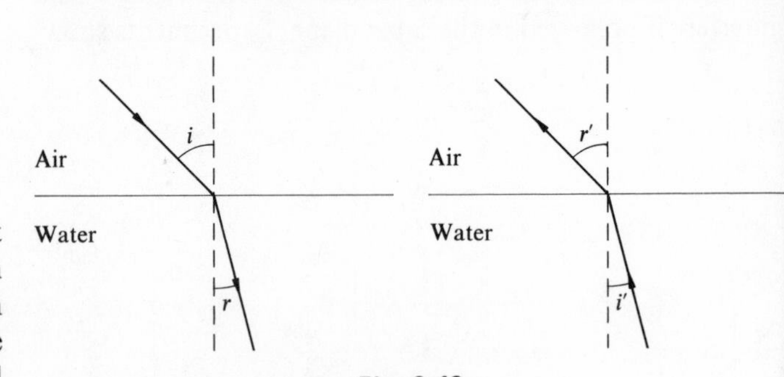

Fig. 6.49

be explained by the path of a ray being completely reversible (Fig. 6.49).

$$\text{Air to water:} \quad \mu = \frac{\sin i}{\sin r}$$

$$\text{Water to air:} \quad \mu' = \frac{\sin i'}{\sin r'}$$

If the ray is reversible then we can have $i = r'$ and $r = i'$. Then

$$\mu = \frac{1}{\mu'}$$

The refractive index of a material depends on the wavelength of the light used. Thus the spectrum produced by Newton passing white light through a prism can be explained in terms of the different colours of light having different refractive indices.

'. . . First the flaming red,
Springs vivid forth; the tawny orange next;
And next delicious yellow; by whose side

Fell the kind beams of all-refreshing green.
Then the pure blue, that swells autumnal skies,
Ethereal played; and then, of sadder hue,
Emerged the deepened indigo, as when
The heavy-skirted evening droops with frost;
While the last gleamings of refracted light
Died in the fainting violet away.'

J. Thomson (1727). *To the memory of Sir Isaac Newton.*

Mirrors and lenses

You look in a mirror and you see an image of yourself staring back at you. You lift your right hand—the image lifts its left hand. In Lewis Carroll's book *Through the looking glass* a little girl, Alice, looks at the image of her room in a mirror and then climbs through the glass into the image room. What are images that we see in a mirror?

Light coming from the object is reflected by the mirror and so to the eye (Fig. 6.50). The eye sees the object by

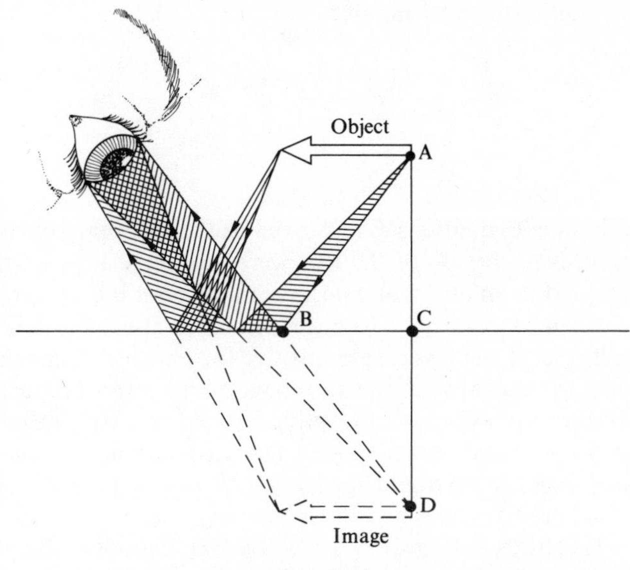

Fig. 6.50

means of rays of light that are bent by the mirror. We call the object we see an image because we see it by bent rays of light. If we think of light as only travelling in straight lines then the object, or rather its image, appears to be behind the mirror. No light actually passes through the mirror to the image. Such an image is called a virtual image. Because of experience we do not expect to be able to touch the image—we recognize it as an illusion. A baby, however, does not have this ability to recognize virtual images as illusions: it considers them to be as real as the objects and is disconcerted when they cannot be touched.

Because the angle of reflection is equal to the angle of incidence, angle ABC equals angle CBD. The line of ADC must cut the mirror surface at right angles—a ray of light going along the path AD will be reflected back along the same path and thus appear to come from the image. The triangles ABC and DBC are identical—thus AC equals CD. The image appears to be as far behind the mirror as the object is in front.

Mirrors do not have to be plane to produce images. The images produced by curved mirrors can, however, be a different size to the object, inverted or the right way up, behind or in front of the mirror. Plane mirrors only produce images of the same size as the objects, only the right way up and only behind the mirror. Images produced in front of a mirror are called real images—light does actually go to those points and the image can be formed on a screen. Figure 6.51 shows the arrangement by which a concave

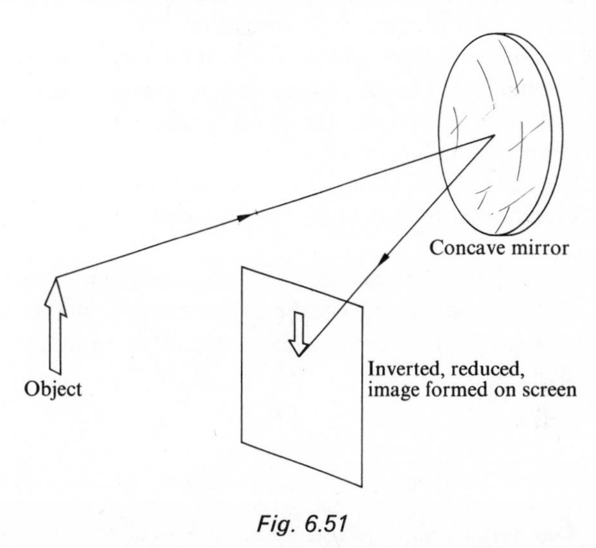

Fig. 6.51

mirror produces a real image. If the object is a long way off, so far that the light reaching the mirror can be considered to be parallel rays, then the image is produced at a distance called the focal length from the mirror. This definition applies regardless of whether the image is real or not.

We can produce images with lenses. Depending on the shape and material of the lens and the distance of the object from the lens, the image can be real or virtual. The lens in a cine projector is used to produce real images on a screen. Such images are inverted and magnified versions of the objects on the cine film.

The following table gives the pattern in which the images occur for both lenses and spherical mirrors. f is the focal length.

	Object distance	*Image*
Concave mirror	Less than f	Erect, magnified, virtual
	Between f and $2f$	Inverted, magnified, real
	Outside f	Inverted, diminished, real
Convex mirror	All distances	Erect, diminished, virtual
Concave (diverging) lens	All distances	Erect, diminished, virtual
Convex (converging) lens	Less than f	Erect, magnified, virtual
	Between f and $2f$	Inverted, magnified, real
	Outside $2f$	Inverted, diminished, real

All rays from a point on an object pass through or appear to diverge from the corresponding point of the image. When they pass through the point the image is real, when they only appear to then the image is virtual.

Ray diagrams

In his book *Compleat system of optics*, published in 1738, R. Smith, who was one of Newton's early successors at Cambridge University, gave a graphical method for determining the image positions given by curved mirrors, curved refracting surfaces and lenses. The method consisted in drawing a number of rays, originating at a point on the object. Such rays will meet at the image. The rays are drawn according to the following rules:

Each ray is assumed to be independent of other rays.

A ray parallel to the axis passes through the focus after refraction (or reflection in the case of mirrors).

A ray directed through the centre of curvature of a surface is undeviated (useful for mirrors and single surfaces).

A ray passing through the focus is after refraction (or reflection) parallel to the axis.

In the case of thin lenses we can add another rule: the ray through the centre of the lens is undeviated.

Figure 6.52 shows such rays.

We will here just consider such rays with lenses. If we take the rules for the rays to be correct, real light rays do follow such paths, then we can establish a number of formulae.

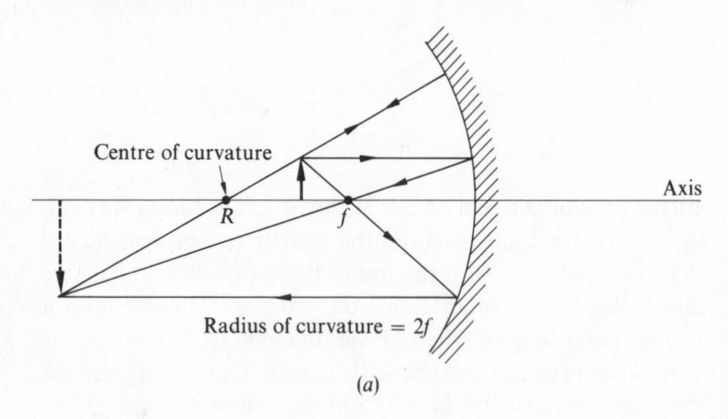

(*a*)

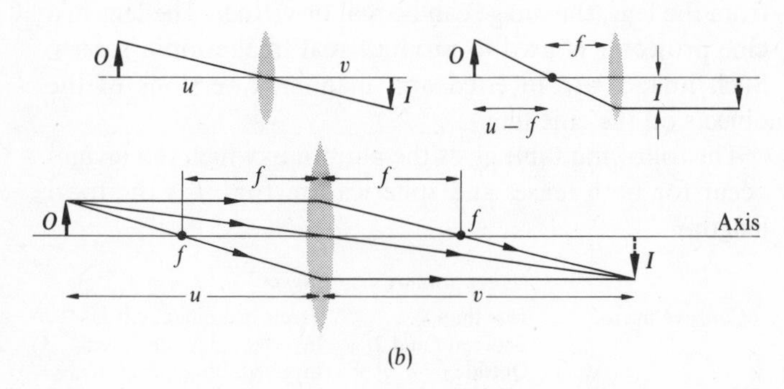

(*b*)

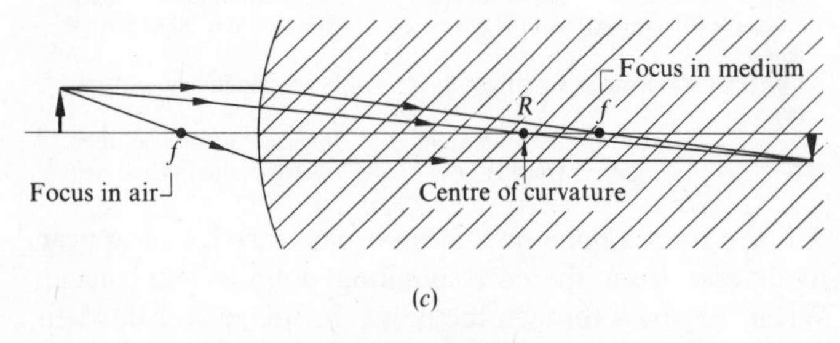

(*c*)

Fig. 6.52

In Fig. 6.52(b) if we consider the ray passing through the centre of the lens

$$\frac{O}{u} = \frac{I}{v}$$

or

$$\text{Magnification} = \frac{I}{O} = \frac{v}{u}$$

If we consider the ray passing from the object through the focus on the object side of the lens

$$\frac{O}{u-f} = \frac{I}{f}$$

Using the magnification equation to eliminate O and I gives

$$\frac{1}{u-f} = \frac{v}{uf}$$

$$uf = vu - f$$

Dividing through by vuf gives

$$\frac{1}{v} = \frac{1}{f} - \frac{1}{u}$$

$$\frac{1}{f} = \frac{1}{v} + \frac{1}{u}$$

The above equations were derived on the assumption that distances to the left of the lens were positive if they were measured to an object and positive to the right if they were measured to an image. Distances measured above the axis, the height of the object, are positive for objects, distances below the axis are positive for images. This gives the 'real is positive' convention (f positive for a convex lens, negative for a concave lens). On the new cartesian convention the object distance would be negative, because it is measured to the left of the lens, and the image height would be negative because it is measured below the axis. On this convention our equations become, if they are to yield the same results,

$$\text{Magnification} = \frac{I}{O} = \frac{v}{u}$$

$$\frac{1}{f} = \frac{1}{v} - \frac{1}{u}$$

For a convex lens f is negative, for a concave lens f is positive.

With the aid of these lens formulae we can 'explain' the pattern of images produced by different object distances for both convex and concave lenses.

Fermat's principle

About 1650 Pierre Fermat discovered a general way of explaining both reflection and refraction, now known as Fermat's principle: out of all the possible paths that light might take to get from one point to another, the path that requires the shortest time is the one taken.

We must qualify this statement if we want the light for example to be reflected at a mirror, Fig. 6.53(a). The path taken would be the shortest time path if the light went directly from source to observer and was not reflected. The path followed is, however, of shorter time than other paths which are closely adjacent to the one followed—the least time path.

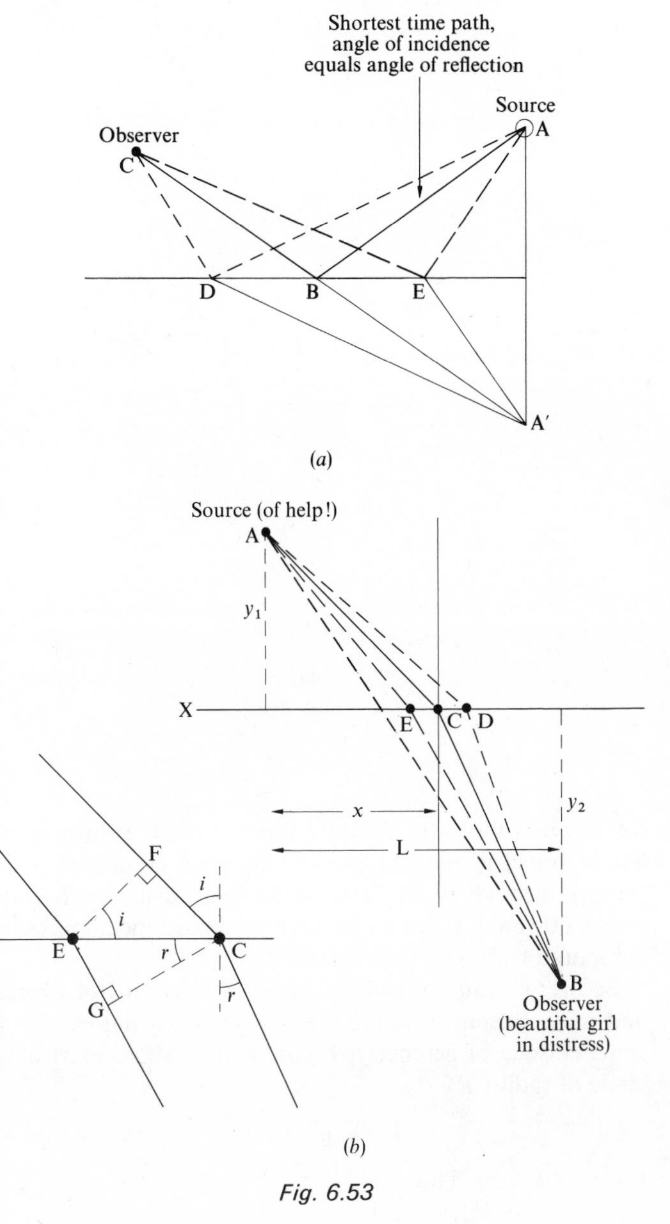

Fig. 6.53

In the case of reflection at a plane surface, Fig. 6.53(a), all the possible paths are in the same medium and thus the minimum time path will be the minimum distance path. The minimum distance path is the one for which the image A′ and the observer are in the same straight line; the path is thus ABC. For this case the angle of incidence is equal to the angle of reflection. Any other path, in which reflection at the mirror occurs, is longer—for instance AEC or ADC.

In the case of refraction we need to find the minimum time path, and as more than one medium is involved this is not the minimum distance path (Fig. 6.53(b)). The following quotation from *The Feynman lectures on physics*, Vol. 1 (by R. P. Feynman, R. B. Leighton, M. Sands; Addison-Wesley, Reading, Mass.), illustrates this point.

> 'To illustrate that the best thing to do is not just to go in a straight line, let us imagine that a beautiful girl has fallen out of a boat, and she is screaming for help in the water at a point B. The line marked X is the shoreline. We are at point A on land, and we see the accident, and we can run and can also swim. But we can run faster than we can swim. What can we do? Do we go in a straight line? (Yes, no doubt!) However, by using a little more intelligence we would realize that it would be advantageous to travel a little greater distance on land in order to decrease the distance in the water, because we go so much slower in the water. . . .'

The time taken to go from A to B via C is

$$\frac{AC}{v_1} + \frac{CB}{v_2}$$

where v_1 is the speed of light in the source medium and v_2 the speed in the observer medium. If this were the minimum time path then the paths on either side, i.e., ADB and AEB will be longer time paths. Moving the point at which the ray of light crosses the interface between the two media in either direction along the interface increases the time. However, for points close to C the time is only slightly different from the time for the ray passing through C; the rate of change of time with distance along the surface is zero at C and very close to zero at points close to C (Fig. 6.54). In a graph of

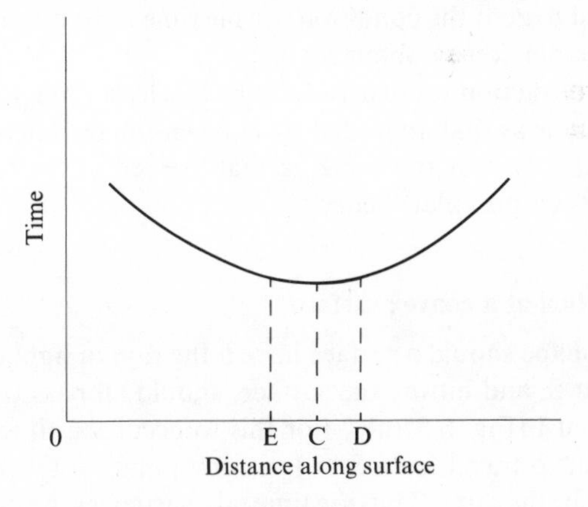

Fig. 6.54

time against distance along the surface of the light crossing point the slope of the graph is zero at C. For small displacements along the surface on either side of C the slope is still very close to zero and thus the time is still about the minimum. A way of finding the minimum time ray is to find the position on the interface for which a small change in position gives essentially no change in time. We will consider EC to be such a small change in position. The ray through C travels further in the source medium than the ray through E; if we consider the rays to be effectively parallel (a good approximation if EC is very small compared with the distance of the source from the interface) the extra distance is FC. The ray through E travels through more of the observer medium than the ray through C; an extra distance EG (essentially same approximation as before). If there is to be no time difference between these rays passing through E and C then the time taken to cover EG must be the same as the time taken to cover FC, i.e.,

$$\frac{\text{EG}}{v_2} = \frac{\text{FC}}{v_1}$$

But

$$\text{EG} = \text{EC} \sin r$$

and

$$\text{FC} = \text{EC} \sin i$$

Thus the condition for the path through C to be the minimum time path is that

$$\frac{v_1}{v_2} = \frac{\sin i}{\sin r}$$

This is the law of refraction.

The above result could have been obtained by differentiating the time equation.

$$\text{time } t = \frac{\text{AC}}{v_1} + \frac{\text{CB}}{v_2}$$

$$\text{AC} = \sqrt{(y_1^2 + x^2)}$$

$$\text{CB} = \sqrt{[y_2^2 + (L - x)^2]}$$

The slope of the time against x graph is a minimum when $dt/dx = 0$. When the above equation is differentiated and equated to zero the condition for the time to be a minimum is as we previously obtained.

The refraction result arrived at by this least time principle is the same as that arrived at by Huygens on the basis of his wave theory, not the same as that arrived at by Newton with his corpuscular theory.

Refraction at a convex surface

What shape should a surface have if the rays of light originating at S, and hitting the surface, should all pass through the point I (Fig. 6.55(a))? For this to occur the time taken for light to travel from S to I via any point on the surface should be the same. Thus the time taken to cover the distance SA + AI must be the same as the time taken to cover SB + BI. If we think in terms of waves—the crest starting at S, and spreading out, must converge on the point I (Fig. 6.55(b)). This can only happen if the time taken to go from S to I via any path passing through the surface is the same.

Thus we must have—

$$\frac{\text{SA}}{v_1} + \frac{\text{AI}}{v_2} = \frac{\text{SB}}{v_1} + \frac{\text{BI}}{v_2}$$

$$\frac{\text{SA} - \text{SB}}{v_1} = \frac{\text{BI} - \text{AI}}{v_2}$$

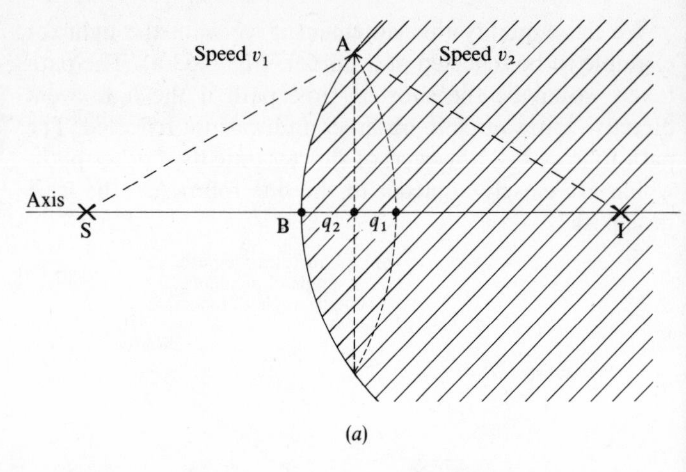

(a)

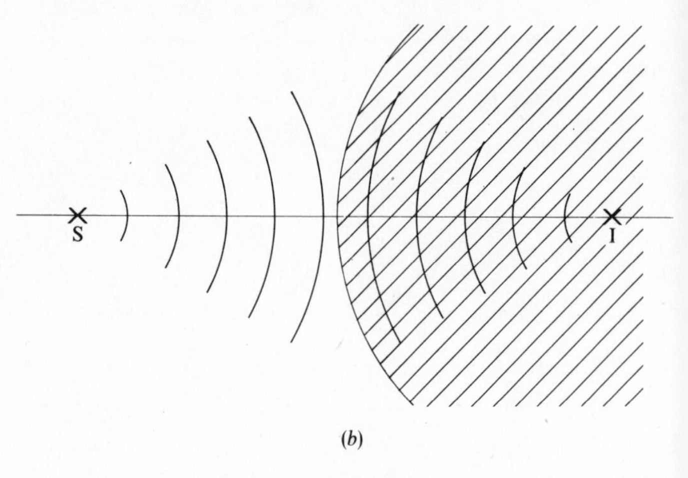

(b)

Fig. 6.55

SA − SB is the extra distance the ray passing through A has to travel in the source medium when compared with the ray through B. BI − AI is the extra distance the ray passing through B has to travel in the image medium when compared with the ray through A.

Before we can write these distances in terms of object and image distances along the axis we must digress for a short episode of geometry. Figure 6.56(a) shows part of a circle of radius R.

$$R^2 = p^2 + h^2$$

But $p = R - q$. Thus

$$R^2 = (R - q)^2 + h^2$$

$$R^2 = R^2 - 2Rq + q^2 + h^2$$

If q^2 is very small compared with the other quantities,

$$q = \frac{h^2}{2R}$$

Now to apply this result to the time equation (Fig. 6.56(b)).

$$\text{SA} - \text{SB} = q_1 + q_2 = \frac{h^2}{2\,\text{SA}} + \frac{h^2}{2R}$$

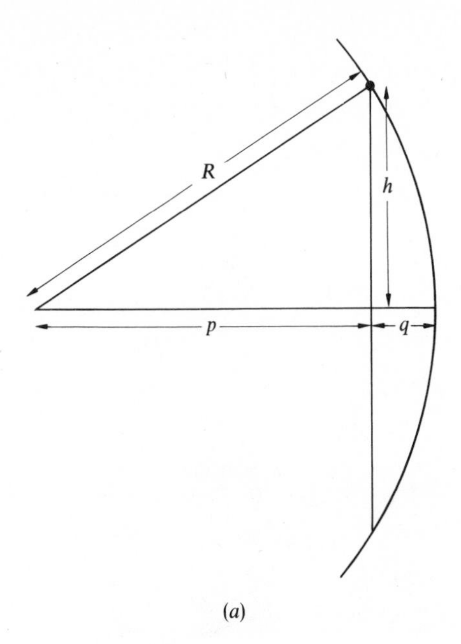

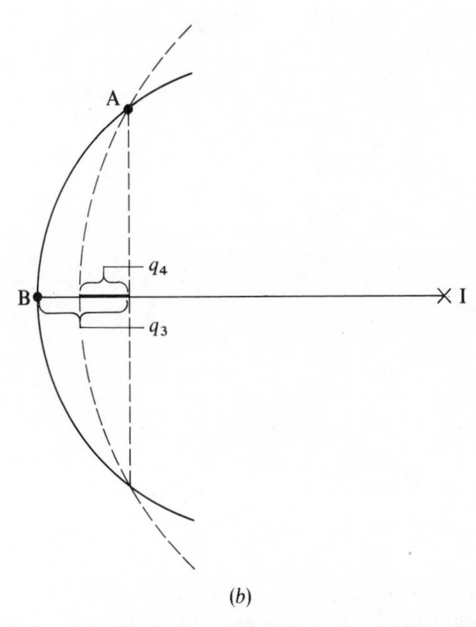

Fig. 6.56

where R is the radius of curvature of the convex surface.

$$\text{BI} - \text{AI} = q_3 - q_4 = \frac{h^2}{2R} - \frac{h^2}{2\,\text{AI}}$$

Thus we have for the time equation—

$$\frac{h^2}{2v_1}\left(\frac{1}{\text{SA}} + \frac{1}{R}\right) = \frac{h^2}{2v_2}\left(\frac{1}{R} - \frac{1}{\text{AI}}\right)$$

If we call SA the object distance u and AI the image distance v then rearranging the equation gives—

$$v_2\frac{1}{u} + v_1\frac{1}{v} = (v_1 - 1)\frac{1}{R}$$

v_1/v_2 is the refractive index on going from the object medium to the image medium, symbol μ.

$$\frac{1}{u} + \frac{\mu}{v} = (\mu - 1)\frac{1}{R}$$

If only rays close to the axis are considered then u and v can be good approximations to distances measured along the axis.

If we have, instead of just one convex surface, two convex surfaces forming a convex lens we can apply the above equation to each surface in turn. At the first surface we have

$$\frac{1}{u} + \frac{\mu}{v'} = (\mu - 1)\frac{1}{R_1}$$

where v' is the image distance due to refraction at the first surface, R_1 is the radius of curvature of the first surface. The image at distance v' from the first surface acts as the object for the refraction at the second surface. The image is, however, on the opposite side of the surface to the direction the light is coming from. If we followed through our previous argument we would find that we must write for the refraction at the second surface object distance $= -v'$; we assume that the lens is so thin that its thickness can be neglected. The other factor which we must take into account is that our 'object' is in a medium where the speed of light is v_2, our final image is in a medium where the speed is v_1, the reverse of the refraction at our first surface. The equation can be written for refraction at the second surface as

$$v_1\frac{1}{-v'} + v_2\frac{1}{v} = (v_1 - 1)\frac{1}{R_2}$$

v is the final image distance, R_2 is the radius of curvature of the second surface. Hence

$$-\frac{\mu}{v'} + \frac{1}{v} = (\mu - 1)\frac{1}{R_2}$$

Combining the two equations for refraction at the two surfaces gives

$$\frac{1}{u} + \frac{1}{v} = (\mu - 1)\left(\frac{1}{R_1} + \frac{1}{R_2}\right)$$

When the object is at infinity the image will be, by definition, at the focal length f. Thus

$$\frac{1}{f} = (\mu - 1)\left(\frac{1}{R_1} + \frac{1}{R_2}\right)$$

In deriving this equation we have taken the object to be on the left side of the lens, the final image to be on the right, the centre of curvature of the first surface to be on the right, the centre of curvature of the second surface to be on the left. If we take distances in opposite directions to these then we must include a minus sign. For example if the first surface was concave then we must write $-R_1$ in place of R_1 in the above equation. This convention of specifying distances is known as 'the real is positive' convention. Another convention, known as 'the new cartesian' convention considers that if the light is incident on a surface from the left then all distances measured to the left of the surface are negative and all distances to the right are positive. On this convention our lens formula becomes, if it is to yield the same results,

$$\frac{1}{f} = \frac{1}{v} - \frac{1}{u} = (\mu - 1)\left(\frac{1}{R_1} - \frac{1}{R_2}\right)$$

Important points to be realized are—our equation is restricted to rays close to the axis and the lens is assumed to be insignificantly thin. If these conditions are not true then the equation will not apply. Also we have assumed that it is possible to have a unique value of refractive index for light. This will be true if the light is of only one wavelength—the refractive index depends on the wavelength of the light used.

The lens equation was first derived by E. Halley in 1693 (*Phil. Trans.*, **17**, 930, 1693). Halley was a friend of Newton.

In the case of a convex lens in air when light comes from a distant object, object distance infinity, the lens refracts the light to a focus on the far side of the lens. If we put the object a distance from the lens equal to the focal length then the image is at infinity. The convex lens has two foci, equal distances on either side of the lens. If, however, we only have refraction at a single convex surface, in air, then when the object is at infinity we again have the light focused to a focal point, by definition, on the far side of the surface. If we put the object at the focal point in front of the surface the image is at infinity, by definition of the focal point. The focal lengths on each side of the surface are, however, not the same. The focal length on the medium side of the surface is μ times that on the air side of the surface. This is because of the μ term occurring above the v term in the equation for refraction at a single surface.

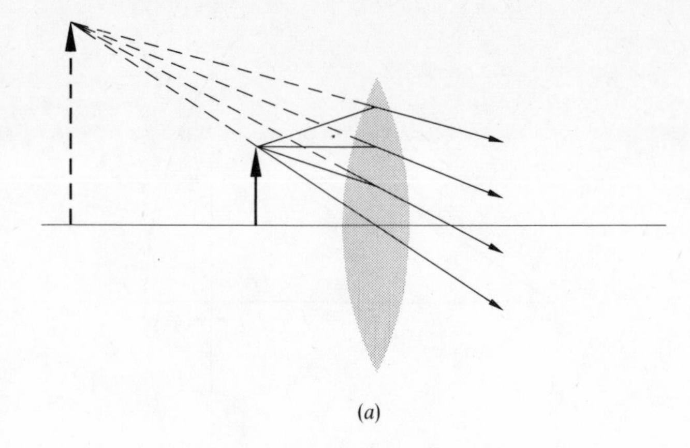

(a)

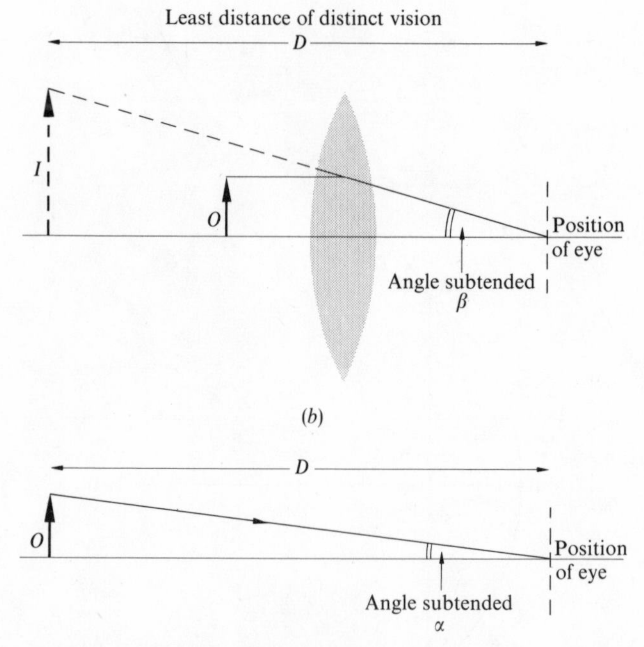

(b)

(c)

Fig. 6.57

The development of the microscope

The following brief outline is intended to show how, over the years, the microscope has developed from a single piece of glass to the complex modern instrument employing many lenses.

The use of a single convex lens as a magnifying glass may possibly date back as far as the Romans, certainly it was known about the year A.D. 1000. An object placed inside the focal length of a convex lens gives rise to an enlarged virtual image (Fig. 6.57(a)). The lens bends the rays of light so that they appear to come from a larger object, the virtual image. (The word virtual is used because the light does not actually pass through the image—the image cannot be put on a screen.) The angle subtended by the rays of light, apparently, from the image is greater than the angle the rays would subtend at the eye if no lens were present (Fig. 6.57(c)). The ratio of these two angles is known as the magnifying power. Without the lens the greatest angle that the object can subtend at the eye is when the object is placed as near to the eye as the eye can give a clear focused image. This distance is known as the least distance of distinct vision and is for the average person about 0·25 m. The virtual image formed by the lens must not be nearer to the eye than this least distance of distinct vision. The bigger the angle subtended at the eye the larger the apparent size of the object. A man 100 m away subtends a much smaller angle than a man 10 m away—at 10 m we may see the buttons on his jacket, at 100 m we may have difficulty in discerning the jacket. A difference in angle subtended at the eye of about one minute of arc is needed if two objects are to appear resolved.

For a single convex lens we have—

(Angle subtended at the eye without instrument) $= \alpha$

(Angle subtended at the eye with instrument) $= \beta$

and thus

$$\text{Magnifying power} = \frac{\beta}{\alpha}$$

$$\tan \alpha = \frac{O}{D}$$

$$\tan \beta = \frac{I}{D}$$

where O is the object size, I the image size, and D the least distance of distinct vision.

If the angles are small, often the case, we can make the assumption that the tangent of the angle is approximately the same as the angle. Then

$$\text{Magnifying power} = \frac{I}{D} \times \frac{D}{O} = \frac{I}{O}$$

If the lens is thin, a very doubtful assumption for most single lens magnifiers, we can use the thin lens formula to obtain I and O in terms of the focal length and D.

$$\frac{1}{f} = \frac{1}{v} - \frac{1}{u} \text{ (New Cartesian convention)}$$

Hence

$$\frac{v}{u} = 1 - \frac{v}{f}$$

and as $I/O = v/u$ we obtain

$$\text{Magnifying power} = 1 - \frac{v}{f}$$

If the eye is fairly close to the lens then v is approximately equal to $-D$ (New Cartesian convention) and we have

$$\text{Magnifying power} = 1 + \frac{D}{f}$$

or to a reasonable approximation, considering the crudity of our assumptions so far, we can neglect the 1 and consider the magnifying power to be D/f. Large magnifying powers are thus produced with small focal length lenses.

Small focal length lenses can be produced by decreasing the radii of curvature of the lens surfaces (Fig. 6.58). For a thin lens with equal radii surfaces the focal length is proportional to the radius.

$$\frac{1}{f} = (\mu - 1)(1/R_1 - 1/R_2) \quad \text{(New Cartesian convention)}$$

If $R_1 = R_2 = R$

$$\frac{1}{f} = \frac{2(\mu - 1)}{R}$$

μ is the refractive index of the lens glass relative to air, it being assumed that the lens is in air.

For a given size lens the limit in decreasing the radii is reached when the lens is spherical. Any further decrease in radii can only be achieved by making the sphere smaller.

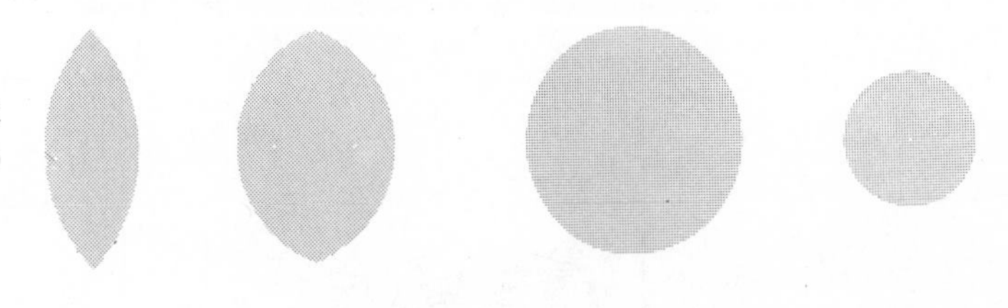

Fig. 6.58

Greater magnifying power can be achieved if two convex lenses are used. The first such compound microscope was probably produced about the end of the sixteenth century. In such an instrument the first convex lens, known as the objective, produces a real image close to the second lens, known as the eyepiece. This lens acts in the same way as the magnifying glass and gives an enlarged virtual image (Fig. 6.59).

(Angle subtended at the eye without instrument) $= \alpha$

(Angle subtended at the eye with instrument) $= \beta$

$$\text{Magnifying power} = \frac{\beta}{\alpha}$$

$$\tan \alpha = \frac{O}{D}$$

$$\tan \beta = \frac{I}{D}$$

where I is the size of the final image. Thus for small angles

$$\text{Magnifying power} = \frac{I}{O}$$

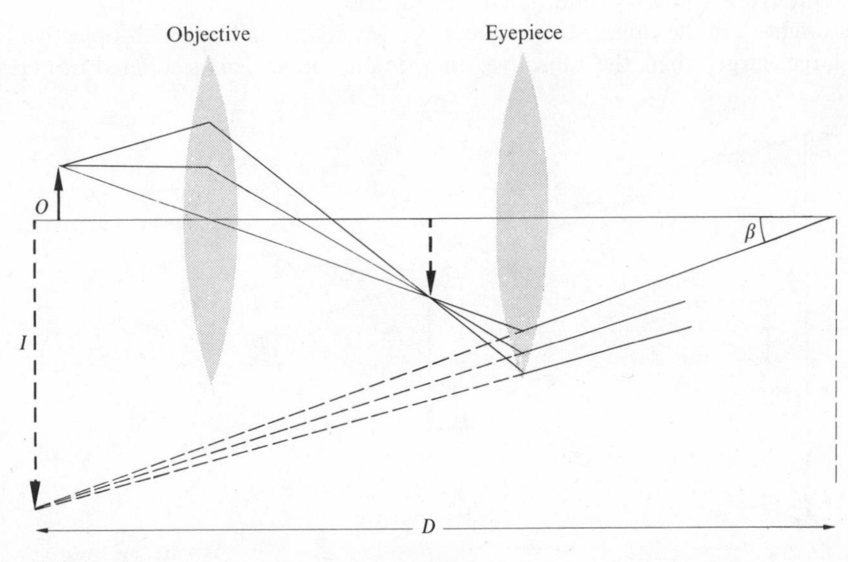

Fig. 6.59

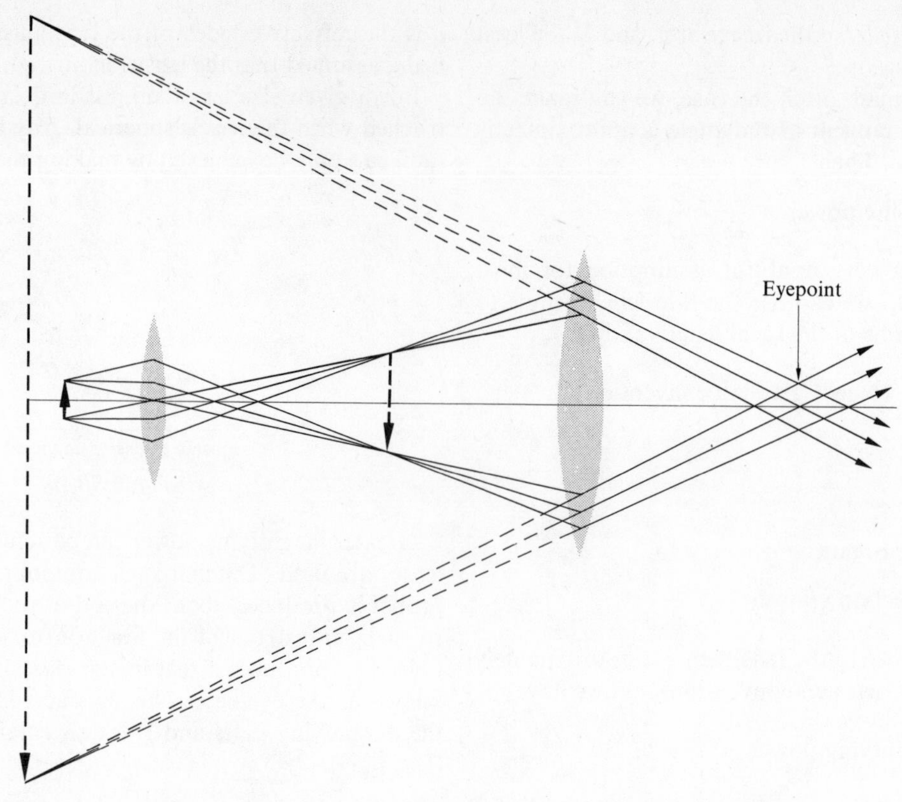

Fig. 6.60

If we multiply the top and the bottom of this equation by the same quantity, I', the size of the intermediate image, we obtain

$$\text{Magnifying power} = \frac{I}{I'}\frac{I'}{O}$$

or

$$\begin{pmatrix}\text{Magnifying}\\\text{power}\end{pmatrix} = \begin{pmatrix}\text{magnification}\\\text{of eyepiece}\end{pmatrix} \times \begin{pmatrix}\text{magnification}\\\text{of objective}\end{pmatrix}$$

In order that the entire cone of rays from the intermediate image can be received at the eye, a necessary condition if the image is not to be less bright than the object, the eyepiece lens has to be quite large, larger than the objective. In addition, all the rays passing through the eyepiece must enter the eye. This is most easily achieved if the eye is placed at the position known as the eyepoint (sometimes called the Ramsden circle), a point some considerable distance from the eyepiece (Fig. 6.60). The difficulties in making large eyepiece lenses and the inconvenience in placing the eye some distance from the eyepiece led to the design of eyepieces using more than one lens. Such eyepieces were invented by Huygens and Kepler, in the early seventeenth century (Fig. 6.61). In the Huygens eyepiece the two lenses are about half the sum of the focal lengths of the two lenses apart.

Microscopes built with objective lenses which were just simple convex lenses suffered from severe distortions of the

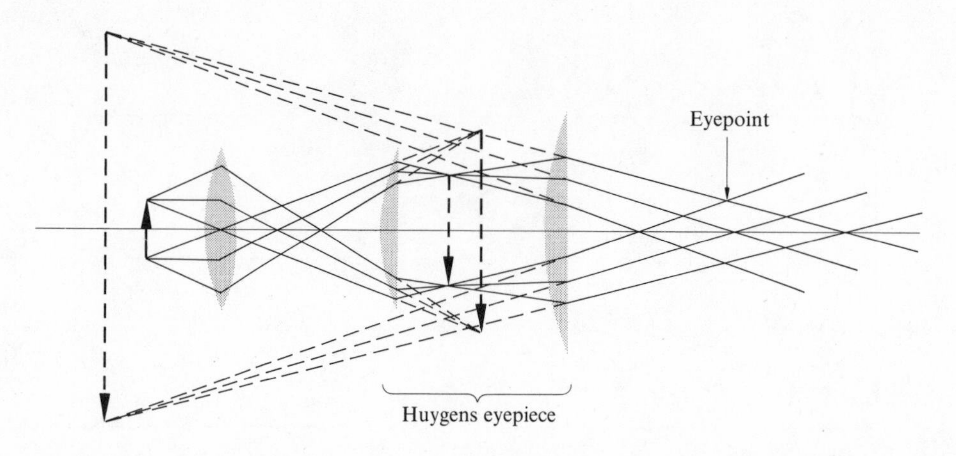

Fig. 6.61

image. The major aberrations were spherical aberration and chromatic aberration. With spherical aberration the rays of light travelling through the outer parts of the lens come to a different focus to that which rays through the inner part of the lens come to (Fig. 6.62(a)). This is a defect

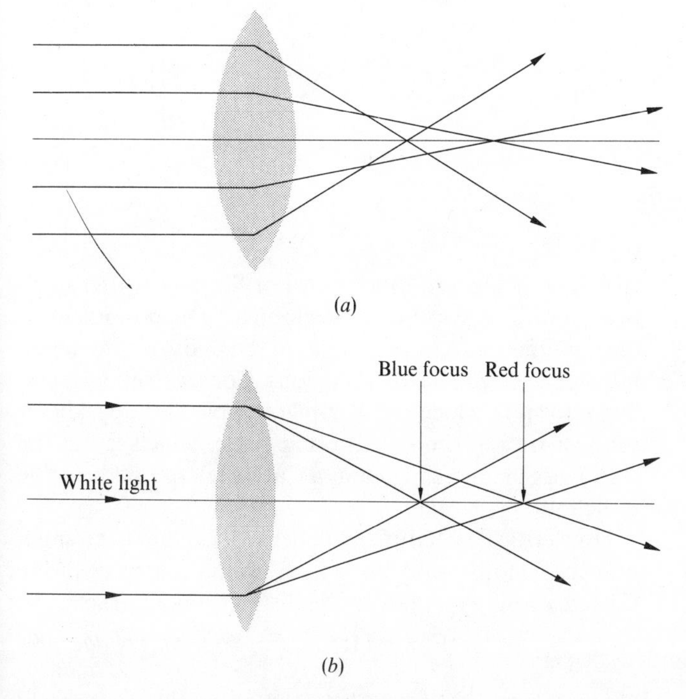

Fig. 6.62 (a) A spherical aberration. (b) Chromatic aberration.

common to all lenses whose surfaces are spherical, i.e., all points on a surface have the same radius of curvature. The effects of this can be reduced by only using the central portion of the lens. With chromatic aberration the different colours of light come to different foci (Fig. 6.62(b)). This occurs because the glass of the lens has different refractive indices for the different colours of light; a lens is rather like two prisms back to back (Fig. 6.63). The result is that the edges of images are blurred and appear coloured. Around about 1740 this problem of chromatic aberration was reduced with the invention of what were called achromatic lenses. Achromatic convex lenses were designed with two components—a convex lens made of crown glass and a

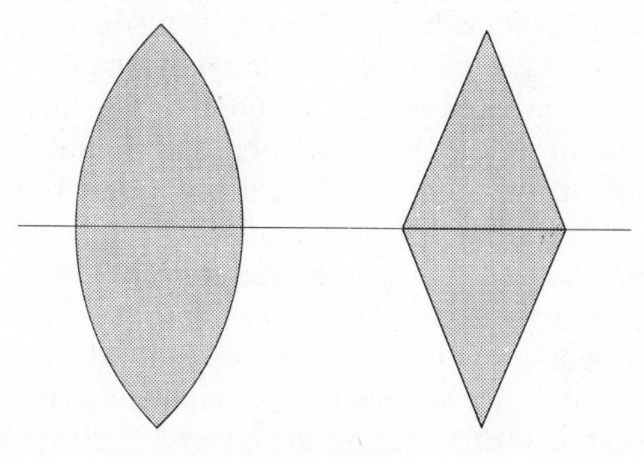

Fig. 6.63

concave lens of flint glass (Fig. 6.64). The focal length of the two lenses together for red light can be made equal to the focal length of the combination for blue light. It was the end of the eighteenth century before such lenses could be satisfactorily made and used in microscopes. Even then

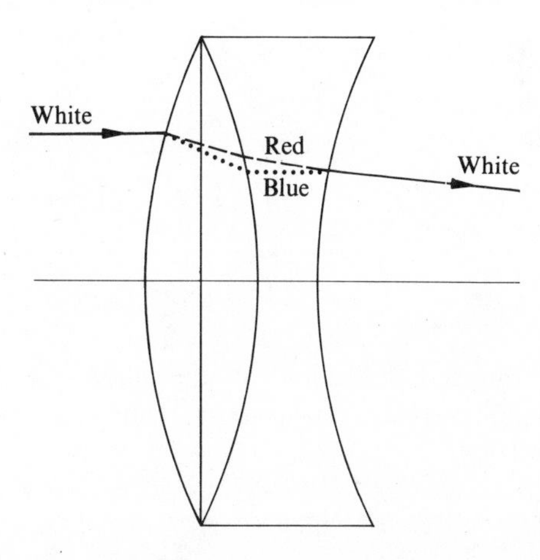

Fig. 6.64 Achromatic lens.

the achromatic lens could only be achieved for long focal lengths. To achieve high magnifying powers short focal length objective lenses were necessary. It was not until about 1830 that short focal length achromatic objective lenses were produced, by using more than one achromatic doublet of lenses. Up to this time empirical methods had been used in combining achromatic lenses. Lister in 1830 showed how to determine theoretically what lenses to use to reduce both chromatic and spherical aberrations to a minimum.

In 1812 Brewster suggested that chromatic aberration could be reduced by immersing the objective lens and the object being examined in a liquid (Fig. 6.65). Effectively

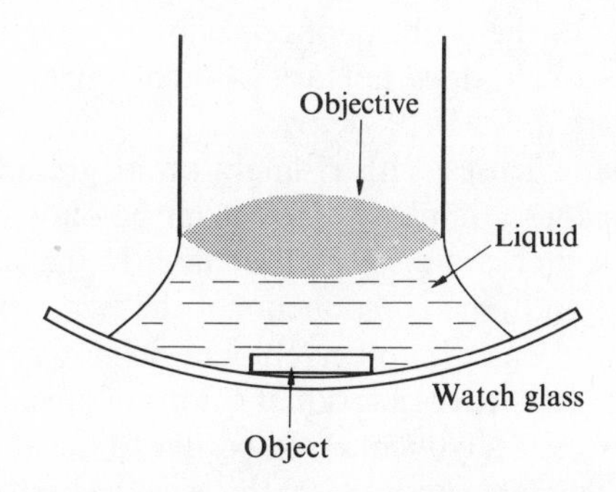

Fig. 6.65 Brewster's immersion lens.

this gave a concave liquid lens in contact with the convex objective lens. Such a combination did reduce chromatic aberration but not as effectively as the achromatic doublets made of glasses. About 1850 Amici developed the idea of immersion objectives for a different purpose—to increase the brightness of the image. With a cover glass over the object, the rays of light entering the objective are limited

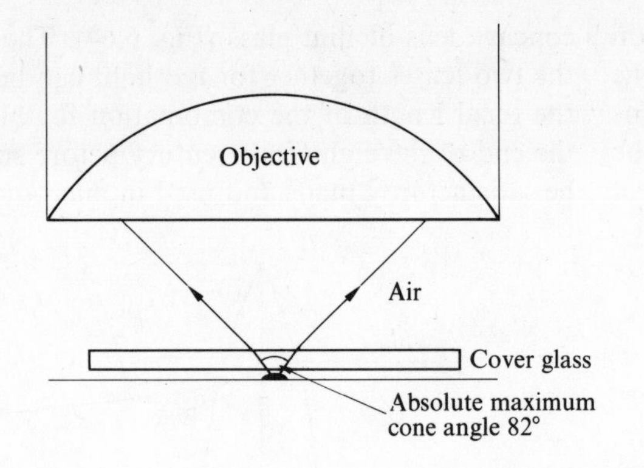

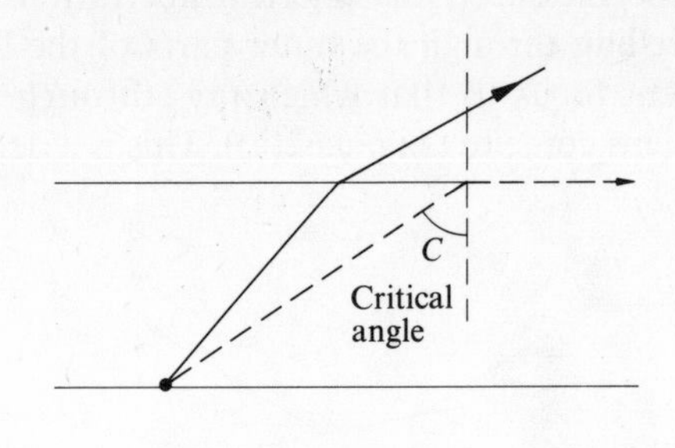

Fig. 6.66

by total internal reflection at the glass-to-air surface (Fig. 6.66). The critical angle occurs when

$$\frac{\sin C}{\sin 90} = \frac{1}{\mu}$$

where μ is the refractive index from air to the glass of the cover glass. Taking μ as 3/2 gives a critical angle of about 41° and thus a maximum cone angle of 82° for light rays from the object to enter the objective. In fact this angle would be less due to the air gap between the cover glass and the objective lens. If the air between the cover glass and the objective is replaced by a liquid the critical angle increases.

$$\frac{\sin C}{\sin 90} = \frac{1}{\mu'}$$

where μ' is the refractive index from liquid to glass. For water as the liquid, μ' is $3/2 \div 4/3$ and the critical angle becomes about 64°. This gives a maximum cone angle of about 128°. By using a liquid whose refractive index is equal to that of the cover glass the cone angle can be increased even further as the problem of reflection at the glass-to-air, or liquid, surface does not arise. Such immersion lenses were realized about 1870.

Achromatic lenses, either singly or in groups, do not produce images completely free from colour—chromatic aberration is not completely eliminated. In the achromatic lens the red and blue components of the light are brought to the same focus, the other colours do not come to the same focus. This leaves a residual colour of greenish-yellow which shows as a slight coloured border to the image. With the glasses available about 1800 this was the best that could be achieved. It was not until near the end of the nineteenth century that new glasses were produced which enabled better colour correction to be achieved. In the new lenses, known as apochromatic lenses, a number of lenses of different glasses are combined to give the red, the blue, and the green components of light all coming to the same focus.

By the end of the nineteenth century the optics of optical microscopes had reached a state from which little improvement has been made so far this century.

'It seems as big as a little Prawn or Shrimp, with a small head, but in it two fair eyes globular and prominent of the circumference of a spangle; in the midst of which you might (through the diaphaneous Cornea) see a round blackish spot, which is the pupil or apple of the eye, beset round with a greenish glittering circle, which is the Iris (as vibrissant and glorious as a Cat's eye), most admirable to behold.

How critical is Nature in all her works! that to so small and contemptile an Animal hath given such an exquisite fabrick of the eye, even to the distinction of parts. . . .'

H. Power (1663). *Experimental philosophy.*

The above is the description of a flea when observed under a microscope.

The development of the telescope

The invention of the compound microscope and the telescope are of the same period of time. The first telescope was probably that of Lippershey in about 1609 and consisted of two spectacle lenses separated by a distance equal to the approximate sum of their focal lengths. Figure 6.67 shows the basic arrangement. The magnifying power is equal to f_o/f_e. Thus long focal length objective lenses and short focal length eyepiece lenses are needed for large magnifying powers.

The Yerkes Observatory in the USA has a telescope with an objective of focal length 20 m and an eyepiece of focal length 2·5 cm, giving a magnifying power of about 800.

Galileo in 1609 made a telescope with a convex objective lens and a concave eyepiece. The magnifying power was 9. He soon, however, was able to improve the magnifying power, and observed that the moon had mountains, that Jupiter had moons and that Saturn had rings. The instruments had magnifying powers of about 20 to 30.

Newton realized that one of the main factors limiting increases in magnifying powers was the lens aberrations. Careful grinding of the lens surfaces enabled him significantly to reduce spherical aberration but he could not reduce chromatic aberration. Seeing no way of reducing this he replaced the objective lens by a concave mirror and invented

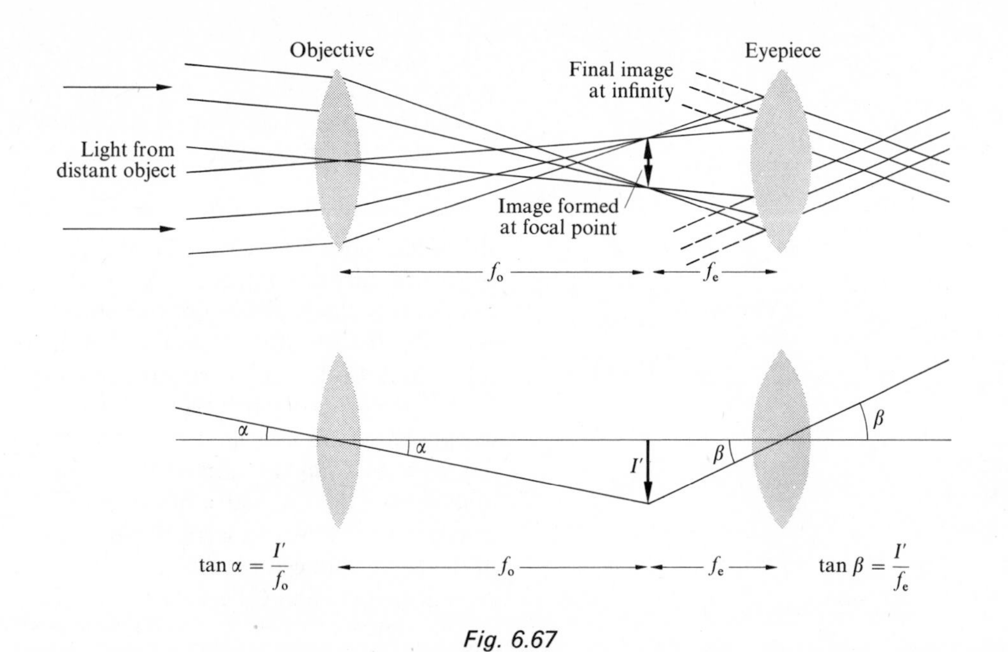

Fig. 6.67

what is known as the reflecting telescope. Figure 6.68(a) shows the basic arrangement.

The reflecting telescope at Mount Palomar (Fig. 6.68(b)) has a mirror 5·5 m in diameter. The observer sits with the eyepiece lens near the focal point of the mirror.

Improvements in lens design (see the previous section on microscopes) have enabled further improvements in telescopes to be realized. The biggest instruments are, however, still reflecting instruments.

Appendix 6A From Galileo to Young

From Galileo to Young, or from the beginning of the seventeenth century to the end of the eighteenth (Fig. 6.69).

The beginning of the seventeenth century saw the death of Elizabeth I of England, and Shakespeare's time. Many people considered that the world was nearing its end. But it was the beginning of a new science that introduced a sceptical attitude to the knowledge of man and nature. The moon had mountains and was not 'perfect', Jupiter had moons and thus we on earth were not the only privileged planet, we were not the centre of the universe.

The middle of the century saw the paintings of Rembrandt, the poetry of Milton:

'. . . What if the sun
Be center to the world, and other stars,
By his attractive virtue and their own
Incited, dance about him various rounds?'

Milton. *Paradise lost*. Book VIII, line 122

the work of Newton, Hooke, and Huygens, civil war in England, the plague (Black death) and the Great Fire of London. It saw the rise of a desire for personal freedom that

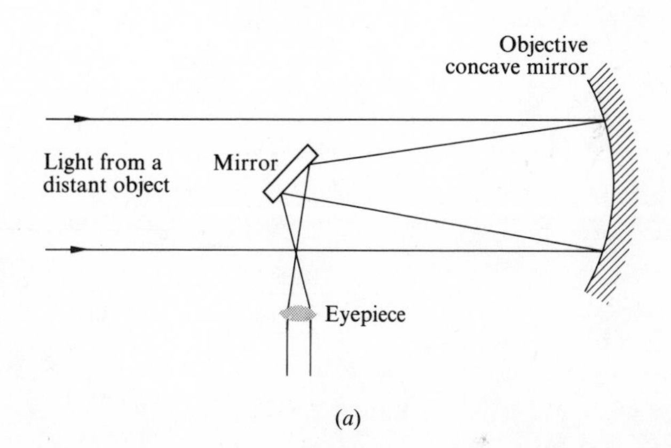

(*a*)

(*b*)

Fig. 6.68 (a) Basic reflecting telescope, (b) 200-inch (5·5 m) Palomar reflecting telescope. The observer is in the prime focus position, the mirror can be seen at the far end of the telescope tube. (Courtesy of the Hale Observatories.)

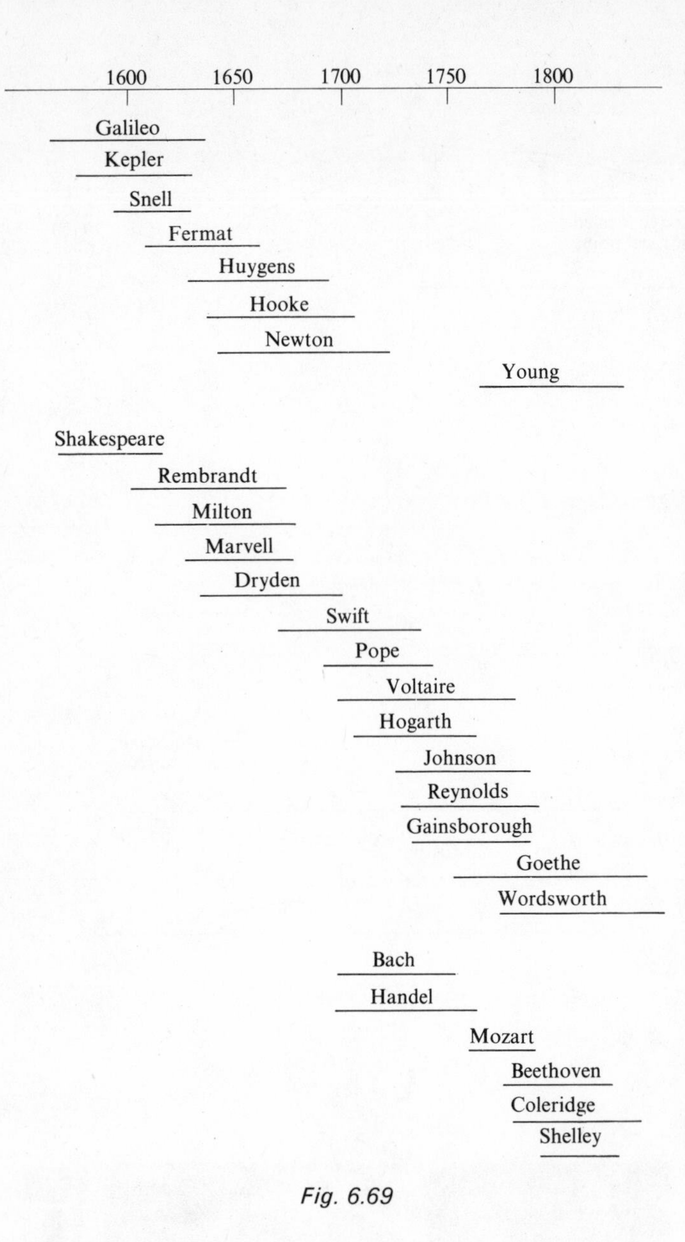

Fig. 6.69

was ultimately to lead to the French Revolution and the American War of Independence near the end of the eighteenth century.

'... For by naturall birth, all men are equally and alike borne to like propriety, liberty and freedome, and as we are delivered of God by the hand of nature into this world, every one with a naturall, inate freedome and propriety (as it were writ in the table of every mans heart, never to be obliterated) even so are we to live, every one equally and alike to enjoy his Birthright and priviledge; even all whereof God by nature hath made him free.'

R. Overton (1645). *An armour against all tyrants and tyrany.*

'Declaration of Independence

In congess July 4th, 1776

The Unanimous Declaration of the Thirteen United States of America.

We hold these truths to be self-evident—that all men are created equal, that they are endowed by their CREATOR, with certain unalienable rights; that among these are life, liberty, and the pursuit of happiness.'

The eighteenth century saw the work of Hogarth, Reynolds, and Gainsborough and the period in which Newton's theories reigned supreme. The middle of the century was the beginning of the British industrial revolution. The industrial revolution thus followed the scientific revolution. Both revolutions probably stem from the same cause—the breakdown of the rigid structure of the medieval world, based on the land, by the growth of trade and industry. In 1700 the population of Britain was about 5 million, by 1800 it had almost doubled. Steam engines, canals, large ironworks were all part of the scene at the end of the eighteenth century. It was also the time of the poetry of Wordsworth and Coleridge.

The eighteenth century could be called the century of revolutions, or the century of enlightenment.

Appendix 6B Human vision

The simplest idea of the human eye is as a camera, i.e., a lens and a light-sensitive surface. The human eye is, however, more complex. The light-sensitive surface, the retina, consists of large numbers, about 130 million, of vision cells, non-uniformly distributed. Animals having eyes with large densities of vision cells have acute vision, e.g., the hawk. Because of the non-uniform distribution the eye tracks backwards and forwards over the field of view so that the image crosses backwards and forwards over the region in the eye where the vision cells are at their densest. Every cell transforms its share of the image into a sequence of electrical discharges, the brighter the light the faster the number of discharges. The output from a vision cell is thus a train of electrical pulses. These signals pass to what might be called sorting stations, cells in the retina along with the vision cells. There is one sorting station for about every 130 vision cells. These sorting stations can exert some control of the eye. For example they can enable you to track a fast moving object; the time taken for the eye to send a signal to the brain and then receive instructions is too long (think of your reaction time to stimuli). From each sorting station a nerve passes to the brain.

When we see an object we are bombarding our brain with large numbers of pulses passing along large numbers of nerves. The cells in the brain respond to these pulses. Some cells respond to the sequence of pulses indicating vertical lines, some respond to horizontal lines. For every type of line, e.g., slit, edge, every position and every direction of motion there is a set of brain cells which respond. The entire image is broken down into lines.

The reaction of the brain to this analysed information depends on its previous experience, though there may be some inherited reactions to some information. Experience has resulted in information, patterns, being lodged in the memory. The information supplied by the eyes is compared with the information in the memory and on the basis of this the brain acts.

How do we perceive colours? Young in 1801 postulated that there were in the eye three different types of light-sensitive cells, one group sensitive to red, one to green, and the other to blue. Any other colour can be broken down into these three colours. Thus if red and green light are mixed the result appears to the eye as yellow light; red, green, and blue lights give white. The colours referred to are not single wavelength colours but regions of the spectrum, the spectrum being considered as three segments—the red end, the green middle, and the blue end. Strong evidence that the Young theory was correct came in 1964 as a result of research in the USA by MacNichol, Marks and Wald. The cone receptors in the eye do have these three different colour sensitivities. There are two types of receptors in the eye, cones and rods. The more numerous cones are responsible for our vision in bright light, the rods come into action in dim light and give a black and white 'picture'. Nocturnal animals have predominantly rod receptors, hence they only see black and white.

Further reading

G. G. Mueller and M. Rudolph. *Light and vision*, Time-Life, 1969.

Problems

1. '... We direct a narrow beam of electrons against the face of a nickel crystal, and observe that under certain conditions a sharply defined stream of electrons leaves the crystal in the direction of regular reflection—angle of reflection equal to the angle of incidence....'

C. J. Davisson (1928). *Franklin Inst. J.*, **205**, 597, 1928.

Davisson was describing the results he and Germer obtained when studying the scattering of electrons by a nickel target.

(a) The law of reflection is obeyed by light—does this prove that electrons have a wave property like light?

(b) Electrons can be thought of as particles about 10^{-15} m, or smaller, in diameter. The distance between the atoms in the nickel target is about $2{\cdot}5 \times 10^{-10}$ m. Can a particle model of reflection, perhaps similar to that of Newton, possibly explain the reflection of electrons?

(c) How can reflection be explained?

2. The recorded voices of people sound 'different' when replayed on a gramophone or tape recorder running at the 'wrong' speed. Why does this occur?

3. 'If a fine beam of homogeneous cathode rays [electrons] is sent nearly normally through a thin celluloid film (of the order of 3×10^{-6} cm thick) and then received on a photographic plate 10 cm away and parallel to the film, we find that the central spot formed by the undeflected rays is surrounded by rings, recalling in appearance the haloes formed by mist round the sun....'.

G. P. Thomson and A. Reid (1927). *Nature*, **119**, 890, 1927.

(a) How can the haloes formed by mist round the sun be explained?

(b) These results were used as an argument in favour of electrons having wave characteristics. Why should the results be characteristic of waves and not particles?

(c) The size of the rings was found to decrease as the energy of the electrons was increased. What does this suggest concerning the wavelength of the electrons?

4. Oxford and Cambridge Schools Examination Board. Nuffield Advanced Physics. Short Answer paper, 1971.

'The map [Fig. 6.70] shows part of a coastline, with two land-based radio navigation stations A and B. Both stations transmit continuous sinusoidal radio waves with the same amplitude and same wavelength (200 m).

A ship X, exactly midway between A and B, detects a signal whose amplitude is twice that of either station alone.

(*a*) What can be said about the signals from the two stations?

(*b*) The ship X travels to a new position by sailing 100 m in the direction shown by the arrow. What signal will it now detect? Explain this.

(*c*) A ship Y also starts at a position equidistant from A and B and then travels in the direction shown by the arrow. Exactly the same changes to the signal received were observed as in the case of ship X in (*b*).

Explain whether Y has sailed 100 m, more than 100 m, or less than 100 m.'

5. Divers who are breathing an atmosphere largely consisting of helium speak in very high squeaky voices, like Donald Duck. Speech normally starts with air being expelled from the lungs, passing through the larynx and out through the mouth and nose. The larynx is in the throat and it is here that the air is broken up into compressions and rarefactions. The throat, mouth, and nose cavities act as resonant cavities which pick out and amplify some of the frequencies.

Why does helium affect the speech?

6. Suggest ways by which sound waves can be focused. Give design details for a short focus system (say a focal length of about 20 cm for short wavelength sound).

7. A wave motion incident on an ordered array of scattering centres is how one might consider light being reflected from some reflector. Can you explain light reflection as an interference effect?

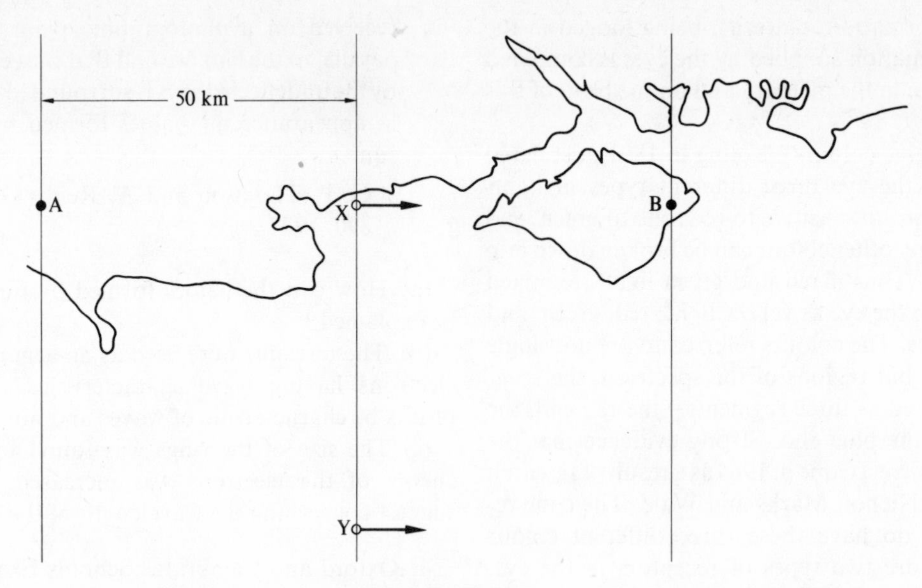

Fig. 6.70

8. 'It has been shown that two equal series of waves, proceeding from centres near each other, may be seen to destroy each other's effects at certain points, and at other points to redouble them; and the beating of two sounds has been explained from a similar interference.'

T. Young. *Course of lectures on natural philosophy and the mechanical arts*, 1807. Lecture 39.

(a) What is meant by the word 'destroy'? Is energy conserved?

(b) Why does Young consider that the waves should proceed from 'centres near each other'?

(c) The 'beating of two sounds' is the result of combining two sounds of slightly different frequencies. Explain what the result would be.

9. Figure 6.71 shows water waves breaking on a shore. What can you deduce from examination of the waves?

10. What wavelength light would you recommend for use with a microscope if the maximum amount of detail was to be seen?

Fig. 6.71 Waves breaking on a shore. (Aerofilms Ltd.)

Teaching note Further problems appropriate to this chapter will be found in:
Nuffield advanced physics. Unit 4 students' book, Penguin, 1972.
Physical Science Study Committee. *Physics*, 2nd ed, chapters 11, 12, 13, 14, 15, 16, 17, 18, Heath, 1965.
Nuffield O-level physics. Question book 3, Longmans/Penguin, 1967.

Practical problems

1. Make a detailed investigation of some musical instrument. Consider the mechanism by which the sound is produced, the characteristics of that particular instrument which distinguish it from other instruments, and the method by which these characteristics are produced.

2. Study the effects occurring when waves meet a beach. Consider the effects of the gradient of the beach on the waves—the wavelength of a wave depends in shallow water on the depth of water. The effects can be studied with a ripple tank and where possible correlated with the behaviour of waves on full-size beaches.

3. When light is incident on the boundary between two media it can be either reflected or refracted or both. The percentage of light that is reflected depends on the angle of incidence and the refractive index between the two media. Devise a refractometer based on the measurement of the light reflected at a glass–sample boundary for a particular angle of incidence.

Suggestions for answers

1. (a) Particles as well as waves give the angle of reflection equal to the angle of incidence.

(b) If the atoms were solid it might. If atoms are mainly empty space a particle model faces difficulty.

(c) As an interference effect. See question 7.

2. The frequencies have been changed.

3. (a) See Fig. 6.40.

(b) The results are characteristic of waves not particles. Particles would not be scattered by regular or irregular scattering centres in any regular manner.

(c) The wavelength is decreasing.

4. (a) They are in phase.

(b) A maximum. The path difference is a whole number of wavelengths.

(c) More than.

5. The speed of sound in helium is 970 m s^{-1}, while that in air is 330 m s^{-1}. The wavelength of the same frequency note in helium will be about three times bigger than that in air. The resonant cavity of the throat, etc., will thus be only able to resonate to the higher frequency notes.

6. A balloon filled with carbon dioxide will act as a convex lens because the speed in carbon dioxide is less than that in air. A concave lens can be made with hydrogen in the balloon; the speed in hydrogen is greater than that in air. A point to consider in the design is the size of the lens and diffraction round the edges.

7. This requires an examination of a reflection diffraction grating when the light is not incident along the normal. Figure 6.72 shows light being scattered by two scattering centres A and D. The path difference is $d \sin \theta' - d \sin \theta$. For a constructive interference we must have

$$d \sin \theta' - d \sin \theta = n\lambda$$

When $n = 0$ we must have $\sin \theta' = \sin \theta$, i.e., the angle of incidence must be equal to the angle of reflection.

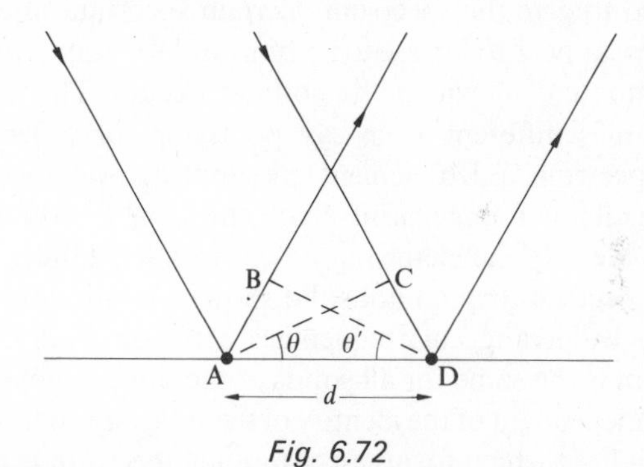

Fig. 6.72

8. (a) Energy is conserved. The energy that is not found at the 'destroy' position is equal to the extra energy found at the 'redouble' position.

(b) The distance between the centres should be of the same order as the wavelength.

(c) The amplitude of the combined wave rises and falls with a frequency equal to the difference in frequency between the two sounds.

9. A decrease in wavelength shows a decrease in speed, a result of the wave moving into shallow water. Note the refraction making the waves bend away from the shore.

10. Short wavelength light. Electron microscopes operate with very short wavelengths. A complication might be that the lenses of the microscope were not corrected for the short wavelength light (violet light).

7 Waves or particles?

Teaching note Practical work appropriate to this chapter will be found in:
W. Bolton. *Physics experiments and projects*, Vol. 2, Pergamon, 1968.
Nuffield advanced physics. Unit 2 and Unit 10 teachers' guide, Penguin, 1972.

7.1 Spectra

If common salt, sodium chloride, is present in a gas flame then the flame has a yellow coloration. If we examine the yellow light with a spectroscope we find that the overall colour is made up by a number of discrete frequencies. A spectroscope could be made from a prism—the amount of bending of light by a prism depends on the wavelength of the light—or a diffraction grating—the angle through which light is diffracted depends on the wavelength of the light.

Strong similarities in spectra occur whatever the sodium salt used to give the spectrum. Certain spectrum lines, frequencies, appear to be due to sodium and to occur whatever the compound in which the sodium occurs. The sodium spectrum is different from say potassium or indeed any other spectrum. Each element gives out its own spectrum, each has its own fingerprint. Such emission spectra can be used to identify the elements present in a substance.

The spectrum from a hot solid shows a continuous spectrum, a wide range of frequencies being emitted. Such a spectrum is the same for all solids at the same temperature, being independent of the identity of the solid (see chapter 12).

When light which forms a continuous spectrum is passed through gases the spectrum is found to contain a number of dark lines. These are absorption lines (see chapter 4) and correspond in position to the frequencies which the gas gives for its emission spectrum. Such absorption lines were first found by Wollaston in 1802 when he examined the spectrum of the sun. Detailed investigation was carried out by Fraunhofer between 1814 and 1824. The sun's spectrum contains a large number of absorption lines due to the continuous spectrum light emitted by the hot solar core having to pass through the hot solar atmospheric gases. These lines are now known as the Fraunhofer lines. Fraunhofer labelled the lines with the letters of the alphabet. The strong yellow sodium lines appear in absorption in the sun's spectrum and are known as the D lines.

Spectroscopy

Spectroscopy is not restricted to the confines of the visible part of the electromagnetic spectrum but spreads over most of the electromagnetic spectrum. Here we concern ourselves with spectrometers which operate in the visible and the ultraviolet and infrared.

Probably the simplest spectroscope for use in the visible is a piece of diffraction grating and a slit, Fig. 7.1. The light

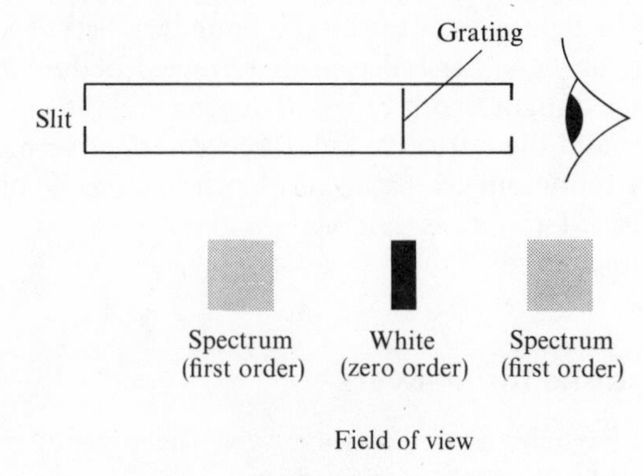

Fig. 7.1

passing through the grating is diffracted through angles which depend on the wavelength. (See chapter 6 for a discussion of the diffraction grating.) This arrangement is restricted to the visible part of the spectrum, only slightly overlapping into the ultraviolet and infrared. The restriction occurs because glass, and many materials which are

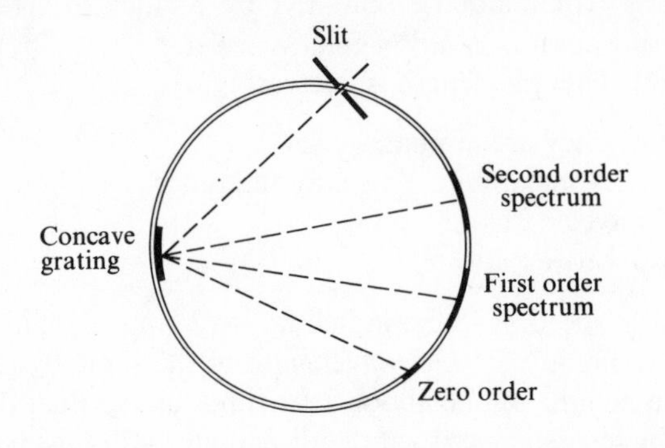

Fig. 7.2

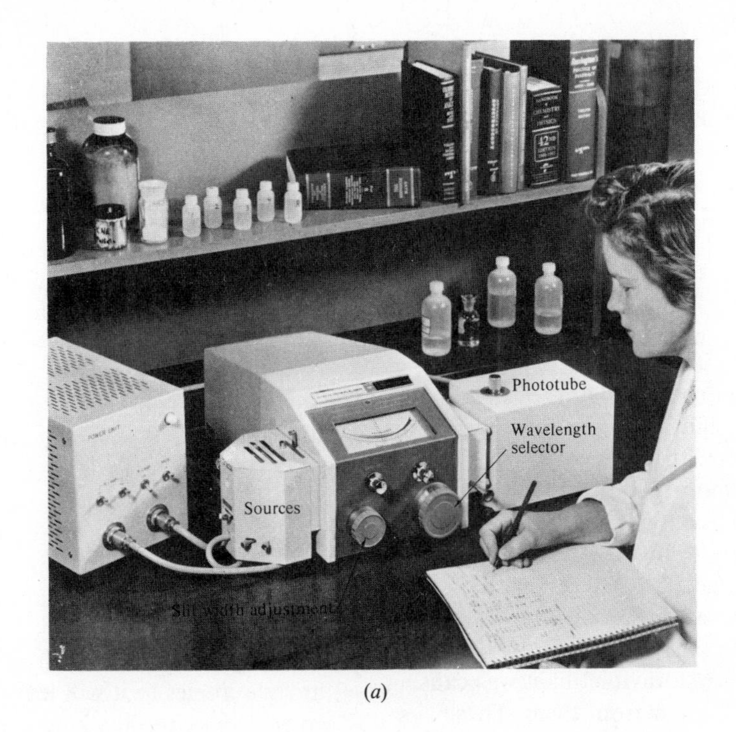

(a)

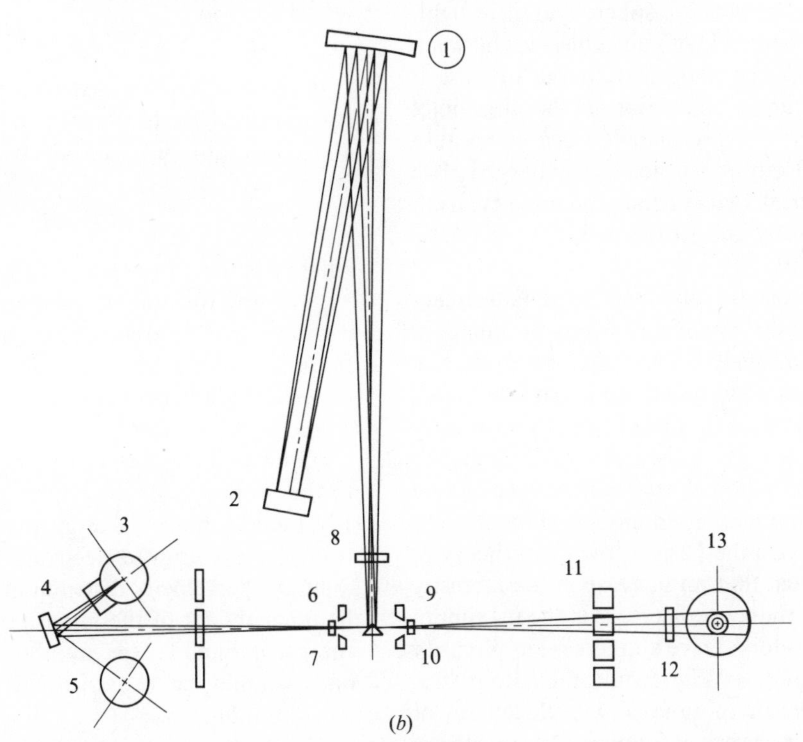

(b)

Fig. 7.3 (a) Hitachi Perkin-Elmer (model 139) Spectrophotometer. (Perkin-Elmer Ltd.) (b) Optical arrangement 1 Collimating mirror; 2 Grating; 3 Hydrogen source lamp; 4 Condensing mirror; 5 Tungsten source lamp; 6 Entrance slit; 7 Slit lens; 8 Window 1; 9 Exit slit; 10 Slit lens; 11 Absorption cell; 12 Window 2; 13 Phototube.

transparent in the visible, is opaque to the far ultraviolet and infrared. One way of overcoming this is to use reflection gratings. If the gratings are ruled on a concave mirror the grating gives both dispersion and focusing. Figure 7.2 shows such an instrument. A strip of photographic film may be used to record the spectrum produced. Photographic emulsion is sensitive to ultraviolet as well as visible radiation.

Figure 7.3 shows a spectrophotometer. Such instruments are used to determine how the absorption of radiation by a substance varies with wavelength. Radiation from a tungsten lamp is used to give, after being incident on a grating, a wide band of wavelengths. A different lamp is used for the ultraviolet region, the tungsten lamp being for the visible region. A narrow band of the wavelengths is selected and passed through a cell containing the sample and the radiation transmitted detected by a phototube. The output of the phototube is shown on the meter.

The absorption at a particular wavelength is determined by the molecules present in the sample (see chapters 4 and 12). The absorption spectrum can thus be used to identify molecules.

7.2 The photoelectric effect

In 1887 Hertz, while doing experiments on electric waves, found that when ultraviolet light from a spark shone on his apparatus, two metal spheres with a high potential difference between them, sparks came more readily. In 1888 Hallwachs found that a negatively charged zinc plate lost its charge when illuminated with ultraviolet light; a positively charged zinc plate did not lose its charge. Later, 1899, Lenard showed that the ultraviolet light was causing electrons to be emitted by metal plates. The sparks passed more readily between Hertz's metal spheres because ultraviolet light was causing the emission of electrons to occur from them. These electrons were attracted to the positive sphere and gave light, i.e., a spark, when they collided with air molecules between the spheres. Hallwachs' zinc plate lost charge because it was emitting electrons under the action of the ultraviolet light. Only the negatively charged plate lost charge because it repelled the emitted electrons, the positively charged plate attracted the electrons back to its surface and none escaped. This emission of electrons due to the action of 'light' is known as photoelectricity.

Lenard in his experiments obtained some surprising results which could not be explained. When the metal is illuminated by monochromatic light the electrons are released with all energies up to a definite maximum value. The value of this maximum energy was found to be directly proportional to the frequency of the light used. Changing the intensity of the illumination at the metal surface had no effect on the value of the maximum energy, though more electrons were released. On the basis of the wave theory of light it would be expected that an increase in the intensity of illumination would produce an increase in the maximum energy of the electrons—for a wave an increase in intensity means an increase in energy. Why should the colour (frequency) of the light have anything to do with the energy of the emitted electrons— a change in frequency means on the wave theory a change of wavelength and not a change of energy. An explanation was given by Einstein in 1905.

'On a Heuristic Point of View about the Creation and Conversion of Light

. . . According to the assumption considered here, when a light ray starting from a point is propagated, the energy is not continuously distributed over an ever increasing volume, but it consists of a finite number of energy quanta, localized in space, which move without being divided and which can be absorbed or emitted as a whole. . . .

The usual idea that the energy of light is continuously distributed over the space through which it travels meets with especially great difficulties when one tries to explain photoelectric phenomena, as was shown in the pioneering paper by Mr. Lenard.

According to the idea that the incident light consists of energy quanta with an energy (hf) . . ., one can picture the production of . . . [electrons] by light as follows. Energy quanta penetrate into a surface layer of the body, and their energy is at least partly transformed into electron kinetic energy. The simplest picture is that a light quantum transfers all of its energy to a single electron; we shall assume that that happens. We must, however, not exclude the possibility that electrons only receive part of the energy from light quanta. An electron obtaining kinetic energy inside the body will have lost part of its kinetic energy when it has reached the surface. Moreover, we must assume that each electron on leaving the body must produce work W, which is characteristic for the body. Electrons which are excited at the surface and at right angles to it will leave the body with the greatest normal velocity. The kinetic energy of such electrons is

$$hf - W$$

If the body is charged to a positive potential V and surrounded by zero potential conductors, and if V is just able to prevent the loss of electricity by the body, we must have

$$Ve = hf - W$$

where e is the . . . (charge) of the electron. . . .

. . . If the formula derived here is correct, V must be, if drawn in Cartesian coordinates as a function of the frequency of the incident light, a straight line, the slope of which is independent of the nature of the substance studied.

As far as I can see, our ideas are not in contradiction to the properties of the photoelectric action observed by Mr. Lenard. If every energy quantum of the incident light transfers its energy to electrons independently of all other quanta, the velocity distribution of the electrons . . . will be independent of the intensity of the incident light; on the other hand . . . the number of electrons leaving the body should be proportional to the intensity of the incident light. . . .'

A. Einstein (1905). *Ann. Physik* **17**, 132. Translation in D. ter Haar (1967). *The old quantum theory*, Pergamon.

According to Einstein light comes in packets, quanta. The energy carried by a quantum is directly proportional to the frequency of the light.

$$E = hf$$

where h is Planck's constant, $6{\cdot}6 \times 10^{-34}$ J s. The intensity of the light is proportional to the number of quanta reaching a surface in unit time. The term photons has been used to describe light quanta.

When Einstein announced his theory no really accurate measurements had been made of the way in which the voltage needed to stop the electrons being emitted varied with different frequencies. Einstein's equation was not accurately verified until the experiments of Millikan in 1915. The following extract is taken from Millikan's book *The electron*, first published in 1917.

'At the time at which it was made this prediction [Einstein's photoelectric equation] was as bold as the hypothesis which suggested it [light quanta], for at that time there were available no experiments whatever for determining anything about how the positive potential V necessary to supply the illuminated electrode to stop the discharge of negative electrons from it under the influence of monochromatic light varied with the frequency of the light, or whether the quantity h to which Planck had already assigned a numerical value appeared at all in connection with photoelectric discharge. We are confronted, however, by the astonishing situation that after ten years of work at the Ryerson Laboratory and elsewhere upon the discharge of electrons by light this equation of Einstein's seems to us to predict accurately all the facts which have been observed.

The testing of Einstein's equation

The method which has been adopted in the Ryerson Laboratory for testing the correctness of Einstein's equation has involved the performance of so many operations upon the highly inflammable alkali metals in a vessel which was freed from the presence of all gases that it is not inappropriate to describe the present experimental arrangement as a machine shop in vacuo. . . .

One of the most vital assertions made in Einstein's theory is that the kinetic energy with which monochromatic light ejects electrons from any metal is proportional to the frequency of the light, i.e., if violet light is of half the wave-length of red light, then the violet light should throw out the electron with twice the energy imparted to it by the red light. In order to test whether any such linear relation exists between the energy of the escaping electron and the light which throws it out it was necessary to use as wide a range of frequencies as possible. This made it necessary to use the alkali metals, sodium, potassium, and lithium, for electrons are thrown from the ordinary metals only by ultra-violet light, while the alkali metals respond in this way to any waves shorter than those of the red, that is, they respond throughout practically the whole visible spectrum as well as the ultra-violet spectrum. Cast cylinders of these metals were therefore placed on the wheel and fresh clean surfaces were obtained by cutting shavings from each metal in an excellent vacuum with the aid of the knife, which was operated by an electromagnet outside the tube. After this the freshly cut surface was turned around by another electromagnet until it was opposite the point O and a beam of monochromatic light from a spectrometer was let in through O and allowed to fall on the new surface. The energy of the electrons ejected by it was measured by applying to the surface a positive potential just strong enough to prevent any of the discharged electrons from reaching the gauze cylinder opposite and thus communicating an observable negative charge to the quadrant electrometer which was attached to the gauze cylinder. . . .'

R. A. R. Millikan (1917). *The electron*. Univ. Chicago Press.

Suppose we arrange a photocell so that the light incident on it just causes the emission of electrons. We then reduce the intensity of the light falling on the cell by placing a piece of glass between the light source and the photocell so that, say, half of the light is reflected and half transmitted through the glass to the photocell (Fig. 7.4). The photocell still has

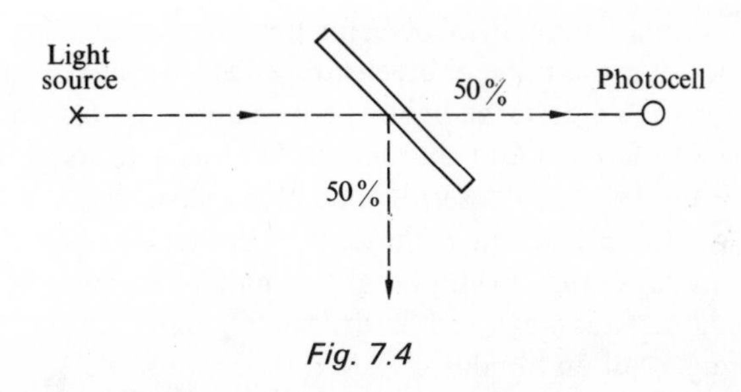

Fig. 7.4

electron emission—but half the number of electrons. Reducing the intensity of the light by half did not affect the energy per photon, it just reduced the number of photons by half.

One of the consequences of Einstein's photoelectric equation is that as the energy is directly proportional to the frequency, high-frequency radiation can do things that low-frequency radiation cannot do. Ultraviolet light can kill bacteria— red light cannot; ultraviolet light has a higher frequency and hence higher energy per quantum than red light. Ultraviolet light has much more effect on a photographic emulsion than red light.

If the energy of a photon is less than W, the energy needed to get an electron out of a metal, then no electron emission can occur. With most metals red light can cause no electron emission, because the energy is less than W. However bright or intense the red light no emission can occur if $W > hf$.

If the photoelectric effect is observed with very small metal particles another piece of evidence in favour of light quanta emerges. Small metal pieces fall between two charged plates and light of the appropriate frequency is shone on them. When an electron is emitted from a piece of metal the balance of charge on the piece is disturbed, it becomes positively or more positively charged. The instant an electron is emitted the piece of metal will thus show a sudden change in its rate of fall between the charged plates. When the experiment is done it is found that on occasions emission of electrons is found to occur as soon as the light is switched on. The emissions of electrons from the metal pieces is found to occur in a random manner with no apparent time lag being necessary before emission can take place. If light came in energy packets then this seems feasible— a piece of metal could take in enough energy for the ejection of an electron almost immediately the light was switched on, if it happened to pick up a quantum. If, however, we think of light as a wave motion we would expect that emission could not occur until a sufficient amount of wave energy had arrived at the metal piece—the wave energy is not localized but spread thinly over an entire wave front. Thus with waves we would expect a time lag which would be the same each time we did the experiment; there would appear to be no random element with waves.

Light, waves or particles?

We face a dilemma if we describe light as a wave motion—we cannot explain the photoelectric effect; we face an equal dilemma if we describe light as particulate—we cannot explain interference and diffraction. We seem to need both models if we are to describe the behaviour of light. As Sir William Bragg has said, in the early 'twenties, 'On Mondays, Wednesdays and Fridays light behaves like waves, on Tuesdays, Thursdays and Saturdays like particles, and like nothing at all on Sundays.'

Monochromatic light incident on two narrow slits placed close together gives an interference pattern (Young's experiment)—a series of fringes—a series of regions where light interferes to give constructive interference, a bright fringe, and where light interferes to give destructive interference, a dark fringe. We use the wave theory of light to calculate where the positions of the constructive and destructive interferences will be. Instead of looking at the interference fringes with the eye, we could use a photocell which we move across the interference pattern. What does the photocell detect?—photons. In the constructive interference fringes the photocell detects a large number of photons, in the destructive interference fringes only a few photons are detected. The photons seem to go to the regions forecast by our wave model. But what happens to a single photon moving up to the double slits? Which slit does it go through? Where does it go—to a constructive or destructive interference region? If it only goes through one slit surely it should be able to land anywhere—for interference we need light from two slits. G. I. Taylor, in 1909, took interference photographs with light so feeble that an exposure time of three months was necessary—the light was so feeble that the chance of there being more than one photon en route to the photographic plate at any instant was rather remote—he still obtained an interference pattern, even though only single photons were involved. Somehow a photon interferes with itself—it goes through both slits!

What is a photon? The following are the opinions expressed by two authors.

'. . . we speak of photons whenever we are thinking of the particle aspect of radiation behaviour.

. . .

. . . we now describe light, X-rays, etc. as "photons", bullets of energy and momentum, with a wave to guide their paths, somewhat like Newton's guess long ago. . . .'

E. M. Rogers (1960). *Physics for the inquiring mind*, Princeton Univ. Press.

'We can conclude with certainty, however, that light has a wave character from the study of interference and diffraction patterns. . . . Evidence just as convincing exists that light consists of small packets of energy which are highly localized, and any one of which can communicate all its energy to a single atom or molecule. These "particles" are known as light quanta or photons. . . .'

F. A. Jenkins and H. E. White (1951). *Fundamentals of optics*, 2nd ed., McGraw-Hill.

The energy necessary for sight

The following extracts are taken from an article 'Energy and vision' by S. Hecht, published in *American Scientist*, **32**, No. 3, 159, July 1944.

'*Meaning of Visual Threshold*

. . . .

How much energy do we need to produce a sensation of light? The first person to ask this question was Langley in 1889, and the reason that he could do so was that he was the first person who could answer it. Eight years before he had invented the bolometer, an instrument for measuring radiant energy. A bolometer is essentially a blackened piece of wire whose electrical resistance varies with temperature. The blackened wire absorbs the light which shines on it, its temperature rises, and its changed resistance is then measured by appropriate means.

With this instrument Langley found that he could see a light when it deposited 3×10^{-16} J on his eye. . . .

Langley's datum achieved interest in an absolute way with the advent of Planck's Quantum Theory in 1900. According to Planck, radiant energy is emitted and absorbed in discrete packets, which act as indivisible units. Each unit of light contains a quantum of energy. One cannot have half a quantum of energy, and this at once sets an absolute unit in terms of which one may express the amount of energy necessary for seeing.

For the light used by Langley, each quantum has a value of $3{\cdot}6 \times 10^{-19}$ J. Therefore, the 3×10^{-16} J found by Langley represents somewhat less than 1000 quanta; and it is this number of absolute energy units which is required to start that complicated train of events beginning at the eye and ending in the cortex of the brain which results in our seeing light.

In 1911 Langley's measurement achieved significance a second time when Einstein formulated what has since become known as the Einstein Photochemical Equivalence Law. This states that for primary photochemical reactions each quantum of energy is absorbed by a single molecule which is then changed chemically by it. In short, one quantum absorbed means one molecule changed. Langley's measurement then says that for us to see a flash of light about 1000 molecules of light-sensitive material in the eye are changed chemically by the light. This is really an astonishingly small chemical change.

We know now that Langley's measurements are wrong. But they are wrong in the right direction. The real value is even smaller than what he found; it is smaller by a factor of 10. This is mainly because Langley worked before the physiology of vision was known sufficiently well to enable him to use the eye most effectively. However, this is of minor importance compared to the establishment of an approximate order of magnitude. We have the fact that a biological process which involves such a high level of organization as a conscious act of seeing can be set in motion by about 1000 quanta of energy and 1000 molecules of matter. This is an exciting thing to contemplate and, if correct, deserves repetition and study. . . .

Apparatus and Manipulations

. . . . [description of apparatus]

Two people are required for the measurements, an operator and an observer. The operator controls the light intensities, keeps the current constant, and gets everything ready for the observer. The observer merely sits comfortably in a curtained cubicle in the dark room with his eye near the exit pupil of the apparatus. When he is informed that all is ready, he looks directly at the little red light, and when he deems the moment just right, he presses the button which permits a flash of light to pass through the apparatus. He then merely records whether or not he has seen a flash of light with his peripheral vision. He has no previous knowledge of whether the light will be bright or dim or whether there will be any light there at all. He merely reports yes or no by ringing a bell, since his mouth is otherwise occupied.

After each flash the operator changes the intensity, and when he is ready he tells the observer to go on. The observer again releases a flash and records his observation. This is continued until the operator is satisfied that he knows what the observer's threshold is, or until he has the requisite number of observations from which to determine it. In general we consider that an observer's threshold has been reached when he sees a given intensity of light 60 per cent of the time.

New Measurements

It is apparent that training and skill are required to be an observer; we used only seven subjects, all of whom were experienced in this kind of experimentation. Over a period of a year and a half we made measurements with them several times, with the results shown in Table II.

TABLE II—MINIMUM ENERGY FOR VISION

Each datum is the result of many measurements during a single experimental period, and is the energy which can be seen with 60 per cent frequency.

$\lambda = 510$ mμ; $h\nu = 3{\cdot}84 \times 10^{-19}$ J.

Observer	*Energy* $/10^{17}$ J	*No. of quanta*	*Observer*	*Energy* $/10^{17}$ J	*No. of quanta*
S. H.	4·83	126	C. D. H.	2·50	65
	5·18	135		2·92	76
	4·11	107		2·23	58
	3·34	87		2·23	58
	3·03	79			
	4·72	123	M. S.	3·31	81
	5·68	148		4·30	112
S. S.	3·03	79	S. R. F.	4·61	120
	2·07	54			
	2·15	56	A. F. B.	3·19	83
	2·38	62			
	3·69	96	M. H. P.	3·03	79
	3·80	99		3·19	83
	3·99	104		5·30	138

Two things come from an examination of this Table. First, it is encouraging that the order of magnitude of our measurements is the same as that of the best three previous determinations already shown in Table I. All our observations lie between 2·1 and $5{\cdot}7 \times 10^{-17}$ J. They are about twice as large as the previous measurements, and there is a certain amount of overlap between them and the others. Nevertheless, one can confidently consider them as confirming one another.

Since the energy of a single quantum of blue-green light is $3{\cdot}84 \times 10^{-19}$ J, our measurements represent thresholds between 54 and 148 quanta. The second thing apparent from Table II is that an individual's threshold varies over a period of time. Mine (S. H.) varies between 79 and 148 quanta; Shlaer's (S. S.) between 54 and 104; and Pirenne's (M. H. P.) between 79 and 138. In other words a person's threshold may vary by a factor of 2, and this agrees with a great deal of previous work on thresholds.

The original question raised at the beginning of our inquiry has now been answered. In order for us to see a light, we must receive from it between 54 and 148 quanta. However, this answer far from exhausts the interest of the measurements. In fact, their real significance has not been even hinted at, and it is to this that we may now turn.

The 54 and 148 quanta necessary for seeing represent the energy incident on the cornea, and we have tacitly made the assumption that they represent the actual energy required for initiating a visual act in the retina. This is not correct. There are at least three corrections that must be made before these numbers acquire meaning in terms of chemical processes in the retina.

Figure [7.5] presents a horizontal section of the human eye with its main landmarks indicated. The drawing shows that light must pass through several structures before it reaches the retina. The first solid surface which the light reaches is the smooth and wet cornea. From such a surface it suffers a reflection of about 4 per cent.

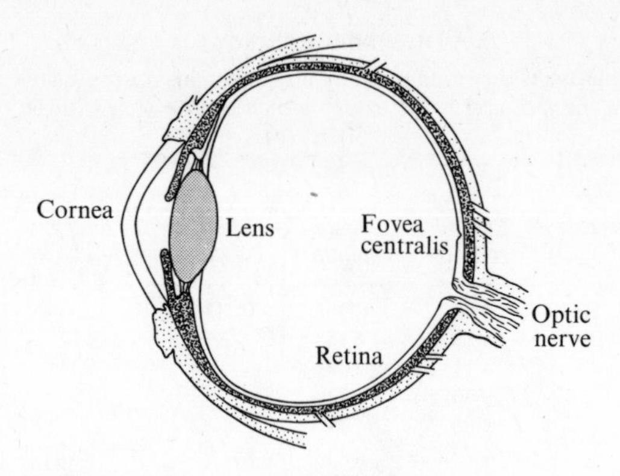

Fig. 7.5 Diagram of a horizontal section through the human eye. It is evident that light in going successively through the cornea, the aqueous humor in the anterior chamber, the lens, and the vitreous humor in the posterior chamber, must suffer some reflections and absorptions before it reaches the retina where it does its photochemical work.

This means that of the minimum 54 quanta 2 are thrown away before entering the eye.

However, after entering the eye the light passes several barriers. As it leaves the cornea it enters the aqueous humor; this is a fluid, undoubtedly less dense than the cornea, and as a result the light will suffer another reflection here. After this it reaches the front surface of the very dense lens, and then the rear surface, to enter and pass through an inch of jelly called the vitreous humor before it impinges on the retina. At all these surfaces it will suffer reflection, and it will also be absorbed and scattered by the ocular media which contain protein.

It used to be thought that the eye media are completely transparent. This must be partly true, because we really do see through it with excellent precision. But it is not completely transparent, as Roggenbau and Wetthauer have shown with cattle eyes. One may then inquire what fraction of the light which enters the cornea finally reaches the retina. This has been measured by Ludvigh and McCarthy with human eyes, and their averaged measurements for the 21-year-old eye are shown in Figure [7.6].

From Figure [7.6] it is evident that for violet light only about 10 per cent of the incident light reaches the retina, while for red the fraction is about 70 per cent. For the

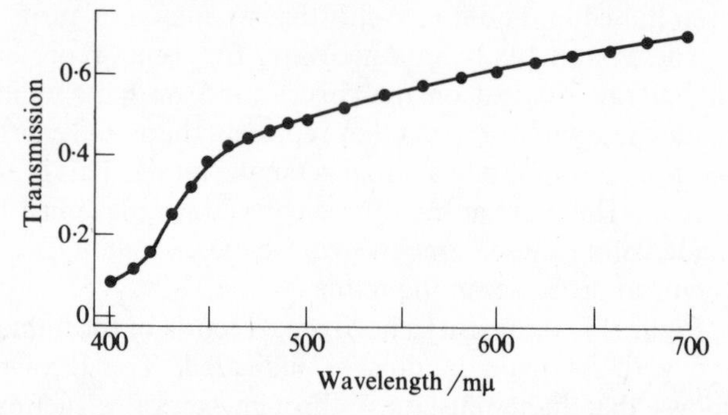

Fig. 7.6 Transmission of light through the human eye. The measurements are by Ludvigh and McCarthy and give the fraction of incident light which finally reaches the retina. This fraction varies with the wavelength of the light.

particular light in which we are interested, namely 510 mμ, almost exactly 50 per cent of the entering light reaches the retina. In an older eye the fraction is even less.

From all this it is clear that of the 54 quanta which fall on the cornea only 26 finally reach the retina. It seems almost profane to be so free with these few precious quanta, yet this is not the end.

Figure [7.7] shows an enlarged diagram of the retina. The light comes from the left and passes through the nerve cells to fall on the layer of sensitive elements to the right. These are the rods and cones, of which three each are drawn, while the rest are indicated by horizontal lines. These retinal elements are closely packed one next to the

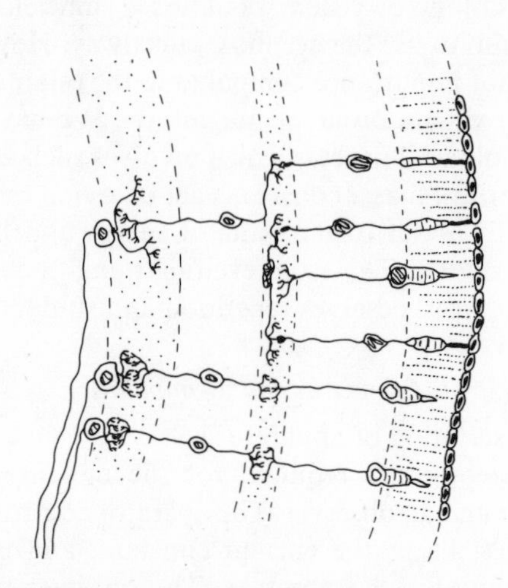

Fig. 7.7 Diagram of the structure of the retina. Light comes from the left, passes through the nerve cells, and impinges on the rods and cones, which are the sensitive elements. The cones are for day vision, and are shown as the stubby elements, while the rods are for night vision, and are drawn as the thinner and longer elements with well-defined terminal segments touching the layer of very black pigment cells at the extreme right. These terminal segments contain visual purple and in them occur the initial photochemical events involved in seeing light at low illuminations.

other, with no spaces between. The cones are of no interest to us here, but the rods—shown by the dark terminal pieces—are the elements concerned in the present problem, and are present to the extent of 150 000 per square millimeter.

In order to be effective, the light must be absorbed by the terminal segments of the rods. Here takes place the photochemical conversion of light into sensory impulses, which pass through the bipolar cells into the ganglion cells and through the brain to the occipital cortex. Any light which is not absorbed by the rods passes through the retina and is absorbed by the densely black pigment cells at the extreme right of Figure [7.7].

In our measurements we have found that when 26 quanta fall on the retina they constitute a minimal visual stimulus. Clearly all of this light cannot be absorbed by the sensitive elements, because they would have to be practically black, and we know they are not. The question then is what fraction of this light is absorbed by the rods for chemical work, and what fraction goes through unabsorbed.

It is the visual purple in the terminal segments of the rods which absorbs the light, and one can determine the amount it absorbs by ascertaining how much visual purple a human retina contains. Such measurements were made by Koenig, who computed that if this quantity were spread over the whole retina, it would absorb only about 5 per cent of light of 510 mμ. This is a small value indeed, and would seem almost unacceptable if it were not that Wald had found similar quantities in the retinas of the rabbit and the rat, namely 4 per cent and 13 per cent respectively.

Even so, these figures are probably too low. . . .

The value of 5 per cent, however, may be considered as setting the lower limit for retinal absorption.

If one accepts this value, or even 10 per cent, as representing retinal absorption, it means that of the 26 quanta falling on the retina no more than 2 or 3 are absorbed by the rods to start the events which end in our seeing light. . . .'

7.3 Energy levels

When a table-tennis ball falls onto a smooth hard surface it bounces and kinetic energy is almost conserved. Collisions in which kinetic energy is conserved are called elastic collisions. When a piece of 'Plasticine' hits the surface rebounding does not occur—kinetic energy is not conserved. Such collisions are called inelastic collisions.

Is it to be expected that kinetic energy will be conserved when electrons hit atoms? If kinetic energy is conserved then the energy of an electron after a collision will be the value it had before the collision. The following are the results obtained by a number of experimenters who fired electrons at atoms.

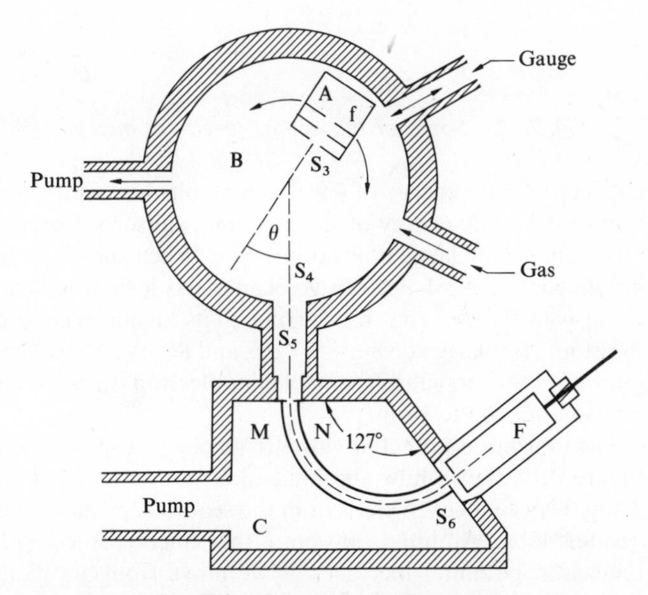

Fig. 7.8 Diagram of apparatus.

'The apparatus used . . . may be briefly described as consisting of an electron projector A, a collision chamber B and an analysing chamber C. A diagram is given in [Fig. 7.8]. Electrons leaving the projector A were scattered by the gas molecules in chamber B. After single collisions in B the electrons passed through the slit S_4 into the analyser C. . . .

In [Fig. 7.9] can be seen an energy loss curve for helium. This curve was taken with electrons of 50 volts energy and indicates the energy losses for those deflected at 10°.

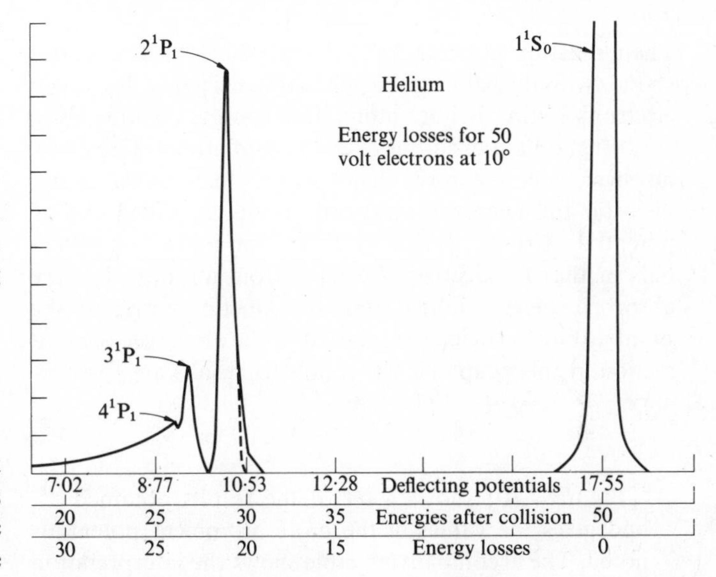

Fig. 7.9 Energy losses in helium for 50 volt electrons simultaneously deflected at 10°.

The abscissa measures the energy of the electrons after collision, while the ordinate measures the number of electrons having that particular energy. The energy loss is also indicated in the diagram. The main peak at 50 volts, comprised of electrons making elastic impacts, is many times higher than the remaining peaks and extends off the figure. A group of peaks is seen near the 30 volts energy region, indicating losses of the order of 20 volts. . . .'

J. H. McMillen (1930). *Phys. Rev.* **36**, 1034.

In the above experiment some of the electrons lost energy, due presumably to collisions with atoms. But there is a strange thing about the results—the electrons do not seem able to lose energy in a random manner but only in well-defined amounts, e.g., a loss of the order of 20 volts energy to helium atoms. (The unit volts is used to indicate the energy an electron would gain, or lose, if accelerated through a potential difference of 1 volt. This is an energy of $1{\cdot}6 \times 10^{-19}$ J.) What is happening when the electrons lose energy?

'It was shown in our experiments on collisions between electrons and molecules of an inert gas or of a metal vapour that the electrons are reflected in such collisions without loss of energy, as long as their kinetic energy does not exceed a certain critical magnitude, but that as soon as their energy becomes equal to the critical value, they lose all of it on collision. The critical velocity is a quantity, characteristic for each gas and is in the cases studied so far equal to the ionization energy. . . .'

Franck and Hertz (1914). *Verh. Dtsch. Phys. Ges. Berlin* **16**, 512.
Translation in D. ter Haar (1967). *The old quantum theory*, Pergamon, 1967.

The ionization potential for helium is 24·58 V. This corresponds with the value, given in McMillen's paper, for which electrons hitting helium atoms lose energy. At ionization electrons are knocked out of the helium atoms. There are, however, energies lower than this at which collisions are inelastic and energy is absorbed. These are called critical potentials. Critical potentials differ from ionization potentials in that no electron is ejected from an atom but the absorbed energy, taken from the kinetic energy of the bombarding particle, is re-emitted by the atom (see the next section in this chapter). The following results are for mercury:

'[Figure 7.10] shows a set of the results obtained . . . and gives the values of the more prominent potentials noted. The accompanying table shows the interpretation placed on the various potentials. The curves obtained . . . could be reproduced without difficulty.

TABLE I

Pot. observed (volts)	Interpreted from Theory	Calculated from Theory	Remarks
4·7	$1^1S - 2^3P_0$	4·66	Metastable state
4·9	$1^1S - 2^3P_1$	4·86	2537
5·2			5·25 obs. by Messenger
5·4	$1^1S - 2^3P_2$	5·43	Metastable state
5·7	2^3P_0	5·73	Ionization of metastable excited atom
6·8	$1^1S - 2^1P$	6·67	
7·6	$1^1S - 2^3S$	7·69	
8·1			8·0 and 8·05 obs. by Messenger
8·4			8·35 obs by F. & E. 8·3 obs. by Messenger
8·6	$1^1S - 3^3P_{0,1}$	8·58	
8·9	$1^1S - 3^1P$	8·79	
	$1^1S - 3^3D_{1,2,3}$	8·82	
	$1^1S - 3D$	8·80	
	$1^1S - 3^3P_2$	8·80	
9·3	$2(1^1S - 2^3P_0)$	9·32	Successive impact
9·8	$2(1^1S - 2^3P_1)$	9·8	Successive impact
10·3	1^1S	10·39	Ionization
	6·7 + 4·9		
11·5	6·7 + 4·7		

J. C. Morris (1928). *Phys. Rev.*, **32**, 447.

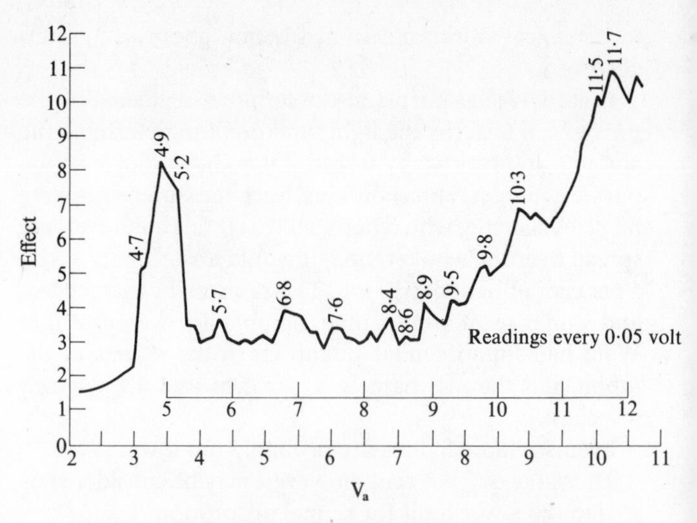

Fig. 7.10 Critical potentials of Hg vapour.

Earlier experiments were too coarse to reveal the detail of the critical potentials. For any atom there exists a whole series of energy levels at which energy can be absorbed by an atom. A convenient way of looking at these levels is as a ladder with rungs at certain energy levels (Fig. 7.11). When

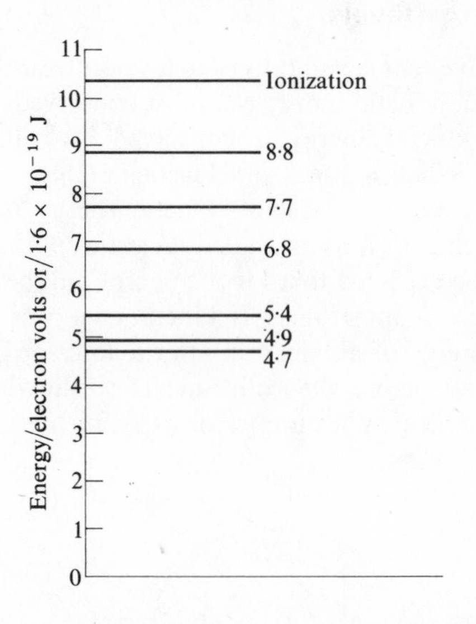

Fig. 7.11 Some of the energy levels for mercury.

an electron with energy of 4·9 electron volts hits a mercury atom the kinetic energy of the electron is absorbed by the atom and an inelastic collision occurs. When the electron has an energy of 5·4 electron volts an inelastic collision can occur with the electron transferring all its kinetic energy to the atom. Similarly at 6·8 eV, 7·7 eV, and 8·8 eV. At 10·4 eV ionization occurs and the incident electron knocks an electron out of the atom.

The ionization potential varies from element to element. Figure 7.12 shows how the ionization potential varies in going from element to element in the sequence given by the periodic table. An important point that emerges is that the ionization potential increases as we move from an alkali element to an inert element, e.g., Li to Ne. The alkali ele-

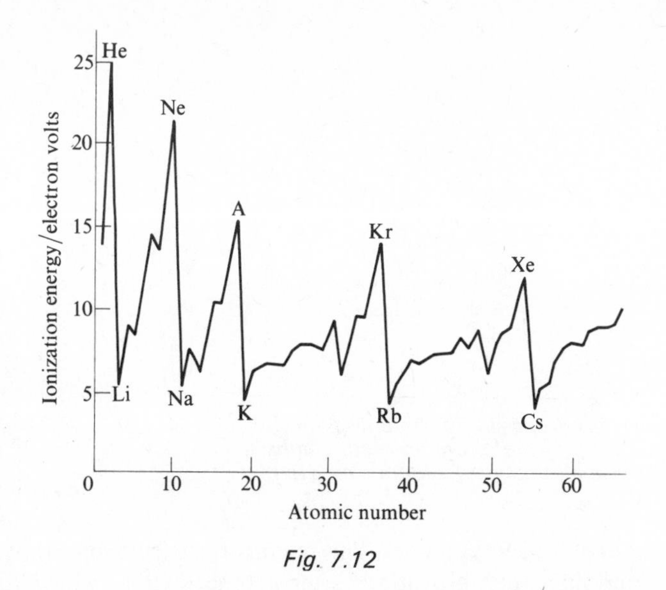

Fig. 7.12

ments are very reactive and have the lowest ionization potentials, i.e., it is very easy to remove one electron from them. The inert elements have high ionization potentials and are very un-reactive.

Light quanta

Atoms can only accept energy in certain well-defined size packets; they have energy levels. Atoms can emit certain well-defined light frequencies, the line emission spectrum. Are these events connected?

> '... As we emphasized in our earlier paper the majority of collisions which transfer to the ... [atom] an energy hf do not lead to ionization.... One should, therefore, expect that such collisions which do not lead to ionization but just to an energy loss hf should be accompanied by an emission of light of frequency f.... This means that if one introduces electrons in mercury vapour and enables them to attain a velocity corresponding to voltage difference of 4·9 volts, one should be able to observe light emission corresponding exclusively to the emission of the mercury 2536 Å... line. Experiments have fully confirmed this expectation.
>
> ... The photographs obtained after exposing for one or two hours showed a continuous spectrum, stretching into the violet, caused by the light emitted by the incandescent wire (the source of the electrons), and then, a long distance away from it, clearly the 2536 Å line: however, in no case was there even a suspicion of the other mercury lines.... The identification of the line was made by comparison with a wavelength scale built for the apparatus and also by imposing upon the spectrum the arc spectrum of mercury as a comparison spectrum....'
>
> J. Franck and G. Hertz (1914). *Verh. Dtsch. Phys. Ges. Berlin*, **16**, 512. Translation in
> D. ter. Haar (1967). *The old quantum theory*, Pergamon.

The smallest energy that a mercury atom can accept and give emission is that corresponding to the acceleration of an electron through 4·9 V. The 4·7 V level is a metastable level from which no emission occurs. At 4·9 V an atom absorbs the energy and then emits it as a single frequency line of wavelength 2536 Å. The ångström unit Å is 10^{-10} m. If we increase the energy that is supplied to the mercury atoms then at other quite specific energy values energy becomes absorbed by the atoms and re-emitted as lines in the mercury spectrum.

The kinetic energy of the bombarding electron has been transformed into light of a particular frequency. What is the relationship between energy and frequency? 4·9 electron volts of energy give a wavelength of 2536 Å, a frequency of $1{\cdot}18 \times 10^{15}$ Hz for mercury atoms. If magnesium is used 2·7 electron volts are needed to give a single line spectrum, wavelength 4571 Å (frequency of $6{\cdot}56 \times 10^{14}$ Hz). The greater energy necessary with the mercury atoms gave a larger frequency than for the magnesium. In fact these two results can be explained if we assume that the frequency of emission is directly proportional to the energy.

For the mercury

$$\frac{\text{energy}}{\text{frequency}} = \frac{4{\cdot}9 \times 1{\cdot}6 \times 10^{-19}}{1{\cdot}18 \times 10^{15}} \frac{\text{joules}}{\text{Hz}}$$

$$= 6{\cdot}6 \times 10^{-34} \text{ J s}$$

For the magnesium

$$\frac{\text{energy}}{\text{frequency}} = \frac{2{\cdot}7 \times 1{\cdot}6 \times 10^{-19}}{6{\cdot}56 \times 10^{14}} \frac{\text{joules}}{\text{Hz}}$$

$$= 6{\cdot}6 \times 10^{-34} \text{ J s}$$

The other lines in the mercury spectrum can fit this relationship energy/frequency $= 6{\cdot}6 \times 10^{-34}$ J s if we assume that the value of energy can be the differences between permitted energy levels. A jump between energy levels gives rise to frequencies given by

$$\frac{E'' - E'}{6{\cdot}6 \times 10^{-34}} = \text{frequency}$$

where E'' is the energy of the level from which the atom drops, E' is the value of the energy level to which the atom drops (this need not be the zero energy level). This relationship can be summmarized as

$$E = hf$$

where h is a constant, known as Planck's constant.

The emission and absorption of radiation

When an atom makes a downward energy jump, from an energy level of more to an energy level of less energy, the energy released appears as a photon. Only a photon whose energy fits the jump between the levels can be emitted (Fig. 7.13(a)). The frequency of the photon is given by

$$E = hf$$

where E is surplus energy, equal to the difference in energy

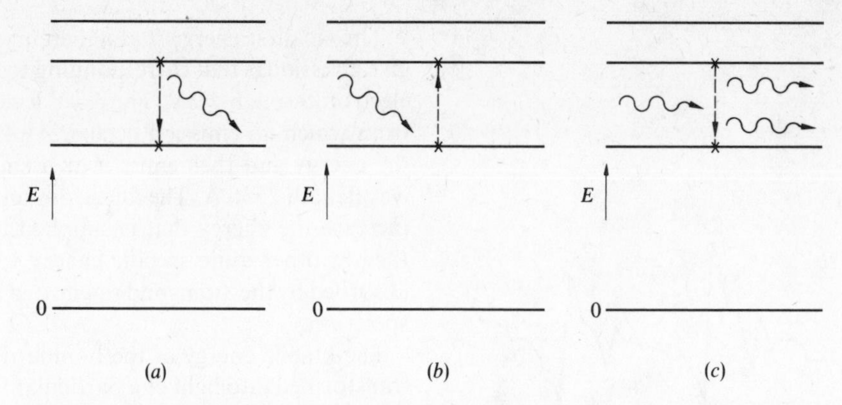

Fig. 7.13 (a) Emission of a photon, an atom drops down the energy level ladder (spontaneous emission). (b) Absorption of a photon, an atom climbs up the energy level ladder. (c) Stimulated emission of a photon, an atom drops down the energy level ladder.

between the two energy levels, h is Planck's constant, and f the frequency of the photons.

In a similar way to emission, an atom can only absorb radiation (photons) whose energy fits one of the possible energy jumps between energy levels (Fig. 7.13(b)).

If we pass white light (i.e., a wide range of frequencies and hence photon energies) through a gas, absorption can only occur for those photons whose energy corresponds to the energy difference between two energy levels in the gas atoms. When white light is passed through sodium vapour the resulting spectrum shows all frequencies with the exception of the yellow sodium lines (Fig. 7.14). The absorption spec-

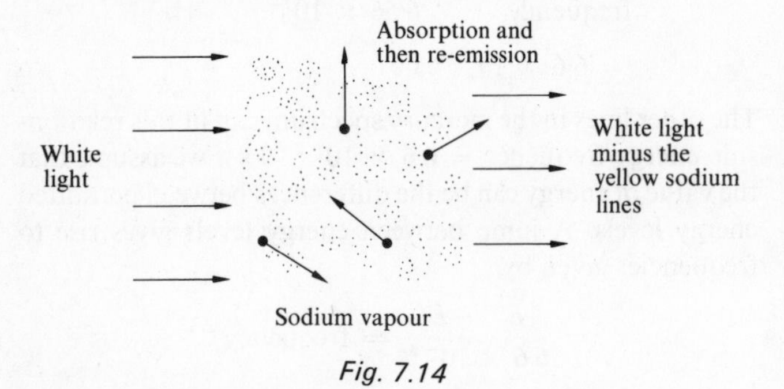

Fig. 7.14

trum of the sodium vapour is at the same frequencies as sodium vapour normally gives in an emission spectrum. The sodium vapour has in fact absorbed the photons with the energy corresponding to that which gives the yellow spectrum lines. The atoms which acquire this surplus energy later re-emit the energy, as the same frequency photons. The photons are, however, not necessarily emitted in the same direction as the incident photons. The result is that the white light has less yellow photons moving in the initial direction.

This absorption and re-emission process can be summarized as—a photon of the right energy hits the atom, the atom becomes excited and moves to a higher energy level, at some time later the atom drops back down the energy level ladder and re-emits the photon. Re-emission can take place in a number of steps as the atom drops back down the steps of the energy ladder. This re-emission occurs some time later than the absorption, after a time interval which is random. This form of emission is known as spontaneous emission.

The absorption and emission process we have just considered assumed that the atom being hit by the photon was able to absorb the photon's energy, i.e., the energy of the photon was just the right value to raise the atom from the energy level it was in to a higher energy level. If, however, the atom was already at this higher level then there is a chance that a process known as stimulated emission can occur (Fig. 7.13(c)). When this occurs the incident photon triggers the emission of an identical photon from the atom, the atom dropping to a lower energy level. Thus the incident photon results in two identical photons moving away from the atom. In this case there is no time lag between the two emerging photons and they both move off in the same direction. This is an increase in the intensity of the light (a factor of two as the number of photons has been doubled). The two photons are said to be coherent.

If the conditions are right for stimulated emission, i.e., a large number of atoms already excited to the correct energy level, then a large build up of intensity occurs. The light is monochromatic, i.e., one frequency, because it is produced by jumps between just one pair of energy levels. A device in which such intense beams of monochromatic radiation is produced is, in the case of visible light, known as a laser (*l*ight *a*mplification by *s*timulated *e*mission of *r*adiation) and in the case of microwaves as a maser (*m*icrowave *a*mplification by *s*timulated *e*mission of *r*adiation). The first such device was produced in 1954 and produced waves of wavelength 0·0125 m. Ammonia molecules were used as the photon target atoms. 1960 saw the first solid state laser, a ruby giving a wavelength of $0{\cdot}694 \times 10^{-6}$ m (a bright red colour). 1961 saw the first gas laser, using a mixture of helium and neon.

All masers and lasers are amplifiers for just one narrow frequency band; a few photons, of the right energy, cause an avalanche of stimulated photons.

The basic requirement for a laser or maser is that there must be a large number of atoms or molecules in the correct

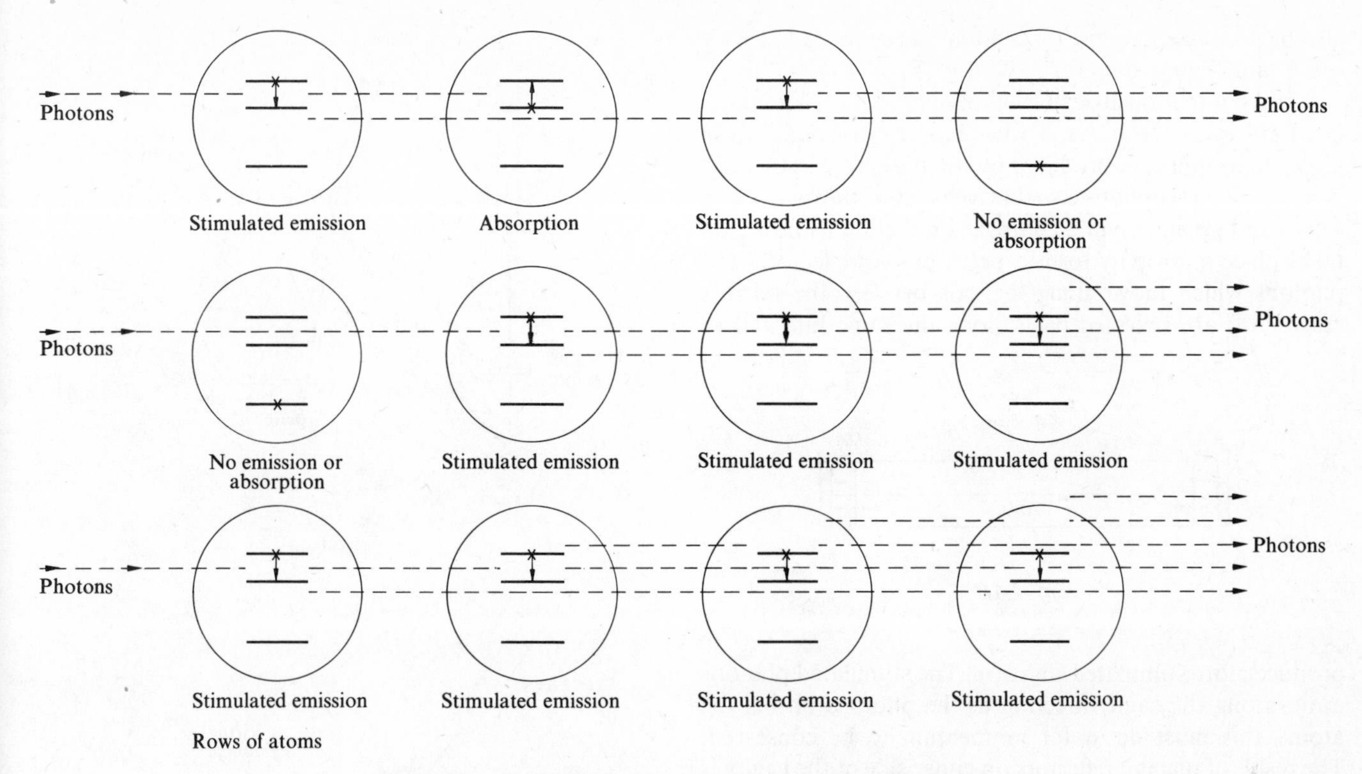

Fig. 7.15 A possible sequence of events for laser action.

excited state for stimulated emission to exceed spontaneous emission (Fig. 7.15). This means that we must have more atoms at the higher energy level than the lower energy level. This is known as population inversion. Under normal conditions more atoms will be at a lower energy level than at a higher one, the population of atoms at a higher level is lower than the population at a lower level. Population inversion can be produced in a number of ways. In the ruby laser bright light from a xenon flashtube is focused on the ruby. The flashlamp only operates for a few microseconds and in that short time interval excites a considerable number of atoms. Because, in this case, the spontaneous decay of the atoms from this excited state takes quite a long time a population inversion is produced—the flash lamp is exciting atoms faster than they can decay. In the helium-neon laser the production of population inversion is achieved in two stages. A high voltage, of the order of 1 kV, is applied across the tube which contains about 85 per cent helium and 15 per cent neon at a combined pressure of about 1 mm of mercury. The high voltage accelerates ions in the tube and so excites the helium atoms to their lowest energy level of 19·8 eV above the ground state. The helium atoms collide with the neon atoms and because neon has an energy level at 19·8 eV energy can be transferred from the excited helium atoms to the neon atoms (Fig. 7.16). This 19·8 eV level is not the lowest level for neon. This results in the neon having more atoms at the 19·8 eV level than its lower levels—a population inversion has been produced. When the neon atoms drop down to the next lower level, at 18·6 eV, photons of wavelength $1{\cdot}15 \times 10^{-6}$ m are produced. Photons of wavelength $0{\cdot}633 \times 10^{-6}$ m (a red colour) can

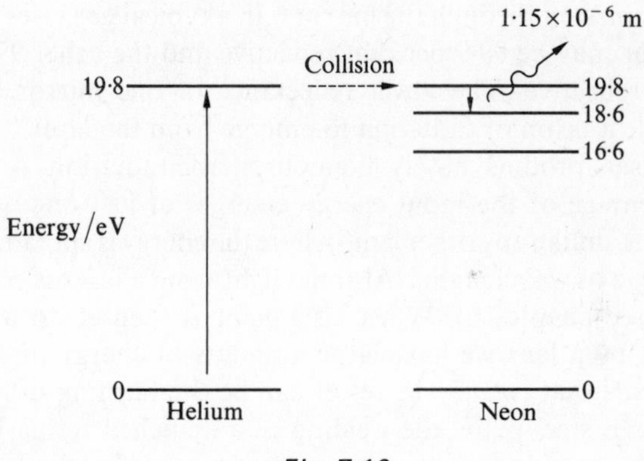

Fig. 7.16

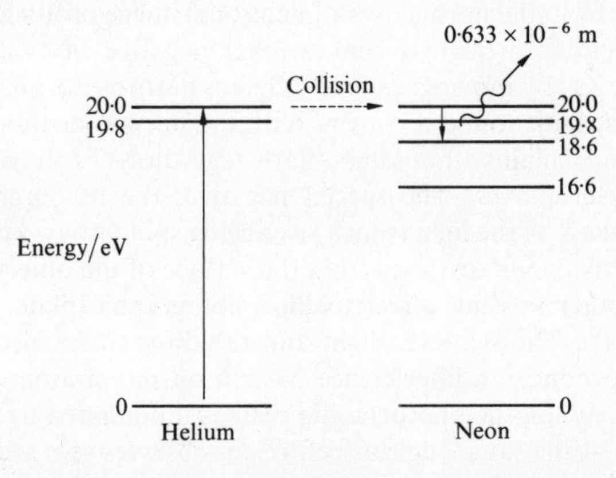

Fig. 7.17

also be produced because the helium energy level at 20·0 eV is the same as a neon level at 20·0 eV (Fig. 7.17).

When population inversion has been produced then stimulated emission can occur. To produce a large build up in intensity the photons produced by stimulated emission must produce yet more photons so that a chain reaction can occur. This build up can be in a controlled direction if the action takes place in a cavity formed between two reflectors. The photons which move along the axis between the mirrors (Fig. 7.18) are reflected back along the same line and so

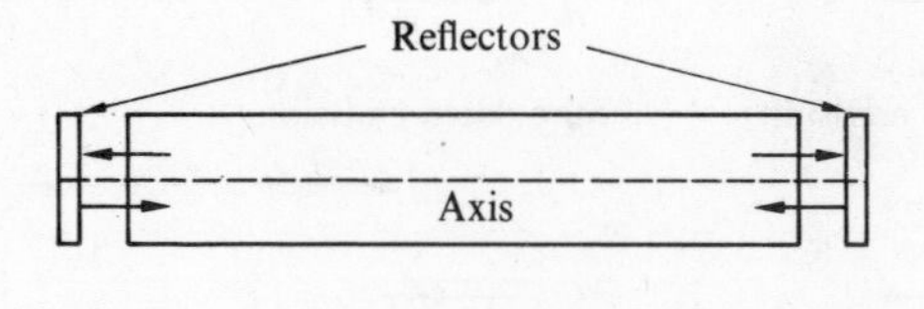

Fig. 7.18

produce more stimulated emission. The stimulated photons move along the same direction as the photons hitting the atom—this must occur for momentum to be conserved. The result of placing reflectors on either side of the cavity is to produce an intense beam along the axis. Since the reflectors are not 100 per cent reflective some photons are lost by transmission through the mirrors. In a typical gas laser, one mirror may be 99·9 per cent reflective and the other 99 per cent reflective. The lower reflectance of one mirror is to enable a beam of radiation to emerge from the laser.

Lasers produce highly monochromatic radiation. A high percentage of the input energy emerges at just one wavelength, unlike any other lamp where the energy is spread over a range of wavelengths. Also the light from a laser is coherent (see chapter 6). When laser light is focused to a fine point by a lens we have large amounts of energy directed towards that point. The result can be the burning of holes through steel plate, the welding of a detached retina back into place in the human eye, or indeed any application where highly localized energy is required.

A technique known as holography permits three-dimensional images of an object to be produced by projection through a photographic negative. Three-dimensional images differ from the normal two-dimensional image produced by projection through a conventional negative in that the observer, by moving, can see different parts of the image—he can look round a corner. With the normal method the image remains the same—flat—regardless of how the observer moves. The special negative, the hologram, is produced by the light from a laser being split into two parts, one part being scattered from the surface of the object and the other passing direct to the photographic plate (Fig. 7.19(a)). The scattered light and the direct light interfere and produce an interference pattern on the photographic plate. When the photographic plate is illuminated by laser light, at the same angle as before, an observer sees a three-dimensional virtual image of the object (Fig. 7.19(b)).

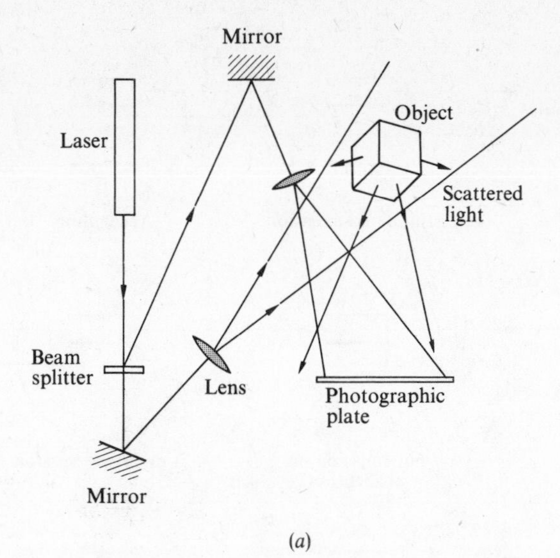

(*a*)

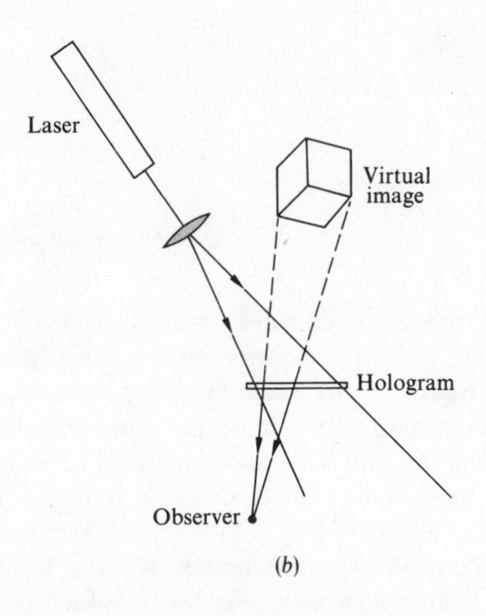

(*b*)

Fig. 7.19

7.4 Electrons, particles or waves?

Charge comes in packets, the minimum size packet being called the electron. This is the evidence given by an experiment such as the Millikan experiment (see chapter 10) where the charge on oil drops was measured and found to be integral multiples of a basic charge. Assuming that the charges on the drops were produced by adding or removing electrons then this gave the charge on an electron. The charge-to-mass ratio for electrons can be measured (see chapter 13). The results of such experiments give for the charge on the electron $1{\cdot}6 \times 10^{-19}$ C and for the mass $9{\cdot}1 \times 10^{-31}$ kg (at speeds not comparable with the speed of light). Electrons would certainly seem to be particles.

In 1925 Davisson, while doing experiments involving the scattering of electrons by a platinum target, had an accident—a liquid air bottle exploded when the target was at a high temperature. The experimental tube was broken and the platinum target became heavily oxidized by the in-rushing air. Davisson managed to rid the platinum target of the oxide by prolonged heating at high temperatures. When, however, he repeated his scattering experiment the results were completely different. The difference was found to be due to the target having previously been composed of small crystals but after the heating it was just a few large crystals. Why should the scattering from a polycrystalline target be different from that of just a few large crystals? Davisson's experiment with electron beams probing into platinum was to give information about the distribution of the atomic electrons in the platinum. Changing the crystalline structure of the platinum would not have been expected to change the distribution of atomic electrons.

A year earlier, 1924, de Broglie in his Ph.D. thesis had put forward a novel suggestion. In his own words:

> '... I assumed that the existence of waves and particles, perceived by Einstein in 1905 in respect of light in his theory of light quanta, should be extended to all types of particle in the form of coexistence of a physical wave with a particle incorporated in it....'
>
> L. de Broglie (1971). *Physics Bull.*, 149.

De Broglie was suggesting that not just light but all matter should have both a wave and a particle characteristic. Other than for light, there was no experimental evidence to justify this suggestion. De Broglie's work was nothing more than speculation. His thesis was, however, to lead to a Nobel prize.

De Broglie suggested that the wavelength of all matter is given by

$$\lambda = \frac{h}{\text{momentum}}$$

This was essentially arrived at by equating $E = hf$ and $E = mc^2$.

How did this fit with the work of Davisson? In 1925 Elsasser suggested that the wave nature of electrons, as forecast by de Broglie, might be found by studying the scattering of electrons from a crystal. The crystal could act as a diffraction grating (see chapter 12 'X-ray diffraction'), and he also suggested that the work of Davisson might yield the results. Davisson, however, doubted this.

> 'The attention of C. J. Davisson was drawn to Elsasser's note of 1925, which he did not think much of because he did not believe that Elsasser's theory of his [Davisson's] prior results was valid. This note had no influence on the course of the experiments. What really started the discovery was the well-known accident with the polycrystalline mass, which suggested that single crystals would exhibit interesting effects. When the decision was taken to experiment with the single crystal, it was anticipated that 'transparent directions' of the lattice would be discovered. In 1926 Davisson had the good fortune to visit England and attend the meeting of the British Association for the Advancement of Science at Oxford. He took with him some curves relating to the single crystal, and they were surprisingly feeble (surprising how rarely beams had been detected!). He showed them to Born, to Hartree and probably to Blackett; Born called in another continental physicist (possibly Franck) to view them, and there was much discussion of them. On the whole of the transatlantic voyage Davisson spent his time trying to understand Schrödinger's papers, as he had then an inkling (probably derived from the Oxford discussion) that the explanation might reside in them. In the autumn of 1926 Davisson calculated where some of the beams ought to be, looked for them and did not find them. He then laid out a programme of thorough research, and on the 6th January 1927 got strong beams due to the line-gratings of the surface atoms, as he showed by calculation in the same month.'
>
> Statement dictated by Davisson in 1937
> (G. P. Thomson (1968). *Contemp. Phys.* **9**, 4).

The papers by Schrödinger referred to above were where Schrödinger introduced the idea that particles were not localized but smeared out over space and their behaviour could be described by a wave equation. Unlike de Broglie who still considered particles to be localized and have a guiding wave, Schrödinger dispensed with the localized particle and just left the wave. (See chapter 13 for a longer discussion of Schrödinger's work.)

Using de Broglie's equation for the wavelength of the electrons $\lambda = h/\text{momentum}$ (the momentum can be calculated from a measurement of the potential difference, V, through which the electrons had been accelerated: $\frac{1}{2}mv^2 = Ve$), Davisson found that the angles at which reflection of electron beams from crystals occurred fitted the idea of the crystal surface being a line grating, spacing d, and the electron a wave. The equation which fitted his results was the diffraction grating equation $n\lambda = d \sin \theta$. The experimental evidence was thus—electrons have wave properties.

Since Davisson's results there have been many more experiments with electrons which reveal the wave nature of electrons, the wavelength being that given by de Broglie's

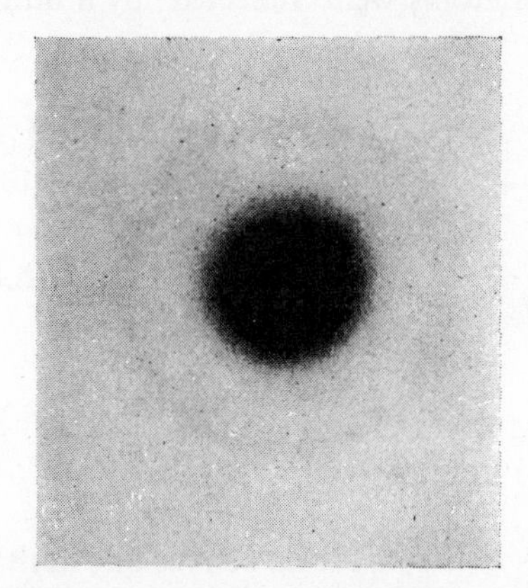

Fig. 7.20 Diffraction haloes for celluloid. (G. P. Thomson. Contemporary Physics, **9**, *8, Fig. 5, 1968.)*

equation. Only a few months after Davisson (and Germer) published their first paper giving evidence of the wave nature of electrons, Thomson and Reid in England fired electrons at a thin sheet of celluloid and found that a photographic plate exposed to the transmitted electrons showed diffraction haloes (Fig. 7.20), a pattern that would be expected for waves diffracted by molecules of definite size orientated at random. Thomson used other materials in place of the celluloid. With aluminium and then gold films he found that the results were again rings; rings whose diameters agreed with what would be expected from waves, wavelength calculated from de Broglie's equation, being diffracted from an array of scattering centres of known spacing (measured by X-rays, see chapter 12). Figure 7.21 shows the results for gold for electrons of two different energies.

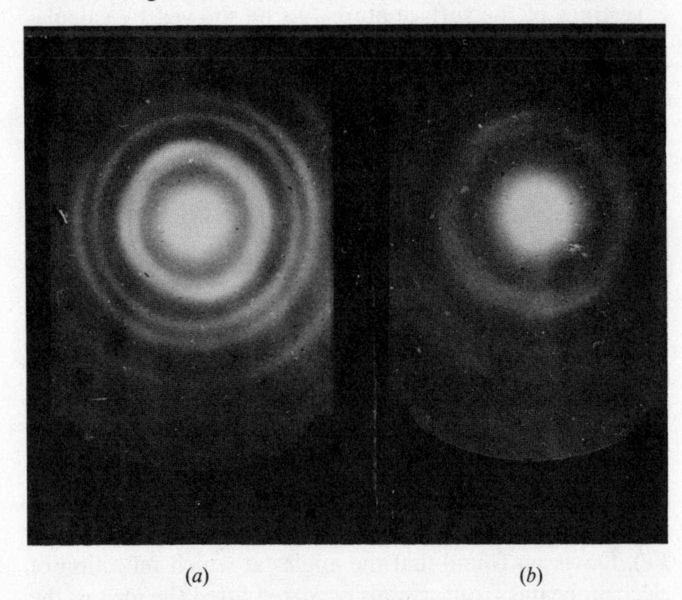

Fig. 7.21 Diffraction rings for gold. (a) Higher energy electrons than (b). (G. P. Thomson. Contemporary Physics, **9**, *9, Fig. 7, 1968.)*

The above evidence is overwhelming in favour of electrons showing a wave characteristic. The de Broglie equation was not, however, restricted to electrons but applied to all matter. Figure 7.22 shows the results of an experiment in which helium atoms were 'reflected' by a lithium fluoride crystal.

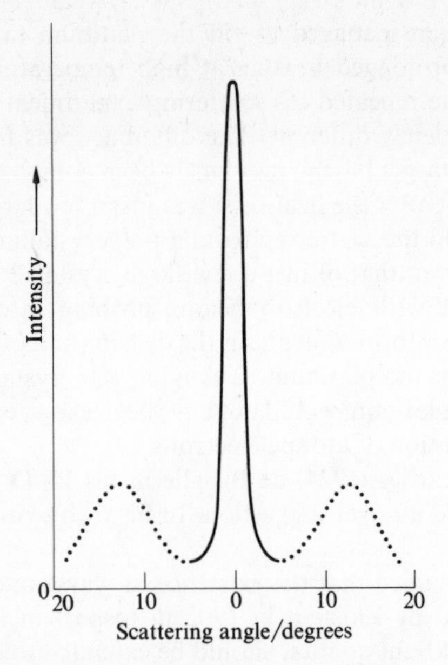

Fig. 7.22 From Estermann and Stern. Zeitschrift für Physik, **61**, *107, 1930.*

> 'My own suspicion is that the universe is not only queerer than we suppose, but queerer than we can suppose.'
>
> J. B. S. Haldane

Quanta

The following quotations serve as a summary of the main points arising from this chapter.

> '... Things on a small scale behave like nothing that you have any direct experience about. They do not behave like waves, they do not behave like particles, they do not behave like clouds, or billiard balls, or weights on springs, or like anything that you have ever seen....
>
> R. P. Feynman, R. B. Leighton, M. Sands (1963). *The Feynman lectures on physics*, Vol. 1, 1st ed. Addison Wesley, Reading, Mass.

> ... many physicists have taken to saying that our models (referring to the wave and particle models) do not describe things as they really are, but act only as frameworks on which to hang the mathematics by which we calculate, correlate and predict the results of specified experiments....
>
> M. B. Hesse (1962). *Quanta and reality. A symposium*, Hutchinson/World Publishing Corp.

> ... You must be prepared to accept that, whenever I use a word like electron or nucleus or particle or wave, it is for the lack of a better one. I am not referring to an object which has all the properties which your commonsense and your knowledge of classical physics leads you to expect. You will see that the quantum particle is not exactly a particle as you know it, nor the wave exactly a wave....'
>
> N. Kemmer (1962). *Quanta and reality. A symposium*, Hutchinson/World Publishing Corp.

Problems

1. By what properties would you identify (a) a wave and (b) a particle?

2. This question is adapted from Berkeley Physics Course, Vol. 4, *Quantum physics*, by E. H. Wichmann, McGraw-Hill and Educational Development Center. Question 7, page 41.

'We all know that stars "twinkle". To see whether this may be a manifestation of the quantum nature of light, estimate the number of photons entering the eye of an observer when he looks at a star of first apparent visual magnitude. Such a star produces a flux at the surface of the earth of about 10^{-6} lumen metre^{-2}. One lumen at the wavelength of maximum visibility, which is about 5560×10^{-10} m, corresponds to 0·0016 W. A star of the first apparent visual magnitude is a fairly bright star, easily visible to the naked eye, although not among the very brightest stars. The star Aldebaran is an example.

Suppose that N photons per second enter the eye of the observer. After you have determined N, decide what is the likely explanation of the twinkling. Why do planets appear to twinkle much less, or not at all?'

3. Photoelectric emission of electrons from copper is not found to occur for incident frequencies lower than $1{\cdot}1 \times 10^{15}$ Hz. What can be deduced from this?

4. How many photons per second are emitted by:

(a) a 5 kW radio transmitter operating at a frequency of 90 MHz.

(b) a 10 W sodium lamp (assume only the yellow frequencies are emitted).

5. What is the wavelength of:

(a) an electron of energy 10 eV,

(b) an electron of energy 1 keV,

(c) a proton of energy 1 keV.

6. What is the momentum of:

(a) a 'yellow' photon,

(b) an X-ray photon,

(c) a radio photon.

7. In 1923 A. H. Compton compared the wavelength of X-rays after scattering by graphite with the incident wavelength. He found that the wavelength of the scattered X-rays was bigger than that of the incident X-rays (a change of about 3 per cent). How can this be accounted for?

8. Many materials fluoresce when exposed to ultraviolet light. Thus natural teeth appear bright in ultraviolet light because light in the visible part of the spectrum is emitted by them. This is one way that detergent powder manufacturers can claim their powders are 'whiter than white'. Included within the detergent packet are fluorescent materials which give off visible light when ultraviolet light falls on them—there is ultraviolet light in natural sunlight. A general law of fluorescence is that the wavelength of the fluorescent light is always longer than that of the absorbed light. This is called Stokes' law.

(a) Offer an explanation for Stokes' law.

(b) Why should ultraviolet light be the main agency for producing fluorescence?

If after the light is switched off the glow of the substances persists the effect is called phosphorescence.

9. What would you expect to happen to a photon when it passes from say air into glass, i.e., it is refracted?

10. In the quotation taken from *The Feynman lectures on physics* on page 186 the following sentence occurs.

'Things on a small scale behave like nothing that you have any direct experience about.'

What do you think is meant by this?

Teaching note Further problems appropriate to this chapter will be found in:
Nuffield advanced physics. Unit 2 student's book and *Unit 10 book*, Penguin, 1972.
Physical science study committee. *Physics*, 2nd ed, chapters 33 and 34, Heath, 1965.

Practical problems

1. When ultraviolet light is incident on certain compounds light is emitted in the visible part of the spectrum. This phenomenon is known as fluorescence and has many uses in industry. For example in fluorescent lamps the ultraviolet radiation produced by the passage of current through mercury vapour is converted into visible light by a layer of fluorescent material on the internal surface of the lamp tube. In this project investigate the factors which affect the light output of a fluorescent material. Consider the thickness and particle size of the material.

2. Examine the visible absorption spectrum of human blood. Then allow a sample to absorb carbon monoxide and re-examine the spectrum. Can the visible absorption spectrum be used to detect the presence of carbon monoxide in blood? Check that your results are not due to the absorption of oxygen in the blood.

3. Design an instrument which can be used to measure the reflection characteristics of a surface at a particular colour. Essentially such an instrument would consist of a white light source directing light through a filter onto the reflecting surface. The light reflected by the surface could be measured by a photoelectric cell.

There are many uses for such an instrument—

(a) Checking for consistency of colour in a product such as linoleum tiles or paper.

(b) Following the course of a chemical reaction in which a colour change occurs.

(c) Determining the ripeness of fruit.

Suggestions for answers

1. A wave—diffraction, interference, energy spread over an area.

A particle—localized energy, momentum, and energy changes in collisions.

2. About 13 000 reach the eye per second. Only about 50 per cent of this light reaches the retina, due to reflection from the eye. Thus about 6500 quanta reach the retina. Fluctuations in this number will occur and could amount to changes of the order of a few per cent in the number of photons reaching the eye per second. This could be an explanation for the twinkling of stars, if changes in the intensity of a few per cent were very noticeable. An alternative explanation would be the movement of dust clouds in the space between the earth and the stars.

If planets were brighter than stars the fluctuations in the number of quanta would be less noticeable and so they would not appear to twinkle. If planets are not as bright as stars then the dust cloud explanation would seem more likely—there would be less dust clouds between us and the planets than between us and the stars and so less chance of variations, also the greater visual area of a planet would mean less chance of the light being significantly reduced by dust cloud variations.

3. The minimum energy that is needed for the ejection of electrons, i.e., the work function, is $7{\cdot}26 \times 10^{-19}$ J or about 4·5 eV.

4. (a) about 9×10^{28}.
(b) about 10^{20}.

5. (a) 400×10^{-10} m.
(b) 40×10^{-10} m.
(c) 3000×10^{-10} m.

$$\lambda = \frac{h}{\sqrt{(2\,mE)}}$$

6. (a) about 10^{-27} kg m s^{-1}.
(b) about 7×10^{-24} kg m s^{-1}.
(c) about 7×10^{-30} kg m s^{-1}.

7. The photons had behaved like particles and lost momentum on collision with graphite atoms.

8. (a) Think of the conservation of energy in a collision between the incident photon and target atom. The atom cannot absorb and then release more energy than it receives. The exception might be the case where the atom is already excited and gives out this energy and the energy it has absorbed from the photon.

(b) It has a higher frequency and thus greater energy per photon.

9. The speed of light in glass is less than that in air. Thus the wavelength in glass is less than that in air. This leads to the conclusion that the mass of the photon is greater in glass than in air. The ratio of the masses is equal to the square of the refractive index. Momentum is still conserved (remember that momentum is a vector if you sum momenta).

10. In discussing this you might mention electrons behaving like particles and like waves. Neither of these two large-scale phenomena, however, is adequate to describe the behaviour of electrons. In an equivalent of the double slit experiment with light we find that apparently an electron goes through both slits simultaneously—just like the single photon.

8 Chance

Teaching note Further reading for this chapter will be found in:
H. A. Bent. *The second law*, Oxford Univ. Press, 1965.
Nuffield advanced physics. Unit 9, Penguin, 1972.

8.1 Scientific laws

What is a scientific law? The *Concise Oxford Dictionary* gives the following explanations of the word 'law'.

> 'Body of enacted or customary rules recognized by a community as binding....
>
> Rule of action or procedure, especially in an art, department of life, or game.
>
> ...correct statement of invariable sequence between specified conditions and specified phenomenon....'
>
> (By permission of the Clarendon Press, Oxford.)

What is a scientific law? Is it different from legal laws?

> 'There was once a government whose leaders decided that it would be nice to relieve the people of the effort of carrying their weight around. They reasoned that life would be much easier without the stresses and strains caused by the force of gravity. Therefore they convened their legislature and repealed the law of gravity....'
>
> Excerpted from *The laws of physics* by Milton A. Rothman. Basic Books Inc. Publishers, New York, 1963.

Scientific laws are different from legal laws or rules of a game in that they describe a relation between specified conditions and specified phenomena and cannot be repealed or altered to suit people—they describe events which happen. For example, Newton's laws of motion are descriptions of the relation between force and motion. Hooke's law describes a relation between the force extending a spring (or length of material) and the amount of extension. A scientific law describes a relationship—a relationship which we expect always to occur (an invariable sequence) for the same specified conditions and phenomena.

> 'Experience...shows him [man] that the natural processes which take place in his environment do not follow one another in an arbitrary, kaleidoscopic manner, but that they present a notable degree of regularity.'
>
> E. Schrödinger (1935, 1957). *Science, theory and man*, Allen & Unwin.

If you bit two, three, ..., ten green apples and they were sour you might conclude that there is a relationship—green apples are sour. True, the experience was limited but if other observers in other places all reported the same result you might conclude that you had a general law.

Causal laws in science

Causal laws can be summarized as—to every effect there is a precise cause. If an object accelerates we say that there is a cause—a force. We do not say—there might be force. When an apple fell off a tree Newton considered there must be a force acting on the apple—the force of gravity. The ancient Greeks did not consider a force but considered the apple to fall because it was the natural thing for apples to fall—no cause was considered.

For example, consider two billiard balls colliding. From a knowledge of their velocities before the collision we can plot their paths and determine their positions at different times. Knowing their masses we can compute their velocities after the collision and hence their positions at different times. To determine the positions of the balls at all times we needed only their velocities, masses, their initial positions, and the laws of conservation of momentum and kinetic energy (Fig. 8.1).

Can we on the basis of causal laws explain the whole of science? The two billiard balls colliding would seem to offer a simpler version of the collisions between molecules of a gas in a container. If we know the initial conditions can we calculate the positions of all the molecules at every future instant of time? The first difficulty is how to determine the initial conditions with sufficient accuracy to permit accurate calculations of the future positions. Only very small errors in the knowledge of the initial positions would con-

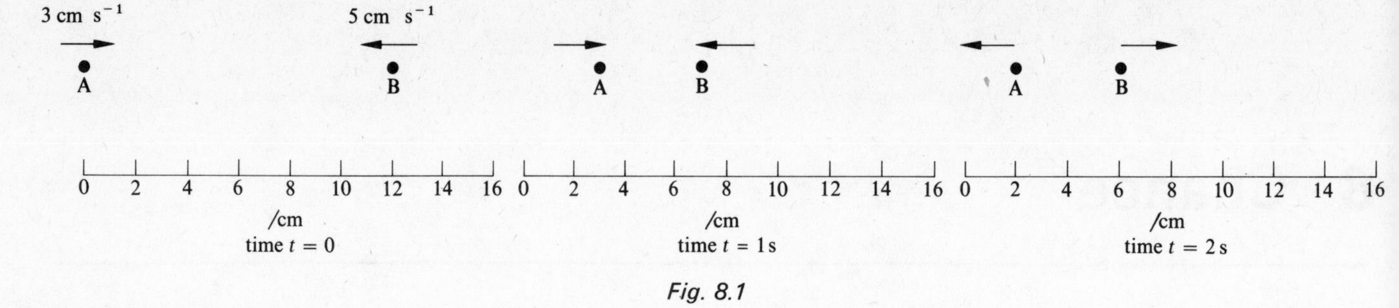

Fig. 8.1

siderably affect the behaviour of a single molecule. If our knowledge of the position of a single molecule was incorrect by say 10^{-10} m we could not calculate its position after just one collision with another molecule. Molecules have a size of about 10^{-10} m and an error in position of this amount might mean that a collision did not take place because the two molecules miss each other. Knowledge of positions to a greater accuracy than this may not be sufficient to determine the angle at which the two molecules collide for their directions after a collision to be accurately calculated. When we think of the large number of molecules in a container then the calculation seems impossible.

> '... for a model corresponding to the kinetic theory of gases, Borel computes that errors of 10^{-100} on initial conditions will enable one to predict molecular conditions for a split second and no more. It is not only 'very difficult,' but actually impossible to predict exactly the future behaviour of such a model.
>
> ... E. Borel ... computed that a displacement of 1 cm, on a mass of 1 gram, located somewhere in a not too distant star (say, Sirius) would change the gravitational field on the earth by a fraction 10^{-100}....'
>
> L. Brillouin (1964). *Scientific uncertainty, and information*, Academic Press Inc., 1964, p. 125.

Causal laws have their uses in the macroscopic sphere, i.e., large-scale effects, but are severely limited in the microscopic sphere, i.e., on an atomic level. If we have just two atoms then we cannot calculate what will happen—they might collide if we somehow aim them at each other but because of the limitations of the accuracy of our aim we cannot forecast their paths. If, however, we fired a large number of atoms at each other then perhaps we could expect some of them to make collisions and perhaps we could expect some of them to make collisions at certain angles. In such a case perhaps we can calculate how such atoms would move after a collision. Large numbers enable us to argue events on the basis of chance—chance that some of the atoms will follow particular paths.

In a rainfall we would expect that the amount of rain falling on each square metre would be the same—we would certainly not expect that only every alternate square metre would have rain, the others remaining dry. We would not, however, feel able to predict the path of a single raindrop.

On an atomic scale we need the idea of chance—is this because of our limitation on the accuracy of our measurements or is it because the events are dictated by chance? The majority opinion is that the events are chance events.

Here are some of the comments of scientists on the place of chance in physics.

> '... We now believe that the ideas of probability are essential to a description of atomic happenings....'
>
> R. P. Feynman, R. B. Leighton and M. Sands (1963). *The Feynman lectures on physics*, Vol. 1, Addison Wesley, Reading, Mass.

> '... As a matter of fact, the most recent development in physics, quantum mechanics ... has shown that we must drop the idea of strict laws, and that all laws of nature are really laws of chance, in disguise.'
>
> M. Born (1951). *The restless universe*, Dover Publications Inc., New York. Reprinted by permission of the publisher.

> 'Within the past four or five decades physical research has clearly and definitely shown—strange discovery—that chance is the common root of all the rigid conformity to Law that has been observed....'
>
> E. Schrödinger (1935, 1957). *Science, theory and man*, Allen & Unwin.

What is chance?

The following appeared in the *Sunday Times* Business News Section for 22 November 1970.

> **'"24 cigarettes a day means 1 in 9 chance of cancer"**
>
> **Doctors call for total ban on tobacco ads**
> *By Gwen Nuttall*
>
> A total ban on all advertising and promotion of cigarettes, cigars and pipe tobacco is the main recommendation of the report of the Royal College of Physicians on the link between smoking and cancer, due early in January.
>
> The shock finding of the Report, I understand, is that anyone smoking 24 cigarettes or more a day has a one in nine chance of contracting lung cancer compared with a possibility of one in 284 for a non-smoker....'

What do we mean by a 1 in 9 chance? If a large number of people, who all smoke 24, or more, cigarettes a day, are considered then 1/9 of them would be expected to contract lung cancer. This does not mean that if just nine people are considered that precisely one will contract lung cancer.

If we throw a six-sided die there are just six ways it can land. We have a 1 in 6 chance of finding one particular die face uppermost, say a five. If we throw 120 dice then we shall expect that we will obtain about 1/6 of them, i.e., 20, showing a five uppermost. We may only get 18 or perhaps 23, but 20 is the most likely result. If we throw the dice a large number of times then we will obtain 20 out of the 120 showing fives more often than any other number.

Chance is the ratio of the number of ways an event, such as a five with a die, can be realized compared with the total number of possible outcomes (with the die—six). We are assuming, in this definition, that each way or outcome is equally likely (our die is not biased). An alternative way of expressing chance is as the ratio of the number of times an event occurs compared with the number of times it could occur.

What is the chance of a coin, when spun, landing heads uppermost? There are just two, equally likely, ways an unbiased coin can land, heads or tails. The total number of possible outcomes is thus two. The number of ways heads can be realized is one. The chance is thus 1/2. A coin spun 1000 times landed with heads uppermost 502 times—this gives the chance of heads as 502/1000 or near enough 1/2.

What is the chance of obtaining two heads when two coins are spun? The possible outcomes are

HH, HT, TH, TT

There are four possible outcomes, one of which gives two heads. The chance is thus 1/4.

What is the chance of obtaining both a head and a tail when two coins are spun? Two ways out of the four give this outcome. The chance is thus 2/4 or 1/2.

If an event can be realized in many ways it is often likely to occur—many ways mean often.

Diffusion

Consider a hypothetical case of a single molecule in a box. What are the chances of finding the molecule in one particular half, say the left-hand half, of the box? The molecule has two alternative positions—in the left-hand half or the right-hand half of the box. There are thus two possible outcomes of which one gives the molecule in the left-hand half. The chance of finding the molecule in the left-hand half is thus 1/2. This assumes that each half is an equally likely place for the molecule to be found.

Now consider two molecules in the box. What are the chances of finding the two molecules in the left-hand half? The possible outcomes are shown in Fig. 8.2. There are four possible outcomes of which one gives all the molecules in the left-hand half. The chance of finding the molecules in the left-hand half is thus 1/4.

With the two molecules, what are the chances of finding one molecule in each half? There are two of the four possible outcomes giving this. The chance is thus 2/4. If our molecules freely wander between the two halves then there is a greater chance of us finding them at any instant with one in each half than with them all in one particular half.

Suppose we have four molecules. Figure 8.3 shows the possible outcomes. There are 16 possible arrangements of the molecules, one of which has all the molecules in the left-hand half. The chance of all the molecules being in the left-hand half is thus 1/16. The chance of there being equal numbers of molecules in each half is 6/16—a much greater chance.

Why do the molecules in a room seem to be reasonably uniformly distributed over the room? The molecules are free to move about the room—why could they not by some fortuitous sequence of collisions all be in one half of the room? What are the chances of finding all the molecules in one particular half of a room? With one molecule in a box we had two possible arrangements (i.e., 2^1), with two molecules four possible arrangements (i.e., 2^2), with four molecules sixteen possible arrangements (i.e., 2^4). If there are two ways a molecule can be located in a box, or room, then with n molecules there are 2^n possible outcomes. In all the previous arrangements only one outcome was with all the molecules in one particular half. The chance of n molecules all being in one half of a room is thus $1/2^n$. How big is n? A room about 5×4 m floor area and 3 m high would under normal conditions contain about $1{\cdot}5 \times 10^{27}$ molecules. Thus the chance of finding them all in one half is

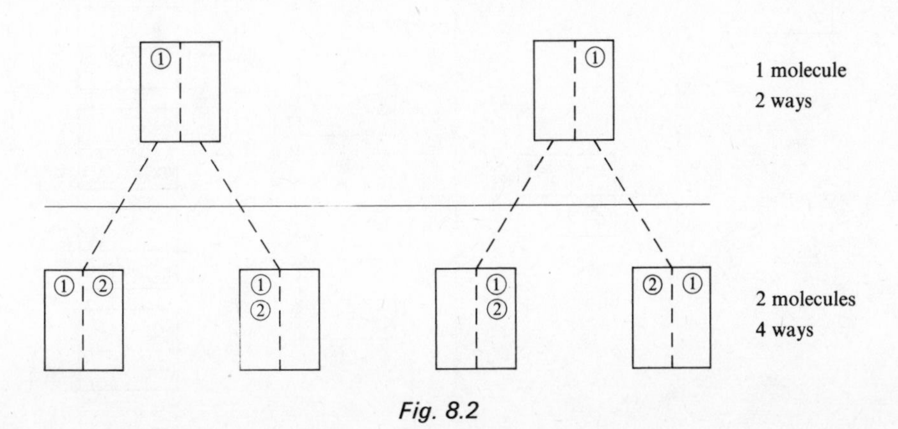

Fig. 8.2

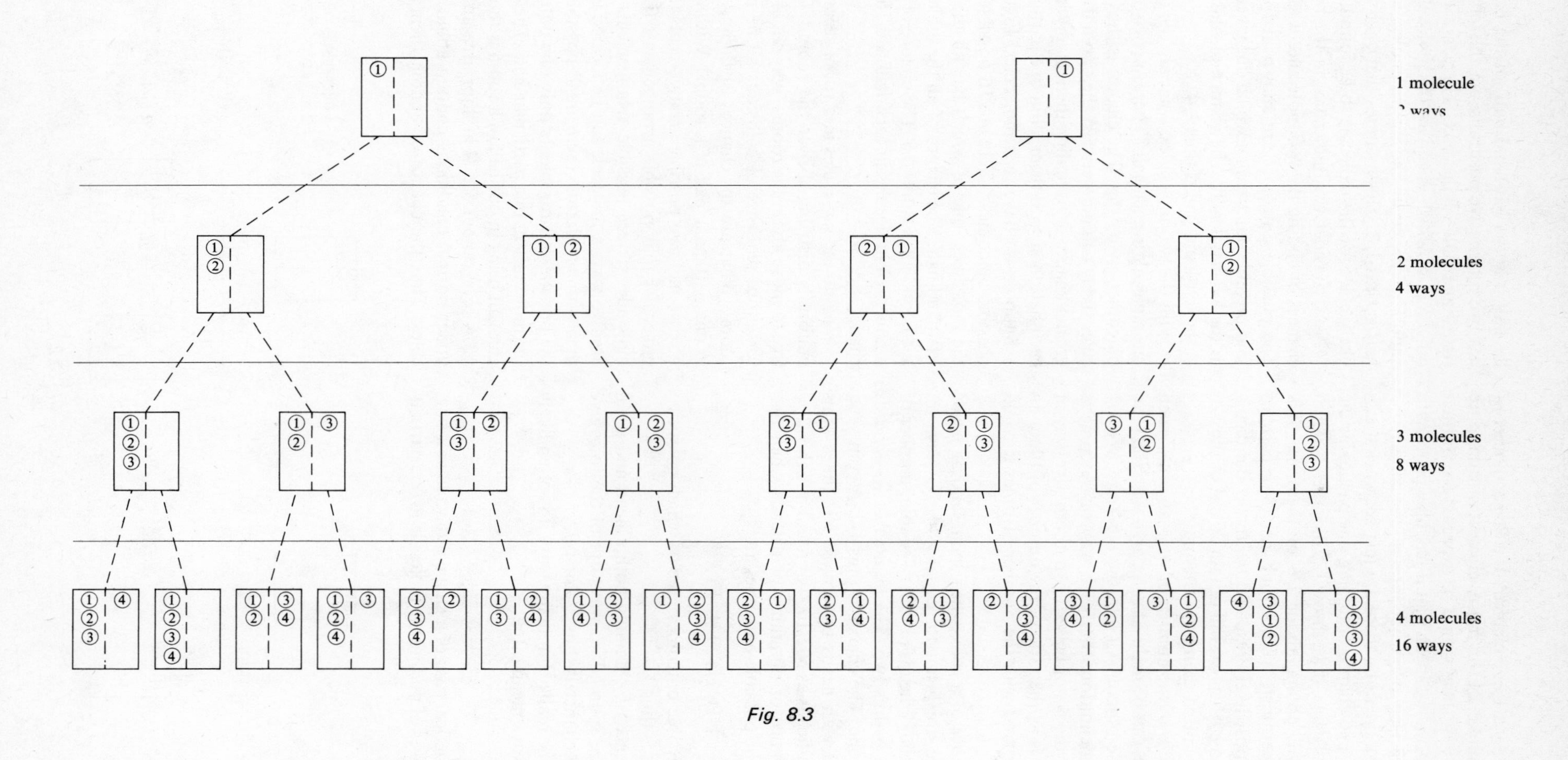
1 molecule
2 molecules
4 ways
3 molecules
8 ways
4 molecules
16 ways

Fig. 8.3

about 1 in $2^{10^{27}}$ (or about 1 in $10^{3\times10^{26}}$; $2^n = 10^{n\log 2}$). If we took a photograph of the molecules, assuming them to be visible to our special camera, once a second for a year we would have $3{\cdot}2 \times 10^6$ pictures. In one year the chance of seeing all the molecules in one half is rather remote. The universe is about 10^{10} years old. If we took photographs every second for the entire life of the universe, about 10^{16}, there is very little chance that we would see all the molecules in one half. We would need to wait about $10^{3\times10^{26}}$ seconds for there to be a chance of one such event occurring. This would seem to be as good as saying that the chance of all the molecules being in one half is extremely improbable.

> 'Not until after a time enormously long compared with $10^{10^{10}}$ years will there be any noticeable unmixing of the gases. One may recognize that this is practically equivalent to never, if one recalls that in this length of time, according to the laws of probability, there will have been many years in which every inhabitant of a large country committed suicide, purely by accident, on the same day, or every building burned down at the same time—yet the insurance companies get along quite well by ignoring the possibility of such events. If a much smaller probability than this is not practically equivalent to impossibility, then no one can be sure that today will be followed by a night and then a day.'
>
> L. Boltzmann (1964). *Lectures on gas theory*, Univ. California Press/Cambridge Univ. Press.

This argument involving chance suggests that the diffusion of a gas is a result of purely the chance movements of molecules. A gas placed in a container will diffuse throughout the container to give a reasonably uniform spread of the gas—purely by the chance movements of the molecules. The greatest chance is for an even distribution of gas within a container; there is, however, a finite chance that the distribution will be non-uniform—but the chance is considerably smaller. This is an essential feature of processes determined by chance—there will be fluctuations from the uniform or most probable state. If we find such fluctuations, for example the gas density varying slightly, then it is strong evidence for a chance-ruled state.

Fluctuations in the distribution of gas molecules in a container would be expected to lead to pressure fluctuations. The pressure of a gas is due to molecular bombardment of a surface. Small particles, of say ash, in a gas show irregular motion, known as Brownian motion. This can be explained as being due to slight differences in the number of molecules bombarding the different sides of a particle; evidence for pressure fluctuations and thus fluctuations in the molecular distribution in space.

Why is the sky blue?

On a clear day the sky is blue—why? Why has it any colour? Air itself does not seem to have any colour. There is a clue in that fine dust, for example chalk dust or smoke, in air gives a bluish colour. Very fine particles in air give the air a bluish colour, a faint colour in any laboratory experiment but when considered for the earth's atmosphere the effect could be significant. This effect can be explained by considering light to be scattered by small particles (Fig. 8.4). The particles have to be very small—at the most of the order of the wavelength of the light. The wavelength of blue light

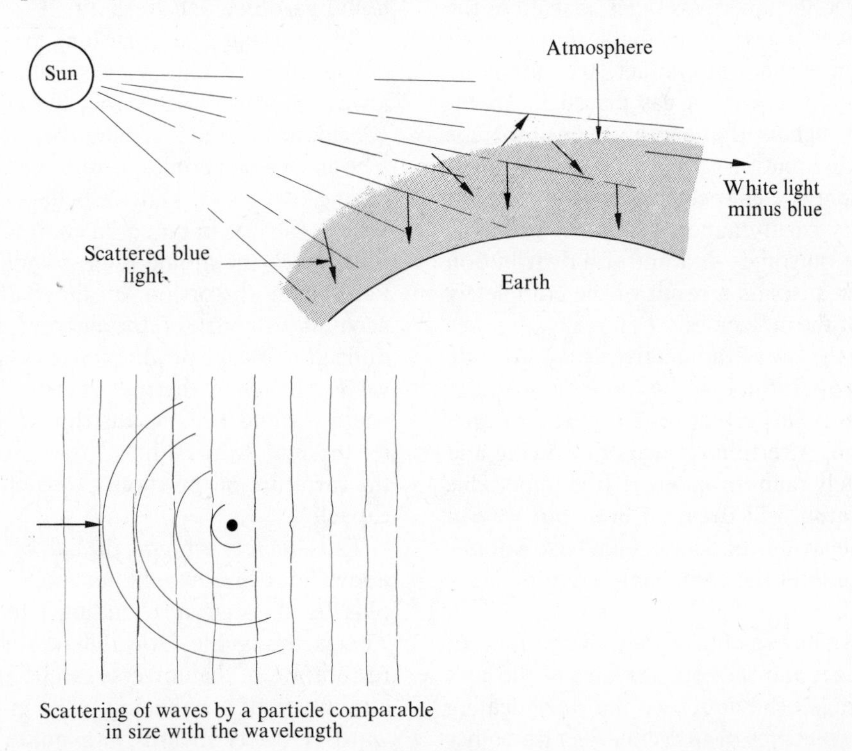

Fig. 8.4

is about half that of red light—this coupled with the fact that the sky is blue would suggest that the particles responsible for the scattering are very small, nearer the wavelength of blue light than red light. This would mean that the scattering particles must have dimensions smaller than 4×10^{-7} m. What particles would be suitable? Air molecules have dimensions of the order of 10^{-9} m—they could be the scattering centres. There is a snag to this idea—if space is uniformly filled with molecules then the waves scattered from one molecule should interfere with the waves scattered from another molecule. If we have sufficient molecules, and we have, and they are uniformly spaced then it is quite easy to argue that for every molecule giving scattering through say 90° there is another molecule giving scattering through the same angle but out of phase with that from the first molecule and so there should be no scattered light. In fact if we do the experiment with water, about 11 m deep to give the same number of molecules as the depth of the atmosphere, we find that for 90° scattering there is almost completely destructive interference and thus no scattered light. Why do we get scattering with the atmosphere? The answer lies in the fluctuations in the distribution of molecules in the atmosphere giving a non-uniform spacing; in water there is a reasonably uniform spacing. The fluctuations are great enough always to give some molecules scattering light which is not subject to destructive interference by light scattered from some other molecule. The fluctuations destroy the uniformity of spacing.

The laws of chance

This could well be called the law of averages, instead of the law of chance. The law states the probable outcome of events which interact in a random manner, i.e., purely by chance. We could state as a law—a gas placed in an enclosure will diffuse throughout that container until there is a reasonably uniform distribution of gas molecules throughout the container. There is a chance that the gas may not diffuse, but it is a very small chance. The most probable outcome—the average outcome—is a uniform distribution of molecules. This law arises as a result of the completely random movements of the molecules.

Another example is the law of radioactive decay: the rate of disintegration is proportional to the amount of radioactive substance present. This law arises as a result of each radioactive atom having a certain chance of decaying and doing so in a completely random manner. It is impossible to say when a single atom will decay, if ever, but we can forecast the average behaviour of a large number of atoms. This law arises as a result of the completely random decay of atoms.

It is considered by many scientists that all the laws of nature are laws of chance and that our versions of the laws only express the average behaviour; we are only dealing with the most probable outcome of an event, other outcomes are possible.

8.2 Reversible and irreversible events

'If, then, the motion of every particle of matter in the universe were precisely reversed at any instant, the course of nature would be simply reversed for ever after. The bursting bubble of foam at the foot of a waterfall would reunite and descend into the water; the thermal motions would reconcentrate their energy, and throw the mass up the fall in drops re-forming into a close column of ascending water. Heat which had been generated by the friction of solids and dissipated by conduction, and radiation with absorption, would come again to the place of contact, and throw the moving body back against the force to which it had previously yielded. Boulders would recover from the mud the materials required to rebuild them into their previous jagged forms, and would become reunited to the mountain peak from which they had formerly broken away. . . .'

W. Thomson (1874). *Proc. Roy. Soc. Edinburgh*, **8**, 325.

'A birch log, for example, cannot be burnt twice. One cannot take the hot flue gases of a wood fire and the warmth of the fire and from these reconstitute an unburnt log, fresh air, and room chilliness. Passage from the burnt state—the hot flue gases and a warm room—to the pre-burnt state—a birch log, fresh air, and a cold room—is impossible, or at most, highly improbable.

Similarly, the onrush of a passing car cannot be reversed in all aspects. Many cars, of course, can be driven backwards. But no car, in an exact reversal of its motion forward, can suck through its tailpipe exhaust fumes and from these produce in its cylinders an ignition spark, liquid gasoline, and fresh air.

Burning logs and onrushing cars are typical examples of irreversible events. Viewed in their entirety, such events always produce unalterable changes in the universe. . . .

Seldom noticed . . . is the thermal energy produced as a bouncing ball comes to rest. Yet if energy is conserved during this event, and we believe energy is always conserved, the loss in potential energy of the system (the ball plus the earth) should appear somewhere in some form. Permanent distortion of the ball or the floor might account for part of the energy; even so, it would be difficult to escape production (through friction and sound waves) of some thermal energy. In time, this thermal energy would become distributed between the ball and its thermal surroundings (the floor, the air, the walls, the furniture of the room), to each according to its heat capacity.

Thus, merely bringing the ball back to its initial position above floor-level would not restore the universe (the ball plus its thermal surroundings) to its initial condition. That is impossible. How remarkable would be a complete restoration of the universe can be appreciated by looking at a movie of a bouncing ball coming to rest, run backwards. Objects obviously are not in the habit of springing spontaneously into the air at the expense of the thermal energy of their surroundings. The idea that they might is absurd. They never do. . . .'

H. A. Bent (1965). *The second law*. Oxford Univ. Press, New York.

There would appear to be no reason why events should not be reversible. Why shouldn't an object spring up from the floor and the floor (and object and surroundings) become cooler? Provided energy is conserved, and there is no reason why it shouldn't be, the event seems possible. There is nothing in the conservation of energy law which says which direction an event should occur. Why is burning, for example the birch log, a one-way process? Energy can be conserved whichever way the event occurs.

Consider a collision between two perfectly elastic balls, A and B. Initially A has a velocity of 4 m s^{-1}, its mass is 30 g. Initially B is at rest, its mass is 10 g. After the collision, the velocity of A becomes 2 m s^{-1}, in the same direction as before, and B is knocked in the same direction with a velocity of 6 m s^{-1}. We could have the exact reverse of this collision, i.e., B moving towards A and colliding with it so that B comes to rest. The velocities would only differ in their directions. B would move with a velocity of 6 m s^{-1} and A 2 m s^{-1} before the collision. After the collision B would be at rest and A would have a velocity of 4 m s^{-1} (Fig. 8.5). All these figures were calculated by the use of the conservation of momentum and the conservation of kinetic energy laws. This theoretical event appears to be perfectly reversible. The event is, however, only theoretical; in practice some kinetic energy would be lost at the collision and a small temperature change produced. If the event were perfectly reversible there should be an increase in temperature at one collision and a drop in temperature for the reverse collision. This does not occur. Our equations of motion are perfectly reversible—real processes are not.

'. . . equations of motion, established by Newton, do not involve the sign of time. If a bullet moves on a certain trajectory from north to south, another similar bullet might move along the same trajectory from south to north. Let us immediately point out that the statement can be correct only if all damping terms happen to be zero. They should be exactly zero, not just negligible!

Air resistance is one of these damping terms, or, if we consider electric charges, damping due to electromagnetic radiation is unavoidable. Physics does not know any motion without damping. . . .'

L. Brillouin (1964). *Scientific uncertainty and information*, Academic Press Inc.

This still does not tell us why the events are not reversible—the bullet moving through the air raises the temperature of the air and so slows down; why doesn't the air temperature fall and increase the speed of the bullet? The bullet moving from north to south slows down due to damping. Why cannot the bullet moving from south to north speed up, by just the right amount to give the same behaviour as the first bullet?

Why are events irreversible? What determines the direction an event will go?

In any frictional effect, e.g., the bullet being slowed down by the air, the organized motion is changed into disorganized motion of atoms. The orderly motion of the bullet is changed into the disorganized motion of air molecules. Whenever an event produces a temperature rise there is an increase in the disorganized motion of atoms. Perhaps the sequence is always order to disorder and this is the clue to the direction an event will go. Consider some of the examples already mentioned in this section: a bouncing ball, the orderly motion is changed into the disorganized motion of atoms (a temperature rise); a burning log, the orderly arrangement of atoms in the log is changed to the disorderly motion of atoms in the flue gas, the energy localized initially in the log is spread around among air molecules; boulders crumbling to mud, an orderly arrangement of atoms in the boulder changed to a disorderly arrangement of atoms in the mud.

Why should the sequence appear to be order to disorder for events?

Number of ways

Consider diffusion as a one-way process. A gas released at one end of a container spreads throughout the container until the gas is uniformly spread throughout the container. A gas spread throughout the container does not reconcentrate itself at one end. Diffusion would seem to be an irreversible process with the direction of events being from concentrated to spread out.

Earlier in this chapter we considered the chance of finding

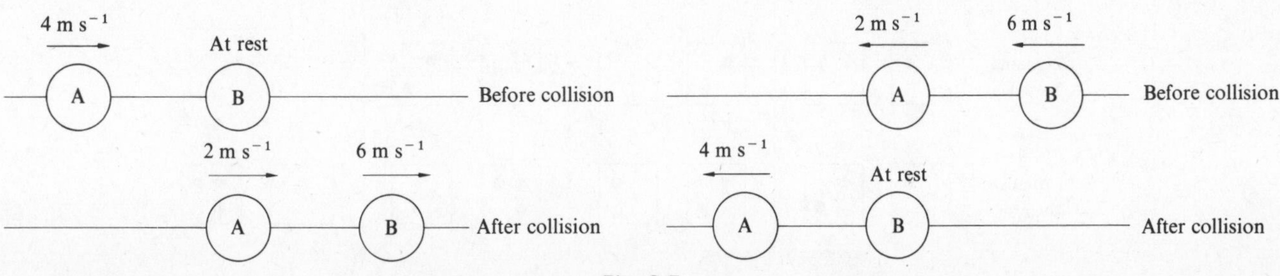

Fig. 8.5

all the molecules in a box in one half of that box. With four molecules there are sixteen possible arrangements or ways of arranging the molecules between the two halves of the box. Only one of these ways has all the molecules in one particular half. Six of these ways have the molecules uniformly spread through the box, i.e., equal numbers in each half. The chance of seeing the molecules all in one particular half is thus only one-sixth of the chance of seeing them uniformly distributed. The greater the chance the more often we expect to find the molecules in a particular arrangement. Thus freely left to itself there is a greater chance that the gas will diffuse throughout the container than that it will become concentrated in one half. This was with just four molecules. If we use a more realistic number of molecules the chance of finding the uniform distribution is considerably greater than finding the molecules all in one half. The word considerable is not strong enough—the factor can be of the order of $10^{10^{10}}$. Diffusion is a one-way process, concentrated to spread out, because the chance of the reverse direction process occurring is considerably less than that of concentrated to spread out. Chance is the factor which decides which way an event will occur. A spread out arrangement is more likely than a concentrated arrangement because there are more ways the spread out arrangement can be realized.

What about events where temperature rises occur? Here instead of being concerned with the spreading out of atoms we are concerned with the spreading out of quanta. Quanta are packets of energy.

A bouncing ball comes to rest and the temperature of the floor (and the ball) rises. The potential energy of the ball has been converted into quanta which have been spread out among the atoms in the floor (and the ball). The reverse event is a drop in temperature of the floor and the ball springing up from the floor. Quanta have been taken from the atoms in the floor and converted to potential energy of the ball. Why is the event with the rise in temperature the event that normally happens?

Consider two quanta shared amongst two atoms (Fig. 8.6)—there are three possible arrangements. If we add another quantum, making three quanta amongst two atoms, we have four possible arrangements. The extra quantum has increased the number of possible arrangements from three to four. If instead of adding a quantum we took one away, making one quantum amongst two atoms, then the number of ways drops from three to two. An arrangement which can be realized in more ways is a more likely arrangement.

The bouncing ball coming to rest increases the temperature of the floor and thus is giving quanta to the floor. The ball springing up from the floor needs to derive its energy from the floor by the floor dropping in temperature; the ball thus needs to take quanta away from the floor. Giving quanta to the floor results in more possible arrangements than taking quanta from the floor and is thus the more likely event. The event that occurs most often is the one that occurs in most ways.

n quanta among *N* atoms

The following is a brief mathematical interlude for those who feel they would like an algebraic equation for the number of ways n quanta can be spread among N atomic sites. We will assume that we cannot distinguish between individual quanta but only count the number occupying any one atomic site.

We have N atomic sites which we can write as

$$\circ\circ\circ\circ\circ\circ\circ\circ\circ\circ\circ\circ \ldots N$$

Each o represents a site. To spread among these sites we have n quanta, represented by x.

$$\times\times\times\times\times\times\times\times\times\times \ldots n$$

Each atom can have 0 or 1 or 2 or 3 or . . . n quanta. Thus the first atom could have say 2 quanta, the second atom 1 quantum, the third atom 4 quanta, etc.

o x x	o x	o x x x x	o x	o x x . . .
first atom	second atom	third atom	fourth atom	

All together, when written in this way, there will be $N + n$ symbols, i.e., o or x, in this line. To find the number of ways n quanta can be arranged among N atoms we have to find the number of ways the symbols in this line can be arranged.

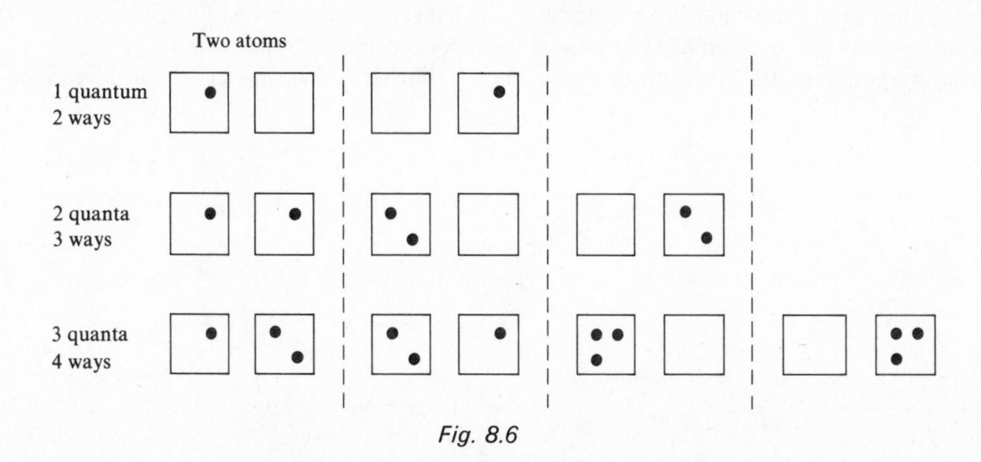

Fig. 8.6

Consider an example, four atoms and two quanta. They can be arranged in the following ways.

o x x o o o
o o x x o o
o o o x x o
o o o o x x
o x o x o o
o x o o x o
o x o o o x
o o x o x o
o o x o o x
o o o x o x

In the first line we have the first site with two quanta and the second, third and fourth sites with zero quanta. In the last line the first two sites have zero quanta and the third and fourth sites have each one quantum. Figure 8.7, solid B, shows this written out as quanta in boxes. With our 4 atom sites and 2 quanta we have 6 symbols in the line. All but the first symbol can be rearranged to give different ways. The total number of ways of arranging the remaining 5 symbols is 5!, i.e., $5 \times 4 \times 3 \times 2 \times 1$. This is $(N - 1 + n)!$. But $5! = 120$ and we have only 10 ways. The error is due to our not being able to distinguish individual atoms or quanta—we do not know after reshuffling whether o is atom two or atom three, or x is quantum one or quantum two. We have thus overestimated by a factor $(N - 1)!$ for the atoms and $n!$ for the quanta. Thus the number of ways is given by

$$W = \frac{(N - 1 + n)!}{(N - 1)!\, n!}$$

Taking the four atom sites and two quanta solid as an example, we have—

$$W = \frac{(4 - 1 + 2)!}{(4 - 1)!\, 2!} = \frac{5 \times 4 \times 3 \times 2 \times 1}{3 \times 2 \times 1 \times 2 \times 1} = 10$$

In this analysis we have assumed that we cannot distinguish between quanta. The following illustration is by Schrödinger.

'...Three schoolboys, Tom, Dick, and Harry, deserve a reward. The teacher has two rewards to distribute among them. Before doing so, he wishes to realize for himself how many different distributions are at all possible. This is the only question we investigate (we are not interested in his eventual decision). It is a statistical question: to count the number of different distributions. The point is that the answer depends on the nature of the rewards. Three different kinds of rewards will illustrate the three kinds of statistics.

(a) The two rewards are two memorial coins with portraits of Newton and Shakespeare respectively. The teacher may give Newton either to Tom or to Dick or to Harry, and Shakespeare either to Tom or to Dick or to Harry. Thus there are three times three, that is nine, different distributions (classical statistics).

(b) The two rewards are two shilling-pieces (which, for our purpose, we must regard as indivisible quantities). They can be given to two different boys, the third going without. In addition to these three possibilities there are three more: either Tom or Dick or Harry receive two shillings. Thus there are six different distributions (Bose-Einstein statistics).

(c) The two rewards are two vacancies in the football team that is to play for the school. In this case two boys can join the team, and one of the three is left out. Thus there are three different distributions (Fermi-Dirac statistics).'

E. Schrödinger (1935, 1957). *Science theory and man*, Allen and Unwin.

The three distributions given by Schrödinger can be written out as—

	Tom	*Dick*	*Harry*
(a)	NS	—	—
	N	S	—
	N	—	S
	—	NS	—
	—	N	S
	S	N	—
	—	—	NS
	S	—	N
	—	S	N

N = Newton, S = Shakespeare

	Tom	*Dick*	*Harry*
(b)	PP	—	—
	—	PP	—
	—	—	PP
	P	P	—
	P	—	P
	—	P	P

P = two shilling-piece

	Tom	*Dick*	*Harry*
(c)	V	V	—
	V	—	V
	—	V	V

V = vacancy in football team

Note: We have considered the distribution of quanta among atomic sites—really we should have considered quanta among localized and independent oscillators. In a solid each atom can have three modes of oscillation and thus N three-dimensional oscillators are equivalent to $3N$ one-dimensional oscillators. In our equation we are referring to the number of one-dimensional oscillators.

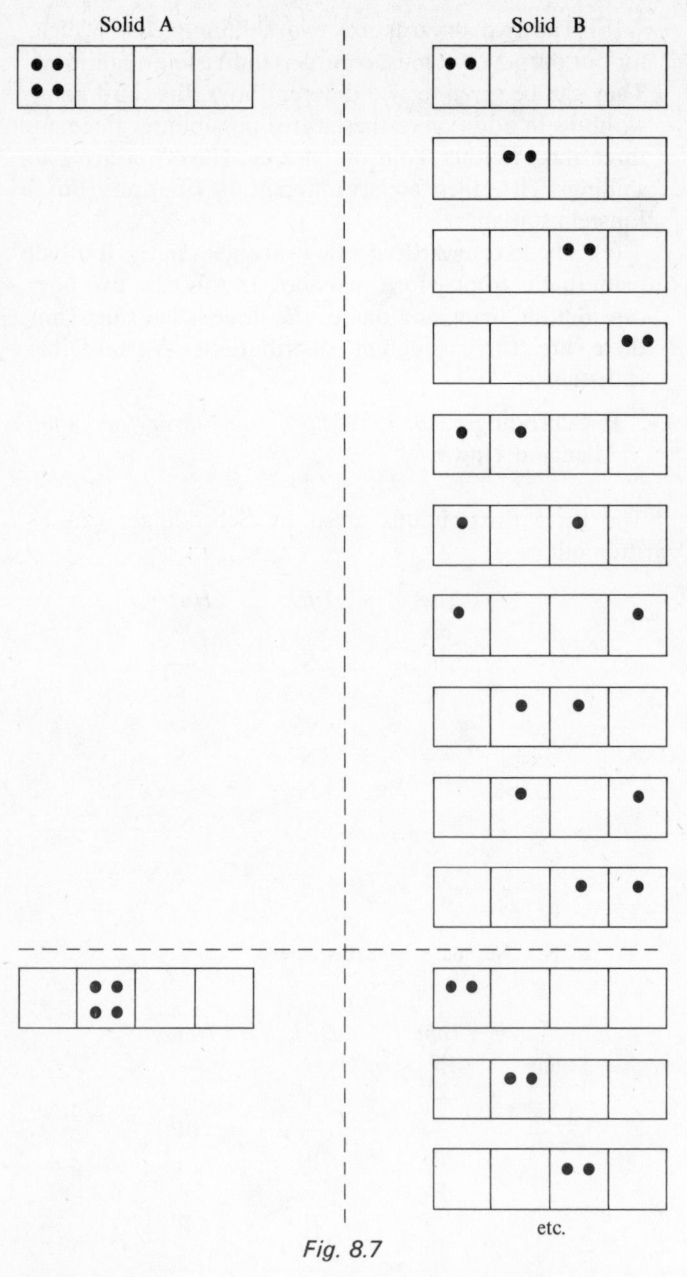

Fig. 8.7

Temperature

When I put an object at say 10°C in thermal contact (i.e., thermal energy can pass between the objects) with another object at say 30°C then they will, if I wait long enough, come to the same temperature (some value between 10 and 30°C). We say that the two objects will reach thermal equilibrium (see section 3.6). Two objects at say 22°C will not of their own accord proceed to change their temperatures so that one ends up at 10°C and the other at 30°C. The event need not violate the conservation of energy principle. There is a one-way process towards thermal equilibrium.

Why should the process be one way? What do we mean by temperature? What is the difference between two objects at different temperatures? What is thermal equilibrium?

Consider two very theoretical solids, solid A has four atoms and four quanta of energy and solid B has four atoms and two quanta. Which solid is at the higher temperature? If we put the two solids in thermal contact we would expect there to be a net flow of energy from the higher temperature solid to the lower temperature solid (we could define our terms high and low temperatures to mean this). Let us make an assumption: chance decides how the quanta move between the two solids.

There are 35 ways of arranging the quanta in solid A and in solid B 10 ways. If A was to give a quantum to B then the number of ways for A would drop to 20 and the number of ways for B would rise to 20. How has this changed the total number of ways of arranging the quanta in the two solids? Initially the number of ways was $35 \times 10 = 350$, after the movement of one quantum from A to B the number of ways became $20 \times 20 = 400$. The total number of ways has increased.

You may rightly query why the numbers of ways for the quanta in A and B were multiplied together to give the total number of arrangements. Each arrangement of quanta in solid A (i.e., way) can be arranged in turn with every one of the arrangements of quanta in solid B (Fig. 8.7). Thus there are initially ten different combinations of B with each arrangement of A. As there are initially 35 possible ways for A then there will be 35×10 total arrangements.

What happens if we take a quantum from B and give it to A? The number of ways for A rises from 35 to 56. The number of ways for B drops from 10 to 4. The total number of ways changes from 350 to $56 \times 4 = 224$, a drop. Taking the quantum from A to B gave an increase in the number of ways and thus this would be the direction the event would tend to go. We conclude that A is at a higher temperature than B.

Consider another example:

Solid A	Solid C
4 quanta, 4 atoms 35 ways	2 quanta, 2 atoms 3 ways

Total of $35 \times 3 = 105$ ways

A gives a quantum to C

3 quanta, 4 atoms 20 ways	3 quanta, 2 atoms 4 ways

Total of $20 \times 4 = 80$ ways

This is a drop in the number of ways and thus is unlikely to occur.

C gives a quantum to A

5 quanta, 4 atoms 56 ways	1 quantum, 2 atoms 2 ways

Total of 112 ways

This is an increase in the number of ways from the initial state and is likely to occur. C is thus at a higher temperature than A.

A and C have the same average number of quanta per atom, i.e., one quantum per atom, but they are at different temperatures. When two objects are at different temperatures and in thermal contact energy flows between the two, giving one a net gain, until the two are at the same temperature. When two objects are at the same temperature energy can still pass between the two but neither object makes a net gain.

Solid A	Solid D
4 atoms, 4 quanta	7 atoms, 9 quanta
35 ways	5005 ways

Total of 35 × 5005 = 175 175 ways

A gives one quantum to D

4 atoms, 3 quanta	7 atoms, 10 quanta
20 ways	8008 ways

Total of 20 × 8008 = 160 160 ways

This is a drop in the number of ways and thus is unlikely to occur.

D gives one quantum to A

4 atoms, 5 quanta	7 atoms, 8 quanta
56 ways	3003 ways

Total of 56 × 3003 = 168 168 ways

This is a drop in the number of ways and thus is unlikely to occur. Both the movement of a quantum from A to D and from D to A give decreases in the number of ways. A and D are thus at the same temperature.

As this last example shows, temperature is not on the microscopic scale the average number of quanta per atom. Two objects at the same temperature can have different numbers of quanta per atom.

How can we distinguish between objects at different temperatures? We can ascertain what happens when we supply energy, i.e., how W, the number of ways, changes.

Let us look at our first example again:

Solid A, higher temperature
4 atoms, 4 quanta
35 ways

Addition of one quantum gives 56 ways. This has increased the number of ways by a factor of 1·6, i.e., 56/35.

Solid B, lower temperature
4 atoms, 2 quanta
10 ways

Addition of one quantum gives 20 ways. This has increased the number of ways by a factor of 2·0, i.e., 20/10.

If A and B had been at the same temperature the effect of adding a quantum to each would be to change the number of ways by the same factor. You can check this with solids A and D which are at the same temperature. Addition of a quantum of energy seems to have a much greater effect on the factor by which the number of ways changes for the lower temperature solid than for the higher temperature solid.

Solid F
4 atoms, 6 quanta
84 ways

Addition of one quantum gives 120 ways. This has increased the number of ways by a factor of 1·4, i.e., 120/84. By putting F in thermal contact with, in turn, A and B and considering the combined numbers of ways we find that F is at a higher temperature than A or B.

Solid G
4 atoms, 8 quanta
165 ways

Addition of one quantum gives 220 ways. This has increased the number of ways by a factor of 1·3, i.e., 220/165. By putting G in thermal contact with A, B, and F we find that G is at a higher temperature than A, B, or F.

Solid	*Atoms*	*Quanta*	*Factor by which ways are changed when one quantum is given to the solid*	
B	4	2	2·0	
A	4	4	1·6	increasing
F	4	6	1·4	temperature
G	4	8	1·3 ↓	

The larger the factor by which the number of ways change when one quantum is supplied the lower the temperature. We could use this factor to specify the temperature of a solid, without recourse to any temperature scale defined by reference to melting or boiling points of substances. Why not

$$\text{factor by which ways are changed when one quantum is given to the solid} \left(\frac{W'}{W}\right) \propto \frac{1}{T}$$

where T is the temperature? We could have said $1/T^2$ or $1/T^3$. The reciprocal has to be used if an increasing temperature is to mean a decreasing factor.

Is this a reasonable temperature scale? The lowest temperature we could envisage would be when our four-atom solid had no energy quanta. With no quanta there is just one way a solid can distribute its energy—all atoms with none. If we give this solid one quantum the number of ways rises from one to four. With one quantum and four atoms there are four ways—each atom in turn having the quantum. The factor W'/W is thus four for the lowest temperature. If we had a real solid with its very large number of atoms the number of ways would have risen from one to a number equal to the number of atoms, when one quantum was added. W'/W would thus be very large for a real solid. As each further quantum was added W'/W would become smaller until it reaches 1·0 for very large number of quanta.

For 4 atoms, 100 quanta $W'/W = 1{\cdot}03$
4 atoms, 1000 quanta $W'/W = 1{\cdot}003$

Suppose the constant of proportionality in the equation relating W'/W and T is one. Our lowest temperature, with a large number of atoms, becomes the reciprocal of a very large number. Our highest temperature becomes the reciprocal of 1·0, i.e., 1·0. The absolute temperature scale goes from zero to infinity. There is no reason why our new scale should do the same but it might be more convenient. We can make our high temperature value infinity by taking the logarithm of W'/W to be proportional to $1/T$. Without the logarithm the highest temperature becomes

$$T \propto \frac{1}{W'/W} \propto \frac{1}{1{\cdot}0}$$

With the logarithm

$$\log\left(\frac{W'}{W}\right) \propto \frac{1}{T}$$

$$T \propto \frac{1}{\log(W'/W)} \propto \frac{1}{\log(1{\cdot}0)} \propto \frac{1}{0}$$

This would seem to be a scale similar to the absolute scale of temperature.

Using the logarithm also has other advantages. For one quantum added:

$$\log\left(\frac{W'}{W}\right) = \log 1{\cdot}5$$

when we have four atoms and 5 quanta. Suppose we add two quanta to give 7 quanta, then:

$$\log\left(\frac{W''}{W}\right) = \log(1{\cdot}5^2) = \log 1{\cdot}5 + \log 1{\cdot}5 = 2\log 1{\cdot}5$$

One quantum changes the number of ways from 56 to 84, a change by a factor of 84/56 or 1·5. Two quanta change the ways from 56 to 120, a change by a factor of 120/56 or 2·1. This is approximately 1·5 × 1·5. Three quanta extra change the factor by approximately 1·5 × 1·5 × 1·5. Thus we have

$$\log\left(\frac{W'''}{W}\right) = 3\log 1{\cdot}5$$

The logarithm of the factor is directly proportional to the number of quanta added, i.e., the total energy supplied to the solid.

$$\log\left(\frac{W'''}{W}\right) = \log W''' - \log W = \Delta \log W$$

The change in the logarithm of the number of ways is directly proportional to the thermal energy supplied (ΔQ)

$$\Delta \log W \propto \Delta Q$$

Incorporating this with our temperature equation we can write

$$\Delta \log W \propto \frac{\Delta Q}{T}$$

or by introducing a constant

$$k\,\Delta \ln W = \frac{\Delta Q}{T}$$

k is known as Boltzmann's constant. We choose the value of the constant to give our temperature scale, generally known as the Kelvin scale, the same size degrees as the absolute temperature scale. The value given is $1{\cdot}38 \times 10^{-23}$ J K^{-1}, if the logarithms are to base e.

The following are the descriptions of temperature used by two authors.

> '... although I say when two things are at the same temperature things get balanced, it does not mean they have the same energy in them; it means that it is just as easy to pick energy off one as to pick it off the other. Temperature is like 'ease of removing energy'.
>
> R. P. Feynman (1965). *The character of physical law*, BBC Publications.

> '... In daily life it is quite easy to define with the help of a thermometer what we mean by stating that a piece of matter has a certain temperature. But when we try to define what the temperature of an atom could mean we are ... in a much more difficult position. Actually we cannot correlate this concept 'temperature of the atom' with a well-defined property of the atom but have to connect it at least partly with our insufficient knowledge of it. We can correlate the value of temperature with certain statistical expectations about the properties of the atom. ...'
>
> W. Heisenberg (1958). *Physics and philosophy*, Allen & Unwin/Harper and Row.

The laws of thermodynamics

Zeroth law (law O)

If any two objects are each in thermal equilibrium with a third, they will be found to be in equilibrium with each other.

First law

If all the energy changes occurring in a process are added up the total energy does not change. This is the law of conservation of energy.

Second law

A closed system will always end up in that condition which can be realized in the greatest number of ways.

Third law

The absolute zero of temperature can never be reached.

There are many other ways of writing these laws—they all, however, amount to the same things.

Entropy

A bouncing ball comes to rest—this is the natural direction for such an event. A ball suddenly springing up from the floor is unnatural. When an event proceeds in the 'natural' direction we say that the entropy is increasing. If the event is considered for the 'un-natural' direction the entropy decreases. At equilibrium any change results in a decrease in entropy. The term change in entropy ΔS is used to denote the quantity $k\,\Delta \ln W$ or $\Delta Q/T$. An increase in the number of ways of arranging energy means an increase in entropy. Thermal energy ΔQ supplied to an object at temperature T produces a change (increase) in entropy for that object of $\Delta Q/T$. We could determine entropy changes by calculating the changes in the number of ways. A more practical method of determining the change is to measure $\Delta Q/T$.

A convention has been adopted that at absolute zero the entropy of a pure substance is zero. (This is another way of writing the Third Law.) By measuring all the $\Delta Q/T$ terms necessary to bring a substance up to a particular temperature we can give that substance an absolute entropy value at that temperature. For example we would measure the energy ΔQ_1 needed to raise the temperature of the substance by, say, 2 K from an initial temperature of 0 K. The entropy change would be

$$\frac{\Delta Q_1}{1}$$

T is taken as the mean temperature, 1 K. The entropy change in going from 2 K to 4 K would be

$$\frac{\Delta Q_2}{3}$$

where ΔQ_2 is the energy used to raise the temperature from 2 to 4 K. By summing all these increments we arrive at the absolute entropy at some temperature T.

$$S = \frac{\Delta Q_1}{1} + \frac{\Delta Q_2}{3} + \frac{\Delta Q_3}{5} + \cdots + \frac{\Delta Q}{T}$$

Typical entropy values are:

1 mole of water, 1 atmosphere pressure, as liquid at 273 K

$$S = 63\ \mathrm{J\,K^{-1}}$$

1 mole of water, 1 atmosphere pressure, as solid at 273 K

$$S = 41\ \mathrm{J\,K^{-1}}$$

The difference between these two values is due to the latent heat needed to make the change from solid to liquid. If it were possible to have water in a vapour state at 273 K and 1 atmosphere pressure then its entropy would (by extrapolation) be about 180 $\mathrm{J\,K^{-1}}$. Liquid and solid water are in equilibrium at 273 K so the liquid should not change into the solid or the solid to liquid. The entropy of the liquid is, however, larger than that of the solid so apparently a change from solid to liquid will give an entropy increase and should be the natural direction of the event. We should not find solid water at 273 K because it has all turned into liquid. This is obviously wrong. The error we have made is not to take into account the energy which must be extracted from the surroundings in order to melt the solid water. The entropy of the surroundings must change, as well as the entropy change for the water. The energy needed to change 1 mole of water from solid to liquid is $5{\cdot}94 \times 10^3\ \mathrm{J\,mol^{-1}}$. The entropy of the surroundings must therefore decrease by $5{\cdot}94 \times 10^3/273$ or 22 $\mathrm{J\,K^{-1}}$. This is in fact the difference between the entropy values for the solid and liquid water, $63 - 41 = 22\ \mathrm{J\,K^{-1}}$. Solid and liquid water are in equilibrium at 273 K because a change from solid to liquid or liquid to solid would not produce any increase in entropy. If we were to increase the temperature to 274 K then the entropy decrease for the surroundings becomes 21 $\mathrm{J\,K^{-1}}$ and there is an entropy increase when solid water changes to liquid water. The natural course of events is for solid water to change into liquid water at 274 K. If we were to decrease the temperature to below 273 K then the entropy decrease for the surroundings becomes more than the difference in entropy values between the solid and liquid.

To summarize: an event will occur if it results in an increase in entropy. Equilibrium exists when any change results in a decrease in entropy.

Engines

An engine converts fuel energy into mechanical energy. The fuel is used to increase the temperature of something, often steam, which is then used to turn perhaps a turbine and in the process becomes cool. The efficiency of even the most modern engines is only about 30 per cent—only 30 per cent of the energy available from the fuel appears as mechanical energy. The rest of the energy is given to the surroundings—perhaps via cooling towers or a river. Power stations have lots of energy to dispose of—lots of energy they 'waste'. Does this have to be so?

Suppose the fuel in a furnace supplies 100 MW to a boiler. If the efficiency is 30 per cent the mechanical energy produced by the steam from the boiler will be 30 MW. The rest of the energy, 70 MW, is dissipated to the surroundings by say a cooling tower. Figure 8.8 illustrates this.

The entropy changes during the above are:

$$\text{Entropy drop at the furnace} = \frac{100 \times 10^6}{T_H}$$

where T_H is the temperature of the furnace;

$$\text{Entropy gain by the surroundings} = \frac{70 \times 10^6}{T_C}$$

where T_C is the temperature of the surroundings. The net entropy change is

$$\frac{70 \times 10^6}{T_C} - \frac{100 \times 10^6}{T_H}$$

For entropy to increase we must have

$$\frac{70 \times 10^6}{T_C} > \frac{100 \times 10^6}{T_H}$$

$$\frac{T_H}{T_C} > \frac{100 \times 10^6}{70 \times 10^6}$$

If $T_C = 300$ K, then T_H must be greater than 430 K.

The higher the temperature of the steam entering the turbine the greater the possible efficiency. For 70 per cent efficiency T_H must be greater than 1000 K. The limitation on the efficiency of an engine is set by the requirement that entropy must increase.

In no event can we convert all our fuel energy into mechanical energy.

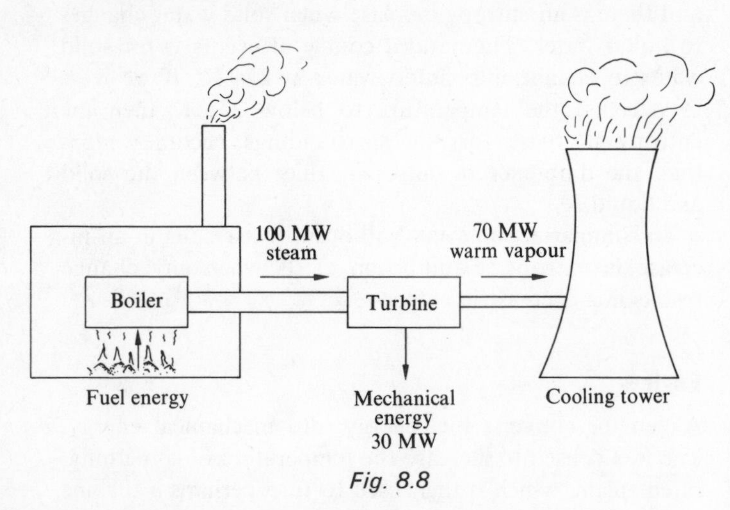

Fig. 8.8

Further reading

J. F. Sandford. *Heat engines*, Heinemann Science Study Series, No. 22, 1962.

Specific heat capacities

We can use the arguments already presented to arrive at the way the temperature of a solid depends on the energy supplied.

$$W = \frac{(N + n - 1)!}{(N - 1)!\, n!}$$

W is the number of ways of arranging n quanta among N atoms. (See page 197.)

When one quantum is added we have

$$W' = \frac{(N + n + 1 - 1)!}{(N - 1)!(n + 1)!}$$

Hence

$$\frac{W'}{W} = \frac{(N + n)}{n + 1}$$

You can check this formula against the earlier results.

$$k\,\Delta \ln W = \frac{\Delta Q}{T}$$

Hence

$$k \ln\left(\frac{N + n}{n + 1}\right) = \frac{\varepsilon}{T}$$

where ε is the energy of one quantum. Thus

$$\frac{N + n}{n + 1} = e^{\varepsilon/kT}$$

Neglecting the 1 on the bottom line as being insignificant in comparison with n gives on rearranging:

$$n = \frac{N}{e^{\varepsilon/kT} - 1}$$

n is the total number of quanta given to the solid. Thus $n\varepsilon$ is the total amount of energy given to the solid.

$$n\varepsilon = \frac{N\varepsilon}{e^{\varepsilon/kT} - 1}$$

The number of atoms in the solid is $N/3$, because N is the number of independent oscillators and each atom can have three independent modes of oscillation. If we consider one mole of atoms and write L for Avogadro's number then

$$\text{Energy supplied to 1 mole} = \frac{3L\varepsilon}{e^{\varepsilon/kT} - 1}$$

As $\varepsilon = hf$, the equation becomes

$$\text{Energy supplied per mole} = \frac{3Lhf}{e^{hf/kT} - 1}$$

This equation tells us how the temperature of one mole of a crystalline solid depends on the energy supplied.

At high temperatures, i.e., when hf/kT is small,

$$e^{hf/kT} \approx 1 + \frac{hf}{kT}$$

$$\text{Thus energy supplied per mole} = \frac{3Lhf}{1 + (hf/kT) - 1}$$

$$= 3LkT$$

Hence the molar heat capacity (at constant volume)

$$C_v = \frac{\Delta(\text{energy})}{\Delta T} = 3Lk$$

As $Lk = R$, the gas constant (see chapter 9), this can be written as

$$C_v = 3R$$

An alternative way of arriving at this equation is given in chapter 9. The other way does not, however, give the equation at temperatures other than high. Chapter 9 gives values of molar heat capacities and compares them with the prediction of this equation.

Energy distribution

n quanta among N atoms—how are the quanta arranged? At some instant how many atoms have 0 quanta, how many 1 quantum, how many 2 quanta, etc.? Suppose we put our quanta into our solid and let chance determine how many an atom will acquire. One way you can represent this is to make a board with six squares by six, i.e., a total of 36 squares, and having put the quanta in some way on the atoms (use counters), then use the chance throws of two dice to pick on a site from which to take a quantum and

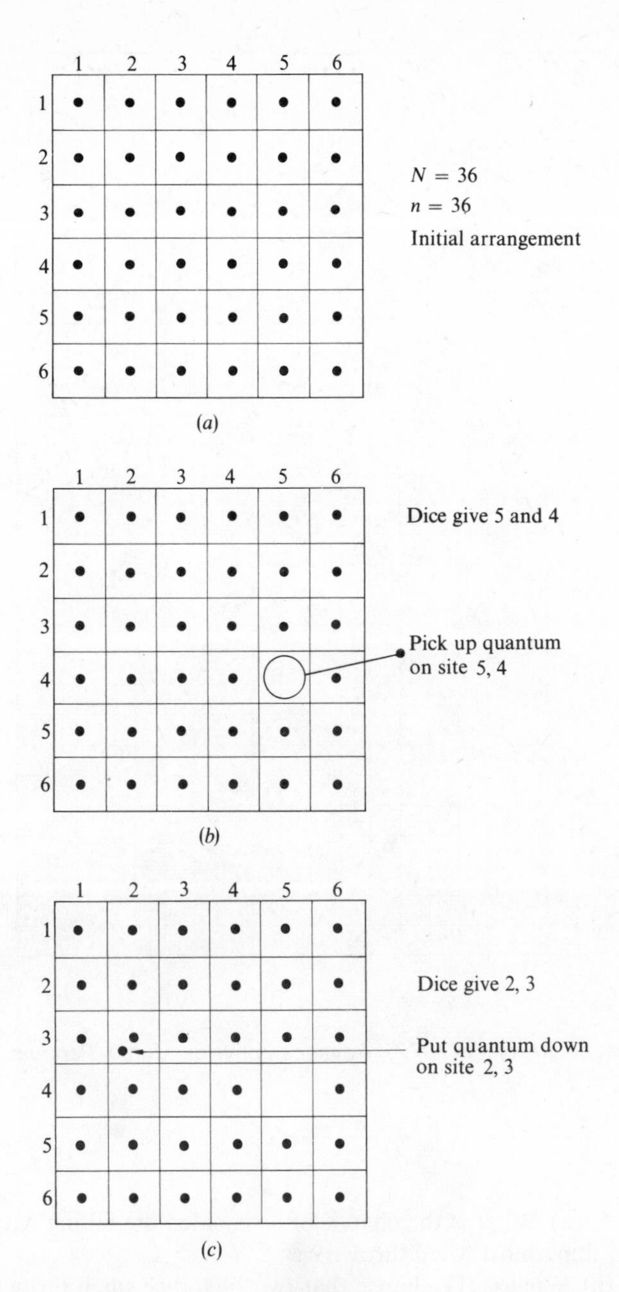

Fig. 8.9

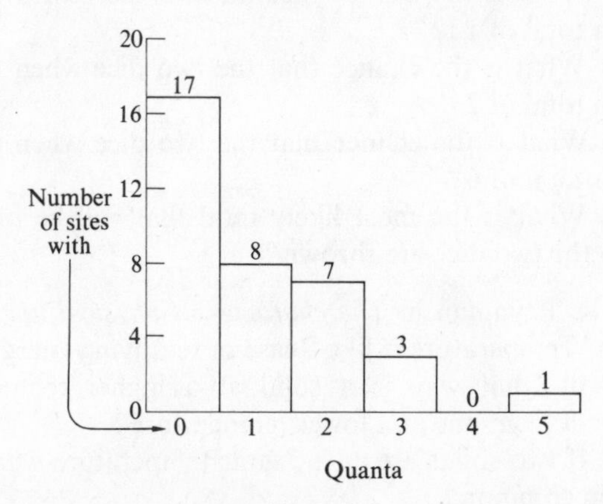

Fig. 8.10

then use the two dice to determine which site to move the quantum to. Figure 8.9 shows a possible first move in such a game. After many moves the distribution becomes more or less static. Figure 8.10 shows the situation after 100 moves. The distribution begins to look as though it may be an exponential. Figure 8.11 shows the situation with 900 sites and 900 quanta after 10 000 moves; the distribution is quite a good exponential. There are twice as many sites with 0 quanta as with 1 quantum, twice as many sites with 1 quantum as with 2 quanta, twice as many with 2 quanta as with 3, etc. Figure 8.12 shows the result when 900 sites shared 300 quanta; there are now four times as many sites with 0 quanta as with 1 quantum, four times as many with 1 quantum as with 2 quanta, etc. With less quanta we have a steeper exponential; this is a lower temperature.

For $N = 900$, $n = 900$

$$\frac{\text{Number of sites with 0 quanta}}{\text{Number of sites with 1 quantum}} = 2$$

But

$$\frac{W'}{W} = \frac{N+n}{n+1} \approx 2$$

For $N = 900$, $n = 300$

$$\frac{\text{Number of sites with 0 quanta}}{\text{Number of sites with 1 quantum}} = 4$$

But

$$\frac{W'}{W} = \frac{N+n}{n+1} \approx 4$$

It would seem that

$$\frac{\text{Number of sites with 0 quanta}}{\text{Number of sites with 1 quantum}} = \frac{W'}{W}$$

$$= e^{\varepsilon/kT}$$

Thus

$$\frac{\text{Number of sites with energy } n\varepsilon}{\text{Number of sites with energy } (n+1)\varepsilon} = e^{\varepsilon/kT}$$

ε is the energy of a quantum.

An application of this equation is given on page 269 for electrical conduction and for evaporation on page 247.

Note: In this chapter it has been assumed that the atoms can accept any number of equally sized quanta, i.e., they have equally spaced energy levels. The arguments can still be applied if this is not the case. See Appendix B of Nuffield Advanced Physics, Unit 9, Penguin 1972.

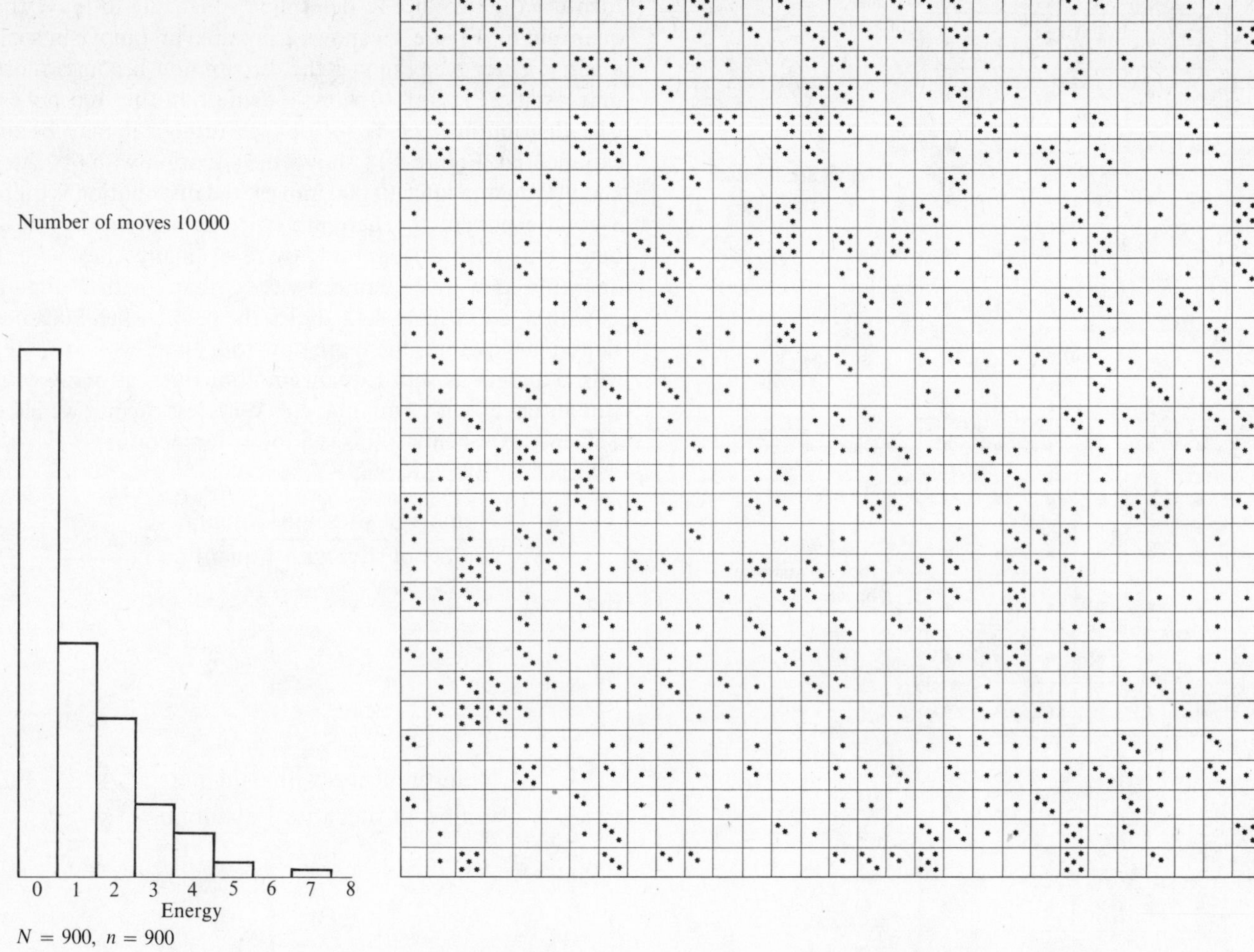

Fig. 8.11 N = *900,* n = *900. (Nuffield Foundation Science Teaching Project (1972).* Advanced physics. Unit 9, *Penguin.)*

Problems

1. How are new laws in physics found?

2. W. Thomson in 1874 wrote: 'Physical processes... are irreversible....' Are physical processes irreversible? Give examples to illustrate your answer.

3. In Darwin's theory of evolution random genetic mutations are considered to be the mechanism of evolution. If the change brought about by such mutations confers on the animal advantages then the effect is passed on—natural selection. Some biologists consider that useful characteristics acquired during an animal's lifetime can be passed on—this effect is not random.

(a) What do we mean by 'random'?

(b) How could tests be devised to distinguish between the random and non-random theories?

(You might like to read about such experiments in the book by A. Koestler, *The case of the midwife toad*, Hutchinson, 1971.)

4. (a) What is the chance of a six-sided die falling with a 5 uppermost when thrown?

(b) What is the chance that two such dice when thrown both give a 5?

(c) What is the chance that the two dice when thrown give a total of 12?

(d) What is the chance that the two dice when thrown give a total of 2?

(e) What is the chance that the two dice when thrown give a total of 6?

(f) What is the most likely total that will be obtained when the two dice are thrown?

5. R. Feynman in *The character of physical law* (BBC) states 'Temperature is like "ease of removing energy".'

(a) In what way is a solid at a higher temperature different from one at a lower temperature?

(b) If two solids are at the same temperature what have they in common?

(c) How valid do you think Feynman's definition of temperature is?

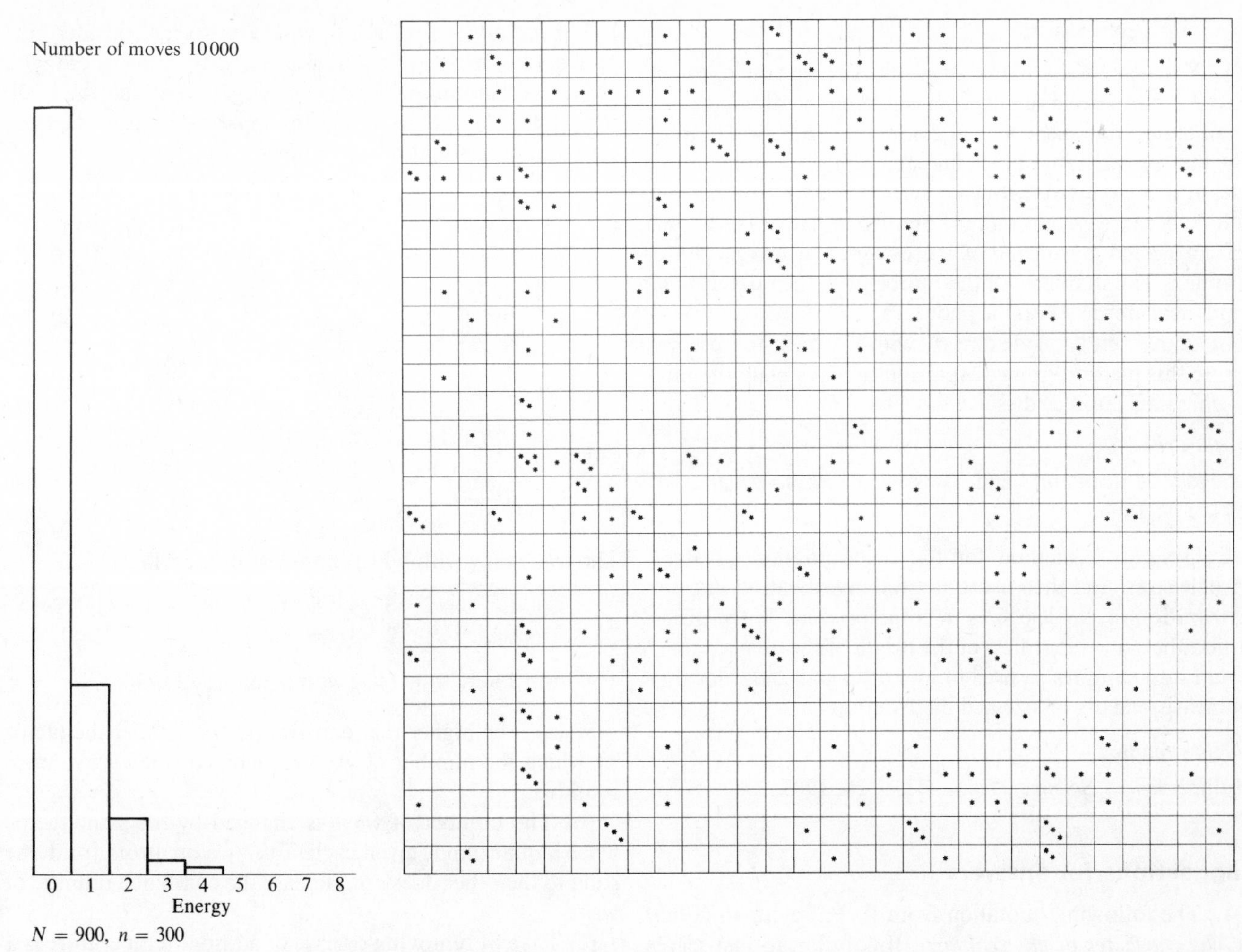

Fig. 8.12 N = *900,* n = *300. (Nuffield Foundation Science Teaching Project (1972)* Advanced physics. Unit 9, *Penguin.)*

6. Hydrogen normally exists as H_2 molecules. But if 1 mole of H_2 changes to two moles of H atoms the entropy of the hydrogen increases from 130 to 230 $J\ K^{-1}$.

(a) Why does hydrogen remain as molecules if an entropy increase would result by it dissociating?

(b) At what temperature should hydrogen spontaneously dissociate?

Energy to break the H—H bond = 430 $kJ\ mole^{-1}$.

7. Why do objects in thermal contact approach a common temperature?

8. I can convert all of a quantity of potential energy into a temperature change. Why cannot I reverse the procedure completely? Why is only a partial reversal possible?

9. Time flows on. The arrow of time. Time is unidirectional. What characterizes the passage of time? Is time reversible?

10. What do you think is meant by the statement that a solid has a negative temperature?

Teaching note Further problems appropriate to this chapter will be found in:

H. A. Bent. *The second law*, Oxford Univ. Press, 1965.

Nuffield advanced physics. Unit 9, Penguin, 1972.

Practical problems

1. Krieger, Mulholland, and Dickey in their article describe how measurements are made of the diameter of a small gas bubble suspended in a liquid as a function of time. The bubble was caught on a fine quartz fibre, a slow bubble stream being ejected through a syringe needle until a single bubble became caught on the fibre. An image of the bubble was projected by means of lenses and the image photographed. The diameter of the bubble was then determined from measurements on the photograph. The measurements could, however, be made directly with a vernier microscope.

Use this method or an adaptation to investigate the rates at which air bubbles dissolve in water.

Further reading

Krieger, Mulholland, and Dickey. *J. Phys. Chem.*, **71**, 1123–9, 1967.

2. Develop a method for the determination of small particle sizes, and then use it in a full investigation into the distribution of particle sizes in common dust. Is the distribution the same regardless of the origin of the dust sample? Could dust samples be used to identify a particular locality by measurements of size and distribution?

Further reading

Mullin. *School Science Review*, **47**, 9–20, 1965.

Suggestions for answers

1. The following quotation from R. P. Feynman (1965) in *The character of physical law* (BBC Publications) serves as an answer.

> 'In general we look for a new law by the following process. First we guess it. Then we compute the consequences of the guess to see what would be implied if this law that we guessed is right. Then we compare the result of the computation to nature, with experiment or experience, compare it directly with observation, to see if it works. If it disagrees with experiment it is wrong. In that simple statement is the key to science. It does not make any difference how beautiful your guess is. It does not make any difference how smart you are, who made the guess, or what his name is—if it disagrees with experiment it is wrong. That is all there is to it. It is true that one has to check a little to make sure it is wrong, because whoever did the experiment may have reported incorrectly, or there may have been some feature in the experiment that was not noticed, some dirt or something; or the man who computed the consequences, even though it may have been the one who made the guesses, could have made some mistakes in the analysis. . . .'

2. Physical processes are all irreversible. See page 195 and the quote from L. Brillouin.

> 'Physics does not know any motion without damping.'

3. I leave this question to you. You might consider how you could decide if a die was biased—how would you recognize random behaviour? I suggest you read the book mentioned in the question to appreciate the difficulties involved in testing for randomness.

4. (a) 1/6
(b) 1/36
(c) 1/36
(d) 1/36
(e)

	1	2	3	4	5	6
1	2	3	4	5	6	7
2	3	4	5	6	7	8
3	4	5	6	7	8	9
4	5	6	7	8	9	10
5	6	7	8	9	10	11
6	7	8	9	10	11	12

The frequency with which numbers occur will be:

Total	2	3	4	5	6	7	8	9	10	11	12
Frequency	1	2	3	4	5	6	5	4	3	2	1

The most likely total is 7, with a chance of 6/36.

5. (a) The higher the temperature the smaller the factor by which the number of ways is changed when a quantum is added.

(b) The number of ways is changed by the same factor when a quantum is given each. Taking a quantum from one solid to the other does not increase the combined number of ways.

(c) Ease of removing or ease of adding; temperature is a measure of the effect on the number of ways of adding or removing quanta—the higher the temperature the less the fractional effect. We are only allowed to take or give energy quanta to a solid if the total number of ways increases.

6. (a) The dissociation results in an increase of 100 J K^{-1} in the entropy of the hydrogen. Breaking the hydrogen molecule apart requires energy—which has to come from the surroundings. The temperature of the surroundings must therefore drop. At 300 K the entropy decrease resulting from this temperature drop will be $430 \times 10^3/300 =$ 1400 J K^{-1}. The net entropy change by the dissociation would be -1300 J K^{-1}, an entropy decrease. For this reason dissociation at 300 K is unlikely.

(b) About 4300 K.

7. The common temperature is when the number of ways of arranging the quanta in the two solids is a maximum.

8. See the section on engines, page 201.

9. The direction of time is the direction of increasing entropy. On a large scale time is not reversible—the chance of the entropy decreasing being very small.

10. It has more atoms with large numbers of quanta than with small numbers of quanta, i.e., the exponential distribution slopes the wrong way.

9 Solids, liquids, and gases

Teaching note Practical work appropriate to this chapter will be found in:
W. Bolton. *Physics experiments and projects*, Vol. 1, Pergamon, 1968.
Nuffield advanced physics. Unit 1 teacher's guide, Penguin, 1972.

9.1 Matter

'From what is to follow... lay aside your cares and lend undistracted ears and an attentive mind to true reason. Do not scornfully reject, before you have understood them, the gifts I have marshalled for you with zealous devotion.... I will reveal those atoms from which nature creates all things....'

Lucretius, about 55 B.C. *On the nature of the universe*. Transl. R. E. Latham. Copyright R. E. Latham (tr) 1951, Penguin.

As far back as the Ancient Greeks there was a suspicion that matter was made up of very minute particles, so minute that they were invisible. The name atoms has since been given to those minute particles. Matter, despite its apparently continuous appearance, was considered to have structure and be discontinuous. It was not, however, until about one and a half centuries ago that the existence of these atoms became more than a suspicion.

'I. *Outline of the Daltonian Theory*

1. With respect to the nature of the ultimate elements of bodies, we have no means of obtaining accurate information; but it is the general opinion that they consist of atoms, or minute solids, incapable of farther division....

They must, I think, be physical points, as minute as you will, but still possessed of length, breadth, and thickness....

2. In cases of the chemical union of one body with another, the substances combined are dispersed everywhere through the whole mass.... However a minute a portion soever of water we take, we shall find it to contain both oxygen and hydrogen.... Now this could not be the case unless the atoms of the combining bodies united with each other....

3. All chemical compounds contain the same constant proportion of constituents with the most rigid accuracy, no variation whatever ever taking place. Water is universally composed of 1 part of hydrogen and 7·5 parts of oxygen; ... by weight....

4. The permanency of chemical compounds cannot be owing to anything else than to the union of a certain determinate number of the atoms of one constituent with certain determinate number of the atoms of the other. Let us suppose water the compound. Let the number of atoms of oxygen which unite be x, and of hydrogen y, then an integrant particle of water will in every case be $x + y$.'

Thomas Thomson (1813) *Annals of Phil.*, **2**, 32–43

Other evidence for the atomic nature of matter was supplied by considerations of the ways in which crystals could be produced from some elementary building block.

'...The substance which most readily admits of division by fracture into these forms is fluorspar....

If we form a plate of uniform thickness by two successive divisions of the spar, parallel to each other, we shall find the plate divisible into prismatic rods, the section of which is a rhomb of 70° 32′ and 109° 28′ nearly; and if we again split these rods transversely, we shall obtain a number of regular acute rhomboids, all similar to each other, having their superficial angles 60° and 120°, and presenting an appearance of primitive molecule, from which all the other modifications of such crystals might very simply be derived....

The theory to which I here allude is this, that, with respect to fluorspar and such other substances as assume the octahedral and tetrahedral forms, all difficulty is removed by supposing the elementary particles to be perfect spheres, which by mutual attraction have assumed that arrangement which brings them as near to each other as possible.

The relative position of any number of equal balls in the same plane, when gently pressed together, forming equilateral triangles with each other [as represented perspectively in Fig. 9.1(a)] is familiar to every one; and it

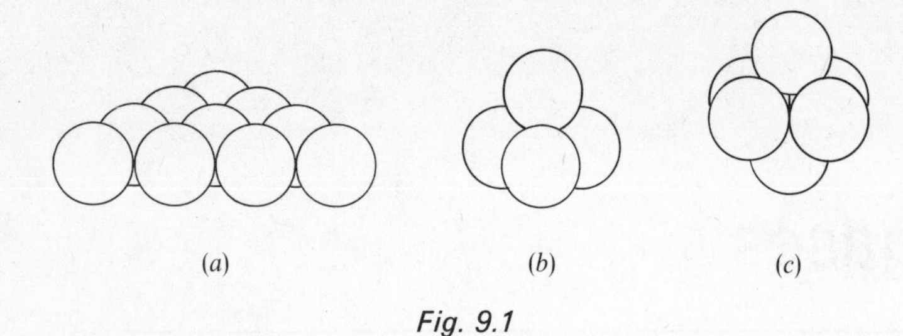

Fig. 9.1

is evident that, if balls so placed were cemented together, and the stratum thus formed were afterwards broken, the straight lines in which they would be disposed to separate would form angles of 60° with each other.

If a single ball were placed any where at rest upon the preceding stratum, it is evident that it would be in contact with three of the lower balls [as in Fig. 9.1(b)], and that the lines joining the centres of four balls so in contact, or the planes touching their surfaces, would include a regular tetrahedron, having all its sides equilateral triangles [Fig. 9.1(c)]....'

Wollaston (1813) *Phil. Trans.*, **2**, 51–63.

Atoms became small spheres, capable of being stacked to give crystal structures and by their moving around in a gas capable of exerting pressure.

'MOLECULES

(Lecture delivered before the British Association at Bradford by Prof. Clerk-Maxwell, F.R.S.)

An atom is a body which cannot be cut in two. A molecule is the smallest possible portion of a particular substance. No one has ever seen or handled a single molecule.

...The old atomic theory, as described by Lucretius and revived in modern times, asserts that the molecules of all bodies are in motion, even when the body itself appears to be at rest. These motions of molecules are in the case of solid bodies confined within so narrow a range that even with our best microscopes we cannot detect that they alter their places at all. In liquids and gases, however, the molecules are not confined within any definite limits, but work their way through the whole mass, even when that mass is not disturbed by any visible motion.

This process of diffusion, as it is called, which goes on in gases and liquids and even in some solids, can be subjected to experiment, and forms one of the most convincing proofs of the motion of molecules.

Now the recent progress of molecular science began with the study of the mechanical effect of the impact of these moving molecules when they strike against any solid body. Of course these flying molecules must beat against whatever is placed amongst them, and the constant succession of these strokes is, according to our theory, the sole cause of what is called the pressure of air and other gases....

If the velocity of the molecules is given, and the number varied, then since each molecule, on an average, strikes the side of the vessel the same number of times, and with an impulse of the same magnitude, each will contribute an equal share to the whole pressure. The pressure in a vessel of given size is therefore proportional to the number of molecules in it, that is to the quantity of gas in it.

This is the complete dynamical explanation of the fact discovered by Robert Boyle, that the pressure of air is proportional to its density. It shows also that of different portions of gas forced into a vessel, each produces its own part of the pressure independently of the rest, and this whether these portions be of the same gas or not.

Let us next suppose that the velocity of the molecules is increased. Each molecule will now strike the sides of the vessel a greater number of times per second, but besides this, the impulse of each blow will be increased in the same proportion, so that the part of the pressure due to each molecule will vary as the square of the velocity. Now the increase of the square of the velocity corresponds, in our theory, to a rise of temperature, and in this way we can explain the effect of warming the gas, and also the law discovered by Charles that the proportional expansion of all gases between given temperatures is the same....'

Nature, **8**, 437–41, 1873.

'BROWNIAN MOVEMENT AND MOLECULAR REALITY

By M. Jean Perrin

...Indeed it would be difficult to examine for long preparations in a liquid medium without observing that all the particles situated in the liquid instead of assuming a regular movement of fall or ascent, according to their density, are, on the contrary, animated with a perfectly irregular movement. They go and come, stop, start again, mount, descend, remount again, without in the least tending toward immobility. This is the Brownian movement, so named in memory of the naturalist Brown, who described it in 1827 (very shortly after the discovery of the achromatic objective), then proved that the movement was not due to living animalculae, and recognised that the particles in suspension are agitated the more briskly the smaller they are.

...it was established by the work of M. Gouy (1888), not only that the hypothesis of molecular agitation gave an admissible explanation of the Brownian movement, but that no other cause of the movement could be imagined, which especially increased the significance of the hypothesis....

In brief the examination of Brownian movement alone suffices to suggest that every fluid is formed of elastic molecules, animated by a perpetual motion.'

By the end of the nineteenth century there was in the main an acceptance of the idea that matter was constituted of atoms. Difficulties existed regarding the nature of the atom, in particular the method by which electrical conduction could take place did not seem to fit the small particle atom. The turn of the century saw, however, great steps forward in the understanding of the atom; the electron and radioactivity were discovered.

> '...Although we know nothing of what an atom is, yet we cannot resist forming some idea of a small particle, which represents it to the mind; and though we are in equal, if not greater, ignorance of electricity, so as to be unable to say whether it is a particular matter or matters, or mere motion of ordinary matter, or some kind of power or agent, yet there is an immensity of facts which justify us in believing that the atoms of matter are in some way endowed or associated with electrical powers, to which they owe their most striking qualities, and amongst them their mutual chemical affinity....'
>
> M. Faraday (1834) *Phil. Trans.*, 77.

Chapter 13 takes up the points made in the quotation by Faraday and looks in more detail at models of atoms and explanations of 'mutual chemical affinity'.

The scientist's view of matter is of a whole made up of numerous small particles. The scientist's view of radiation is of a composition of small quanta. To the modern scientist nature is discontinuous. Figure 9.2 shows a painting by

Fig. 9.2 Bridge at Courbevoie. (G. P. Seurat. Photograph, Courtauld Institute Galleries London.)

Seurat, 1888, of a scene in which the composition has been built up from a multitude of small dots of paint. At a distance the dots blend together to give a continuous appearance; close to, the dots show. Nature appears like that to a scientist.

Further reading

M. P. Crosland. *The science of matter*, Penguin, 1971.
S. Toulmin and J. Goodfield. *The architecture of matter*, Penguin, 1965.

9.2 Solids

In the earlier part of this chapter we quoted from a paper by Wollaston (1813) in which he suggests that by stacking spheres he could build up the crystalline form of fluorspar. Suppose we examine the forms of piles of stacked spheres. Figure 9.3 (a) and (b) show two possible sequences. In (i) we start off with the spheres packed, on one plane, as close together as possible. In (ii) we place on top of the first layer, called layer A, another layer of spheres. These are placed so that their centres rest over the voids in the lower layer. This gives a stable arrangement. This second layer we call layer B. In (iii) we have two different ways of stacking the next layer. In the (a) part of the diagram we have placed another layer identical in placing to the first layer, layer A. This thus gives a layer sequence of ABA. In part (b) of the diagram the spheres have been placed in a different stable arrangement. This gives a layer called C. The arrangement is thus ABC. These two arrangements can be repeated to give large structures,

ABABABABABABABABA...
and ABCABCABCABCABC...

The first gives a structure known as hexagonal close-packing (Fig. 9.4) and the second a face-centred cubic structure (Fig. 9.5).

How can we check whether these arrangements exist in solids? X-ray, or electron, diffraction methods (see chapter 12) enable us first of all to determine that there is an orderly structure in a particular solid and then to measure the various interatomic spacing distances. X-rays have wavelengths comparable with the interatomic spacings in crystals and reflect from layers of atoms like light reflecting from a mirror. However, X-rays reflect from each layer of atoms in a crystal and so there are many reflected beams (Fig. 9.6). These reflected beams interfere with each other and only at certain angles do the reflected beams constructively interfere and give a reflected beam, for a particular wavelength. At other angles destructive interference occurs and there is no reflection.

The planes of atoms act like mirrors behave with light. Thus for each plane we have an image of the X-ray source. Interference occurs between the waves which apparently originate at these images. The path difference between the waves from successive images is $2d \sin \theta$ (see Fig. 9.6). Thus the condition for constructive interference is

$$2d \sin \theta = n\lambda \qquad \text{(Bragg equation)}$$

where d is the spacing between the planes of atoms, θ the angle between the atomic plane and the incident (or reflected) X-rays, λ the wavelength of the X-rays and n an integer, i.e., 0, 1, 2, 3, etc.

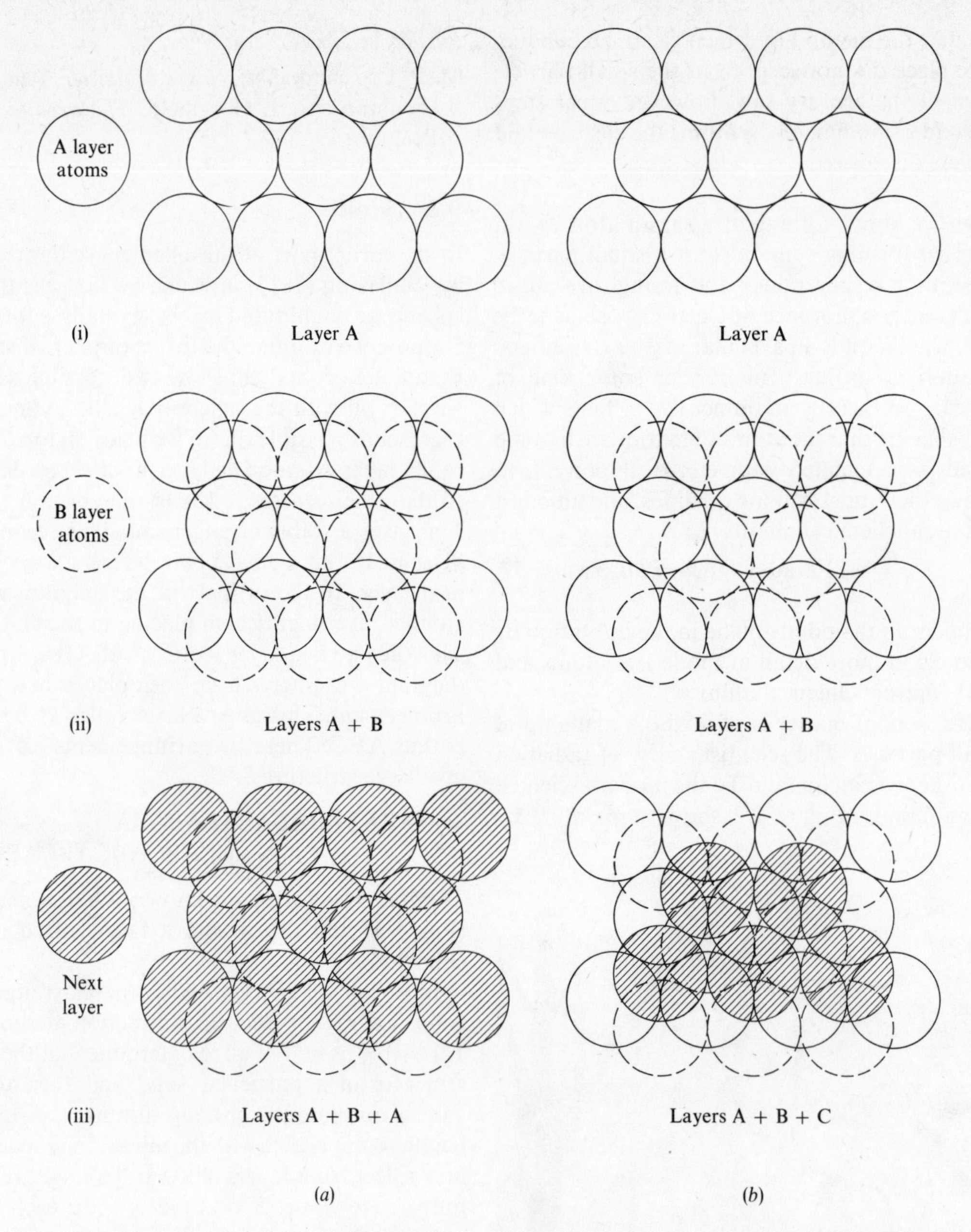

Fig. 9.3 (a) Hexagonal close-packed structure. (b) Face-centred cubic structure.

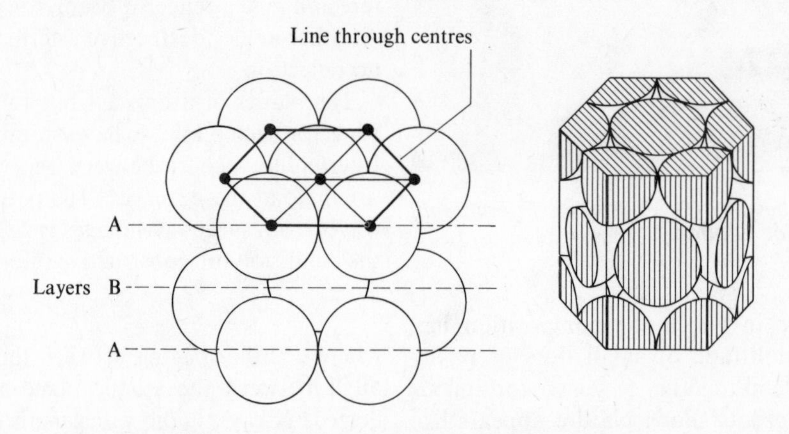

Fig. 9.4 Hexagonal close-packed structure.

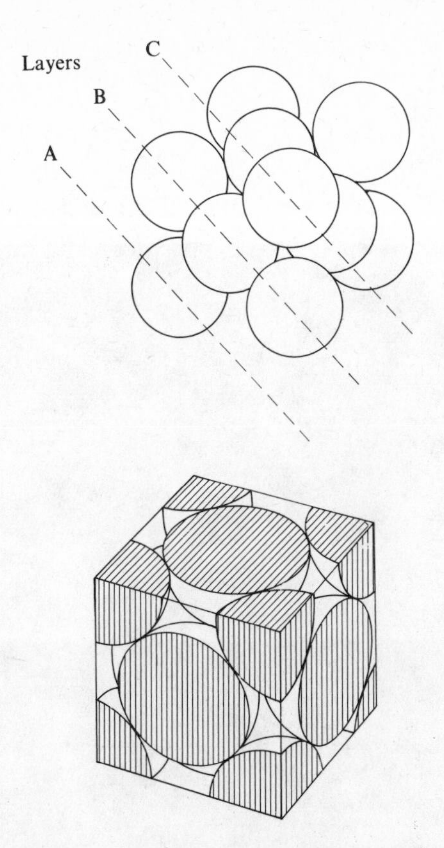

Fig. 9.5 Face-centred cubic structure.

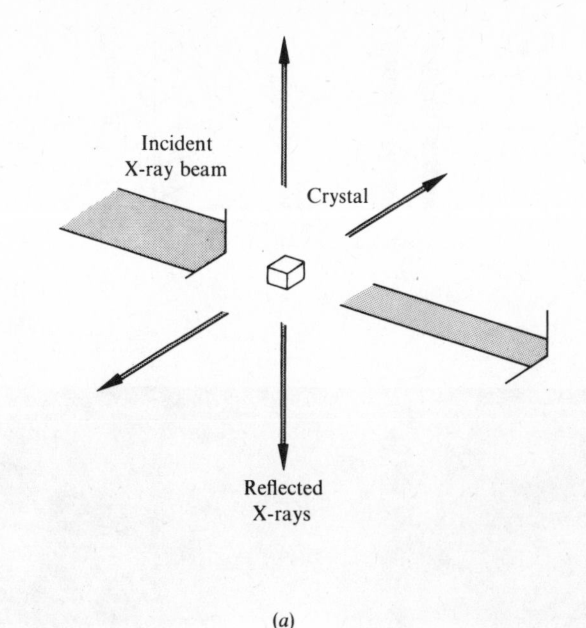

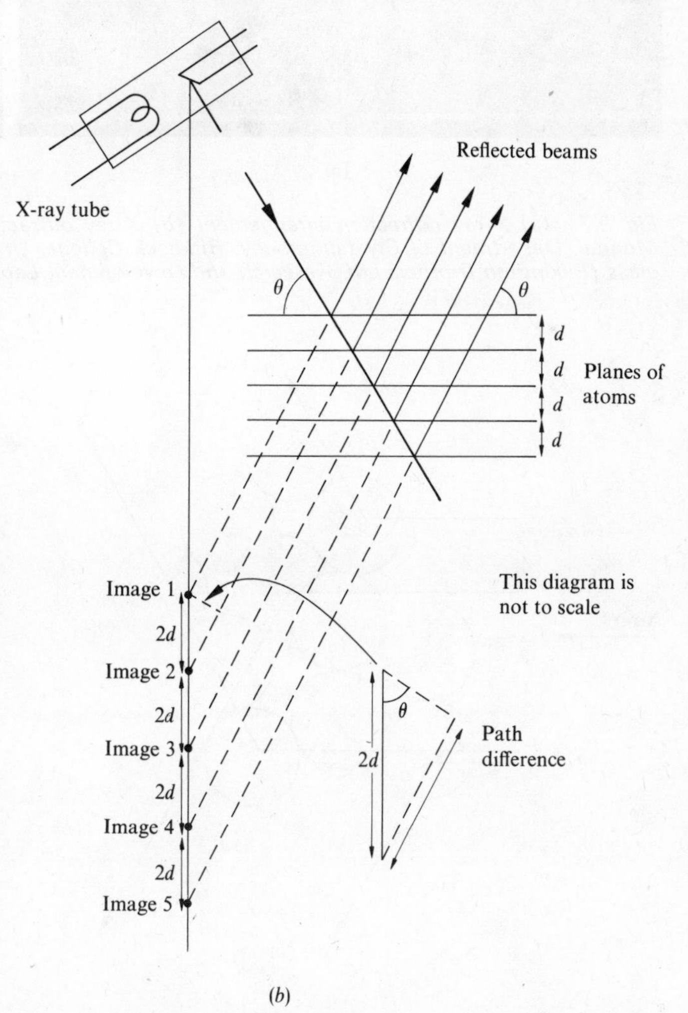

Fig. 9.6

The path difference between the X-rays reflected from any one plane must be a whole number of wavelengths different from the X-rays reflected from the next plane. If a solid has no regular atomic planes there are no regular reflections. Compare the diffraction photographs produced by X-rays (not monochromatic) being reflected off the planes of a single sodium chloride crystal with those produced by 'Pyrex' glass (Fig. 9.7). Many diffraction photographs are, however, of fine powders containing many small crystals. In the multitude of crystals some are at the correct angles to give reflections (Fig. 9.8) and a cone of reflected X-rays results. Figure 9.9 shows the diffraction patterns produced with white tin for different size crystals. The larger the crystal size in the sample the more the diffraction pattern takes on the appearance of dots, the smaller the crystal size the more the pattern becomes a series of continuous rings. Such rings still denote a crystalline structure for the sample.

Measurements of the angles at which reflections occur, θ, and use of the Bragg equation enable values for the spacings between the layers of atoms in a solid to be determined. All the ways of packing atoms give rise to many planes of atoms. Figure 9.10(a) shows some of the possible planes for the hexagonal close-packed structure. We can calculate the various spacings between the planes in terms of the distance between atom centres. For a hexagonal close-packed structure we have planes with spacings of

$$a, \frac{a}{\sqrt{4/3}}, \frac{a}{2}, \frac{a}{\sqrt{8}}, \frac{a}{\sqrt{28/3}}, \text{etc.}$$

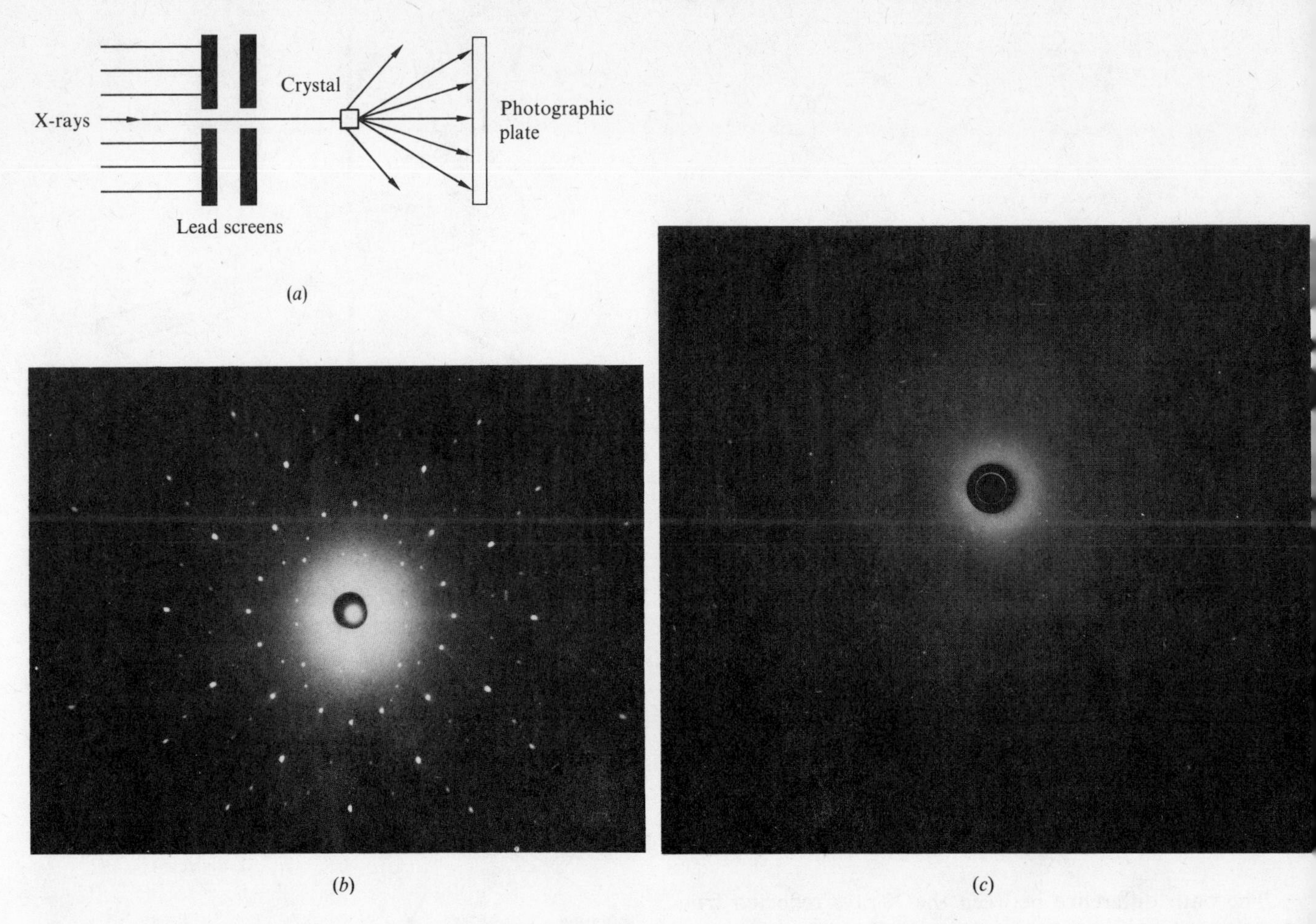

Fig. 9.7 (a) X-ray diffraction arrangement (b) X-ray diffraction pattern with a single crystal of sodium chloride (Miss P. Mondal. Department of Crystallography, Birkbeck College, University of London). (c) X-ray diffraction pattern with 'Pyrex' glass (Pilkington Brothers Ltd, Research and Development Laboratories).

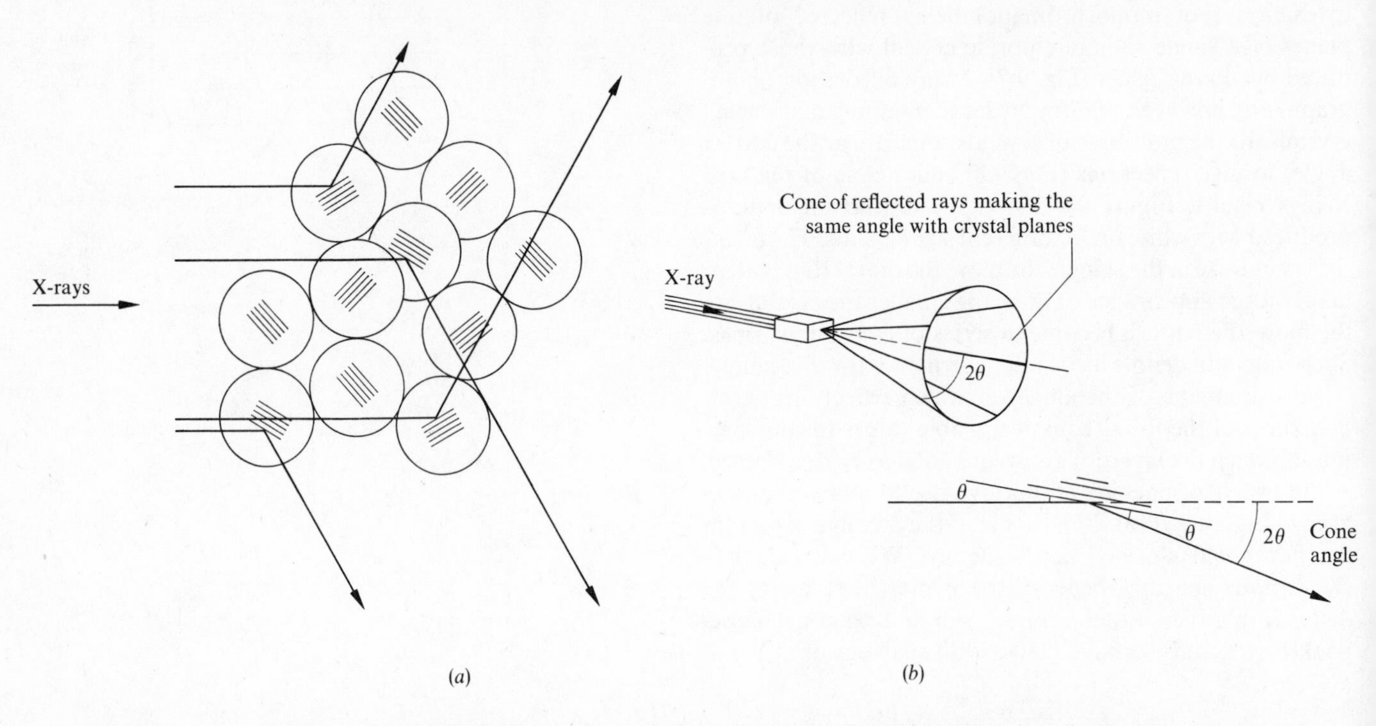

Fig. 9.8 X-ray diffraction. (a) Two-dimensional view. (b) Three-dimensional view.

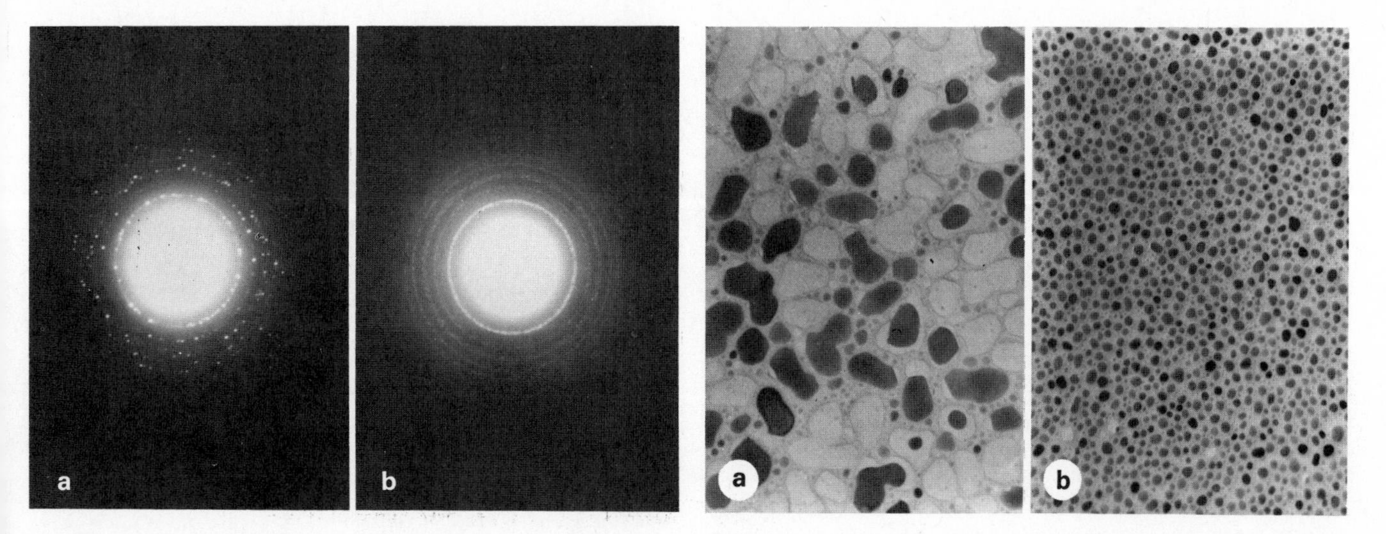

Fig. 9.9 Electron diffraction patterns for white tin crystals deposited on a film of silicon monoxide. The wavelength was $0{\cdot}04 \times 10^{-10}$ m (100 keV electrons). (a) Average crystal size 600×10^{-10} m. (b) Average crystal size 200×10^{-10} m. (Berkeley Physics Course Vol. 4. Quantum physics, *E. H. Wichman. McGraw-Hill Book Company. Copyright Educational Development Center.)*

For a simple cubic array the spacings are

$$a, \frac{a}{\sqrt{2}}, \frac{a}{\sqrt{3}}, \frac{a}{\sqrt{4}}, \frac{a}{\sqrt{5}}, \text{etc.}$$

For a face-centred cubic array we have

$$\frac{a}{\sqrt{3}}, \frac{a}{\sqrt{4}}, \frac{a}{\sqrt{8}}, \frac{a}{\sqrt{11}}, \frac{a}{\sqrt{12}}, \text{etc.}$$

By examining the spacing results from the X-ray diffraction picture we can establish the nature of the atomic packing.

For example, sodium chloride is a face-centred cubic; as also are copper, gold, and lead. Solid helium is hexagonal close-packed; as also are magnesium, zinc, and calcite.

If we assume that atoms are really like our spheres in contact then we can find the radius of the spheres from the values of the spacings between the atomic planes. Figure 9.11 shows a face-centred cubic, a section of the earlier sketch (Fig. 9.5). As can be seen from the diagram, the radius, r, of the sphere is related to a, the minimum distance between atom centres in the structure, by

$$r = \frac{a\sqrt{2}}{4}$$

For a simple cubic $r = a/2$.

Figure 9.12 shows the results for the different elements. A significant point about these results is that the larger radii atoms are those of the alkali elements, the most reactive elements. The radii of ions of some of the elements can be found as in some crystals the elements are present as ions.

For example:

	Atom radius/ 10^{-10} m	*Ion radius/* 10^{-10} m	
Lithium	1·52	0·60	Li^+
Sodium	1·86	0·95	Na^+
Potassium	2·31	1·33	K^+
Chlorine	0·99	1·81	Cl^-
Bromine	1·14	1·95	Br^-

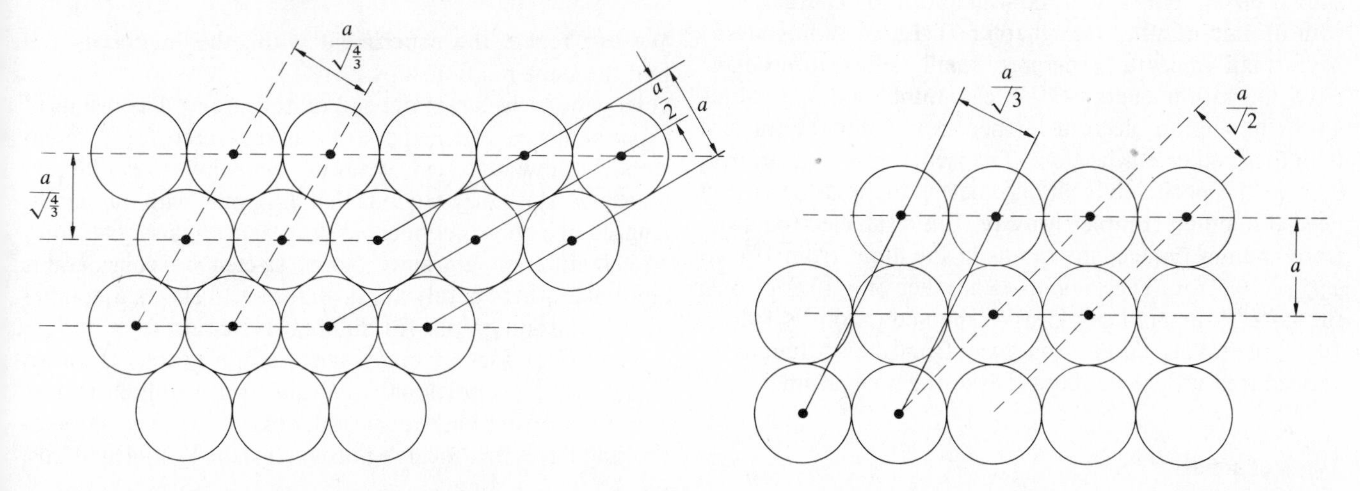

Fig. 9.10 (a) Some of the possible planes of atoms in a hexagonal close-packed structure. (b) Some of the possible planes of atoms in a simple cubic array.

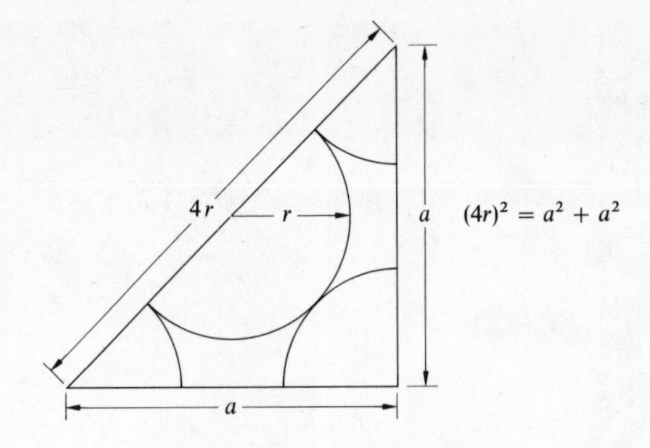

Fig. 9.11 Face-centred cubic.

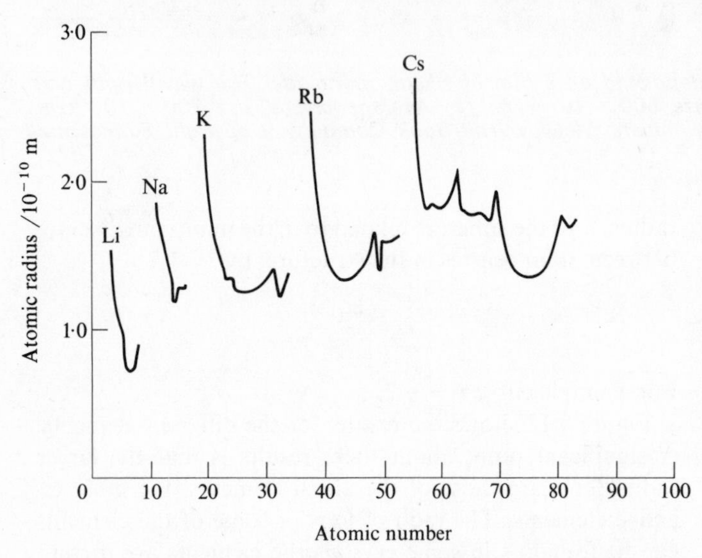

Fig. 9.12 Atomic radii of the elements.

Taking an electron away from alkali elements makes their radii smaller, giving an electron to chlorine and bromine makes their radii bigger.

The variation of atomic radius with atomic number should be compared with the way ionization energies vary with atomic number (see chapter 7). Large radius atoms have small ionization energies; small radius atoms have large ionization energies. We can think of large radius atoms having an electron farther away from the nucleus than the smaller radius atom. The farther away an electron is from the positively charged nucleus the less the energy needed to pull it completely away. Taking an electron away from sodium reduces the radius of the atom from 1·86 to $0{\cdot}95 \times 10^{-10}$ m. The sodium ion has then only 10 electrons, the same as neon. The radius of the neon atom is $1{\cdot}12 \times 10^{-10}$ m; taking an electron away from the sodium atom has reduced its size to about that of the neon atom.

Types of solids

Solids can be divided into two groups: crystalline and amorphous. In the crystalline solids the atoms are arranged in a regular way, in an amorphous solid there is little regularity—certainly not the long-range regularity that occurs in crystalline solids. Figures 9.7(c) and 9.9 show diffraction photographs for a glass and a metal.

Metals are crystalline, generally consisting of a large number of crystals, polycrystalline. A particularly easy surface in which to see crystals is that of galvanized buckets or pipes.

Glasses, polymers, and rubbers all are generally amorphous. They are made up of long molecules which during solidification become tangled. X-ray diffraction photographs do not show the many sharp rings or dots which are characteristic of crystals and long-range order but show only a few diffuse rings or spots, indicating only short-range order.

Elasticity

When you pull a length of rubber it gets longer. On what factors does the change in length depend? Figure 9.13 illustrates a simple experiment for investigating the possible variables. A length of rubber, of uniform cross-sectional area, is held vertical with the upper end in a clamp and the lower end attached to weights. With a constant weight, doubling the length of the rubber doubles the extension; the extension seems to be proportional to the length of the sample. A constant force, load, produces a constant value of extension/original length. This is called the strain.

$$\text{Strain} = \frac{\text{extension}}{\text{length}}$$

Doubling the cross-sectional area of the rubber, with both the length and weight constant, halves the extension. The extension is inversely proportional to the area. With double the area, $2A$, we can obtain the same extension as for area A if we double the weight. A constant extension is produced by a constant value of force/area. This is called stress.

$$\text{Stress} = \frac{\text{force}}{\text{area}}$$

We can repeat the experiment with other materials and find the same relationships.

How does the strain depend on the stress? The results of experiments in which the strain is measured for different stresses show different types of behaviour for different materials. It is fairly obvious that if we pull, with our hands, lengths of rubber, copper, steel, plastics, glass, etc., they stretch different amounts. Some stretch by considerable amounts, others barely at all. Figure 9.14 shows a number of stress–strain graphs for different materials.

The initial part of many stress–strain graphs is linear, i.e., stress is proportional to strain over a limited region. This relationship is known as Hooke's law. The slope of the graph over this linear region is called the Young modulus.

$$\text{Young modulus} = \frac{\text{stress}}{\text{strain}}$$

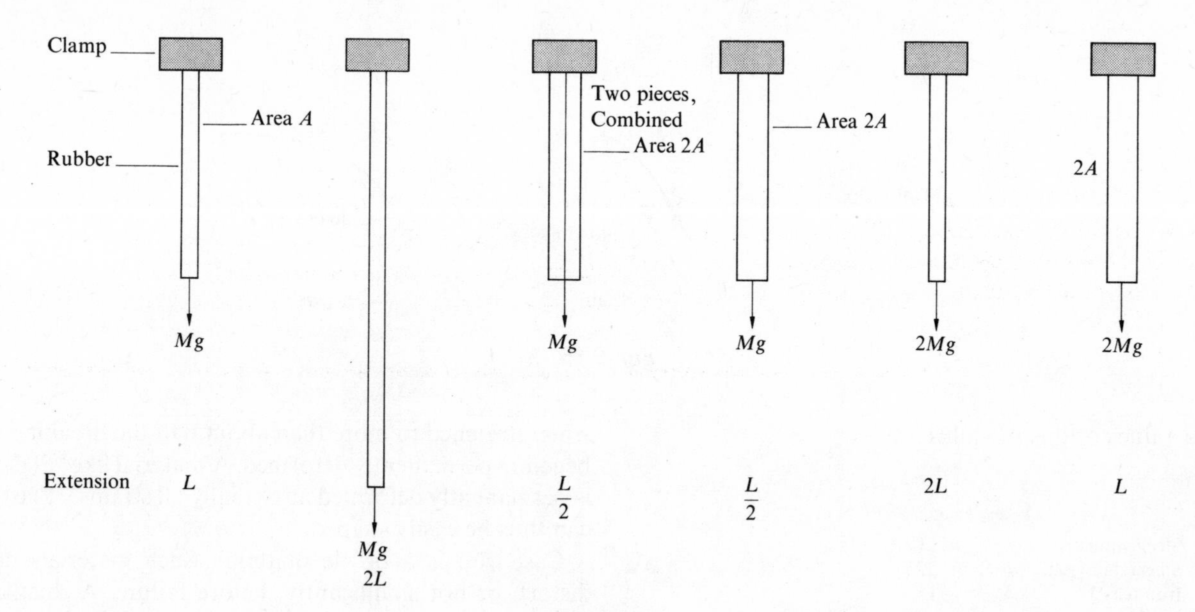

Fig. 9.13

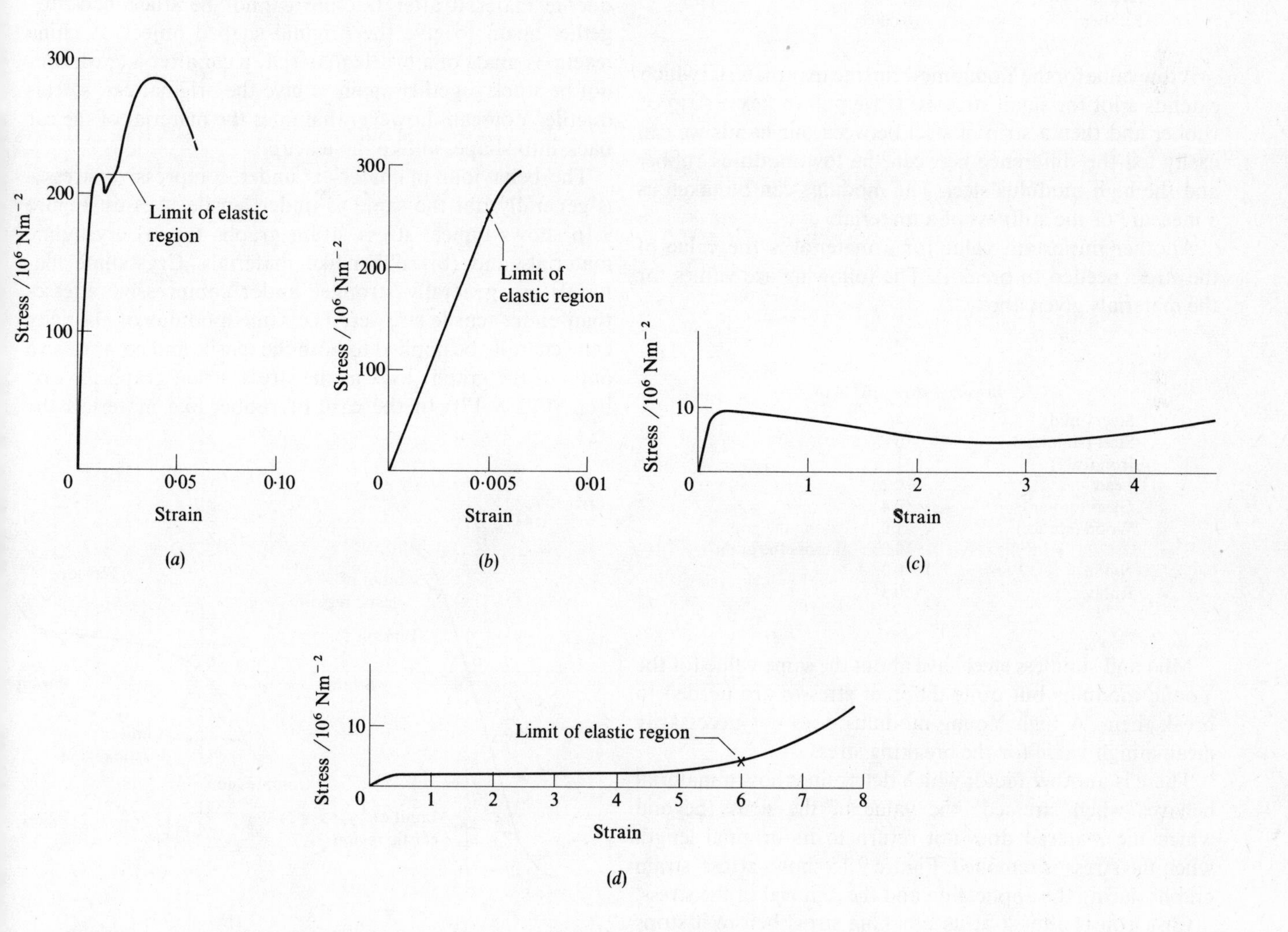

Fig. 9.14 Stress–strain graphs (the axes are not the same for each graph). (a) Steel; (b) cast iron; (c) polythene; (d) rubber.

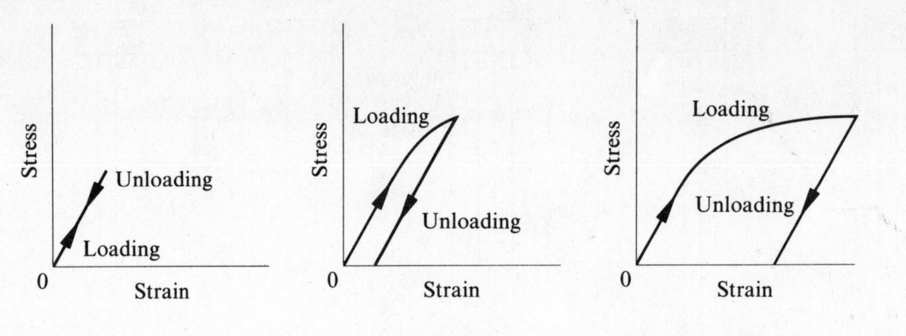

Fig. 9.15

Typical values of this modulus are:

	Young modulus/10^{10} N m^{-2}	
Steel (mild)	21	
Steel (stainless)	22	
Iron (cast)	15	
Lead	2	
Glass (crown)	7	
Wood (spruce)	1·2	along the grain
	0·06	at right angles to grain
Nylon	0·2	
Rubber	0·0007	

A low value for the Young modulus means a material which extends a lot for small stresses. If we pull or flex a strip of rubber and then a strip of steel between our hands we can easily tell the difference between the low modulus rubber and the high modulus steel. The modulus can be taken as a measure of the stiffness of a material.

Another important value for a material is the value of the stress needed to break it. The following are values for the materials given above.

	Breaking stress/10^{8} N m^{-2}	
Steel (mild)	4	
Steel (stainless)	10	
Iron (cast)	1	
Lead	0·1	
Glass (crown)	0·4	
Wood (spruce)	1·0	along the grain
	0·03	across the grain
Nylon	0·3	
Rubber	0·3	

Mild and stainless steel have about the same value for the Young modulus but quite different stresses are needed to break them. A high Young modulus does not necessarily mean a high value for the breaking stress.

There is another factor which determines how a material behaves when stressed—the value of the stress beyond which the material does not return to its original length when the stress is removed. Figure 9.15 shows stress–strain graphs during the application and the removal of the stress.

Cast iron is almost at its breaking stress before it stops coming back to its original length when the stress is removed. Materials like cast iron are difficult to form into different shapes, when bent they do not retain the bent shape. Steel when stretched to more than about half the breaking strain becomes permanently deformed. A material like 'Plasticine' is permanently deformed at virtually all strains. 'Plasticine' can thus be easily shaped.

Cast iron is a brittle material. Such materials do not distort, or not significantly, before failure. A ductile material, like 'Plasticine', suffers considerable distortion before failure. A brittle material when broken can have the pieces all stuck back together again and they will fit. A ductile material after fracture cannot be stuck back together again to give the original shaped object. A china teacup is made of a brittle material; a car after a crash cannot be stuck together again to give the original car, steel is ductile. You can, however, hammer the material of the car back into shape, unlike the teacup.

The behaviour of materials under compressive stresses is generally not the same as under tensile stresses. Figure 9.16 shows typical stress–strain graphs for (a) crystalline materials and (b) rubber-like materials. Crystalline materials are generally stronger under compressive stresses than under tensile stresses. The same modulus of elasticity can generally be applied to both the tensile and compressive parts of the graph; look at the stress–strain graph for cast iron (Fig. 9.17). In the case of rubber-like materials the

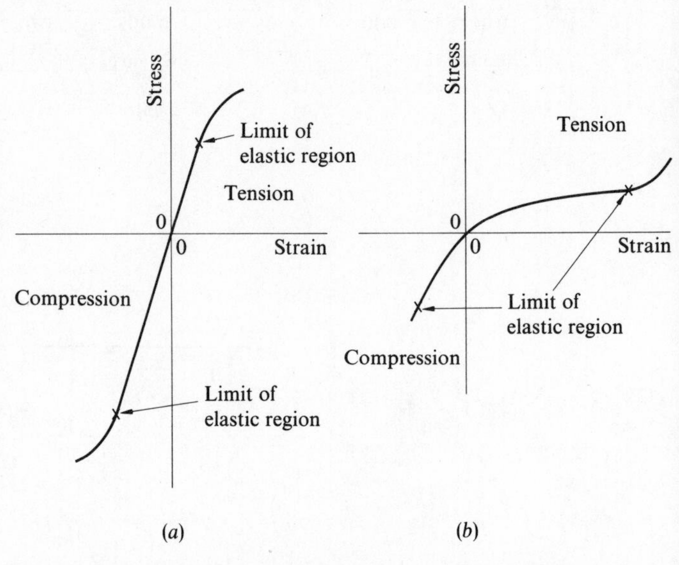

Fig. 9.16 General form of stress–strain graphs for (a) crystalline materials, (b) rubber-like material.

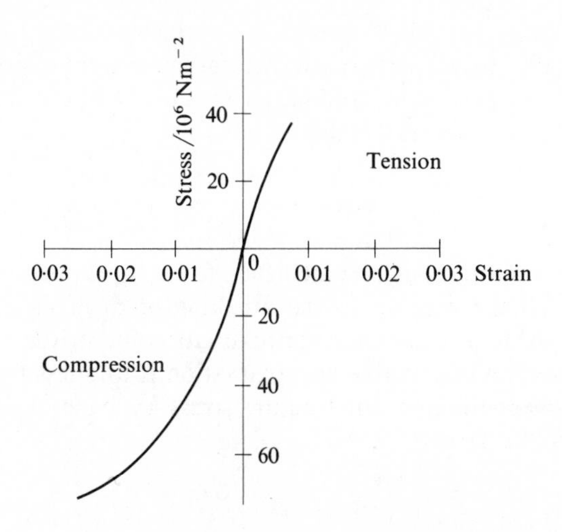

Fig. 9.17 Stress–strain graph for cast iron.

shape of the stress–strain graph for compression is completely different to that for tension. The compressive stress–strain graph is much steeper, i.e., higher modulus, and more linear. Though the stress at which the material breaks is much the same in compression and in tension there is considerable difference in the strains at which breaking occurs.

If a piece of rubber, or any other material, is stretched or compressed and then released it springs back, provided it is stretched within its elastic region. Applying a stress to the material stores energy in the material, which is released when the material is allowed to spring back to its original size. The energy needed to stretch, or compress, a length of material by a distance Δx is the product of the average force acting on the specimen and the distance Δx. The force, however, depends on the extension of the specimen. If the stress is proportional to the strain we have

$$F \propto \Delta x$$

and a graph of force against extension looks like Fig. 9.18 To find the total energy needed to extend a specimen by x we must sum all the $F\,\Delta x$ terms. But $F\,\Delta x$ is the area of a strip below the force–extension graph. Therefore the total energy will be the total area under the graph between extension being equal to zero and equal to x.

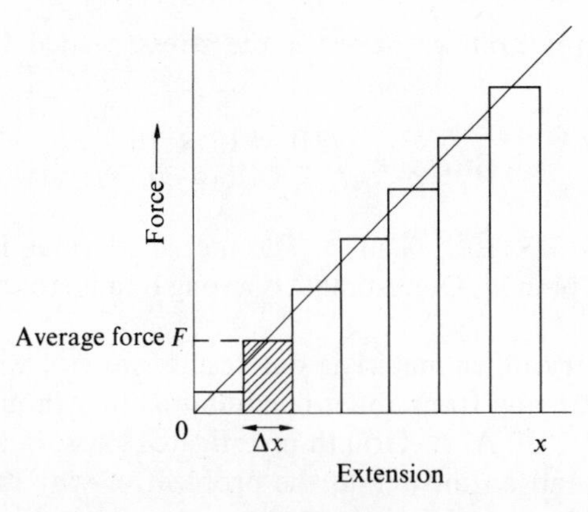

Fig. 9.18

$$\text{Energy} = \tfrac{1}{2}Fx$$

If the cross-sectional area of the specimen is A and its length L then we can write the equation as

$$\frac{\text{Energy}}{AL} = \frac{1}{2}\frac{F}{A}\frac{x}{L}$$

or

$$\text{Energy per unit volume} = \tfrac{1}{2}(\text{stress})(\text{strain})$$

This is the energy needed to extend or compress the specimen and if we assume that all this energy is stored in the sample, then the energy stored in unit volume of the specimen is half the product of the stress and the strain. As stress/strain equals the Young modulus, E, the energy stored per unit volume can be expressed as

$$\text{Energy/volume} = \tfrac{1}{2}(\text{stress})^2/E$$

Further reading

J. E. Gordon. *The new science of strong materials*, Penguin, 1968.

Interatomic forces

The atoms in a solid are held together by interatomic forces. We can picture them as small springs linking the atoms. If we stretch a spring we find a linear relationship between the force and the extension, provided the spring does not become overstretched and deformed. We can write

$$F = k\,\Delta x$$

where F is the force, Δx the extension and k a constant we can call the stiffness or force constant. Suppose we assume that we can apply the same relationship to atoms in a solid: if we displace an atom by a distance Δx from another atom then a force F is needed and F is directly proportional to Δx.

For simplicity we will consider a solid in which the atoms are in a simple cubic array, Fig. 9.19. Stretching increases the separation between two atoms by Δx, thus the force necessary to cause the stretching (also the force trying to restore the atoms to their original sites) is $k\,\Delta x$.

If the cross-sectional area of the sample is A, there will be A/x^2 atoms with springs being stretched. Thus the total force involved is

$$\frac{A}{x^2}\,k\,\Delta x.$$

The stress is therefore

$$\frac{k}{x^2}\,\Delta x.$$

The strain is $\Delta x/x$.
This is the strain for the entire specimen: we are assuming that every pair of planes has been separated by the same distance Δx.

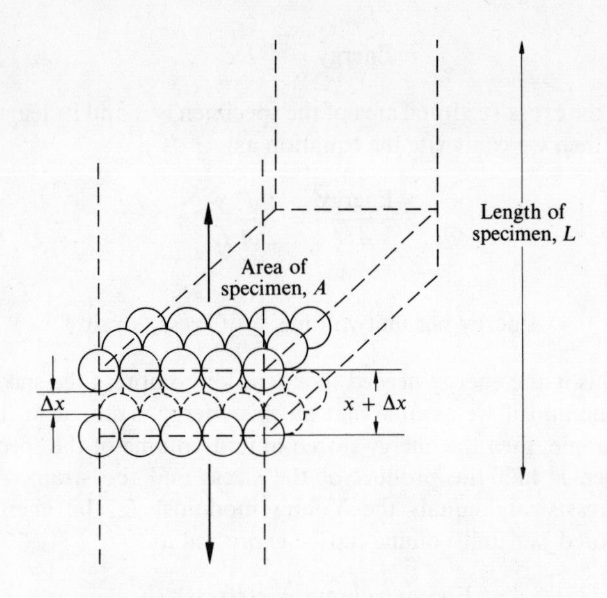

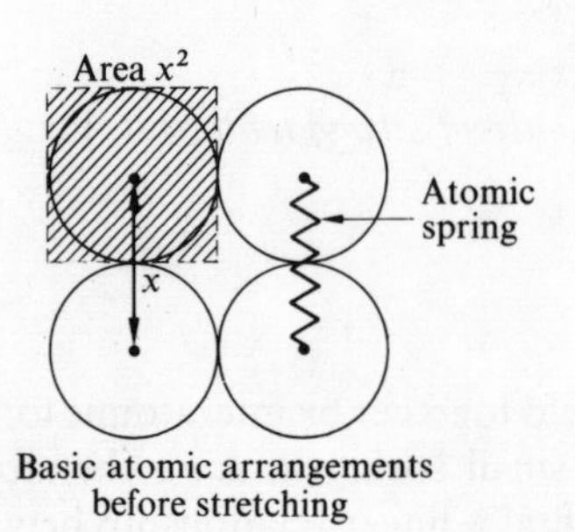

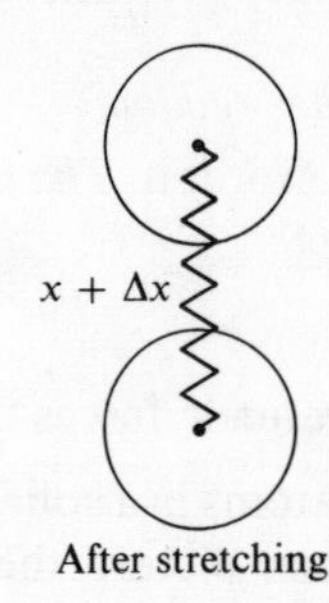

Fig. 9.19

Thus Young modulus $E = \dfrac{\text{stress}}{\text{strain}}$

$$E = \frac{k}{x}$$

If we assume that iron atoms are arranged in the cubic array then, as $E = 15 \times 10^{10}$ N m^{-2} and the spacing of the atoms is about 3×10^{-10} m, the value of k is about 45 N m^{-1}.

Cracks and dislocations

Pulling two layers of atoms apart needs energy. The energy per unit volume will be given by

$$\frac{\text{energy}}{\text{volume}} = \frac{(\text{stress})^2}{2E}$$

Assuming we are just, in this case, pulling two layers apart, i.e., producing a fracture, then the volume will be xA.

Hence

$$\frac{\text{energy}}{A} = \frac{x}{2E}(\text{stress})^2$$

Thus the stress needed to facture a material is given by

$$\text{Stress} = \sqrt{\frac{2E(\text{energy})}{xA}}$$

The energy needed to produce a surface of area 1 m^2 is called the surface energy, G. The energy in our equation is the energy needed to produce an area of $2A$ (there are two sides to the fracture). Hence

$$\text{Stress} = 2\sqrt{\frac{GE}{x}}$$

Our calculation has assumed that stress is proportional to strain all the way up to the breaking point. This is only likely to be a reasonable approximation for brittle solids. In general a reasonable approximation is to guess that we have overestimated our fracture stress by a factor of two (Fig. 9.20). Hence

$$\text{Stress} = \sqrt{\frac{GE}{x}}$$

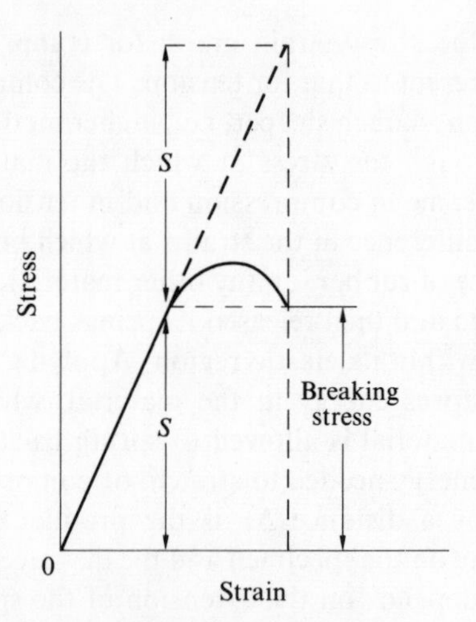

Fig. 9.20

The following are the results of experimental determinations of G.

	G/J m^{-2}
Lead	0·5
Iron	2·0
Tungsten	2·9

(Converted from data given in H. R. Tipler. *Metallurgical Reviews*, **8**, 105, 1963)

Thus for iron we have for the stress needed to cause fracture

$$\text{Stress} = \sqrt{\frac{2{\cdot}0 \times 15 \times 10^{10}}{3 \times 10^{-10}}}$$

or about 3×10^{10} N m^{-2}. The measured value is about 1×10^{8} N m^{-2}. Our estimate is wrong by a factor of about 300.

If we try other materials our results are still wrong by factors ranging from about one hundred to a thousand.

About 1920 A. A. Griffith investigated glass, in order to try and find an answer to the problem of why materials were much weaker than the theoretical reasoning indicated.

He used glass because he could measure G for liquid glass and assume that it was much the same in the solid (see the section on glass in this chapter) and also because it was a convenient brittle substance. From his G measurement and the theory he estimated that glass should have a breaking stress of about 10^{10} N m^{-2}. The glass he was using had a breaking stress of only about 10^8 N m^{-2}. However, he found that when he made very thin fibres he obtained much higher breaking stresses; fibres about 10^{-3} mm diameter having breaking stresses of about 3×10^9 N m^{-2}. Thin fibres were much nearer the theoretical value than bulk glass. The reason for this strange behaviour was cracks, very fine, almost invisible cracks. Thin fibres were less prone to damage and so had less cracks. It is now possible to produce very thin fibres of many materials and obtain very high breaking stresses. Herring and Galt in 1952 produced the first high strength metal fibres—they were called whiskers and were found growing from the surface of tin.

The low strength of metals is now ascribed to the presence of dislocations inside the metal. Figure 9.22 shows a diagram of a dislocation in a simple cubic crystal. G. I. Taylor in 1934 first thought of the idea but it was not until the 'fifties that convincing evidence for their existence was found. Figure 9.21(a) shows the first electron microscope photograph of a dislocation.

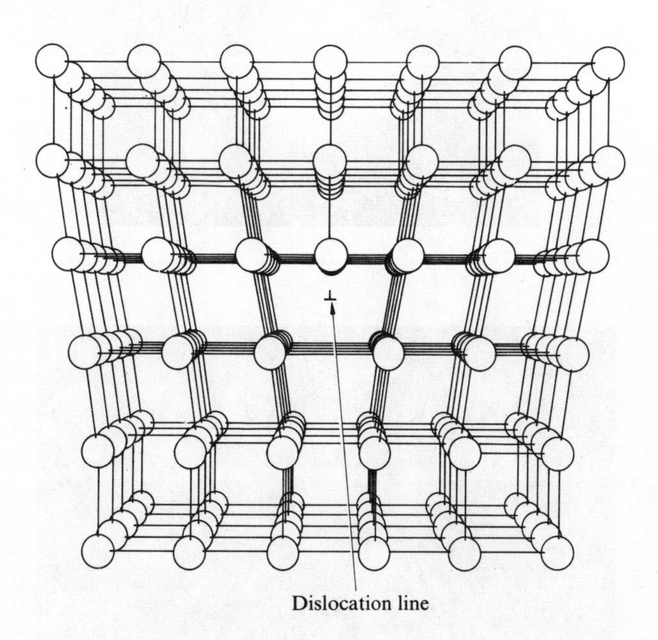

Fig. 9.22

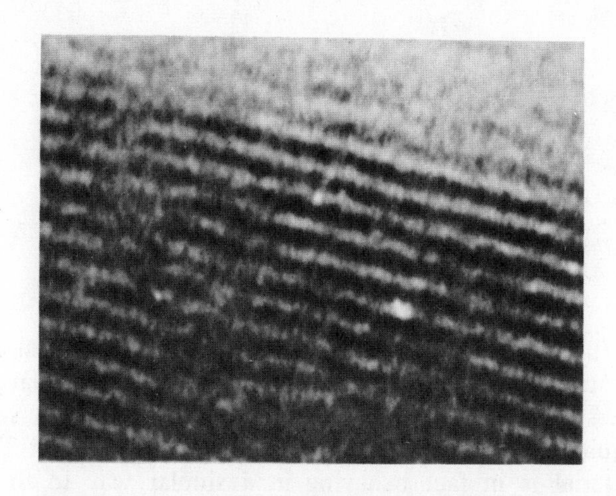

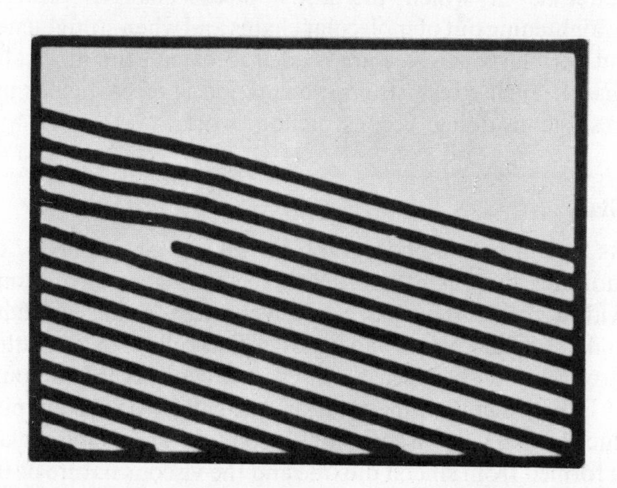

Fig. 9.21 The first photograph of a dislocation. (J. W. Menter. Proc. Roy. Soc., A**236**, *119, 1956.)*

Polymers

How are the atoms, or molecules, arranged in polymers such as rubber or nylon? Figure 9.23 shows diffraction photographs for (a) stretched rubber and (b) nylon. Both show some signs of crystallinity.

The diffraction photograph for rubber changes quite significantly when the rubber is stretched. Stretched rubber shows a significant amount of crystallinity; unstretched rubber shows very little, giving a pattern more like that of glass. The atomic arrangement in rubber is thought to change from a liquid like state to a crystal state, i.e., a disordered state to an ordered state. The stress–strain graph for rubber (Fig. 9.14(d)) is much steeper at high strains than at low strains and corresponds more to the stiffness of a crystalline solid.

Rubber molecules are long chain molecules, in natural rubber basic units of C_5H_8 linked together in very long chains. In the unstretched state these molecules are considered to be randomly orientated. When rubber is stretched these long molecules uncoil and so straighten. This is completely different to metals and other crystalline substances where stretching involves lengthening atomic bonds. It is because of this difference that we have the characteristic rubberlike elasticity—easily stretched, can be stretched to large strains, elastic behaviour. When the rubber molecule chains have been straightened then the stretching force has to extend interatomic bonds. The rubber is then like a crystal (Fig. 9.24).

Nylon consists of molecules of $NH_2(CH_2)_6NH.CO.(CH_2)_4CO$ which link together in long chains containing

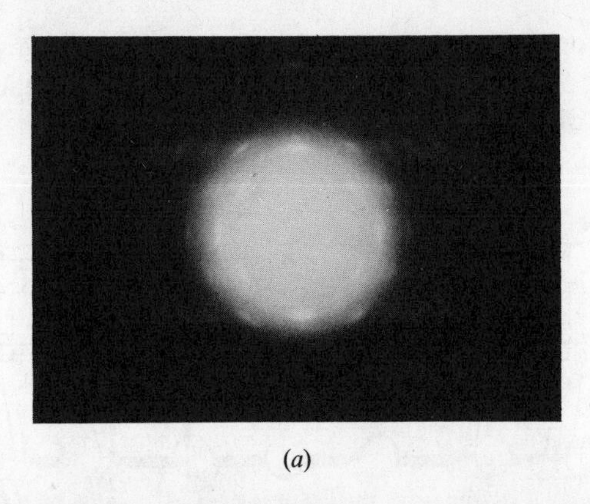

(a)

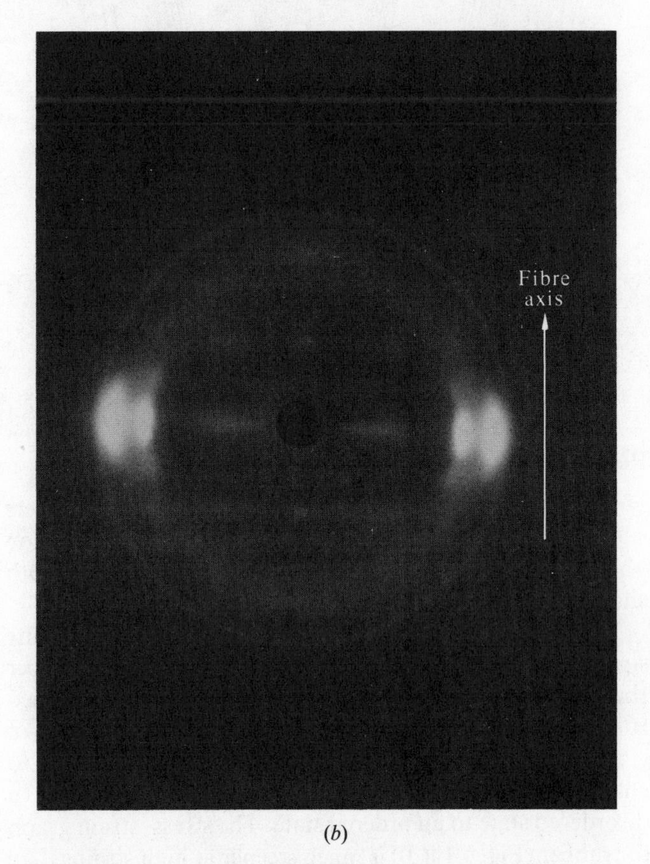

(b)

Fig. 9.23 Diffraction photographs. (a) Stretched rubber (Professor E. H. Andrews). (b) Nylon (Courtaulds Ltd, Synthetic Fibres Laboratory, Coventry).

as many as 100 or more of these units. An important part of the process of producing nylon is a stretching of the material to line up the molecular chains to give increased strength. This lining up of molecular chains is confirmed by the X-ray diffraction photograph.

Polythene, whose stress–strain graph is given in Fig. 9.14(c), is a long chain molecule consisting of the CH_2 unit repeated many times, of the order of a thousand times. Polythene in the unstressed condition is partially crystalline and it is thought that it has long chains folded in a regular manner (Fig. 9.25). This gives the polythene the useful property of strength and flexibility, hence its wide use for bags and wrapping material.

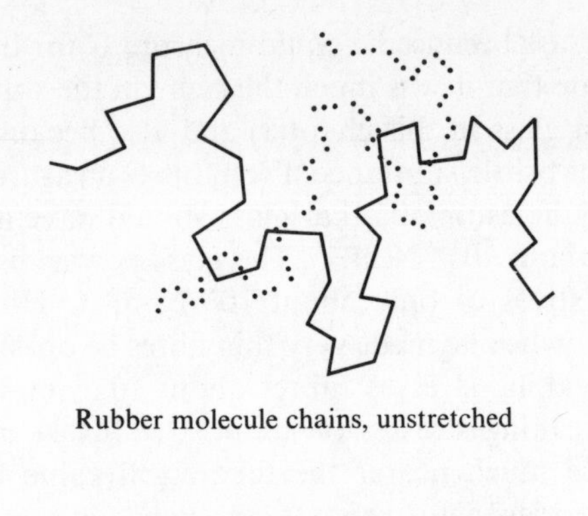

Fig. 9.24

The properties of polymers can be changed if cross links, extra bonds, are established between the different molecular chains. Thus natural rubber can be made much stiffer by the production of cross links; the rubber is said to be vulcanized. A very rigid material can be produced if sufficient cross links are established.

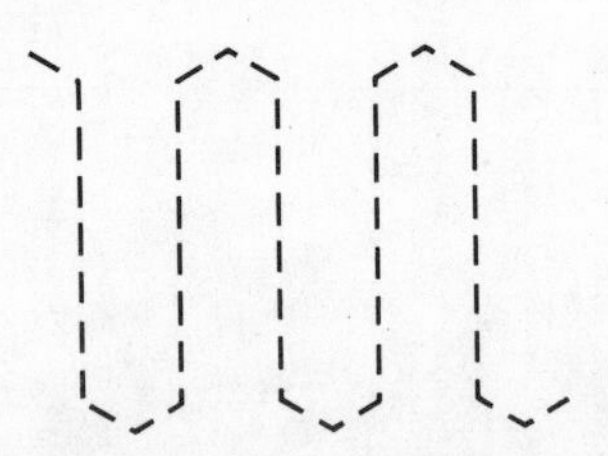

Fig. 9.25 Polythene molecular chain.

Human skin stretches with virtually little change in stress for strains up to about 20 per cent, beyond this much larger stresses are needed and beyond 40 per cent strain the skin requires very large stresses to produce higher strains. The material is in fact behaving in a similar way to those materials in which the initial stress changes produce straightening out of molecular chains and when straightened out very large stresses are needed to extend the molecular chains. Such stress–strain information is of particular use to surgeons doing 'reconstruction' work.

Glass

As the diffraction photograph for 'Pyrex' glass, Fig. 9.7(c), indicates there is little order in the arrangement of atoms within glass. Glass has a relatively high Young modulus and is a rather brittle substance. This would indicate rather strong bonds between atoms and a considerable amount of linkage between the atoms in the solid. Glasses are produced when very viscous liquids solidify. The simplest glass is formed from silicon dioxide and the viscous nature of the liquid glass near solidification would seem to indicate that strong bonds are being formed between the silicon dioxide

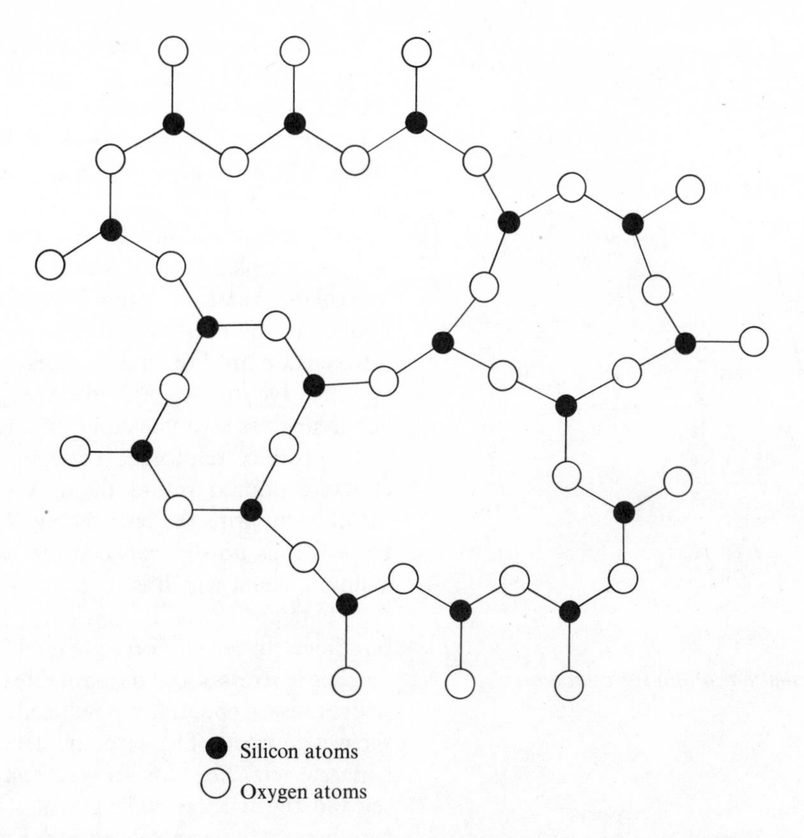

Fig. 9.26 Silicon dioxide network in glass. This is a two-dimensional drawing of what is really a three-dimensional network.

molecules even before the solid is formed. These bonds prevent the atoms in the liquid arranging themselves into the orderly crystal arrangement. Thus glass can be considered a network of linked silicon dioxide molecules (Fig. 9.26).

Wood

Wood is a natural polymer with long chains of cellulose molecules, $C_6H_{10}O_5$; perhaps three to four thousand such molecules in a chain. The long chains are generally parallel and thus X-ray diffraction photographs of wood show a crystalline structure. Because of the alignment of the molecules the properties of wood are very much dependent on the orientation of the wood when used. Thus for spruce, the tensile stress at which fracture occurs is $1{\cdot}0 \times 10^8$ N m^{-2} when the extension is along the grain direction and only $0{\cdot}03 \times 10^8$ N m^{-2} when it is across the grain.

Wood is a brittle substance and unlike cast iron, which is also brittle, it is weaker in compression than in tension. Along the grain the stress at which fracture occurs in compression is $0{\cdot}3 \times 10^8$ N m^{-2}, considerably less than the $1{\cdot}0 \times 10^8$ N m^{-2} value for tension.

Composites

Glass fibre laminated small boats are now quite common. Glass fibre laminates are just layers of a glass fibre woven cloth stuck together by a resin. The tensile stress at fracture of such a composite may be as high as or higher than that of mild steel and it may have a Young modulus about half that of mild steel.

	Young modulus	*Fracture stress*
Glass fibre laminate	2×10^{10} N m^{-2}	5×10^8 N m^{-2}

A higher Young modulus is necessary if large boats are to be built from laminated glass fibre—the material is not stiff enough. The lower the Young modulus of a material the more it will stretch or bend for a given force. The use of other fibres which have higher moduli is a possibility currently being explored. Carbon fibres can give much higher values for the Young modulus and the fracture stress.

	Young modulus	*Fracture stress*
Carbon fibre laminate	50×10^{10} N m^2	10×10^8 N m^{-2}

Wood is in fact a natural composite, consisting of cellulose fibres in a matrix of lignin. Because the fibres are laid all in the same direction the composite has stiffness and strength predominantly in the one direction. With fibre glass laminates this is avoided by running the fibres in different directions in the laminate. Plywood is the wood structure in which this is done. Plywood is made by gluing together thin sheets of wood with the grain crossed.

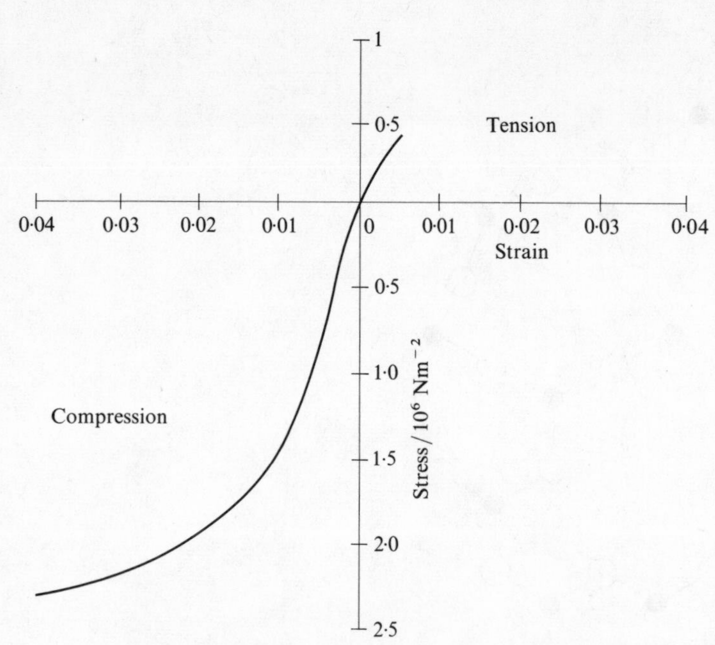

Fig. 9.27 Stress–strain graph for concrete.

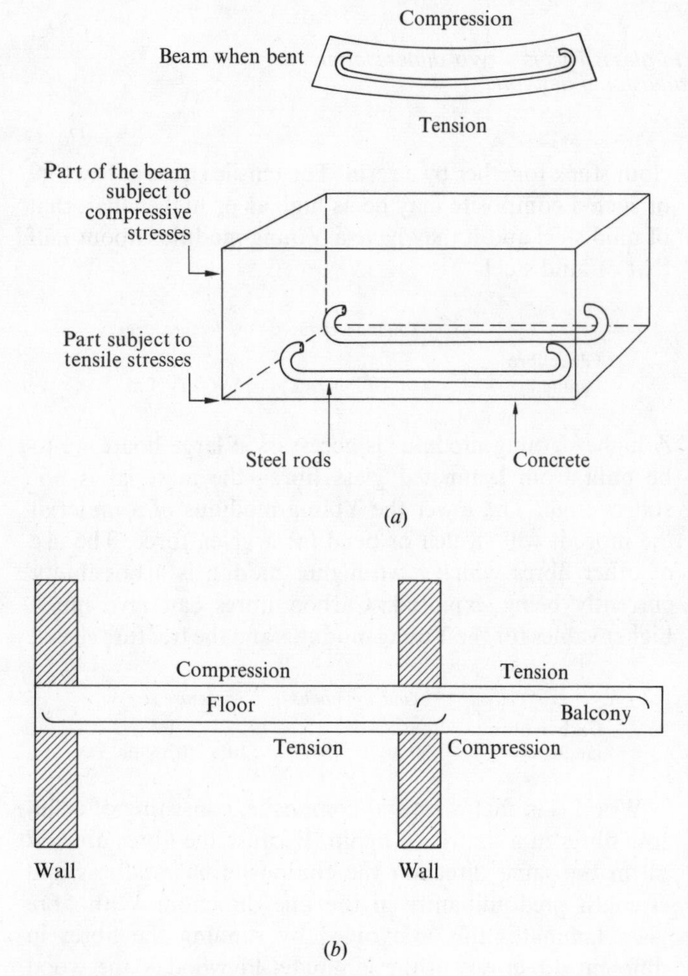

Fig. 9.28 (a) Reinforced concrete beam. (b) Reinforced concrete floor with cantilevered balcony.

Concrete has a very low tensile fracture stress, but a high compressive fracture stress (Fig. 9.27), and thus has to be used in situations where the only stresses the concrete will experience are compressive. Reinforcement of concrete by steel rods does enable the concrete composite to carry greater tensile loads than would otherwise be possible with just the concrete alone. If, however, the reinforcement bars are put in tension in the concrete they put the concrete into a permanent state of compression. When tensile stresses are applied to the pre-stressed concrete the concrete is not put into tension until the tensile stresses exceed the value of the compressive stresses built into the system. Such pre-stressed concrete gives a very useful building material.

To produce reinforced concrete steel bars have the wet concrete poured round them. When the concrete sets it shrinks and grips the bars firmly. When the beam is loaded there is thus no slip between the bars and the concrete. In a simple beam which is to be subject to bending loads the steel bars are placed in that part of the beam which would be subject to tensile forces (Fig. 9.28). The steel thus takes the tensile stresses and the concrete the compressive stresses.

Prestressed concrete is produced by running wires through the wet concrete. The wires are held under tension while the concrete sets. The tension is released after the concrete has set and the wire in endeavouring to revert to its original length puts the concrete in compression. Another way of achieving the same result is to place the wires in ducts in the concrete and apply the tension only when the concrete is set. In this last method the tension has to be maintained during the life of the beam and thus careful anchoring of the ends of the tensioned wire is necessary.

Bridges

The materials available to man for constructional purposes have changed over the years. The early materials were the naturally occurring ones, timber and stone. The properties of these materials dictated the way in which they could be used in structures. Timber is weaker in compression than in tension, stone is strong in compression and weak in tension.

The simplest form of bridge is a horizontal beam resting on columns, Fig. 9.29. The columns are in compression and the beam when loaded has the lower surface in tension and the upper surface in compression. Stone is thus a suitable material for the columns but if used for the horizontal beam has to be quite thick to keep the tensile stresses low. Because of the weight of stone beams they were only able to be used, even in considerable thicknesses, to bridge small gaps; the weight of the stone itself caused bending. Use of wood for the beam enabled the thicknesses to be reduced.

One way of overcoming the weakness of stone in tension was to build arches (Fig. 9.30). The action of the forces on the arch is to endeavour to straighten it out, i.e., shorten both the upper and lower surfaces of the arch, and this puts the arch material in compression. The supporting columns must, however, be able to withstand the sideways push of

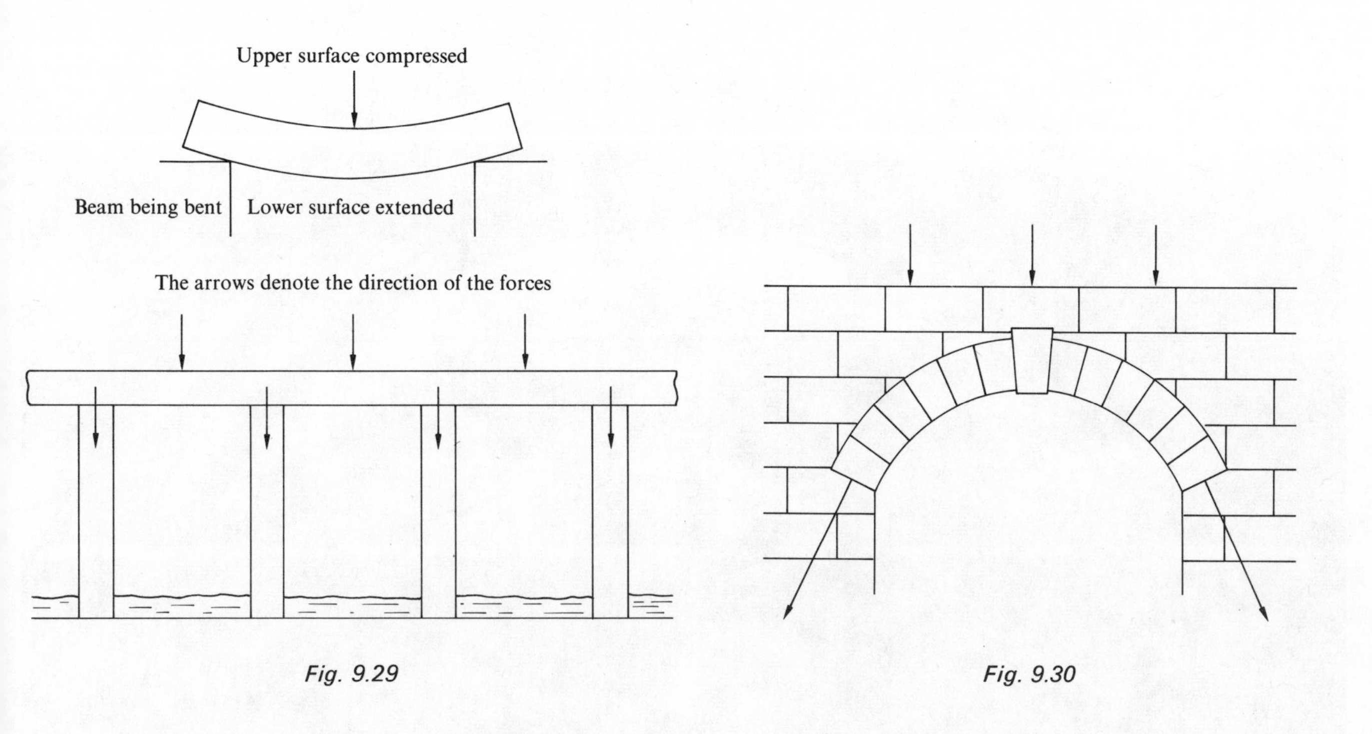

Fig. 9.29

Fig. 9.30

Fig. 9.31 Roman bridge at Rimini Italy, built by Augustus and Tiberius. (J. Allan Cash.)

Fig. 9.32 The World's first iron bridge 1779. At Ironbridge over the River Severn, built by Abraham Darby of cast iron. About 8 m wide and 100 m long (still standing). (J. Allan Cash.)

the arch, also the foundations must be secure enough to stop the base of a column being displaced. Figure 9.31 shows the Roman bridge at Rimini in Italy; the maximum span of an arch being about 9 metres.

The end of the eighteenth century saw the introduction into building of a new material—iron. Cast iron was used to build the first iron bridge in 1779 (Fig. 9.32). Cast iron is strong in compression and weak in tension (Fig. 9.17(c)), like stone. Thus the iron bridge followed much the same arch design as a stone bridge.

The beginning of the nineteenth century saw the introduction of suspension bridges (Fig. 9.33). A suspension bridge is the inverse of an arch—the cable being in tension. The cable supports the bridge deck. The cables have to be anchored on the land sides of the towers so that the forces on the cables do not pull the cable supporting towers out of the vertical. Figure 9.34 shows one of the early suspension bridges, built by Brunel over the Avon Gorge at Clifton, Bristol. The chains were made of wrought iron. Wrought iron differs from cast iron in being much purer iron; it is ductile and much stronger in tension than cast iron, having properties much closer to those of mild steel. Later suspension bridges used steel for the cables. A recent suspension bridge is that over the river Severn, near Bristol. This bridge (Fig. 9.35) was completed in 1966. The supporting cables consist of 440 parallel high-tensile steel wires about 5 mm

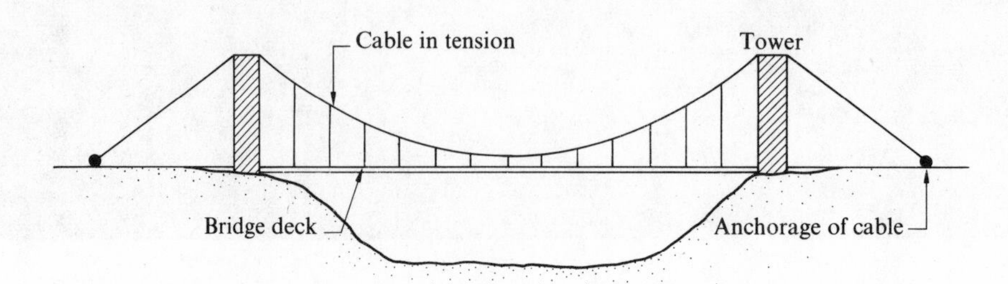

Fig. 9.33

Fig. 9.34 The Clifton Suspension Bridge. (J. Allan Cash.)

in diameter. The fracture stress for the cable material is about 14×10^8 N m^{-2}. It was only with the development of materials that were strong in tension that suspension bridges became feasible.

Wrought iron began to displace cast iron as a bridge-building material about the middle of the nineteenth century. The end of the nineteenth century saw mild steel replacing wrought iron. The strength of these new materials in tension led to the use of what is almost the old 'beam resting on columns' bridge. Instead of using a continuous beam, a beam is built up of separate members jointed together. Figure 9.36 shows such an arrangement. In the same way as the loaded beam, the top of this structure is in compression and the lower part in tension. The diagonal struts are hinged, some of them carrying tensile and some compressive stresses. Many different forms of such trusses were used in bridge building. Figure 9.37 shows a truss bridge built in 1864 by Eiffel, the same man who was responsible for the Eiffel tower in Paris (that is also based on the use of trusses).

Reinforced and prestressed concrete have become the materials for twentieth century bridge building. The Gladesville bridge in Sydney, Australia (Fig. 9.38) shows a modern prestressed concrete arch bridge, very similar in design to the old Roman bridges. Bridges of a completely different appearance are used to carry some modern high speed traffic roads up high out of the way of other local traffic. Figure 9.39 shows part of the flyover linking the M4 motorway to the centre of London. The roadway is made of a number of precast box sections linked together.

Fig. 9.35 The Severn Suspension Bridge. (J. Allan Cash.)

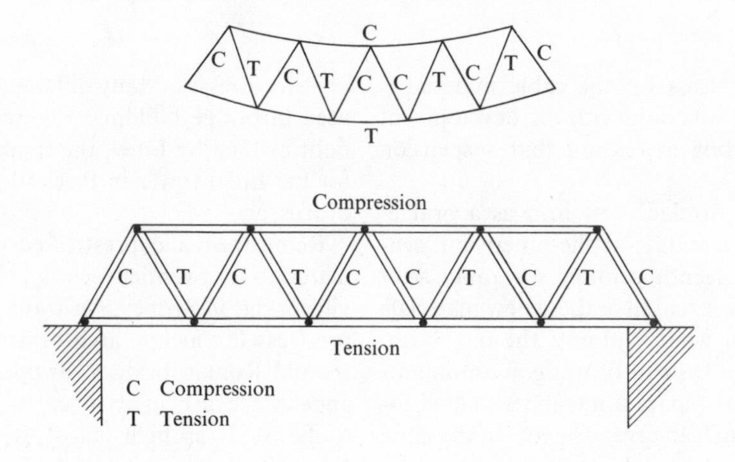

Fig. 9.36

Fig. 9.37 The Viaduct by Eiffel at Busseau sur Creuse. The distance between the piers is about 60 metres. (Prof. Arch. Italo Insolera.)

Fig. 9.38 The Gladesville Bridge, span about 330 metres. (Australian News and Information Bureau.)

Fig. 9.39 The Hammersmith Flyover, London. (J. Allan Cash.)

9.3 Liquids

X-ray diffraction photographs for liquids (Fig. 9.40) show some diffuse maxima indicative of some degree of order amongst the molecules in a liquid. Suppose instead of packing spheres together in a careful well-ordered array, as with solids, we toss the spheres into a container and let them take up their positions completely by chance; does this give sufficient order to conform with the X-ray diffraction evidence? In a hexagonal close-packed solid each sphere is in contact with twelve other spheres. In the random array each sphere is in contact with between four and eleven spheres, the number is not constant as is the case for the hexagonal packing. If we take a number of spheres and measure the different distances neighbouring spheres make from each of the spheres we can obtain a graph of the radial distribution of spheres round a sphere. Figure 9.41 shows results of such measurements made by Bernal and Scott.

Fig. 9.40 X-ray diffraction photograph for water. (Pilkington Brothers Ltd, Research and Development Laboratory.)

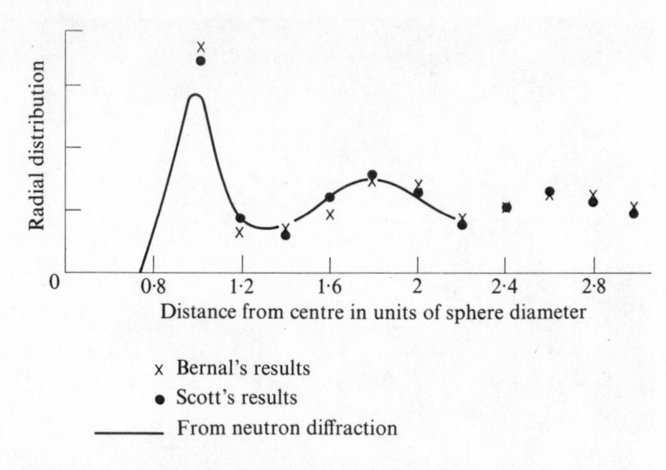

Fig. 9.41 Radial distribution for liquid argon. (D. Tabor (1969). Gases, liquids and solids, *Penguin.)*

The results are compared with the distribution forecast by the result of neutron diffraction; remarkably good agreement is obtained. The density of liquid argon is about 12 per cent less than that of solid argon; the random packing of the spheres gives a volume about 10 per cent greater than that obtained with close packing—again reasonable agreement.

In the solid we consider the atoms to remain in their positions, oscillating but not moving to new positions. In the liquid we must, however, consider our random arrangement to be just a snapshot of the arrangement at some instant, the molecules constantly shifting around but retaining the same radial distribution.

For a solid in a close-packed arrangement an atom has twelve near neighbours, if in the liquid form that same substance has only ten near neighbours to each atom then bonds must have been broken when the solid turned into a liquid. When the liquid turns to vapour the atoms lose these ten near neighbours. On this basis we would roughly expect that the energy needed to form unit mass of a liquid from the solid would be about 2/10 times that needed to convert the same mass from liquid to vapour, i.e., the ratio of the number of bonds broken. This is in fact the ratio of the molar latent heats of fusion compared to those of evaporation for some elements.

Element	*Molar latent heat of fusion* L_F/kJ mole^{-1}	*Molar latent heat of vaporization* L_V/kJ mole^{-1}	*Ratio* L_F/L_V
Argon	1·18	6·5	0·18
Neon	0·33	1·8	0·18
Krypton	1·64	9·0	0·18
Xenon	2·29	12·6	0·18
Chlorine	3·20	10·2	0·32
Bromine	5·29	15·0	0·35
Iodine	7·89	20·9	0·39
Sodium	2·60	89·0	0·029
Potassium	2·30	77·5	0·030
Rubidium	2·36	69·2	0·029

The ratio of the molar latent heat of fusion to that of vaporization for the inert gases would seem to be about the ratio 2/10, i.e., 0·2. The group of chlorine, bromine, and iodine would seem to fit a ratio of 3/9 while the alkali elements form another group with what would appear to be atoms in the liquid having almost the same number of neighbours as those in the solid.

In one mole we have 6×10^{23} atoms and thus if we take each atom in the liquid to have 10 near neighbours we have to break

$$\frac{10}{2} \times 6 \times 10^{23}$$

bonds to evaporate the liquid. The 2 occurs in the expression because each bond is between two atoms, and must be only counted once. Argon has a latent heat of vaporization of 6·5 kJ mole^{-1}; this must be equal to

$$\frac{10}{2} \times 6 \times 10^{23} \times \text{(energy to break one bond)}$$

Thus the energy necessary to break a bond in liquid argon is about $2{\cdot}2 \times 10^{-21}$ J, about 0·014 eV. (Most molecules need energies of this order to break their bonds.)

We can write this as

$$\text{Molar latent heat} = 5L\varepsilon$$

where L is Avogadro's number and ε the energy to break a bond.

Water

The density of water changes from 917 kg m^{-3} for the solid, ice, to 1000 kg m^{-3} for the liquid at 0°C. There is a 10 per cent increase in density in going from the solid to the liquid; this is not typical of most substances which decrease in density by about 10 per cent in going from the solid to liquid.

Snow flakes show a hexagonal symmetry, see Fig. 9.42(a). This is an indication of the hexagonal structure of ice, confirmed by X-ray diffraction. Figure 9.42(b) shows the atomic arrangement in an ice crystal. The structure is very open, far from being close packed. Every oxygen atom has four immediate neighbours; in a close-packed hexagonal arrangement each atom, or molecule, has twelve near neighbours. When ice melts the molecular separation increases

(*a*)

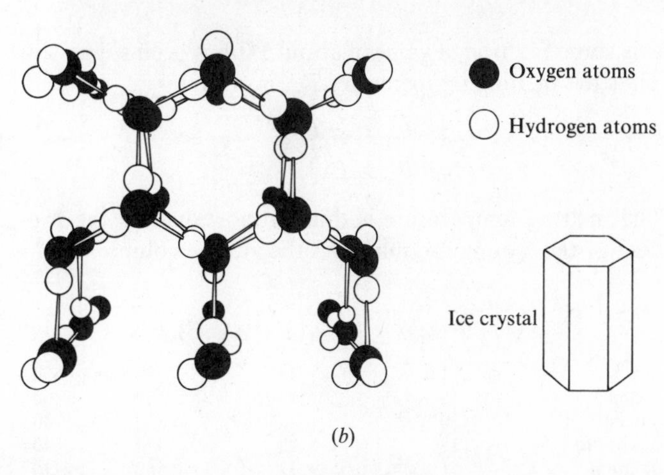

(*b*)

Fig. 9.42

but the packing becomes more dense; in the liquid a water molecule has an average of five near neighbours. The effect of this improvement in packing is an increase in density.

Melting

What happens when a solid melts? We imagine that in the solid the atoms are vibrating about their mean positions and that at the melting temperature the vibration becomes so great as to break the atoms free of the bonds holding them in their fixed positions. They are then free to wander within the confines of the liquid, bonds between atoms being constantly made and then broken.

If the oscillatory motion of an atom in a solid is simple harmonic, the energy of the atom will be

$$\text{Energy} = \tfrac{1}{2}kA^2$$

where k is the force constant of the bond and A the amplitude (see chapter 4). If the solid has the atoms in a simple cubic array we can write

$$\text{Young modulus } E = \frac{k}{x}$$

(see earlier this chapter). Thus, eliminating the force constant between the two equations gives

$$\text{Energy} = \tfrac{1}{2}EA^2x$$

x is the interatomic spacing.

We will assume that the temperature of the solid is directly proportional to the energy of the vibrating atom (see later this chapter and chapter 8).

$$\text{Energy} \propto T$$

Thus we have

$$T \propto \tfrac{1}{2}EA^2x$$

At melting we will assume that the amplitude A is some constant fraction of x, the interatomic spacing.

At melting

$$A = Cx$$

C is some fraction, a value of about 1/7 has been suggested. Thus the melting temperature, T_m, is given by

$$T_m \propto \tfrac{1}{2}EC^2x^3$$
$$T_m \propto Ex^3$$

The melting temperature is thus proportional to the product of the Young modulus and the atomic volume.

	$E/10^{10}$ N m^{-2}	$x^3/10^{-30}$ m^3	Ex^3/N m	T_m/K
Lead	1·6	43	69	600
Copper	13	13	169	1360
Nickel	20	16	320	1730
Platinum	17	22	374	2040
Iridium	51	13	660	2730
Tungsten	41	20	820	3650

Figure 9.43 shows the above data plotted on a graph. The result is approximate confirmation of the proportionality of the melting temperature to Ex^3. Considering the approximations made in the derivation this is quite good agreement (not all the solids have a cubic array).

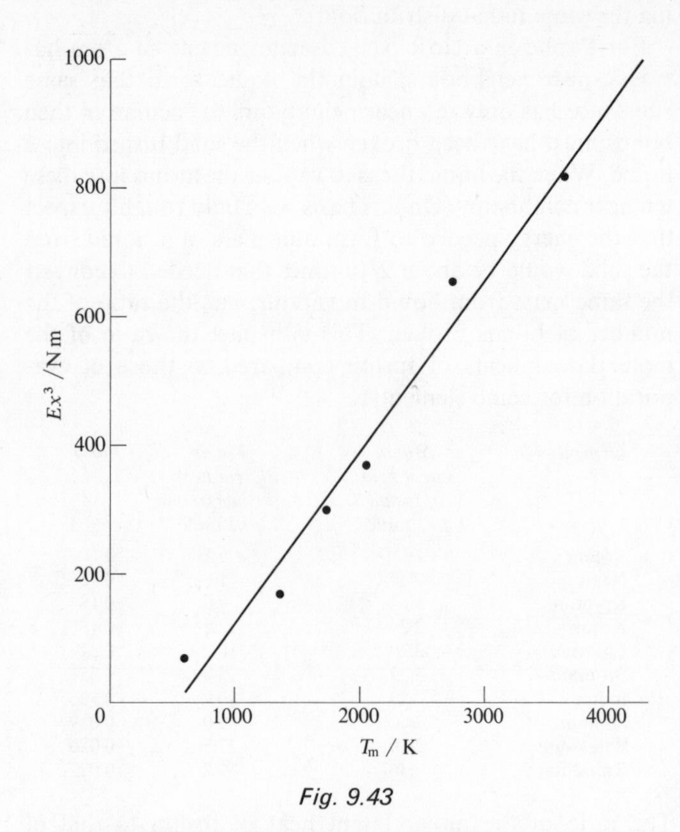

Fig. 9.43

The following table shows the molar latent heats of fusion of substances, at their melting points. A mole is used for the mass so that equal numbers of molecules are compared.

	Molar latent heat of fusion/kJ mole^{-1}
Lead	4·77
Copper	13·05
Nickel	17·61
Platinum	19·66
Iridium	26·36
Tungsten	35·23

Correlation would seem to occur between the molar latent heats of fusion and the melting points, a high melting point indicating a high value for the latent heat.

Surfaces

In a liquid we consider the molecules to be moving around, constantly interchanging positions. Why do those molecules moving near the edges of the liquid not all leave? Some leave (evaporation) but not generally all. The obvious answer would be to postulate a force of attraction between the molecules in the liquid. A drop of mercury on a clean glass surface, drops of water falling from a tap, all show a tendency to assume a spherical shape. A sphere has the minimum surface area—this would be consistent with an attractive force between molecules. The molecules in the surface are acted on by a force directed towards the molecules in the bulk of the liquid, the molecules in the bulk of the liquid are on the average uniformly surrounded by other molecules and thus experience no net force. Thus the surface assumes the minimum area. A force acting on a molecule in the surface would tend therefore to pull the molecule in to the bulk of the liquid—liquids, however, do not shrink to a vanishingly small point, there must therefore be another force which stops the molecules moving too close to one another. Liquids are not easily compressed, there are therefore strong repulsive forces resisting the movement of molecules closer together. Figure 9.44 shows how the force between two molecules varies with the separation; it fits these observations (see chapter 3 for similar curves).

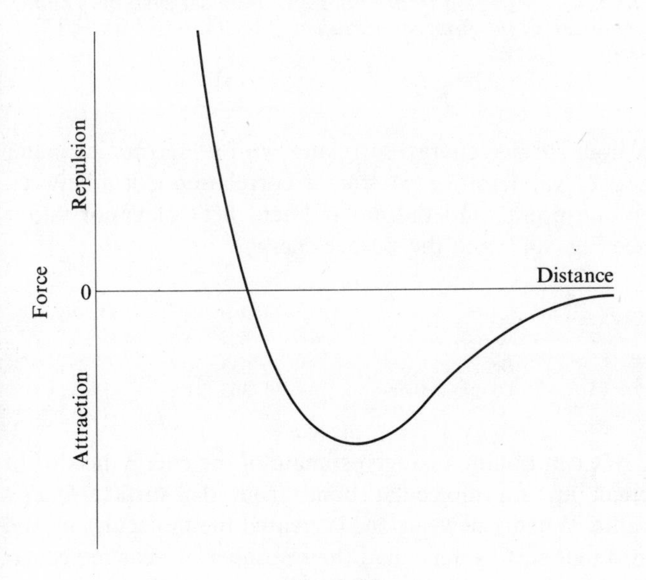

Fig. 9.44

Suppose we want to increase the surface of liquid—we must push more molecules into the surface against the attractive forces between molecules. Energy is necessary to do this. A simple measurement of this energy is possible with liquids, like soap solutions, which form thin films. Figure 9.45 shows a simple wire frame containing a soap film. To increase the area a force must be applied. The energy needed, with no change in temperature, to produce unit area of a surface is called the surface energy, G.

$$G = \frac{\text{energy to form a surface}}{\text{area of the surface formed}}$$

With the soap film the area produced is $2WL$, there are two surfaces to the film. The energy needed is FL. Thus

$$G = \frac{FL}{2WL} = \frac{F}{2W}$$

F is the force parallel to the film surface which is necessary to keep the film extended at the increased area. We can think of this force being balanced by an opposite and equal force, a tension, within the liquid and parallel to its surface.

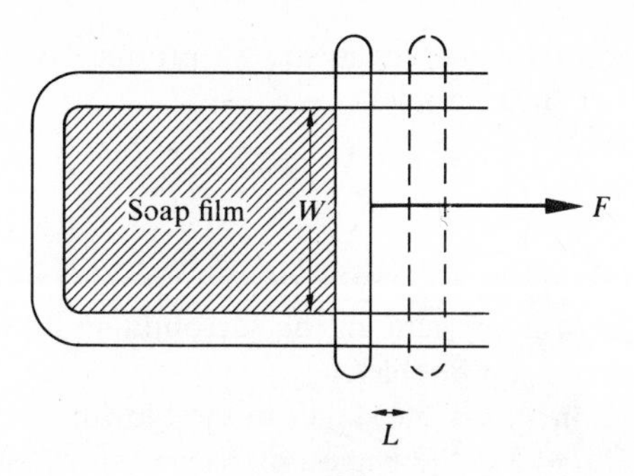

Fig. 9.45

The value of this force per unit length is called the surface tension, γ.

$$\gamma = \frac{F}{2W}$$

γ is thus equivalent to G. This is not an accurate derivation of the relationship. In order to keep the temperature constant during the area increase, some extra energy must be supplied. An accurate relationship is

$$G = \gamma - T\frac{\mathrm{d}\gamma}{\mathrm{d}T}$$

The last term is negative and so G is larger than γ. The term free surface energy is often used for that part of the energy given by γ and the two terms together give the total surface energy.

Blowing bubbles is a way of determining surface energies. When we blow a bubble under the surface of a liquid we are increasing the surface area between the liquid and air (plus the liquid's own vapour). Suppose the radius of the bubble increases by ΔR, the surface area will have increased from $4\pi R^2$ to $4\pi(R + \Delta R)^2$, the initial radius of the bubble being R.

Change in area $= 4\pi(R^2 + 2R\,\Delta R + \Delta R^2) - 4\pi R^2$. If we neglect ΔR^2,

$$\text{Change in area} = 8\pi R\,\Delta R$$

The energy needed to produce this increase in area will be

$$\text{Energy needed} = 8\pi RG\,\Delta R$$

This energy must be supplied by pressure inside the bubble. If the pressure necessary for the bubble to exist at this radius is p, then as pressure is force per unit area, the force will be

$$\text{Force} = 4\pi R^2 p$$

This force pushes out the surface of the bubble by a distance ΔR, therefore the energy supplied is

$$\text{Energy supplied} = 4\pi R^2 p\,\Delta R$$

Thus if we neglect temperature changes produced by the change in area

$$4\pi R^2 p\,\Delta R = 8\pi R\,\Delta RG$$

$$p = \frac{2G}{R}$$

If we blow soap bubbles in air we produce two surfaces and thus for such bubbles

$$p = \frac{4G}{R}$$

p is the amount by which the pressure inside the bubble is greater than the pressure in the surrounding liquid, or air in the case of soap bubbles.

Figure 9.46 shows the stages in the blowing of a bubble within a liquid. At the beginning the radius of the bubble is very large (a). Increasing the pressure decreases the radius of the surface (b). In (c) the bubble has become a hemisphere, its radius is a minimum at this point and thus the pressure a maximum. Any increase in pressure beyond this point causes the bubble to increase its radius; the pressure needed to maintain this larger radius is, however, less than the value for the hemispherical bubble. The condition is thus unstable and the bubble expands further, becomes more unstable, and either bursts or breaks away.

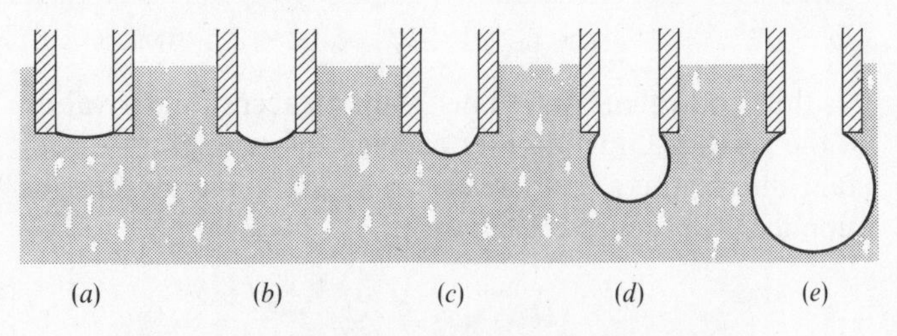

Fig. 9.46 Stages in the blowing of a bubble.

If we use a small tube through which to blow and produce bubbles we would find that much greater pressure was necessary to form the bubble. The maximum pressure is determined by the radius of the bubble and this in turn is determined by the diameter of the tube.

The following are some surface energies for a few liquids in contact with air, at 20°C.

	Free surface energy/J m^{-2} (*or surface tension*/N m^{-1})
Water	0·073
Benzene	0·029
Di-ethyl ether	0·017

The energy needed to produce a liquid–air surface is less for ether than benzene and less for benzene than water. The surface energy is a measure of intermolecular forces and thus at 20°C we can say that the forces between water molecules are stronger than those between benzene molecules and these in turn are stronger than those between ether molecules. At 20°C ether evaporates faster than benzene and benzene faster than water. The following are the values at 20°C for the molar latent heat, i.e., for equal numbers of molecules.

	Molar latent heat of vaporization at 20°C/kJ mole^{-1}
Water	44
Benzene	33
Di-ethyl ether	27

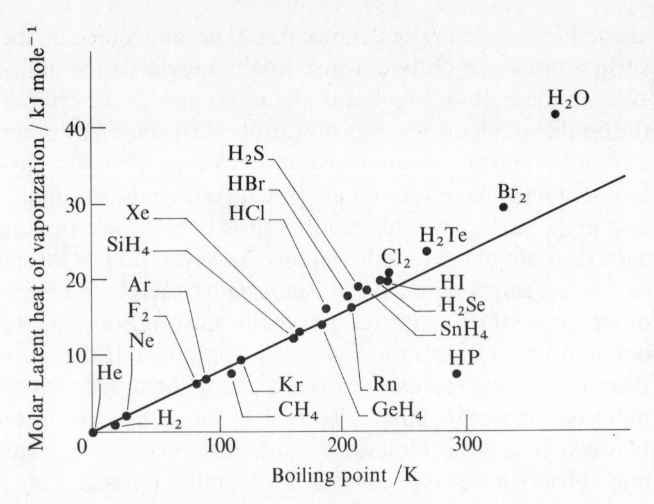

Fig. 9.47 Adapted from Fig. 8.25 Chemical systems *(1964), Chemical Bond Approach Project. McGraw-Hill Book Company.*

A high surface energy correlates with a high molar latent heat of vaporization. A similar correlation exists between boiling points and the molar latent heat of vaporization (see Fig. 9.47) and the surface energy.

	Boiling point/K
Water	373
Benzene	353
Di-ethyl ether	308

We can obtain a rough estimate of the energy needed to break an intermolecular bond from the surface energy value. When a new surface is created the molecules moved into that surface have had their number of near molecular neighbours reduced by a factor of two. Typically the number of near neighbours may be reduced from ten to five. If n is the number of molecules per unit area of surface then $5n$ bonds are broken for each two unit areas of surface produced. If ε is the energy needed to break one bond then $5n\varepsilon/2$ is the energy needed to produce unit area of surface. Thus

$$G = \frac{5n\varepsilon}{2}$$

If the effective radius of a molecule is r then

$$n = \frac{1}{\pi r^2}$$

Thus

$$G = \frac{5\varepsilon}{2\pi r^2}$$

Thus for benzene: $G = 0{\cdot}029$ J m^{-2}, $r = 2{\cdot}6 \times 10^{-10}$ m and we obtain for ε a value of $3{\cdot}1 \times 10^{-21}$ J or about 2×10^{-2} eV.

Earlier we obtained the equation

$$\text{Molar latent heat} = 5L\varepsilon$$

Thus we can write a relationship between the molar latent heat of vaporization and the surface energy, G

$$\text{Molar latent heat} = \frac{2LG}{n}$$

A high surface energy should thus correlate with a high molar latent heat—reasonably confirmed by experimental results.

The pressure difference across a curved liquid surface has many consequences. If a narrow bore tube is dipped into water, the water rises up the tube (Fig. 9.48). The water

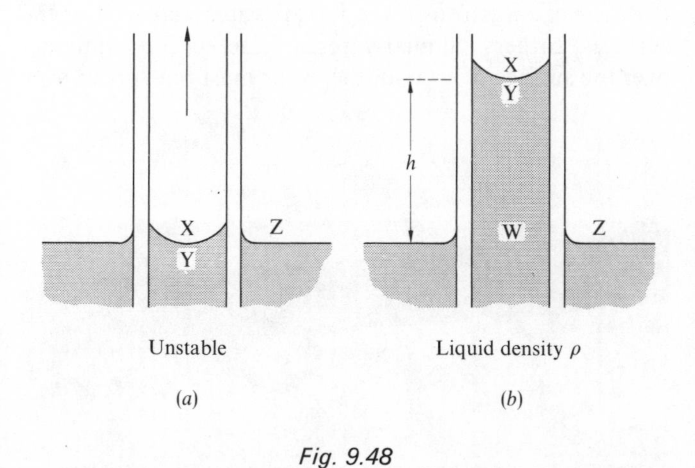

Fig. 9.48

surface inside the tube is almost a hemisphere, concave to the air. Thus the situation is like the bubble in the liquid—a curved surface means a pressure difference, the pressure being greater on the air side of this surface. The pressure at X is thus greater than that at Y by $2G/R$, where R is the radius of curvature of the liquid surface and the tube radius. But the pressure at X must be the same as that at Z, i.e., the atmospheric pressure. Therefore the pressure at Y must be below atmospheric pressure. Y cannot remain on the same level as Z, Fig. 9.48(a), because the pressure difference between Z and Y will force the liquid up the tube. The liquid will go up the tube until the pressure at W is the same as that at Z.

$$\text{Pressure at W due to the liquid column} = h\rho g$$

This must be equal to the pressure difference produced by the curvature of the liquid surface.

$$h\rho g = \frac{2G}{R}$$

Why does the surface of the liquid become curved in the first instance—if there was no curvature there would be no rise of liquid up the tube? For simplicity we will consider a drop of liquid resting on a horizontal surface (Fig. 9.49). If the drop spreads a little farther over the solid surface the area of liquid–solid interface has been increased. Suppose the increase in area to be ΔA. Solids have surface energies (see 'cracks and dislocations' in this chapter) and thus as we have reduced the area of exposed solid surface by ΔA we have decreased (released) the energy associated with the solid surface by $G_{SV}\,\Delta A$. The symbol G_{SV} is used for the surface energy of the solid when in contact with its own vapour. The spreading liquid has increased the solid–liquid interface area by ΔA and thus if we consider G_{SL} to be the energy needed to form unit area of such an interface, the energy increase needed is $G_{SL}\,\Delta A$. The spreading drop will also have increased the surface area of the liquid–vapour surface. The increase in area is $\Delta A \cos\theta$, where θ is the contact angle. Thus the energy increase needed is $G_{LV}\,\Delta A \cos\theta$. The net change in energy is thus

$$G_{SL}\,\Delta A + G_{LV}\,\Delta A \cos\theta - G_{SV}\,\Delta A$$

If spreading the liquid farther over the surface produces a net energy release then spreading will occur. The spreading will continue until no change in energy occurs, i.e., when

$$G_{SL} + G_{LV}\cos\theta = G_{SV}$$

or in the case of liquid rising up a tube the energy release equals the change in potential energy.

This equation can be derived by considering the surface tension forces and considering when they result in no net force (Fig. 9.50).

$$\gamma_{SL} + \gamma_{LV}\cos\theta = \gamma_{SV}$$

The small net resultant force in the vertical direction has

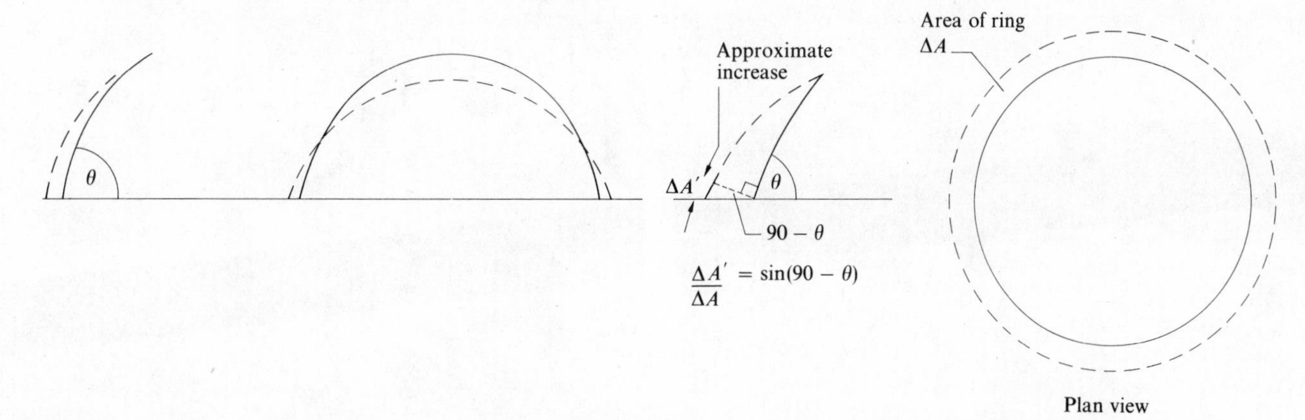

Fig. 9.49

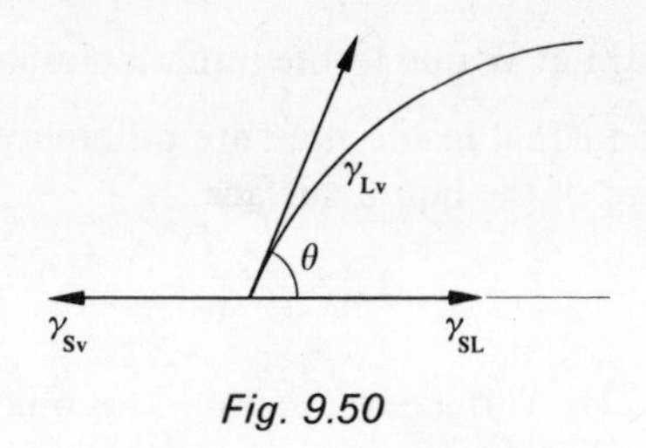

Fig. 9.50

been shown to produce some slight deformation of the solid surface.

When a narrow bore tube is dipped into water, the liquid creeps up the inside of the tube (and the outside) in accordance with the above argument. This in a narrow tube produces a sufficient curvature of the liquid surface to produce the pressure difference.

The equation for the spreading of liquids over solid surfaces has many implications. An adhesive if it is to be of any use must spread over the surfaces to be glued together. For spreading to occur we must have

$$G_{SV} > G_{SL} + G_{LV} \cos \theta$$

Solids with low surface energies are difficult to stick together. For example, 'Teflon', with a surface energy of about 0·018 J m^{-2} presents considerable difficulties. Most metals have surface energies of the order of 1 J m^{-2}, about 60 times larger than that of Teflon.

Water runs off a duck's back. Here we have a case where it would be disastrous if the liquid, water, spread over the surface, feathers. Similarly for a water beetle skimming over the surface of a pond, the water must not spread over

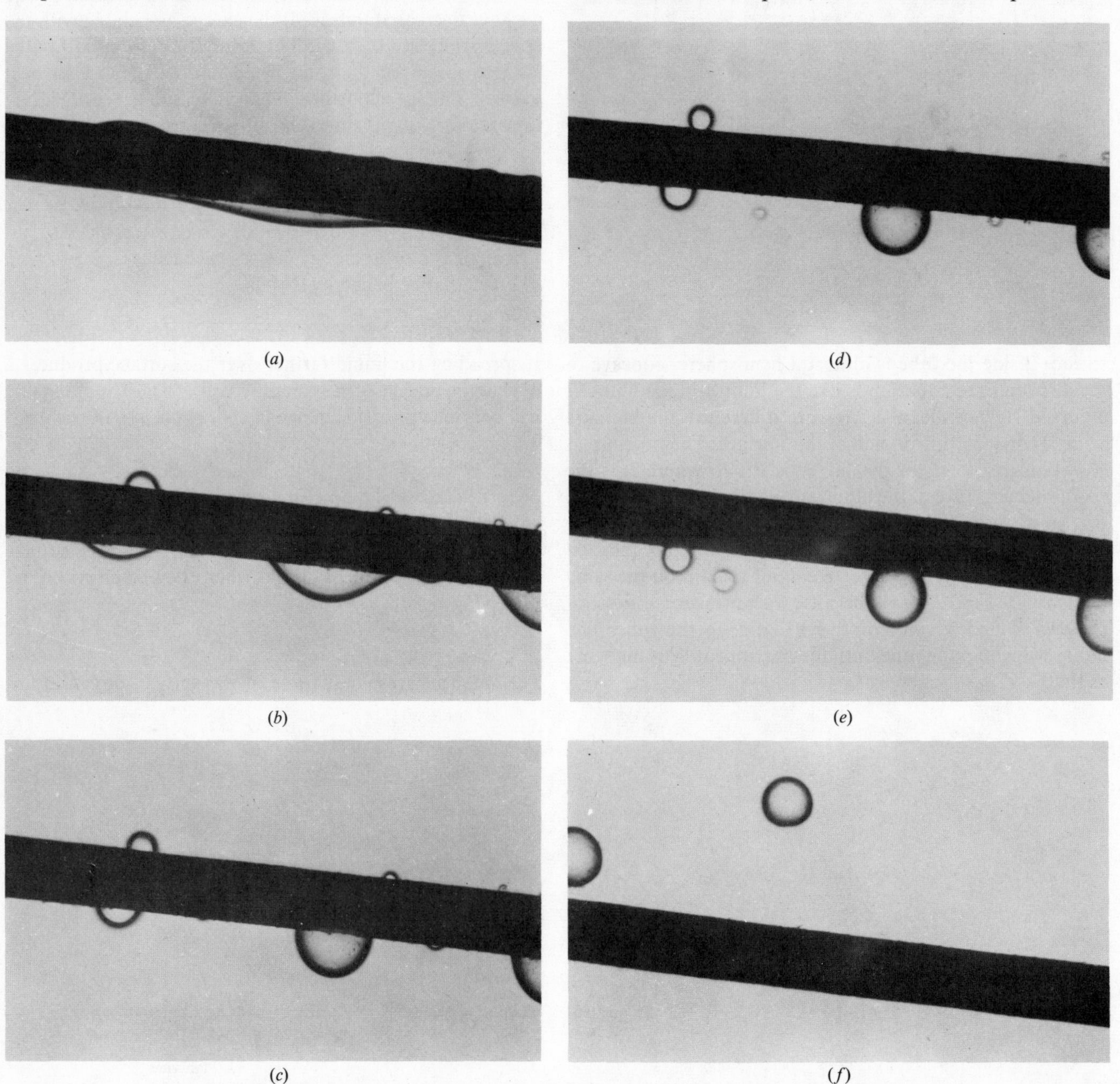

Fig. 9.51 Removal of stearic acid from a strand of hair. (Unilever Educational Publications.)

the surface of the beetle. To remove grease from clothes or dishes we add a detergent to the washing water. The effect of the detergent is to increase the angle of contact between the grease and the material (Fig. 9.51) so that it no longer spreads over it but rolls up into globules.

Viscosity

Knock over a cup of water and the water soon flows out of the overturned cup. Knock over a cup of treacle and quite an appreciable time elapses before the treacle all flows out. We say that treacle is more viscous than water. We can, however, regard the pouring as a sliding of layers of liquid molecules over one another—it seems easier to slide water molecules over one another than treacle molecules over one another.

Figure 9.52 shows the type of results obtained when a liquid flows near a wall. The layer of liquid at the surface of the wall does not move. The layers of liquid well away from the wall move with constant speed. A velocity gradient

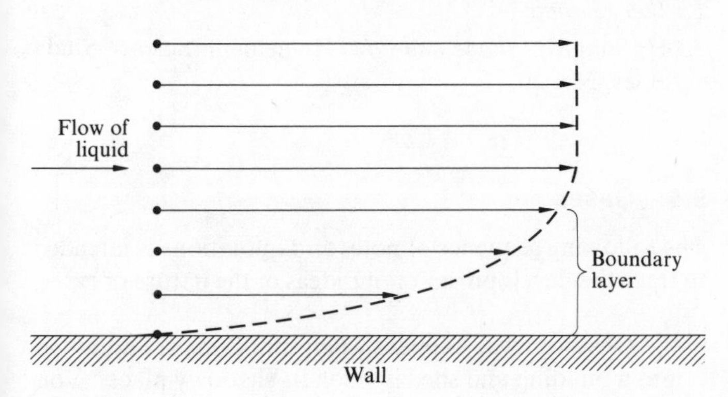

Fig. 9.52 Distances covered by particles of fluid in the same time.

is created by the presence of the wall; this gradient exists in what is called the boundary layer. The velocity gradient is almost constant within the boundary layer. The bigger the velocity gradient the bigger the drag experienced by the wall. The liquid flowing past the wall tries to pull the wall with it. The force necessary to keep the wall stationary depends on the size of the velocity gradient. The drag force also depends on the area of the wall in contact with the liquid. The drag will differ from liquid to liquid. Newton defined a coefficient of viscosity η in terms of the velocity gradient.

$$F = \eta A \frac{dv}{dx}$$

F is the drag force, A the surface area and dv/dx the velocity gradient. Fluids which obey this are known as Newtonian fluids. Many liquids encountered in daily life do not obey this relationship, the viscosity depending on the applied stress, e.g., a paint which can be brushed on easily but does not drip.

The following table gives values for the coefficient of viscosity for some liquids.

Liquid	*Coefficient of viscosity at* 20°C /10^{-4} N s m^{-2}
Water	10·0
Benzene	6·5
Di-ethyl ether	2·4

The results would seem to confirm the trend given by the surface energy, boiling points, and latent heat values of stronger intermolecular forces for water than benzene and benzene in turn having stronger forces than ether. Ether has a viscosity about four times smaller than that for water. An object moving through water thus needs about four times the force it would in ether to achieve the same velocity. Treacle has a large coefficient of viscosity and thus does not flow as easily as water or ether.

Liquid	*Formula*	*at* 35°C /10^{-4} N s m^{-2}
Pentane	$CH_3(CH_2)_3CH_3$	2·1
Hexane	$CH_3(CH_2)_4CH_3$	2·7
Heptane	$CH_3(CH_2)_3CH_3$	3·6
Octane	$CH_3(CH_2)_6CH_3$	4·6
Nonane	$CH_3(CH_2)_7CH_2$	5·9
Decane	$CH_3(CH_2)_2CH_3$	7·5

The above table gives the viscosities for a series of alcohols. Successive alcohols in each row have had one CH_2 group added. Increasing the length of molecular chain has increased the viscosity and so the longer length alcohols flow slower. Treacle has long molecules and so a high viscosity.

An air bubble rising through water, dust particles falling through air, objects falling through fluids, are all experiencing viscous drag forces. If there is a relative velocity between an object and a fluid then there will be viscous drag. In 1846 Sir George Stokes derived a relationship for the drag on an object in a fluid:

$$\text{Drag force} \propto (\text{velocity}) \times (\text{viscosity}) \times (\text{size})$$

In the case of a sphere this becomes

$$\text{Drag force} = 6\pi\eta r v$$

where r is the sphere radius, v the velocity of the sphere and η the coefficient of viscosity of the medium in which the sphere is moving. Figure 9.53 shows a distance–time graph for a steel ball bearing falling through glycerine.

After the initial acceleration the ball reaches a constant velocity. This is called the terminal velocity. An object falling freely in a vacuum accelerates, with a constant acceleration, over its entire path (see chapter 1). The force causing such a ball to accelerate is mg, where m is the mass and g the acceleration due to gravity. In a fluid another force will act on the object, viscous drag. As the viscous drag is proportional to the

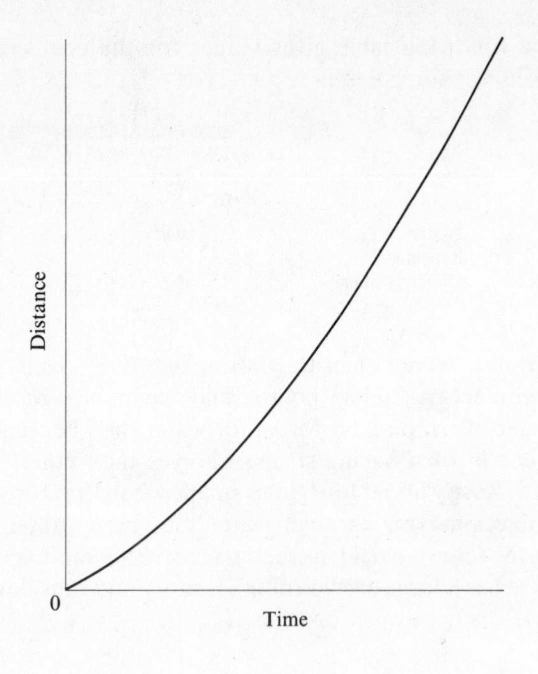

Fig. 9.53

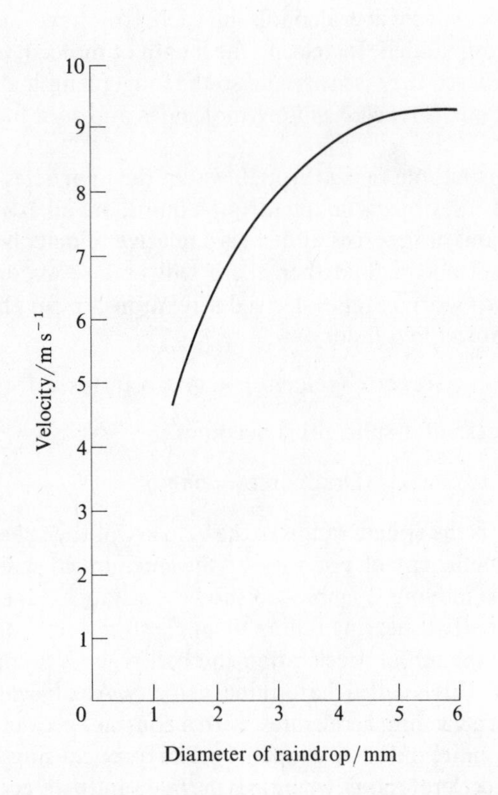

Fig. 9.54 Adapted from J. O. Laws. Trans. Amer. Geophys. Un., **22**, *719, 1941.*

velocity its value will increase as the object accelerates from rest. At some value the force on the object will balance and there will be no net force to cause any further acceleration. The object will then continue its motion with a constant velocity. For the ball bearing this will occur when

$$m'g = 6\pi\eta rv$$

m' is the mass of the sphere in the fluid. This will be less than its value in a vacuum due to the upthrust exerted on the sphere due to its immersion in a fluid.

$$m' = m - V\rho$$

where V is the volume of the object and ρ the density of the fluid.

Raindrops falling in air or air bubbles rising in liquid columns might be expected to obey Stokes' law. Because the raindrops and air bubbles are deformed by their motion through the fluid the law is only valid for very small diameter drops or bubbles, where the deformation is very small. Figure 9.54 shows some results for raindrops.

Further reading

A. H. Shapiro. *Shape and flow*, Heinemann Science Study Series, No. 20.

9.4 Gases

The following sequence of notes and quotations is intended to trace the development of our ideas of the nature of gases.

> '...Observe what happens when sunbeams are admitted into a building and shed light on its shadowy places. You will see a multitude of tiny particles mingling in a multitude of ways in the empty space within the light of the beam ... their dancing is an actual indication of underlying movements of matter that are hidden from our sight. There you will see many particles under the impact of invisible blows changing their course and driven back upon their tracks, this way and that, in all directions. You must understand that they all derive this restlessness from the atoms. It originates with the atoms, which move of themselves.'
>
> Lucretius, about 55 B.C. *On the nature of the universe*. Transl. R. Latham, Penguin, 1951.

Small dust particles buffeted around by the incessant bombardment of invisible atoms is the surprisingly modern view advanced by Lucretius. The dust particle motions so easily seen in air are, however, the results of convection currents. In 1827 Brown used the idea of atomic bombardment to explain the movements of small particles in liquids (see earlier in this chapter). The motion of small particles due to atomic bombardment is quite small, even in air.

The seventeenth century saw the beginnings of a theory of the nature of gases. The barometer was invented by Torricelli (about 1643), a student of Galileo, and the first 'vacuum' pump was invented by Otto von Guericke (about 1650). It was Boyle, in 1660, who first began to find laws governing gases. Boyle compressed a column of air and found how the volume of the air depended on the pressure. The apparatus he used was of the form shown in Fig. 9.55.

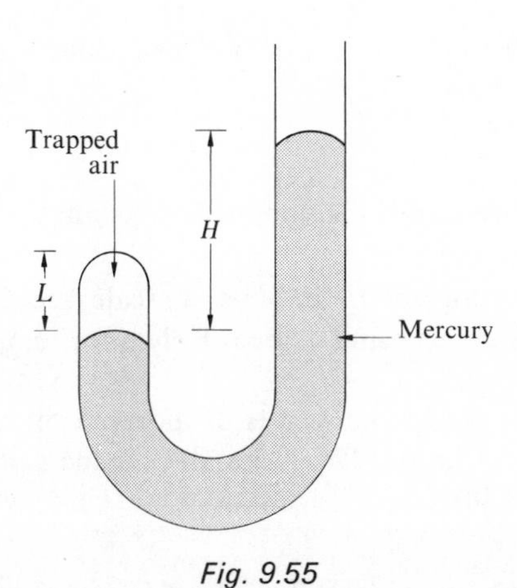

Fig. 9.55

When he added mercury to the J-tube the trapped air was compressed into a smaller volume. The pressure acting on the trapped air was, in terms of the length of a mercury column, H plus the atmospheric pressure. The volume of the trapped air was proportional to the length L. Boyle tabulated the data and left it to others to obtain the relationship. The relationship, which is known as Boyle's law, was that the volume of the air was inversely proportional to the pressure (provided the temperature remains constant).

$$V \propto \frac{1}{P}$$

or $PV =$ a constant.

The same result was arrived at by Mariotte in France, in 1679.

Boyle imagined the molecules in air to be—

> '... the air near the earth to be such a heap of little bodies, lying one upon another, as may be resembled to a fleece of wool. For this ... consists of many slender and flexible hairs; each of which may indeed, like a little spring, be easily bent or rolled up; but will also, like a spring, be still endeavouring to stretch itself out again....'
>
> Boyle (1660) *New experiments Physico-Mechanical, touching the spring of the air, and its effects; made, for the most part in a new pneumatical engine.*

The idea of air molecules being like small springs was criticized by many.

> '... [this] ... vast contraction and expansion seems unintelligible by feigning the particles of air to be springy or ramous, or rolled up like hoops or by any other means than a repulsive force....'
>
> Newton (1704) *Opticks.*

Newton's idea of air was that it was composed of particles between which existed repulsive forces. These repulsive forces kept the particles apart.

> 'If a fluid be composed of particles fleeing from each other, and the density be as the compression, the centrifugal forces [by centrifugal he means repulsive] of the particles will be inversely proportional to the distances of their centres. And, conversely, particles fleeing from each other, with forces that are inversely proportional to the distances of their centres, compose an elastic fluid, whose density is as the compression.'
>
> Newton (1687). *Philosophiae Naturalis Principia Mathematica.*

Newton argued that if the density of a fluid is directly proportional to the pressure then there will be a repulsive force between particles proportional to the reciprocal of the distance between the particle centres. As the density is inversely proportional to the volume of a fluid Newton was, by his repulsive forces, arriving at Boyle's law.

$$P \propto \rho$$

where ρ is the density.

$$\rho = \frac{m}{V}$$

and thus

$$P \propto \frac{1}{V}$$

D. Tabor in a book published in 1969 (*Gases, liquids and solids*, p. 43, Penguin) uses this idea of Newton to show that results exactly the same as those given by kinetic theory can be obtained.

1738 saw the beginnings of the kinetic theory of gases. Unfortunately it was too big a step for the scientists of the time and it was not until the end of the nineteenth century that such ideas were acceptable.

> '... Consider a cylindrical vessel ACDB [Fig. 9.56] set vertically, and a movable piston EF in it, on which is placed a weight P: let the cavity ECDF contain very minute corpuscles, which are driven hither and thither with a very rapid motion; so that these corpuscles, when they strike against the piston EF and sustain it by their repeated impacts, form an elastic fluid which will expand of itself if the weight P is removed or diminished, which will be condensed if the weight is increased.... Such therefore is the fluid which we shall substitute for air....'
>
> D. Bernoulli (1738) *Hydrodynamica.*

The end of the seventeenth century, 1699, had seen another law for gases to add to that of Boyle. This was the relationship between pressure and temperature for a constant volume gas, originating with the work of Amontons. In modern language, Amontons found that, at constant

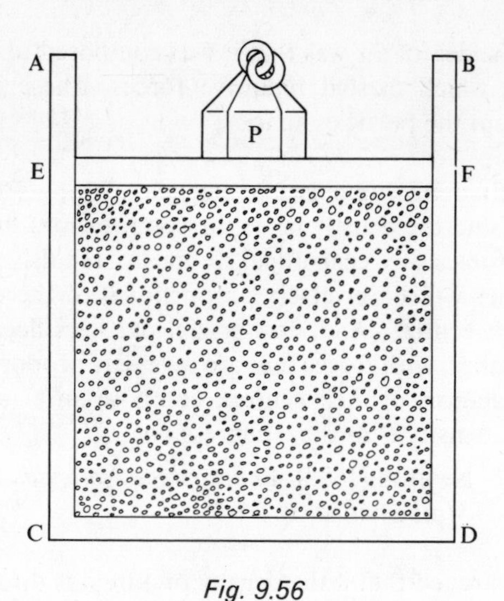

Fig. 9.56

volume, a change in temperature produced a change in pressure such that

$$\frac{\Delta P}{P} \propto \Delta\theta,$$

i.e., the fractional change in pressure was proportional to the change in temperature $\Delta\theta$. This can be written as

$$\frac{P_2 - P_1}{P_1} = \beta\,\Delta\theta$$

where β is a constant.

$$P_2 = P_1(1 + \beta\,\Delta\theta)$$

If temperature is measured from 0°C, then

$$P = P_0(1 + \beta\theta)$$

where P_0 is the pressure at 0°C. For the 'permanent' gases β has the value 1/273 deg C^{-1}. In 1703 Amontons realized that if, as he reduced the temperature below 0°C, the gases still followed his law then at a temperature of −273°C the pressure would become zero. This temperature became known as the Absolute zero. In fact the pressure law does not apply at very low temperatures; the gases liquefy.

About 1787 J. A. Charles discovered the relationship between the volume and temperature, at constant pressure. He did not publish his results and it was left to J. L. Gay-Lussac, who repeated the experiments, to express the law

> 'that all gases, speaking generally, expand to the same extent through equal ranges of heat; provided all are subject to the same conditions.'
>
> J. L. Gay-Lussac (1802).

This gives a similar relationship to that given for the variation of pressure with temperature.

$$V = V_0(1 + \beta\theta)$$

The constant β has the same value, 1/273 deg C^{-1}. Thus we might infer, if we assumed the law to hold at low temperatures, that the volume of a gas became zero at −273°C.

If the temperatures are reckoned from the absolute zero value, that is 0°C is given the value of 273 degrees on the absolute scale, the volume and pressure laws become

$$\frac{V}{T} = \text{a constant, if the pressure is constant,}$$

and

$$\frac{P}{T} = \text{a constant, if the volume is constant.}$$

T is the temperature on the absolute scale (see chapter 8 for the reasons for us using degrees Kelvin for temperature on this scale).

The table shows the results of more recent experiments with different gases. The gases all had the same constant volume of 1 litre.

Gas	*Pressure at* 0°C/ atm	*Pressure at* 100°C/ atm
Argon	1·0000	1·3675
Hydrogen	1·0000	1·3663
Helium	1·0000	1·3661
Oxygen	1·0000	1·3676
Air	1·0000	1·3675
Nitrogen	1·0000	1·3672

(Table 8-4. *Chemical Systems*, McGraw-Hill, 1964 Chemical bond approach project)

As can be seen, over the 100 degrees change in temperature the change in pressure is almost the same for the gases considered (only 'permanent' gases were used). The unit of pressure, the atmosphere, is the pressure given by a column of mercury 760 mm high. Extrapolating these results back to the zero pressure value gives virtually the same temperature point for all the gases, −273°C.

The three gas laws can be combined to give the equation

$$\frac{PV}{T} = \text{a constant}$$

That this equation is reasonably valid can be seen from the experimental results plotted in Fig. 9.57 for helium. A fixed mass of gas was used.

The pressure law and Charles' law concern pressure and volume changes with temperature. In the early work the temperatures were defined by means of mercury-in-glass thermometers and thus the laws were obtained by taking a temperature scale defined as 'equal changes in temperature produce equal changes in the length of a mercury column', i.e., a linear relationship between temperature and length of mercury column. Later it was decided to use the gas laws to define the temperature scale. On this basis the laws become definitions.

In 1811 Amedo Avogadro followed up the work of J. L. Gay-Lussac, published 1808, on the volumes of gases in chemical reactions with a deduction of what is often called Avogadro's law.

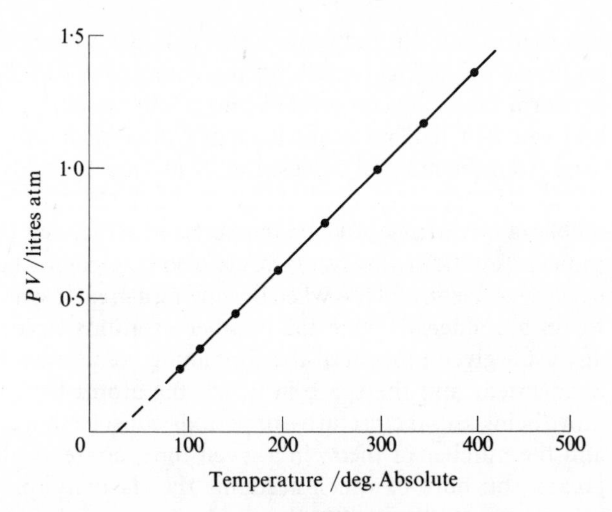

Fig. 9.57 Fig. 8-14 Chemical systems. *Chemical Bond Approach Project, 1964. McGraw-Hill Book Company.*

'M. Gay-Lussac has shown in an interesting memoir that gases always unite in a very simple proportion by volume, and that when the result of the union is a gas, its volume also is very simply related to those of its components. But the quantitative proportions of substances in compounds seem only to depend on the relative number of molecules which combine, and on the number of composite molecules which result. It must then be admitted that very simple relations also exist between the volumes of gaseous substances and the numbers of simple or compound molecules which form them. The first hypothesis to present itself in this connection, and apparently even the only admissible one, is the supposition that the number of integral molecules in gases is always the same for equal volumes, or always proportional to the volumes. . . .'

A. Avagadro (1811).
Alembic Club reprint No. 4, 1890.

The work of Gay-Lussac can be considered in terms of some simple examples: one volume of oxygen always needs two volumes of hydrogen in order that all the gases should be used to give water; one volume of hydrogen combines with one volume of chlorine to give two volumes of hydrogen chloride. Thus if we have one molecule of hydrogen combining with one molecule of chlorine we obtain two molecules of hydrogen chloride (Fig. 9.58). Thus equal volumes of gases contain the same number of molecules. It is assumed that all the volumes are at the same temperature and pressure. It was, however, not until Cannizzaro resurrected Avogadro's law in 1858 that it became generally accepted.

If equal volumes contain the same number of molecules then the mass of a molecule must be proportional to the density of the gas. The masses of the same volume of different gases will be directly proportional to the molecular masses of the gases. The following are some of the results quoted by Cannizzaro.

'*Name of substance*	*Weight of one volume, i.e., weight of the molecule referred to the weight of half a molecule of Hydrogen* = 1
Hydrogen	2
Oxygen	32
Chlorine	71
Bromine	160
Nitrogen	28
. . .'	

Cannizzaro (1858).

Thus the molecular mass of hydrogen is taken as 2, that of oxygen as 32, chlorine 71, etc. By comparing these results with those of compounds containing atoms of these substances Cannizzaro was able to arrive at the atomic masses, in terms of the hydrogen atom having a mass of 1. He obtained for oxygen 16, two atoms of oxygen forming the molecule; for chlorine 35·5, again two atoms needed to give the molecule; and so on. Atomic masses are now measured with respect to C^{12}, its atoms being considered to have mass 12.

The amount of a substance containing the molecular mass in grams is called one mole of that substance. Thus one mole of hydrogen has a mass of 2 g, one mole of oxygen has a mass of 32 g, one mole of chlorine has a mass of 71 g, etc. One mole of any substance contains the same number

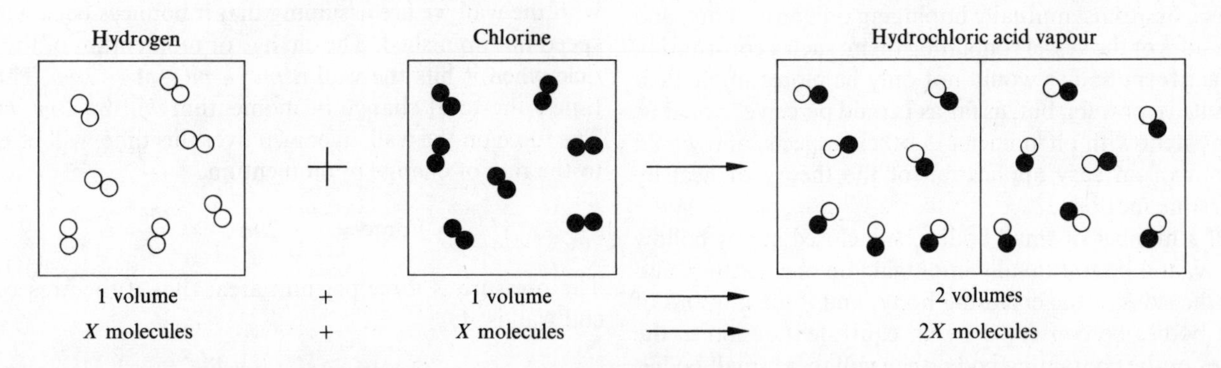

Fig. 9.58

of molecules. The number of molecules in one mole is $6{\cdot}02 \times 10^{23}$. This is known as Avogadro's constant or number. At standard temperature and pressure (called s.t.p.) one mole of a gas occupies 22·4 cubic decimetres. The standard temperature is 0°C, the standard pressure 1 atmosphere, i.e., the pressure given by a column of mercury 760 mm high, 10^5 N m^{-3}.

Many methods have been used to determine Avogadro's constant. Many radioactive elements decay by the emission of alpha particles. The rate at which the alpha particles are produced can be measured by perhaps a Geiger counter. Alpha particles when they pick up stray electrons become helium atoms, a gas. Thus if the helium gas is collected over a measured time and the rate of production of the alpha particles is measured, we can find the number of helium atoms in a volume of helium gas. For radium decaying we have 0·043 cm^3 of gas produced per year by 1 g of radium. The number of alpha particles produced per year by 1 g of radium is $11{\cdot}6 \times 10^{17}$. Thus we have for the number of atoms in a mole

$$\frac{11{\cdot}6 \times 10^{17} \times 22\,400}{0{\cdot}043} = 6{\cdot}04 \times 10^{23}$$

22 400 is the number of cm^3 in a mole.

As 1 mole of any gas occupies the same volume at the same temperature and pressure the constant in the combined gas equation is the same for 1 mole of any gas.

$$\frac{PV}{T} = R$$

Thus for one mole of a gas R has the value 8·31 J deg^{-1} mole^{-1}. If two moles of gas are used the value of the constant is $2R$, if three moles $3R$. This constant is independent of the gas used (we are, however, only applying the equation to the 'permanent' gases).

The idea that gas pressure could be explained in terms of the molecular bombardment of surfaces reappeared in the work of Herapath in 1821.

> 'Philosophers, since the time of Newton, have taught us that the elasticity of gases is owing to a mutual repulsion between their particles, by which they endeavour to fly from one another; ... if gases ... were made up of particles, or atoms, mutually impinging on one another, and the sides of the vessel containing them, such a constitution of aeriform bodies would not only be more simple than repulsive powers, but, as far as I could perceive, would be consistent with phenomena in other respects, and would admit of an easy application of the theory of heat by intestine motion....
>
> If a number of small bodies be inclosed in any hollow body, and be continually impinging on one another, and on the sides of the enclosing body; and if the motions of the bodies be conserved by an equivalent action in the sides of the containing body, then will these small bodies compose a medium, whose elastic force (pressure) will be like that of our air and other gaseous bodies; for if the bodies be exceedingly small, the medium might, like any aeriform body, be compressed into a very small space; and yet, if it had no other tendency than what would arise from the internal collision of its atoms, it would, if left to itself, extend to the occupation of a space of almost indefinite greatness. And its temperature remaining the same, its elasticity (pressure) would also be greater when occupying a less, and less when occupying a greater space; for in a condensed state the number of atoms striking against a given portion of the containing vessel must be augmented; and the space in which the atoms have to move being less, their returns, or periods, must be shorter; and the number of them, in a given time, consequently greater, on both of which accounts the elasticity (pressure) is greater the greater the condensation....'
>
> J. Herapath. *Annals of Philosophy* (1821), **2**, 343.

Suppose we have, as in Fig. 9.59, a single molecule moving backwards and forwards along the length of a box.

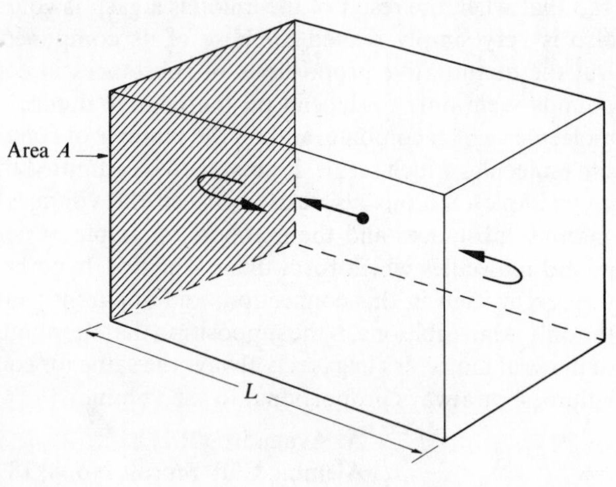

Fig. 9.59

When it hits the walls at the ends it bounces back. If the speed of the molecule is v and the length of the box L then the number of collisions it will make with one end wall in time t will be $vt/2L$, i.e., the total distance the molecule travels in time t divided by the distance between collisions with the particular end wall. Each time the molecule collides with the wall we are assuming that it bounces back with its speed undiminished. The change of momentum of the particle when it hits the wall is $mv - m(-v) = 2mv$. Thus in time t the total change of momentum will be $2mv.\ vt/2L$. The force on the wall, averaged over this time, will be equal to the rate of change of momentum.

$$\text{Force} = \frac{vt}{2L}.2mv.\frac{1}{t} = \frac{mv^2}{L}$$

The pressure is force per unit area, thus if the area of the end wall is A,

$$\text{Pressure P} = \frac{mv^2}{AL}$$

But AL is the volume v of the box. Thus

$$PV = mv^2$$

This was for just one molecule. Suppose we have n molecules in the box. Not all the molecules will hit the end face, some will be moving towards faces of the box which are perpendicular to the end face. The velocities of all the molecules can be resolved into three mutually perpendicular directions and thus we can consider that $n/3$ molecules will move towards the particular end face. Hence

$$PV = \tfrac{1}{3}nmv^2$$

If the molecular speed is constant then PV is a constant. Temperature would seem to be related to v^2. A possible connection would be to link temperature with the kinetic energy $\tfrac{1}{2}mv^2$ of a molecule. Herapath made an error and linked temperature with the momentum mv.

'Mr. Herapath unfortunately assumed heat or temperature to be represented by the simple ratio of the velocity instead of the square of the velocity—being in this apparently led astray by the definition of motion generally received—and thus was baffled in his attempts to reconcile his theory with observation. If we make this change in Mr. Herapath's definition of heat or temperature . . . we can without much difficulty deduce . . . the primary laws of elastic fluids. . . .'

J. W. Waterson (1858) *Phil. Mag.*

Waterson in 1846 derived the equation much in the same form as given above. However, when he submitted the paper to the Royal Society it received the comment 'The paper is nothing but nonsense'. It was not published until 1892 and then only to set the record straight.

In 1848 Joule used the ideas of Herapath to arrive at a value for the speed of gas molecules. The density ρ of a gas is equal to mn/V and so

$$P = \tfrac{1}{3}\rho v^2$$

The density of hydrogen at 0°C and 760 mm pressure is $0{\cdot}0899\ \text{kg m}^{-3}$.

760 mm pressure is $1{\cdot}013 \times 10^5\ \text{N m}^{-2}$; it is $0{\cdot}76 \times$ density of mercury in $\text{kg m}^{-3} \times 9{\cdot}8$ (i.e., $h\sigma g$). Thus

$$v^2 = \frac{3 \times 10^5 \times 1{\cdot}013}{0{\cdot}0899}$$

and

$$v = 1840\ \text{m s}^{-1}$$

Joule obtained a value of 6055 feet per second at 0°C, this is $2042\ \text{m s}^{-1}$.

R. Clausius in 1857 calculated the speed for hydrogen as $1844\ \text{m s}^{-1}$, for oxygen $461\ \text{m s}^{-1}$ and for nitrogen $492\ \text{m s}^{-1}$. For 1 mole of a gas at constant pressure we have P, V, and n constant and thus

$$mv^2 = \text{a constant}$$

where m is the mass of a molecule. Thus the product of the atomic mass and the square of the speed is a constant, hence the speed of molecules in a gas is inversely proportional to the square root of the atomic mass of the gas.

Clausius rederived the gas equation and in a later paper, 1858, calculated the mean length of the path of molecules between collisions. This distance is known as the mean free path. The following is a simple form of his derivation.

Two spheres collide when their centres are a distance d apart, both spheres having the same diameter d. We will consider all except one of the molecules in the gas to be at rest. This molecule moves with a velocity v. In a time t it will cover a distance vt (Fig. 9.60). It will collide with all the

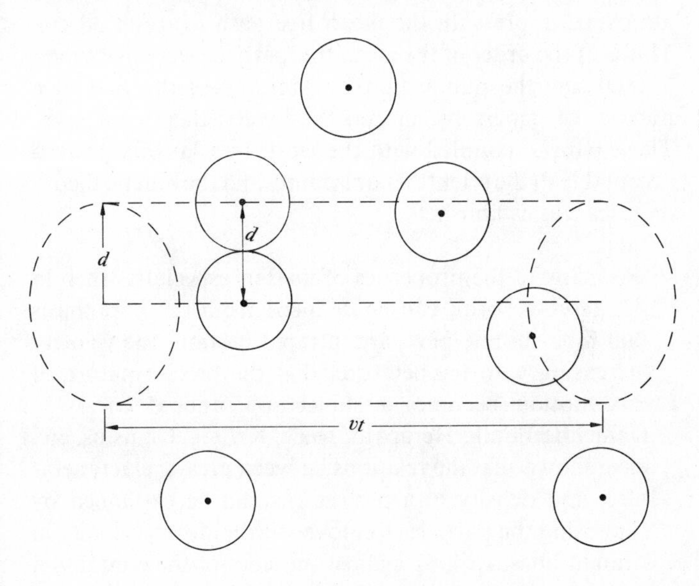

Fig. 9.60

other molecules whose centres come within a distance d of the path its centre follows. This is all the molecules whose centres come within a cylinder of volume $vt\pi d^2$. If there are n particles per unit volume then there will be $nvt\pi d^2$ molecules whose centres fall within the collision cylinder. Thus the mean distance between collisions is

$$\lambda = \frac{\text{distance travelled}}{\text{number of collisions}}$$

$$\lambda = \frac{vt}{nvt\pi d^2}$$

$$\lambda = \frac{1}{n\pi d^2}$$

If all the molecules are considered to be moving with the velocity v the equation becomes

$$\lambda = \frac{1}{n\pi d^2\sqrt{2}}$$

In air at normal atmospheric pressure and at 0°C one

mole occupies $22 \cdot 4 \times 10^{-3}$ m^3 and contains 6×10^{23} molecules. Thus

$$n = \frac{6 \times 10^{23}}{22 \cdot 4 \times 10^{-3}}$$

If we take d as 2×10^{-10} m, then we have a value for the mean free path of about 3×10^{-7} m. As the mean free path is inversely proportional to n and n is directly proportional to the pressure, we have

$$\lambda \propto \frac{1}{P}$$

thus at a pressure about one millionth that of normal atmospheric pressure the mean free path is about 30 cm. This is of the order of the mean free path in electronic tubes.

1860 saw the publication by Maxwell of the first of a number of papers by him on the kinetic theory of gases. These papers, coupled with the work by Clausius, proved acceptable to the scientific community and the kinetic theory of gases was established.

'So many of the properties of matter, especially when in the gaseous form, can be deduced from the hypothesis that their minute parts are in rapid motion, the velocity increasing with temperature, that the precise nature of this motion becomes a subject of rational curiosity. Daniel Bernoulli, Herapath, Joule, Krönig, Clausius, etc. have shown that the relations between pressure, temperature, and density in a perfect gas can be explained by supposing the particles to move with uniform velocity in straight lines, striking against the sides of the containing vessel and thus producing pressure. . . . I shall demonstrate the laws of motion of an indefinite number of small, hard, and perfectly elastic spheres acting on one another only during impact.

If the properties of such a system of bodies are found to correspond to those of gases, an important physical analogy will be established, which may lead to more accurate knowledge of the properties of matter. . . .

Instead of saying that the particles are hard, spherical, and elastic, we may if we please say that the particles are centres of force, of which the action is insensible except at a certain small distance, when it suddenly appears as a repulsive force of very great intensity. It is evident that either assumption will lead to the same results. . . .'

J. C. Maxwell (1860) *Phil. Mag.*, **19**, 19.

Maxwell derived an equation for the distribution of velocities amongst the molecules (he did not assume as our earlier derivation did that they all had the same velocity), and then proceeded to derive the gas equation. He applied his results to the process of diffusion, viscosity of gases, and heat conduction in gases. Maxwell in deriving the velocity distribution equation applied the laws of chance (probability), considering the chances of collisions occurring. Boltzmann followed this to introduce what is now known as statistical mechanics. (See chapter 8 for a discussion of chance and kinetic theory.)

The following extracts are from a paper by I. F. Zartman entitled 'A direct measurement of molecular velocities', published in *Physical Review*, **37**, 383, 1931.

'One of the fundamental assumptions of the kinetic theory of gases is the continual heat motion of the molecules in the gas. Even though there was no direct experimental verification of this assumption, the theoretical development progressed rapidly, resulting in the derivation of an expression for the distribution of molecular velocities by Maxwell in 1859 and, later by Boltzmann. This law has found many applications and until 1920 was only indirectly verified. . . .

Consider a gas issuing through a rectangular slit, g_1 [Fig. 9.61] in a side of the enclosure E which can be maintained at a constant temperature T. . . . With the aid of the slit g_2 a sharply defined rectangular beam is formed. Let a cylinder D, having a slit g_3 in its periphery

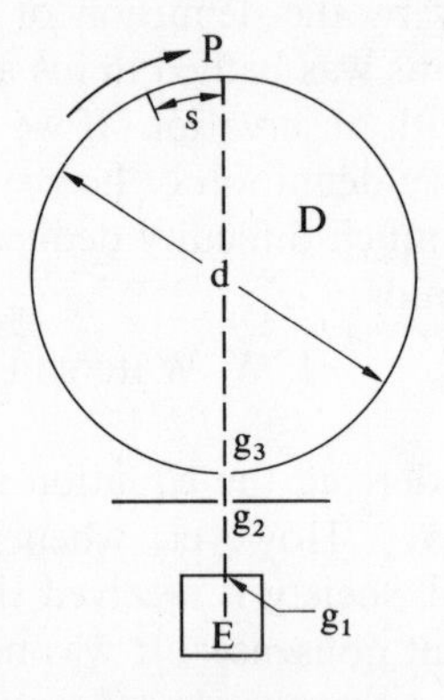

Fig. 9.61

and capable of being rotated, be placed in the path of this beam. If the cylinder is at rest, and slit g_3 is in the path of the molecular beam, a deposit forms on the rim at P diametrically opposite the slit. If the cylinder is rotated, the molecules entering g_3 require a finite time to traverse the diameter and consequently strike the rim at a point s to the left of P. The displacement s of a molecule moving with a speed c is given by the equation

$$s = \pi d^2 n/c = A/c \qquad (1)$$

where d is the diameter of the cylinder and n is the number of revolutions per second. The equation also shows that, by making d and n large, one may secure large displacements of the molecules, i.e., high resolving power.

The molecules in the beam possess not one definite velocity, but a distribution of velocities and consequently a "velocity spectrum" is formed on the rim of the cylinder. The density of this deposit should bear some relation to the velocity distribution in the enclosure E. According to the Maxwell-Boltzmann distribution law, the number of molecules dn out of a total number N in equilibrium within the enclosure possessing speeds between c and $c + dc$, is given by the expression

$$dn = \frac{4N}{\alpha^3 \pi^{1/2}} e^{-c^2/\alpha^2} c^2 dc$$

where α is the most probable speed of the molecules within the enclosure....

Apparatus

The substance whose velocity distribution is to be determined is vaporized in a steel crucible A [Fig. 9.62]. The vapor then passes through the channel in the neck to the slit S_1, there escaping as a molecular beam,... Directly above the neck of the crucible is a second slit S_2 used to define sharply the molecular beam. . . .

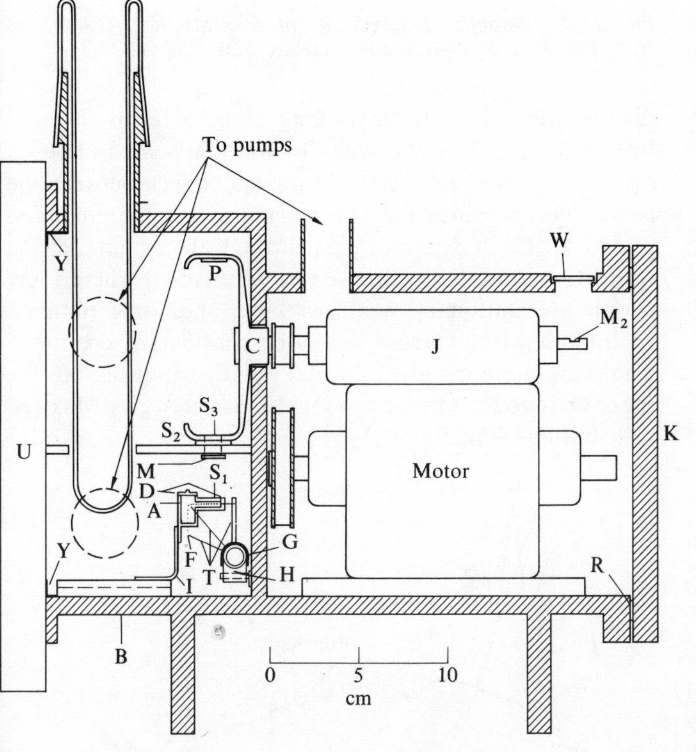

Fig. 9.62 Diagram of apparatus.

A cylinder C, 10 cm in diameter, machined out of a solid piece of machine steel, is placed above the slit system. This cylinder is designed to withstand the forces due to high speeds of rotation with the least possible deformation. A knife edge slit S_3, 0·6 mm wide, is cut into the rim of the cylinder. Directly opposite on the inner rim a device is fastened to hold a curved glass plate P. The cylinder is carefully balanced and mounted on the high speed spindle J of an internal grinder. Its speed is determined by the stroboscopic method.... The entire system is evacuated....

Experimental Procedure

...A bent glass plate is thoroughly cleaned and dried, then placed in the "initial depositing" apparatus and given a thin uniform coating of bismuth. The plate is fastened to the cylinder. The cylinder is adjusted with the aid of a traveling microscope until the slit S_3 is observed to be directly above the defining slit S_2. It is then fixed in this position. The system is evacuated, the crucible heated to about 800°C and the plate exposed to the molecular beam for about 20 seconds. This operation gives a visible deposit on the plate and forms the zero mark from which the displacements are measured. After admitting air to the system, the plate is removed and a microphotometer record obtained of the density of the initial deposit and location of the zero mark.

The plate is again fastened to the cylinder, the crucible filled with bismuth and the system evacuated. A slow increase in the temperature is necessary to avoid the violent expansion of gas bubbles in the molten metal which results in a "spitting" of the metal through the slit. When the temperature of the furnace reaches equilibrium and the motor speed is adjusted by varying the resistance in the line to the absence of beats, the shutter is opened and the beam allowed to pass to the plate. The maximum of the deposit first becomes visible in from three to six hours depending upon the temperature and speed of the run. The run is continued for periods ranging from eight to twenty-two hours. The plate is removed from the cylinder and again photometered. Finally, the two photometer curves are superimposed and the density due to the velocity distribution is measured.

Results

... In [Fig. 9.63] the solid lines represent the theoretical distributions for the conditions of operation given in Table I. The circles represent the experimental results obtained from Run 1, and the crosses those from Run 2. The experimental points are in close agreement (within the experimental error) with the theoretically derived curve except for a few points on the high velocity side.'

Because of the variation of speed amongst the molecules in a gas we need to find out what the mean pressure will be. The mean will depend on the mean value of mnv^2. As m and n are constant, we need the mean value of v^2. The mean value of this term is written as $\overline{c^2}$. The square root of this quantity is called the root mean square velocity.

$$\text{Root mean square velocity} = \sqrt{\overline{c^2}}$$

Incidentally, the average velocity will be zero as there are as many molecules going in one direction as there are in the opposite direction and both sets will have the same velocity distribution.

To summarize: by making some simple assumptions about the behaviour of gas molecules it is possible to derive a gas equation which will tally with the experimentally

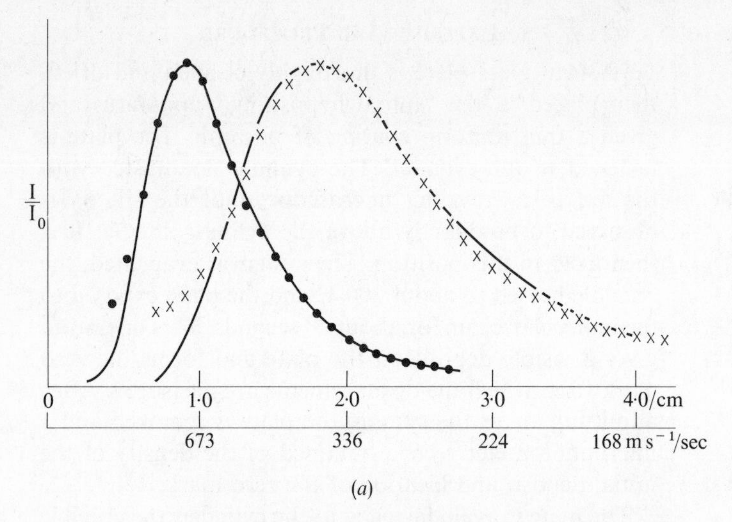

(a)

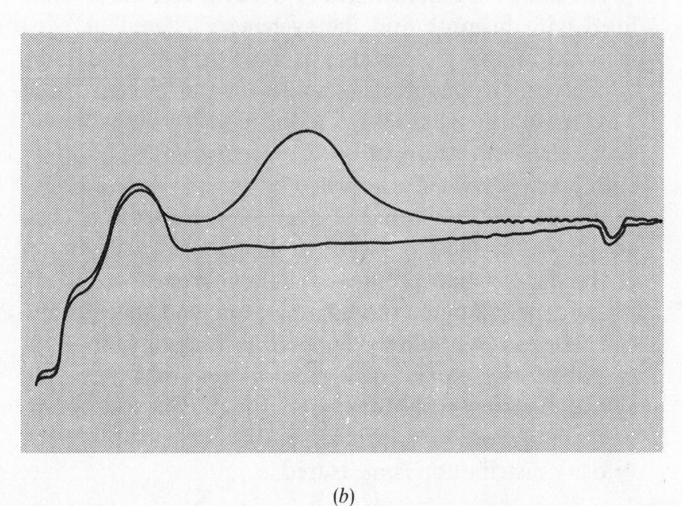

(b)

Fig. 9.63 (a) Theoretical and experimental intensity distributions assuming a vapour composition of 40 percent Bi and 60 percent Bi_2. T equals 851°C. Curve 1, n = 120·7 r.p.s., curve 2 n = 241·4 r.p.s. Bottom line gives the molecular velocity corresponding to several displacements at a cylinder speed of 241·4 r.p.s. (b) Contact print of photometer record of Run 1 n = 120·7 r.p.s. Abscissas twice actual deflections.

obtained PV/T equation. An important assumption is that temperature can be identified with the mean kinetic energy of the molecules (see chapter 8). The experimental equation is, however, only an approximation and thus we must regard our kinetic model of a gas as an approximation.

The behaviour of crowds

The following account is based on an article by L. F. Henderson 'The statistics of crowd fluids' in *Nature*, **229**, 381, 1971.

In a gas each molecule has a velocity and these velocities fall into a general pattern—the Maxwell distribution. Each person in a crowd will likewise have a velocity—does the same distribution occur? A gas is a crowd of molecules at a relatively low density—our crowd of people should therefore be also low density. Figure 9.64 shows the velocity

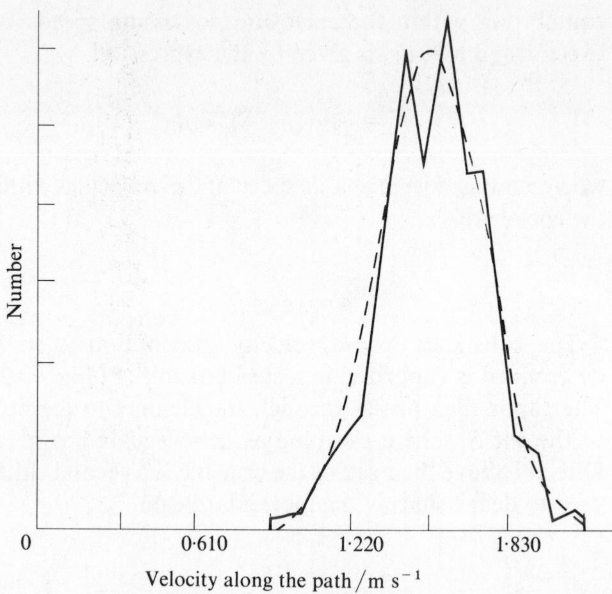

Fig. 9.64 Velocity distribution for 693 students. (Adapted from Fig. 1, L. F. Henderson. Nature, **229**, *382, 1971.)*

distribution for students walking along a footpath away from a library. The Maxwell distribution is superimposed on the graph and close correlation will be seen between the two curves, provided the Maxwell distribution curve is displaced so that it does not pass through the origin. All the students were walking. In the case of a crowd where some people are stationary, some walking, and some running, each mode of movement has to be considered separately. The standing mode gives no Maxwell distribution, but the other two modes when considered separately give Maxwell distributions (Fig. 9.65).

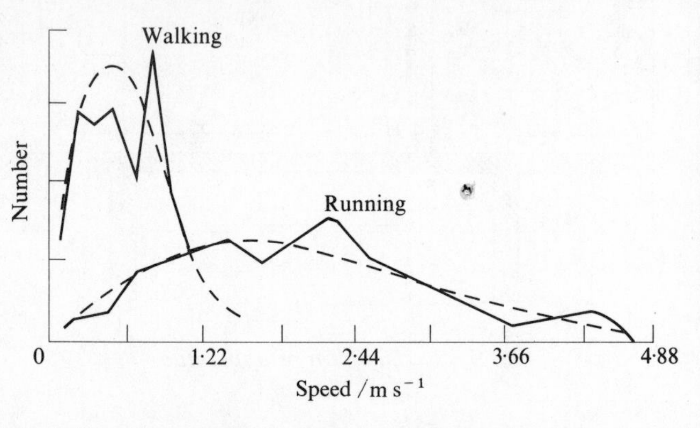

Fig. 9.65 Speed distribution for children in a playground, 377 walking and 217 running. (Adapted from Fig. 4. L. F. Henderson. Nature, **229**, *382, 1971.)*

Equipartition of energy

For one mole of gas we can write the gas equation as

$$PV = \tfrac{1}{3}mL\overline{c^2}$$

where L is Avogadro's number, and

$$PV = RT$$

Thus

$$m\overline{c^2} = \frac{3RT}{L}$$

and thus the mean kinetic energy of a molecule is

$$\text{Mean kinetic energy} = \tfrac{1}{2}m\overline{c^2} = \frac{3}{2}\frac{RT}{L}$$

R/L is a constant with the value $1{\cdot}38 \times 10^{-23}$ J K^{-1}. But this is the value of the Boltzmann constant k, defined in chapter 8. R, L, and k are not independent constants.

$$\text{Mean kinetic energy} = \tfrac{3}{2}\,kT$$

The mean kinetic energy of a molecule thus only depends on the temperature of the gas.

Suppose a gas was composed of two different mass molecules, m and m'. Because the gas was all at the same temperature we must have the mean kinetic energy the same for both sets of molecules. The energy is shared out so that the average kinetic energy of one set of molecules is equal to the average kinetic energy of the other set.

We can think of all the molecules in a gas as being divided into groups moving in three independent directions at right angles to each other, the x, y, and z directions. As there is no reason to believe that the molecules moving in any one direction have more average kinetic energy than those in another direction, the mean kinetic energy in the x direction will be equal to the mean kinetic energy in the y direction, will be equal to the mean kinetic energy in the z direction. The total mean kinetic energy is thus divided into three shares. Hence the mean kinetic energy in any one of these directions will be $\frac{1}{2}kT$. These directions are called degrees of freedom.

We can generalize this result to: the mean kinetic energy per degree of freedom is $\frac{1}{2}kT$. This is called the equipartition of energy.

Suppose we take a gas at absolute zero (!) and supply energy to it and so raise its temperature. The energy needed to raise its temperature to T will be the product of the mean kinetic energy and the number of molecules in the gas. If we consider one mole of gas, then

$$\text{Energy needed} = \tfrac{3}{2}kTL$$

where L is Avogadro's number. As $Lk = R$ we can write

$$\text{Energy needed} = \tfrac{3}{2}RT$$

But we also have a quantity called the specific heat capacity. This is the energy to raise unit mass of a substance by one degree. In the case of a gas the value of the specific heat capacity is determined by whether the gas is allowed to expand (we will look at this point shortly). Here we will assume that the gas is at constant volume and all the energy supplied goes into producing the temperature change. Hence

$$C_v = \frac{\text{energy}}{\text{mass} \times \text{temp. change}}$$

C_v is the specific heat capacity at constant volume. Thus for one mole

$$C_v T = \tfrac{3}{2}RT$$

$$C_v = \tfrac{3}{2}R$$

This is a prediction which can be compared with experimental results.

The theoretical value is about 12·5 J K^{-1} mole^{-1}. At -180°C helium and argon have specific heat capacities, at constant volume, of about 12·6 J K^{-1}. The agreement is very good. Argon and helium are monatomic gases, i.e., in the gaseous state they exist as single atoms. If we look at the results for diatomic gases the results are all close to 21 J K^{-1}, at room temperature.

Gas	C_v/J K^{-1} mole^{-1}
Hydrogen	20·2
Nitrogen	20·8
Oxygen	21·0
Carbon monoxide	20·8

This is, however, quite different from the 12·5 J K^{-1} mole^{-1} forecast. We have a clue to a possible answer in that all these gases are diatomic and all have virtually the same specific heat capacity. If we imagine diatomic molecules to be like two spheres joined together by a spring then when energy is supplied to such a molecule it could result in the molecule as a whole moving or it could cause the molecule to rotate (Fig. 9.66), or the atoms to vibrate. 21 is about 5/3 of 12·5 and we could obtain agreement between our theory and the experimental results if we had

$$C_v = \tfrac{5}{2}R$$

This would mean five degrees of freedom for the molecule, instead of the three we used for monatomic gases. Suppose we allow diatomic gases to have the degrees of freedom represented by (a), (b), (c), (d), and (e) in Fig. 9.66 for room

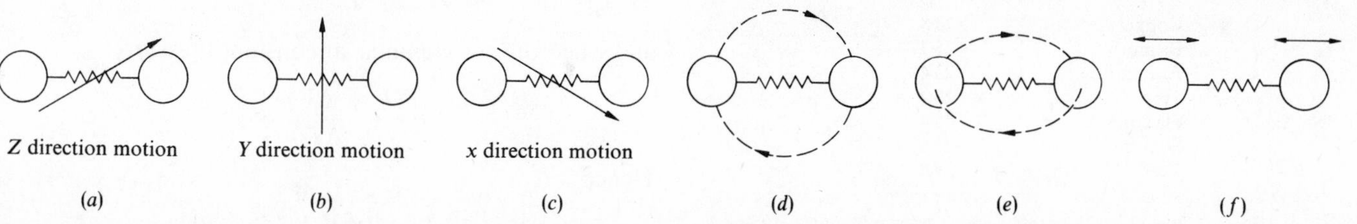

Fig. 9.66 (a), (b), (c) Translational degrees of freedom. (d), (e) Rotational degrees of freedom. (f) Vibrational degree of freedom.

temperature conditions. At very low temperatures hydrogen has a specific heat capacity of about 12 J K^{-1} $mole^{-1}$; it would appear that rotation of the molecule does not start to occur until the energy supplied has reached a certain value (in the case of hydrogen this is when the temperature is about 100 K). Figure 9.67 shows how the specific heat capacities of hydrogen and chlorine change with temperature. Above about 1000 K the specific heat capacity begins

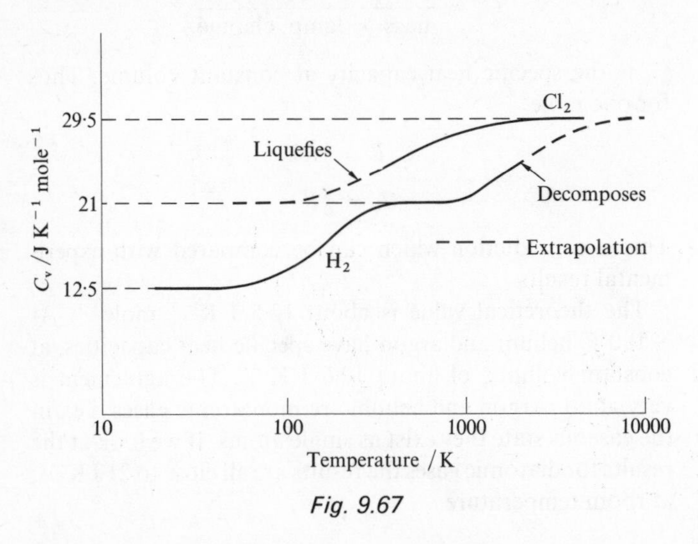

Fig. 9.67

to rise above the five degrees of freedom value. We can consider that the molecule is at higher temperatures beginning to vibrate and vibrational degrees of freedom must be considered. Chlorine at room temperature has a specific heat capacity of about 25 J K^{-1} $mole^{-1}$ and would appear to have translational, rotational, and some vibrational degrees of freedom. At about 2000 K the specific heat capacity value for chlorine indicates seven degrees of freedom; three translational, two rotational, and two vibrational. At high temperatures we thus have the atoms vibrating backwards and forwards along their axes. Such a mode of oscillation is considered to have two degrees of freedom, i.e., need two shares of energy, because vibration has both potential and kinetic energies.

We can try to apply these ideas to solids. The atoms in a solid can only vibrate and if we consider there to be three different modes of vibration (along the x axis, the y axis, and the z axis) there will be six degrees of freedom and thus the specific heat capacity will be $3R$ or 25 J K^{-1} $mole^{-1}$.

Solid	*Specific heat capacity, at constant volume,* /J K^{-1} $mole^{-1}$
Copper	23·8
Bismuth	25·3
Aluminium	23·4
Tin	25·4
Platinum	25·4
Sodium	25·6
Lead	24·8

For many solids the predicted result appears to be very close to the experimental result. That solids should have this value of specific heat capacity was arrived at empirically by Dulong and Petit in 1819 and is known as the Dulong and Petit law. The law, however, only holds above a certain temperature. Figure 9.68 shows how the specific heat capacity of lead varies with temperature. For lead the law holds above 200 K.

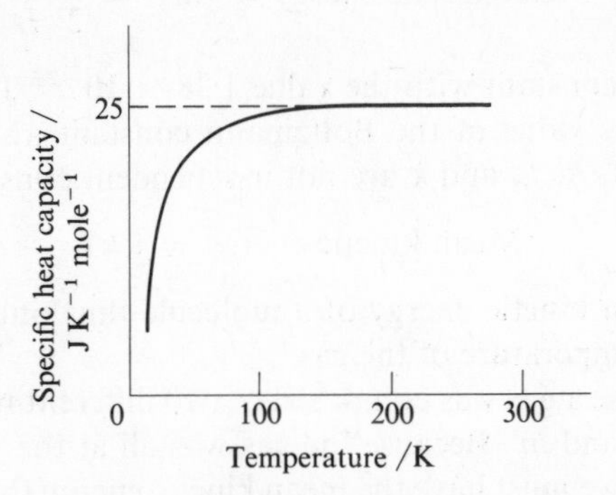

Fig. 9.68 Specific heat capacity of lead.

As has been mentioned earlier the specific heat capacity of a gas depends on whether the gas is allowed to expand. The specific heat capacity at constant volume is the one we have used. We can relate this to the specific heat capacity at constant pressure (C_p). At constant volume all the energy supplied to the gas is used to produce higher molecular velocities and so higher temperatures. At constant pressure the gas expands and so some of the energy supplied to the gas is used to produce this expansion. This energy is equal to $P\,\Delta V$, where ΔV is the change in volume produced when the pressure is kept constant at P (see chapter 3). We write the energy supplied as equal to

$$C_p\,\Delta T$$

where ΔT is the temperature rise produced. We are considering one mole of gas. We can produce the same temperature rise by supplying energy to the gas at constant volume. This energy will be equal to $C_v\,\Delta T$. This energy will be less than that supplied at constant pressure, because of the expansion by $P\,\Delta V$. Hence

$$C_v\,\Delta T + P\,\Delta V = C_p\,\Delta T$$

$$C_p - C_v = P\frac{\Delta V}{\Delta T}$$

But we already have a relationship between P, V, and T.

$$\frac{PV}{T} = R$$

and when the gas expands, at constant pressure,

$$\frac{P(V + \Delta V)}{T + \Delta T} = R$$

Hence

$$\frac{P\,\Delta V}{\Delta T} = R$$

and thus

$$C_p - C_v = R$$

Thus when for a monatomic gas we have $C_v = 3/2R$, we have for the constant-pressure condition $C_p = 5/2R$.

By making enough assumptions about possible degrees of freedom it has been possible to account for the specific heats of many substances. A difficulty, however, lies in explaining why the specific heat capacities vary with temperature. Polyatomic gases, such as H_2O and CO_2, have specific heat capacities which are even more difficult to explain. The solution of all these problems lies with quantum theory—see chapter 8 for a discussion of a solid.

Evaporation

Evaporation is the escape of molecules from a liquid. If no energy is supplied to the liquid the evaporation causes a drop in temperature of the liquid. The mean energy of the molecules in the liquid must have fallen. This would suggest that the molecules escaping are the high energy ones.

We can estimate the energy a molecule must have to escape from a liquid. If the latent heat is measured for one mole of liquid, then as there are 6×10^{23} molecules in a mole the energy needed for one molecule to escape is

$$\frac{\text{Molar latent heat}}{6 \times 10^{23}} = \frac{\Delta H_{vap}}{6 \times 10^{23}}$$

At 20°C the molar latent heat of vaporization for water is 34 000 J mole^{-1}. Hence the escape energy for a molecule is $5{\cdot}7 \times 10^{-20}$ J.

When a vapour condenses energy is released. When in a cloud water droplets grow due to condensation there is a rise in temperature of the surroundings.

With a liquid equilibrium conditions will occur when the rate of evaporation equals the rate of condensation, i.e., the rate at which molecules leave the surface is equal to the rate at which the molecules enter the surface. At this condition the pressure due to the molecules of the liquid in the vapour state is called the saturation vapour pressure. A simple method that has been used to measure this vapour pressure is to inject the liquid into the vacuum space above the mercury in a simple barometer (Fig. 9.69). The evaporating liquid gives a vapour which produces a pressure which depresses the mercury level. At saturation no further drop in mercury level occurs. The total drop in mercury level is the saturation vapour pressure.

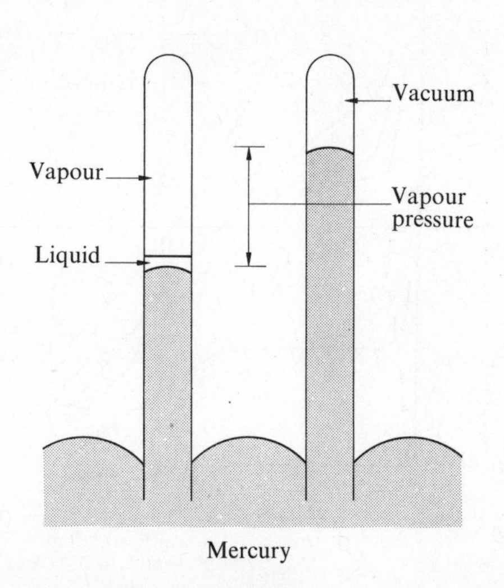

Fig. 9.69

The saturation vapour pressure depends on the temperature. Figure 9.70 shows how the saturation vapour pressure of water changes with temperature. The ratio of the number

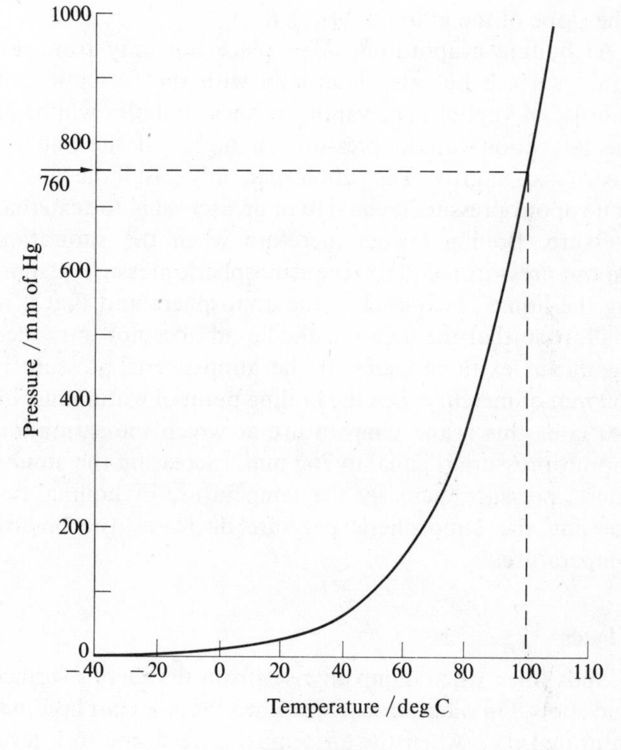

Fig. 9.70

of molecules with an energy E greater than that of another group of molecules is given by (see chapter 8)

$$\frac{n_1}{n_2} = e^{-E/kT}$$

If n_1 is the number of vapour molecules and n_2 the number of liquid molecules, then E is of the order of the escape energy.

$$E = \frac{\Delta H_{vap}}{L}$$

where L is Avogadro's number. Thus

$$n_1 = n_2 \, e^{-\Delta H_{vap}/LkT}$$

But the pressure of the vapour will be proportional to the number of vapour molecules ($PV = \frac{1}{3}mn\overline{c^2}$), thus

$$\text{vapour pressure } p \propto n_2 \, e^{-\Delta H_{vap}/LkT}$$

n_2 will change only slightly with temperature and thus a reasonable approximation is

$$p = A\,e^{-\Delta H_{vap}/LkT}$$

or as $Lk = R$

$$p = A\,e^{-\Delta H_{vap}/RT}$$

This equation is a reasonable description of the graph. If the vapour pressure results are replotted as log p against $1/T$ a straight line is produced.

$$\ln p = -\frac{\Delta H_{vap}}{RT} + \ln A$$

The slope of the graph is $\Delta H_{vap}/R$.

At boiling evaporation takes place not only from the liquid surface but also internally with the formation of bubbles of vapour. The vapour in these bubbles will be at the saturation vapour pressure, or higher, if the bubbles are to exist, or grow. The bubbles can only exist if the saturation vapour pressure is equal to or greater than, the external pressure. Boiling occurs therefore when the saturation vapour pressure is equal to the atmospheric pressure, assuming the liquid is exposed to the atmosphere and that it is shallow so that the depth of the liquid does not introduce significant extra pressure. If the atmospheric pressure is 760 mm of mercury then the boiling point of water must be 100°C as this is the temperature at which the saturation vapour pressure is equal to 760 mm. Increasing the atmospheric pressure increases the temperature of boiling, decreasing the atmospheric pressure decreases the boiling temperature.

Clouds

Clouds form when damp air rises from the earth's surface and cools. Do water drops form when the damp air becomes saturated, i.e., when the air temperature drops to a level where the vapour pressure becomes equal to the saturation vapour pressure?

In 1897 C. T. R. Wilson wrote a paper in which he describes experiments where he measured the vapour pressure at which water drops were formed. He used dust-free air and found that the air needed to be supersaturated before condensation in the air (excluding that on the walls of the container) occurred. The degree of supersaturation was measured by taking air at its saturation vapour pressure and finding by what factor the volume of the air had to be reduced before condensation occurred. His results show that regardless of the initial volume of saturated air the ratio of the vapour pressure for condensation to the saturation vapour pressure was 1·25 at about 20°C. At this supersaturation sufficient water molecules come together to form a water drop which grows.

If the air is not dust free condensation occurs at much lower vapour pressures, pressures only slightly greater than the saturation vapour pressure. The water condenses on the dust particles. If the dust particles are hygroscopic, for example sodium chloride particles, then condensation can occur below the saturation vapour pressure.

To form a liquid surface, or increase its area, energy is needed (see the section in this chapter on 'Surfaces'). To keep a group of water molecules together as a drop the vapour pressure above its surface needs to be greater than the saturation vapour pressure. For a drop of radius R the pressure difference is about $2G/R$, where G is the surface energy. This pressure difference is greatest when the drop radius is small—hence the high degree of supersaturation necessary in the absence of dust. Increasing the size of dust particles lowers the degree of supersaturation necessary as the drop can start with a higher initial radius. This assumes that the dust is insoluble in water. If this is not the case the water molecules form a solution with the particles and supersaturation is not necessary.

Wilson in his 1897 paper found another factor which could cause clouds—X-rays. With X-rays passing through the air the degrees of supersaturation needed for condensation was much less than with dust-free air. In 1911 and 1912 Wilson utilized this fact to design what is now known as the Wilson cloud chamber. Ions can act as centres for condensation and thus any radiation which produces ions when it passes through air, or some other gas, will give trails of condensation along the ion path it produces. (See chapter 2 for some cloud chamber photographs.)

Liquefying gases

In 1823 Faraday liquefied chlorine by applying pressure. By 1908 the last of the so-called 'permanent' gases had been liquefied (this was helium). How good are the gas laws in

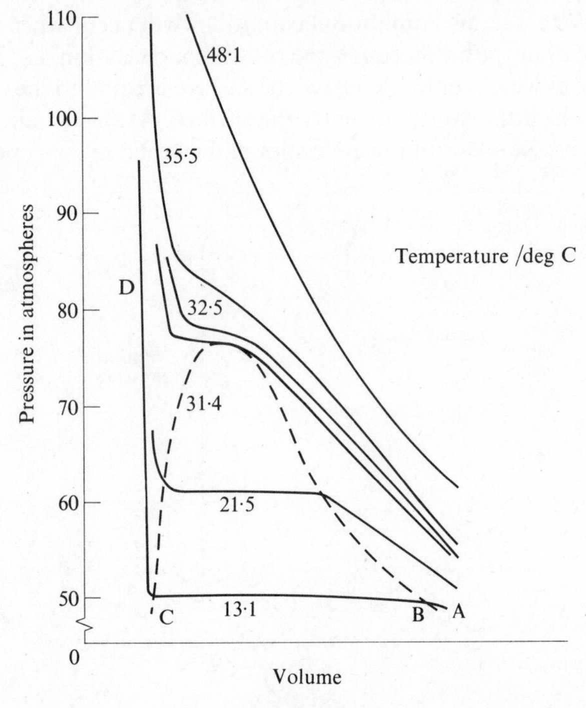

Fig. 9.71

describing the pressure, volume, temperature relationships when a gas is near liquefaction?

The classic work on this was that by T. Andrew in 1863 on the liquefaction of carbon dioxide. He measured pressures and volumes for a sample of carbon dioxide at different temperatures. Figure 9.71 shows his results. Above about 50°C his results gave reasonable agreement with the gas laws. At temperatures below 31·4°C pressure was able to produce liquefaction, above that temperature liquefaction could not be produced however great the pressure. This temperature is called the critical temperature. At say a temperature of 13·1°C we have the following sequence of events as the volume of the gas is decreased: from A to B a large decrease in volume produces only a small change in pressure; at B liquefaction commences; from B to C liquefaction is occurring and the large change in volume produces no change in pressure; at C liquefaction is complete; from C to D the liquid is being compressed and only small changes in volume occur for large changes in pressure.

The following are the values for some materials of the critical temperatures and the pressures needed at those temperatures to produce liquefaction.

	Critical temp./ K	*Critical pressure/* 10^5 N m^{-2}
Carbon dioxide	304·2	73·8
Oxygen	154·8	50·8
Nitrogen	126·2	33·9
Hydrogen	32·99	12·94
Helium	5·2	2·29

(One atmosphere pressure is about 10^5 N m^{-2}.) The reason for helium being one of the last gases to be liquefied becomes obvious from the fact of its very low critical temperature; it cannot be liquefied at temperatures greater than 5·2 K.

There have been many attempts to modify the gas equation to fit the behaviour of gases. A useful equation was put forward by van der Waals in 1873.

$$\left(P + \frac{a}{V^2}\right)(V - b) = RT$$

a and b are constants. The volume term in the gas equation is reduced by b to account for the finite size of the molecules. This is because the volume of a container which is available for the molecules to move in is significantly reduced by the volumes of the molecules themselves. In our kinetic theory derivation we were assuming that a molecule could move to any other part of the container. The pressure term is modified to take into account intermolecular forces. These would become significant when the gas molecules become closely packed together. Because of these forces the molecules spend more time near each other than in the absence of such forces and thus the number of impacts on the wall in a given time is reduced, also the attractive forces reduce the momentum with which a molecule hits the wall. Both these effects will depend on the number of molecules per unit volume, n/V, and thus we have a term proportional to $1/V^2$. Figure 9.72 shows the curves for carbon dioxide given by van der

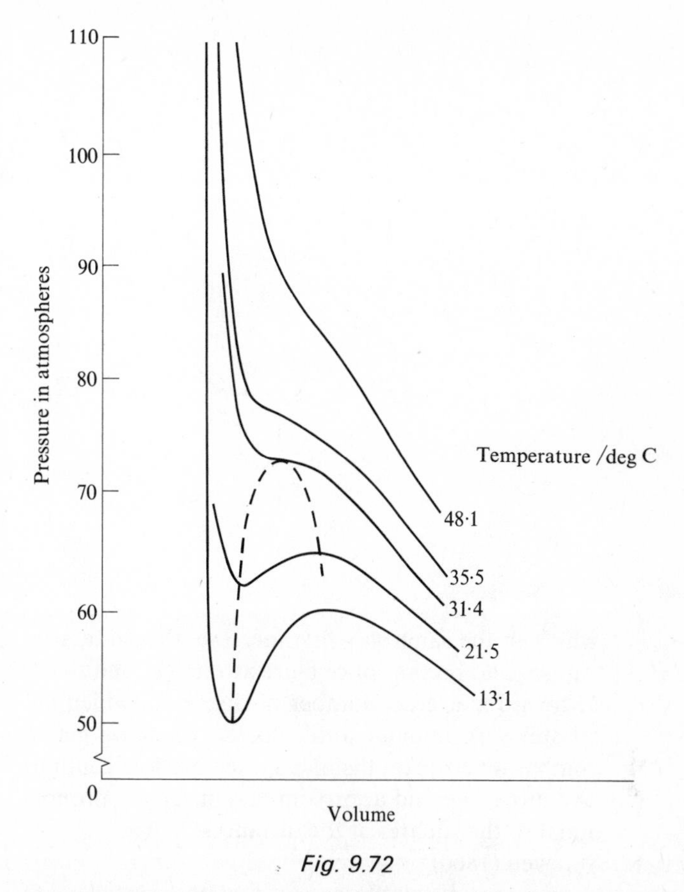

Fig. 9.72

Waals' equation. The agreement is only approximate and there is considerable divergence in the liquefaction region. The equation is, however, a useful approximation.

Problems

1. The following quotations give the views that have been held concerning the way that the force between molecules in a gas varies with the separation of the molecules. Draw a possible force–distance graph for each of the views.

(a) Newton (1687)

'. . . the [repulsive] forces of the particles will be inversely proportional to the distances [between] their centres.'

(b) Boscovich (1763)

'. . . the forces are repulsive at very small distances, and become indefinitely greater and greater, as the distances are diminished indefinitely, in such a manner that they are capable of destroying any velocity, no matter how large it may be, with which one point may approach another, before ever the distance between them vanishes. When the distance between them is increased, they [the forces] are diminished in such a way that at a certain distance, which is extremely small, the force becomes nothing. Then as the distance is still further increased, the forces are changed to attractive forces; these at first increase, then diminish, vanish, and become repulsive forces,

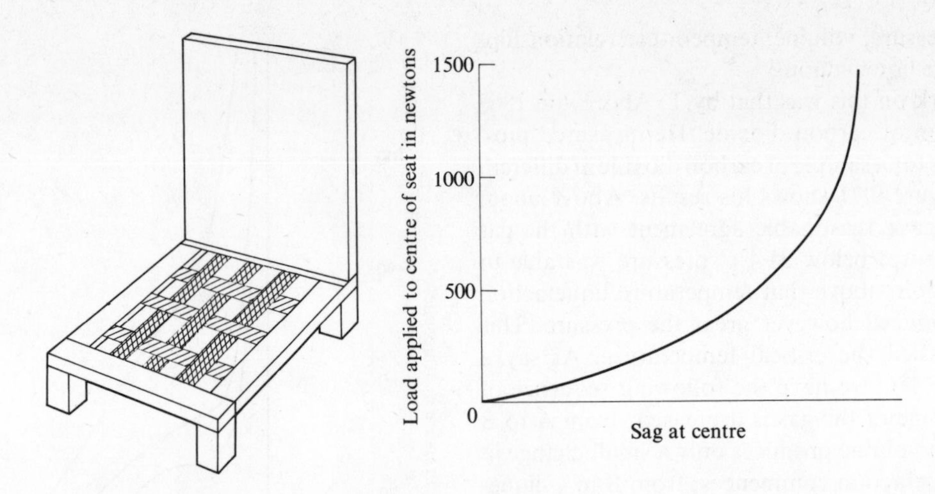

Fig. 9.73

which in the same way first increase, then diminish, vanish, and become once more attractive; and so on, in turn, for a great number of distances, which are all still very minute: until, finally, when we get to comparatively great distances, they become continually attractive and approximately inversely proportional to the squares of the distances....'

(c) Maxwell (1860)

'...we may if we please say that the particles are centres of force, of which the action is insensible except at a certain small distance, when it suddenly appears as a repulsive force of very great intensity....'

(d) Now draw your idea of the force–distance graph.

2. Oxford and Cambridge Schools Examination Board. Nuffield Advanced Physics. 1971. Short Answer Paper Q2.

'Suppose that the graph [Fig. 9.73] is the load-extension curve for rubber webbing used as the base of a chair:

(a) The following are predictions about what will happen when a person sits on the chair.

(i) He will sink only a little, unless he is very heavy, when he will suddenly sink a lot.

(ii) He will sink a moderate distance and then be supported at about the same level whether he is heavy or light.

Explain which is the correct prediction.

(*b*) Suppose that in a test on the chair a weight of 250 newtons is set oscillating freely on the seat. Explain under what circumstances simple harmonic motion will be performed.'

3. Graphite consists of plates of carbon atoms. Between atoms in the plates there are strong interatomic forces. Between the atoms in the different plates there are only very weak forces.

What physical properties would you expect a sample of graphite to have?

4. Bismuth has a Young modulus of 3×10^{10} N m^{-2}, chromium has a modulus of 28×10^{10} N m^{-2}. Which material would be most easily bent?

5. The interatomic spacing in aluminium is $2{\cdot}86 \times 10^{-10}$ m, the mass of an aluminium atom is $4{\cdot}5 \times 10^{-26}$ kg, and the value of Young modulus for aluminium is $7{\cdot}0 \times 10^{10}$ N m^{-2}.

(a) What is the force constant for interatomic bonds in aluminium?

(b) At what frequency would you expect aluminium atoms to resonate?

6. Deuterium, a naturally occurring isotope of hydrogen, was discovered by H. Urey, F. Brickwedde, and G. Murphy in 1932. Normal hydrogen has an atomic mass of 1, deuterium has an atomic mass of 2. The method used to obtain a deuterium-rich sample of hydrogen was to allow four litres of hydrogen to evaporate until only a few cubic centimetres was left. This was found to give spectrum lines which indicated that deuterium was present.

Why should evaporation appear to increase the concentration of deuterium in a sample of hydrogen?

7. Write an essay on the properties of water. The following information may be of help.

Density of water, liquid, at 0°C = 1000 kg m^{-3}
Density of water, ice, at 0°C = 917 kg m^{-3}
Free surface energy at 20°C = 0·073 J m^{-2}
Latent heat of vaporization at 20°C = 44 kJ mole^{-1}
Boiling point = 373 K
Freezing point = 273 K
Coefficient of viscosity at 20°C = 10×10^{-4} N s m^{-2}
Latent heat of fusion at 0°C = 6·0 kJ mole^{-1}
Latent heat of vaporization at 100°C = 41 kJ mole^{-1}

8. Equal volumes of gases contain the same numbers of molecules.

(a) How could you convince another student that this statement was correct?

(b) How much of a substance is a mole?

(c) Avogadro's number is 6×10^{23}. What is this the number of?

(d) The atomic mass of lead is about 207 and its density 1134 kg m^{-3}. What is the effective volume of a lead atom?

9. (a) Why are small raindrops almost spherical?

(b) Why does water rise up capilliary tubes?

(c) How does water behave on a 'waterproof' surface?

(d) How would you expect the surface energy of a liquid to be changed by an increase in temperature?

10. Sugar is a crystalline substance but when it is made into toffee it behaves like a glass.

By what properties can you distinguish a crystalline substance from a glass-like substance?

Teaching note Further problems appropriate to this chapter will be found in:
Nuffield advanced physics. Student's book 1, Penguin, 1972.
Nuffield O-level physics. Questions books 3 and 4, Longmans/Penguin, 1966.
Physical Science Study Committee. *Physics*, 2nd ed., chapters 8 and 9, Heath, 1965.

Practical problems

1. Investigate the motion of air bubbles in liquid filled tubes. Consider the case where the diameter of the bubble is comparable with the diameter of the tube. Such bubbles can be produced by filling a tube apart from a certain amount which determines the bubble size; the tube can then be sealed and the motion studied with the tube vertical, inverting the tube causing the bubble to move. Consider the factors which determine the velocity of the bubble (is there a terminal velocity?), e.g., viscosity of the fluid, size of bubble, size of tube, etc.

2. A corrugated strip of paper between two sheets of paper is a commonly used method of producing a strong structure from relatively weak materials. The product is known as corrugated cardboard. Other possible forms of reinforcement can be devised, e.g., a series of tube sections, a honeycomb structure, a square framework, etc., between the two sheets. Construct different reinforced cardboard structures and compare their elastic properties.

3. Determine how the rate of evaporation of water from drops depends on the drop size. A possible method of doing this at room temperature is to measure the diameter of a suspended drop with the aid of a vernier microscope. The rate of evaporation can then be calculated from the density of the water and the rate at which the drop diameter changes. Care must be taken to ensure that the air conditions round the drop remain constant during the experiment.

The result of this investigation is of significance in the study of clouds.

Suggestions for answers

1. (a) (b) (c) See Fig. 9.74

(d) You might consider something like Fig. 9.44.

2. (a) (ii).

(b) The amplitude of the motion must not be so great that the displacement of the seat moves off the lower linear part of the graph.

3. Strong in one direction, tension along the length of the plates, and weak in a direction at right angles to the plates. In fact the plates slide easily over one another—hence graphite pencils.

4. The lower modulus material bends most easily, i.e., bismuth.

5. (a) $20\ \text{N m}^{-1}$.

(b) $3{\cdot}4 \times 10^{12}$ Hz.

6. Equipartition of energy would suggest that the deuterium molecules would have a root mean square velocity smaller than that of normal hydrogen, by a factor of $1/\sqrt{2}$. This would mean less high velocity deuterium molecules than normal hydrogen molecules and thus less would evaporate.

7. You might care to discuss the reason for the difference between the density in the liquid and in the solid, the effects of water having a high surface energy (for a liquid), perhaps estimate the energy needed to break the intermolecular bond. Try reading *Water the mirror of science*, by K. S. Davis and J. A. Day (Heinemann Science Study Series, No. 21).

8. (a) You could discuss chemical reactions along the lines discussed on page 239.

(b) The amount of the substance containing the molecular mass in grams.

(c) The number of particles, molecules, in a mole.

(d) Mass of a lead atom $= \dfrac{207}{6 \times 10^{23}}$ g

Volume $= 3{\cdot}0 \times 10^{-30}\ \text{m}^{-3}$

9. (a) A sphere has the smallest surface area for its volume and thus is the minimum energy shape.

(b) See page 233.

(c) The water has a large angle of contact. See E. M. Rogers *Physics for the inquiring mind* (OUP) page 98.

(d) For water:

Surface energy/ J m^{-2}	0·076	0·073	0·070	0·066
Temperature/ °C	0	20	40	60

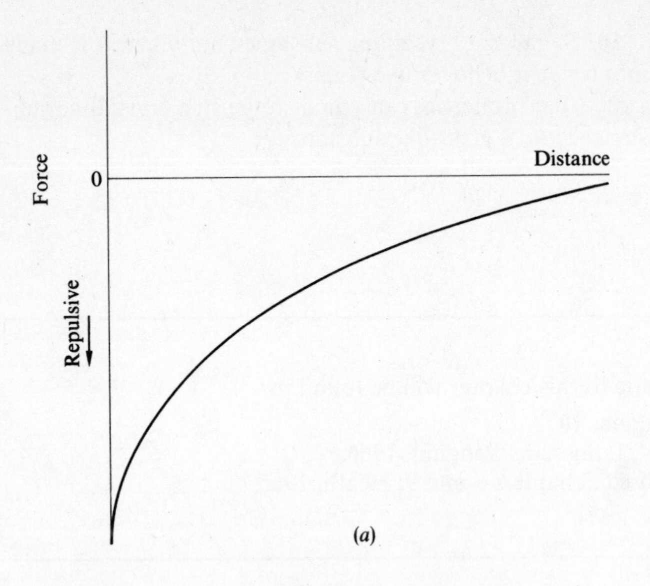

(a)

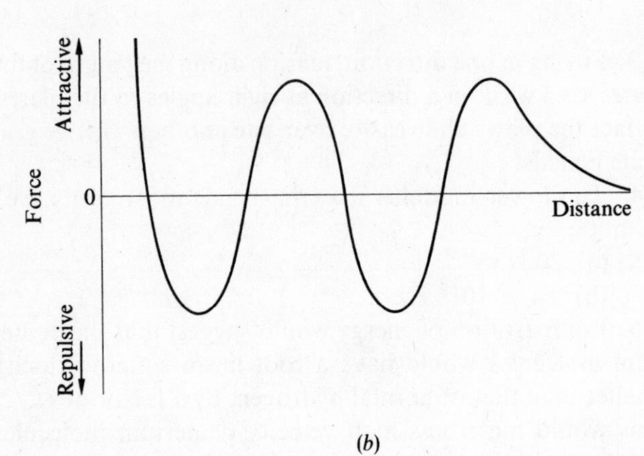

(b)

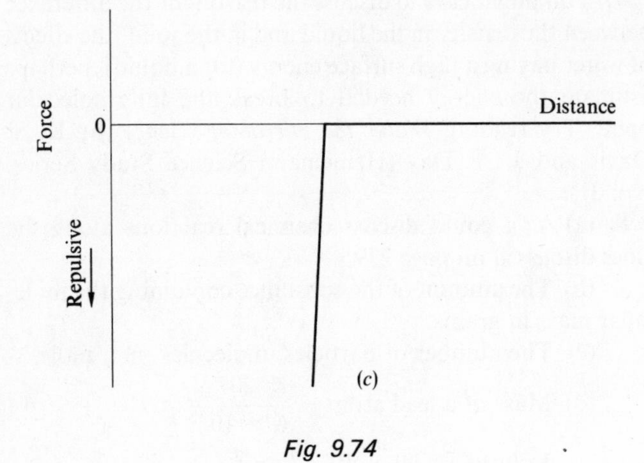

(c)

Fig. 9.74

Increasing the temperature puts the water molecules farther apart and thus less energy is needed to separate them completely.

10. Think of sugar and toffee. You might like to try some experiments with toffee. Crystalline substances have well defined melting points, glasses do not. Crystalline substances involve regular packing of atoms, glasses are a disorderly tangle.

10 Electric circuits

Teaching note Practical work appropriate to this chapter will be found in:
W. Bolton. *Physics experiments and projects*, Vol. 4, Pergamon, 1968.
Nuffield advanced physics. Units 2 and 6, teacher's guides, Penguin, 1972.

10.1 Charge and its movement

Charge in any isolated system is conserved, that is, if we add up algebraically the sum of the positive and negative charge in an isolated system we always obtain the same number. Another characteristic of charge is that it is quantized, that is, charge comes in packets. We can treat it like packets of sugar, one packet plus another packet makes two packets—twice the quantity of sugar. One coulomb of charge plus another coulomb makes two coulombs—twice the amount of charge. This additive property is not a characteristic of all quantities in physics—temperature is not additive: if we have one item at 10°C and we add another item at 10°C we do not get 20°C but 10°C. The size of the basic charge packet is fixed—no matter what object or particle is charged the charge is always an integral multiple of this one basic packet size. The electron and proton carry the basic charge. The charges on the electron and proton are equal to within at least 1 part in 10^{20} (J. G. King. *Phys. Rev. Letters*, **5**, 562, 1960). The above is a summary of our knowledge about charges. We now take a look at the evidence for the above statements.

Whenever we produce a negative charge on an object we always produce a positive charge on another part of the object or on some other object. The total positive charge is equal to the total negative charge. We rub a plastic pen against a strip of wool—equal positive and negative charges are produced. Gamma ray photons can end their existence by the production of a negative and a positive electron, the positron. Two charged particles have been produced from an uncharged photon. The net charge is, however, still zero as the charge on the negative electron is opposite in sign and equal in magnitude to that carried by the positive electron. In all events where we have charge we have charge conservation. Non-conservation of charge is incompatible with physics as we now know it.

The charge on any object is always made up of an integral number of basic charges. The classic experiment on this was the oil drop experiment by Millikan in 1911. Millikan measured the charge on a large number of drops of oil and other liquids. Changing the liquid or the size of the drop made no difference to his result—'in not one single instance has there been any change which did not represent the advent upon the drop of one definite invariable quantity of electricity, or a very small multiple of that quantity . . .'.

'. . . [Fig. 10.1]. By means of a commercial "atomizer" A a cloud of fine droplets of oil is blown with the aid of dust-free air into the dust-free chamber C. One or more of the droplets of this cloud is allowed to fall through a pin hole p into the space between the plates m, n of a horizontal air condenser and the pinhole is then closed by means of an electromagnetically operated cover not shown in the diagram. If the pin hole is left open air currents are likely to pass through it and produce irregularities. The plates m, n are heavy, circular, ribbed brass

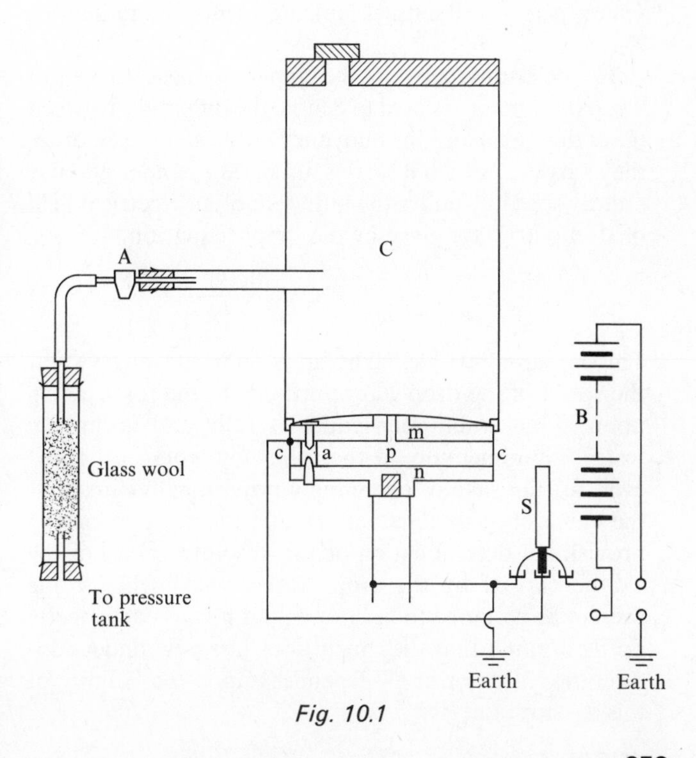

Fig. 10.1

castings 22 cm in diameter having surfaces which are ground so nearly to true planes that the error is nowhere more than ·02 mm. These planes are held exactly 16 mm apart by means of three small sheet ebonite posts a held firmly in place by ebonite screws. A strip of thin sheet ebonite c passes entirely around the plates, thus forming a completely enclosed air space. Three glass windows, 1·5 cm square, are placed in this ebonite strip at the angular positions 0°, 165° and 180°. A narrow parallel beam of light from an arc lamp enters the condenser through the first window and emerges through the last. The other window serves for observing, with the aid of a short focus telescope placed about 2 feet distant, the illuminated oil droplet as it floats in the air between the plates. The appearance of this drop is that of a brilliant star against a black background. It falls, of course, under the action of gravity, towards the lower plate; but before it reaches it, an electrical field of strength between 3,000 volts and 8,000 volts per centimeter is created between the plates by means of the battery B, and, if the droplet had received a frictional charge of the proper sign and strength as it was blown out through the atomizer, it is pulled up by this field against gravity, towards the upper plate. Before it strikes it the plates are short-circuited by means of the switch S and the time required by the drop to fall under gravity the distance corresponding to the space between the cross hairs of the observing telescope is accurately determined. Then the rate at which the droplet moves up under the influence of the field is measured by timing it through the same distance when the field is on. This operation is repeated and the speeds checked an indefinite number of times, or until the droplet catches an ion from among those which exist normally in air, or which have been produced in the space between the plates by any of the usual ionizing agents like radium or X-rays. . . .

The relations between the apparent mass (the term "apparent mass" is used to denote the difference between the actual mass and the buoyancy of the air) m of a drop, the charge e_n, which it carries, its speed v_1 under gravity, and its speed v_2 under the influence of an electrical field of strength F_1 are given by the simple equation

$$\frac{v_1}{v_2} = \frac{mg}{Fe_n - mg} \text{ or } e_n = \frac{mg(v_1 + v_2)}{Fv_1} \qquad (1)$$

This equation involves no assumption whatever save that the speed of the drop is proportional to the force acting upon it, an assumption which is fully and accurately tested experimentally in the following work. . . . However, for the sake of obtaining a provisional estimate of the value of m in equation (1), and therefore at once a provisional determination of the absolute values of the charge carried by the drop, Stokes's law will for the present be assumed to be correct, but it is to be distinctly borne in mind that the conclusions just now under consideration are not at all dependent upon the validity of this assumption.

The law in its simplest form states that if η is the coefficient of viscosity of a medium, x the force acting upon a spherical drop of radius a in that medium, and v the velocity with which the drop moves under the influence of the force, then

$$x = 6\pi\eta a v \qquad (2)$$

The substitution in this equation of the resulting gravitational force acting on a spherical drop of density σ in a medium of density ρ gives the usual expression for the rate of fall, according to Stokes, of a drop under gravity, viz.,

$$v_1 = \frac{2}{9}\frac{ga^2}{\eta}(\sigma - \rho) \qquad (3)$$

The elimination of m from (1) by means of (3), and the further relation $m = \frac{4}{3}\pi a^3(\sigma - \rho)$ gives the charge e_n in the form

$$e_n = \frac{4}{3}\pi\left(\frac{9\eta}{2}\right)^{3/2}\left(\frac{1}{g(\sigma - \rho)}\right)^{1/2}\frac{(v_1 + v_2)v_1^{1/2}}{F}$$

It is from this equation that the values of e_n in tables I–XII are obtained. . . .

TABLE I
Negative drop

Distance between cross-hairs = 1·010 cm
Distance between plates = 1·600 cm
Temperature = 24·6°C
Density of oil at 25°C = ·8960
Viscosity of air at 25·2°C = ·0001836

	G sec.	F sec.	n	$e_n \times 10^{10}$	$e_1 \times 10^{10}$
G = 22·28, V = 7950	22·8	29·0	7	34·47	4·923
	22·0	21·8	8	39·45	4·931
	22·3	17·2			
	22·4	—			
	22·0	17·3	9	44·42	4·936
	22·0	17·3			
	22·0	14·2	10	49·41	4·941
	22·7	21·5	8	39·45	

. . . In the interval between December, 1909, and May, 1910, Mr. Harvey Fletcher and myself took observations in this way upon hundreds of drops which had initial charges varying between the limits 1 and 150, and which were upon as diverse substances as oil, mercury and glycerine and found in every case the original charge on the drop an exact multiple of the smallest charge which we found that the drop caught from the air. The total number of changes which we have observed would be between one and two thousand, and in not one single instance has there been any change which did not represent the advent upon the drop of one definite invariable quantity of electricity, or a very small multiple of that quantity. . . .

TABLE II

n	$4{\cdot}917 \times n$	*Observed charge*	n	$4{\cdot}917 \times n$	*Observed charge*
1	4·917	—	10	49·17	49·41
2	9·934	—	11	54·09	53·92
3	14·75	—	12	59·00	59·12
4	19·66	19·66	13	63·92	63·68
5	24·59	24·60	14	68·84	68·65
6	29·50	26·62	15	73·75	—
7	34·42	34·47	16	78·67	78·34
8	39·34	39·38	17	83·59	83·22
9	44·25	44·42	18	88·51	—

Millikan (1911) *Phys. Rev.*, **32**, 349–387.

The charge unit we use in this book is different to that used by Millikan. In our system the basic charge unit is $1{\cdot}6 \times 10^{-19}$ C.

Though there are now many particles we consider to be fundamental particles, i.e., they are not thought of as being made up of other particles, they all carry either no charge or one positive or one negative charge unit. The unit in all cases is $1{\cdot}6 \times 10^{-19}$ C.
For example:
neutrinos, no charge.
muon, either one positive or one negative charge.
electron, either one positive or one negative charge.
pion, either no charge, one positive or one negative charge.
proton, either one positive or one negative charge.
neutron, no charge.

The following is a report of the situation existing in July 1971 regarding the search for particles with charges less than the charge on the electron.

'*The elusive quark*

Recent experimental studies at Liverpool and Berkeley of more than 100 000 cosmic ray showers have failed to reveal any tracks made by fractionally charged particles and so another search for quarks has proved fruitless.

...

The hunt for quarks has been going on since their existence was independently proposed by Murray Gell-Mann and George Zweig in 1964. They have been looked for since then with particle accelerators, in cosmic radiation, and in investigations of macroscopic material, for example seawater: most of these searches rely on the fact that the quarks should possess non-integral electric charges of one-third, or two-thirds of that of the electron, and should therefore yield tracks in particle detectors with substantially lower ionizations than conventionally charged particles.

Although no substantial evidence for the quark has been revealed by these investigations, there have been a few false hopes. In 1969, an Australian group, led by C. B. A. McCusker of the Cornell-Sydney University Astronomy Centre, claimed that certain anomalies, observed in cloud chamber studies of cosmic ray ionizations, were most likely to have been caused by quarks of charge $2/3e$....

Five candidates for such quark events were reported from a total of some 6000 cloud chamber photographs. This result was greeted with some scepticism by the physics community and was soon under severe attack: an article by R. K. Adair of Yale University and H. Kasha of Brookhaven ... indicated that the anomalous low ionizations reported by McCusker et al. were probably due to low energy electrons and muons associated with air showers. A more recent report of a quark sighting by a group from Ohio State University, and based on a cosmic ray bubble chamber photograph, has also received very little support. . . .'

Nature, **232**, 298, 1971.

Current and voltage

We may look at an electrical circuit and say—the current is 5 A, the battery has an e.m.f. of 6 V, the potential difference across that resistor is 3 V. What do these terms mean?

Current is the term used to describe the rate of movement of charge past some point.

$$I = \frac{dQ}{dt}$$

If 5 coulombs pass a point in a circuit every second we call the current 5 coulombs per second, or 5 amps.

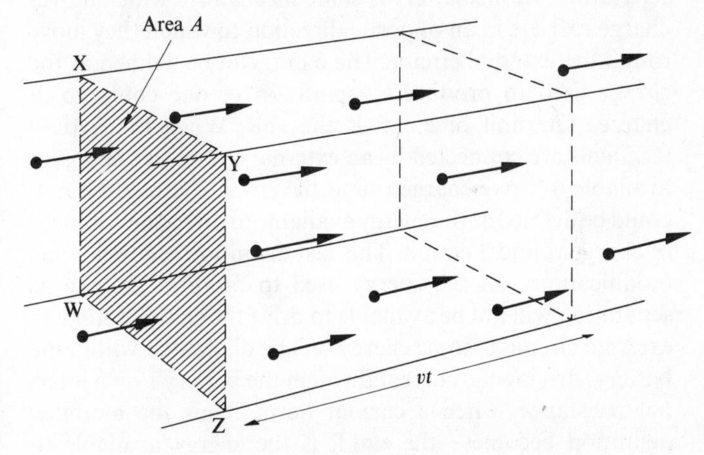

Fig. 10.2

Suppose we have a conductor of cross-sectional area A carrying a current I (Fig. 10.2). Also we suppose that there are n charge carriers per unit volume, each carrying a charge q and each moving with a velocity v along the conductor. In time t a charge carrier would cover a distance vt. In time t all the charge carriers in a volume Avt will have crossed through an area WXYZ. Thus the charge passing through the area in time t is $Avtnq$. Current is the rate of movement of charge, thus

$$I = Avnq$$

How big is v? If we consider a copper wire of cross-sectional area 1 mm², i.e., 10^{-6} m², and assume that each copper atom donates one electron for conduction purposes, n will be $8{\cdot}5 \times 10^{28}$ per m^3.

(Atomic mass of copper 63·6, density 8930 kg m^{-3}, 6×10^{23} atoms per mole. Volume of 1 mole of copper $63{\cdot}6 \times 10^{-3}/8930$.)

For a current of 1 A, we have

$$v = \frac{1}{10^{-6} \times 8{\cdot}5 \times 10^{28} \times 1{\cdot}6 \times 10^{-19}}$$

$$v = 0{\cdot}74 \times 10^{-3} \text{ m s}^{-1}$$

or about $0{\cdot}7 \text{ mm s}^{-1}$. On our theory the charge carriers are moving very slowly. (The high speed with which a lamp comes on when you press the electric light switch can be explained in terms of the high speed with which an electric field travels through the circuit.)

What causes the current? What causes the electrons to have even this low velocity v? A battery could be the origin of the force causing the electrons to move. We say the battery supplies an electromotive force, i.e., an e.m.f. A battery is a device in which by chemical action one electrode, or terminal, becomes positively charged and the other negatively charged. When an external connection is made between the two battery terminals charge flows, i.e., there is a current through the connecting wire, and the chemical reaction in the battery endeavours to maintain the charge separation. In the battery is some mechanism which moves charge carriers in an opposite direction to which they move round the external circuit. The e.m.f. can be defined as the energy used to produce a separation of one coulomb of charge. The unit of e.m.f. is the volt. When the battery terminals are connected by an external circuit this energy is available to drive charge round the circuit. Thus the e.m.f. could be defined as the energy available to drive one coulomb of charge round a circuit. This last definition requires some modification—all the energy used to produce the charge separation will not be available to drive the charge round an external circuit as some energy will be dissipated within the battery, driving the current through the battery's own internal resistance, when a current flows. Thus the modified definition becomes—the e.m.f. is the energy available to drive one coulomb round a circuit when no current is being taken.

The energy needed to drive one coulomb through a circuit component such as a resistor is called the voltage or potential difference, V.

$$V = \frac{\text{energy}}{\text{charge}}$$

The unit of V is volts if energy is in joules and charge in coulombs. A voltmeter placed in parallel with the resistor is a means of measuring the energy per coulomb needed. Many voltmeters are current-measuring instruments and depend on the fact that there is a linear relationship between the current flowing through the voltmeter coil (resistor) and the potential difference across the coil.

The energy dissipated by a charge q passing through a circuit component is the same as the energy needed to drive the charge through the component.

$$\text{Energy dissipated} = Vq$$

where V is the potential difference across the component. If the energy dissipated in time $t = Vq$, then the energy dissipated per second $= Vq/t$. This is known as the power—rate of energy dissipation.

$$\text{Power} = \frac{Vq}{t} = VI \quad \text{as} \quad I = \frac{q}{t}$$

The unit of power is the watt. In the above equation this is when V is in volts and I in amps.

Measurements of the power due to a current passing through a coil can be used to give a measure of the potential difference without the use of a voltmeter. If the coil is mounted inside a metal cylinder the rise in temperature of the cylinder per second can be measured, for a known current. The energy needed to produce this temperature change can be estimated if the same temperature change is produced by the expenditure of a measured amount of mechanical energy.

The ratio of the potential difference V across a resistor to the current I is called the resistance R.

$$R = \frac{V}{I}$$

For some materials R is a constant, independent of the value of the current. The relationship with R as a constant is known as Ohm's law.

In terms of electron movements we can write

$$R = \frac{V}{I} = \frac{V}{Avnq}$$

For Ohm's law to be obeyed we must have

$$v \propto V$$

Doubling the potential difference across a resistor doubles the velocity of the electrons along the length of the resistor.

Some resistors are known as voltage-dependent resistors. For them the voltage–current relationship can be written as

$$V = CI^{\beta}$$

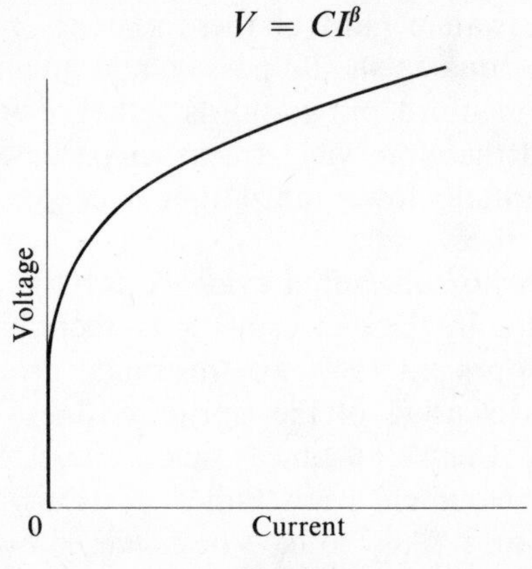

Fig. 10.3

where C and β are constants. Figure 10.3 shows a voltage–current graph for such a resistor.

$$\text{Resistance} = \frac{V}{I} = \frac{CI^{\beta}}{I} = \frac{C}{I^{1-\beta}}$$

β has values generally between 0·15 and 0·30. C has a value between 14 and a few thousand.

Electrical circuits

Circuits can be quite complex; here we look at some rules which enable simple relationships to be established.

Figure 10.4 shows a circuit junction with meters measuring the current flowing into the junction and out of it. A simple rule emerges from examination of the meter readings.

$$I_1 = I_2 + I_3$$

The current entering the junction is equal to the current leaving the junction. This rule is known as Kirchoff's first law. All the rule is really stating is that the rate at which charge enters the junction is equal to the rate at which charge leaves the junction. This is what we would expect if charge does not accumulate at the junction—and assuming that charge is conserved.

Figure 10.5 shows a simple circuit in which we measure the potential differences across every component. A simple rule emerges from examination of the meter readings.

$$V_1 = V_2 + V_3 + V_4$$

The sum of all the potential differences is zero. This is known as Kirchoff's second law. We can express this law in a different way: if we carry a charge around any circuit loop, and back to where we started, the net work is zero. Figure 10.6 shows the arrangement with a piece of circuit in which there is no source of e.m.f. For the closed loop

$$V_1 + V_2 = V_3$$

The same work is done in taking a charge from A to B via one arm of the circuit as via the other or any other arm. This is all that is being said for the earlier circuit with the source of e.m.f.: the same work is done in taking a charge between any two points regardless of the path, whether it be just through a resistor or via a source of e.m.f.

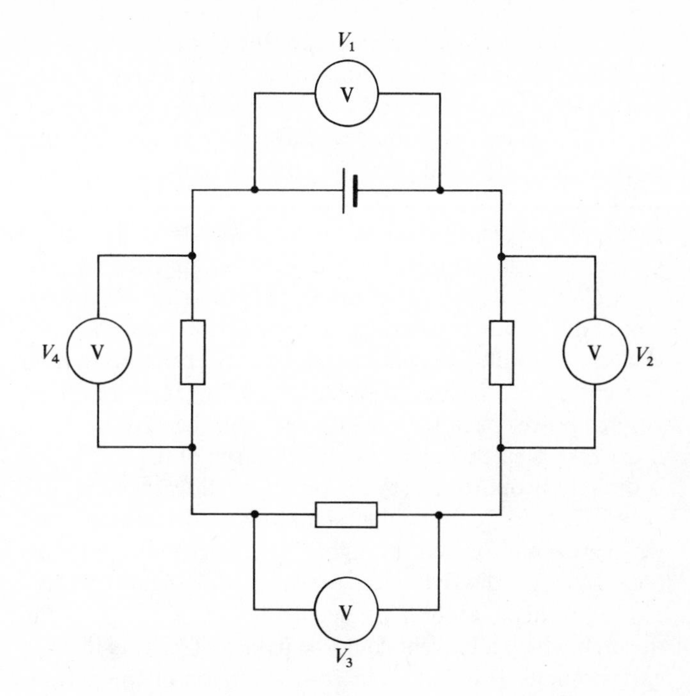

Fig. 10.5

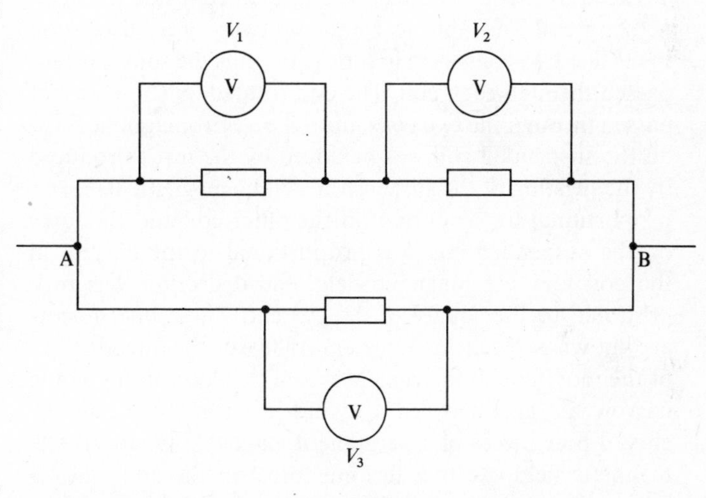

Fig. 10.6

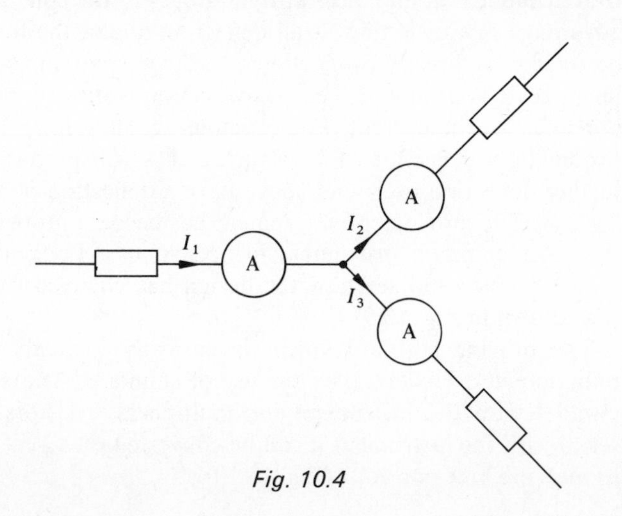

Fig. 10.4

Measurement instruments

The first electrical measurement instrument was probably the electroscope, invented by Rev. A. Bennet in 1787. The original electroscope was essentially just a loop of very thin gold leaf suspended over an insulating rod. When the gold leaf was put in contact with a charged object the two parts of the gold leaf diverged (Fig. 10.7). The charge on the charged object becomes shared with the gold leaf and as both parts of the leaf receive the same charge they move apart due to the repulsive force between equally charged objects, and the attractive force between the oppositely charged leaf and the walls of the electroscope. Bringing a charged object, a gold leaf, near an object such as the wall of the electroscope induces an opposite charge on that part of the wall nearest the leaf and so attraction occurs, rather like picking up a piece of paper by bringing a charged rod near it. If the wall is earthed we have in essence a capacitor consisting of a central plate, the leaf, surrounded by a concentric plate, the wall of the electroscope, each plate being charged. The deflection of the leaves depends on the charge on the leaf, and that on the wall. As for a capacitor we have $Q = CV$; the deflection of the leaf is a measure of the potential difference between the two capacitor plates, i.e., the leaf and the wall. The gold leaf electroscope can thus be used as a voltmeter. Connecting a battery, a large one, or a high voltage power pack between the leaf and the case, or earth if the case is earthed, will give a deflection of the leaf which is directly proportional to the potential difference applied.

In 1820 Oersted found that a compass needle was deflected when a current passed through a wire placed alongside the compass. This discovery was the basis of a current measuring instrument, a compass needle at the centre of a coil through which a current could be passed. This was the first galvanometer (Fig. 10.8). In 1841 a version of the galvanometer was produced in which the compass needle was replaced by a coil. This coil was suspended inside the magnetizing coil and able to rotate by twisting a suspension wire. The two coils were in series and thus the same current passed through each coil. The coil rotated, when a current passed through the two coils, until the electromagnetic force on the suspended coil was balanced by the force produced by the twisting of the suspension. As the magnetic field was proportional to the current in the outer coil and the force on the suspended coil was proportional to the current in the coil and the magnetic field, the deflection was proportional to the square of the current. These instruments are known as electrodynameters. 1888 saw the introduction of the moving-coil instrument where the coil rotated in the narrow air gap between a cylindrical iron core and the curved pole pieces of a permanent magnet (Fig. 10.9). The magnetic field exerts a turning force on the coil when a current flows in the coil. The coil rotates against a spring or twists a suspension, reaching a steady deflection when the

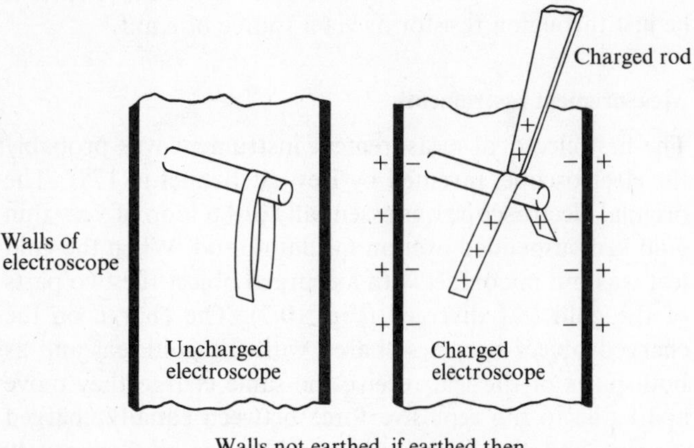

Fig. 10.7

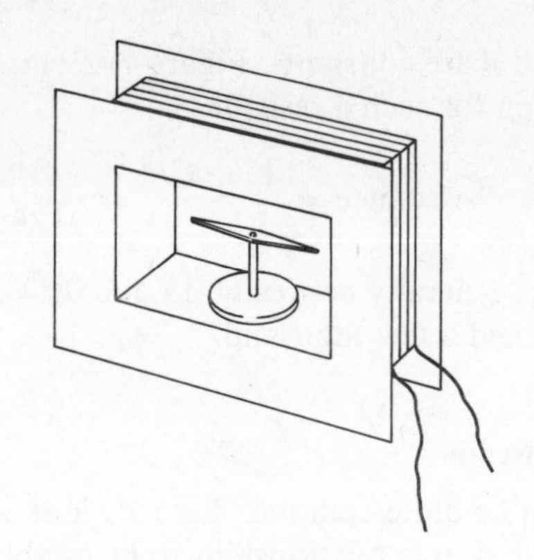
Fig. 10.8

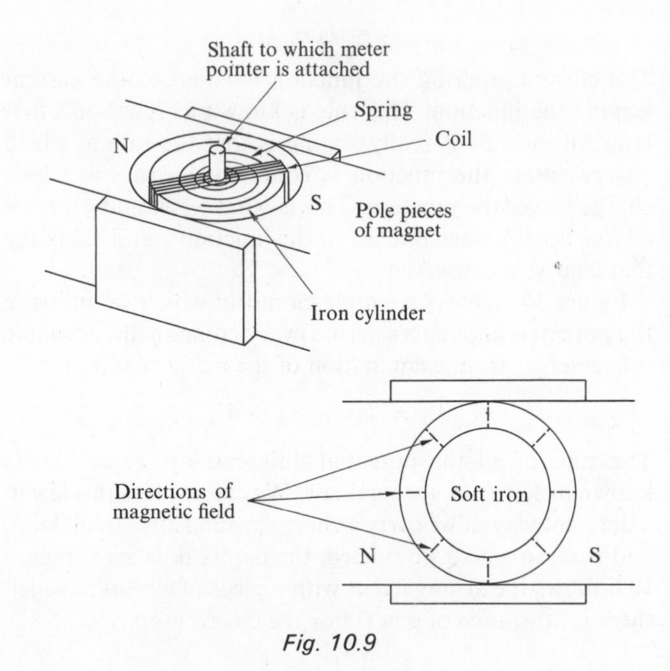

Fig. 10.9

turning force due to the magnetic field is balanced by the turning force exerted by the spring or suspension. The magnetic field in the air gap between the pole pieces and the iron cylinder is at all places at right angles to the coil. The advantage of this is that at all deflection angles the force on the coil is directly proportional to the current and so a linear scale is produced, i.e., the deflection is directly proportional to the current. For example, we may have 1 A producing a deflection of 5°, a further 1 A will produce a further deflection of 5° and so we have a deflection of 10° for 2 A. The moving coil instrument has become probably the most common instrument in present use. Look at a modern meter and see how the design has changed from that shown in Fig. 10.9.

The moving coil instrument is basically a micro- or milliammeter. However, by the use of shunts, resistors in parallel with the instrument, or multipliers, resistors in series with the instrument it can be converted into a meter to measure amps or volts (Fig. 10.10).

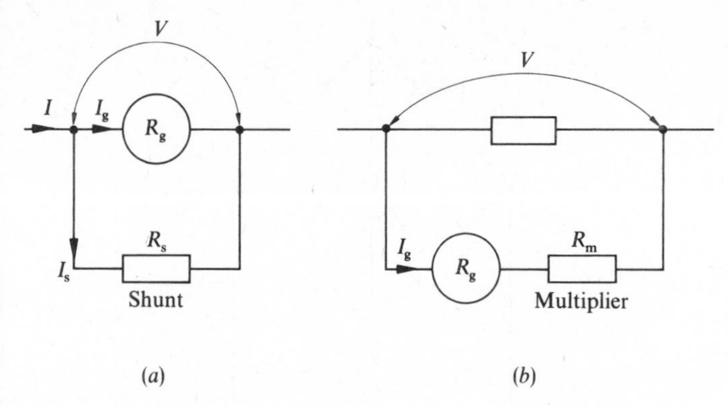

Fig. 10.10 Moving-coil meter as (a) an ammeter, (b) a voltmeter.

As an ammeter:

$$I = I_g + I_s$$

where I is the current to be measured, I_g is the current indicated by the meter, and I_s is the current passing through the shunt. The shunt takes a constant fraction of the current, leaving only a small fraction to be detected by the meter. The fraction taken depends on the resistances of the meter R_g and the shunt R_s;

$$V = I_g R_g = I_s R_s$$

Thus

$$\frac{I_g}{I_s} = \frac{R_s}{R_g}$$

and therefore

$$I = I_g + \frac{R_g}{R_s} I_g = \left(1 + \frac{R_g}{R_s}\right) I_g$$

The resistance of the shunt is less than the resistance of the meter. To convert a microammeter, say full-scale deflection 100 μA, to an ammeter, say full-scale deflection 10 A, we must have

$$10 = \left(1 + \frac{R_g}{R_s}\right) 100 \times 10^{-6}$$

or approximately

$$\frac{R_g}{R_s} = 10^5$$

If the meter has a resistance of 100 Ω the shunt will need to have a resistance of 10^{-3} Ω.

As a voltmeter:

$$V = I_g(R_g + R_m)$$

For a 100 μA meter of resistance 100 Ω to be converted to read 10 V at full-scale deflection, we must have

$$10 = 100 \times 10^{-6}(100 + R_m)$$

$$R_m = 99\,900\ \Omega$$

Multipliers have high resistances. In both the case of the shunt and the multiplier it is assumed that the resistors and the meter obey Ohm's law.

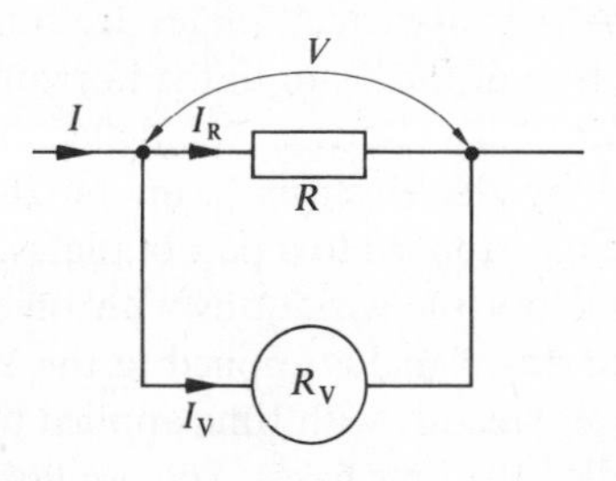

Fig. 10.11

The resistance of a voltmeter affects the voltage reading given by the instrument when it is placed across a circuit component (Fig. 10.11). The potential difference across a resistor, resistance R, is IR when no voltmeter is in parallel with the resistor. When the voltmeter is in parallel with the resistor the potential difference drops to $I_R R$ as some current flows through the voltmeter, I_R is the current through the resistor.

$$I = I_R + I_V$$

$$V = I_R R = I_V R_V$$

Thus

$$I = I_R\left(1 + \frac{R}{R_V}\right)$$

$$IR = I_R R\left(1 + \frac{R}{R_V}\right)$$

$$\text{p.d. with no voltmeter} = (\text{p.d. with meter})\left(1 + \frac{R}{R_V}\right)$$

The p.d. with the voltmeter approaches the value without the voltmeter when R_V becomes considerably greater than R. Thus the higher the voltmeter resistance the more 'correct' the value.

The cathode ray oscilloscope evolved from the work of many scientists about the beginning of this century. Essentially it consists of an electron gun which produces a high speed beam of electrons which is focused on a fluorescent screen (Fig. 10.12). A pair of horizontal plates are

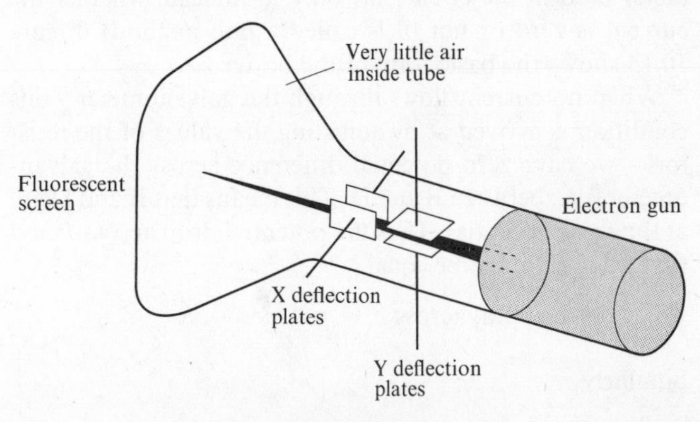

Fig. 10.12

placed one on each side of the beam. These are called the Y deflection plates because a potential difference placed across the plates causes the beam to be deflected in the vertical, Y, plane. A pair of vertical plates are similarly used to produce deflection of the beam in the horizontal, X, plane. The oscilloscope is a voltmeter; the deflection of the light spot produced by the electron beam on the screen is a measure of the p.d. applied to a pair of plates. The spot can be used to trace out the variations with time of a voltage signal if the varying signal is applied to the Y plates and a p.d. which varies steadily with time applied to the X plates (this is then called the time base). The oscilloscope is a very high resistance voltmeter.

Very high resistance voltmeters, called valve voltmeters, use the p.d. to control electrons moving within a valve (see details of the triode valve in From valves to transistors, later in this chapter). Changes in the numbers of electrons moving across the valve produce changes in current in an external circuit which can be registered on a meter.

Mention must be made of the way in which the measurement of resistance has changed over the years. Ohm's law was discovered in 1826 and was used to define resistance. The measurement of the p.d. across a resistor and the current flowing through it gives a value for the resistance. If moving coil instruments are used the accuracy of the result is limited due to the limited accuracy with which such meters can be read. The meters used by Ohm were of the type compass needle at the centre of a coil. Incidentally, Ohm's work was not at first received with enthusiasm, as the following review of Ohm's book shows:

> 'He who looks on the world with the eye of reverence must turn aside from the book as the result of an incurable delusion, whose sole effort is to detract from the dignity of nature.'

In 1843 Sir Charles Wheatstone gave a lecture, published in *Philosophical Transactions*, **133**, 323, 1843, in which he described what has since become known as Wheatstone's bridge. In fact it had previously been discovered by S. H. Christie in 1833 (*Phil. Trans.*, **123**, 95, 1833); Wheatstone did acknowledge the earlier work. The bridge compares resistances and is capable of high accuracy because the meter used in the circuit has only to indicate whether the current is zero or not (it is called a null method). Figure 10.13 shows the basic form of the bridge.

When no current flows through the galvanometer—this condition is arrived at by adjusting the values of the resistors—we have zero potential difference across the galvanometer, i.e., between B and D. This means that B and D are at the same potential. Thus the potential drop across P and that across R must be equal.

$$\text{p.d. across } P = \text{p.d. across } R$$

Similarly

$$\text{p.d. across } Q = \text{p.d. across } S$$

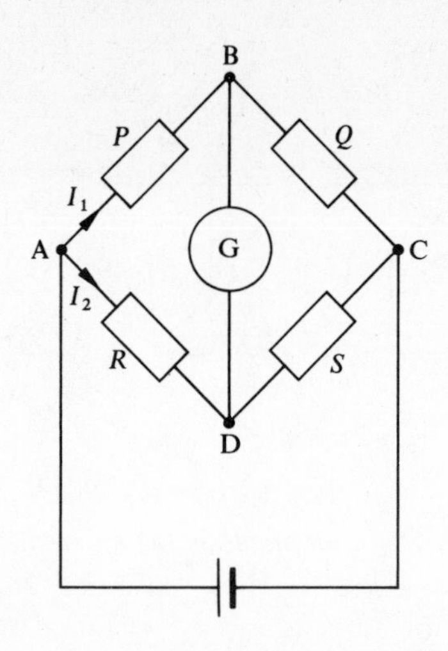

Fig. 10.13

The current through P must be equal to the current through Q as no current flows through the galvanometer. Similarly the current through R must equal that through S. Hence using the above equations

$$I_1P = I_2R$$
$$I_1Q = I_2S$$

and thus

$$\frac{P}{Q} = \frac{R}{S}$$

Knowing the ratio R/S and the value of Q, the value of P can be found. The accuracy of this result depends on the accuracy of these resistance values and the accuracy with which the zero current condition can be determined.

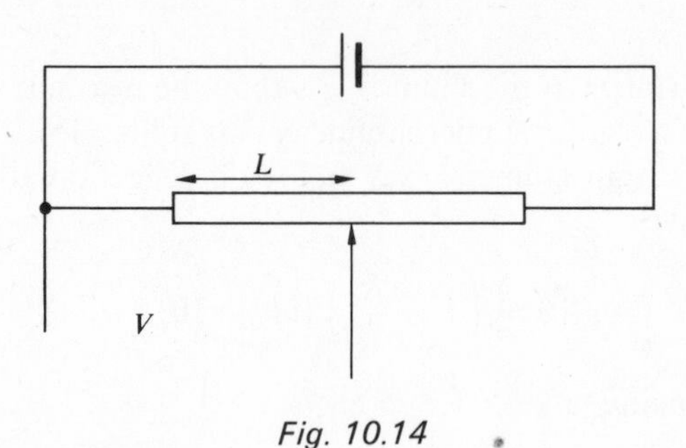

Fig. 10.14

1841 saw the invention by Poggendorff of the potentiometer. This is a method of comparing potential differences. A battery connected across a tapped resistor, see Fig. 10.14, enables a variable potential difference to be obtained between the tapping point and one end of the resistor. The resistor may be of the form of a strip of carbon or a wire along which a sliding contact, the tapping point, can move. If the resistor is perfectly uniform along its length, the distance the tapping point is from one end is proportional to the p.d. between the tapping point and that end.

$$V \propto L$$

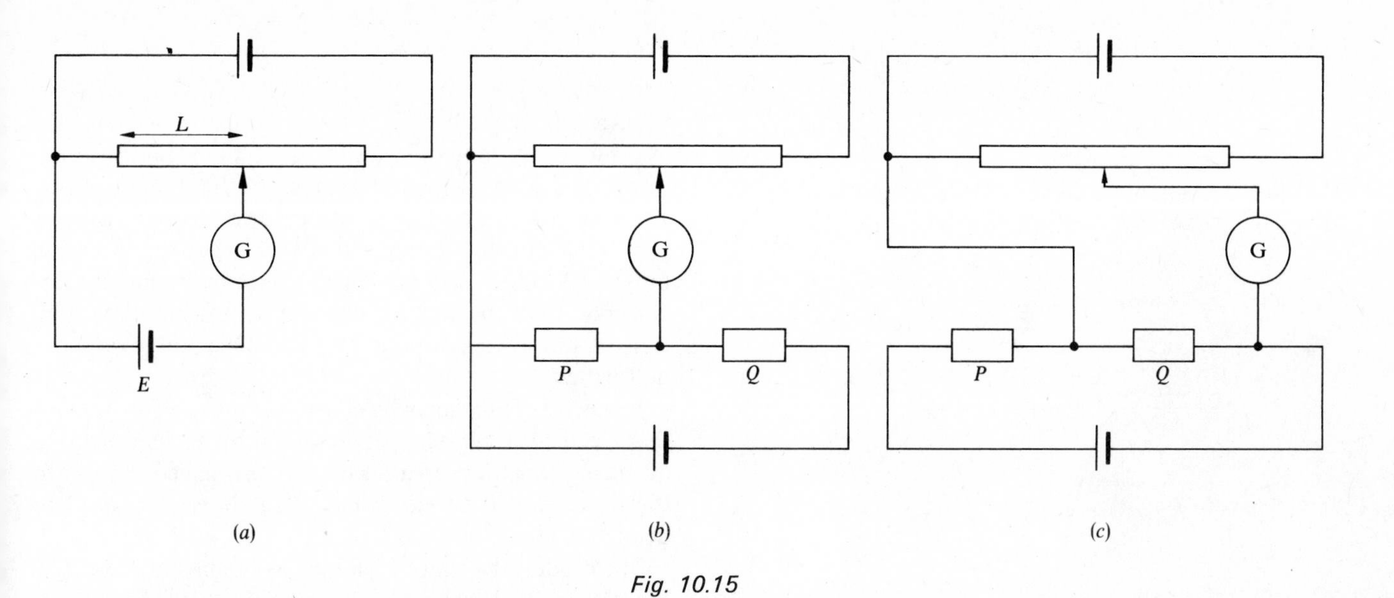

Fig. 10.15

Figure 10.15(a) shows the circuit used for the comparison of the e.m.f.s of cells. The sliding contact is moved along the resistor until there is no current detected by the galvanometer. When this occurs

$$E \propto L$$

The cell is then replaced with a standard cell and a new no-current position of the contact found.

$$E_S \propto L_S$$

Hence

$$\frac{E}{E_S} = \frac{L}{L_S}$$

Two measurements of length, or angle in the case of the slider moving round a circular track, and a knowledge of the e.m.f. of one cell thus enable an e.m.f. to be measured. Figure 10.15(b) shows the circuit when a potential difference is being measured. The p.d. across one resistor is compared with that across another resistor when the same current flows through each resistor.

$$\frac{V_P}{V_Q} = \frac{L_P}{L_Q}$$

As the same current flows through each resistor

$$\frac{P}{Q} = \frac{L_P}{L_Q}$$

and the resistance of P can be found if that of Q is known.

Alternating current

When a direct current, d.c., is passed through a resistor power is dissipated.

$$\text{Power} = IV$$

If Ohm's law is obeyed, $V = IR$, and we can write

$$\text{Power} = I^2R$$

or

$$\text{Power} = \frac{V^2}{R}$$

What is the power dissipated when an alternating current, a.c., passes through a resistor? The power dissipated at any instant will be that due to the values of I and V at those instants. Suppose we have an alternating current where the voltage, and the current, vary as $\sin\theta$, where θ is some function of time, say $V = V_{max}\sin\theta$. Figure 10.16(a) shows how the voltage varies with time. We can find how the power varies with time by taking the square of the sine curve.

$$\text{Power} = \frac{V_{max}^2}{R}\sin^2\theta$$

Figure 10.16(b) shows the square of the sine curve. The average power dissipated will be the average value of $\sin^2\theta$ multiplied by V_{max}^2/R. The average value of $\sin^2\theta$ is half the maximum value. (The $\sin^2\theta$ curve is symmetrical about the 0·5 line.) Thus the average power is

$$\text{Average power} = \frac{V_{max}^2}{2R}$$

If we compare this average power equation with the power equation for d.c. we can see that for the same power dissipation in the same resistor we must have

$$V_{d.c.}^2 = \frac{V_{max}^2}{2}$$

$$V_{d.c.} = \frac{V_{max}}{\sqrt{2}}$$

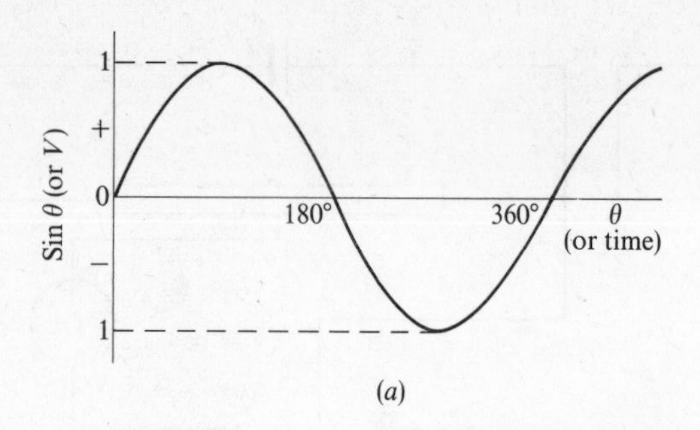

(a)

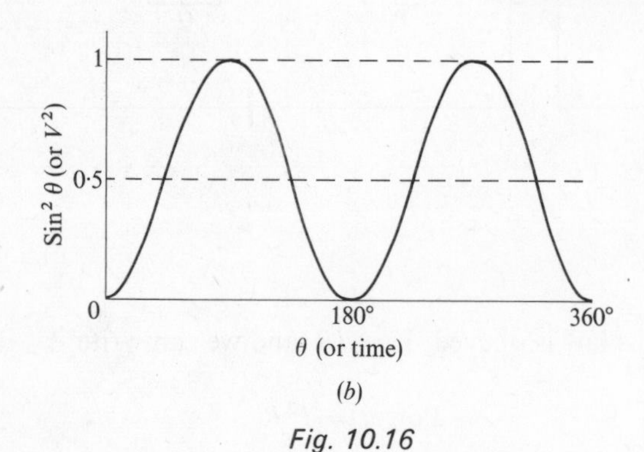

(b)

Fig. 10.16

The value of the d.c. voltage which gives the same power dissipation as that given by the a.c. is called the RMS or root mean square voltage.

$$V_{\mathrm{RMS}} = \frac{V_{\max}}{\sqrt{2}}$$

In a similar way we can arrive at the root mean square current.

$$I_{\mathrm{RMS}} = \frac{I_{\max}}{\sqrt{2}}$$

Thus a root mean square current of 4 A will have a maximum value of 5·6 A, a root mean square voltage of 240 V will have a maximum of 336 V.

From valves to transistors

In 1883 Edison while studying the blackening which occurred inside the early electric lamps sealed an extra electrode in the lamp. Figure 10.17 shows the basic arrangement, a heated filament to give the light and another electrode. He found that if this extra electrode was made positive a current flowed, if it was negative no current flowed. We now call this effect thermionic emission and consider that the heated filament is emitting electrons. Edison, however, saw no use for the effect and investigated no further.

It was J. J. Thomson who elucidated the effect by showing that electrons were being emitted and by his investigations of the nature and properties of electrons paved the way for the development of electronic valves (see chapter 13 for details of Thomson's work).

The diode valve can be claimed as the invention of J. A. Fleming in 1904. Fleming was an associate of Marconi and concerned with the development of 'wireless' (see chapter 12). In this work he saw the need for a device to rectify high frequency alternating current, i.e., to convert currents alternating from positive to negative values to currents which alternated only from positive to zero values (Fig. 10.18(a)). The diode valve is essentially the light bulb of Edison with the extra electrode. Current can only flow in one direction through such valve—when the extra electrode (called the anode) is positive with respect to the filament (Fig. 10.18(b)).

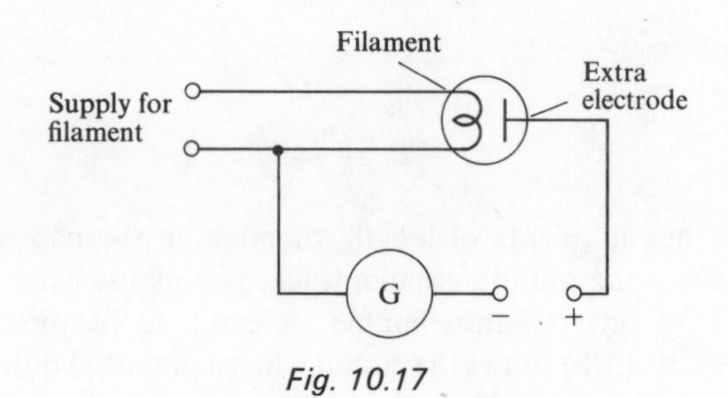

Fig. 10.17

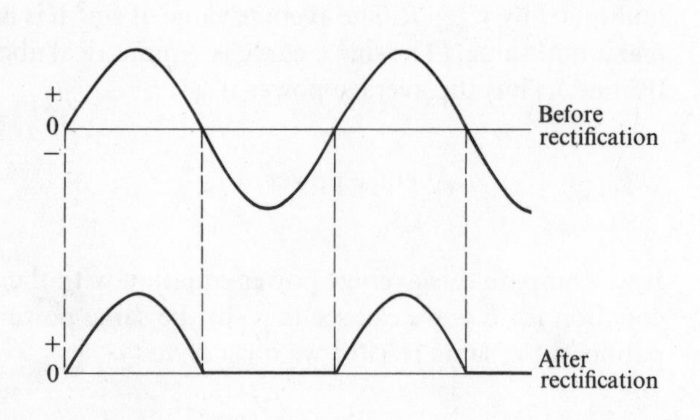

(a)

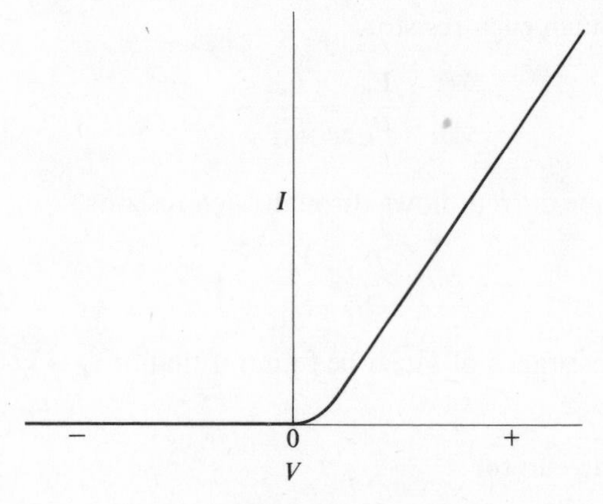

(b)

Fig. 10.18

In 1906 Lee de Forest modified the valve by adding a third electrode, a grid of wires, between the anode and the filament. This valve is known as the triode (Fig. 10.19). Small changes in the potential on the grid produce large changes in the current flow between filament and anode. This was the invention which enabled radio communication to develop from a short distance operation to long distance. It gave amplification of small aerial signals.

Later years saw the improvement of the triode by the addition of more electrodes; the tetrode in 1916, the pentode in 1926.

In the early days of wireless a very simple detector (see chapter 12) was used, called the crystal with a 'cat's whisker' wire making contact with it. Certain crystals in contact with one another are found to possess a rectifying action—the resistance in one direction is much greater than in the other direction. Copper oxide rectifiers, copper in contact with cuprous oxide, and selenium rectifiers, selenium in contact with aluminium, were produced about 1925 (Fig. 10.20).

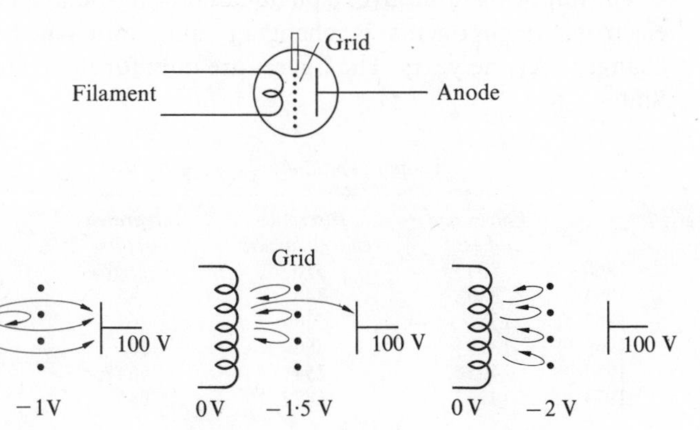

Typical voltages and the effect of grid potential on electron paths

Fig. 10.19

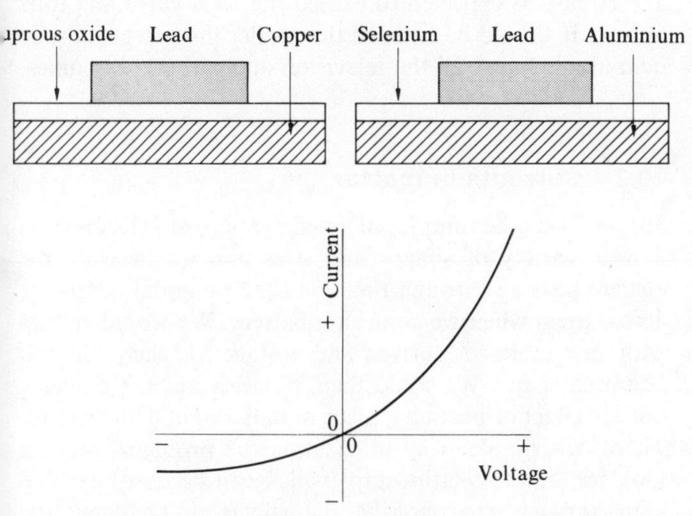

Typical current–voltage graph

Fig. 10.20

In 1948 the transistor was invented by Bardeen and Brattain. This was the equivalent of the triode valve and consisted of a sandwich of three materials (Fig. 10.21); rectifiers were essentially sandwiches of just two materials. Semiconductors such as germanium and silicon have their conduction properties markedly changed by the inclusion in them of certain impurities—the process is called 'doping'. Phosphorus as an impurity in silicon increases the supply of electrons which take part in the conduction process. Boron as the impurity increases the conductivity by the introduction of 'holes'—vacant sites into which electrons can move. The phosphorus impurity gives what is called an n-type semiconductor, conduction is by *n*egative charge carriers. The boron impurity gives a p-type semiconductor, conduction is by holes which are equivalent to *p*ositive charge carriers. p- and n-type semiconductors in contact give a rectifier. A transistor is two such junctions placed back to back. Transistors will do most of the jobs which the triode did and are cheaper, smaller, lighter, and more efficient.

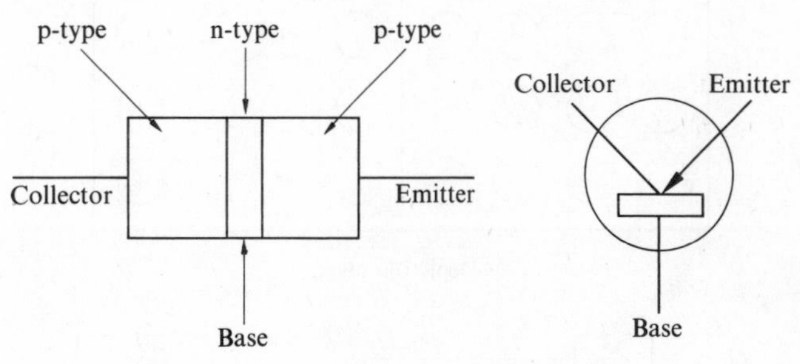

Fig. 10.21 One form of transistor.

In a p-type material we have mobile holes, in n-type material mobile electrons. When a junction between such materials is formed diffusion of holes and electrons occurs across the junction (Fig. 10.22). The holes and electrons combine with the result that the layer of material on either side of the boundary becomes depleted of charge carriers. This narrow layer at the junction is called the depletion layer. Because both the n- and p-type materials were initially neutral there is a negative charge on the p-type material side of the junction and a positive charge on the n-type material side. The movement of holes and electrons continues across the junction until the n side is sufficiently positive to stop further movement of holes and the p side sufficiently negative to stop the electrons. The potential difference that now exists between the two sides acts as a potential barrier to further charge movement.

If a battery is connected across the junction a significant current is only found to flow when the applied p.d. is sufficient to overcome the effect of the potential barrier. Thus when the p side of the junction is made sufficiently positive a current will flow, limited only by the resistance of the material and that of the external circuit; when the p side is made negative virtually no current flows. The junction is a rectifier, giving only a current in one direction.

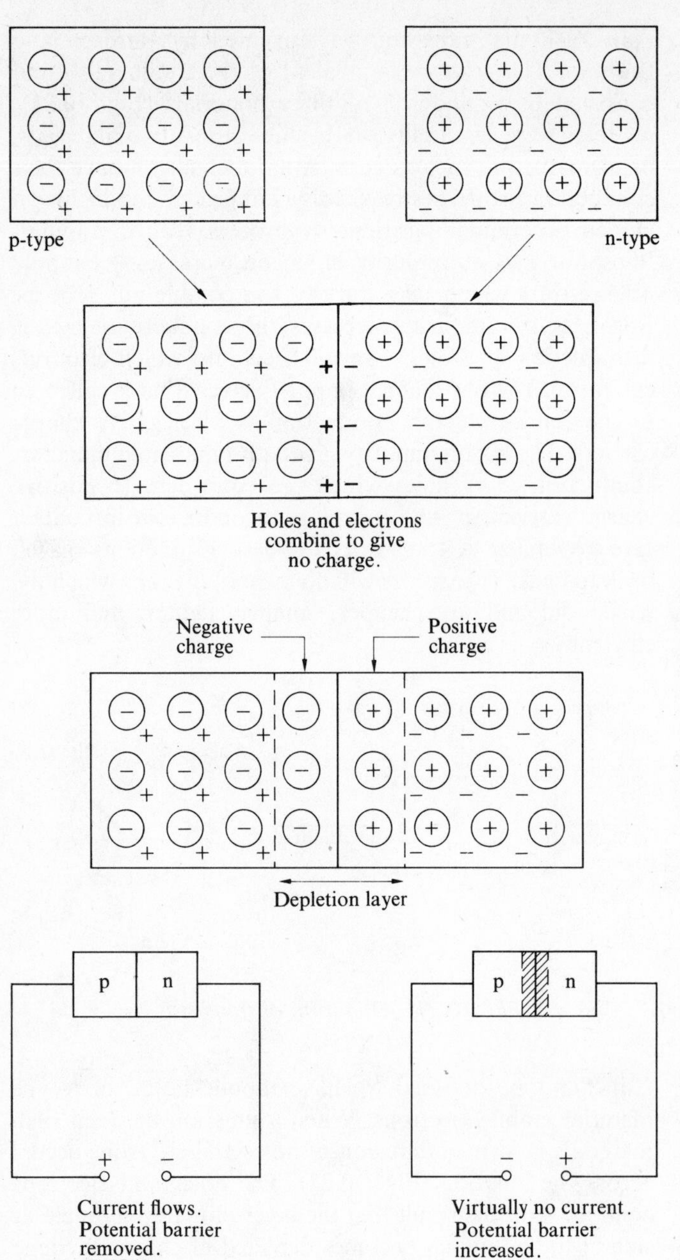

Fig. 10.22

In a transistor there are two p–n junctions, depletion layers forming at each junction (Fig. 10.23). The output circuit has a potential difference externally applied between the collector and emitter so that no current flows. If, however, a sufficiently high p.d. is applied between the base and the emitter so that the base is made negative the depletion layers vanish due to the movement of holes from the emitter to the base. In practice a small current in the input circuit can produce a very much larger current in the output circuit.

The next development was the integrated circuit. Instead of making individual transistors, resistors, capacitors or diodes, the entire circuit is made in one piece of material by suitably doping the various parts. The transistor was much smaller than the triode valve, the integrated circuit gives

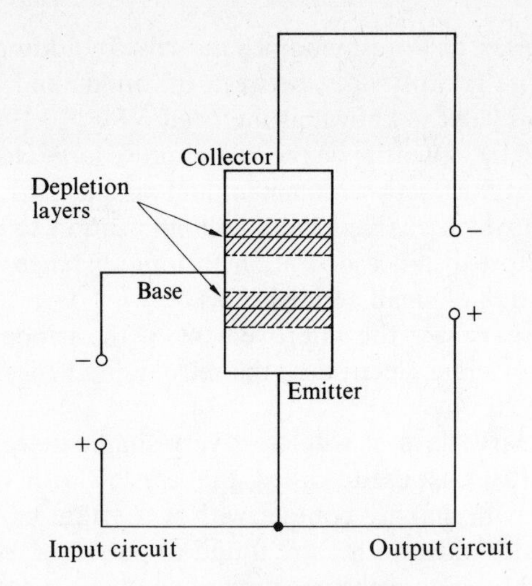

Fig. 10.23

us entire circuits in very little volume; modern electronics is microelectronics (see later in this chapter).

The following table gives a picture of how the demand for electronic components is changing, and anticipated to change, over the years. The figures are only for the United States.

Output in millions of dollars

	Valves and tubes	*Discrete semiconductors*	*Integrated circuits*
1960	800	550	negligible
1965	880	840	85
1967	1361	730	269
1969	1388	780	503
1970	1405?	754?	564?
1973	1474?	687?	945?

(D. Forsyth. *New Scientist*, 30 April 1970, 223.)

By 1973 the combined total of semiconductor components and circuits is expected to exceed the total valve and tube output. It should be realized that under the valve and tube item are included all the television and cathode ray tubes.

10.2 Currents in matter

Suppose we take samples of a wide variety of substances in a wide variety of shapes and sizes and we measure the current passing through them and the potential difference across them when we connect a battery. We would end up with vast tables of current and voltage for each material and each shape. We could then, by using our tables, work out the effect of putting a piece of material in a d.c. circuit. There is a simpler way of tackling the problem—we can look for simple relationships which can be used, perhaps approximately, to describe the effects of changing the material and the shape.

Experiments show that the ratio of the potential difference to the current, i.e., the resistance, is directly proportional to the length of a uniform specimen of a material,

$$R \propto L$$

and is inversely proportional to the cross-sectional area of a uniform length of material,

$$R \propto \frac{1}{A}$$

Thus we can write

$$R \propto \frac{L}{A}$$

and if we introduce a constant ρ

$$R = \frac{\rho L}{A}$$

The value of ρ, called the resistivity, depends on the nature of the material.

This relationship can be deduced if we assume that charge is conserved and the current entering a wire must equal the current leaving the wire. Suppose we consider two equal lengths, equal cross-sectional area, of the same material. If we have the same current in each length we expect the same potential difference across each length. Suppose we now put the two lengths together to give a wire of double the length, between the ends we must have, for the same current double the potential difference (Fig. 10.24(a)). The ratio V/I has been doubled by doubling the length of a wire. The resistance R is thus proportional to the length. Suppose we now put the two wires together to give one wire of double the cross-sectional area (Fig. 10.24(b)). For the same potential difference we have double the current. Thus doubling the area gives half the value for the ratio V/I, the resistance is inversely proportional to the area.

The reciprocal of resistance is called conductance and the reciprocal of resistivity the conductivity. Figure 10.25 shows the values of conductivity, and resistivity, for some materials. Note that the scale is logarithmic. The figure shows also how the values change with temperature. Metals have, at room temperature, resistivities of the order of 10^{-8} to 10^{-7} ohm m, insulators have resistivities of the order of 10^{12} ohm m—about 10^{20} times greater than that of a metal. Roughly mid-way between these we have resistivities of the order of 10^{2} ohm m; materials with resistivities in this region are called semiconductors. The resistivity of pure metals increases with an increase in temperature, that of semiconductors and insulators decreases with an increase in temperature.

Earlier in this chapter we obtained the expression for resistance of

$$R = \frac{V}{I} = \frac{V}{Avnq}$$

Rewriting this in terms of resistivity gives

$$\rho = \frac{V}{vnqL}$$

V is the potential difference between the ends of a wire of length L, this can be written in terms of the electric field E

$$E = \frac{V}{L}$$

Thus

$$\rho = \frac{E}{vnq}$$

If the resistivity increases with temperature (Fig. 10.25) then if V/L, i.e., E, remains constant the term vnq must decrease. There seems no reason to expect that q will change so either n or v or both must change. The Hall effect, discussed in chapter 11, however, enables us to measure n and the results show that for metals there is no significant change in n with temperature. There must therefore be a decrease in v with temperature for metals.

For the semiconductors and insulators the resistivity decreases with temperature. There is no reason to expect that v will not still decrease with temperature—thus there must be a large increase in n with an increase in temperature.

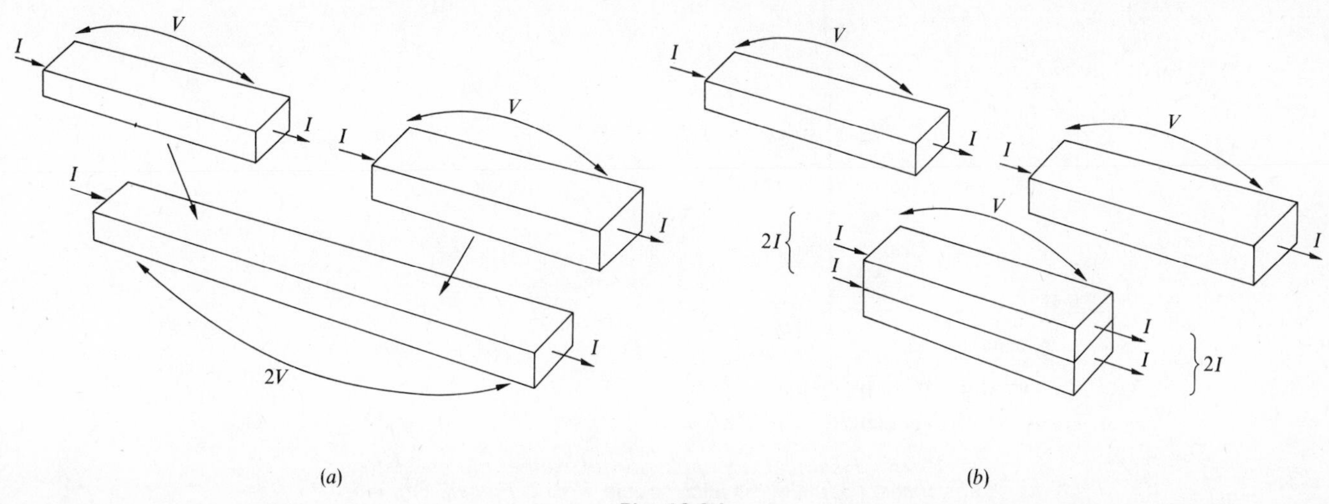

Fig. 10.24

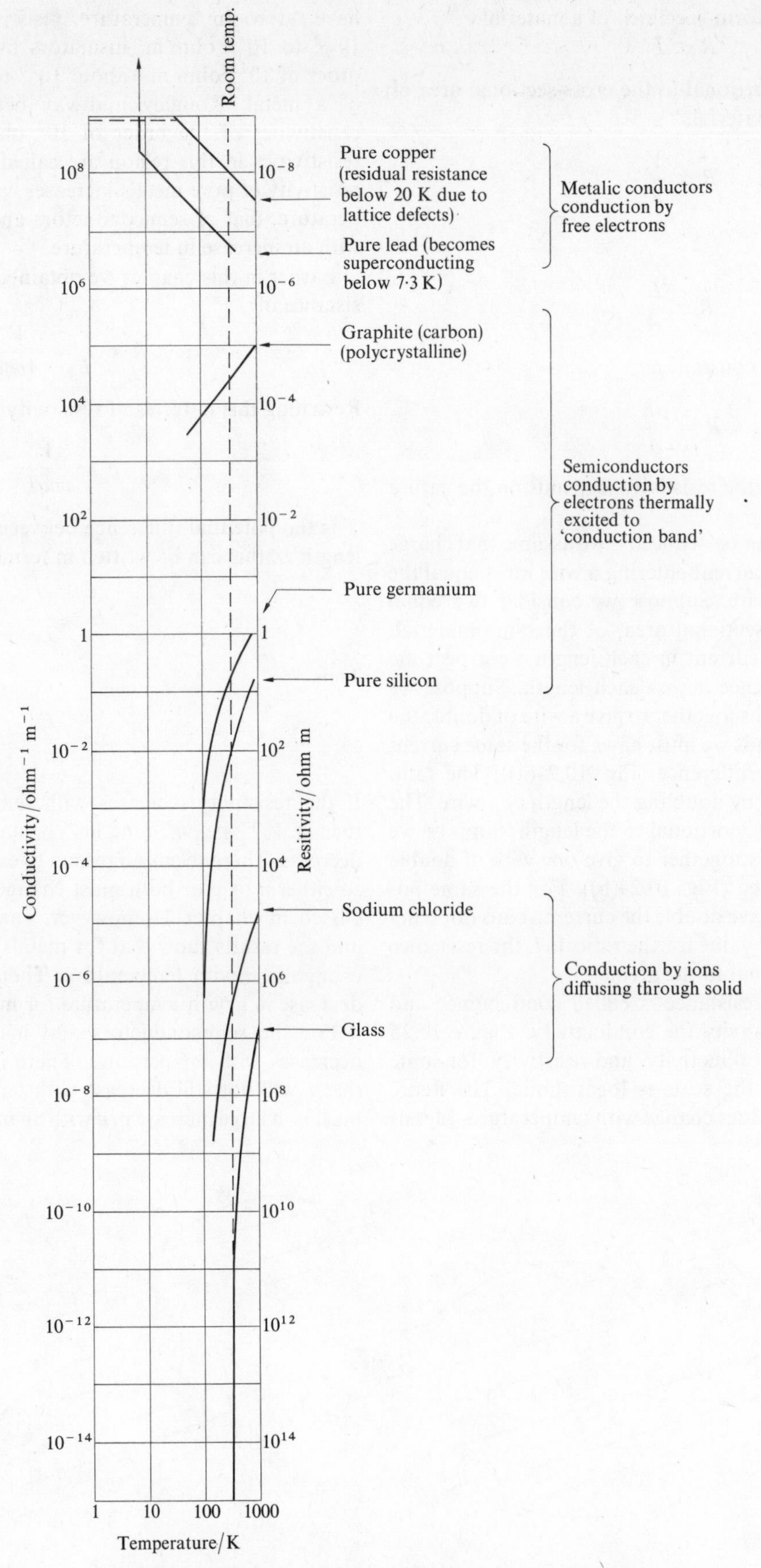

*Fig. 10.25 Electrical conductivity of some representative substances. (*Berkeley physics course, *Vol. 2. Purcell, Fig. 4.6. McGraw-Hill Book Company.)*

This is confirmed by measurements of the Hall effect. Increasing the temperature releases more charge carriers.

Figure 10.26 shows, across part of the periodic table, how the energy needed to ionize an atom correlates with its electrical property in the solid. Good conductors have atoms with low ionization energies, insulators have atoms with high ionization energies.

Electrons in metals

A potential difference between the ends of a wire causes a current which we have considered as being due to the p.d. causing electrons to move with a low velocity along the wire. The electrons in a metal can be considered to be free, within the confines of the metal, and able to move about within the metal like molecules in a gas. At room temperature they will be about 300 K above absolute zero and would therefore be expected to be moving about because of this thermal energy. This motion would, however, be random with no particular direction—unlike the motion produced by the potential difference. We can make a rough estimate of this random velocity—

$$\text{roughly } \tfrac{1}{2}mv^2 = kT$$

k is Boltzmann's constant, $1{\cdot}4 \times 10^{23}$ J K^{-1}. This gives v as about 3×10^5 m s^{-1}—much higher than the drift velocity of about a millimetre per second produced by a potential difference giving a current of 1 A.

Under the action of a p.d. the electrons which are moving around with a high velocity between collisions with atoms are slowly drifting along the wire making, it might be assumed, numerous collisions en route.

The energy needed to move a charged particle, charge q, along the length of a wire, length L, is Vq. V is the potential difference between the ends of this wire. This energy must be equal to FL, where F is the force component acting along the length of the wire.

$$FL = Vq$$

The acceleration of the charged particle will be

$$a = \frac{F}{m} = \frac{Vq}{mL}$$

where m is the mass of the charged particle. Suppose the particle accelerates from rest. In time t the particle will have acquired a velocity v where

$$v = at = \frac{Vqt}{mL}$$

If t is the time over which the particle accelerates before colliding with some atom, v will be the maximum velocity. It is assumed that after the collision the particle has no velocity in this force direction. The average velocity will be $\frac{1}{2}v$.

$$\text{Average velocity} = \frac{Vqt}{2mL}$$

But $I = Avnq$, thus by eliminating the velocity term (the v in the current expression is the average velocity) we have

$$\frac{I}{Anq} = \frac{Vqt}{2mL}$$

$$I = \left(\frac{A}{L}\right)\left(\frac{nq^2t}{2m}\right)V$$

We assume here that all the electrons have the same value of t, regardless of the value of V. This is only a reasonable approximation if the velocity change produced by the p.d. is very small compared with the random velocity. Then the random velocity determines the time between collisions and this depends only on the temperature. In this case the terms in brackets are constant, at a particular temperature, and thus I is proportional to V. This is Ohm's law. Thus we have the resistance R given by

$$R = \frac{L}{A}\frac{2m}{nq^2t}$$

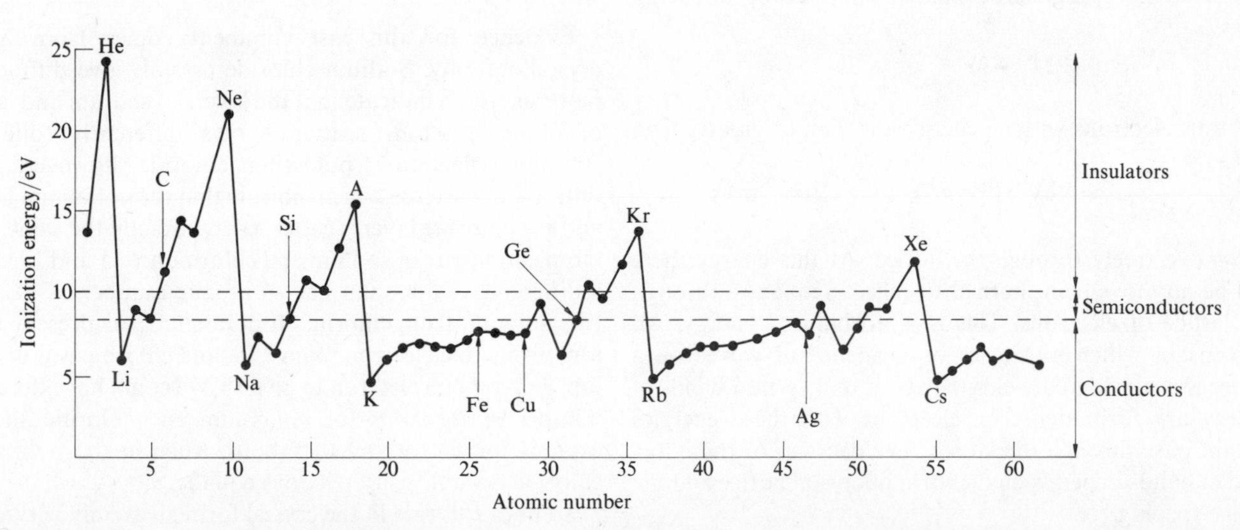

Fig. 10.26 Correlation between the energy needed to ionize an atom and its electrical property in the solid.

But

$$R = \frac{L}{A}\rho$$

where ρ is the resistivity. Thus

$$\rho = \frac{2m}{nq^2t}$$

For copper: ρ is about $1{\cdot}6 \times 10^{-8}\ \Omega$ m, n about $8{\cdot}5 \times 10^{28}\ \text{m}^{-3}$, the mass of an electron, m, about $9{\cdot}1 \times 10^{-31}$ kg, q $1{\cdot}6 \times 10^{-19}$ C. Thus we obtain for t a value of about 5×10^{-14} s. With the electrons having random velocities of the order of $3 \times 10^5\ \text{m s}^{-1}$, the electrons will travel about 15×10^{-9} m between collisions. This means electrons travelling about 100 atom lengths before suffering a collision. This seems highly improbable in the densely packed structure of a metal. We must consider the electron as a wave and its progress along a wire as the passage of a wave past scattering centres if we are to develop a better explanation. Some idea of this approach can be gained by considering an electron wave passing through an array of scattering centres. The condition for the 'reflection' of a wave from an array is the Bragg equation (see chapter 6).

$$2d \sin \theta = n\lambda$$

d is the spacing between the scattering centres, θ is the angle between the direction of the incident wave and the 'reflecting' planes of scattering centres, n is an integer 0, 1, 2, 3, etc., λ is the wavelength. Thus for waves incident normally on the scattering planes, reflection will occur for those wavelengths which satisfy the Bragg equation, i.e., $2d = \lambda_1$, $2d = 2\lambda_2$, $2d = 3\lambda_3$, etc. For electrons with a wavelength not equal to $2d$ no reflection occurs and the electrons can move freely through the lattice.

The wavelength of an electron is given by

$$\lambda = \frac{h}{mv} \quad \text{(see chapters 7 and 13)}$$

and as the energy is given by $\frac{1}{2}mv^2$ we can write for the energy

$$\text{Energy} = \frac{h^2}{2m\lambda^2}$$

and thus electrons with an energy less than or greater than

$$\frac{h^2}{8md^2} \approx 6\ \text{eV} \quad (\text{for } d = 2{\cdot}5 \times 10^{-10}\ \text{m})$$

will move freely through the lattice. At this energy there will be no movement, normal to reflecting planes, through the lattice of electrons. This is a 'forbidden' energy. An analysis in which all angles of incidence of waves on a lattice are considered yields the result that certain bands of energy are forbidden, i.e., electrons with these energies cannot pass through the lattice. The energies of these forbidden bands depends on the form of crystal lattice and the lattice spacing.

The electrical behaviour of a solid depends on these energy bands and the energies of the electrons in the solid. In an insulator the addition of energy to electrons would, if possible, result in the electrons in an allowable band of energies increasing their energies and having energies in the forbidden band. In a good conductor the electrons can accept the energy and remain within an allowable band of energies.

Conduction by ions

If a potential difference is applied between two inert (perhaps carbon) electrodes immersed, but not touching, in very pure water (Fig. 10.27) there is only a very minute current. If, however, a small amount of sodium chloride, or some other 'salt', is dropped in the water there is a large increase in current. We can explain this increase by an increase in the number of charge carriers. Sodium chloride breaks up into its separate ions when it is put into water (see chapter 11, polar molecules, for further information), and these provide charge carriers additional to the few ions that are already present in the water. In fact we believe that the sodium chloride is already in the form of ions in the crystal and that the effect of the water is to pull the ions apart from each other.

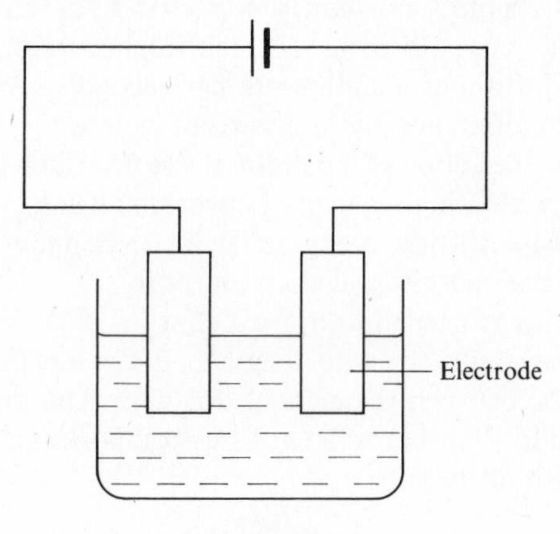

Fig. 10.27

Evidence for this last comment comes from X-ray crystallography. Sodium chloride crystals give diffraction patterns which indicate that the layers of sodium and layers of chlorine atoms scatter X-rays differently (different strength reflections); potassium chloride, however, gives diffraction patterns which indicate that the potassium layers and the chlorine layers scatter X-rays exactly the same. The atomic numbers of sodium and chlorine are 11 and 17, quite a difference. Potassium has an atomic number of 19, only two different from chlorine. If potassium was present as an ion, losing one electron to give 18, and chlorine was also an ion, gaining one electron to give 18, we could have the same number of electrons for potassium and chlorine and an explanation as to why they both, when in the potassium chloride crystal, scatter X-rays equally.

Sodium chloride in the crystal form gives only very small currents when a potential difference is applied across a block of the material, the ions are not free to move. Sodium chloride is a good insulator.

Semiconductors

Figure 10.28 shows how the conductivity of germanium changes with temperature. The conductivity increases as the temperature increases, the resistivity decreases as the temperature increases. The conduction process is often called intrinsic if it refers to pure specimens.

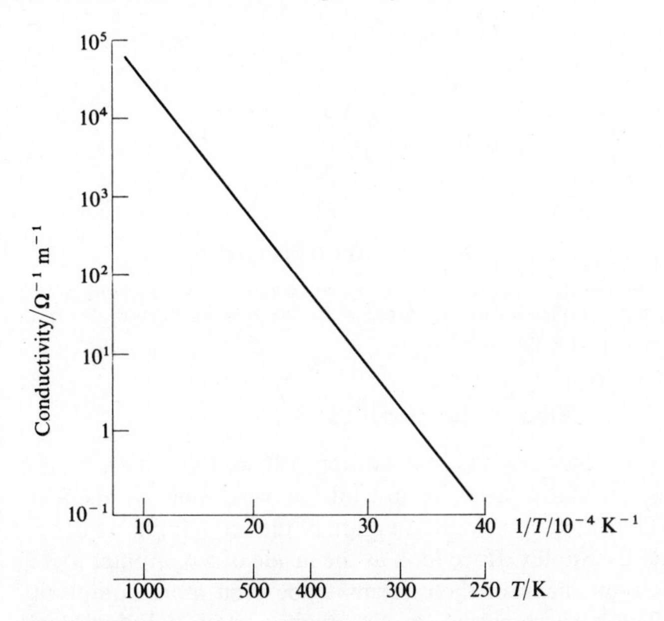

Fig. 10.28 Intrinsic conductivity of germanium. (After F. J. Morin and J. P. Maita. Phys. Rev., **94**, *1525, 1954.)*

The increase in conductivity with temperature can be explained by an increase in temperature producing an increase in the number of charge carriers. This is confirmed by measurements of the Hall effect (see chapter 11). Figure 10.29 shows how the number of charge carriers in silicon increases as the temperature increases. This is typical of semiconductors. As the graph indicates the relationship between the number density of charge carriers and temperature is of the form

$$\log n = -\frac{m}{T} + c$$

where m and c are constants. c is the value of $\log n$ when m/T is zero; we can simplify the equation by writing $\log n_0$ for c.

$$\log n = -\frac{m}{T} + \log n_0$$

or

$$\log\left(\frac{n}{n_0}\right) = -\frac{m}{T}$$

m is the gradient of the graph. What physical quantity, or quantities, does m represent?

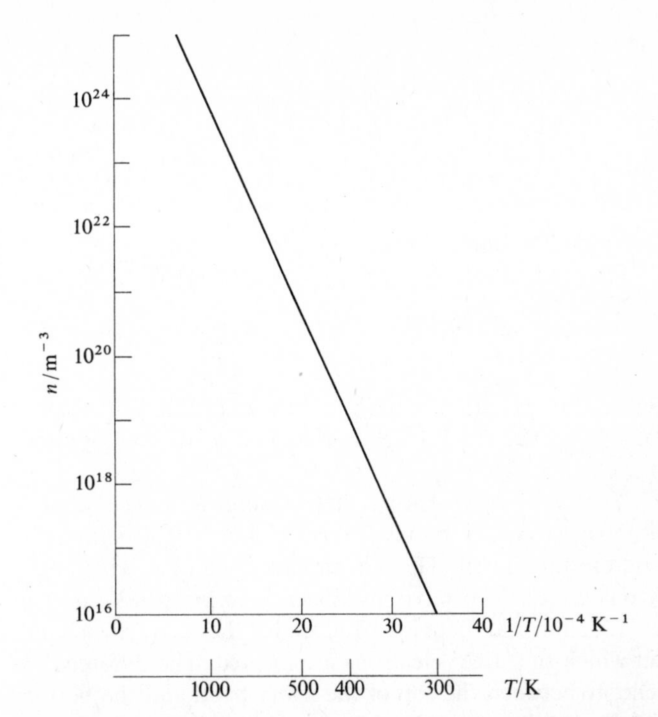

Fig. 10.29 Intrinsic carrier concentration for silicon. (After F. J. Morin and J. P. Maita. Phys. Rev., **94**., *1525, 1954.)*

When we increase the temperature we are supplying energy to the solid. If the solid is a metal then there is no change in the number of conduction electrons when we supply energy. The number of conduction electrons is, however, not the total number of electrons in a metal; for example in silver there is about one conduction electron per silver atom (see chapter 11, the Hall effect), a silver atom has, however, 47 electrons. Thus when silver atoms are bound together into a solid we appear to release some electrons from the atoms to such an extent that they can take part in the conduction process. The remaining electrons appear to remain tightly bound to the silver atoms. With a semiconductor the number of charge carriers changes when we supply energy and raise the temperature. At room temperature we may have released about one charge carrier per million of semiconductor atoms (see Hall effect, chapter 11). Energy is needed to pull electrons away from atoms. n/n_0 is the ratio of the number of atoms with energy greater than this release energy value to the total number of atoms in the same volume of solid.

In chapter 8 we arrived at the equation

$$\frac{n_2}{n_1} = e^{-E/kT}$$

where n_2/n_1 is the ratio of the number of particles in energy levels which differ by E. k is Boltzmann's constant. This is not quite the same as our ratio n/n_0 but if $n_0 \gg n$ it may be a reasonable approximation. Thus

$$\frac{n}{n_0} = e^{-E/kT}$$

or

$$\ln\left(\frac{n}{n_0}\right) = -\frac{E}{kT}$$

The gradient m becomes E/k, where E is the energy needed for an electron to escape from an atom, still, however, remaining within the solid.

The gradient of the graph for silicon, Fig. 10.29, is about

$$\frac{\ln 10^{24} - \ln 10^{17}}{20 \times 10^{-4}}\left(=\frac{E}{k}\right)$$

when the logarithm is used for the vertical axis. Taking k as $1{\cdot}38 \times 10^{-23}$ J K^{-1} gives E as $13{\cdot}2 \times 10^{-20}$ J or about 0·7 eV.

A way of representing these results is to use what is known as energy band pictures. Figure 10.30 shows the picture for silicon. There is an energy band in which electrons exist which have insufficient energy to take part in conduction and there is a band, called the conduction band, in which the 'free' electrons are located. The difference in energy between the top of the lower band and the bottom of the conduction band is the value of E. An energy gap exists between the two bands.

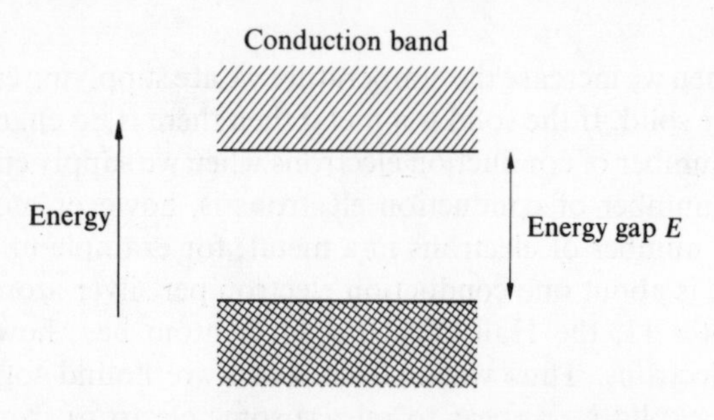

Fig. 10.30 Energy band picture for silicon.

When an electron becomes 'free' it leaves an atom short of an electron. An electron from another atom may move into this 'hole'. The 'hole' moves from atom to atom. In a pure conductor we will have as many 'holes' as we have 'free' electrons. By suitably doping the material, i.e., adding very small amounts of impurity elements, we can increase the number of 'holes' or the number of 'free' electrons. Increasing either increases the conductivity, i.e., lowers the resistivity. Figure 10.31 shows this for p-type conduction, i.e., conduction predominantly by holes, and n-type conduction, i.e., conduction predominantly by electrons.

In a metal the conduction is considered to be due entirely to the movement of electrons.

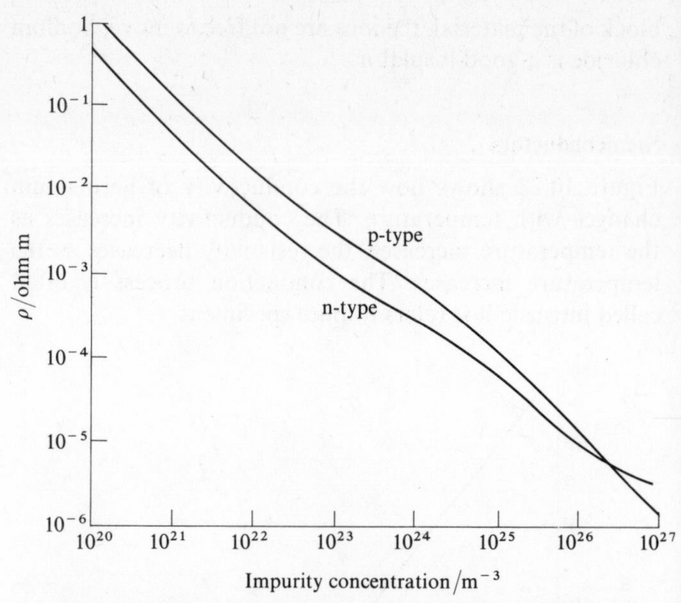

Fig. 10.31 Silicon resistivity at 300 K as a function of impurity concentration. (After J. C. Irwin, Bell System Tech. J., *41, 387. Mar. 1962.)*

10.3 Electronic circuits

Electronic circuits look complex. If we look in say a transistor radio set there are lots of wires and components. If we look at a circuit diagram of the set, it looks neater but still complex. If we look at the inside of a computer and its circuit diagram there seems to be even more complexity. There are, however, in all circuits certain building blocks with simple functions. We can by looking for these building blocks discern the method by which a circuit performs its function. We can also do the reverse job and by putting together appropriate building blocks assemble a system to do a job.

Figure 10.32 shows the breakdown into building blocks of a simple radio set. Each block in some way modifies the input signal to give the required output. The tuner has as input a large number of radio frequencies, modulated by the audio signals (see chapter 12), and gives as output just one narrow radio frequency band, modulated by the audio signal. We are thus able to select the radio station we wish to listen to.

Figure 10.33 shows the breakdown of a nuclear ratemeter. From an input to the Geiger tube of random radiation packets, random electrical pulses of various sizes are produced. The next block chops all the pulses down to the same size. After amplification the pulses reach the integrator unit. Here the pulses are summed over some definite

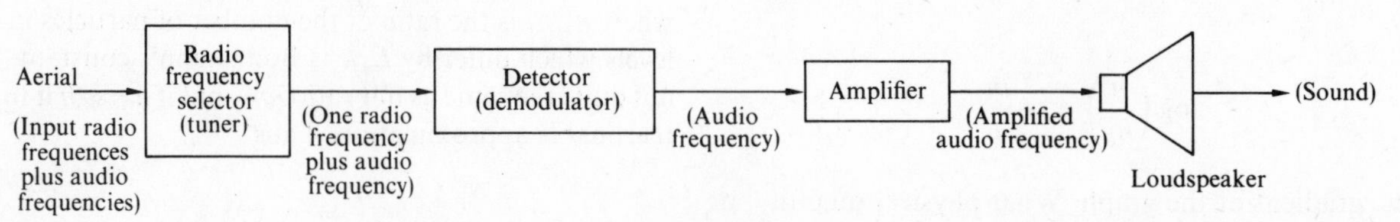

Fig. 10.32

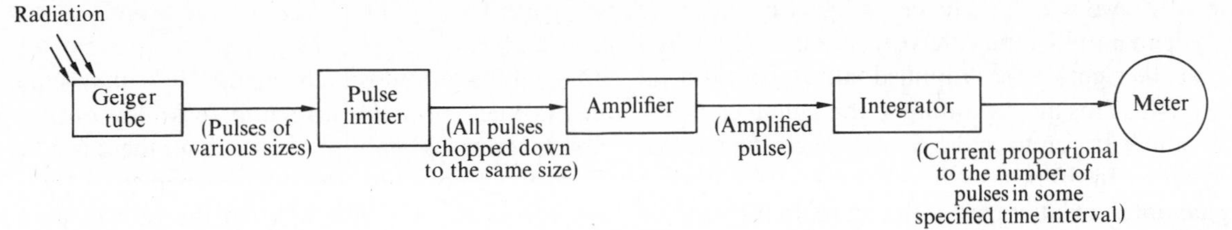

Fig. 10.33

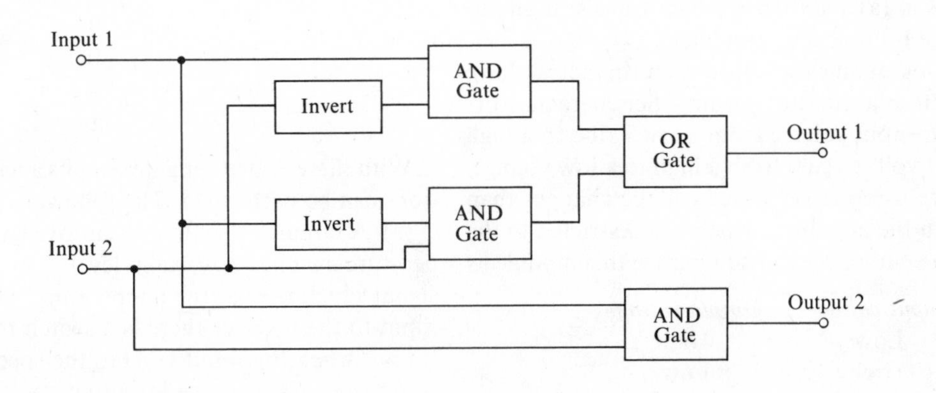

Input 1	Input 2	Output 1	Output 2
0	0	0	0
1	0	1	0
0	1	1	0
1	1	0	1

Fig. 10.34 A Binary adder circuit.

time interval and the output current, proportional to the number of pulses, fed to a meter.

Figure 10.34 shows a binary adder circuit. High (1) or low (0) pulses are supplied to the two inputs and the binary sum displayed at the two outputs. For example, suppose we have a high input at input 1 and a low one at input 2. The upper AND gate receives from input 1 a high signal—this is a direct line—and from input 2 a high signal—the low input has been inverted to give a high input to the gate. Thus because both inputs to the gate are high there is a high output. The second AND gate receives only low inputs and so gives a low output. Because one of the inputs to the OR gate is high there is a high output and so output 1 registers a high output. The third AND gate receives a high input from input 1 and a low input from input 2. As both 1 and 2 are not high there is a low output. For a high output from an AND gate both inputs, 1 *and* 2 must be high. For a high output from an OR gate either input, 1 *or* 2, must be high.

Let us take a closer look at some of these building blocks. If we note how the output from an amplifier varies as the input is changed we can determine the characteristic of an amplifier, i.e., its amplification and any conditions which have to be met to achieve amplification. Figure 10.35 shows some typical amplifier characteristics, Fig. 10.35(d) showing the basic module. In all the cases (a), (b), and (c) there are portions of the characteristic where a small change in input voltage gives rise to a large change in output voltage. In (a)

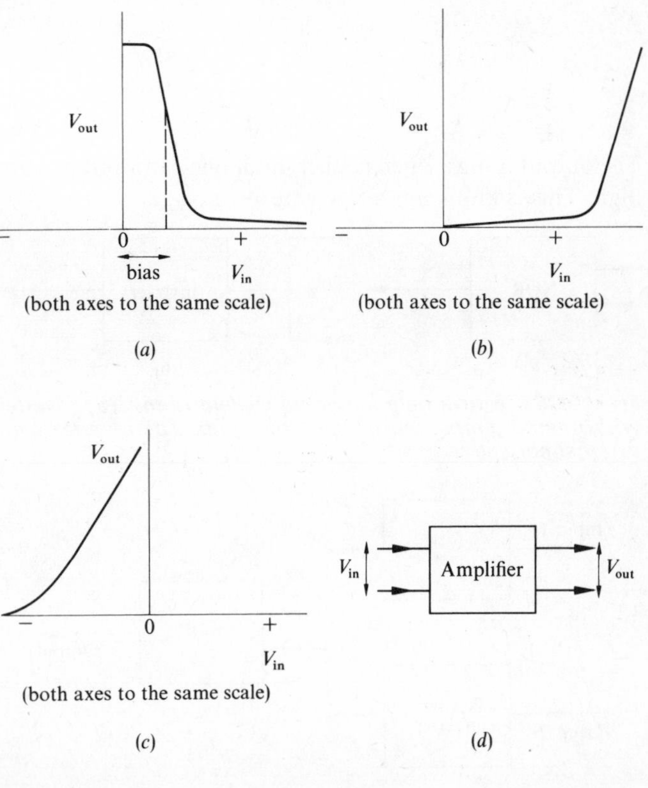

Fig. 10.35

for amplification we must use the central part of the curve and thus such an amplifier must have its input signal biased in order that the signal to be amplified falls on this central part of the curve. This means adding to the signal a positive voltage of sufficient size to move the signal to this central part of the curve. In (b) positive bias again must be used to get the signal onto the latter part of the curve. In (c) negative bias must be used to get the signal on the appropriate part of the curve.

The amplifiers in (a) and (b) are in fact transistor amplifiers while that of (c) is a valve amplifier.

Now take a look at an electronic switch. In fact we have already met one in Fig. 10.35(a), an amplifier characteristic. If we change our input voltage from a low value to a high value our output voltage goes from a high to a low voltage. This time we are using larger input voltage changes than was the case with the amplifier, which was restricted to the centre part of the curve. We can summarize this module as

Input voltage	*Output voltage*
Low	High
High	Low

If we signify a low signal by 0 and a high signal by 1 we have

Input	*Output*
0	1
1	0

This characteristic is that of an inverter.

Consider a version of the inverter where we have two inputs and just one output (Fig. 10.36) giving the following

Input one	*Input two*	*Output*
0	0	1
1	0	0
0	1	0
1	1	0

The output is high when neither input one *nor* input two are high. This is known as a NOR gate.

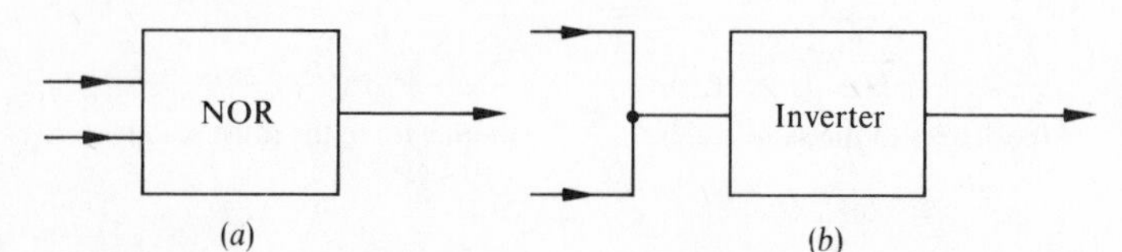

Fig. 10.36 A NOR gate in two equivalent forms. (a) Inverter with internal correction of the two inputs. (b) Inverter with external correction.

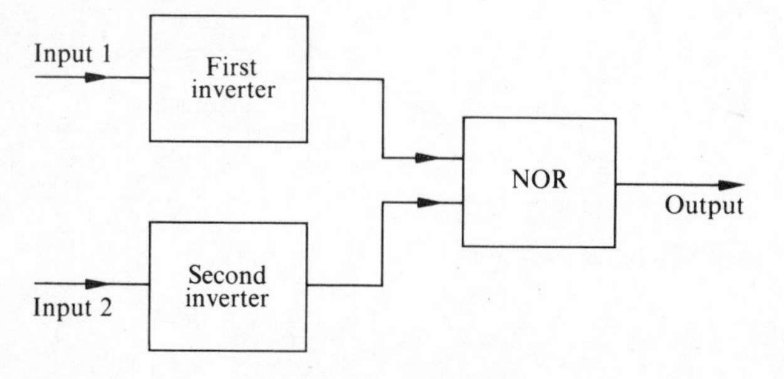

Fig. 10.37 AND gate.

By using two of our inverter circuits and the NOR gate we can make an AND gate, i.e., when input one *and* input two are high the output is high. Figure 10.37 shows the arrangement. When input one is high there is a low output from the first inverter. When input two is high there is a low output from the second inverter. Only when both these outputs are low is there a low input to the NOR gate and a high output.

Input one	*Input two*	*Output*
0	0	0
1	0	0
0	1	0
1	1	1

With these and other gate circuits a wide variety of operations can be performed. The following are examples.

(a) A circuit to switch off an oven when the oven temperature reaches a certain level. This requires an input signal which rises as the temperature rises. When this is the input to the inverter there is a switch from a high to a low output when the input reaches the specified high temperature value. This change in output could be used to control the current to the oven heating element.

(b) A NOR gate can be used to indicate when two safety locks are both operating. If the making of a safety lock gives a low signal then only when both give low signals will there be a high output.

(c) If in (b) the making of a safety lock gave a high signal then an AND gate would give a high output when both inputs are high, i.e., both safety locks made.

Combining gate circuits widens the operations that can be performed. The binary adder illustrated in Fig. 10.34 is an example. With the aid of gates we can build digital computers.

R, C, L circuits

What happens when we pass electrical pulses through circuits containing resistors, capacitors, and inductors? (See chapter 11 for details of inductors and capacitors.) Consider a generator which produces square pulses (Fig. 10.38) and the results when such pulses are fed into circuits containing these different elements. With a resistor, Fig. 10.38, the oscilloscope trace has the same shape as when the output from the generator is fed directly to the oscilloscope. If, however, we have a capacitor as well as the resistor in the circuit and examine the p.d. across the capacitor as a function of time, we find a change in the shape of the pulse (Fig. 10.39). The pulse across the resistor is also changed (Fig. 10.40).

We can explain the form of these pulses by considering the rise in the voltage pulse of the generator charging up the capacitor and the drop in the generator voltage pulse as discharging the capacitor. The potential difference across the capacitor plates is directly proportional to the charge on the plates ($Q = CV$) and thus the oscilloscope trace given by the oscilloscope when connected across the capa-

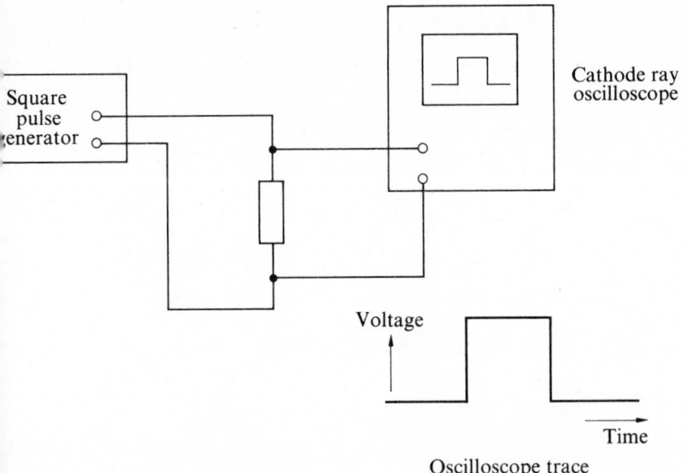

Fig. 10.38 Diagram of square pulse generator directly feeding oscilloscope.

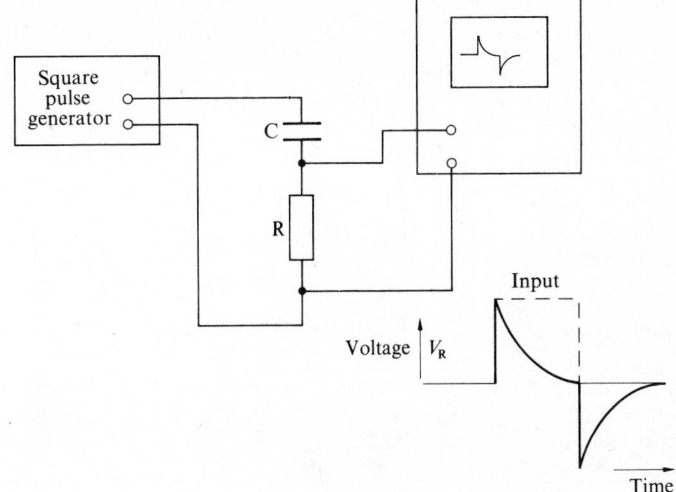

Fig. 10.40

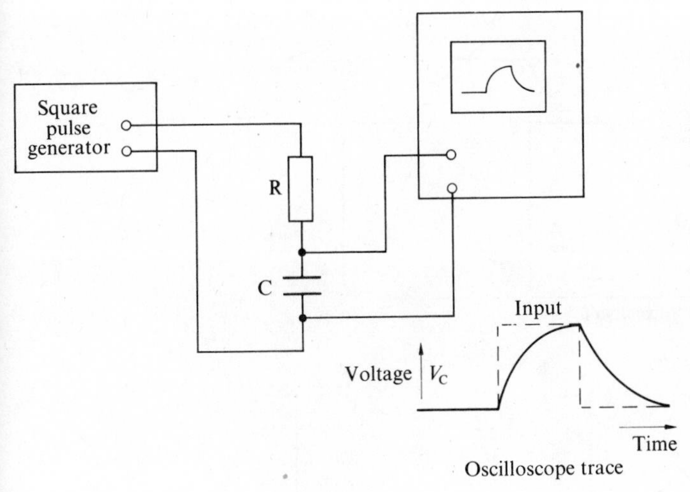

Fig. 10.39

citor shows the charge and discharge process. The sum of the voltages across the resistor and across the capacitor gives the voltage supplied by the generator (Fig. 10.41).

The product of the capacitance C and the resistance R is known as the time constant of the circuit.

$$\text{Time constant} = CR$$

If C is in farads and R in ohms the unit of the time constant will be farads × ohms, or seconds.

$$\left(CR = \frac{QR}{V} = \frac{Q}{I} = \frac{Qt}{Q} = t\right)$$

The time constant is the time taken for the voltage. the current, and the charge on the capacitor to fall to 0·3679 of its initial value or grow to 0·6321 of its final value.

$$0{\cdot}3679 = \frac{1}{e}; \quad 0{\cdot}6321 = 1 - \frac{1}{e}$$

Chapter 5 discusses growth and decay and introduces the constant e.

Figure 10.42 shows how the pulse shapes differ for different time constants. With a small time constant the pulse across the resistor consists of two sharp voltage spikes, the pulse across the capacitor consists of only a slight change from the generator pulse. The pulse across the resistor is, if Ohm's law is obeyed, proportional to the circuit current.

$$V_R = RI$$

But

$$I = \frac{dQ}{dt}$$

Thus

$$V_R = R\frac{dQ}{dt}$$

But

$$Q = CV_C$$

Thus

$$V_R = RC\frac{dV_C}{dt}$$

The pulse seen on the oscilloscope when it is connected across the resistor is the differential of the pulse seen when the oscilloscope is connected across the capacitor. With a small value of CR the pulse across the capacitor is almost the same as that given by the generator. Thus with a small time constant the output pulse across the resistor is approximately equal to the differential, i.e., the gradient, of the generator pulse. Such a circuit is known as a differentiating circuit.

With a large time constant the pulse across the resistor is almost the same shape as that supplied by the generator, that across the capacitor is almost a steady rise with time followed by a steady fall.

$$\frac{dV_C}{dt} = \frac{1}{RC}V_R$$

$$V_C = \frac{1}{RC}\int V_R\,dt$$

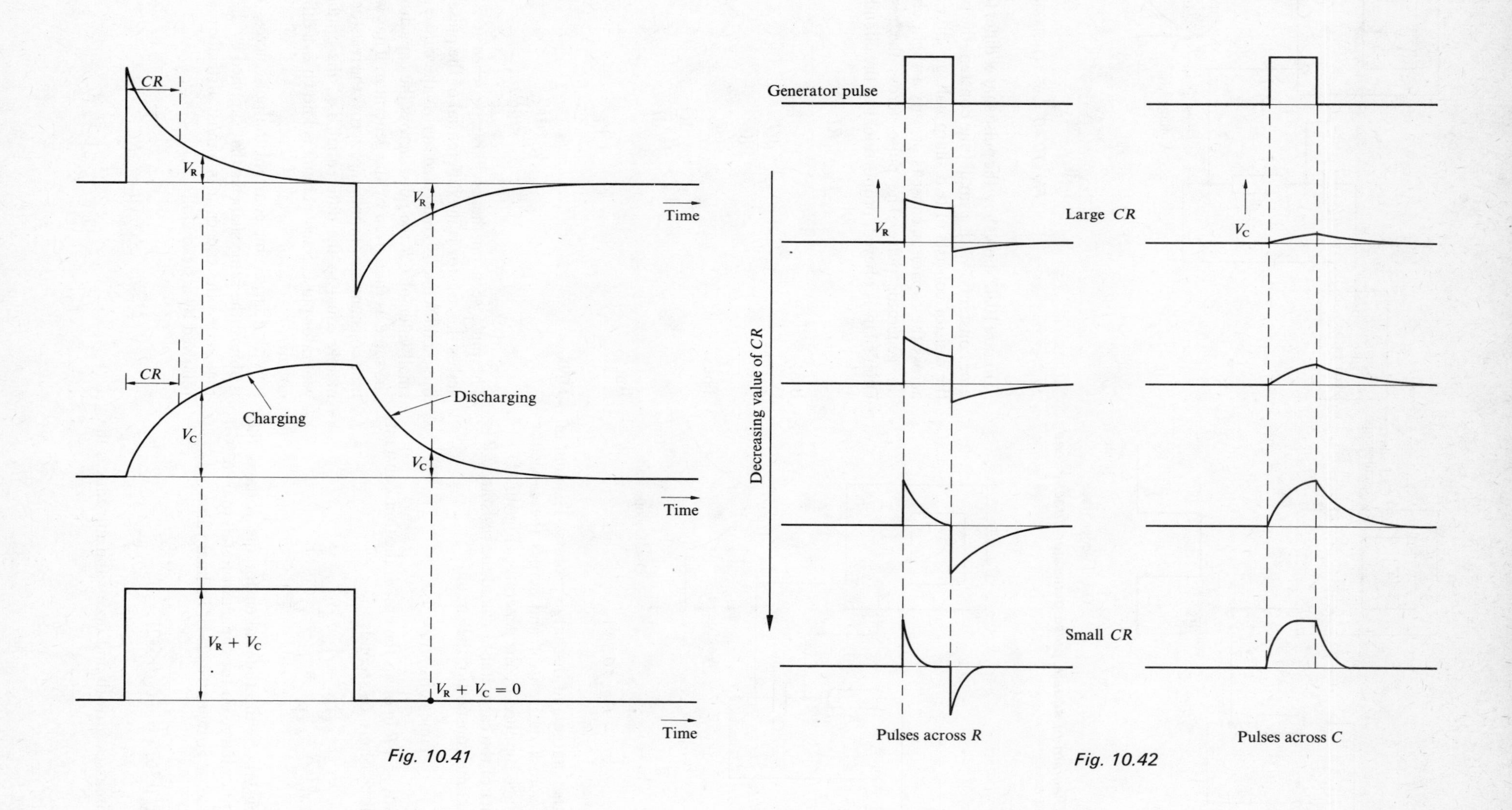

Fig. 10.41

Fig. 10.42

V_C is the integral of the variation with time of V_R. As with a large time constant V_R varies with time in almost the same way as the generator pulse, the output across the capacitor is roughly the integral, i.e., the area under the voltage–time curve, of the generator pulse. Such a circuit is known as an integrating circuit. With the pulse the integration is about the mean voltage level of the pulse, see Fig. 10.43.

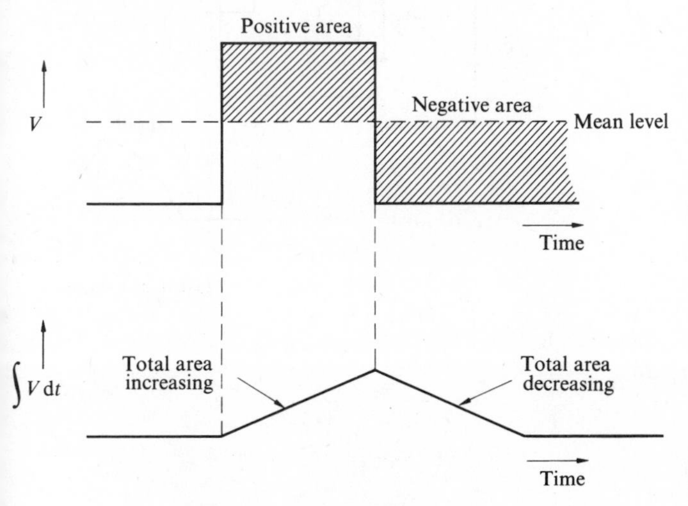

Fig. 10.43

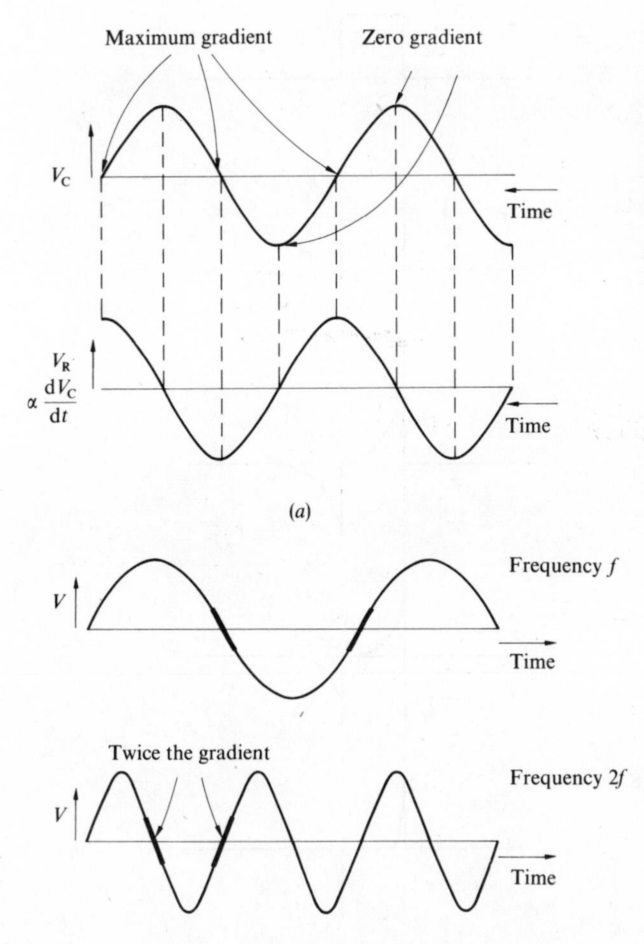

Fig. 10.44

An application of the integrator circuit has already been met in the ratemeter circuit. Another application is to produce a steadily varying voltage to drive the time base of an oscilloscope. Another application is in analogue computers where a differential equation is to be solved by being integrated.

Figure 10.44 shows the results of supplying a sine wave input voltage, instead of a square pulse, to the resistor and capacitor circuit. The output across R is proportional to dV_C/dt, i.e., the gradient of the V_C variation with time. The potential difference across the capacitor is a maximum when the p.d. across the resistor is zero (as $V = IR$ the current is zero) and zero when the p.d. across the resistor is a maximum (current a maximum). The size of the maximum p.d. across the resistor depends on the maximum rate of change of voltage across the capacitor. This in turn depends on the frequency of the voltage, see Fig. 10.44(b). Twice the frequency means that the voltage change takes place in half the time. Thus twice the frequency means twice the gradient and thus twice the maximum value for V_R. As V_R is proportional to the current in the circuit this means that doubling the frequency doubles the maximum current. The maximum current in the circuit is thus proportional to the frequency. A circuit containing a resistor and a capacitor in series thus gives bigger currents at high frequencies than it does at low frequencies.

Figure 10.45 shows the circuit of what is called a 'low pass' filter. The resistor and capacitor arm of the circuit takes more of the high frequency current than the load arm, the load arm takes more of the low frequency current than the resistor and capacitor arm. The low frequencies thus pass through the load. Such a circuit is used in a radio, being placed across the loudspeaker, so that the loudspeaker responds only to the low frequency audio signal. Another application is to 'smooth' a mains rectified voltage. The low pass filter is placed in parallel with the output terminals of the mains rectifier and so filters out of the signal the high frequency components—hence giving a smoother, more steady, d.c. output.

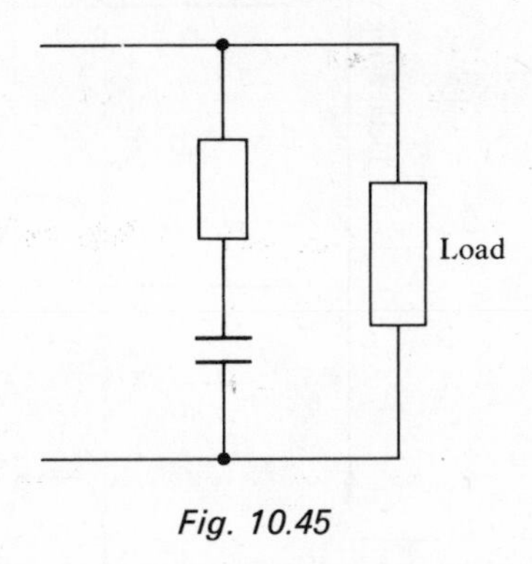

Fig. 10.45

Instead of combining a capacitor with a resistor we can use an inductor with a resistor. Figure 10.46 shows what happens when a square pulse is fed into such a circuit. The sum of the voltages across the resistor and across the inductor gives the voltage supplied by the generator. The

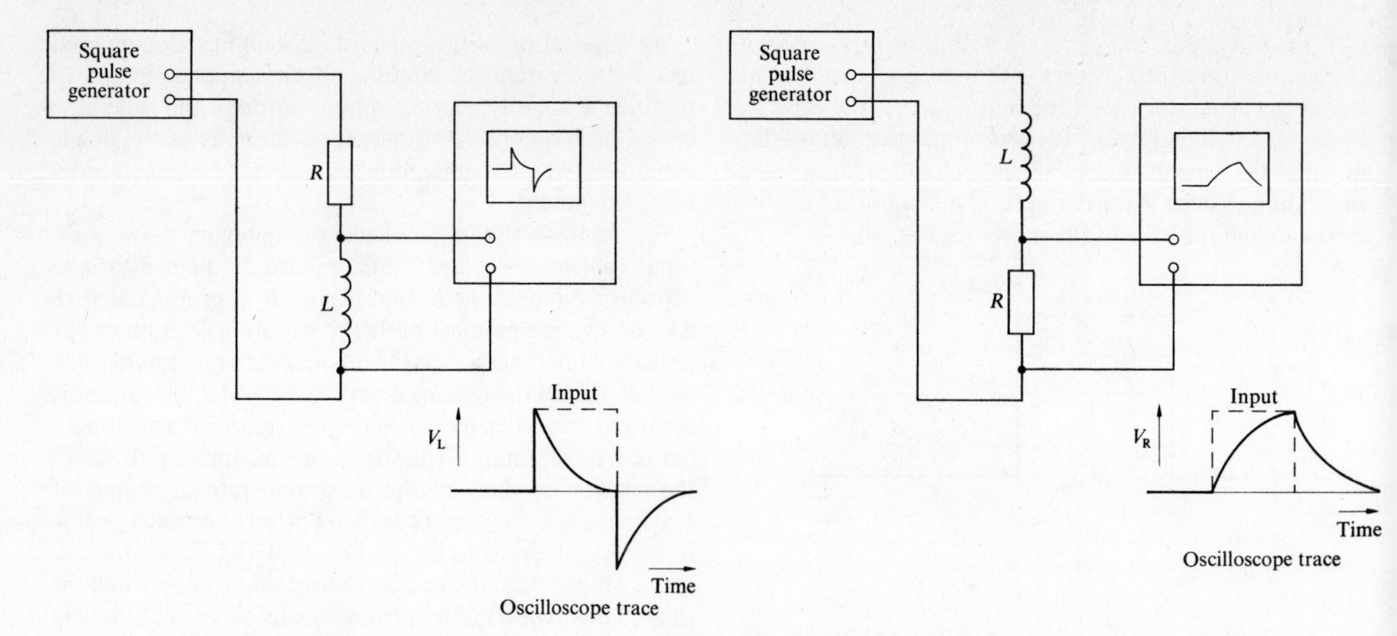

Fig. 10.46

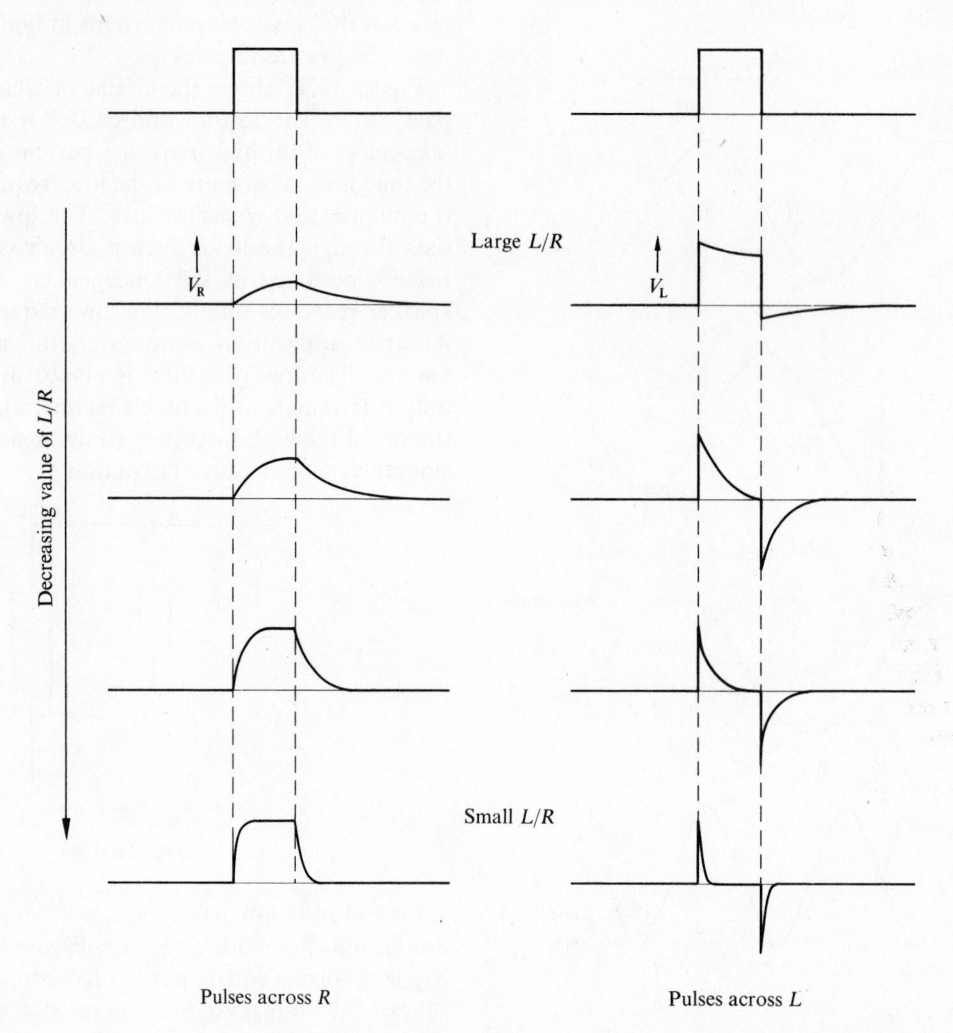

Fig. 10.47

voltage across the resistor shows that the current in the circuit first grows with time and then decays exponentially. The time constant for this circuit is L/R, where L is the value of the inductance. If L is in henries and R in ohms, the time constant is in seconds. Figure 10.47 shows the effects on the pulse of using different time constants. For a large time constant the voltage across the resistor rises almost linearly with time and then decays almost linearly, the voltage across the inductor is almost the same as that supplied by the generator. With a small time constant the voltage across the inductor is in the form of two sharp spikes, that across the resistor is almost the same as that supplied by the generator. Figure 10.48 shows the results when a sinusoidal signal is used instead of the square pulse. The potential difference across the inductor is a maximum when the gradient of the p.d. across the resistor is greatest. When the gradient of the V_R graph is zero the p.d. across the inductor is zero. The p.d. across the resistor is proportional to the current in the circuit, thus we have

$$V_L \propto \frac{dI}{dt}$$

V_L

Zero gradient

V_R

Maximum gradient

Fig. 10.48

(or $V_L = L(dI/dt)$, where L is called the inductance), or in terms of V_R

$$V_L \propto \frac{dV_R}{dt}$$

Thus when we have a small time constant and the voltage across the resistor is almost the same as that at the generator, the p.d. across the inductor is almost the differential of the input. This is another form of a differentiation circuit. The voltage across the resistor is the integral of that across the inductor. This applies regardless of the form of the input.

$$V_R \propto \int V_L \, dt$$

Thus when the time constant is large and the voltage across the inductor is almost the same as that supplied by the generator, the voltage across R is almost the integral of the input—another integration circuit.

What happens if we use a sinusoidal input voltage and increase the frequency? The current in the circuit is proportional to V_R and thus

$$I \propto \int V_L \, dt$$

The integral is the area under the V_L variation with time curve. The maximum area will be that of one entire half wave, see Fig. 10.49. Doubling the frequency halves the maximum area. Thus increasing the frequency decreases the maximum current. The maximum current is inversely proportional to the frequency. A circuit containing a resistor and an inductor in series thus gives bigger currents at low frequencies.

What happens if we put a capacitor in parallel with an inductor and apply a sinusoidal voltage signal (Fig. 10.50)? At low frequencies we will have a large current in the inductor arm and a small current in the capacitor arm. At high frequencies we will have the reverse, a small current in the inductor arm and a large current in the capacitor arm. At some intermediate frequency we will obviously have equal currents in each arm. The currents in the two arms are half a cycle out of phase, see Fig. 10.51. (This is only completely true if there is no resistance in either arm. Provided the resistance is low this is a reasonable approximation.)

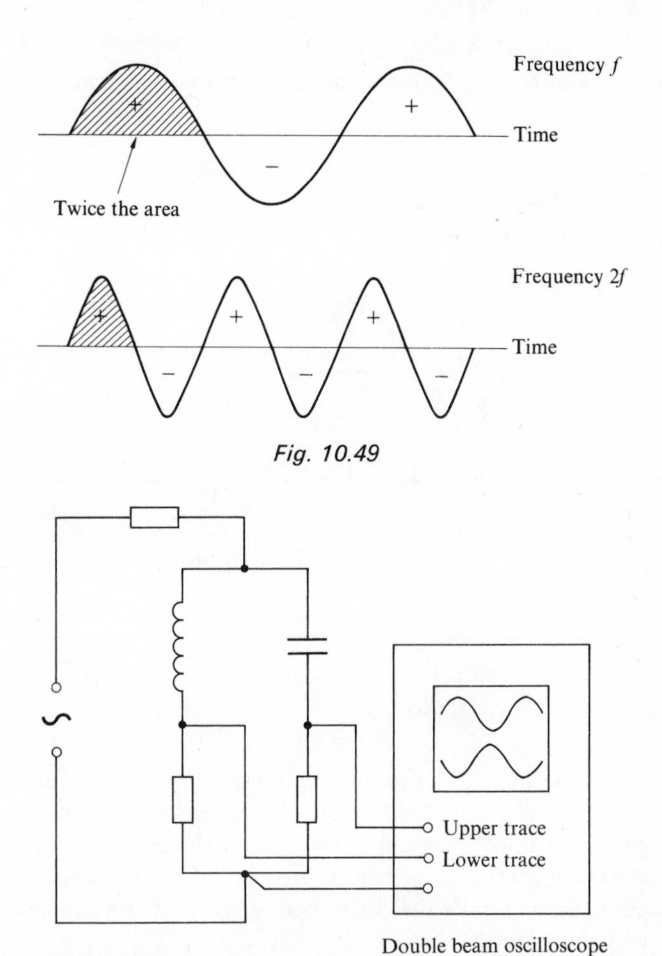

Fig. 10.49

Fig. 10.50

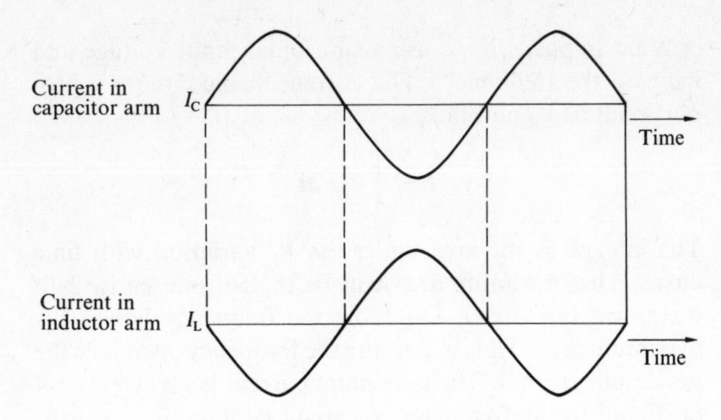

Fig. 10.51 Out of phase currents.

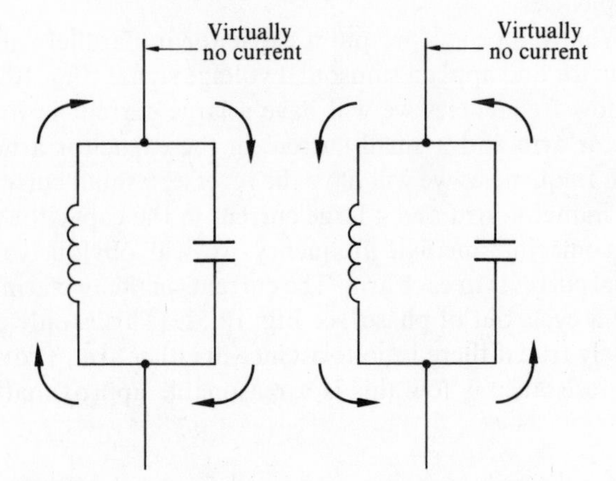

Fig. 10.52 Current directions in resonant circuit.

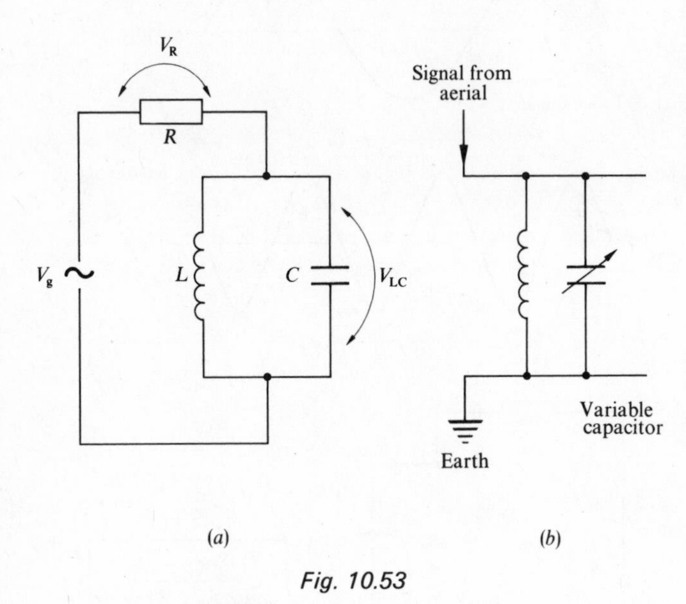

Fig. 10.53

When the current is a maximum in one arm of the circuit it is a minimum in the other arm. The current supplied by the generator must be equal to the sum of the currents in the capacitor and inductor arms at any instant. If these currents are opposite and equal there is no current being supplied by the generator.

$$I = I_L + I_C$$

If $-I_L = I_C$ then $I = 0$. With little resistance in either circuit arm these two out of phase currents are almost equal and very little current is taken from the generator. The frequency at which this occurs is called the resonant frequency.

How can we have large currents in the capacitor and in the inductor arms of the circuit and yet have very little current being supplied by the generator? The clue to this problem is in the out of phase nature of the currents in the two circuit arms. When the current is going one way in the capacitor arm it is going in the opposite direction in the inductor arm (Fig. 10.52). The current seems to be circulating within the capacitor–inductor circuit and not passing out or passing in from the rest of the circuit. The charge is just passing backwards and forwards between the capacitor and the inductor.

For the potential differences in the circuit, Fig. 10.53(a), we must have

$$V_g = V_R + V_{LC}$$

At resonance the current through R is very small, practically zero, thus V_R must be very low ($V = IR$). Thus at resonance V_{LC} must be almost equal to V_g. This only occurs at one frequency. At all other frequencies there is a larger current through R and hence V_{LC} is much smaller. This is in fact the basic tuning circuit of a radio (Fig. 10.53(b)). Only at the selected frequency is there a high potential difference across the rest of the radio circuit.

We can calculate this frequency from a knowledge of the capacitance and the inductance. At resonance we have

$$V_C = -V_L$$

But

$$V_C = \frac{Q}{C}$$

and

$$V_L = L\frac{dI}{dt}$$

Thus

$$\frac{Q}{C} = -L\frac{dI}{dt}$$

But

$$I = \frac{dQ}{dt}$$

Thus

$$\frac{Q}{C} = -L\frac{d^2Q}{dt^2}$$

$$\frac{d^2Q}{dt^2} = -\frac{1}{LC}Q$$

We can solve this equation by the techniques used in chapters 2 and 4. In fact the equation is of exactly the same form as the simple harmonic motion equation

$$\frac{d^2x}{dt^2} = -\omega^2 x$$

The charge equation can be written as

$$\frac{\Delta(\text{slope})}{\Delta t} = -\frac{1}{LC}Q$$

The slope referred to is the slope of the Q against t graph. The solution of this is, as the simple harmonic equation, a cosine curve. The periodic time of the oscillation is $2\pi\sqrt{(LC)}$.

$$\text{Frequency} = \frac{1}{2\pi\sqrt{(LC)}}$$

In the radio tuner the required frequency is obtained by varying the capacitance.

If we pass a square pulse into our capacitor and inductor circuit (Fig. 10.54) oscillations are produced at a frequency given by the above equation. Every time we give the circuit a 'bump' it oscillates. The oscillations are, however, damped. This is because of the resistance in the circuit. If we increase the resistance the oscillations die away more rapidly. With resistance R in the capacitor–inductor circuit our equation becomes

$$V_L + V_R + V_C = 0$$

There is no source of p.d. in the circuit at resonance. Hence we have

$$L\frac{d^2Q}{dt^2} + R\frac{dQ}{dt} + \frac{Q}{C} = 0$$

Chapter 4 gives a solution of a similar damped oscillation equation.

The circuit of Fig. 10.55 can be used to convert a sinusoidal variation of voltage with time into a square wave variation. This depends on the characteristic of the unit being like that of Fig. 10.35, i.e., a step-like characteristic.

We can use the inverter unit, already referred to in this section, to produce a pulse output when an input is switched.

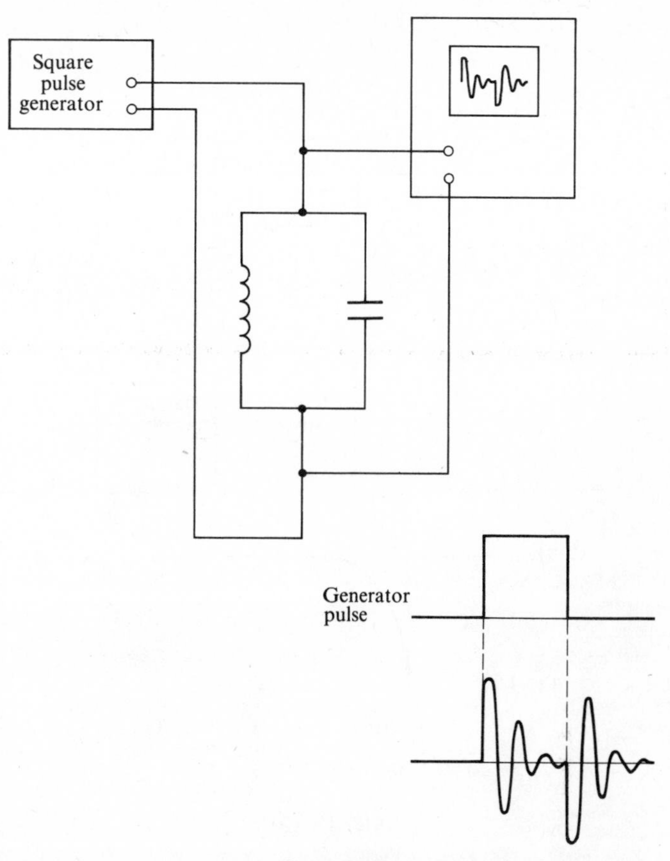

Fig. 10.54

We need a resistor and a capacitor circuit to give a voltage input to the inverter which changes from high to low and then back to high, giving an output which goes low to high to low. Figure 10.56 shows the circuit. (a) is the initial condition, both the inputs being high and thus there being a low output. When the capacitor input is switched to low, (b), the input drops to low and then rises exponentially to high. The output thus rises to high and remains at high until the input has sufficiently risen to switch the output back to low. The output is thus a pulse, the size of which depends on the value of CR.

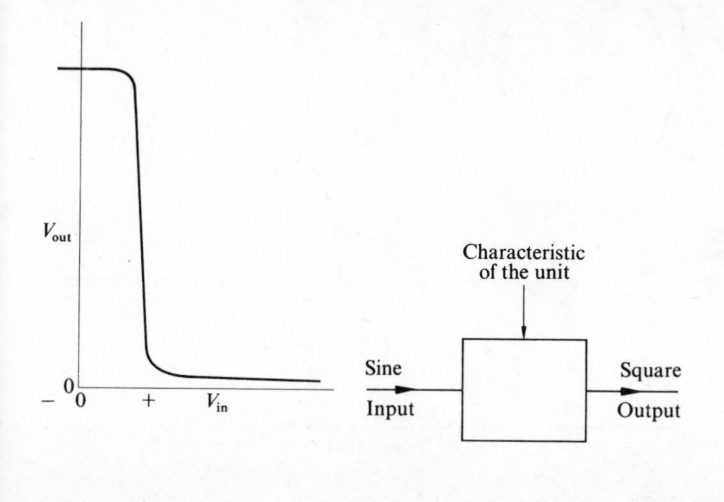

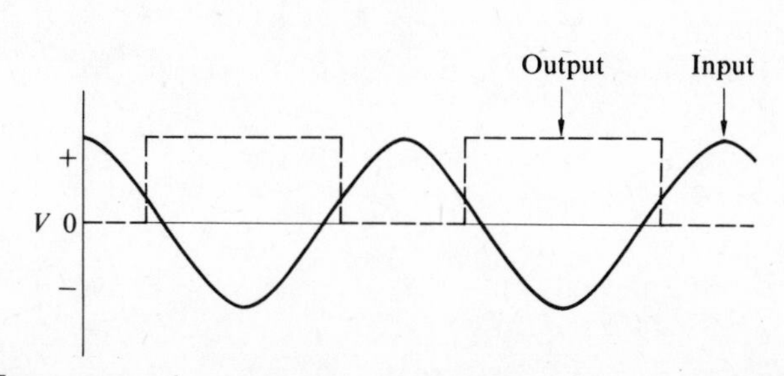

Fig. 10.55

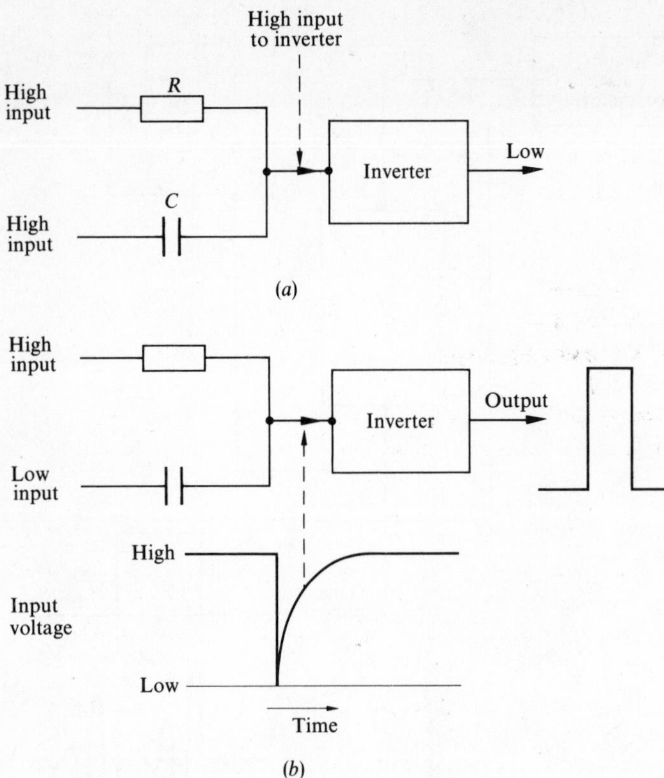

Fig. 10.56

Electrical pulses are the means by which we as human beings function. You prick a finger—electrical pulses race along the nerves, some perhaps taking a short cut to your muscles to give a quick reflex action, others going on to the brain. From the brain electrical pulses are sent out along nerves to the appropriate muscles—the result is for you to pull your finger away from the point. The greater the intensity of the stimulus to a nerve ending the higher the pulse frequency. Your reaction time is the time taken for the pulses to race along the nerves, be processed, and then race to the appropriate muscles—a time of the order of 0·1 s.

Telecommunications

The first transatlantic cable was laid in 1857–8; only some 700 messages were, however, transmitted before it broke. 1866 saw the next completed cable across the Atlantic. These early cables were for the transmission of pulse signals, not speech. The first telephone cable was not laid across the Atlantic until 1956. Since that time more long-distance cables have been laid—Fig. 10.57 shows the situation. Telephones were first used, for speech transmission over a very short distance, in 1876—why the delay before speech could be transmitted over a transatlantic cable? What were the difficulties?

The first transatlantic cable was made of seven copper

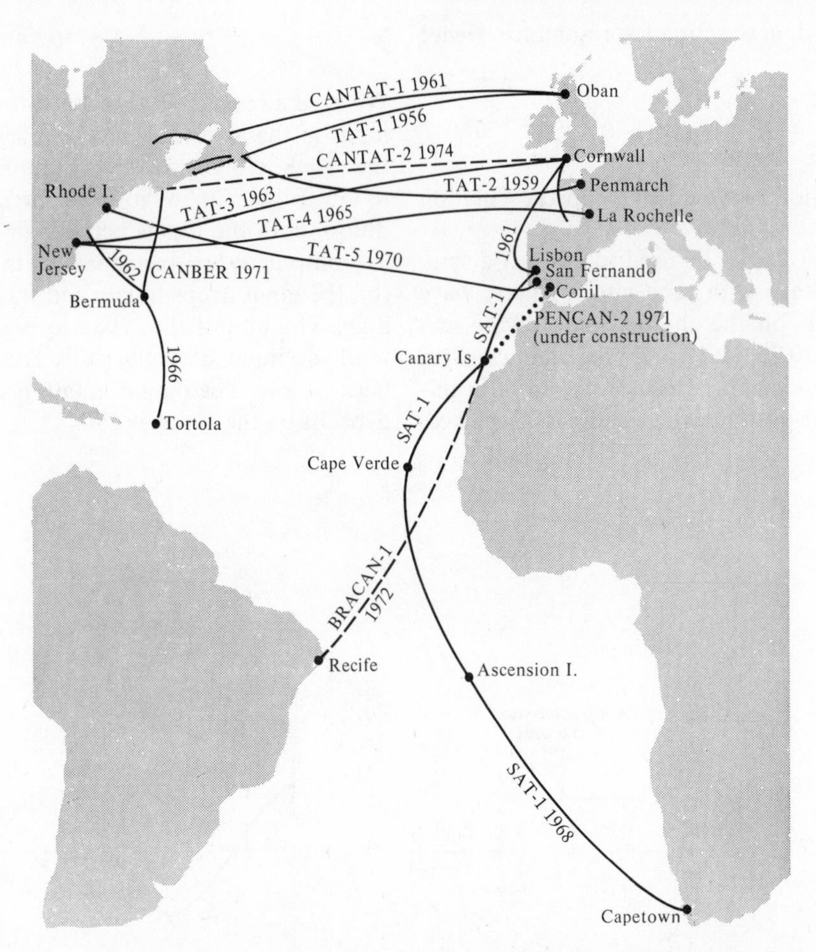

Fig. 10.57 R. Brown. This map first appeared in 1971 in New Scientist, *London.*

wires, about 1 mm in diameter, covered in a mixture of hemp, tar, and wax and enclosed in an armouring of eighteen strand iron wires. The distance the cable covered was about three thousand kilometres. The copper wire in such a length would have a resistance of about $9 \times 10^3\ \Omega$. The capacitance of such an arrangement would be of the order of 20 μ F. This gives a time constant CR of about 0·2 s. (The inductance would be about 1 H and L/R about 5×10^{-6} s and thus insignificant by comparison with 0·2 s).

The cable behaves like a capacitor which needs about 4 s for a pulse to grow, or decay, by about one-third. For a pulse passing along such a cable there would have to be a growth time at the beginning of the pulse of the order of 3×4 s and a similar time at the end of the pulse for it to decay. The minimum length of a pulse would thus have to be about 24 s. Thus only a small number of pulses could be transmitted in any one minute. To pass a message, in the form of pulses of differing lengths, along the cable would take many minutes. The transmission of audio frequency signals or audio modulated signals was therefore not possible—the time interval between wave crests would be much less than CR.

The capacitance of the cable is proportional to the length D of the cable.

$$C \propto D$$

The resistance of the cable is proportional to the length.

$$R \propto D$$

Thus the time constant CR is proportional to D^2.

$$CR \propto D^2$$

The improvement which enabled the 1956 telephone cable to operate successfully was the insertion in the cable of repeaters at regular intervals. Fifty-one repeaters were used in the transatlantic cable. The repeaters, as the word suggests, repeat the signal with some amplification. This effectively reduces the length of the cable and so the time constant decreases by a factor of about 5^2 to about $1{\cdot}5 \times 10^{-3}$ s. The 1956 cable was also of different construction; the copper had three times the diameter of the first cable and thus only 1/9 of the resistance. The cable is thus able to operate at a frequency of 144 kHz. The audio signals are used to modulate this high frequency signal.

The future of telecommunications by cable may, however, lie with the use of pulse code modulation. This is the conversion of the audio signal to a sequence of pulses for transmission over the cable, converting back to the audio signal at the receiver. The first lines using this were introduced in 1960.

Feedback

A rise in the general cost of goods causes demand for an increase in wages, which if met causes an increase in manufacturing costs, which in turn results in an increase in the cost of goods. The cost of the goods thus spirals upwards—there is inflation.

Oil is fed into a central heating furnace. This results in an increase in the air temperature in a room. A thermostat in that room responds to the temperature change and controls the oil flow. Changing the oil flow changes the room temperature and so causes the thermostat to change its response. The air temperature oscillates about the value set by the thermostat.

Power stations in a city produce air pollution, which means that more people buy air conditioners, which need more power, which means more power stations which means more air pollution, which means more air conditioners, which means yet more power, and so on.

These are all examples of feedback. We have a stimulus which produces a response which feeds back a signal to modify the stimulus. In some cases the return signal reinforces the stimulus, as in the price of the goods, and is known as positive feedback. If the return signal produces a decrease in the stimulus it is known as negative feedback.

Feedback has many applications in electronics. For example, consider a circuit made up of two of the inverters referred to earlier in this chapter (Fig. 10.58). A high input to the first inverter produces a low input to the second inverter and a high output. A low input to the first inverter produces a high input to the second inverter and a low output. What happens if we take the output from the second inverter and use it as the input to the first inverter (Fig. 10.59)? A high input to the first inverter produces a low input to the second inverter which produces a high input to the first inverter which produces a low input to the second inverter, and so on. Once the first input has been made high the feedback will maintain the high state. This is positive feedback. This arrangement is known as a bistable circuit: it 'remembers' what was last done to it. If the input to the first inverter was last made high it remains high, if it was last made low it remains low. The circuit has two stable

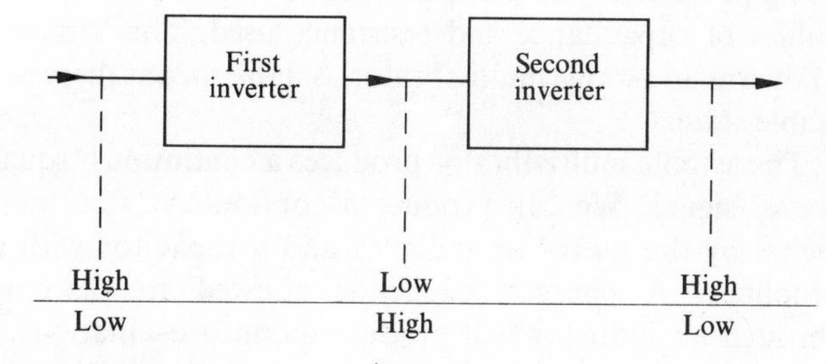

Fig. 10.58

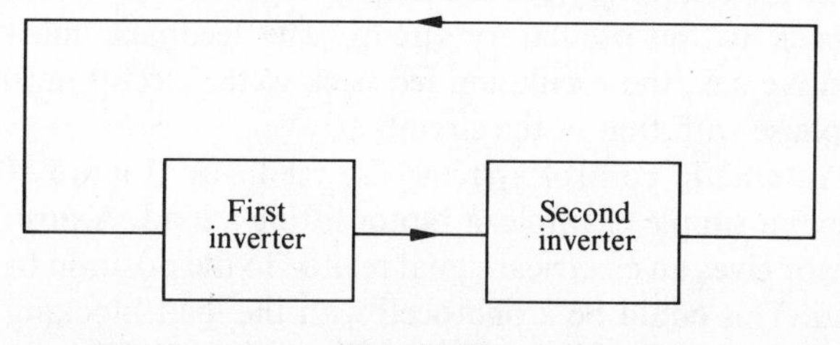

Fig. 10.59

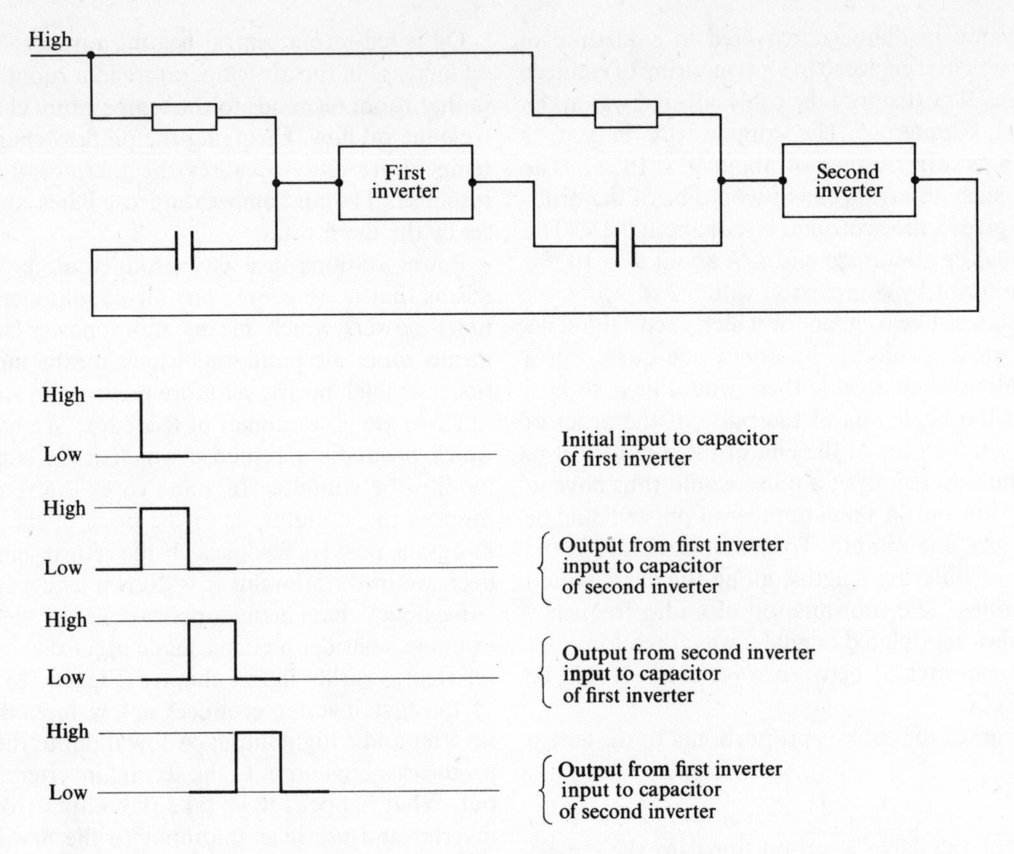

Fig. 10.60

states, the output from inverter one low and from inverter two high or that from one high and two low. This is one form of a memory unit used in computers.

If instead of just using two inverters we use the modified inverters which give pulses when the input to a capacitor drops from high to low we can produce a continuous train of pulses (Fig. 10.60). Once the first pulse has been produced the process continues, a series of equally spaced pulses being produced. The duration of the pulses depends on the values of capacitance and resistance used. This circuit is known as an astable multivibrator. Astable means there is no stable state.

The astable multivibrator produces a continuous 'square wave' signal. We can produce a continuous 'sine wave' signal by the use of an inductor and a capacitor with an amplifier. A charged capacitor allowed to discharge through an inductor will produce damped oscillations. If the oscillations are to be continuous energy must be fed into the oscillatory circuit to make good the losses. This can be achieved (Fig. 10.61) by extracting some of the energy from the oscillatory circuit, amplifying it and then feeding it back to the oscillatory circuit. The feedback must be positive, i.e., the oscillation fed back to the circuit must be in phase with that in the circuit.

Automatic control systems use feedback. Figure 10.62 shows a simple example, a motor lifting a load. A position sensor gives an electrical signal related to the position of the load. This could be a photocell with the load blocking off part of a beam of light which falls on the cell. The output signal from the sensor is then subtracted from a reference input signal. The difference, in these signals is a measure of how far off position the load is. When the load is in the correct position there is no difference between the two signals. This difference is amplified and then used to control the motor and so move the load that the difference is reduced. The position of the load is thus controlled and can be made to follow a prearranged programme. The input reference signal is made to change and so move the load through the

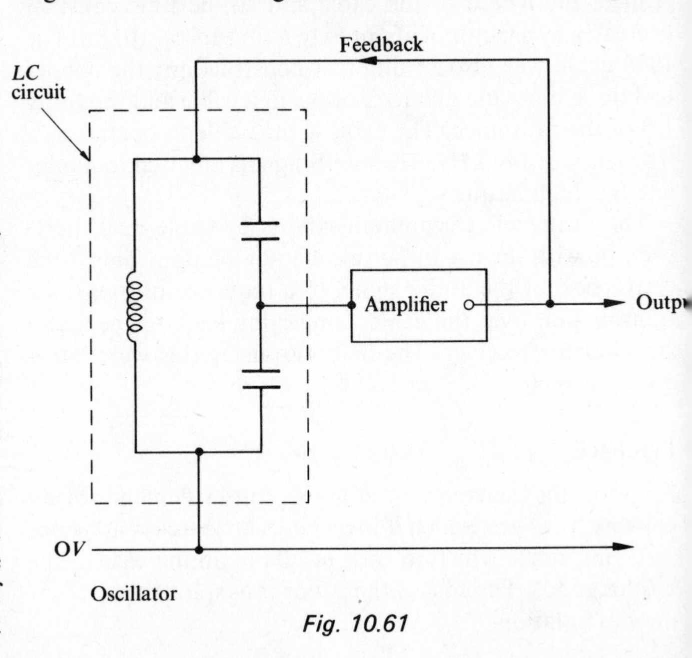

Fig. 10.61

required moves. Similar systems are used to control tools cutting or machining materials, often to intricate patterns.

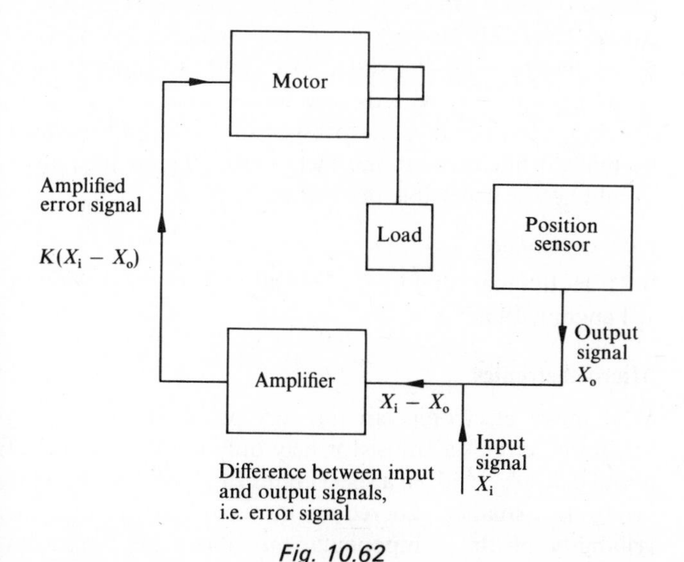

Fig. 10.62

A stimulus producing some response which then supplies a signal to modify the stimulus is a description of a feedback-controlled situation. Such situations are not restricted to physics. Close your eyes and then try to pick up an object. Now try it with a friend giving you instructions—his voice supplies the feedback loop. With your eyes open the feedback loop is supplied by light entering your eye, signals passing from your eye retina to the brain, the brain sending signals to the appropriate muscles. The action of trying to pick up an object requires an almost continuous supply of information to produce feedback to modify the action so that the required aim occurs, i.e., the object is picked up.

The following quotations are taken from an article by E. de Bono in the *Times Educational Supplement* for 30 April 1971. They illustrate the concept of feedback and were intended to present the concept to people who had not necessarily a scientific background.

'Leaves blown by the autumn wind pile up behind a stone. As the pile gets bigger it catches more leaves and gets bigger still. And so it catches still more leaves . . .

(The size of the pile affects the way it collects more leaves and this affects the size of the pile.)

* * *

A little boy pulls his sister's pigtails until she screams and then he knows that he has pulled hard enough.

(The strength of the pull affects the girl whose cry then affects the boy and his pull.)

* * *

More wages are needed to pay higher prices. But increased wages to people who make the product increase its price, so wages have to rise again.

(Wages affect price which in turn affects wages.)

* * *

More and more cars drive into the city until the traffic is so bad that no one can move and then people think twice about bringing their cars into the city.

(Cars affect traffic congestion which in turn affects the number of cars coming into the city.)

* * *

The more the horse bolted the harder he pulled on the reins which only made the horse bolt more.

(The bolting horse affected the pull on the reins which in turn affected the bolting horse.)

* * *

The smallest puppy could not fight his way past his bigger brothers to get to the food so he did not grow so he found it more difficult to fight his way to the food.

(The puppy's size affected the amount he ate which in turn affected his size.)

* * *

The frontiersman chopped some wood to see him through the winter. He then went to ask the wise old Red Indian what sort of winter it would be.

"Cold" said the Indian.

So the frontiersman chopped some more wood and went back to the Indian.

"Very cold" said the Indian.

So the man chopped still more wood.

"Very very cold" said the Indian.

"Why don't you make up your mind" said the man who was tired of chopping wood.

"Well," said the Indian, "I am a wise old Indian and when I see frontiersman chopping wood I know it is going to be a cold winter. When I see more wood chopped then it means a colder winter." '

Computers

At a rough estimate there were in 1968 about 70 000 computers in the world, about two-thirds being in the United States (Great Britain had about 3000). The beginning of electronic computers was probably a digital computer built at Harvard University and first operating in 1944; the first British computer started operation in 1948.

There are two main classes of computer, digital and analogue. Digital computers manipulate electrical signals which represent digits—the signals are not continuously variable but go up in steps. Analogue computers, however, use electrical signals which are directly proportional to some other physical quantity—they give electrical analogues. In the case of, say, the oscillation of a building a differential equation may be proposed as a description of the oscillation. The analogue computer is then set the problem of solving the analogous electrical oscillation equation. This it will do by using integrators which operate on the continuously varying input. A digital computer solving a differential equation would have to break it down into a number of digits which can be individually operated on.

The method used in chapter 4 for the solution of the oscillation differential equation can be adopted by a digital computer.

Essentially a digital computer can be represented as a sequence of units:

input, memory, arithmetic, output

Figure 10.63 shows the basic arrangement of the units.

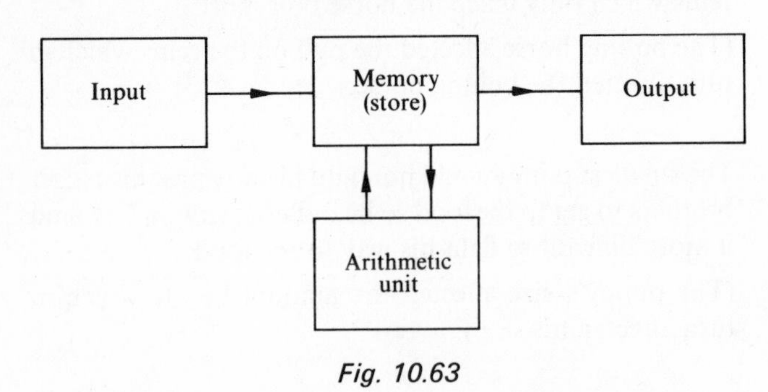

Fig. 10.63

The sequence of operations with a digital computer can be roughly summarized as:

(a) A programmer (human) breaks the problem down into a sequence of simple arithmetic operations that can be tackled by the machine. A program is written.

For example suppose we require the sum $5x + 3y$ to be evaluated, the sequence of operations might be: read x; multiply x by 5; store the result; read y; multiply by 3; take the earlier result from store and add; print out the result.

(b) The program is fed into the computer memory.

(c) The data are fed into the computer memory.

(d) The control section of the computer then processes the data according to the program, passing information backwards and forwards between the memory and the arithmetic sections.

(e) The answers are supplied at the output.

Many of the jobs tackled by digital computers are what one might call stock control. Thus in a motor factory a computer can keep stock of the numbers of each different part available and issue orders to keep the stock up to adequate levels. This is a complex task when it is realized that in just one car alone there may be more than one thousand different parts, the factory will produce many models, and may be producing a new car every minute.

Your account with a bank may be kept by a computer; the computer deducting all the money you draw out or pay out by cheque, adding all the money you pay in, paying out direct to, perhaps, another computer sums of money for which you have issued a standing order (perhaps the mortgage on a house), and informing your bank manager when you run into the red.

Computers are good at extracting information from large collections of data. They act as 'data bases', holding files of information. In the motor factory it can be details of the stock of components—if a strike occurs in a component factory it could be useful to ask the computer how long the factory can run on existing stocks. The files of information could, however, be about humans, about you and me. Whenever and wherever you earn money the information could be directly fed to a computer—you could not avoid paying your full income tax. Many branches of life lend themselves to computer surveillance and as computers can keep very detailed records and make them accessible on demand, very rapidly, the prospect is viewed with alarm by many who feel that the facts pertaining to their lives should not be accessible but private.

Further reading

S. H. Hollingdale and G. C. Toothill. *Electronic computers* Penguin, 1965.

Microelectronics

Why make electronic circuits very small? In a modern integrated circuit a transistor may only occupy an area of about 6×10^{-12} m^2 in a sheet about 10^{-6} m thick. Making electronics smaller can reduce the cost and increase the reliability of the components, but above all it reduces signal propagation times. Signals take a finite time to pass through a circuit, a component, or even just along a lead. Modern computers require operations to be performed by circuits in a matter of nanoseconds (10^{-9} s). If we take the velocity with which a signal travels through the circuit to be of the order of 10^8 m s^{-1}, then for the travel time to be 10^{-9} s the length of circuit path must only be about 10^{-1} m or 10 cm. Pulses travelling through the circuits would be out of step if they were produced at 10^{-9} s intervals and followed paths which differed by 10 cm or more. The limiting pulse frequency in such a case would be $1/10^{-9}$ or 10^9 Hz.

The cost of a circuit is the sum of the costs of the materials, fabrication, and assembly. Making circuits smaller can reduce the materials cost term and as in an integrated circuit the entire circuit can be manufactured as one item the fabrication cost of an entire circuit is not too much more than the cost of fabricating say a single transistor. Assembly costs are reduced as individual components do not have to be assembled to form the circuit. This reduction of the human assembly item increases the reliability of the circuit. The main cause of failures with discrete component circuits is not in the active device but the interconnections, particularly soldered joints.

High speed circuits require very close packing of minute components. One of the difficult problems associated with this packing is the dissipation of heat. Currents passing through devices produce temperature rises. If the temperature rises too high the device may break down or have its characteristic modified to such an extent as to change the function of the circuit. Putting components close together reduces the amount of air that can circulate round a component. Another factor limiting close packing is the interaction that can occur between neighbouring components. A changing current in one component can induce currents in other components (electromagnetic induction—see chapter 11).

With integrated circuits the entire circuit, or group of circuits, is produced on one chip of semiconductor material. The process involved is one of etching away layers in some parts, depositing layers of other materials in some parts, diffusing into some parts different elements to form p and n regions. Figure 10.64 shows an integrated circuit, considerably enlarged, the circuit diagram and the way the circuit is grown.

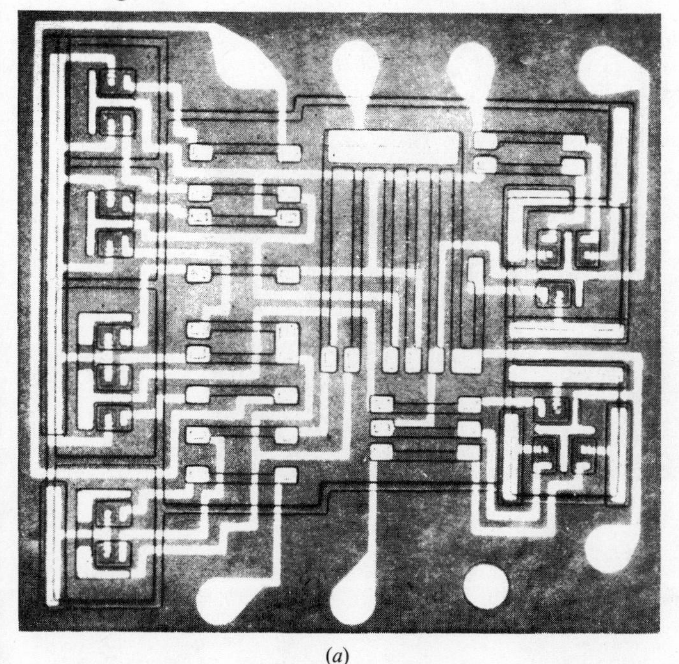

(*a*)

Fig. 10.64 From Microelectronics *edited by Edward Keonjian. Chapter 5 by G. E. Moore. Copyright 1963 by McGraw-Hill, Inc. Used with permission of McGraw-Hill Book Company. (a) Photomicrograph of a circuit with 15 transistors and 21 resistors (Fig. 5.52). The area is about 4 mm square.*

Appendix 10A Communication

Somewhere out in space there is life on another planet orbiting some other sun. How can we communicate with them? How can we tell if they are trying to communicate with us?

This is not science fiction but real life. Astronomers at the National Radio Astronomy Observatory in West Virginia, USA, have used their 28 m diameter radio telescope to look for communications from some distant planet, conferences have been held, papers have been written in serious scientific journals, the Russians even at one stage in 1965 thought they had detected a communication—at the present, however, 1973, there is no firm evidence that anybody is trying to communicate with us here on Earth.

What form of communication are we expecting from outer space? How could we distinguish such a communication from the general background 'noise' that is picked up by radio telescopes? How can you tell that your doorbell is ringing—how can you distinguish its sound from all the other sounds, from the background noise? How can you distinguish my communication to you, the words on this page, from all the other ways ink can be splattered over paper? Background noise, ink splattered over paper, we expect to fall into no regular pattern. The astronomers are searching the sky for some well-defined and repeated pattern of radio signals, some sequence that could not have arisen purely by chance.

The following is part of a communication by G. Cocconi and P. Morrison to scientists, published in the scientific journal *Nature* in September 1959 (**184**, 844).

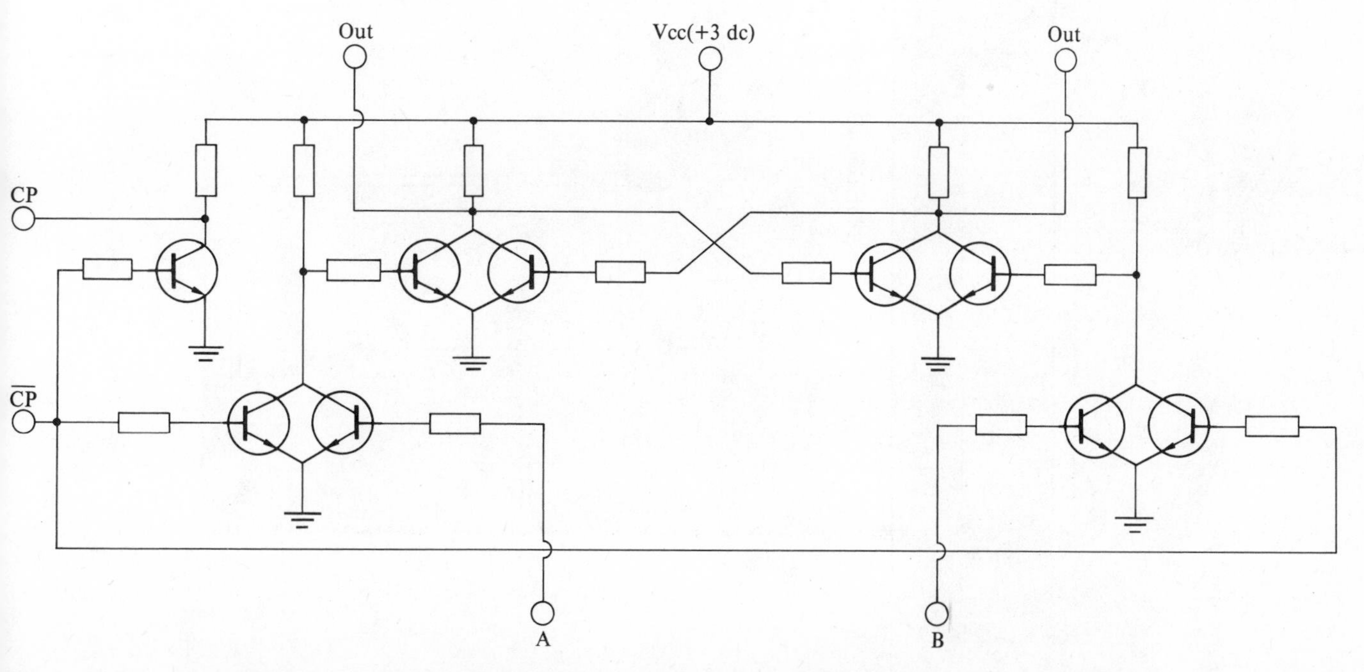

Fig. 10.64 (b) Circuit diagram of the circuit whose assembly is shown in (c). (Fig. 5.52.)

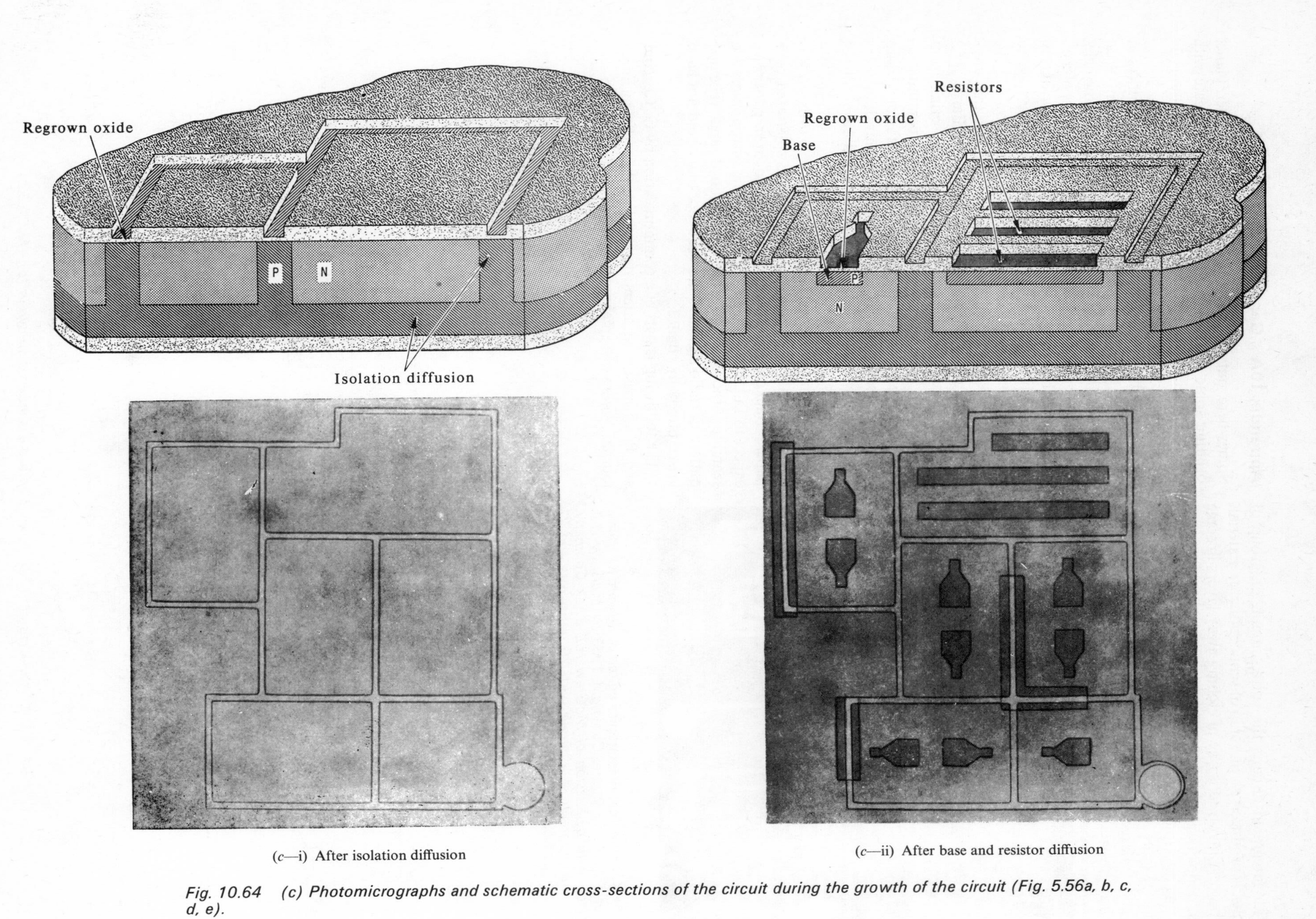

(*c*—i) After isolation diffusion

(*c*—ii) After base and resistor diffusion

Fig. 10.64 (c) Photomicrographs and schematic cross-sections of the circuit during the growth of the circuit (Fig. 5.56a, b, c, d, e).

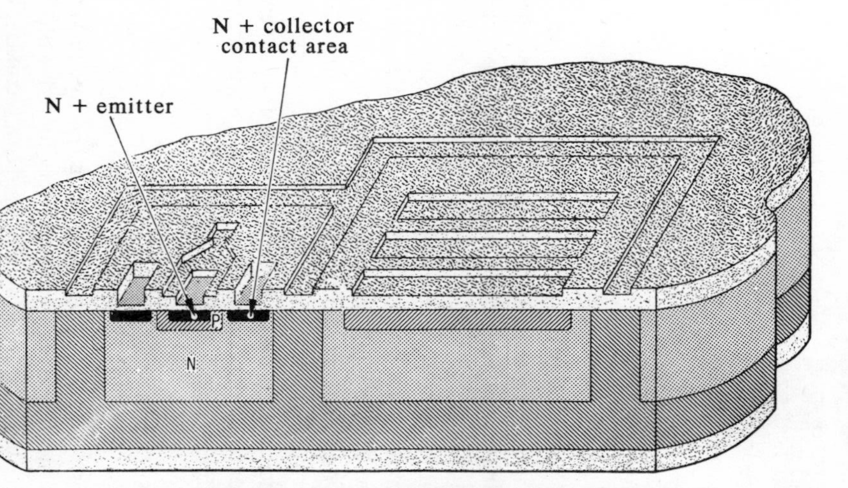

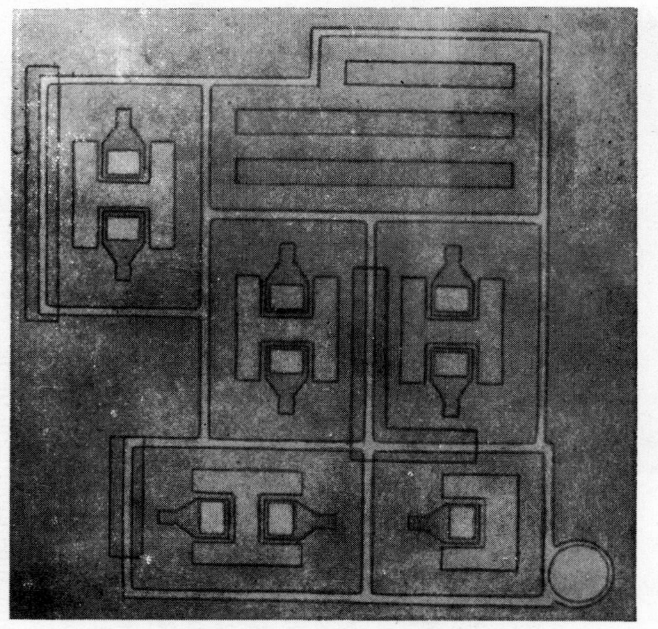

(c—iii) After emitter diffusion

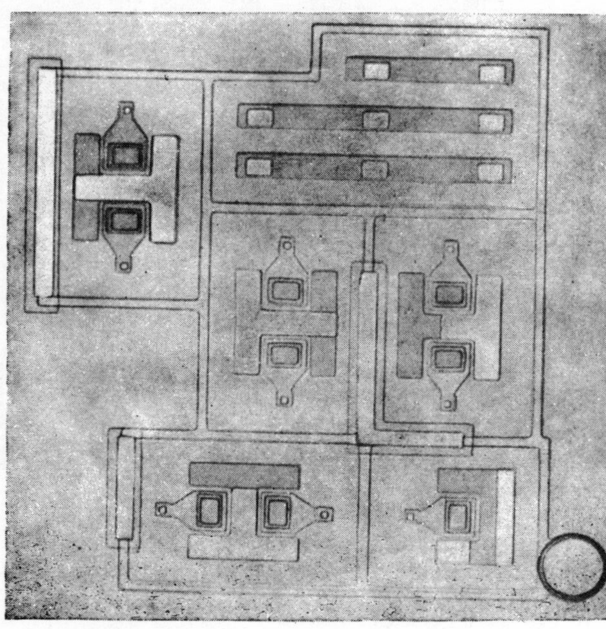
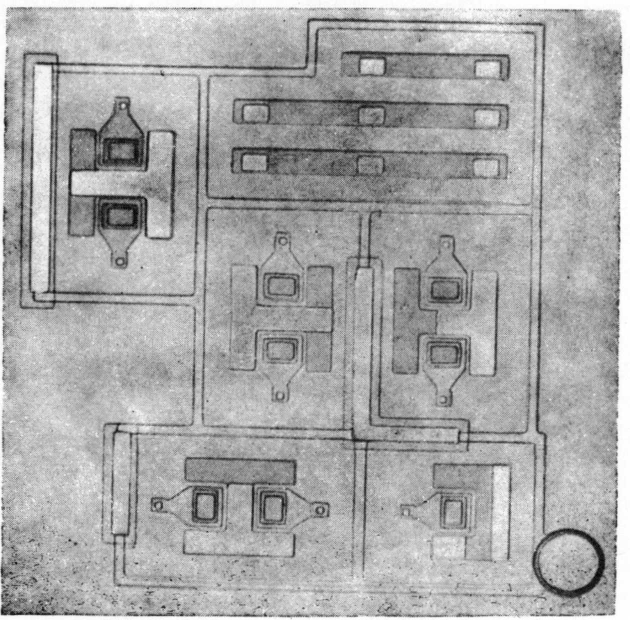

(c—iv) After oxide removal for contacting

Fig. 10.64 (c) Photomicrographs and schematic cross-sections of the circuit during the growth of the circuit (Fig. 5.56a, b, c, d, e).

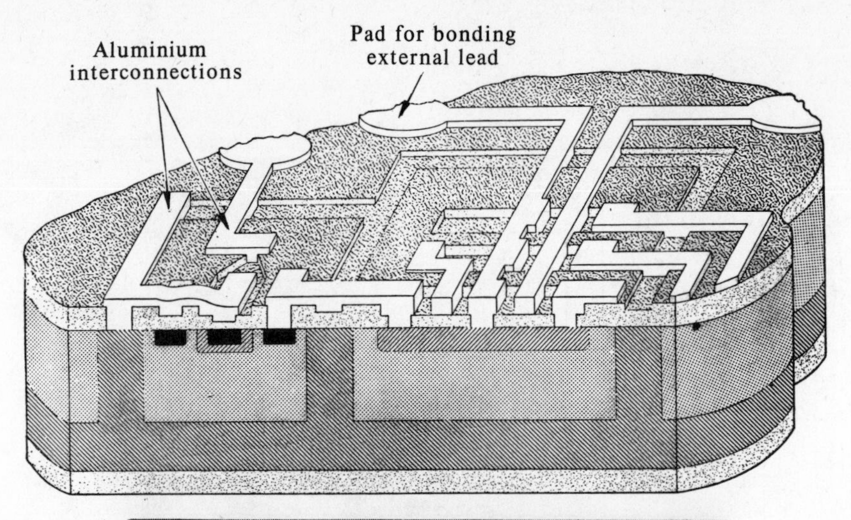

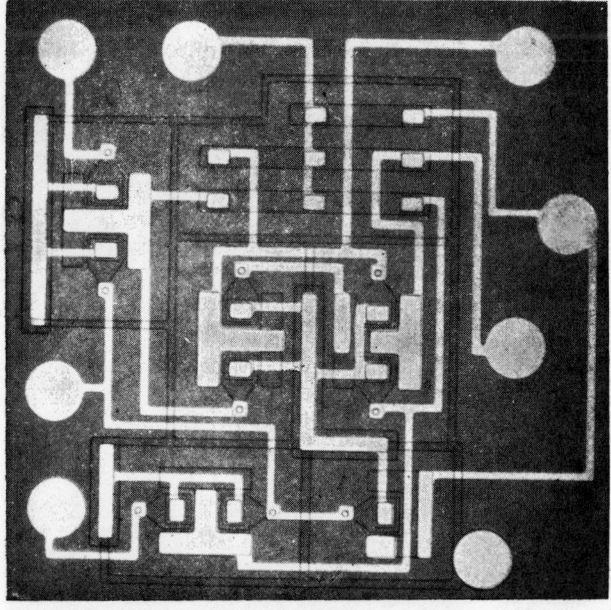

(c—v) After intraconnection metallization

Fig. 10.64 (c) Photomicrographs and schematic cross-sections of the circuit during the growth of the circuit (Fig. 5.56a, b, c, d, e).

'Searching for interstellar communications

By Giuseppe Cocconi and Philip Morrison
Cornell University– Ithaca– New York

No theories yet exist which enable a reliable estimate of the probabilities of (1) planet formation; (2) origin of life; (3) evolution of societies possessing advanced scientific capabilities. In the absence of such theories, our environment suggests that stars of the main sequence with a lifetime of many billions of years can possess planets, that of a small set of such planets two (Earth and very probably Mars) support life, that life on one such planet includes a society recently capable of considerable scientific investigation. The lifetime of such societies is not known; but it seems unwarranted to deny that among such societies some might maintain themselves for times very long compared to the time of human history, perhaps for times comparable with geological time. It follows, then, that near some star rather like the Sun there are civilizations with scientific interests and with technical possibilities much greater than those now available to us.

To the beings of such a society, our Sun must appear as a likely site for the evolution of a new society. It is highly probable that for a long time they will have been expecting the development of science near the Sun. We shall assume that long ago they established a channel of communication that would one day become known to us, and that they look forward patiently to the answering signals from the Sun which would make known to them that a new society has entered the community of intelligence. What sort of a channel would it be?

. . .

Nature of the Signal and Possible Sources

No guesswork here is as good as finding the signal. We expect that the signal will be pulse-modulated with a speed not very fast or very slow compared to a second. . . . A message is likely to continue for a time measured in years, since no answer can return in any event for some ten years. It will then repeat, from the beginning. Possibly it will contain different types of signals alternating throughout the years. For indisputable identification as an artificial signal, one signal might contain, for example, a sequence of small prime numbers of pulses, or simple arithmetical sums.

...

The reader may seek to consign these speculations wholly to the domain of science-fiction. We submit, rather, that the foregoing line of argument demonstrates that the presence of interstellar signals is entirely consistent with all we now know, and that if signals are present the means of detecting them is now at hand. Few will deny the profound importance, practical and philosophical, which the detection of interstellar communications would have. We therefore feel that a discriminating search for signals deserves a considerable effort. The probability of success is difficult to estimate; but if we never search, the chance of success is zero.'

Problems

1. Oxford and Cambridge Schools Examination Board. *Nuffield advanced physics*. Short Answer Paper, 1970.

'The diagram [Fig. 10.65] is of a very simple radio receiver which can be used for broadcasts from one station only.

(a) The tuning circuit selects one station (one frequency). Is the output of energy of the selected signal coming from the tuning circuit greater than the energy of the signal collected by the aerial? Explain your answer briefly.

(b) For which of the other three boxes is the energy of the output signal of the box larger than the energy of the input signal to the box? Give... the name of the box or boxes for which there is energy increase.

(c) How could the tuning circuit be altered so that it could select other stations?

(d) Which box or boxes is/are designed to transform energy from one form to another?

(e) The frequency of the signal received at the aerial is about 10^6 Hz, and the frequency of the speaker output is about 10^3 Hz. Would it matter if the amplifier could only amplify signals of frequency less than 10^5 Hz and so failed to amplify signals of 10^6 Hz? Explain your answer.'

2. Estimate the leakage current through the insulator of a parallel plate capacitor. The plates have an area of 10^{-2} m^2, are at a p.d. of 1 kV, and the insulator has a thickness of 1 mm and a resistivity of 10^{13} ohm m.

3. Sketch a graph showing how the resistance of the component described by Fig. 10.3 changes with the applied voltage. Is Ohm's law valid for this component?

4. The 'lie detector' works on the principle that when a person lies he perspires more and his surface resistance changes. Devise a 'lie detector'. Consider both the problem of finding a suitable circuit and that of attaching the electrodes so that movements of the body are not responsible for changes in reading.

5. When the current through a length of metal wire is doubled what happens to (a) the number of electrons in the wire, (b) their drift velocity, and (c) the number of collisions made per unit time by the electrons?

6. Suggest systems which could be used to:

(a) keep the temperature of a furnace at a fixed level,

(b) keep a telescope 'locked on to' the moon for many hours,

(c) keep the moisture content of paper in a paper drying machine at a constant level.

(d) keep the level of liquid in a tank at a constant level.

7. Figure 10.66 shows a current which varies with time according to a sine function.

(a) When is dI/dt a maximum?

(b) When is dI/dt zero?

(c) When is $\int_0^t I\,dt$ a maximum?

(d) When is the integral zero?

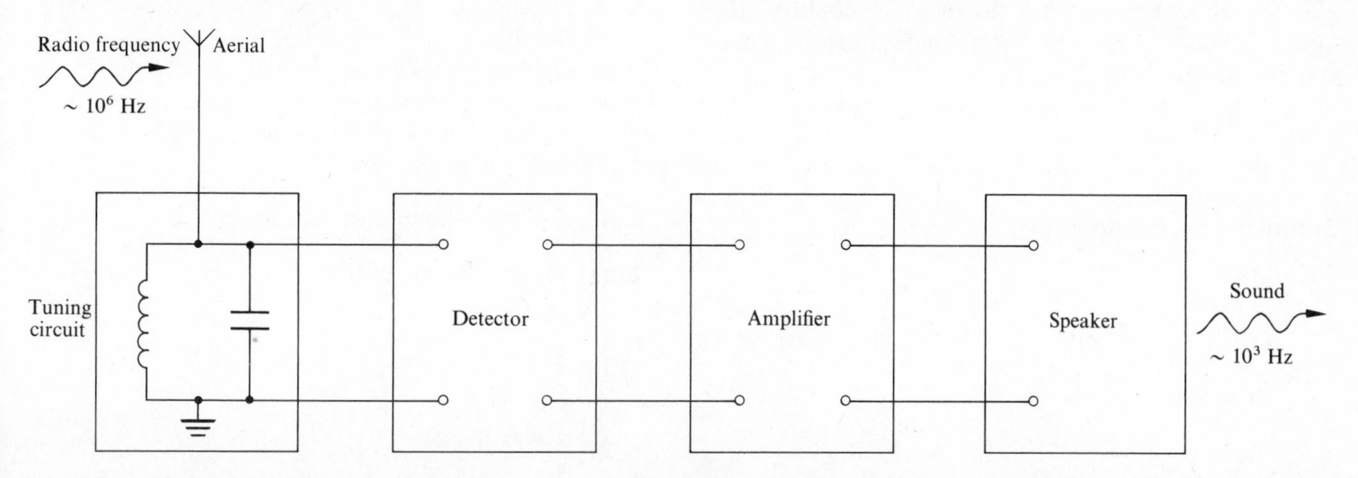

Fig. 10.65

(e) How would doubling the frequency of the alternating current affect the maximum value of dI/dt?

(f) How would doubling the frequency affect the maximum value of the integral?

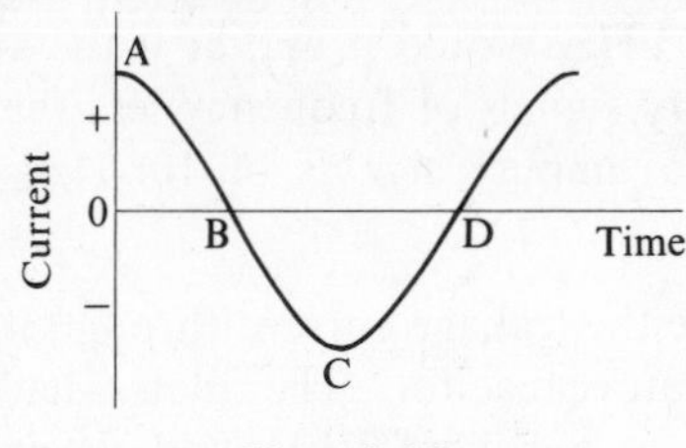

Fig. 10.66

8. Explain in simple language, perhaps to a student who does not study physics, the functions of the following circuits:

(a) an amplifier,
(b) an OR gate,
(c) an AND gate.

9. This question is about the changing of the shapes of voltage pulses. How would you change:

(a) a sinusoidal voltage signal to a square wave signal,
(b) a square wave to a series of sharp blips,
(c) a sinusoidal signal to a series of sharp blips?

10. Is man a machine or a self?

This was the question posed by J. Bronowski in his book *The identity of man* (Penguin, 1967). In discussing the question you might try to explain what is meant by the term 'machine'.

Teaching note Further problems appropriate to this chapter will be found in:
Nuffield advanced physics. Units 2 and 6, student's books, Penguin, 1972.
Physical Science Study Committee. *Physics*, 2nd ed, chapters 28 and 29, Heath, 1956.

Practical problems

1. Determine the electrical conductivity of pure ice and hence consider whether ice is an insulator, semiconductor or good conductor.

Reference
Pounder. *Physics of ice* (Pergamon, 1965).

2. Investigate the factors affecting the electrical conductivity of flames. An obvious factor is the temperature. Consider the possible mechanism by which a flame conducts, i.e., are the charge carriers ions or electrons or both? Does the conductivity depend on the materials present in the flame; possibly their ionization potentials could affect the result?

3. Design and construct a circuit which could be used to control the exposure time for bromide paper in a photographic enlarger.

Suggestions for answers

1. (a) No.
(b) Amplifier.
(c) Change the value of either the inductance or the capacitance.
(d) Speaker.
(e) No.

2. 10^{-9} A. Ohm's law is assumed to apply.

3. Your graph should show how the slope of the V against I graph changes with V. Ohm's law does not apply. For Ohm's law to apply the resistance should be independent of the voltage.

4. The Wheatstone bridge, Fig. 10.13, offers one possibility. Tests to find out whether the resistance changes will be large or small and whether the changes will be superimposed on a high or low resistance are necessary. Do all people have the same initial resistance? Must allowances be made for variations in initial resistance? Contact with the skin via a low resistance grease or liquid would seem to be one way of eliminating resistance variations due to movements of the body. This problem requires a number of tests before a satisfactory answer can be produced.

5. (a) No change.
(b) Doubled.
(c) Unchanged. This is virtually unaffected by the changes in the drift velocity, being determined largely by the velocity of the electrons given by their being at some temperature above absolute zero.

6. Figure 10.67 shows possible answers.

7. (a) D; a minimum at B.
(b) A, C.
(c) B.
(d) C.
(e) Double it.
(f) Halve it.

8. (a) You might introduce the terms input and output

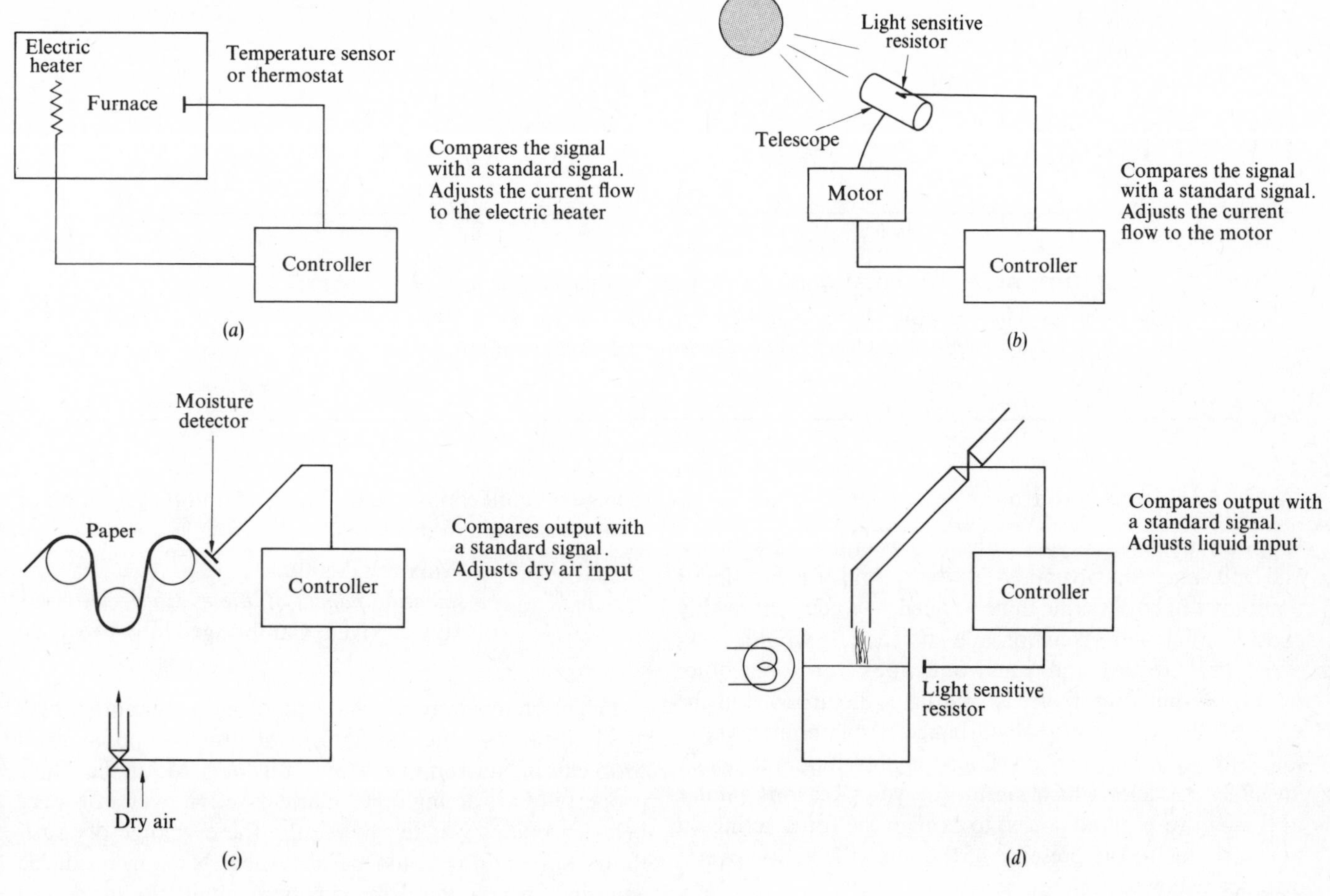

Fig. 10.67

and talk of the output being a magnified version of the input. You might have to explain how you appear to have 'got something for nothing'.

(b) Figure 10.68(a) might help.

(c) Figure 10.68(b) might help.

9. (a) A device with a characteristic like Fig. 10.35(a) can be used. See Fig. 10.55.

(b) See Fig. 10.42 or 10.47.

(c) You could use (a) followed by (b).

10. I would suggest you read the book by Bronowski. He considers that 'Man is a machine by birth but a self by experience'.

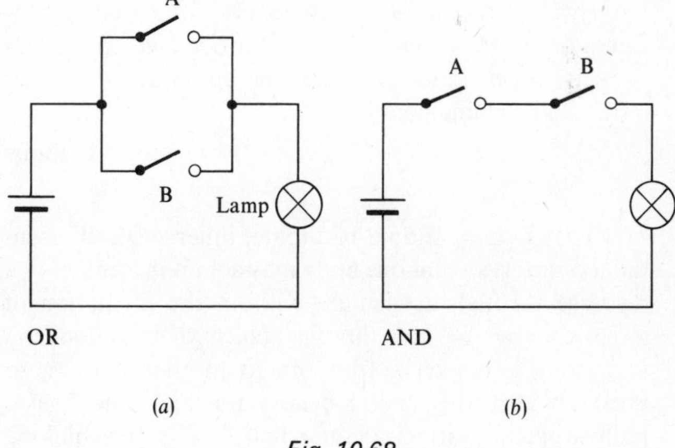

Fig. 10.68

11 Fields

Teaching note Practical work appropriate to this chapter will be found in:
W. Bolton. *Physics experiments and projects*, Vol. 4, Pergamon, 1968.
Nuffield advanced physics. Units 3 and 7, teacher's guides, Penguin, 1972.

11.1 Action at a distance

If I point my finger at you, without any contact between me and you, and you promptly fall over—and if every time I point my finger the same thing happens—you might readily consider that there is action at a distance. It would, however, seem absurd and you would look for some other reason. Perhaps my finger is creating a disturbance in the air between us and this disturbance is communicating a force to you. Perhaps my finger is shooting out small, invisible, particles which are hitting you. Reasons similar to these have been advanced to explain the force acting on the earth due to the presence of the sun—the action over a distance which we call gravity.

> 'A body is never pushed naturally, except by another body which touches it and pushes it; after that it continues until it is prevented by another body which touches it. Any other kind of operation on bodies is either miraculous or imaginary.'
>
> Leibniz.

> '... That gravity should be innate, inherent, and essential to matter, so that one body may act upon another at a distance through a vacuum, without the mediation of anything else, by and through which their action and force may be conveyed from one to another, is to me so great an absurdity that I believe no man who has in philosophical matters a competent facility for thinking can ever fall into it...'
>
> Newton (1692). Letter to Bentley.

> 'When we observe one body acting on another at a distance, before we assume that this action is direct and immediate, we generally inquire whether there is any material connection between the two bodies; and if we find strings, or rods, or mechanism of any kind, capable of accounting for the observed action between the bodies, we prefer to explain the action by means of these intermediate connections, rather than to admit the notion of direct action at a distance.'
>
> J. C. Maxwell (1890).
> *The scientific papers of James Clerk Maxwell.*
> Vol. II. ed. Niven. Cambridge Univ. Press.

The mechanism by which one body attracts another over great distances—the mechanism of gravity—posed great problems in Newton's time (and still does). Matter has some property of attracting other matter. Action over a distance between bodies was felt to require the exchange of something, but nothing could be discerned. Newton produced equations which enabled predictions about the motion of matter to be made but he did not produce any 'mechanism' responsible for his equations. In this chapter we look at the equations without knowledge of the mechanism.

> 'Physics is mathematical not because we know so much about the physical world, but because we know so little: it is only its mathematical properties that we can discover.
>
> B. Russell. *An outline of philosophy.*
> Allen & Unwin.

One piece of matter does affect another piece of matter and we can arrive at laws, based on observations of matter, which can be used to describe the effects of such an interaction. A useful concept—a useful language—by which we can describe such interactions is that of fields. Around one piece of matter we construct (mathematically) a field—a region of influence—and say that this is produced by our piece of matter. When another piece of matter comes within that field we have a force acting on the matter due to its interaction with the field.

> '"If we pick up a stone and then let it go, why does it fall to the ground?" The usual answer to this question is: "Because it is attracted by the earth." Modern physics formulates the answer rather differently.
>
> The action of the earth on the stone takes place in-

directly. The earth produces in its surroundings a gravitational field, which acts on the stone and produces its motion of fall.'

A. Einstein (1920). *The theory of relativity*, 3rd ed. transl. R. W. Lawson, Methuen.

The concept of a field is not just restricted to gravitation but is useful in electricity and magnetism. A charged particle can be thought of as producing an electric field and if another charged particle is placed in the field an interaction occurs between the particle and the field.

'How the magnetic force is transferred through bodies or through space we know not:—whether the result is merely action at a distance, as in the case of gravity; or by some intermediate agency, as in the case of light, heat, the electric current, and (as I believe) static electric action.'

M. Faraday (1852). Item 3075, *Phil. Trans.*, pl.

Despite Faraday's not knowing the mechanism, he was able by means of the field concept to develop an understanding of the behaviour of magnets and charges.

The field concept is abstract; many physicists have tried to devise some 'machinery' to avoid this abstraction, but it is a useful concept enabling considerable simplifications to be made.

11.2 Gravitational fields

Two pieces of matter attract each other with a force which varies inversely with the square of their distance apart (d) and directly in proportion to the masses of the two bodies (m and M).

$$F = \frac{GmM}{d^2}$$

G is the constant of proportionality, known as the Universal constant of Gravitation. The experimentally determined value of this constant is $6{\cdot}67 \times 10^{-11}$ m^3 kg^{-1} s^{-1}. This paragraph summarizes the situation considered in chapter 2.

Suppose we take a mass m and place it near a mass M, it experiences a force F. If we put another mass m' at the same distance from M then the force is F' where

$$\frac{F'}{F} = \frac{m'}{m}$$

It does not matter what position we choose to put our masses, m and m', the ratio of the forces is the same (provided both are put in turn at the same point). Thus if $m' = 2m$ then $F' = 2F$. Thus if I measure the force on a 2 kg mass at a particular point above the earth's surface I will measure twice the force that I would have measured with a 1 kg mass. Thus if I took a 1 kg mass to every point in space and measured the forces at these points all I would need to do to find the force on another mass would be to multiply the force value by the mass. This value of force per unit mass is called the gravitational field strength.

$$\text{Gravitational field strength} = \frac{\text{force}}{\text{mass}}$$

At the surface of the earth the force on a 1 kg mass is about 9·8 N and thus the gravitational field strength is 9·8 N kg^{-1}. This is identical in magnitude with the acceleration due to gravity, in m s^{-2}, at the point concerned. The gravitational field strength is always equal to the acceleration due to gravity at the point concerned.

By putting a 1 kg mass at different positions near another mass we can plot the contour lines of gravitational field strength. These lines tell us what the force would be if we put a 1 kg mass at particular points. These contour lines are said to give a map of the gravitational field. Is the field there when we have no test mass at the point concerned? The field is not radiated by a mass but is only a measure of the force that would occur when a mass is put in the vicinity of another mass.

'There was a young man who said, "God
Must think it exceedingly odd
If he finds that this tree
Continues to be
When there's no one about in the Quad."

REPLY

Dear Sir:
Your astonishment's odd:
I am always about in the Quad.
And that's why the tree
Will continue to be,
Since observed by
Yours faithfully,
GOD'

R. Knox.

B. Russell (1961). *History of western philosophy*, Allen & Unwin, Simon and Schuster.

For a point mass (!) M, the force on a 1 kg mass a distance, d, away is

$$F = G\frac{M \times 1}{d^2}$$

The gravitational field strength is the force acting on unit mass and thus we have

$$\text{Gravitational field strength } g = \frac{GM}{d^2}$$

The direction of the field is the direction of the force. The gravitational field strength thus varies as the inverse square of the distance of the point concerned from the point mass.

What if we do not consider point masses but masses of finite size, say spheres? Do we measure the distance d from the surface of the mass M or the centre? Figure 11.1 shows a situation where we calculate the field strength at a point P close to the surface of a sphere. We can consider the sphere to be made up of a large number of small masses. These masses will be different distances from P and the forces experienced by a test mass placed at P will not all be in the same direction. To find the field strength at P we need to sum all the forces due to all these small masses. The direction of the field at P will be the direction of the resultant force.

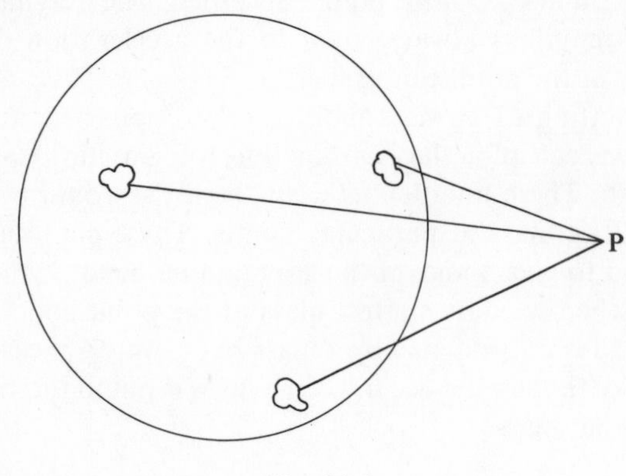

Fig. 11.1

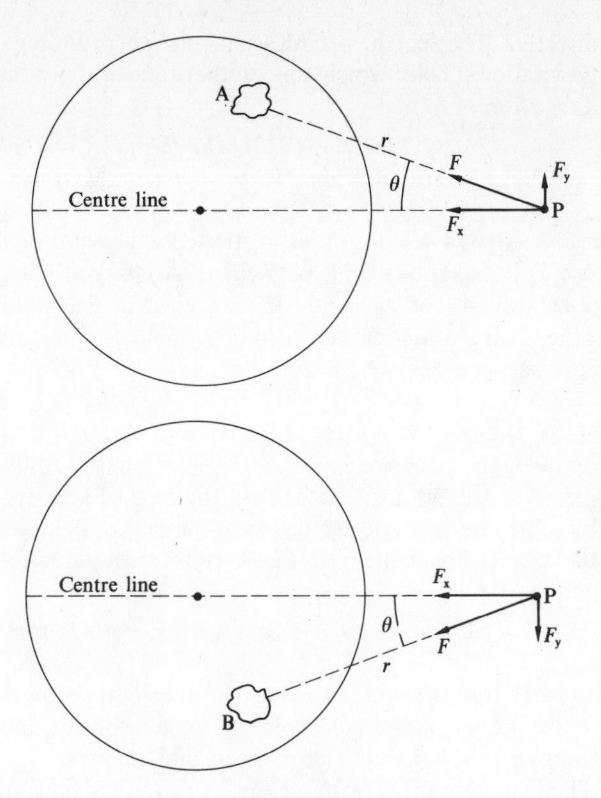

Fig. 11.2

We can see that the resultant force, and hence the field, must be along a line joining P to the centre of the sphere if we consider the small masses in pairs, one below the centre line (B) and a corresponding one (A) above the centre line. If these masses are equal and the same distance from P then the force F acting at P due to each mass will be the same. The forces can be resolved into components along the centre line and at right angles to the centre line (Fig. 11.2). As the forces F due to each mass make the same angle with the centre line, we have chosen corresponding pairs of masses, then the magnitudes of the resolved components will be the same, i.e., F_X and F_Y. Both the F_X components will be in the same direction—towards the centre. The F_Y components are, however, in opposite directions and as their magnitudes are equal they cancel. The resultant force is thus $2F_X$, along the line to the centre. To find the field at P due to all the small masses making up the sphere we need to find the sum of all their F_X components.

$$F_X = F \cos \theta$$

The force ΔF due to any small mass Δm a distance r away is

$$\Delta F = G \frac{\Delta m}{r^2}$$

Thus

$$\Delta F_X = G \frac{\Delta m}{r^2} \cos \theta$$

Consider all the components due to masses like A and B, i.e., all those at a distance r and making an angle θ with the centre line (Fig. 11.3). We can think of the small masses as being part of a thin ring, each mass having an area on the

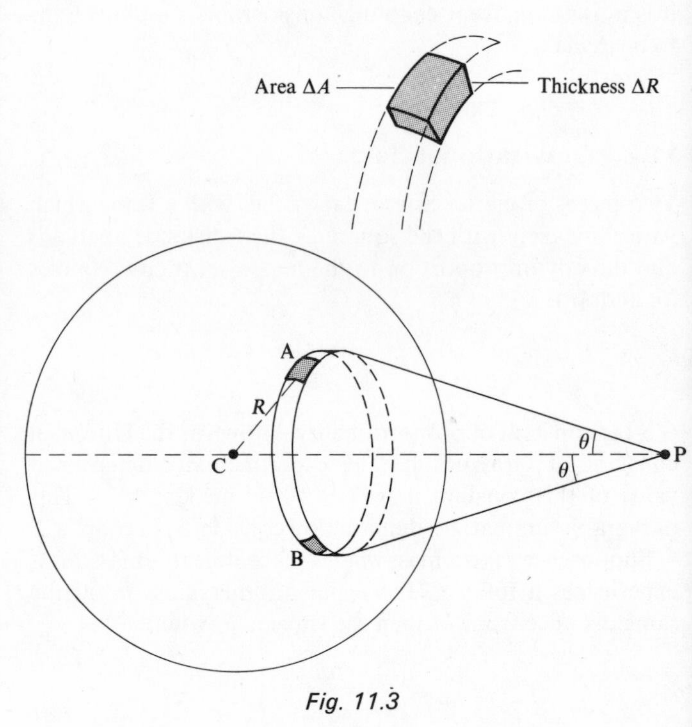

Fig. 11.3

ring surface of ΔA. If the ring thickness is ΔR and the density of material in the sphere ρ

$$\Delta m = \rho \, \Delta A \, \Delta R$$

Thus for the F_X components due to any of the masses in this ring we have

$$\Delta F_X = G \frac{\rho \, \Delta A \, \Delta R}{r^2} \cos \theta$$

We can simplify this expression. Take a point T on the centre line so that the triangle CAT is similar to the triangle CAP, i.e., corresponding angles are the same and the sides of one triangle are just scaled-up versions of the corresponding ones in the other triangle. Then we have (Fig. 11.4)

$$\frac{\text{PA}}{\text{PC}} = \frac{\text{AT}}{\text{AC}}$$

$$\frac{r}{d} = \frac{\text{AT}}{R}$$

$$r^2 = \text{AT}^2 \frac{d^2}{R^2}$$

Also we have

$$\Delta A \cos \theta = \text{AT}^2 \, \Delta\omega$$

where $\Delta\omega$ is the angle subtended at T by the area ΔA. Hence

$$\Delta F_{\text{X}} = G\rho \text{AT}^2 \, \Delta\omega \, \Delta R \frac{R^2}{\text{AT}^2 d^2}$$

$$\Delta F_{\text{X}} = G \frac{\rho \, \Delta\omega \, \Delta R \, R^2}{d^2}$$

Summing all the small masses which make up this ring gives

$$F_{\text{X}} = G \frac{\rho 4\pi \, \Delta R \, R^2}{d^2}$$

But the total mass of the spherical shell of which this ring is a part is $4\pi R^2 \rho \, \Delta R$. Hence for a spherical shell we have

$$F_{\text{X}} = \frac{Gm}{d^2}$$

where m is the mass of the shell. The entire sphere can be considered to be made up of shells and thus the force at P due to all the shells will be

$$F_{\text{X}} = \frac{GM}{d^2}$$

where M is the total mass of the sphere. d is the distance of the centre of a shell from P and is the same for all shells. This result is of importance because it means that we can in all field calculations involving spheres consider the mass to be effectively located at the centre.

What happens if d is less than the radius of the sphere? Consider a uniform spherical shell of matter, as in Fig. 11.5, and the field at a point P somewhere inside the spherical shell. We can divide the interior of the shell into a number of small areas and compute the field at P due to each of these areas. Consider an area ΔA_1. It makes a cone with the apex at P and we can construct an exactly symmetrical cone on the other side of P. This other cone has a base area on the interior of the shell of ΔA_2. The squares of the distances of the two areas from the point P are in the same ratio as the areas.

$$\frac{\Delta A_1}{\Delta A_2} = \frac{r_1^2}{r_1^2}$$

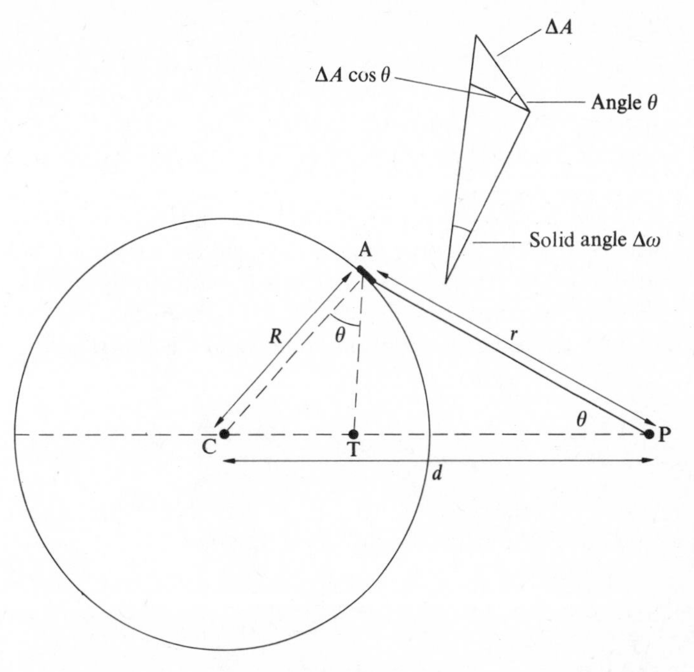

Fig. 11.4

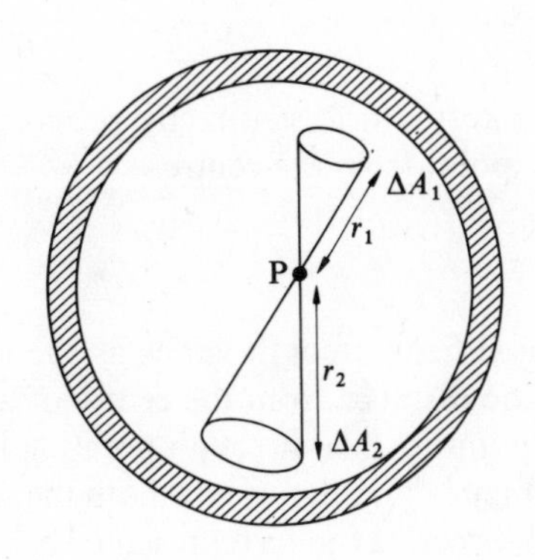

Fig. 11.5

(If the area was a square, each side would be proportional to r and thus the area proportional to r^2.) The mass of each element of the shell covered by the areas is proportional to the area.

$$\frac{\Delta m_1}{\Delta m_2} = \frac{\Delta A_1}{\Delta A_2}$$

Thus

$$\frac{\Delta m_1}{\Delta m_2} = \frac{r_1^2}{r_2^2}$$

The force on a test mass m placed at P will be

$$F_1 = Gm \frac{\Delta m_1}{r_1^2}$$

for area ΔA_1 and for area ΔA_2

$$F_2 = Gm \frac{\Delta m_2}{r_2^2}$$

The forces will be in opposite directions. But as the masses are in the same ratio as the squares of the distances these two forces F_1 and F_2 must cancel. The same thing will happen with all the cones coming from the small areas which make up the inside area of the spherical shell. There is no field inside a shell.

If we now consider a sphere made up of concentric spherical shells the only mass which will contribute to the field at a point P within the sphere is that due to the matter between the point P and the centre. If d is the distance of the point P from the centre of the sphere and d is less than the radius of the sphere we have

$$\text{Field at P} = \frac{GM}{d^2}$$

where M is the mass of that part of the sphere with radius equal to d. If ρ is the density of the sphere

$$M = \frac{4}{3}\pi d^2 \rho$$

and thus

$$\text{Field at P} = \frac{4G\pi d^2 \rho}{3d^2} = \frac{4\pi G\rho d}{3}$$

Within the sphere the field is directly proportional to the distance of the point from the centre.

Field maps

The gravitational field strength varies as the reciprocal of the square of the distance from the centre of a sphere and the field at all points is in a direction pointing to the centre of the sphere. Figure 11.6 shows how both the field strength magnitude and direction can be represented by diagrams for the earth. The direction lines are often called 'lines of force'.

The figure has assumed a perfectly spherical earth. This is not quite true. The departure from the sphere is, however, not large enough significantly to alter the results of Fig. 11.6 which shows the field on a relatively coarse scale. On a more local scale irregularities in the earth's field are used to establish the presence of underground deposits of coal or salt, etc. Figure 11.7 shows some typical field values. The irregularities arise because of local variations in the earth density.

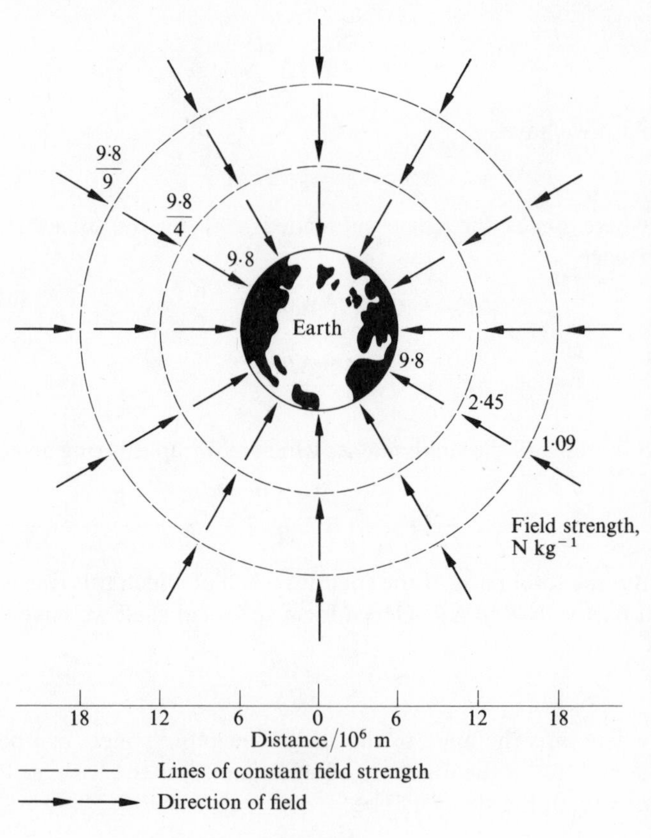

Fig. 11.6

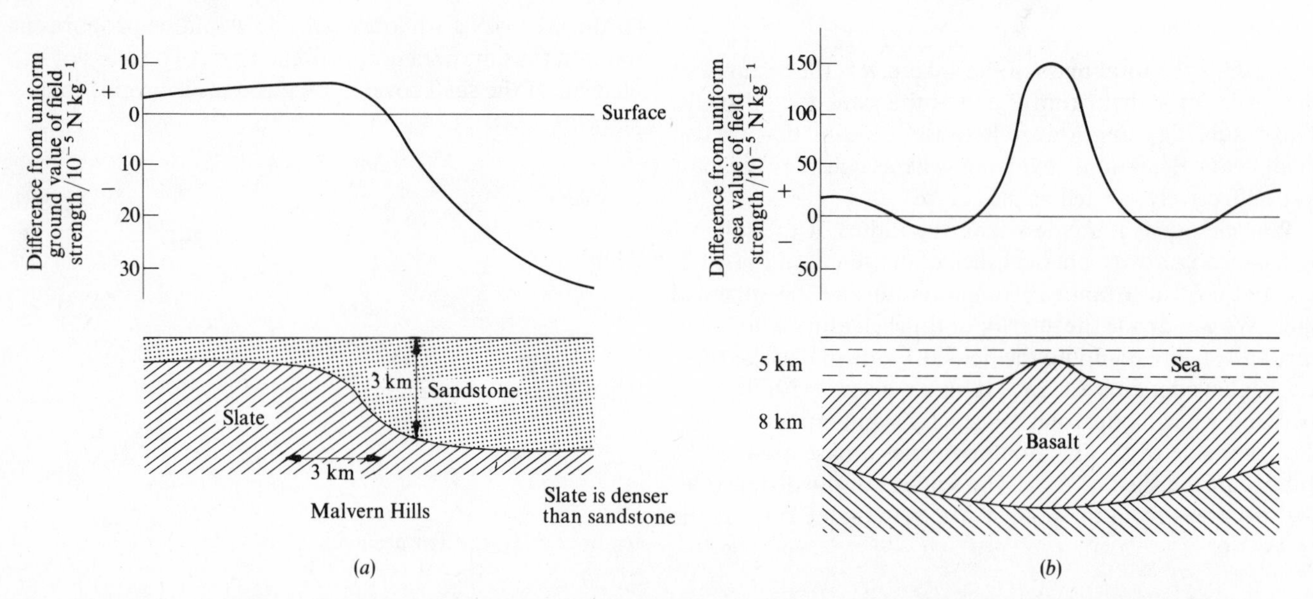

Fig. 11.7 (a) Field strength variations over rock discontinuity. (b) Field strength variations over an oceanic island (A. H. Cook and V. Saunders (1969). Gravity and the earth, *Wykeham Publications (London) Ltd.)*

Even if the earth were perfectly uniform there would be changes in the field strength due to the earth's surface being undulating. The top of a mountain is farther from the centre of the earth than sea level and as the value of the field strength varies as the reciprocal of the square of the distance from the earth's centre we would expect the field strength to be lower at the top of the mountain.

$$\frac{\text{Field strength at height } h}{\text{Field strength at sea level}} = \frac{R^2}{(R + h)^2}$$

R is the radius of the earth, i.e., the distance of sea level from the centre of the earth. A mountain of height 1 km would give a difference in field strength between the top and the base of about 3×10^{-3} N kg^{-1}.

Gravitational potential

If we move a mass m away from another mass M then energy must be supplied to move it against the force of attraction.

$$\begin{matrix}\text{Energy needed} \\ \text{to move } m \text{ in} \\ \text{a direction}\end{matrix} = \begin{matrix}\text{force} \\ \text{component in} \\ \text{that direction}\end{matrix} \times \begin{matrix}\text{distance} \\ \text{moved along} \\ \text{that direction}\end{matrix}$$

But the gravitational field strength at a point is force/mass. If the distance moved ΔR is very small so that we can assume that the field strength does not change,

$$\begin{aligned}\text{Energy} &= \text{field strength} \times \text{mass} \times \text{distance} \\ &= gm\,\Delta R\end{aligned}$$

This energy must be supplied to the mass to move it through this distance—if the mass moves back to its original position then it will give up this amount of energy, perhaps as kinetic energy. The new position has thus an energy associated with it when the mass is present—this is known as potential energy (see chapter 3).

The potential energy per unit mass is called the potential. The potential difference between the two points a distance ΔR apart is thus $g\,\Delta R$.

If the field is not constant over the distance the mass is moved then we must consider the movement of the mass in a large number of small steps—the field strength being considered constant over the distance of each step (Fig. 11.8(a)).

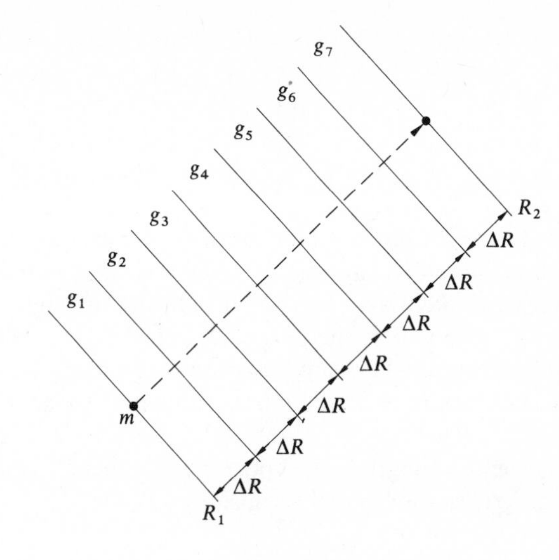

(a)

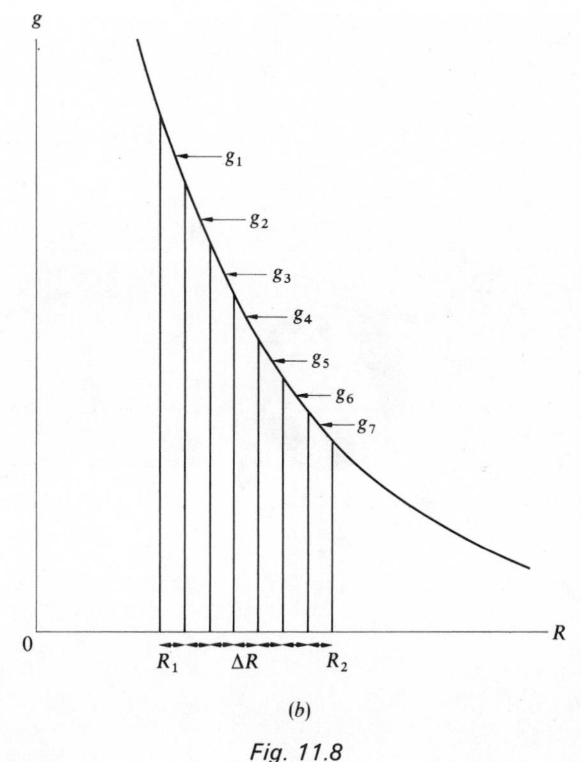

(b)

Fig. 11.8

$$\begin{aligned}\text{Potential difference} &= g_1\,\Delta R + g_2\,\Delta R + g_3\,\Delta R \\ &\quad + g_4\,\Delta R + g_5\,\Delta R + g_6\,\Delta R + g_7\,\Delta R\end{aligned}$$

We can write this summation as

$$\text{Potential difference} = \int_{R_1}^{R_2} g\,\mathrm{d}R$$

If we have a graph showing how the field strength changes with distance the summation is the area under the graph between R_1 and R_2 (Fig. 11.8(b)). Chapter 3 gives an example of this summation.

The integral can be written as

$$\begin{aligned}\text{Potential difference} &= \int_{R_1}^{R_2} \frac{GM}{R^2}.\mathrm{d}R \\ &= \left[-\frac{GM}{R}\right]_{R_1}^{R_2} \\ &= \frac{GM}{R_1} - \frac{GM}{R_2}\end{aligned}$$

It is customary to measure potential differences between infinity, our zero, and a particular point if we want to specify the absolute potential at the point. Thus if we fetch a mass from infinity we have $R_1 = \infty$ and thus

$$\text{Potential} = -\frac{GM}{R_2}$$

where R_2 is the distance from the centre of mass M at the point we are considering. For a spherical mass the potential varies as the reciprocal of the distance. We can plot lines of constant potential around a spherical mass—they are called equipotentials (Fig. 11.9). The minus sign in the equation means that the potential increases, i.e., becomes more positive, as the distance increases. This sign is a result of our convention of taking the potential to be zero at infinity.

The field strength is the potential gradient. Rewriting the integral in differential form shows this.

$$\text{Potential difference} = \int_{R_1}^{R_2} g\,\mathrm{d}R$$

$$\mathrm{d}V = g\,\mathrm{d}R$$

$$g = \frac{\mathrm{d}V}{\mathrm{d}R}$$

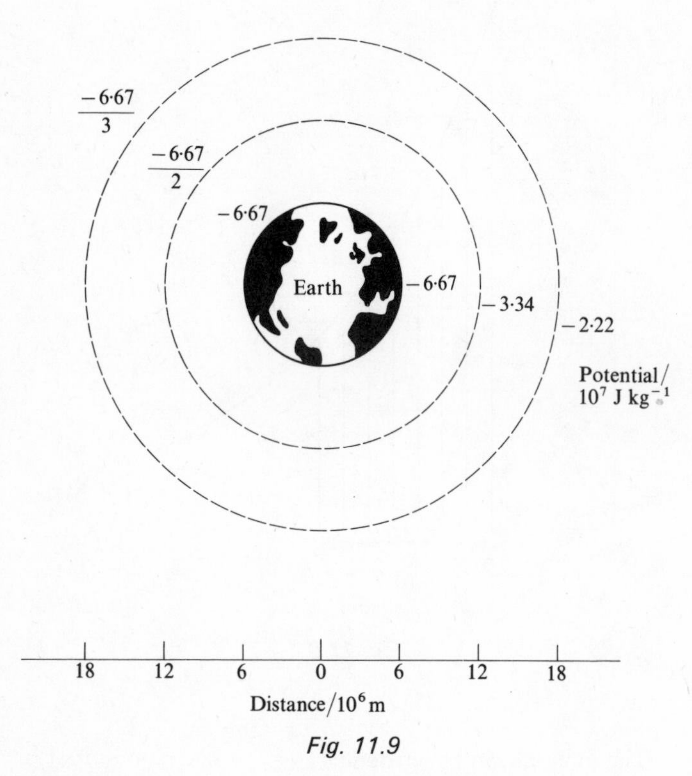

Fig. 11.9

Matter

What is matter? What is the property of matter called mass? We can say that two pieces of matter attract each other and that the force of attraction depends on their masses. The only observable event is however, motion—we can say that this is the result of the force of attraction which arises because mass is a source of force.

> '... What has been thought of as a particle will have to be thought of as a series of events. The series of events that replaces a particle has certain important physical properties, and therefore demands our attention; but it has no more substantiality than any other series of events that we might arbitrarily single out. Thus "matter" is not part of the ultimate material of the world, but merely a convenient way of collecting events into bundles.'
>
> B. Russell (1961). *History of western philosophy*, p. 786, Allen & Unwin.

> '... We need not regard matter as a foreign entity causing a disturbance in the gravitational field; the disturbance is matter. In the same way we do not regard light as an intruder in the electromagnetic field, causing the electromagnetic field to oscillate along its path; the oscillations constitute light. Nor is heat a fluid causing agitation of the molecules of a body; the agitation is heat.
>
> This view, that matter is a symptom and not a cause, seems so natural....'
>
> A. Eddington (1960). *Space, time and gravitation*, Cambridge Univ. Press.

11.3 Electric fields

Suppose we take a charged particle and place it near another charged particle, it experiences a force. If our charged particle has twice the charge the force is doubled. Coulomb's law in chapter 2 considered the effects of changing the charges on two spheres.

$$F = \frac{q_1 q_2}{4\pi\varepsilon_0 r^2}$$

where q_1 and q_2 are the charges on the two bodies, r their distance apart and ε_0 a constant. With all the points in the space around a charge we can associate a force—the force that a test charge would experience if placed there. We can imagine that there is a field of force surrounding the charge and assign to each point a field strength, the field strength being the force per unit charge. The direction of the electric field is taken as the direction of the force on a positive charge.

$$\text{Electric field strength} = E = \frac{\text{force}}{\text{charge}}$$

At a distance r from a point charge Q the force on a test charge q will be

$$F = \frac{Qq}{4\pi\varepsilon_0 r^2}$$

and thus as $E = F/q$ we have

$$E = \frac{Q}{4\pi\varepsilon_0 r^2}$$

The electric field strength varies as the reciprocal of the square of the distance from the charge. The same equation applies to a spherical charge, provided r is greater than the radius of the sphere. This can be shown in the same way as for the gravitational field in 11.2. Within a charged shell the electric field is zero—this can be derived in the same way as the gravitational case.

The Coulomb's law equation and the electric field concept have implicit in them the idea that electric charge is additive in its effect. Doubling the charge at one point doubles the force on another charge, i.e., doubles the field at the point where the second charge is situated. But also the force acting on the second charge due to the first one is not influenced by the presence of another, third, charge. Coulomb's law can be used to calculate the interaction force for any pair of charges, regardless of whether other charges are present. We can calculate the force on one charge due to the presence of say two other charges by calculating the force due to each charge separately and then finding the vector sum.

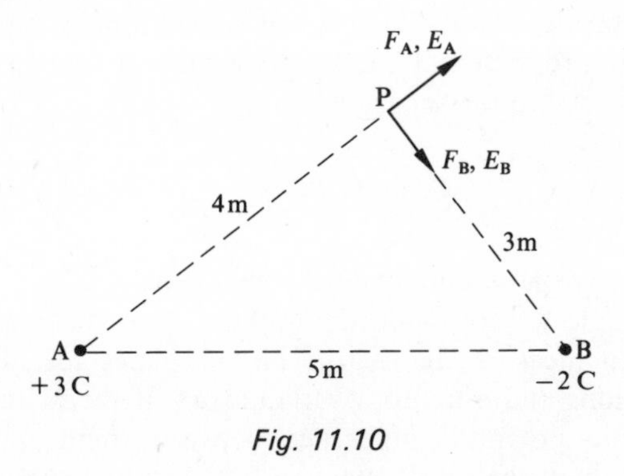

Fig. 11.10

Consider two charges, A with +3 C and B with −2 C (Fig. 11.10) and the field produced by them at a point P. P is 4 m from A and 3 m from B. A and B are 5 m apart. To calculate the field we consider the force on a +1 C charge placed at P. The field at P due to A will be

$$E_A = \frac{3}{4\pi\varepsilon_0 4^2}$$

The field at P due to B will be

$$E_B = -\frac{2}{4\pi\varepsilon_0 3^2}$$

To find the resultant field at P we must take into account the directions of these two fields. In this particular case the two fields are at right angles and thus the resultant field E is given by

$$E^2 = E_A^2 + E_B^2$$

$$E = \frac{1}{4\pi\varepsilon_0}\sqrt{\left[\left(\frac{3}{16}\right)^2 + \left(-\frac{2}{9}\right)^2\right]}$$

$$E = \frac{0{\cdot}5}{4\pi\varepsilon_0}$$

The tangent of the angle between this resultant and the E_B direction is given by

$$\tan\theta = \frac{E_A}{E_B}$$

The directions of the electric field can be plotted in the space around the charge to give a field map. The lines marking these directions are known as lines of force (Fig. 11.11).

Electric field patterns can be 'seen' if charged conductors are in an insulating liquid and grass seed is sprinkled on the surface of the liquid. The grass seed lines up along the direction of the field.

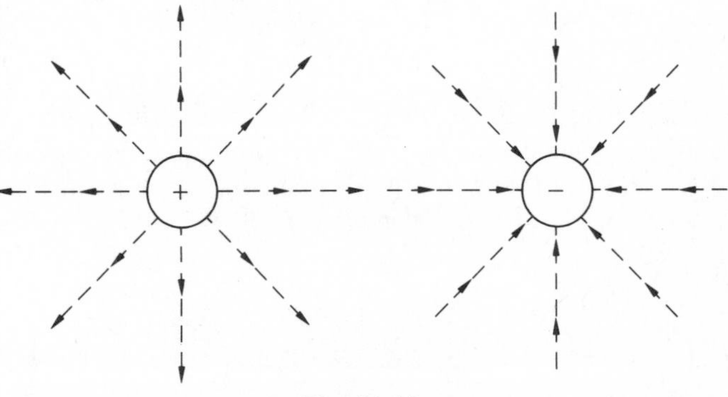

Fig. 11.11

A case which is of particular interest is the one involving two parallel plates. The electric field is everywhere between the plates at right angles to the plates. Indeed for any conducting surface the electric field is always at right angles to the surface. If the field were not at right angles there would be a component of the field parallel to the conducting surface—this would cause electrons to move within the surface and hence a current flow. This current would continue until the electrons have so arranged themselves that the electric field is at right angles to the surface. The parallel plate arrangement is a capacitor and we would normally charge the plates by connecting a potential difference between the plates. This would result in one plate having a positive charge and the other a negative charge. The two charges would have the same magnitude. The charges would, however, be spread over the plates—two carpets of charge facing each other.

A voltmeter connected between the two parallel plates gives a reading V when there is an electric field between the plates. When $V = 0$ there is no field. The voltmeter reading tells us the energy transformed when a unit charge moves between the plates.

$$\text{Energy} = Vq$$

There is nothing in this equation which suggests that it matters between which two points on the plates the charge moves. The field between the plates is uniform. This can be tested by putting a charged oil drop between the plates (the Millikan experiment)—the drop experiences the same force, the same acceleration is measured, wherever the drop is between the plates. Some peculiarities occur near the edges but the greater part of the volume between the plates has the same field strength. The energy transformed by moving a charge between the plates must be the product of the force acting on the drop and the distance through which the drop is moved.

$$\text{Energy} = Fd$$

where d is the separation of the plates and F the force acting on the charge due to the electric field. Hence

$$Fd = Vq$$

or as $E = F/q$ we must have, for the field strength between the plates,

$$E = \frac{F}{q} = \frac{V}{d}$$

The units of E can be N kg^{-1} or V m^{-1}. E is the potential gradient. Our definitions (Fig. 11.12) of the directions of the field and the potential gradient are such as to require us to write $E = -$(potential gradient).

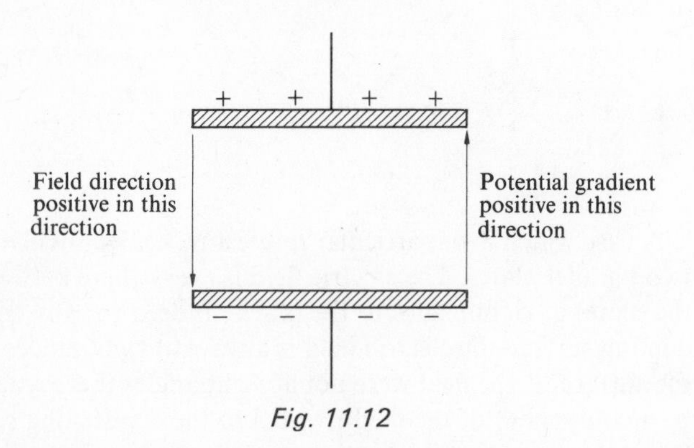

Fig. 11.12

Suppose we move the charge in one case along the direction of the field, i.e., perpendicular to the plates, and in the second case at an angle to the field—would this affect the above analysis? Consider moving the charge along lines BC and RS in Fig. 11.13. Along the line RS we have a force F acting and the distance over which the charge is moved is d—the energy transformed is thus Fd. Along the line BC we have a force component $F/\sin\theta$ and the distance the charge is moved is $d \sin\theta$—the energy transformed is thus

$$\frac{F}{\sin\theta} d \sin\theta = Fd$$

The result is precisely the same. It does not matter what path is chosen between the two plates the energy transformed is the same. The result is more general than this—the energy transformed is independent of the path followed by a charge moving between any two equipotentials. A field with this property is called conservative.

The electric field between parallel plates thus depends on the potential difference between the plates and their separation.

$$E = \frac{V}{d}$$

But the charge on the plates Q is related to the potential difference V

$$Q = CV$$

where C is the capacitance. Thus we have

$$E = \frac{Q}{Cd}$$

The electric field between the plates is proportional to the charge on the plates. The capacitance of a pair of parallel plates is proportional to the plate area A and inversely proportional to the separation

$$C \propto \frac{A}{d}$$

These are experimentally determined facts.

A simple way of determining these is to use an electrometer to measure the charge on the plates for different separations and different overlap areas. If the plates have the same potential difference between them for each measurement then the capacitance is directly proportional to the charge ($Q = CV$). The electrometer measures the charge by being connected to one of the parallel plates—after it has been charged by the application of a potential difference. The charge on the plate becomes shared between the plate and a capacitor which is across the electrometer terminals. If the parallel plate arrangement has a small capacitance compared with that of the electrometer capacitor then most of the charge moves to the electrometer capacitor. The electrometer is really a very high impedance voltmeter and measures the potential difference across the electrometer capacitor and as its capacitance is known the charge on it can be calculated. Some electrometer instruments have built-in capacitors and voltmeter scales directly calibrated in coulombs.

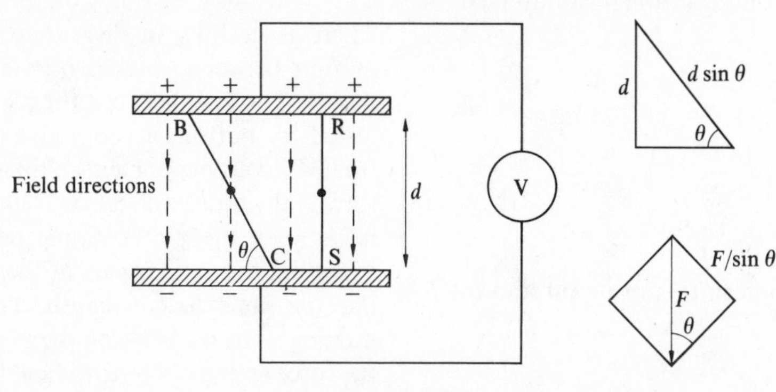

Fig. 11.13

Hence the electric field between the parallel plates is given by

$$E = \frac{Q}{Cd}$$

and as we have

$$C \propto \frac{A}{d}$$

we must have

$$E \propto \frac{Q}{A}$$

The electric field is thus proportional to the charge per unit area of plate. This is known as the charge density σ.

$$E \propto \sigma$$

The constant of proportionality relating the electric field and the charge density is the reciprocal of the same constant as relates the capacitance of a parallel plate capacitor and the plate area divided by the plate separation. It is possible to measure the capacitance of a parallel plate capacitor, perhaps by the use of the electrometer, and measure the plate area and separation. Hence the constant can be experimentally determined. It turns out to be a constant already met—the constant ε_0 which occurs for the force between spherical charges. Why is this?

$$E = \frac{\sigma}{\varepsilon_0}$$

(The constant depends on the medium. We are considering a vacuum, though air is a good approximation.)

The electric field at a point between parallel charged plates does not depend on the distance of the point from a charged plate. The charges on a plate can be thought of as a carpet of point charges. Each individual charge will give an electric field which depends on the reciprocal of the square of the distance—the net result does, however, seem to be an electric field independent of distance when we have plates or carpets of charge. How can this be?

Consider the field at a point P above a carpet of charge. We will divide the carpet into small segments of charge and consider the effects due to these separate small charges, for which the inverse square law will apply. Firstly we will consider the field at a point which is height h above the carpet.

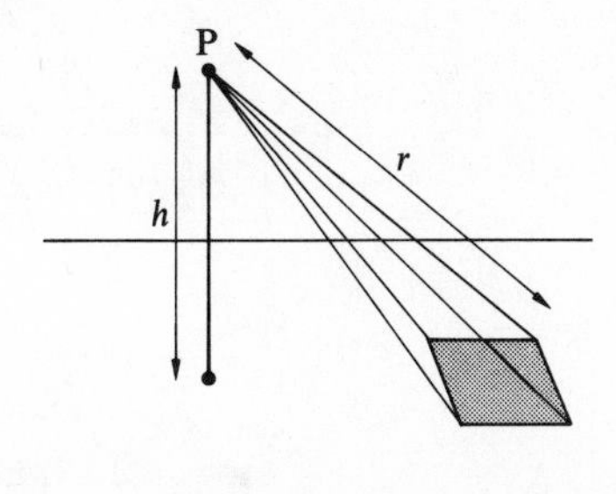

Fig. 11.14

The field at this point P (Fig. 11.14) due to an area of the carpet ΔA will be

$$\frac{\text{charge on } \Delta A}{4\pi\varepsilon_0 r^2}$$

where r is the distance of P from the area. If the charge density is σ then the charge on area ΔA will be $\sigma\ \Delta A$ and thus the field is

$$\frac{\sigma\ \Delta A}{4\pi\varepsilon_0 r^2}$$

To find the total field at P we must sum the effects due to all the separate areas, like ΔA which make up the carpet of charge. Before looking at this we will consider the problem of the field being independent of distance.

Consider the field at a point P which is at a height $2h$ above the carpet of charge. The field at this point due to the entire carpet of charge would be obtained by summing all the effects due to a large number of small areas of charge. We will keep the same number of small areas as when we considered P to be at a height h. To find the area of one of these new patches we merely need to extend the bundle of lines from point P which outlined the earlier area ΔA until they meet the plane distance $2h$ from P (see Fig. 11.15). Doubling the height doubles each side of the rectangle marking out the charge area and thus we now have an area of $4\ \Delta A$. The charge on this area is thus $4\sigma\ \Delta A$. This area is, however, a distance $2r$ from P. Thus the field due to this area is

$$\frac{4\sigma\ \Delta A}{4\pi\varepsilon_0 (2r)^2}$$

This becomes

$$\frac{\sigma\ \Delta A}{4\pi\varepsilon_0 r^2}$$

the same value as we obtained for a height h. This thus confirms the experimental result that the field above a charged plate is independent of the distance away from the plate.

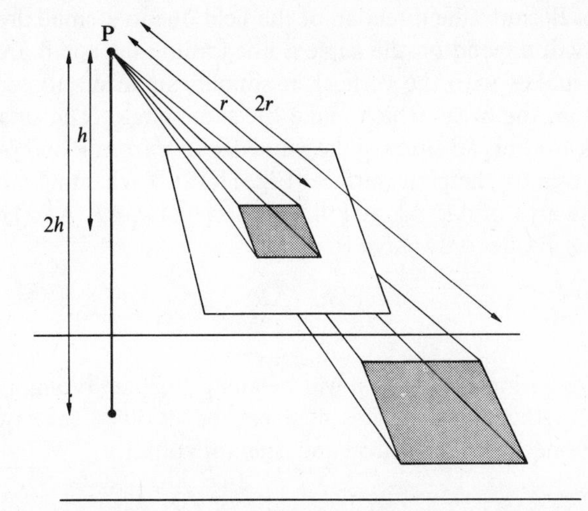

Fig. 11.15

The sum of all these terms will be the same in both cases if we have the same number of areas. This will be the case as long as the bundles of lines marking out the areas do not go beyond the edge of the plate. Figure 11.16(a) shows this. The result is thus valid provided point P is close enough to a large plate for virtually all the bundles of lines to fall on the plate (Fig. 11.16(b)).

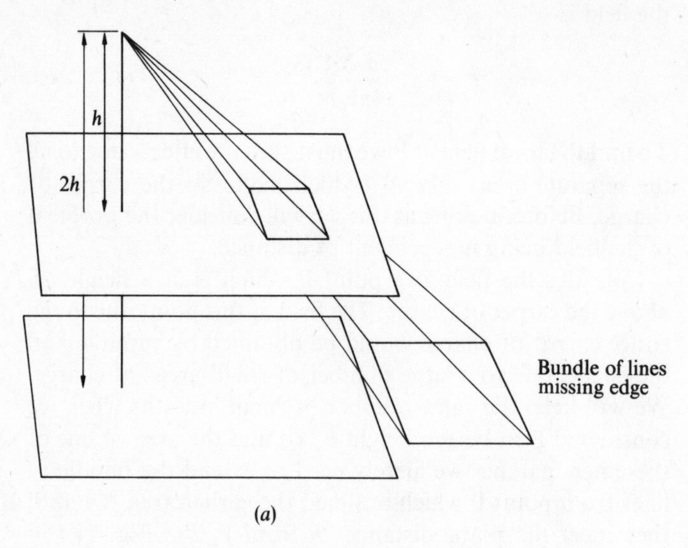

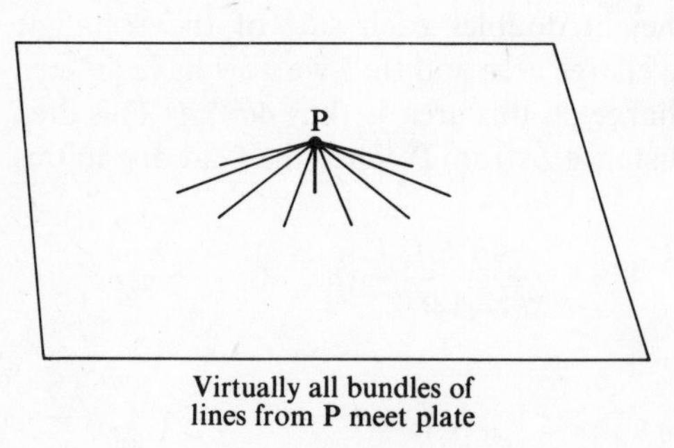

Fig. 11.16

Now consider the field at P due to all the small areas of charge which make up the carpet of charge on a large flat plate. Because the direction of the field due to a small area at P will depend on the angle a line joining the small area to P makes with the vertical, it simplifies matters to consider all the areas which make the same angle. The small areas are thus all situated the same distance from P and fall on a ring on the plate surface (Fig. 11.17). Each small area has an area of $\Delta L\ \Delta R$ and thus a charge of $\sigma\ \Delta L\ \Delta R$. The field at P due to this area is

$$\frac{\sigma\ \Delta L\ \Delta R}{4\pi\varepsilon_0 r^2}$$

The direction of this field will be along the line joining the area to the point P. This field can be resolved into two components—one vertical and one horizontal.

$$\text{Vertical component of field} = \frac{\sigma\ \Delta L\ \Delta R}{4\pi\varepsilon_0 r^2}\cos\theta$$

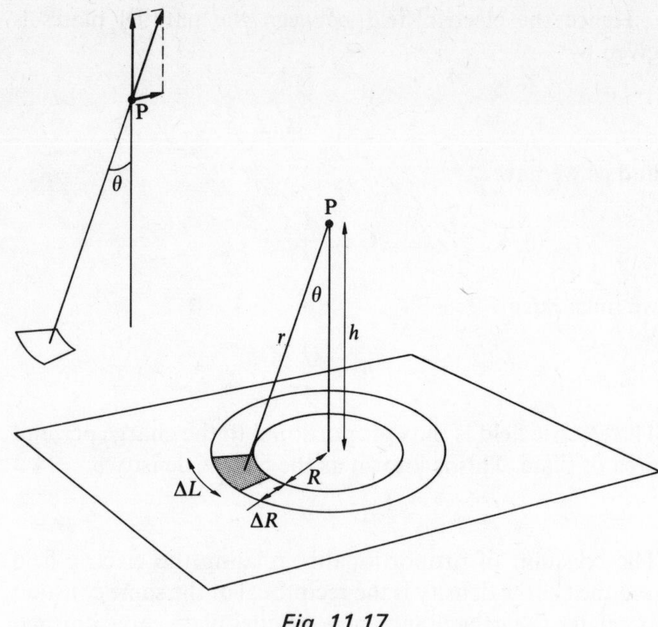

Fig. 11.17

$$\text{Horizontal component of field} = \frac{\sigma\ \Delta L\ \Delta R}{4\pi\varepsilon_0 r^2}\sin\theta$$

If we now take into account all the areas which make up the ring, the horizontal components will all cancel out. Areas diametrically opposite each other give equal and opposite horizontal components. The vertical components are, however, all in the same direction. The number of small areas making up the ring will be the circumference of the ring divided by the length of one area, i.e., $2\pi R/\Delta L$. Thus the total vertical field due to all the areas which make up the ring will be

$$\frac{2\pi R}{\Delta L}\cdot\frac{\sigma\ \Delta L\ \Delta R}{4\pi\varepsilon_0 r^2}\cos\theta$$

$$= \frac{\sigma R\ \Delta R}{2\varepsilon_0 r^2}\cos\theta$$

The ring only covers a small sweep of the total angle which we must sum over in order to find the total field due to all

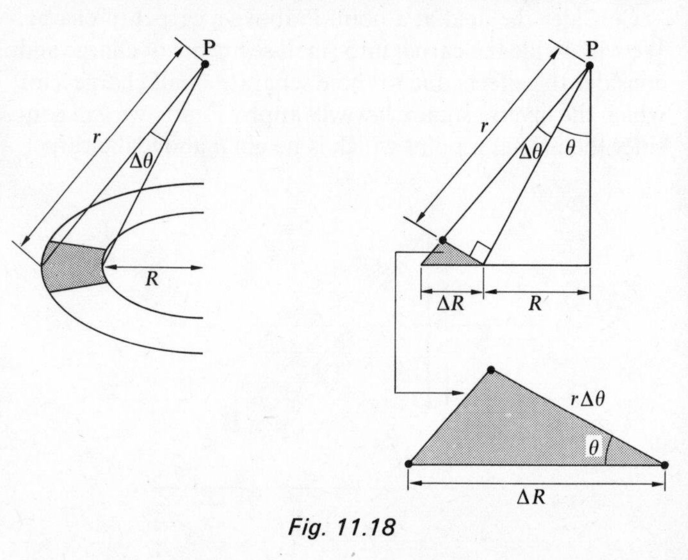

Fig. 11.18

the charge on the plate. The ring covers an angle of $\Delta\theta$ (Fig. 11.18). From the diagram we have

$$\cos\theta = \frac{r\,\Delta\theta}{\Delta R}$$

The same result could have been obtained by differentiating $R/r = \sin\theta$. Substituting this in our field equation gives

$$\text{Vertical component of field due to ring} = \frac{\sigma R\,\Delta\theta}{2\varepsilon_0 r}$$

But $R/r = \sin\theta$, thus

$$\text{Vertical component} = \frac{\sigma}{2\varepsilon_0}\sin\theta\,\Delta\theta$$

To find the total field due to all the rings we need to sum this result for all angles from 0° to 90° or 0 to $\pi/2$.

$$\text{Field at P} = \int_0^{\pi/2} \frac{\sigma}{2\varepsilon_0}\sin\theta\,d\theta$$

The integral $\int_0^{\pi/2}\sin\theta\,d\theta$ has the value 1. Hence the field at P due to a plane of charge is

$$\text{Field} = \frac{\sigma}{2\varepsilon_0}$$

If we have two parallel and oppositely charged planes of charge then the field at any point between them becomes (Fig. 11.19)

$$2 \times \frac{\sigma}{2\varepsilon_0} \text{ or } \frac{\sigma}{\varepsilon_0}$$

Thus we have

$$E = \frac{\sigma}{\varepsilon_0}$$

At points above or below the two plates the fields due to each plate cancel. The field is almost entirely located between the plates.

We can think of matter as made up of sheets of positive and negative charge—between the sheets large electric fields, outside the sheets virtually no field. There will thus be large forces between the charges, holding them together, and very low forces between the charges on one block of matter and the charges in another block of matter.

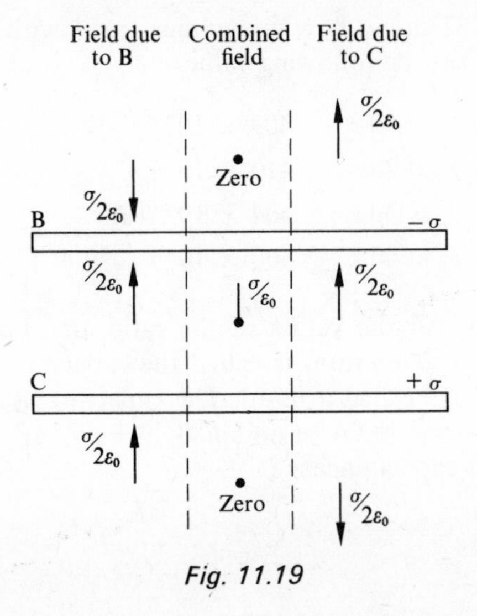

Fig. 11.19

Capacitors

When a potential difference is applied across two plates they become charged, one positive and the other negative. A galvanometer in the charging circuit shows a current which decreases exponentially with time (see Chapter 5). The potential difference across the two plates rises exponentially with time. How does the charge on a plate vary with time?

$$I = \frac{\Delta Q}{\Delta t}$$

Thus the change in charge on a plate in time Δt is

$$\Delta Q = I\,\Delta t$$

We can find the charge at any instant by summing all the $I\,\Delta t$ values that have occurred since switching on. Figure 11.20 shows results obtained with the arrangement shown. Consider time intervals Δt of 10 s. In the first 10 s the value of $I\,\Delta t$ is about 90 × 10. Thus the charge that flows onto the plate in the first 10 s is 900 μC; I is in microamps. In the next 10 s the charge flowing onto the plate is about 70 × 10 or 700 μC. The total charge now on the plate is thus 900 + 700 = 1600 μC. In the next 10 s the charge flowing onto the plate is about 54 × 10 or 540 μC and thus the total charge is 2140 μC. In the next 10 s the charge

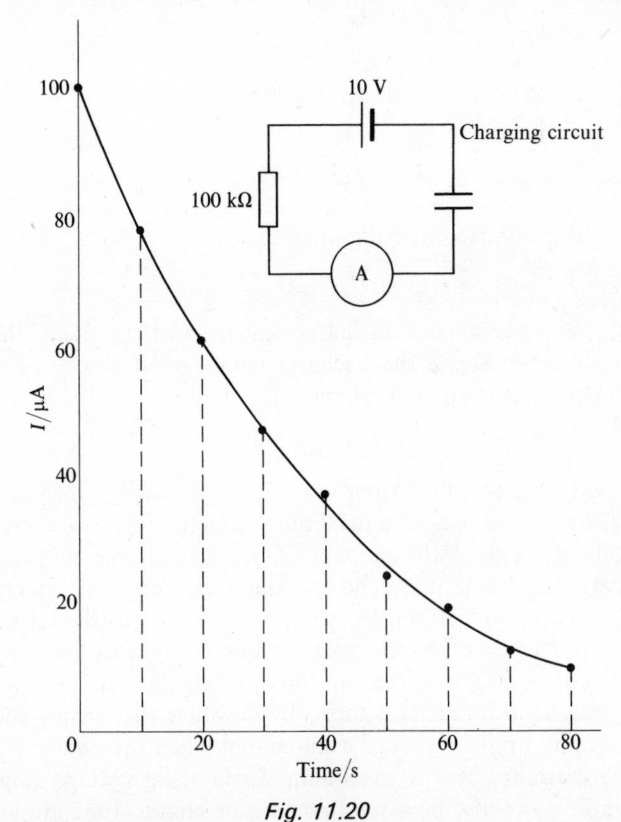

Fig. 11.20

flowing onto the plate is about 420 μC; in the next 10 s about 300 μC, in the next 10 s about 210 μC, in the next 10 s about 150 μC, the next 10 s about 110 μC. Figure 11.21 is a plot of these charge results.

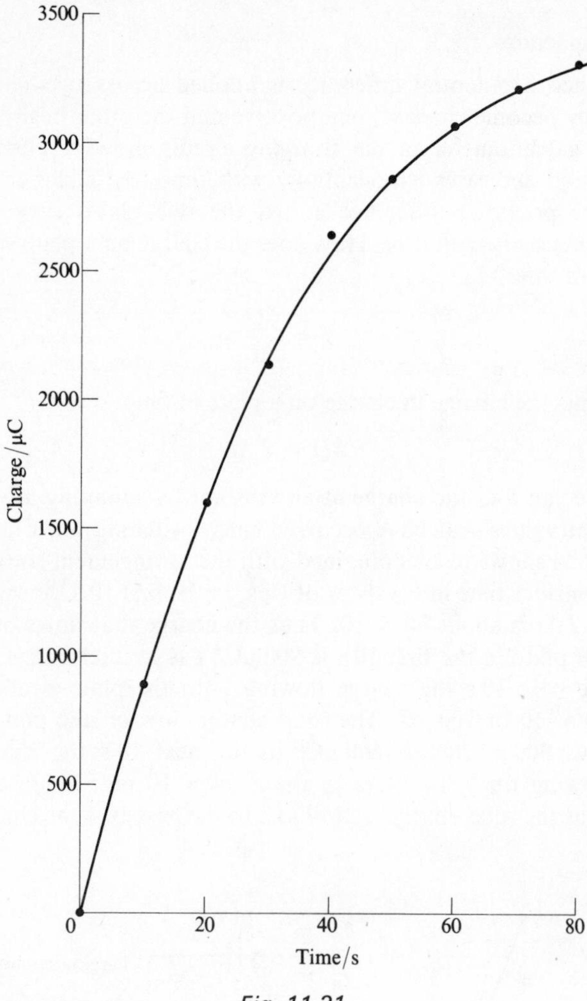

Fig. 11.21

The graph has the equation

$$Q = Q_m - Q_m e^{-t/CR}$$

Q_m being the maximum charge reached by the plate, C the capacitance, and R the discharge circuit resistance.

The current–time graph has the equation

$$I = I_m e^{-t/CR}$$

I_m is the maximum current.

The charge rises to a maximum value, in this case about 3500 μC. This is the amount of negative charge that has flowed onto one plate and the amount of negative charge that has flowed off the other plate, leaving it positive. If we had put a very high impedance voltmeter, say a cathode ray oscilloscope, across the capacitor during the charging we would have found that the voltage would rise during the charging until it reached a maximum when the charge on the capacitor was a maximum. In fact the voltage–time graph is exactly the same shape as the charge–time graph. The ratio of the charge on a plate to the voltage at any instant is a constant.

$$\frac{Q}{V} = \text{a constant}$$

We could construct the voltage–time graph from the current–time graph. The potential difference across the capacitor is equal to the potential difference across the battery terminals (10 V) minus the potential drop across the resistor in the charging circuit.

$$V = 10 - IR$$

R has the nominal value of 100 kΩ. Thus the voltage across the capacitor at time $t = 0$ is

$$V = 10 - 100 \times 10^{-6} \times 100 \times 10^3 = 0$$

After 10 s the voltage is

$$V = 10 - 78 \times 10^{-6} \times 100 \times 10^3 = 2{\cdot}2 \text{ V}$$

After 20 s

$$V = 10 - 61 \times 10^{-6} \times 100 \times 10^3 = 3{\cdot}9 \text{ V}$$

After 30 s

$$V = 10 - 47 \times 10^{-6} \times 100 \times 10^3 = 5{\cdot}3 \text{ V}$$

After 40 s

$$V = 10 - 36 \times 10^{-6} \times 100 \times 10^3 = 6{\cdot}4 \text{ V}$$

After 50 s

$$V = 10 - 26 \times 10^{-6} \times 100 \times 10^3 = 7{\cdot}4 \text{ V}$$

After 60 s

$$V = 10 - 18 \times 10^{-6} \times 100 \times 10^3 = 8{\cdot}2 \text{ V}$$

After 70 s

$$V = 10 - 13 \times 10^{-6} \times 100 \times 10^3 = 8{\cdot}7 \text{ V}$$

After 80 s

$$V = 10 - 10 \times 10^{-6} \times 100 \times 10^3 = 9{\cdot}0 \text{ V}$$

Figure 11.22 shows how the voltage varies with time. The ratio Q/V has the following values:

at 10 s	409×10^{-6} C V^{-1}
at 20 s	410×10^{-6} C V^{-1}
at 30 s	404×10^{-6} C V^{-1}
at 40 s	400×10^{-6} C V^{-1}

and so on. All the values of the ratio are about 400×10^{-6} C V^{-1}. This ratio is called the capacitance and the unit C V^{-1} is called a farad. The capacitor used for this experiment was in fact a nominally 400 μF capacitor. The symbol for capacitance is C.

$$\frac{Q}{V} = C$$

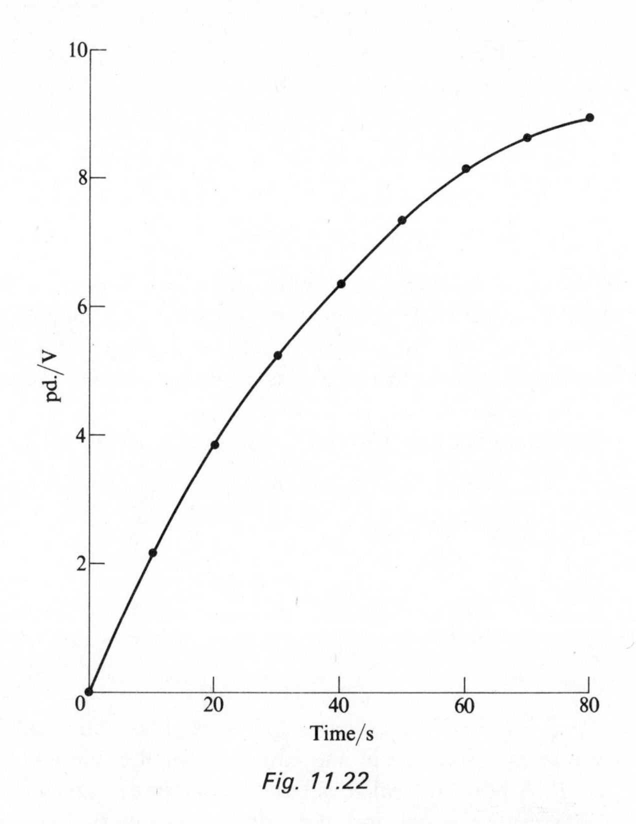

Fig. 11.22

The discharge of a capacitor through a resistor is considered in chapter 5.

Energy is needed to move a charge through a potential difference. To charge a capacitor we bring charges and put them on one of its plates. For the first charge brought up no energy is needed as the potential of the plate is zero (we assume the plate is initially uncharged). As we keep on putting charges on the plate so its potential increases and so for each new charge brought up we have to supply more energy to get the charge up the potential gradient: the gradient keeps on growing steeper. We can think of the charge on the plate exerting bigger and bigger repulsive forces on the new charge being brought up, the forces grow because of the increase of charge on the plate.

If at some instant the potential on the plate is V and we bring up a new charge ΔQ then the energy needed will be $V\,\Delta Q$. This in fact the area under a graph of V against Q (Fig. 11.23) which corresponds to the potential and size of ΔQ. The total energy used in charging up the capacitor to a potential V will be the area under the V against Q graph from zero potential to V. The area is $\frac{1}{2}VQ$.

$$\text{Energy} = \tfrac{1}{2}VQ$$

As $Q = CV$ we can write this as

$$\text{Energy} = \tfrac{1}{2}CV^2$$

If we charge a capacitor up to 4 V we supply 16 times the energy we did when charging it up to 1 V. When we discharge a capacitor the energy used in putting the charge onto the capacitor plates is released. Capacitors are convenient ways of storing energy. In an electric flashgun used

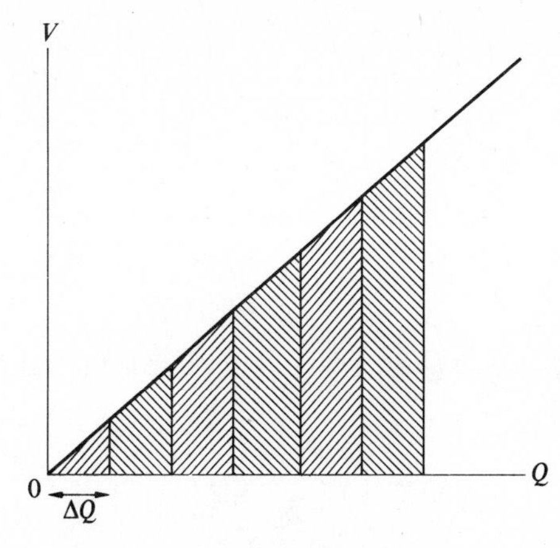

Fig. 11.23

in photography we charge a capacitor up by slowly supplying energy and then release it, all in one quick lump, by discharging it through the flashbulb.

In a d.c. supplied circuit a capacitor only gives a current when the supply is switched on and when it is switched off, i.e., during charging and discharging. In an a.c. circuit we can think of the current as being switched on, off, reversed, off, on, etc. What is the effect of a capacitor in such a circuit? With a.c. and a capacitor in the circuit (Fig. 11.24) we continue to have a current, an alternating current, after switching on. The capacitor is continually charging and discharging.

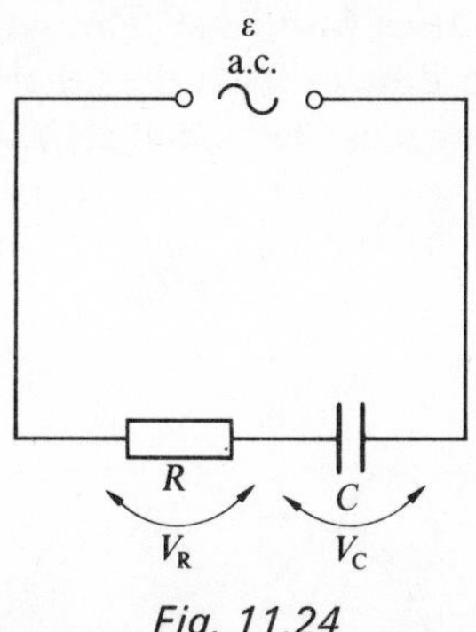

Fig. 11.24

If ε is the e.m.f. of the supply, V_R the potential drop across the resistance in the circuit and V_C the potential drop across the capacitor, we must have

$$\varepsilon = V_R + V_C$$

For the capacitor

$$V_C = \frac{Q}{C}$$

where Q is the charge at some instant on a capacitor plate, C is the capacitance. Thus V_C will vary with time as the charge on the capacitor changes with time. For the resistor

$$V_R = RI$$

where R is the resistance and I the current in the circuit at some instant. Hence

$$\varepsilon = RI + \frac{Q}{C}$$

But

$$I = \frac{dQ}{dt}$$

Thus the charge Q on a capacitor plate is the sum of all the $I\,dt$ terms

$$Q = \int_{Q \text{ at } t=0}^{Q \text{ at time } t} dQ = \int_0^t I\,dt$$

Hence

$$\varepsilon = RI + \int_0^t \frac{I}{C}\,dt$$

If the variation of I with time is given by

$$I = I_m \sin 2\pi ft$$

where I_m is the maximum current, we have

$$\varepsilon = RI_m \sin 2\pi ft + \int_0^t \frac{I_m}{C} \sin 2\pi ft$$

$$\varepsilon = RI_m \sin 2\pi ft - \frac{I_m}{2\pi fC} \cos 2\pi ft$$

Figure 11.25 shows how these terms vary with time and how the sum, ε, varies with time. The sum can be arrived at by adding the ordinates from the other two curves. The term $1/2\pi fC$ is known as the reactance X_C of the capacitor.

$$X_C = \frac{1}{2\pi fC}$$

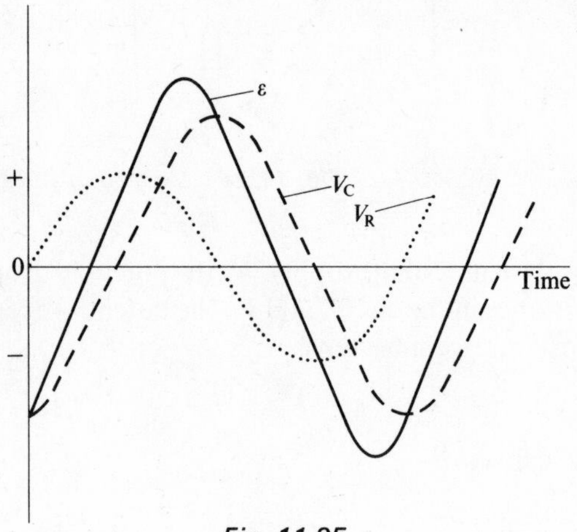

Fig. 11.25

The equation, of the variation of ε with time shown in Fig. 11.25, can be simplified to give

$$\varepsilon = \sqrt{[(RI_m)^2 + (I_m/2\pi fC)^2]} \sin (2\pi ft - \theta)$$

or

$$\varepsilon = \sqrt{[(RI_m)^2 + (X_C I_m)^2]} \sin (2\pi ft - \theta)$$

θ is the angle by which the e.m.f., ε, is behind the V_R term, or the current in the circuit which is in phase with the potential difference across the resistor. This angle is known as the phase difference.

$$\tan \theta = \frac{1}{2\pi fRC}$$

The e.m.f., ε, reaches a maximum an angle θ later than the potential difference across the resistor. The potential difference across the capacitor reaches a maximum 90° after that across the resistor, it is a negative cosine graph, whereas the p.d. across R is a sine graph.

The maximum e.m.f. in the circuit occurs when

$$\sin (2\pi ft - \theta) = 1$$

Then we have

$$\varepsilon_m = I_m\sqrt{(R^2 + X_C^2)}$$

We can rewrite this as

$$\varepsilon_m = I_m Z$$

where Z is called the circuit impedance. It is comparable to the term resistance in the equation for d.c. circuits of $V = IR$. A point to realize is that the maximum e.m.f. and the maximum current do not occur at the same times.

$$\text{Impedance } Z = \sqrt{(R^2 + X_C^2)}$$

The reactance term represents the effect of the capacitor on the overall circuit impedance. The units of Z, R, and X_C are all ohms.

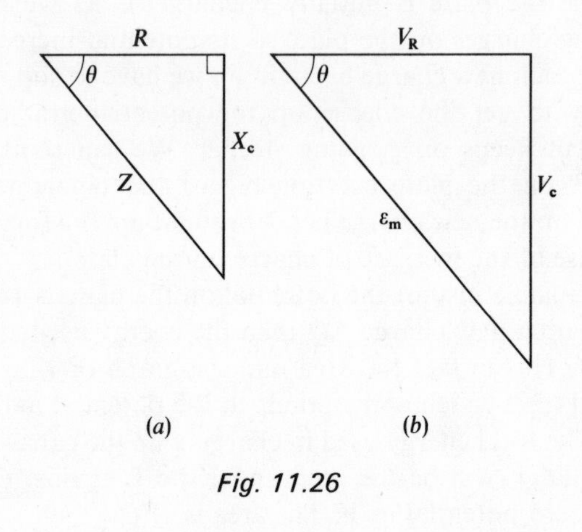

Fig. 11.26

The above equation is similar to the equation relating the lengths of the sides for a right angled triangle,

$$\text{Hypotenuse} = \sqrt{[(\text{base})^2 + (\text{vertical})^2]}$$

We can represent Z by the hypotenuse of a triangle, R by the base, and X_C by the vertical (Fig. 11.26(a)). The hypotenuse makes an angle with the base of

$$\tan \theta = \frac{X_C}{R}$$

This is the same as the angle θ in our earlier expression, i.e., the phase angle.

$$\tan\theta = \frac{1}{2\pi fRC}$$

As $V_R = I_m R$ and $V_C = I_m X_C$ the triangle also represents the relationship between the e.m.f. and the potential differences in the circuit (Fig. 11.26(b)).

$$\varepsilon_m^2 = (RI_m^2) + (X_C I_m)^2$$

Lightning

'. . . It is possible that by the dashing of two clouds the lightning may flash out as is the case when two stones are struck against each other. . . .'

Pliny (2nd century A.D.).

'I have been much amused by ye singular phenomena resulting from bringing a needle into contact with a piece of amber or resin fricated on a silk cloth. Ye flame putteth me in mind of sheet lightning on a small—how very small—scale.

I. Newton (1716).

Small sparks can be observed when some charged objects discharge. The effect is most pronounced when conditions are dry. Newton observed the spark (ye flame) when he brought a needle close to a piece of amber or resin which had been rubbed (fricated—same word stem as friction) against a silk cloth. Lightning is a discharge on a much grander scale—not clouds dashing against each other. That clouds carry charge, was first shown, it is said, by Franklin's experiment in which he flew a kite up into a thundercloud. Rain made his kite string conducting and charge flowed down his kite string. A knuckle of his hand placed close to the end of the wet kite string produced a spark between his hand and the string.

What is the cause of such sparks, the small ones and lightning?

A charged capacitor will not maintain its charge indefinitely—the charge slowly leaks away. This is due to the

Fig. 11.27 Lightning. (R. F. Addie.)

ions that are always present in normal air. These ions are in the main produced by cosmic radiation. Thus if we think of two parallel charged capacitor plates the ions will move between the plates, the positive ions moving to the negative plate and the negative ions to the positive plate. The plates are charged and thus there is a potential difference between the plates and so an electric field. This accelerates the ions.

$$\tfrac{1}{2}mv^2 = Vq$$

v is the velocity acquired by an ion of mass m and charge q moving through a potential difference V. The greater the potential difference V the larger we would expect the kinetic energy to be. There is, however, air between the plates and the ions will collide with air molecules many times in their movement between plates. The vital factor in thus deciding the kinetic energy that can be acquired by an ion between collisions is therefore the potential difference per unit length of ion path. The greater the value of this the greater the energy of an ion immediately prior to a collision. The potential difference per unit length has been already met—it is the electric field.

$$E = \frac{V}{d}$$

If the kinetic energy acquired by an ion is large enough it can ionize an air molecule when they collide. The energy needed for this will be about 10 electron volts, i.e., about 10^{-18} J. For air molecules moving about among air molecules at atmospheric pressure and normal room temperature the mean distance between collisions is about 5×10^{-7} m. An ionized air molecule could be expected to have the same mean distance between collisions. Electrons are considerably smaller than ionized air molecules and thus collide less with air molecules, their mean distance between collisions would be about 2×10^{-6} m. This mean distance is called the mean free path (see chapter 9). Thus the ions or electrons must acquire an energy equivalent to being accelerated through 10 volts in a distance ranging from 5×10^{-7} to 2×10^{-6} m. For capacitor plates 1 cm apart these would mean a voltage of 2×10^5 V for ionized air molecules to produce ionization and 5×10^4 V for electrons to produce ionization.

In dry air, at atmospheric pressure and normal room temperature, with plates 1 cm apart a spark will jump between the plates when they have a potential difference between them of 3×10^4 V. This is of the same order as our calculated value for electrons to produce ionization.

When ionization occurs there is an increase in the number of ions and electrons, these in turn can become accelerated and produce more ions and electrons, these in turn produce more. Thus when ionization starts there is a very rapid build-up in the number of charge carriers which can cause leakage of the charge from the capacitor plates. The effect is known as an avalanche and the voltage at which it occurs as the breakdown voltage. All these ions provide a conducting path of low resistance between the two plates. The resistance thus suddenly drops from a very high value (air is a reasonable insulator) to almost a short circuit. The effect of this is for the charge on the plates to become very quickly discharged along the leakage path provided by the ions. It is this surge which appears as the spark. The ions and electrons rushing between the plates excite many air molecules and visible light is produced.

The presence of water droplets considerably reduces the breakdown voltage: water molecules act as small dipoles. Drops about 1 mm radius reduce the breakdown voltage to about 10 000 V for a 1 cm path. Bigger drops reduce the breakdown value even more.

The discharge of charge occurs more readily from an object which has points or sharp edges than one that is perfectly smooth. Lightning conductors are points for this very reason. Newton put a needle close to his charged piece of amber and obtained a spark—leakage of the charge from the amber occurred rapidly when the needle was brought close. For a charged conducting sphere we can think of all the charge Q as being at the centre, the potential at the surface of the sphere, radius R, is thus

$$V = \frac{Q}{4\pi\varepsilon_0 R}$$

Suppose we have two spheres connected together by a conducting wire. Both spheres will have surfaces at the same potential because they are connected together. If the second sphere has a radius r then the charge q must be given by

$$V = \frac{Q}{4\pi\varepsilon_0 R} = \frac{q}{4\pi\varepsilon_0 r}$$

$$\frac{Q}{R} = \frac{q}{r}$$

The electric field at the surface for the first sphere is given by

$$E = \frac{Q}{4\pi\varepsilon_0 R^2}$$

$4\pi R^2$ is the surface area of a sphere. For the second sphere we have

$$E' = \frac{q}{4\pi\varepsilon_0 r^2}$$

The ratio of the electric fields is thus

$$\frac{E}{E'} = \frac{Qr^2}{R^2 q}$$

But we have $Q/R = q/r$, hence

$$\frac{E}{E'} = \frac{r}{R}$$

If R is bigger than r we have the result that the field is greatest near the smaller radius sphere. The fields are inversely proportional to the radii. A point has a small radius and thus near a point the field will be much greater than near a flat or larger radius of curvature surface. It is thus easier for discharge to occur from points because for the same potential as any other surface they have larger fields in their vicinity.

Charges are produced on sheet insulators running over rollers. Dust particles are attracted to charged surfaces—think of how you probably sometime have rubbed a plastic pen or comb against your jacket and used it to pick up small pieces of paper. The dust particles become dipoles in the field of the charged surface. If the charges are produced on fabrics running over rollers in a factory then this dust can mark the fabric. There are two main methods of reducing this effect—both involve providing a leakage path for the charge to escape from the fabric by supplying ions. One method uses a radioactive source to ionize the air in the vicinity, other methods employ points placed close to the fabric. Sometimes fine wires are used.

The earth as a whole carries a negative charge and thus near the earth's surface there is an electric field. The electric field can be measured; in fine weather its value is about 130 N C^{-1}. Hence the charge density will be given by $E = \sigma/\varepsilon_0$ as

$$\sigma = 130 \times 8{\cdot}85 \times 10^{-12}$$
$$= 1{\cdot}2 \times 10^{-9}\ \mathrm{C\,m^{-2}}$$

As the earth has a surface area of $5{\cdot}1 \times 10^{14}\ \mathrm{m}^2$, this would mean a total charge of $6{\cdot}1 \times 10^5$ C, about half a million coulombs. The charge is negative. Thunderclouds are generally positive in the upper part of the cloud and negative in the lower part. Balloons or aircraft flying through the clouds give this information—Franklin with his kite was able to determine the sign of the charge in the base of thunderclouds. The base of a thundercloud is, however, more negative than the immediate ground area under the cloud. We have effectively a parallel plate capacitor, the base of the cloud being one plate and the surface of the earth the other plate. The base of the cloud will be about 2 km above the surface of the earth. The breakdown voltage for wet air is about 10 000 volts per centimetre of path ($E = 10^6$ N C^{-1}). Thus for breakdown of the air between the cloud and the earth we would need about 2×10^9 V.

The sequences occurring in a lightning flash can be followed by high speed photography. The avalanche of ionization moves out from the cloud with a speed about 1/6 that of light. This gives a moving spot of light which progresses down to the earth. It does not move down to the earth in one uninterrupted step but comes down in a series of steps with short pauses between steps. This is called the step leader—it is not the main lightning flash. The progress of the step leader to the ground is irregular. When the step leader reaches the ground the negative charge in the base of the cloud is rapidly discharged to the earth. This discharge gives the bright lightning flash, and the corresponding thunder. The lightning, however, travels from the ground up to the cloud: it is known as the return stroke. The light is produced when the charges can move rapidly and it is those charges in the step leader near to the ground which first move rapidly. When they shoot into the ground then those higher up the leader can move rapidly—then those higher up, and so on up to the cloud. After a short interval of time a further leader comes down the path of ions left by the first leader and again there is a return stroke. Near pointed or tall objects on the earth there will be higher electric fields and so the step leader is more likely to strike them than the flat ground. The high electric field will already have produced some ions and thus there is a lower resistance path for the step leader in their vicinity. Lightning conductors on tall buildings provide a lower resistance path to earth than the fabric of the building and thus the discharge current, perhaps of the order of 10^4 A, bypasses the building.

Lightning is the means by which the earth acquires its negative charge. During fine weather the earth is steadily loosing charge, lightning replenishes it.

Dielectrics

The capacitance of a parallel plate capacitor, or indeed any capacitor, depends on the medium between the plates. So far the medium has been assumed to play no part—it is a vacuum, or air which seems to play no part. If we measure the capacitance of a capacitor both with air and with an insulator such as polythene or glass or paper between the plates we find that the effect of the insulator is to increase the capacitance. If the entire volume between the plates is occupied by the insulator then the ratio of the capacitance with the insulator to that without is called the dielectric constant (the term relative permittivity is sometimes used).

The following table gives dielectric constants for some materials.

Substance	*Dielectric constant*
Air, 0°C, 1 atm.	1·00059
Water, liquid at 20°C	80
Paraffin wax at 20°C	about 2·3
Glass, soda	7·5
Nylon	about 4

The difference between using air between capacitor plates and having a vacuum is very slight. Water has a very high dielectric constant.

The effect of the insulator is thus to increase the capacitance. Now can we explain this? When a charged rod, perhaps a plastic pen which has been rubbed against your coat, is brought near to a small piece of paper (dry) the paper adheres to the rod. We say that charges have been induced in the paper—one end of the paper becoming positive and the other negative (the section in this chapter on Van der Waals' forces looks more closely at this effect). Perhaps when we put an insulator between charged plates charges become induced on it. Figure 11.28 shows the possible effect. The plates produce a field E, the charges on the dielectric produce a field $-E'$ and the resultant field is $E - E'$. If the field between the plates is reduced, the net charge density at the plates is reduced by the presence of the dielectric. Thus an applied potential difference between the plates produces a lower electric field and effectively a lower

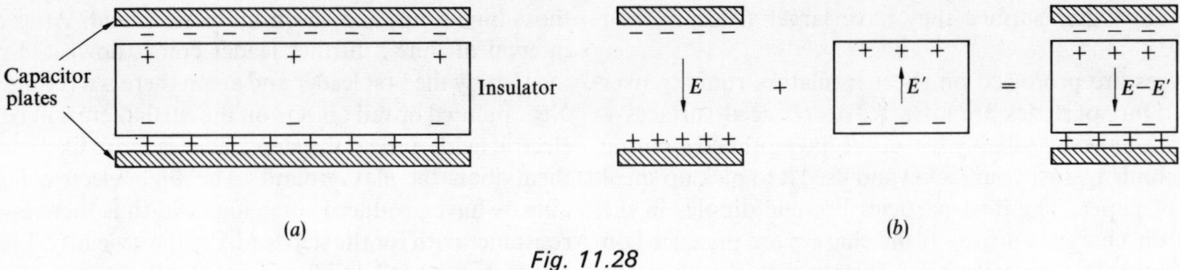

Fig. 11.28

charge density than it would if there was a vacuum between the plates. The dielectric has increased the capacitance.

Bringing a dielectric or a conducting object near a capacitor, whether it be a parallel plate arrangement or just an isolated object, changes its capacitance and hence the electric field. Some fishes produce electric fields around themselves and sense the disturbance of the field produced by objects in their vicinity. Field changes as low as 10^{-6} V m^{-1} can be detected.

Van der Waals forces

Though an atom or molecule may be completely neutral, i.e., equal numbers of positive and negative charges, electrical forces can still exist between such atoms or molecules. The charge may not be uniformly distributed within the atom or molecule; or if it is uniformly distributed there may be some momentary fluctuation in the distribution which results in asymmetry. We can think of such an asymmetrical molecule as being two spheres of charge, one positive and the other negative, with centres slightly displaced from each other. Because a charged sphere behaves as if all the charge were located at its centre, the picture becomes one of two point charges separated by a distance. This is known as a dipole.

We will consider here the forces occurring between atoms and molecules which have no permanent dipoles. How does the electric field produced by a dipole vary with distance from the dipole centre? Consider a dipole with charges $+q$ and $-q$ separated by a distance $2L$. We can calculate the field at a point P, along the axis of the dipole, by considering the forces on a test charge due to each of the dipole charges (Fig. 11.29).

$$F = \frac{qQ}{4\pi\varepsilon_0(x-L)^2} - \frac{qQ}{4\pi\varepsilon_0(x+L)^2}$$

Q is the test charge placed at point P. The electric field at P is F/Q.

$$E = \frac{q}{4\pi\varepsilon_0}\left[\frac{1}{(x-L)^2} - \frac{1}{(x+L)^2}\right]$$

$$E = \frac{q}{4\pi\varepsilon_0}\frac{2xL}{(x-L)^2(x+L)^2}$$

$$E = \frac{q}{4\pi\varepsilon_0}\frac{2xL}{(x^2-L^2)^2}$$

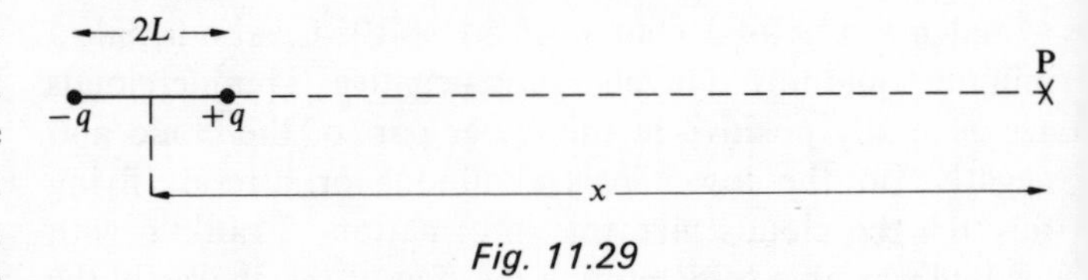

Fig. 11.29

If x is considerably greater than L, we have the following approximation,

$$E = \frac{2qL}{4\pi\varepsilon_0 x^3}$$

The electric field strength along the axis of a dipole varies as the reciprocal of the distance cubed. The length of the dipole multiplied by the charge, $2Lq$, is known as the dipole moment, μ.

$$E = \frac{\mu}{4\pi\varepsilon_0 x^3}$$

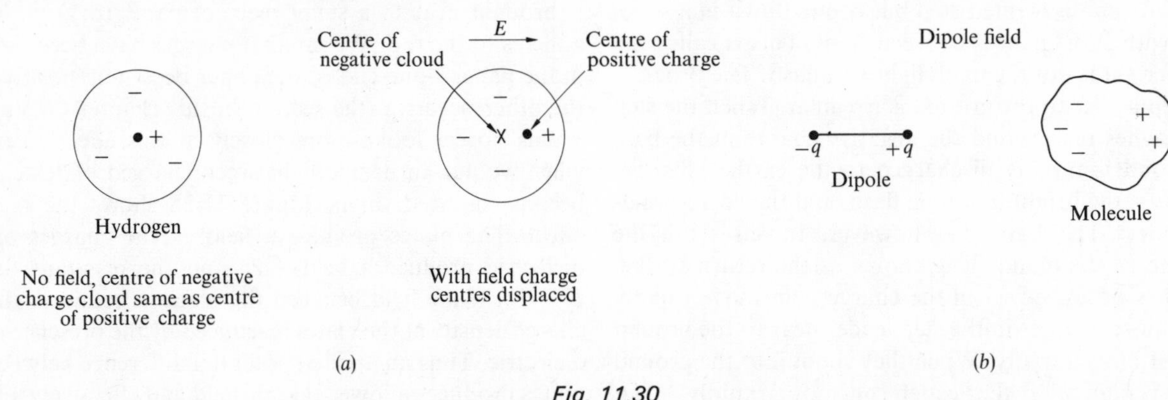

Fig. 11.30

The effect of this electric field on another molecule is to distort its charge distribution (Fig. 11.30) and so make it into a dipole. The molecule will thus acquire a dipole moment. It seems reasonable to assume that this new dipole moment μ' will be proportional to the electric field acting in the region where the molecule is situated.

$$\mu' \propto E$$

or in terms of charge q' and dipole length $2L'$

$$(q'2L') \propto E$$

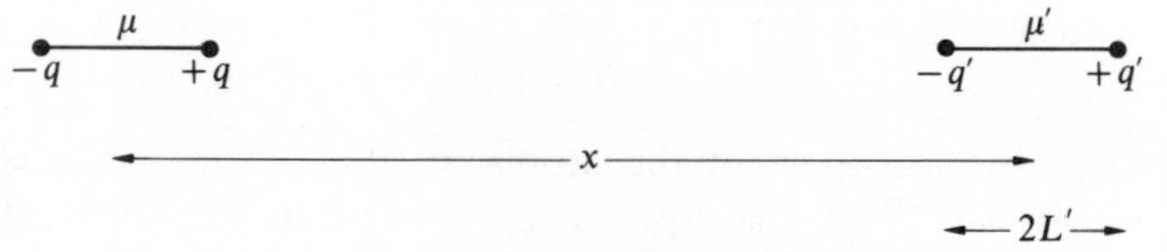

μ $-q$ $+q$ μ' $-q'$ $+q'$ x $2L'$

Fig. 11.31

The force acting on this dipole because it is in an electric field is (using $F = qE$) (Fig. 11.31)

$$F = -\frac{\mu q'}{4\pi\varepsilon_0(x - L')^3} + \frac{\mu q'}{4\pi\varepsilon_0(x + L')^3}$$

$$= \frac{\mu q'}{4\pi\varepsilon_0}\left[-\frac{1}{(x - L')^3} + \frac{1}{(x + L')^3}\right]$$

$$= -\frac{\mu q'}{4\pi\varepsilon_0}\frac{6L'x^2}{(x^2 - L'^2)^3}$$

For $x \gg L'$ the equation approximates to

$$F = \frac{\mu q'}{4\pi\varepsilon_0}\frac{6L'}{x^4}$$

But $q'2L' \propto E$, therefore

$$F \propto -\frac{3\mu E}{4\pi\varepsilon_0 x^4}$$

But $E = \mu/4\pi\varepsilon_0 x^3$, thus we have

$$F \propto -\frac{3\mu^2}{(4\pi\varepsilon_0)^2}\cdot\frac{1}{x^7}$$

The force between the molecules is proportional to the reciprocal of the seventh power of their separation. Such a force is known as a Van der Waals' force. The minus sign is because the force is attractive. The same $1/x^7$ variation applies to permanent dipole interactions as well as induced dipole interactions.

It is this type of force which is responsible for the bonding in solids such as solid hydrogen or neon, where atoms are involved, or solid methane or pentane, where molecules are involved. These are the forces responsible for the condensation of gases into liquids. These forces act between all molecules and atoms; in some cases there are additional forces due to other forms of bonding, e.g., ionic bonds (see chapter 3). The Van der Waals force is a weak force by comparison with other atomic and molecular bonding forces and solids which depend only on this force need little energy to be melted or broken apart (low mechanical strength). Only about 10^{-21} J is needed to separate a pair of hydrogen atoms in solid hydrogen; the melting point of the solid is only 14 K. Sodium chloride, which melts at 1074 K, needs $12{\cdot}7 \times 10^{-19}$ J for a pair of sodium and chlorine ions to be torn apart. The energy needed for sodium chloride is about 1000 times greater than that needed for solid hydrogen.

How big are these temporary dipole moments which are responsible for these Van der Waals forces? Measurements of the capacitance of a capacitor between whose plates the substance has been placed, i.e., measurement of the dielectric constant, can yield results for the dipole moment of the atoms or molecules in the substance. Hydrogen in an electric field of 1 V m^{-1} has a dipole moment of $0{\cdot}8 \times 10^{-40}$ C m. The dipole moment is proportional to the size of the electric field. As the dipole moment is the product of the dipole charge and the distance b between the dipole charges we can calculate this distance if we assume that the charge on, say, the hydrogen dipole is $1{\cdot}6 \times 10^{-19}$ C, the charge on an electron.

$$1{\cdot}6 \times 10^{-19} \times b = 0{\cdot}8 \times 10^{-40}$$

$$b = 5 \times 10^{-22} \text{ m}$$

With a larger electric field of, say, 10^6 V m^{-1} we obtain $b = 5 \times 10^{-16}$ m. The effect of this field is to displace the hydrogen charge centres by only about one millionth of the atomic diameter.

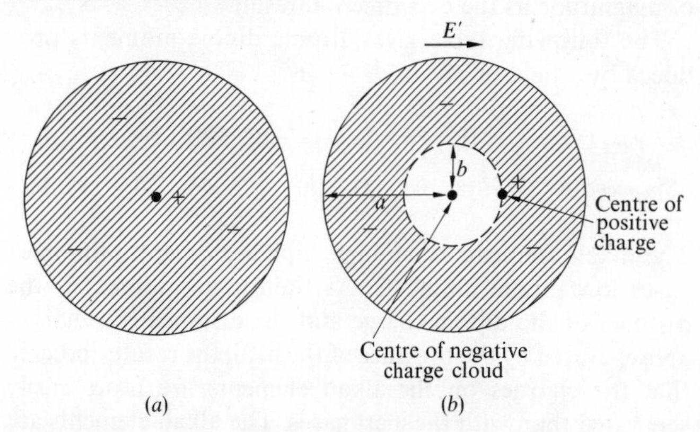

Fig. 11.32 (a) Hydrogen with no field. (b) Hydrogen with field.

We can estimate this displacement, and hence the dipole moment. Figure 11.32(a) shows the hydrogen atom in no field, the centres of the negative charge cloud and the positive nuclear charge coincide as the atom is symmetrical. When a field is applied the two centres become displaced, Fig. 11.32(b). Call this displacement distance b. The electric field acting on the nucleus due to the negative charge cloud will be that due to the negative charge inside a sphere of radius b. All the charge outside that sphere will have no effect on the nuclear charge, provided the negative charge is uniformly spread over the sphere. The electric field inside a charged shell is zero. The charge inside the sphere of radius b will be the fraction b^3/a^3 of the total negative charge. a is the radius of the negative charge cloud, i.e. the radius of the atom. Thus the field, E, acting on the nucleus due to the negative charge is

$$\frac{b^3}{a^3}\frac{q}{4\pi\varepsilon_0 b^2}$$

q is total negative charge. The force acting on the nucleus due to this field will be given by

$$F = qE$$

The charge q on the nucleus is the same as that of the total negative charge cloud.

$$F = \frac{b^3}{a^3}\frac{q^2}{4\pi\varepsilon_0 b^2}$$

This force must be opposite and equal to the force acting on the nucleus due to the applied field E'. This force is qE'.

Hence

$$qE' = \frac{b^3}{a^3}\frac{q^2}{4\pi\varepsilon_0 b^2}$$

and thus

$$b = 4\pi\varepsilon_0 a^3 \frac{E'}{q}$$

If we take for hydrogen a value of $0{\cdot}5 \times 10^{-10}$ m for a and $1{\cdot}6 \times 10^{-19}$ C for q we obtain a value for b of about 8×10^{-22} m for a field of 1 V m^{-1}. This is of the same order of magnitude as the experimental result.

The following table gives atomic dipole moments produced by a field of 1 V m^{-1}

Element	H	He	Li	Be	C	Ne	Na	A	K
Dipole moment/10^{40} C m	0·8	0·2	13	10	1·7	0·4	30	1·8	38

The alkali elements have high dipole moments, the inert gases low dipole moments. As the dipole moment is the product of the dipole charge and the distance the charges are separated by the presence of the field, the results indicate that the charges on the alkali elements are more easily separated than with the inert gases. The alkali elements are more easily distorted than the inert elements.

Polar molecules

In the previous section we considered the forces occurring between atoms and molecules which have no permanent dipole, i.e., no permanent asymmetry. In a molecule such as HCl with a permanent dipole moment we can consider that the electron of hydrogen has shifted partially over to the chlorine atom. This gives a dipole with the hydrogen end of the molecule having a net positive charge and the chlorine end a net negative charge (Fig. 11.33). The dipole moment for HCl has been measured as $3{\cdot}4 \times 10^{-30}$ C m. The dipole moment is the product of the charge on the ends of the dipole and the distance separating them. If we take the charge to be that of one electron, the distance apart b of the charges for HCl is given by

$$1{\cdot}6 \times 10^{-19} \times b = 3{\cdot}4 \times 10^{-30}$$
$$b = 0{\cdot}21 \times 10^{-10}\ \text{m}$$

The distance apart of the atoms in the HCl molecule is $1{\cdot}27 \times 10^{-10}$ m; the dipole charges are thus effectively at some positions between the atom centres and separated by only 0·21/1·27, i.e., 1/6, of the interatomic spacing.

An alternative and completely equivalent way of considering the HCl dipole would be charges of $e/6$ at the interatomic spacing. A complete electron has not been transferred between the atoms, unlike an ionic bond where the complete transfer of an electron occurs.

Water has a dipole moment of $6{\cdot}3 \times 10^{-30}$ C m. It is this high dipole moment which makes water such a good solvent for ionic solids or substances with polar molecules. The polar water molecules break up the solids by exerting stronger forces on, say, the sodium and chlorine ions in sodium chloride than the interatomic forces between the ions. The sodium and chlorine ions then become attached to the water molecules (Fig. 11.34). The sodium and chlorine ions are thus in solution not as free ions but attached to water molecules.

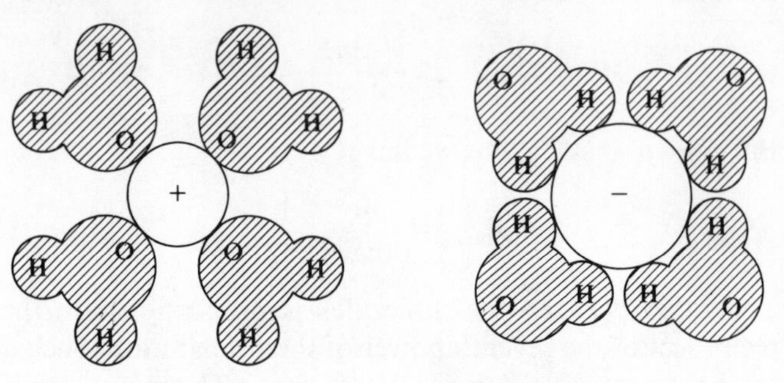

Fig. 11.34

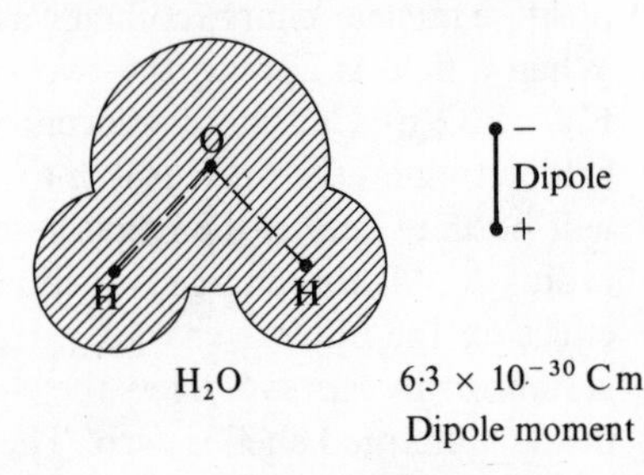

H_2O $6{\cdot}3 \times 10^{-30}$ Cm Dipole moment

HCl $3{\cdot}4 \times 10^{-30}$ Cm Dipole movement

H Cl + Dipole −

Fig. 11.33 Polar molecules.

11.4 Magnetic fields

Iron filings sprinkled round a permanent magnet or round a wire carrying a current show the 'existence' of what is called a magnetic field. We say there is a field because without there being any contact between either the magnet or the current-carrying wire and the iron filings forces act on the filings, forcing them into patterns.

Consider two parallel and perfectly free wires—when there is no current in the wires there seems to be no noticeable force acting on the wires due to the presence of each other. The wires must contain both positive and negative charges and as the electrical forces between the two wires seem to cancel out there must be equal numbers of positive

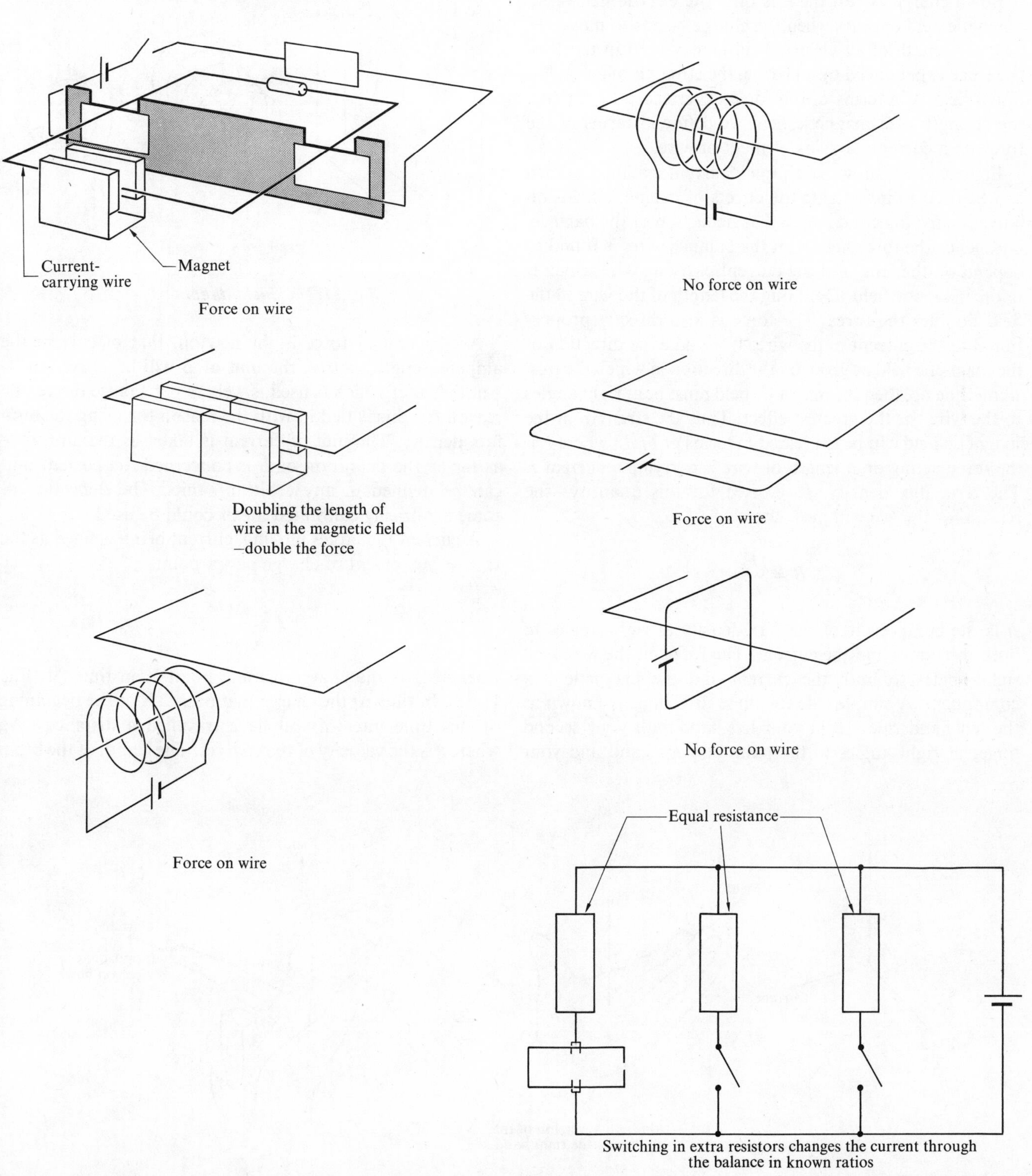

Fig. 11.35

and negative charges in each wire. If, however, the charges are made to move, i.e., currents flow, forces are experienced by the wires. Moving charges in each wire give forces where previously no forces were apparent. We call these forces magnetic forces.

Motion of charge leads to a current and an associated magnetic field. Without motion there is no magnetic field.

For a charge at rest there is only the electric field—the magnetic field appears when the charge begins to move.

The strength of an electric field is measured in terms of the force experienced by a charge, the strength of a gravitational field is in terms of the force experienced by a mass, the strength of a magnetic field we define in terms of the force on a current-carrying element of wire.

Figure 11.35 shows a simple 'current balance' which can be used to investigate the effects of magnetic fields on current-carrying conductors. The deflection of the balance, and hence the force acting on the balance wire, is found to depend on the length of the current-carrying wire which is in the magnetic field. Doubling the length of the wire in the field doubles the force. The force is also directly proportional to the current in the wire. If we take the direction of the magnetic field as given by the direction in which the iron filings line up, then the magnetic field must be at right angles to the wire for the greatest effect. Thus the strength of the magnetic field can be expressed in terms of F/IL, where F is the force acting on a length of wire L carrying a current I. The term flux density, B, is used for this quantity—the reason for this we will meet shortly.

$$B = \frac{F}{IL}$$

B is the component of the flux density at right angles to both the forces and the current. The force on the wire is at right angles to both the current and the magnetic flux component. A simple rule for these directions is known as the left hand rule: Hold your left hand with your second finger at right angles to the palm of your hand and your thumb in the plane of your hand but at right angles to the first finger, the first finger then gives the field direction, the second finger the current (opposite to electron flow) direction and the thumb the force direction (Fig. 11.36). Figure 11.37 shows the accepted directions of some fields.

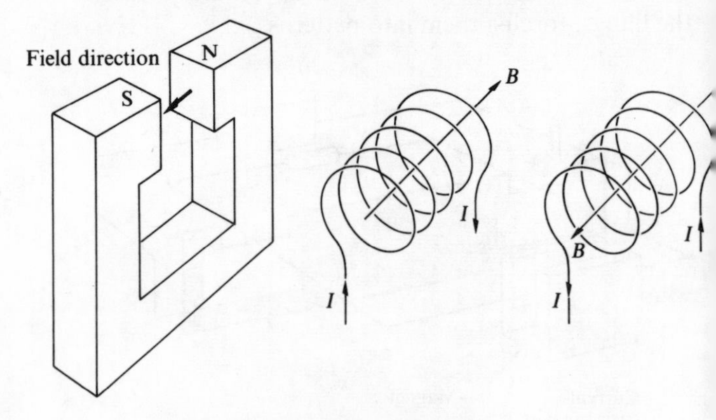

Fig. 11.37 Field directions.

As the unit of force is the newton, that of current the ampere, length metres, the unit of B will be N A^{-1} m^{-1}. Another unit which is used is webers per square metre, the reason for this is tied in with the reason for using the term flux density. The unit of current is taken as the ampere—as far as the definition of B is concerned the current unit can be defined in any arbitrary units. The deflection of some arbitrarily calibrated meter could be used.

A current is charges moving, current being defined as the rate of movement of charge past a point.

$$I = \frac{\Delta Q}{\Delta t}$$

where ΔQ is the charge passing point X in time Δt (Fig. 11.38). In time Δt the charge that passed X at the beginning of this time interval will have travelled a distance $v\,\Delta t$, where v is the velocity of the charges. The length of the beam

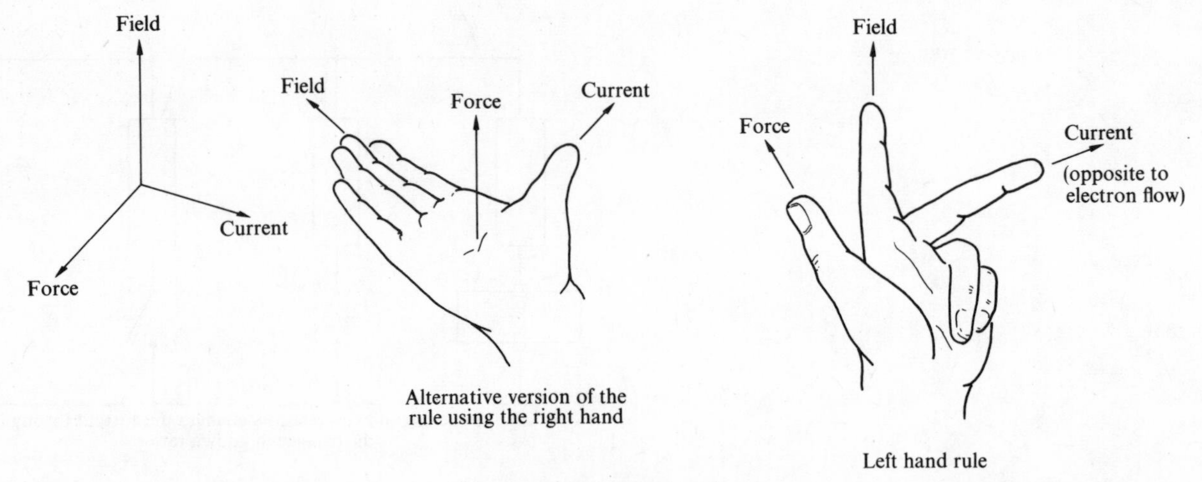

Fig. 11.36 Force direction.

of charges produced in time Δt is thus $v \, \Delta t$. If this beam of charge is in a magnetic field then the force experienced will be

$$F = BIL$$

$$F = B\frac{\Delta Q}{\Delta t}v \, \Delta t$$

$$F = Bv \, \Delta Q$$

Fig. 11.38

The charge ΔQ is made up of N charges each carrying a charge q, hence

$$F = BvNq$$

This is the force on N charges, the force on one charge will thus be

$$F = Bvq$$

The charges need not be restricted to motion in a wire for a force to act on them when in a magnetic field. Figure 11.39(a) shows how a beam of charge becomes deflected in a magnetic field which is everywhere at right angles to the 'current'. The beam of charge was made visible by being fired through hydrogen gas under low pressure. The electrons excite the gas and cause the emission of visible light. The beam is deflected into a circular path by the magnetic field because the force produced is everywhere at right angles to the motion of the charges. The force is thus equal to mv^2/r when the charges move in a circular path of radius r. The mass of a charge is m. Hence

$$\frac{mv^2}{r} = Bvq$$

The radius of the path is thus

$$r = \frac{mv}{Bq}$$

If the magnetic flux density B is measured by using a current balance, the velocity of the charges calculated from a measurement of the potential difference, V, through which they were accelerated ($\frac{1}{2}mv^2 = Vq$), and the radius measured, then the charge-to-mass ratio of the charged particles can be calculated. Chapter 13 gives an example of how this method was used to measure the charge-to-mass ratio of electrons.

(*a*)

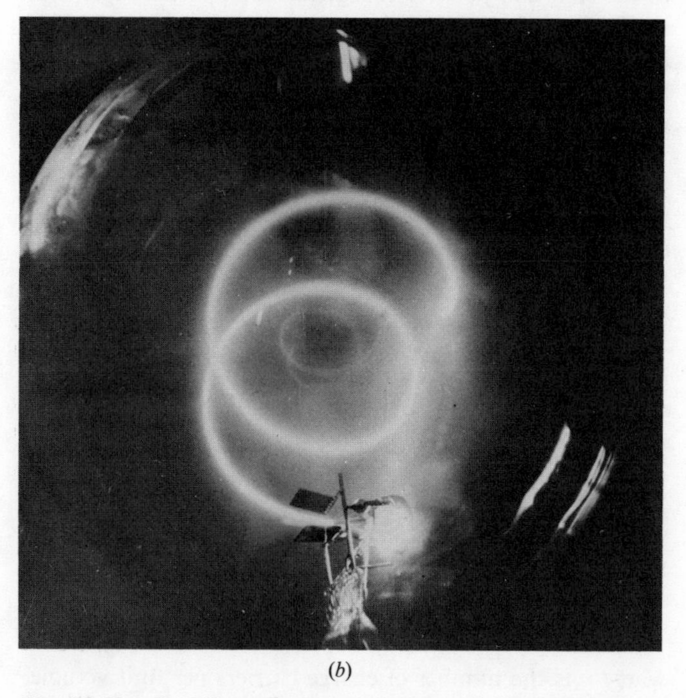

(*b*)

Fig. 11.39 (a) Magnetic field everywhere at right angles to the electron beam. (b) Magnetic field not quite at right angles to the electron beam. The electron beam has been made visible by its collision with hydrogen atoms.

Hall effect

Suppose we have a current passing through a slab of material, as in Fig. 11.40. If we apply a magnetic field normal to the surface of this material, the charge carriers giving the current should experience a force.

$$F = Bqv$$

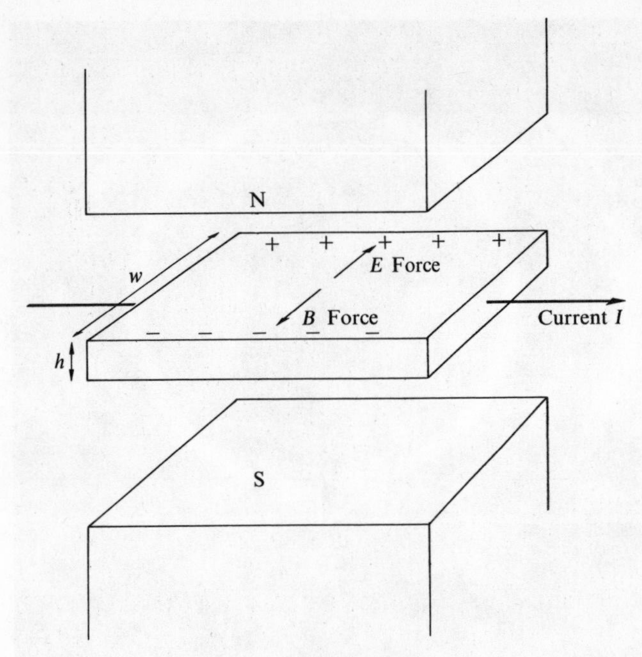

Fig. 11.40

This force will cause negative charge to drift to one edge of the slab and positive charge to the opposite edge. This separation of charge will produce an electric field. This field will give another force acting on the charged particles, in the opposite direction to the force produced by the magnetic field.

$$F = Eq$$

The charge separation will grow until the force due to the electric field becomes equal to the force exerted by the magnetic field.

$$Eq = Bqv$$

Because there is an electric field there will be a potential difference between the opposite edges of the slab.

$$E = \frac{V}{w}$$

Thus

$$V = Bvw$$

But

$$I = nAqv \quad \text{(see chapter 10)}$$

where n is the number of charge carriers per unit volume, A is the cross-sectional area, and I the current.

$$A = wh$$

Thus

$$I = nwhqv$$

and

$$V = Bw\frac{I}{nwhq}$$

$$V = \frac{BI}{nhq}$$

This potential difference occurs when no current is drawn off the slab in a transverse direction, i.e., the charge separation is not affected. When a current is drawn off, Fig. 11.41, then

$$V = I'R$$

where I' is the transverse current and R the total resistance of the transverse circuit. This current can be measured.

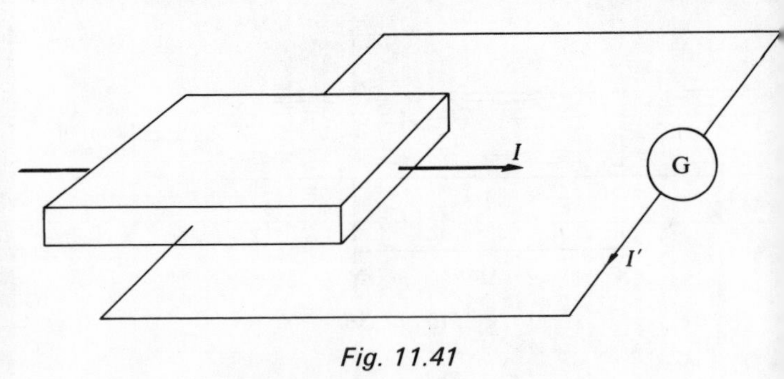

Fig. 11.41

For the same material and current I we have

$$I' \propto B$$

We can thus use such a slab for the measurement of magnetic flux densities. The slab is known as a Hall probe and is generally made of a semiconductor material, for this purpose.

We can use the measurements, with known magnetic flux density, for the measurement of the number of charge carriers per unit volume in a material.

For example, the following data refer to silver. $B = 0{\cdot}1\ \text{N A}^{-1}\ \text{m}^{-1}$, $I = 18$ A, $h = 0{\cdot}1$ mm, $V = 1{\cdot}6 \times 10^{-6}$ V. This gives for n, if we take q as $1{\cdot}6 \times 10^{-19}$ C, a value of $6{\cdot}9 \times 10^{28}\ \text{m}^{-3}$.

The density of silver is 10 500 kg m^{-3}, its atomic mass 107·87 g. In $107{\cdot}87 \times 10^{-3}$ kg there will be 6×10^{23} atoms. This mass has a volume of $107{\cdot}87 \times 10^{-3}/10\,500\ \text{m}^3$. Thus the number of atoms per unit volume is

$$6 \times 10^{23} \times \frac{10\,500}{107{\cdot}87 \times 10^{-3}} = 5{\cdot}9 \times 10^{28}\ \text{m}^{-3}$$

The number of charge carriers in silver would thus seem to be about one per atom, each charge carrier having the charge of an electron.

Hall effect measurements with germanium show that the number of charge carriers increases rapidly as the temperature is increased (Fig. 11.42). The increase is a good approximation to an exponential

$$n = n_0\, e^{-\varepsilon/kT}$$

ε is the energy needed to make a charge carrier available for conduction: k is Boltzmann's constant. The results also show that the number of charge carriers is only a very small fraction of the number of germanium atoms, about one charge carrier per million germanium atoms. This is typical of semiconductors. Metals have roughly equal numbers of charge carriers and atoms.

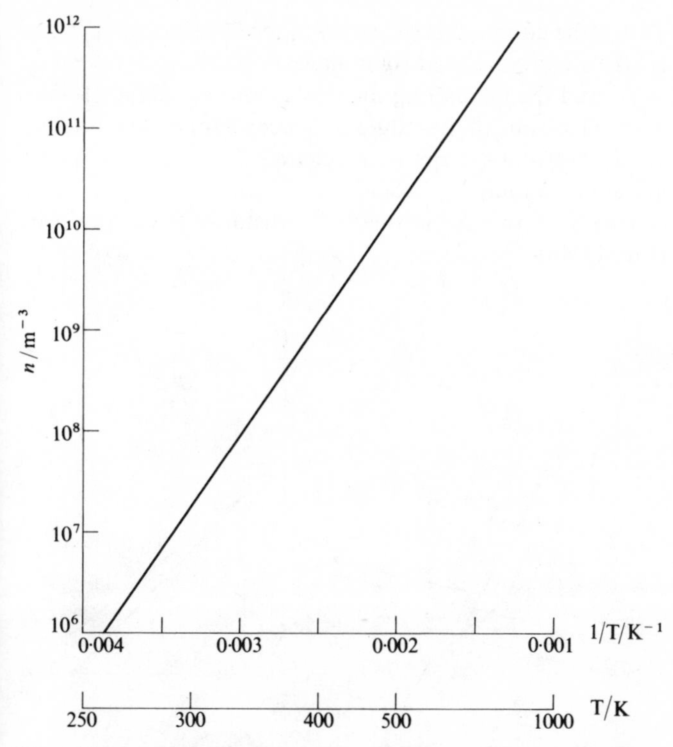

Fig. 11.42 Hall effect with germanium. (After F. J. Morin and J. P. Maita. Phys. Rev., **94**, *1525, 1954.)*

With some metals, zinc and cadmium, the direction of the Hall voltage is in the opposite direction to that given by silver or copper. As there seems to be no reason why these metals should differ from others it would seem that our basic theory is at fault.

Magnetic fields produced by currents

Exploration of the field inside a solenoid shows that the field within the greater part of the internal volume is a constant. It only departs from the constant value near the ends. The magnetic field is at virtually all points parallel to the axis of the solenoid, the exceptions again being near the ends. The field can be measured by means of a current balance, a search coil, or a Hall probe. (Details of all these appear in this section.)

On what factors does the magnetic field inside a solenoid depend? We can do experiments to find out. The results—the magnetic flux density, B, is directly proportional to the current carried by the solenoid:

$$B \propto I$$

B is independent of the cross-sectional area of the solenoid; B is independent of the length of the solenoid provided the number of turns of wire per unit length is constant; B is directly proportional to the number of turns of wire per unit length (n);

$$B \propto n$$

B is independent of the nature of the wire or its cross-section provided the current is constant; B is independent of the shape of the solenoid. B seems to depend on just two quantities, the current and the number of turns per unit length.

$$B \propto nI$$

As B, n, and I can be measured we can arrive at a value for the constant of proportionality μ_0. This constant has been called 'the magnetic space constant' or 'permeability of free space'.

$$B = \mu_0 nI$$

The experimentally determined value for μ_0 is $4\pi \times 10^{-7}$ N A^{-2}. The reason for putting the constant in this form is that it is more general than just a constant relating the flux density B to the current and number of turns per metre for a solenoid. n is assumed to be in turns per metre and I in amperes (specified in the same way as for our definition of B).

If we explore the magnetic field near a long straight current-carrying wire we find—at a particular distance from the wire B is directly proportional to the current I carried by the wire;

$$B \propto I$$

B varies as the reciprocal of the distance r from the wire;

$$B \propto \frac{1}{r}$$

B is independent of the cross-section of the wire, its material or shape, provided the current is constant. The directions of B are in circles, centred on the wire, which lie in planes at right angles to the wire (Fig. 11.43).

$$B \propto \frac{I}{r}$$

In the same way as for the solenoid we can experimentally determine the value of the constant of proportionality.

$$B = k\frac{I}{r}$$

$$k = 2 \times 10^{-7} \text{ N A}^{-2}$$

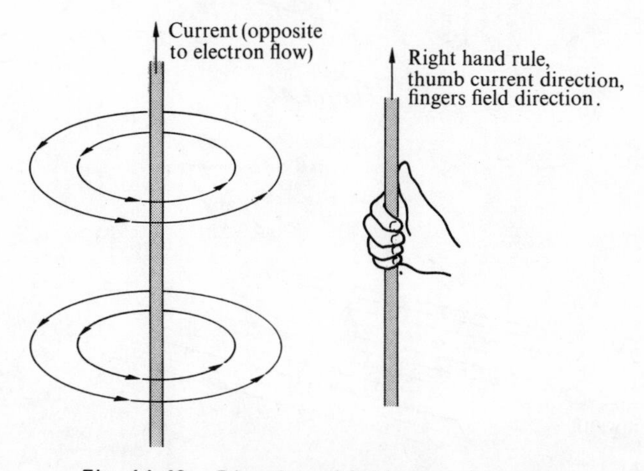

Fig. 11.43 Direction of field near a long wire.

which gives us

$$k = \frac{\mu_0}{2\pi}$$

Thus we have

$$B = \frac{\mu_0 I}{2\pi r}$$

We do not, however, have to rely on experiment to show the relationship between k and μ_0. A solenoid is only a lot of parallel wires. If we sum the effects due to each single wire in a solenoid we should end up with the solenoid formula.

One of the experimental facts found with the solenoid was that the field inside the solenoid was independent of the shape of the solenoid. We will therefore consider a solenoid which has a rectangular cross-section (Fig. 11.44). We can experimentally check that such a solenoid behaves like the 'normal' solenoid—it does.

The advantage of considering a solenoid of this shape is that we can consider it to be two sets of long straight wires. Let us calculate the flux density at a point P above one such set of long straight wires (Fig. 11.45). Each wire will make a contribution to the field at P. Each contribution will be different as the distance of the wires from P are different. The direction of the field at P will also be different for each wire. The flux density due to the wire at R will be

$$\Delta B = \frac{kI}{r}$$

the distance PR $= r$.

The direction of this field is at right angles to the line PR. The horizontal component of this field will be

$$\Delta B \cos\theta = \frac{kI}{r}\cos\theta$$

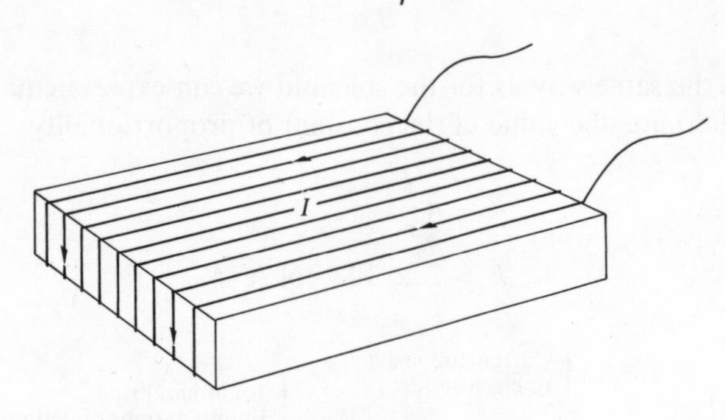

Fig. 11.44

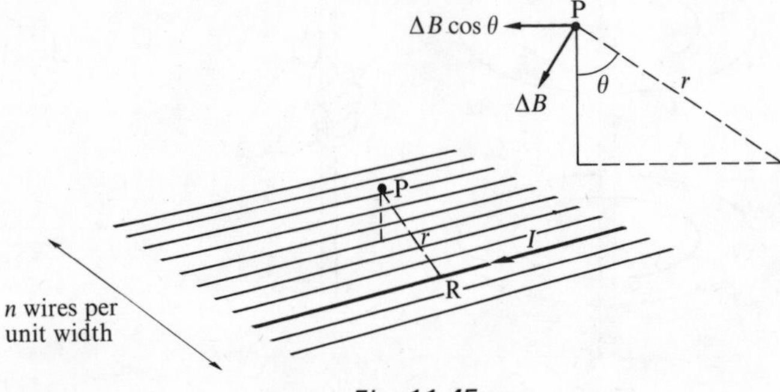

Fig. 11.45

This is the component along the axis of the solenoid. There is also a component at right angles to the axis.

To find the field along the axis due to the plane of wires we need to sum all the values of $\Delta B \cos\theta$ for all the different values of θ and r. Suppose we change the angle θ by $\Delta\theta$, we are now summing the field due to all the wires between R and S. If this $\Delta\theta$ is small, the distance RS is given by (Fig. 11.46)

$$\text{RS} = \frac{r\,\Delta\theta}{\cos\theta}$$

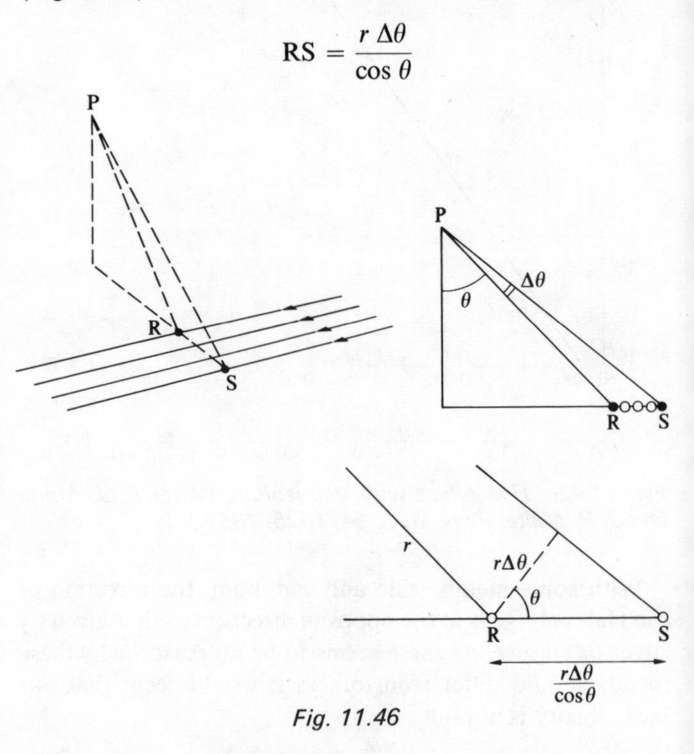

Fig. 11.46

If there are n wires per unit width, we have in the distance RS

$$\text{Number of wires} = \frac{nr\,\Delta\theta}{\cos\theta}$$

As our value of $\Delta\theta$ is small we will assume that all these wires are the same distance, r, from P. Thus the horizontal field component due to these wires is

$$\frac{nr\,\Delta\theta}{\cos\theta}\frac{kI}{r}\cos\theta = knI\,\Delta\theta$$

If the length of the solenoid is infinite (Fig. 11.47) our plane of wires stretches to infinity on either side of our point P. As k, n, and I are independent of angle our summation of the field due to all the wires means considering all the values of $\Delta\theta$ which make up the total angle subtended by the entire plane of wires. This total angle is 180° or π.

$$\text{Horizontal field at P} = knI\pi$$

The vertical components of B cancel because there are as many wires on one side of P as on the other. The wires on one side of P produce vertical components in one direction and the wires on the other side produce them in the opposite direction—in all cases the horizontal components are in the same direction (Fig. 11.48).

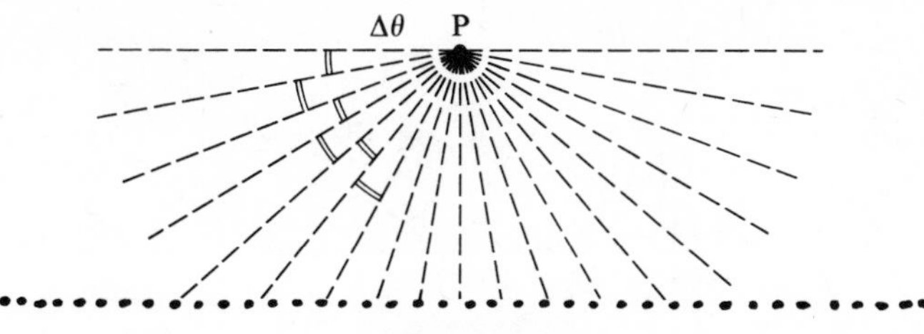

Fig. 11.47

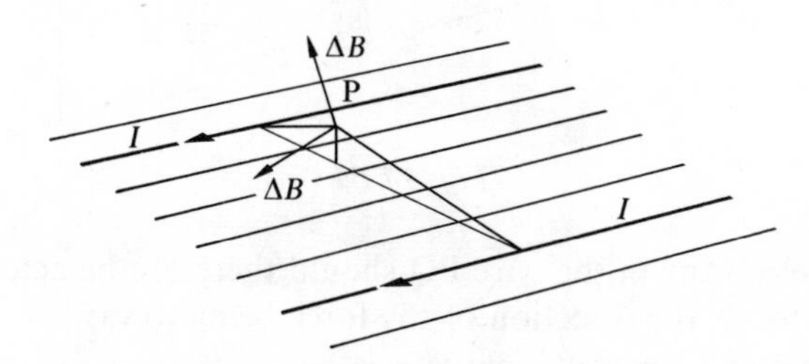

Fig. 11.48

Our solenoid is made up of two such planes and as they both give the horizontal component in the same direction, the field due to the two planes is

$$B = 2\pi knI$$

The vertical components for each plane cancel each other. This is thus the field along the axis between two planes of wires. The short lengths of wire making up the ends of the solenoid are considered to be at infinity by virtue of our summing the field at P over 180° for the horizontal planes. Their effect is thus zero. Hence the flux density in a solenoid is

$$B = 2\pi knI$$

Experimentally we have

$$B = \mu_0 nI$$

Thus $\mu_0 = 2\pi k$.

The unit of current

The flux density B a distance r from a long straight wire is

$$B = \frac{\mu_0 I}{2\pi r}$$

If at this distance we place another long straight wire, everywhere at right angles to B, a force will act on this wire when a current passes through it.

$$F = BIL$$

Thus if the current in both the wires is the same,

$$F = \frac{\mu_0 I}{2\pi r} IL$$

$$= I^2 \frac{\mu_0 L}{2\pi r} = I^2 \times 2 \times 10^{-7} \frac{L}{r}$$

The force depends on the square of the current, the value of μ_0, and the geometrical arrangement of the wires, i.e., r and L. If the wire, it doesn't matter which, is part of a current balance (Fig. 11.49) the force can be measured. The force will cause the balance frame to be deflected—the force necessary to restore balance, perhaps putting weights on the frame, can be measured. Thus we can measure the current. This is the basis of the definition of the ampere.

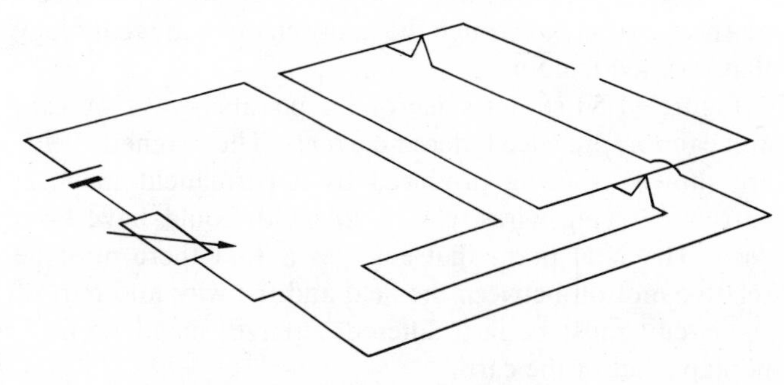

Fig. 11.49

The definition of the ampere is that constant current which if maintained in two straight parallel conductors of infinite length, of negligible circular cross-section, and placed 1 metre apart in a vacuum, would produce between these conductors a force equal to 2×10^{-7} N per metre length.

It is our definition of the ampere which determines the value of μ_0, the ampere is the primary unit. The values of the constants we have obtained in this section on magnetic fields have been fixed by our fixing our unit of current. Fixing the current unit not only fixes the value of μ_0 but also the value of ε_0. As we shall later (chapter 12) see, μ_0 and ε_0 are related by the equation

$$\text{Speed of light} = \frac{1}{\sqrt{(\mu_0 \varepsilon_0)}}$$

Electromagnetic induction

'Two hundred and three feet [about 65 m] of copper wire in one length were coiled round a large block of wood; other two hundred and three feet of similar wire were interposed as a spiral between the turns of the first coil, and metallic contact everywhere prevented by twine. One of these helices was connected with a galvanometer, and the other with a battery of one hundred pairs of plates four inches [10 cm] square, with double coppers, and well charged. When the contact was made, there was a sudden and very slight effect at the galvanometer, and there was a sudden and very slight effect when the contact with the battery was broken. But while the voltaic current was continuing to pass through the one helix, no galvanometrical appearances nor any effect like induction upon the other helix could be perceived. . . .'

M. Faraday (1831).
Experimental researches in electricity.
First series, Item 10.

When a current in one coil of wire changes a current can be observed in another isolated coil of wire. The current is only produced when there is a changing current in the first coil. Currents produce magnetic fields; thus changing magnetic fields can produce currents. This work of Faraday was the beginning of that part of physics known as electromagnetic induction. It was to have profound effects on the life of man through the many changes in technology that resulted from it.

Figure 11.50 shows some of the possible ways we can, and cannot, produce induced currents. The magnetic fields are shown as being produced by a permanent magnet; current-carrying wires, e.g., a solenoid, could have been used. The vital point that emerges is that there must be relative motion between the field and the wire and part of the circuit must be in a different strength magnetic field, perhaps that of the earth.

Consider the vertical part of the wire, PQ, which is moving in the field (Fig. 11.51). Suppose it is moving with a velocity v. The wire contains electrons which are being dragged in the wire through a magnetic field. Charges moving in a magnetic field experience a force if there is a field component at right angles to their motion.

$$F = Bqv$$

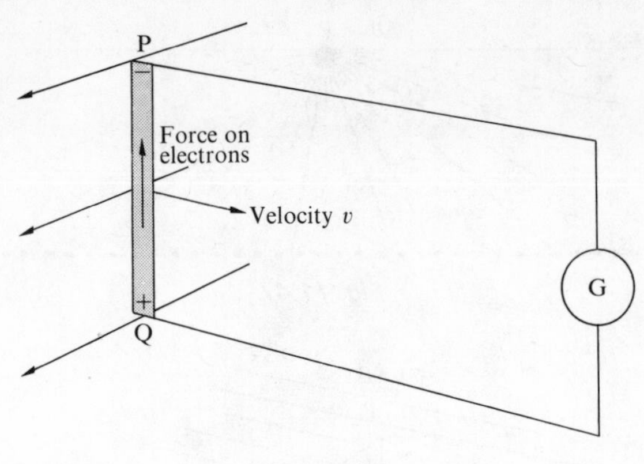

Fig. 11.51

The electrons in the wire PQ should therefore be acted on by a force, the direction of the force being towards P. The force on an electron, charge e, is

$$F = Bev$$

This will cause the electrons to move along the wire. The result of this will be to produce an electric field E, which will produce an opposing force on the electrons.

$$F = Ee$$

The field is produced by a potential difference V between the ends P and Q of the wire distance, L.

$$E = \frac{V}{L}$$

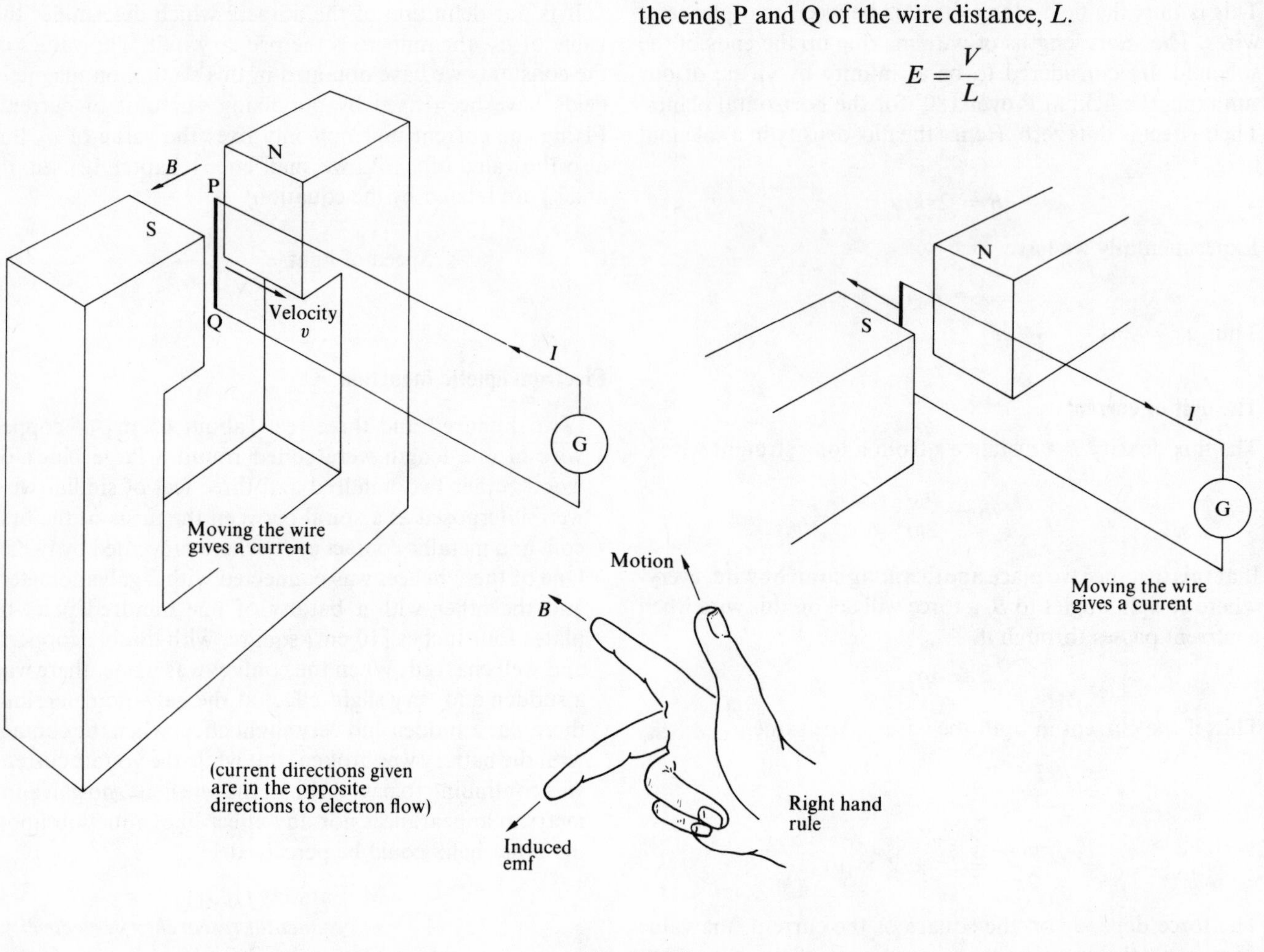

Fig. 11.50 (a)

Thus

$$F = \frac{Ve}{L}$$

The net force acting on an electron is thus

$$Bev - \frac{Ve}{L}$$

It is this force which is responsible for driving the current round the circuit. The maximum value of the potential difference V will occur when

$$Bev = \frac{Ve}{L}$$

This gives for V

$$V = BLv$$

The maximum potential difference occurs when there is no current taken by the circuit. The higher the resistance of a voltmeter connected across the ends of the wire the nearer will be the measured potential difference to the value given by this equation. This potential difference with no current drain is called the electromotive force (e.m.f.), symbol ε.

$$\varepsilon = BLv$$

The e.m.f. is thus proportional to the magnetic flux density component normal to the wire, the length of the wire, and the velocity with which the wire sweeps through the field.

In a time t the distance moved by the wire PQ is vt (Fig. 11.52). The area A of field swept out by the wire in time t is Lvt, the area being at right angles to the field. Thus

$$\varepsilon = B\frac{Lvt}{t} = \frac{BA}{t}$$

The quantity BA is called the flux, ϕ.

$$\varepsilon = \frac{\phi}{t}$$

The e.m.f. is thus proportional to the amount of flux the wire sweeps through, links, in time t.

B is generally called the flux density because it is the flux

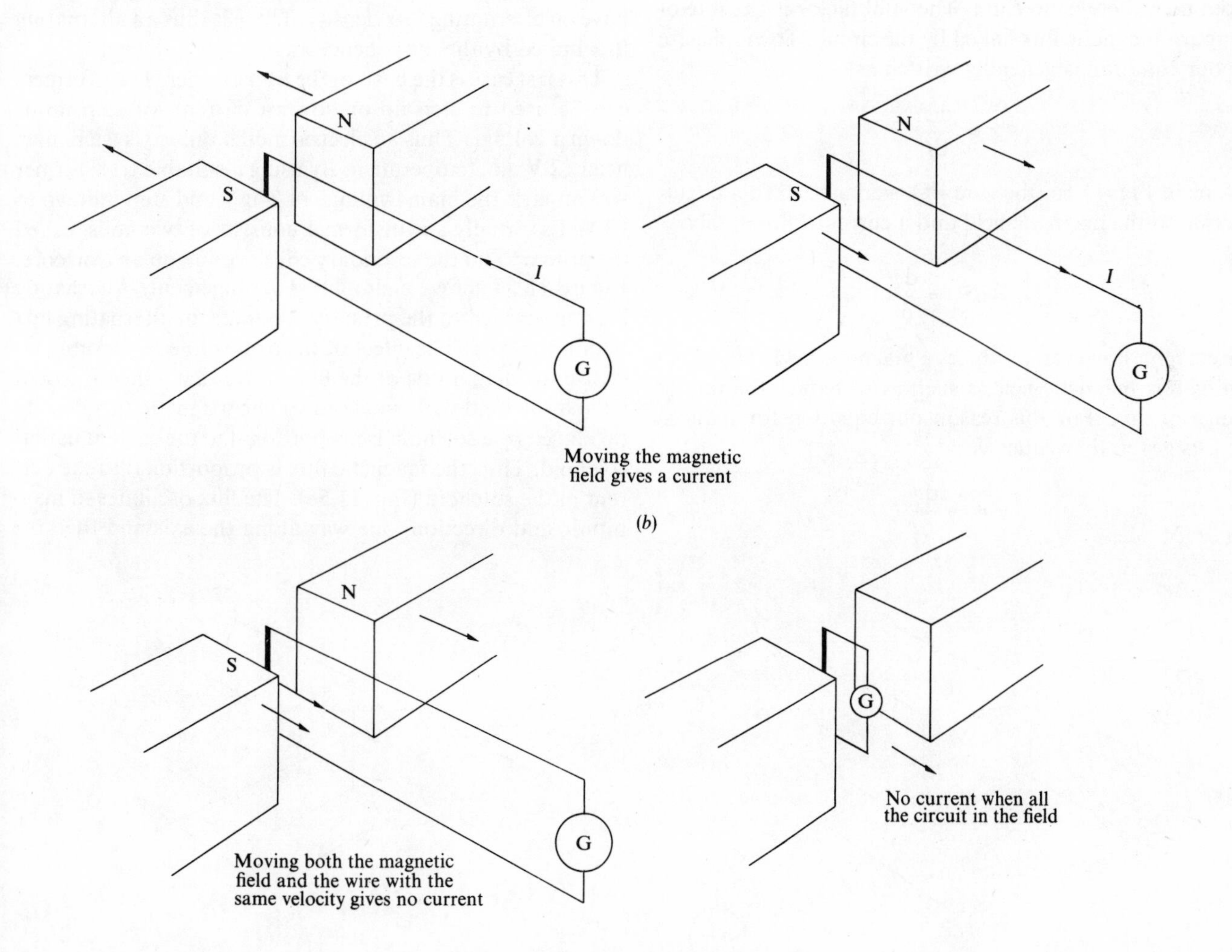

Fig. 11.50 (c)

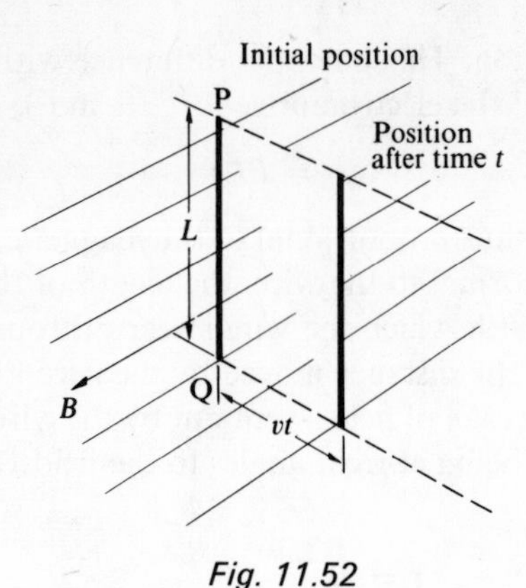

Fig. 11.52

per unit area. The unit of flux is the weber and hence B has the units webers per square metre.

$$B = \frac{\phi}{A}$$

The induced e.m.f. is zero if there is no change in magnetic flux linked by the circuit. Thus if the coil is moving in a magnetic field such that there is no change in the magnetic flux linked then there is no e.m.f. The vital factor is the rate of change of magnetic flux linked by the circuit. To emphasize this our equation is generally written as

$$\varepsilon = \frac{d\phi}{dt}$$

If, as in Fig. 11.53, the wire PQ is connected to a circuit external to the magnetic field and a current I flows, then

$$\varepsilon = IR = \frac{d\phi}{dt}$$

The current, however, produces a magnetic field. The direction of this magnetic field is such as to reduce the rate of change of flux. For this reason our equation for induced e.m.f. is generally written as

$$\varepsilon = -\frac{d\phi}{dt}$$

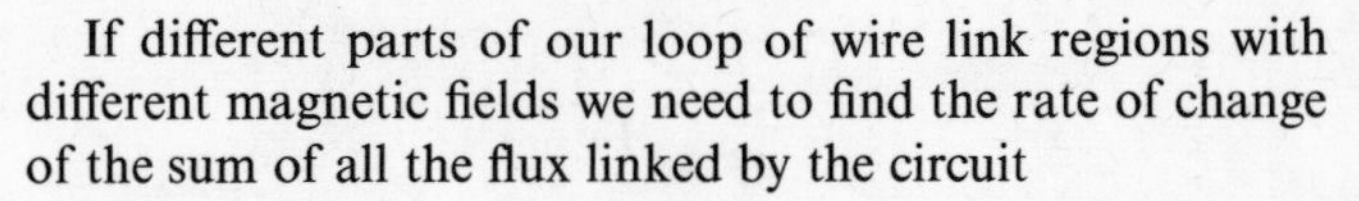

If different parts of our loop of wire link regions with different magnetic fields we need to find the rate of change of the sum of all the flux linked by the circuit

$$\varepsilon = -\frac{d}{dt}\int B\,dA$$

Figure 11.54 shows some arrangements in which loops of wire move in magnetic fields. In (a) the flux linked by the loop is changing (the flux is passing through the shaded area) and thus an e.m.f. is induced. In (b) although the loop is moving in a magnetic field there is no change in the flux linked by the loop, the flux linked remains a constant. There is therefore no induced e.m.f. In (c) the loop is rotating. The flux through the loop is changing because the flux is the product of the area and the component of the flux density at right angles to the area, the component is changing. In (d) there is no flux linked by the loop and thus no induced e.m.f. In (e) there is flux linked and moving the loop changes the flux linked—hence an e.m.f. is induced. In (f) the area of the loop in the magnetic field is not changing but the magnetic field is. The magnetic field is proportional to the current in the solenoid and as this is alternating current we have an alternating flux density. There is thus an alternating flux linked by the loop, hence an e.m.f. is induced.

This last case is the basis of the transformer. Transformers can be used to step up or down a current, or step up or down a voltage. Thus an electric model railway system may need 12 V a.c. to operate it. By using a suitable transformer we can take the mains voltage of 240 V and step it down to 12 V. Essentially a transformer consists of two coils, called the primary and the secondary coils, wound on an iron core. Figure 11.55 shows a simplified arrangement. Alternating current applied to the primary produces an alternating flux within the core. The effect of the iron is to considerably increase the magnitude of the flux above that which it would be in a solenoid with an air core. The magnetic flux density produced by a solenoid is proportional to the current in that solenoid. Thus the magnetic flux is proportional to the current in the solenoid (Fig. 11.56). The flux oscillates in magnitude and direction, one way along the axis and then the

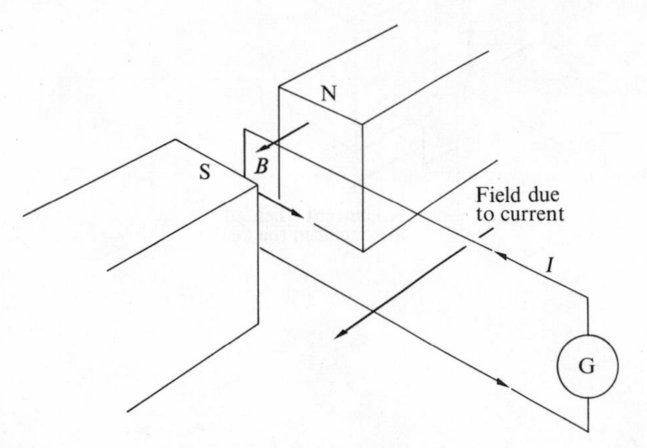

N S B G

Fig. 11.53

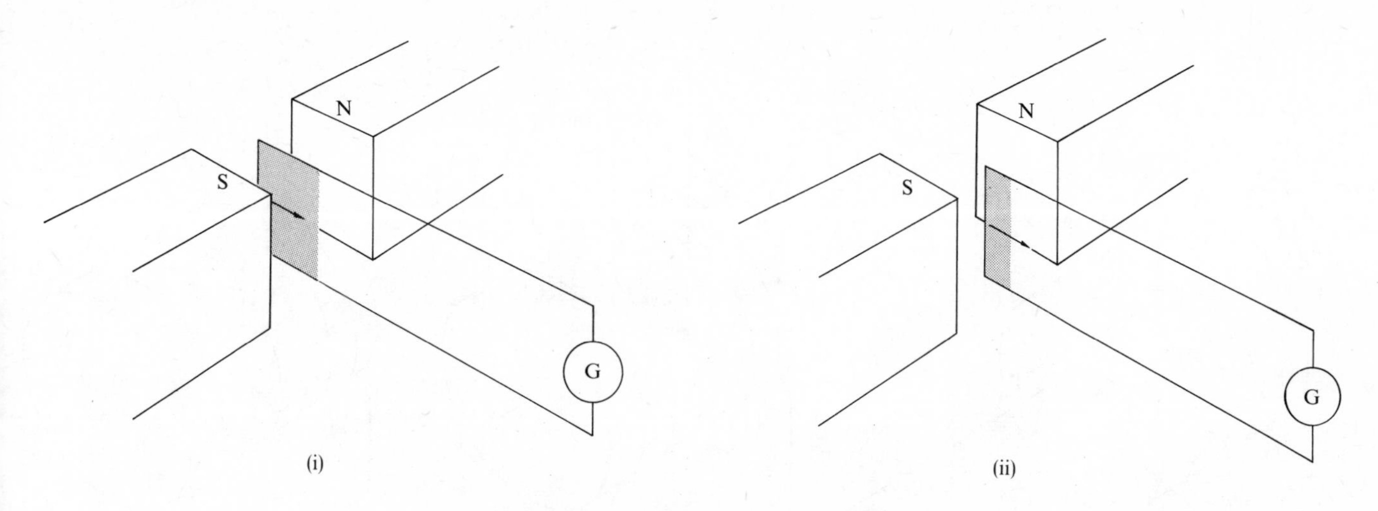

(a) Flux changes, hence e.m.f.

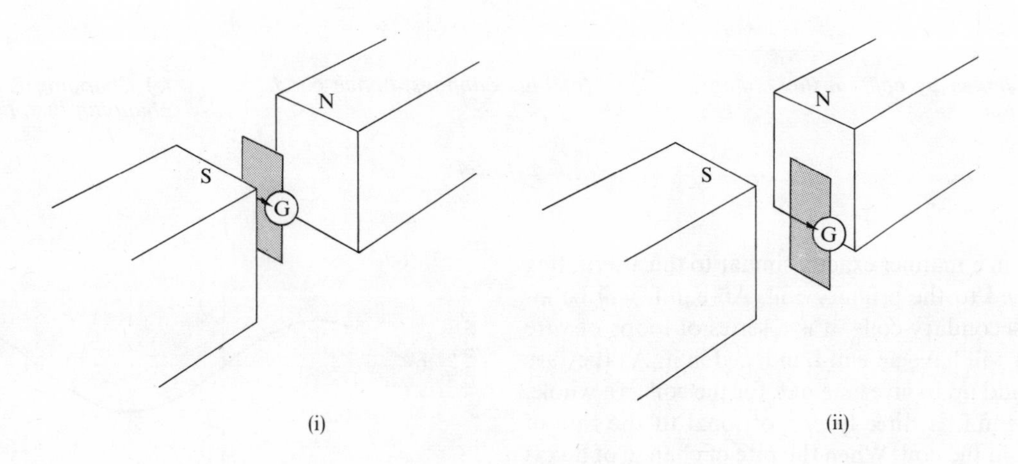

(b) No flux change, hence no e.m.f.

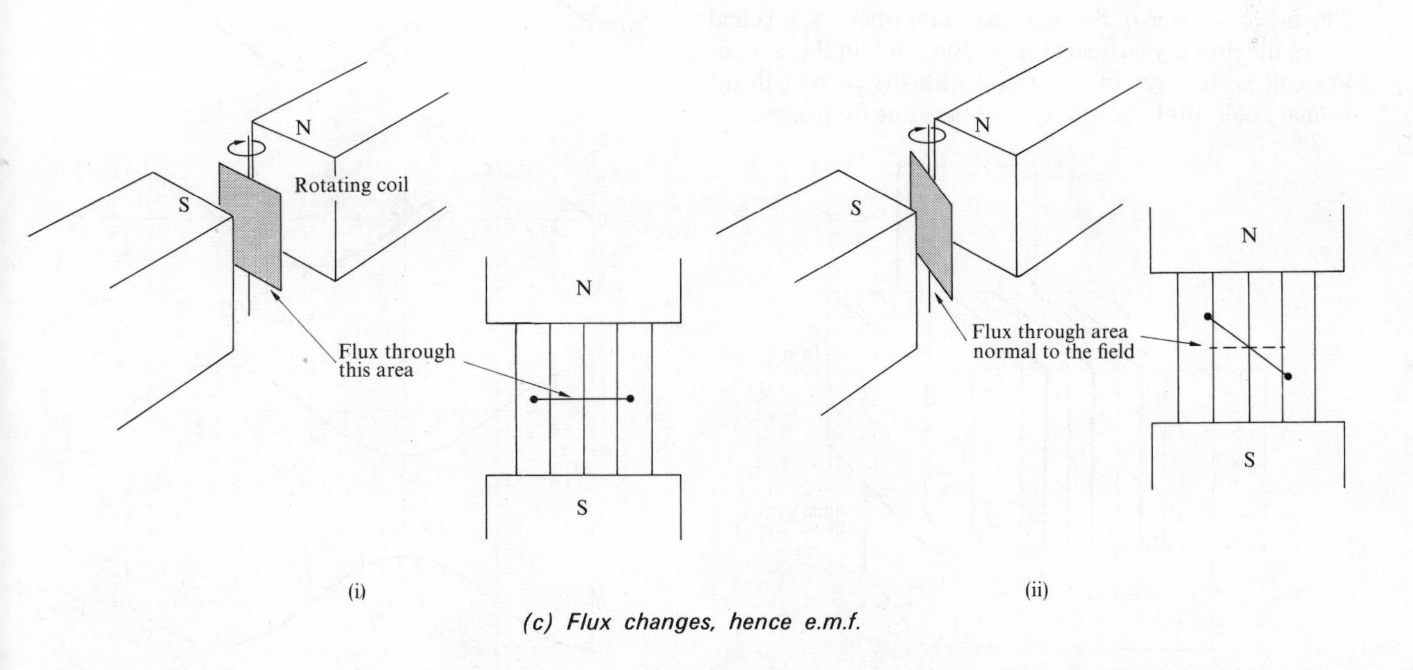

(c) Flux changes, hence e.m.f.

Fig. 11.54

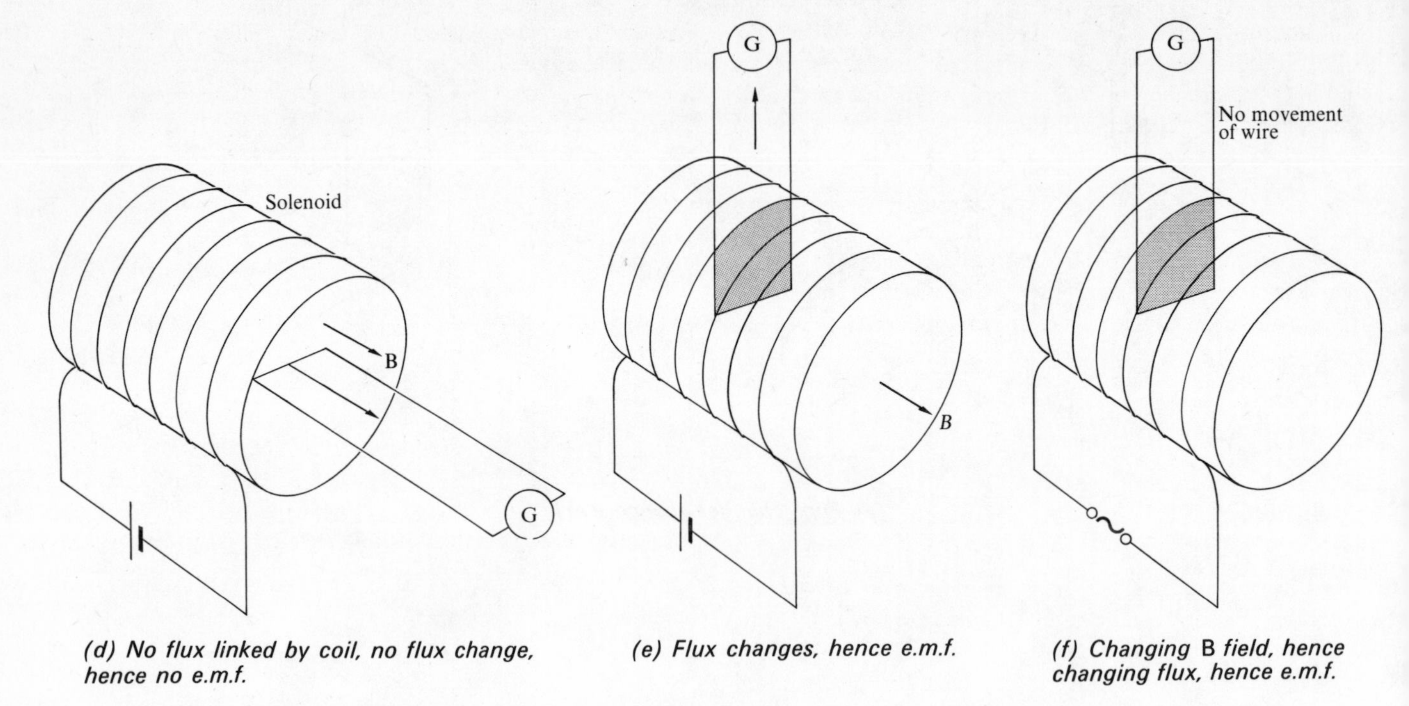

(d) No flux linked by coil, no flux change, hence no e.m.f.

(e) Flux changes, hence e.m.f.

(f) Changing B field, hence changing flux, hence e.m.f.

Fig. 11.54

opposite way, in a manner exactly similar to the alternating current supplied to the primary coil. An e.m.f. will be induced in the secondary coil—it is a series of loops of wire and each loop will have an e.m.f. induced in it. As they are in series they add up to give an e.m.f. for the coil as a whole. The induced e.m.f. is directly proportional to the rate of change of flux in the coil. When the rate of change of flux is zero the secondary e.m.f. is zero—this occurs when the flux and hence the primary current is a maximum or a minimum. The rate of change of flux is a maximum when the flux and hence the primary current is zero. The e.m.f. in the secondary coil is thus 90° out of phase with the current in the primary coil. Both, however, have the same frequency.

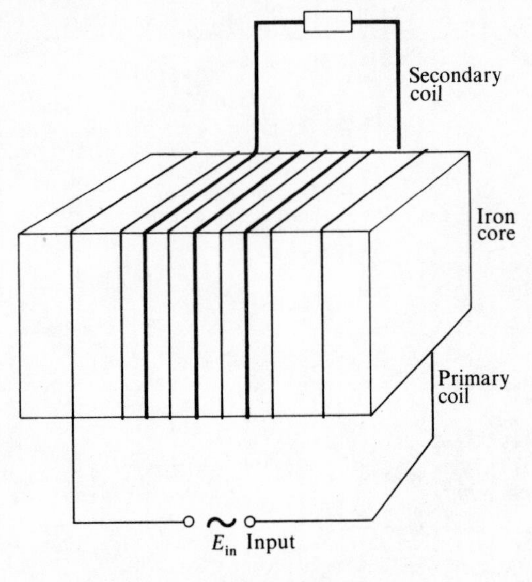

Fig. 11.55

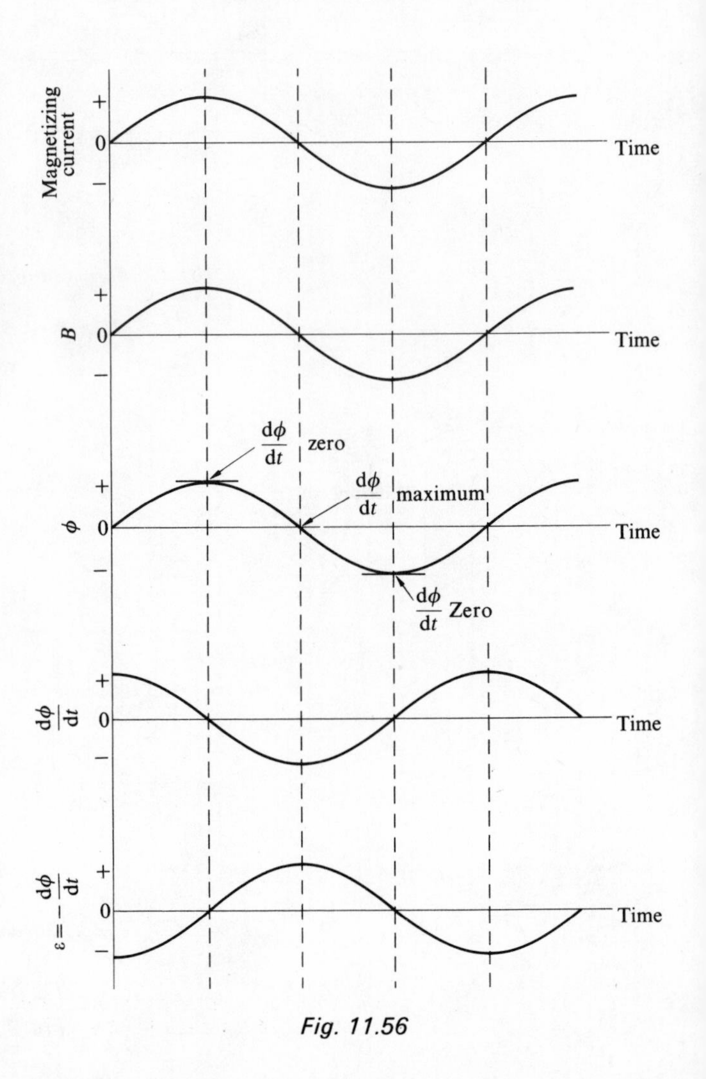

Fig. 11.56

The amount of flux linked by the secondary coil is directly proportional to the number of coil turns of the secondary. Doubling the number of secondary coil turns will double the induced secondary e.m.f.

$$\varepsilon_s = -N_s \frac{d\phi}{dt}$$

Two coils within which there is a changing magnetic flux—we have only considered that there will be an induced e.m.f. in the secondary coil. There must be an induced e.m.f. in the primary coil as well as the one in the secondary coil.

$$\varepsilon_p = -N_p \frac{d\phi}{dt}$$

Thus

$$\frac{\varepsilon_p}{\varepsilon_s} = \frac{N_p}{N_s}$$

The ratio of the induced e.m.f.s is the same as the ratio of turns numbers.

The induced primary e.m.f. is 90° out of phase with the magnetizing current supplied by the input source. If no current is taken from the secondary and we assume conservation of electrical energy, e.g., no temperature changes produced, then no energy is taken from the secondary and none is dissipated in the primary. This can only be the case if the input e.m.f. is equal to the induced primary e.m.f. and in opposite direction, i.e., out of phase.

$$\varepsilon_{in} - \varepsilon_p = 0$$

Thus for this condition of no secondary current

$$\frac{\varepsilon_{in}}{\varepsilon_s} = \frac{N_p}{N_s}$$

This condition is approximately true for large resistance in the secondary circuit.

Thus if we have an input of 240 V a.c. and require an output of 12 V a.c. across a high resistance load, a turns ratio of 240:12 or 20:1, primary to secondary is required.

If the secondary coil passes a current through a resistance power is being dissipated.

$$\text{Power} = \varepsilon_s I_s$$

where I_s is the secondary current. The energy to supply this power must be provided by the primary circuit, i.e., taken from the input. Thus ε_{in} must become larger than ε_p. Thus in the above example, a 20:1 turns ratio would give less than 12 V output for a 240 V input if the secondary current was significant.

If we have conservation of electrical energy we must have the following relationship between the primary and secondary currents

$$\varepsilon_{in} I_p = \varepsilon_s I_s$$

The power drawn from the input source must be equal to the power dissipated in the secondary. Hence

$$\frac{\varepsilon_{in}}{\varepsilon_s} = \frac{I_s}{I_p} \simeq \frac{N_p}{N_s}$$

The transmission of power

The mains voltage in Britain is nominally 240 V. The voltage is alternating. The power is transmitted from the generator to your home area at voltages of either 400 kV or 132 kV (through what is called the 'grid', Fig. 11.57). The power lines carrying these voltages are to be seen strung between pylons which straggle across the country.

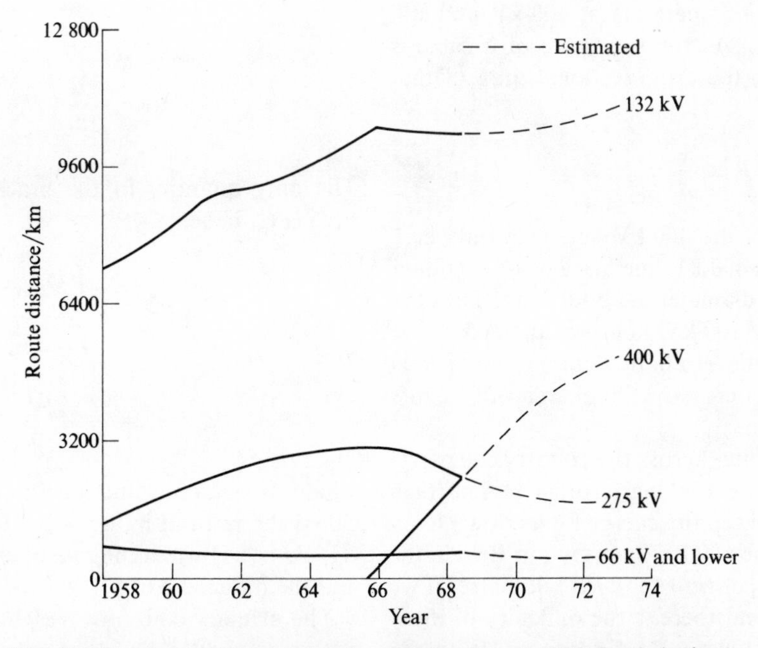

Fig. 11.57 Central Electricity Generating Board 1968 Statistical year book. *Route distance for the 'grid' in Britain.*

The generator produces the power at a voltage of about 10 to 20 kV. The step up of the voltage from the generator to that of the pylon wires is done with a transformer, the step down from the voltage of the pylon wires to that of the area supply and then down again to that of your mains supply is again done with transformers. Why the high voltage for transmission?

The power lost by passing a current along a power cable, or indeed any wire, is proportional to the square of the current.

$$\text{Power loss} \propto I^2$$

For a fixed power input, the current along a line is inversely proportional to the potential difference between the ends of the wire.

$$I \propto \frac{1}{V}$$

$$\text{Thus power loss} \propto \frac{1}{V^2}$$

For a 400 kV power line the power lost is $\frac{1}{16}$ that of a power line operating at 100 kV. A 400 kV line loses only 1 part in 4 millionth of that a 200 V line would lose. Operating at a high voltage reduces the power loss. If we wanted to transmit power at a low voltage we would need much thicker cables, and hence more pylons to take the load, in order to reduce the power loss.

For the same power loss at all voltages, how must the resistance of a cable be changed for the different transmission voltages? For a constant current, i.e., constant power loss, we have

$$R \propto V$$

Thus the resistance per kilometre for a 400 kV line can be four times that of a line operating at 100 kV and still give the same power loss. As the resistance of a cable is inversely proportional to the cross-sectional area of that cable

$$R \propto \frac{1}{A}$$

the cross-sectional area of the 400 kV line need only be $\frac{1}{4}$ that of the 100 kV line. A 400 kV line has a cross-sectional area of $2{\cdot}6 \times 10^{-4}$ m^2, a diameter of about 2 cm. Thus the cross-sectional area of a 100 kV line would need to be $10{\cdot}4 \times 10^{-4}$ m^2, a diameter of about 4 cm. A line operating at 200 V would need a cross-sectional area of $0{\cdot}52$ m^2, a diameter of about 90 cm.

We have pylons straggling across the countryside carrying the power because we use high voltages. The high voltages are necessary to keep the energy losses down to an acceptable value. With the 400 kV lines used in Britain the energy loss is about 0·02 per cent for each kilometre. If we put the cables underground there is the difficulty of dissipating the heat that this energy loss produces. With the cables in the air cooling is by the air flowing over them; in the ground cooling can only occur by the flow of heat through the soil—soil is a bad conductor of heat. Thus underground cables have to be thicker in order to keep the energy loss low enough for the cable temperature to keep within acceptable limits. Also there is the problem of electrical insulation when the cables are in the ground, perhaps wet ground which may be a reasonable conductor of electricity. The cost of underground cables has been estimated to be about ten times greater than that for overhead cables. If you dislike the sight of pylons, do you want to pay the higher cost for electricity?

Inductance

Consider two coils wound on the same former. If we put alternating current in the first coil, the primary, an alternating e.m.f. is induced in the second coil. The flux density produced by the current I_p in the primary coil is in air

$$B = \mu_0 \frac{N_p}{l} I_p$$

where l is the length of the primary coil and N_p the total number of turns on that coil. If we assume that this is the flux density in the secondary coil then the flux linked by the secondary coil will be

$$\phi = N_s BA$$

where N_s is the total number of secondary turns, A the cross-sectional area of the secondary coil. Thus

$$\phi = N_s \mu_0 \frac{N_p}{l} I_p A$$

The e.m.f. induced in the secondary coil is equal to the rate of change of flux linked by the coil.

$$\varepsilon_s = -\frac{d\phi}{dt}$$

$$\varepsilon_s = -\frac{d}{dt}\left(N_s \mu_0 \frac{N_p}{l} I_p A\right)$$

The only quantity in the bracket which is varying with time is I_p, hence

$$\varepsilon_s = -\left(N_s \mu_0 \frac{N_p}{l} A\right)\frac{dI_p}{dt}$$

or

$$\varepsilon_s = -M \frac{dI_p}{dt}$$

where M is a constant for a particular pair of coils. M is called the mutual inductance. The unit of M is the henry (H). M is 1 H when the rate of change of current is 1 A s^{-1} and the induced e.m.f. 1 V.

The primary coil, however, links changing flux and thus an e.m.f. should be induced in the primary, due to the alternating current supplied to the primary. The flux linked by the primary is

$$\phi = N_p BA$$

$$\phi = N_p \mu_0 \frac{N_p}{l} I_p A$$

The e.m.f. induced in the primary is equal to the rate of change of flux linked by the primary.

$$\varepsilon_p = -\frac{N_p^2}{l} \mu_0 A \frac{dI_p}{dt}$$

$$\varepsilon_p = -L \frac{dI_p}{dt}$$

where L is a constant known as the self inductance.

A single wire has self inductance. It does not need to be in the form of a coil or even be near any other coil. A changing current in a wire produces a changing flux and so an induced e.m.f. Because this induced e.m.f. opposes the change in current (hence the minus sign in the equation) it is sometimes called the 'back e.m.f'. The value of L depends on the arrangement of the wire and is thus a constant for a particular arrangement. The induced e.m.f. depends only on the rate of change of current in the circuit. If there is no change of current, i.e., d.c. is used, there is no opposing e.m.f. If, however, a.c. is used there is an opposing e.m.f. The opposing e.m.f. is proportional to the rate of change of current and hence to the frequency of the a.c.

The effect of this opposing e.m.f. is to reduce the current in the circuit; suppose it to be an a.c. source and a coil of wire.

$$\varepsilon - \varepsilon' = IR$$

ε is the e.m.f. supplied by the a.c. source, ε' is the induced e.m.f., I is the current in the circuit, and R the resistance. As ε' varies with the frequency of the a.c. so must the current in the circuit.

If the current variation with time can be represented by

$$I = I_m \sin 2\pi ft$$

where I_m is the maximum current and f the frequency, then

$$\frac{dI}{dt} = 2\pi f I_m \cos 2\pi ft$$

and we have

$$\varepsilon - 2\pi f L I_m \cos 2\pi ft = IR$$

The quantity $2\pi fL$ is known as the reactance, X_L, of the inductor, i.e., the thing having the inductance. Hence by rearranging the equation

$$\varepsilon = RI_m \sin 2\pi ft + X_L I_m \cos 2\pi ft$$

Figure 11.58 shows graphs for these two terms on the right-hand side of the equation, one being a sine curve and the other a cosine curve. The sum of these two graphs gives the e.m.f., ε. The above equation can be simplified to give the equation of this variation of ε with time.

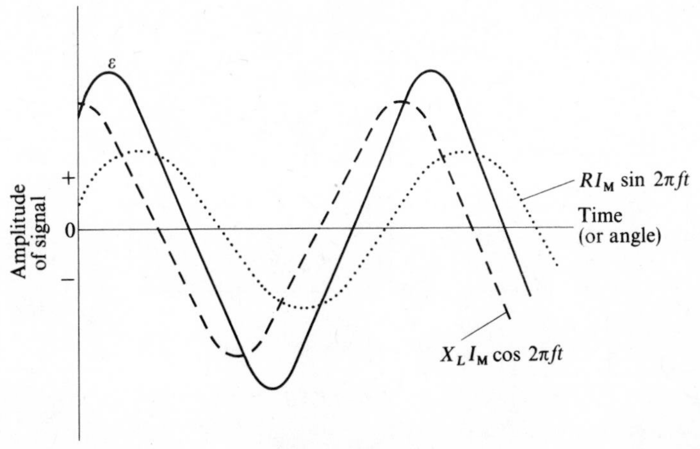

Fig. 11.58

$$\varepsilon = \sqrt{[(RI_m)^2 + (X_L I_m)^2]} \sin (2\pi ft + \theta)$$

or

$$\varepsilon = \varepsilon_m \sin (2\pi ft + \theta)$$

where the maximum e.m.f. is

$$\varepsilon_m = \sqrt{[(RI_m)^2 + (X_L I_m)^2]}$$

$$\varepsilon_m = I_m \sqrt{(R^2 + X_L^2)}$$

and

$$\tan \theta = \frac{X_L}{R}$$

θ is called the phase angle of the circuit; it is the angle by which the source voltage is ahead of the current.

We can rewrite the e.m.f. equation as

$$\varepsilon_m = I_m Z$$

where Z is called the circuit impedance. It is comparable to the term resistance in the equation for d.c. circuits of $V = IR$. The maximum e.m.f. and the maximum current will not, however, be occurring at the same times.

$$\text{Impedance } Z = \sqrt{(R^2 + X_L^2)}$$

The reactance term represents the effect of the inductor on the overall circuit impedance. The units of Z, R, and X_L are all in ohms.

The above equation is similar to the equation relating the lengths of the sides in a right-angled triangle.

$$\text{Hypotenuse} = \sqrt{[(\text{base})^2 + (\text{vertical})^2]}$$

We can represent Z by the hypotenuse, R by the base, and X_L by the vertical (Fig. 11.59(a)). The angle between the base, R, and the hypotenuse, Z, is the same as the phase angle θ.

$$\tan \theta = \frac{X_L}{R}$$

As $V_R = I_m R$ and $V_L = I_m X_L$, the sides of the triangle are also proportional to the potential differences in the circuit, the hypotenuse being proportional to the e.m.f., ε_m (Fig. 11.59(b)).

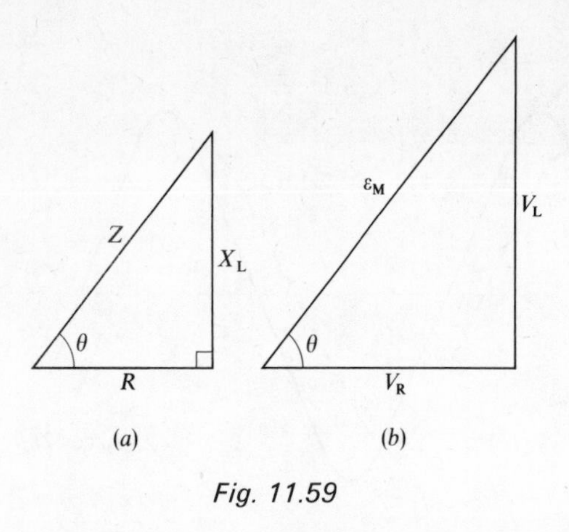

Fig. 11.59

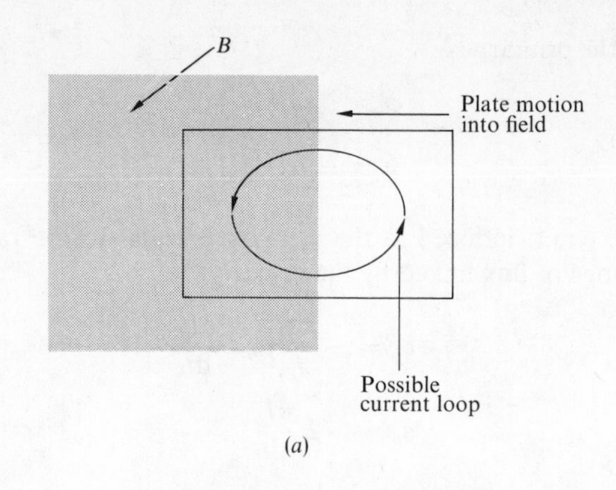

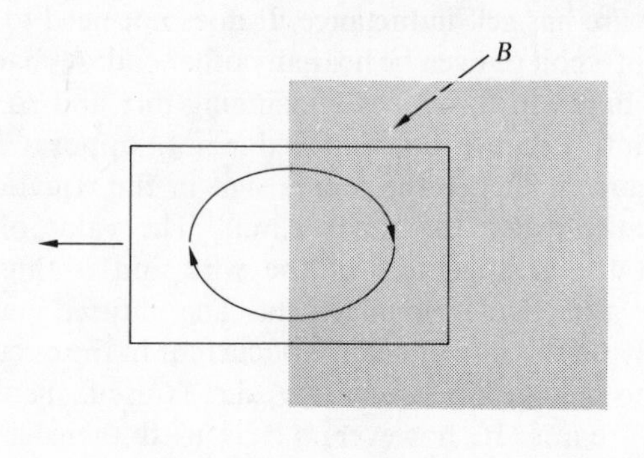

Fig. 11.61

Eddy currents

What happens when a metal plate, e.g., an aluminium plate, moves into a magnetic field? (see Fig. 11.60). The plate slows down on entering the field. Why? An e.m.f. is induced in the plate—this means a force acts on the 'free' electrons within the plate due to its moving in a magnetic field—and because there is a possible low resistance path currents flow in the plate. Closed loops of currents are produced (Fig. 11.61). These produce magnetic fields which oppose the field producing them. The current acts in such a way as to oppose the change in flux through the plate. Thus if, as in this case, the flux linked by the plate is increasing then the current produces a field in a direction which tries to reduce this increase in flux. If the plate is moving out of the field the eddy currents are in such a direction as to oppose the flux change—in this case the flux linkage is

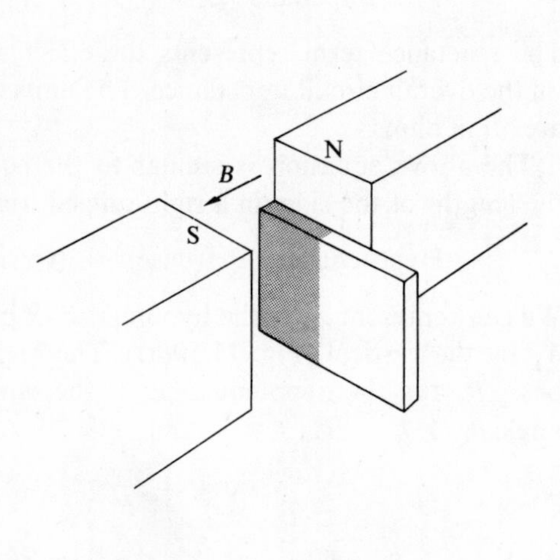

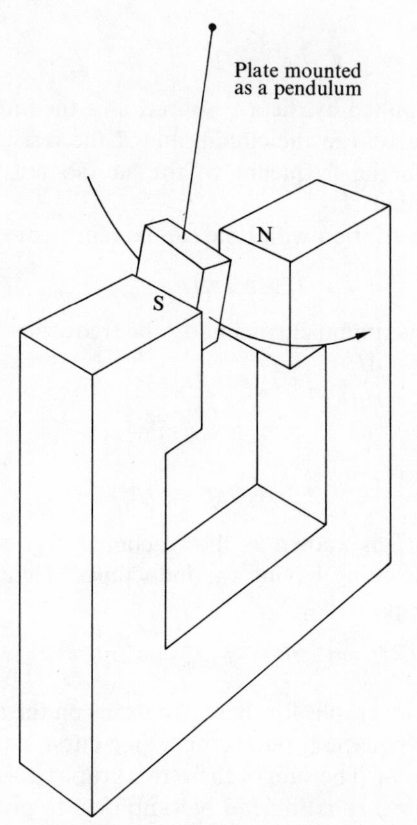

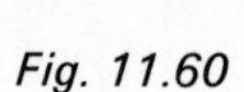

Fig. 11.60

being reduced so the field produced is in the same direction as the magnet's field, to stop the flux linked becoming smaller. The eddy currents in this case have the effect of opposing the motion out of the field. To sum up, eddy currents are in such a direction as to oppose motion into a field or out of a field, i.e., they oppose motion relative to the field. If we move a magnet past an aluminium plate, the plate will endeavour to follow; the eddy currents oppose relative motion between the field and the plate.

If we have a metal plate stationary in a constant magnetic field there is no changing magnetic flux and thus no eddy currents. If, however, the magnetic field is changing there is changing flux linked by the plate and thus eddy currents. The core of a transformer will have eddy currents induced in it when a.c. is supplied to the primary coil.

One effect of currents flowing in a metal sheet is to produce an increase in temperature. Transformer cores get hot. The size of these eddy currents depends on the size of the induced e.m.f., the nature of the material, and its shape.

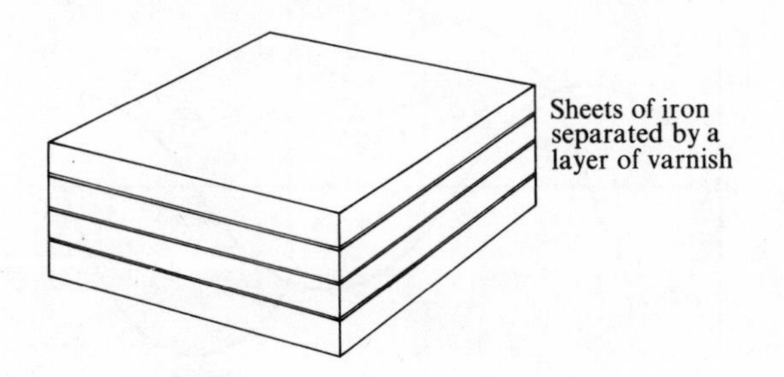

Fig. 11.62 Laminated core.

The lower the resistance of the material the larger the eddy currents. If the material is in small segments less current occurs. Each segment links less flux than the material in bulk and thus the e.m.f. produced is less. In addition the resistance is higher. Transformer cores are laminated (Fig. 11.62) in order to reduce the size and effects of eddy currents. Reducing the size of eddy currents reduces both the temperature rise and the 'wastage' of energy.

Motors and generators

A current-carrying wire in a magnetic field experiences a force if there is a magnetic field component at right angles to the wire. A pivoted current-carrying coil in a magnetic field can experience forces which will cause it to rotate (Fig. 11.63). With the coil VWXY, all the wires experience forces. It is, however, only the forces on the wires VY and WX which can cause the coil to rotate. In diagram (b) the forces are causing the coil to rotate in an anti-clockwise direction, in (c) the forces are not causing rotation, in (d) they produce a clockwise rotation, in (e) a clockwise rotation. With the arrangement shown the coil will rotate to the vertical position and there stop. If it overshoots this position there is a force returning it to that position. The coil can be made to rotate continuously if the current is reversed every time the coil reaches the vertical position. All the forces will then be acting to cause rotation in the same direction. Figure 11.64 shows an arrangement, known as a commutator, by which this reversal of current can be achieved. The wires from the coil are connected to split

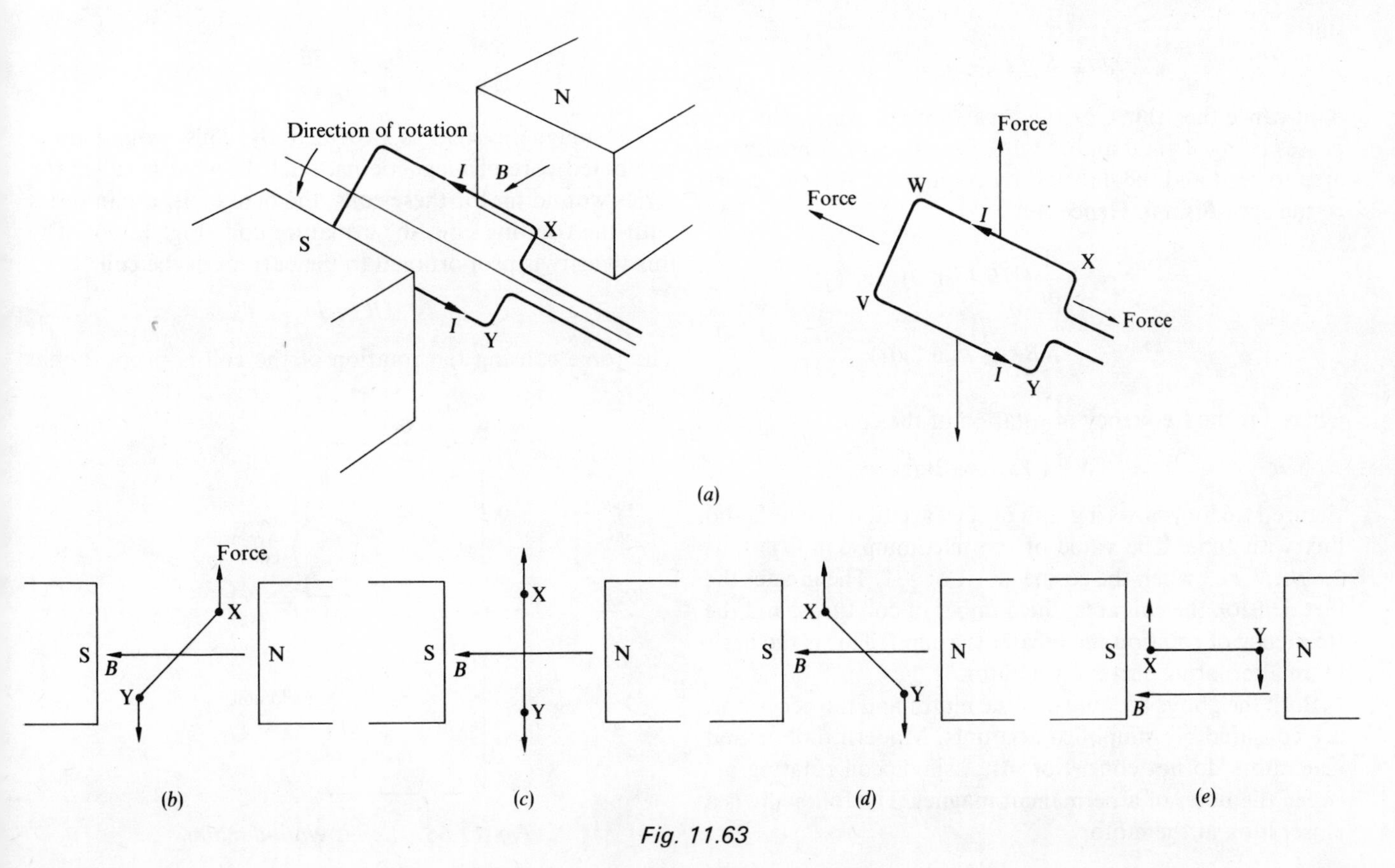

Fig. 11.63

rings which rotate with the coil. Contact with the current source is made by these split rings being in contact with fixed brushes. As the coil rotates these brushes make contact with the rings, changing the rings they are in contact with every time the coil passes through the vertical. This is the basis of the d.c. motor.

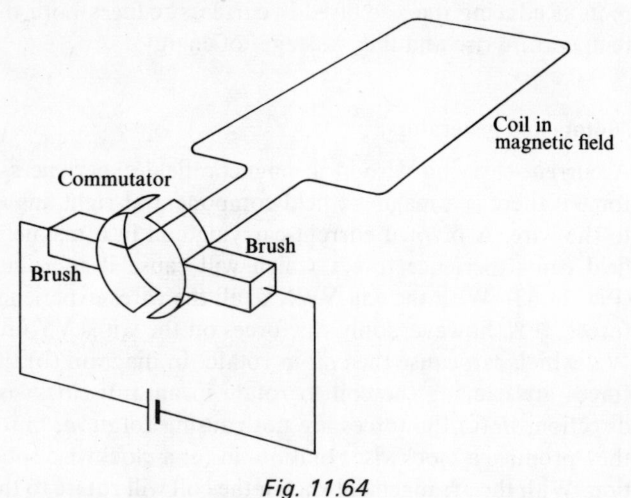

Fig. 11.64

This arrangement will produce an alternating current if the coil is rotated in the magnetic field. The flux linked by the coil keeps changing. If the coil is connected to simple slip rings, not split, the alternating e.m.f. can be used to produce an alternating current in a circuit (Fig. 11.65).

$$\varepsilon = -\frac{d\phi}{dt}$$

But

$$\phi = NBA \sin \theta$$

You can either think of the area through which the flux passes being $A \sin \theta$ and the flux density B or consider the area to be A and the flux density component at right angles to the area $B \sin \theta$. Hence

$$\varepsilon = -\frac{d}{dt}(NBA \sin \theta)$$

$$\varepsilon = -NBA \frac{d}{dt}(\sin 2\pi ft)$$

where f is the frequency of rotation of the coil.

$$\varepsilon = -NBA\, 2\pi f \cos 2\pi ft$$

Figure 11.65(c) shows a graph of the variation of e.m.f. and flux with time. The value of the maximum e.m.f. is thus $2\pi fBAN$, i.e., when the cosine is $+1$ or -1. The greater the flux density, the coil area, the number of coil turns, and the frequency of rotation the greater the e.m.f. This is the basis of an alternating current generator.

Both the above accounts, of the motor and the generator, are considerably simplified accounts. Modern motors and generators do not consist of just a single coil rotating between the poles of a permanent magnet. The following is a closer look at the motor.

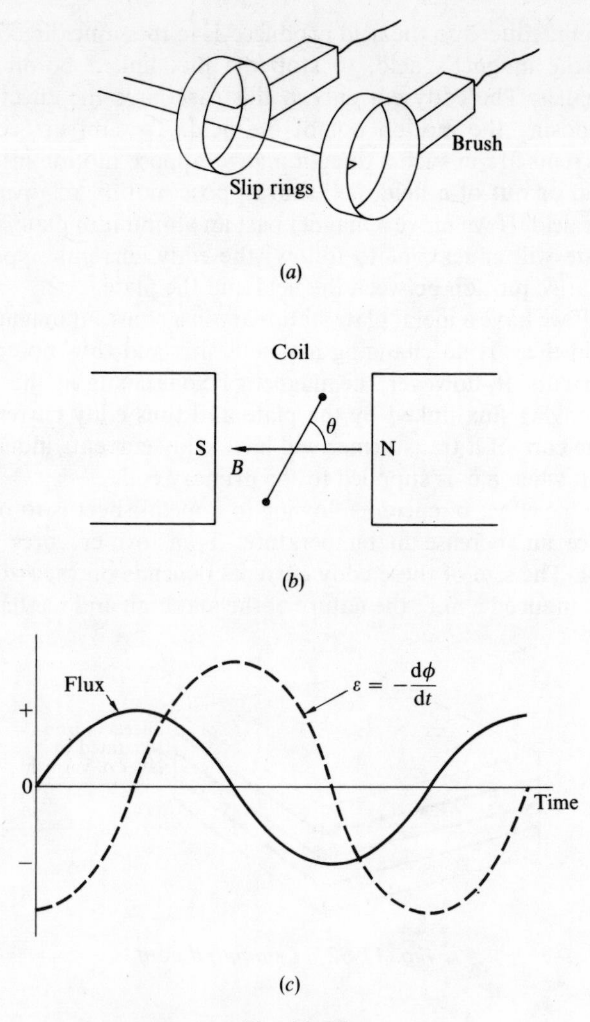

Fig. 11.65

The magnetic field is provided by coils wound on a laminated core of magnetic material. In what is called the series wound motor these coils, the field coils, are in series with the rotating coil, the armature coil (Fig. 11.66). The flux density is proportional to the current in the coil.

$$B \propto I$$

The force causing the rotation of the coil is proportional

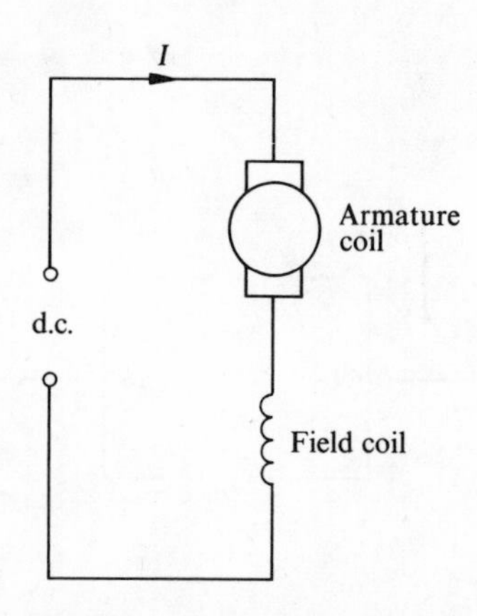

Fig. 11.66 Series wound motor.

to the flux density and the current carried by the armature coil,

$$F = NBIL$$

where N is the number of armature coil turns, L the length of the coil, I the armature coil current, and B the flux density component at right angles to the coil (Fig. 11.67(a)).

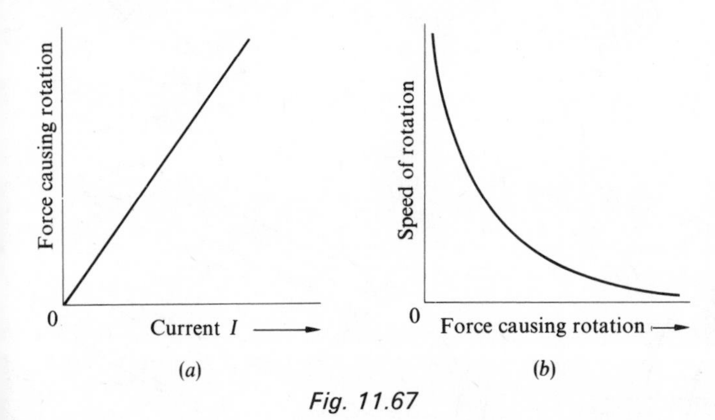

Fig. 11.67

Thus

$$F \propto I^2$$

The current I will, however, depend on the frequency of rotation of the armature coil. A coil rotating in a magnetic field will have an e.m.f. induced in it, this induced e.m.f., ε' being in the opposite direction to the e.m.f. supplied, ε.

$$\varepsilon - \varepsilon' = I(R_a + R_f)$$

R_A is the resistance of the armature coil, R_f that of the field coil. Hence

$$I = \frac{\varepsilon - \varepsilon'}{R_A + R_F}$$

The induced e.m.f. depends on the rate of change of flux linked by the coil and hence the frequency of rotation of the coil. Thus when a series wound motor is switched on and begins to rotate ε' is small and hence I large. As the motor picks up speed and its frequency of rotation increases so ε' increases and the current I decreases. The forces acting on the armature coil thus decrease as the speed of rotation increases (Fig. 11.67(b)).

When a series wound motor is used to drive something, e.g., the wheels of an electric locomotive, the forces acting on the armature coil are being used to rotate both the coil and the 'something'. High speed rotation only occurs with low forces; only at low speeds are the forces high. This is in fact suitable for a locomotive where large forces are needed to get it moving and much lower forces needed to keep it going when it has reached speed.

A difficulty with the d.c. motor is the need to use a split ring commutator to periodically reverse the current to the armature coil in order to keep the coil rotating. In the induction motor no connections are made to the armature and so a considerable simplification occurs. The induction motor depends for its rotation on the fact that a conductor will follow a moving magnetic field (see the section on eddy currents).

Figure 11.68 shows a section of an induction motor. The rotating part is the 'squirrel-cage' (Fig. 11.68(g)). This is a number of conducting bars of copper or aluminium connected together at the ends by conducting rings. The whole rotor is generally sunk into slots in a laminated iron cylinder. The cylinder is attached to the shaft which is to rotate. The magnetic field is provided by two sets of coils, opposite pairs being connected in series. Each of these pairs is supplied with an alternating current—there is, however, a phase difference between the currents of 90° (Fig. 11.68(f)). The effect of this is to give a rotating field. The resulting induced currents in the rotor drag the rotor after the field.

Appendix 11A Machines*

Machines have an input, a processing part, and an output. They need an energy supply. With their aid we can with only a small expenditure of energy on our part command other energy resources and do jobs which perhaps would not have been feasible in the absence of the machine. How much energy do you command? If you live in England the answer is probably about 4×10^{11} J per year. You by the action of your own muscles only directly produce about 4×10^9 J per year: you command 100 times more than you can produce. If you lived in India you might only command about 2 times more energy than you can produce.

Our economic welfare has changed; we can now expect an average life span of almost 70 years. At the beginning of this century the life expectation was not quite 50 years, in the mid-nineteenth century it was only 40 years. In India it is now only 32 years. Quality of diet, housing, clothing, medical care, education have all improved as our exploitation of energy increased. This has occurred through the discovery of new sources of energy and improvements in the efficiency with which we make use of the energy.

The year 1785 saw the first commercial use of the steam engine. This gave improved transport and made possible an increased production of coal (by pumping water out of mines, lifting and moving coal), so increasing the supply of energy. The early steam engine had an efficiency of only a few per cent. Over the years the efficiency has been improved until now the modern steam turbine has an efficiency of nearly 40 per cent.

A new source of energy was discovered in about 1850—petroleum. The industry now supplies about $0{\cdot}6 \times 10^{20}$ J per year, about half the amount supplied by coal. The year 1876 saw the birth of the car industry. There are now more than 128 million passenger cars.

In 1831 Faraday discovered the electromagnetic generation of electricity. The capacity of the electrical power stations in Britain is now about 4×10^{10} W. In 1960 about 8×10^{17} J were used in Britain. Our consumption of energy increases each year (Fig. 11.69). In 1956 nuclear energy

* The correct title for this appendix is engines but common usage makes machines the more acceptable title.

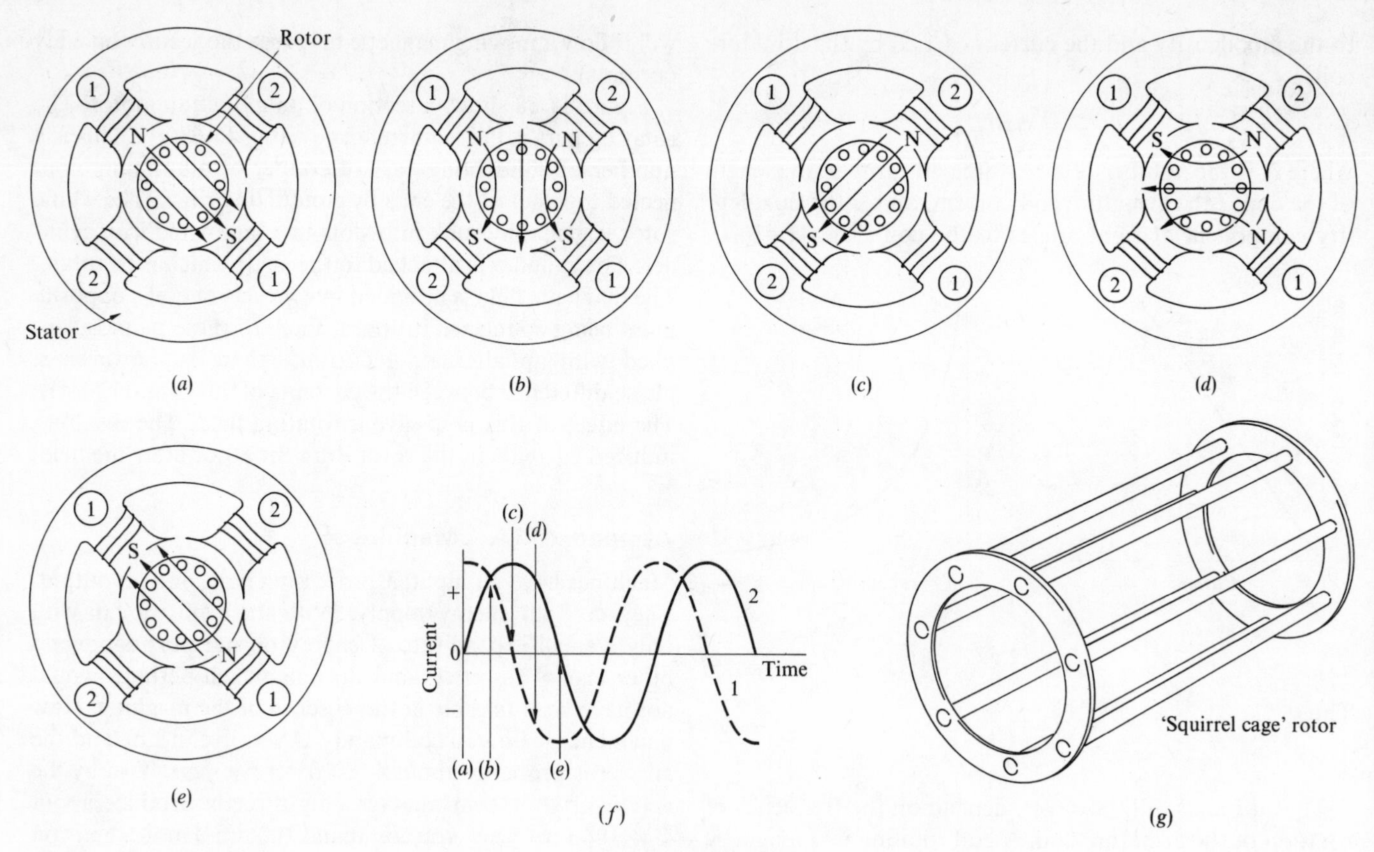

Fig. 11.68

became harnessed as a source of energy (the first nuclear power station).

In 1831 34 per cent of the British population were employed in agriculture and fishing, in 1951 only 5 per cent were. In India 74 per cent of the population were in 1950 employed in agriculture. The end of the eighteenth century saw the beginning of the industrial revolution in Britain—the mid-twentieth century is seeing the industrial revolution occurring in India.

Are all these developments for the 'good'? The development of the coal industry has laid waste many areas of England, has destroyed many men with its working conditions; is the cost worthwhile? The menace of cars with their deadly fumes. The pollution of the environment by industry. The waste heat from power stations changing the environment. The drain on the world's energy resources (see chapter 3). Is our present organization with its emphasis on using machines and energy resources almost regardless of the consequences the right one for society?

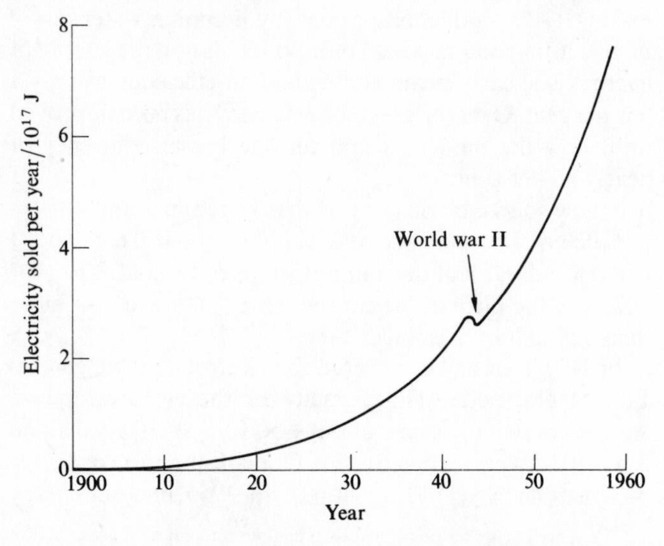

Fig. 11.69 Consumption of electricity in Britain.

> 'Machines are taking the upper hand in a way that makes me anxious and apprehensive.'
>
> Goethe. 1794–6. *Wilhelm Meisters Wanderjahre.*

> '...it is a crisis of values. What is at issue is two diametrically opposed views of how man ought to live. Up to now it has seemed that the materialist view—supported equally by parties to the right and the left—must triumph. Suddenly it begins to appear that it cannot triumph. The choice then becomes whether to make an intelligent selection from the gifts of technology and use them to enrich an humane existence, or whether to commit suicide....'
>
> G. R. Taylor (1970). *The doomsday book*, Thames and Hudson.

Further reading

C. M. Cipolla. *The economic history of world population*, Penguin, 1962.

Problems

1. F. Hoyle and J. V. Narlikar in an article in *Nature* (3 Sept. 1971, **223**, 41) put forward the idea that the gravitational constant G may vary with time and suggested that the variation might be of the form

$$G \propto \frac{1}{t}$$

What effects do you consider such a change might have?

2. Oxford and Cambridge Schools Examination Board. *Nuffield advanced physics*. Long answer paper, 1970.

'[Figure 11.70(a)] shows a section through a type of microphone called a "capacitor microphone". [Figure 11.70(b)] is a circuit which shows how the microphone may be connected for use. (The output from the microphone could be taken from terminals B and C, which would be connected to a suitable amplifier).

(*a*) If the switch S were closed for a few seconds, then opened again, and the diaphragm then pushed slightly inwards, explain:

(i) what would happen to the capacitance of the microphone.

(ii) what would happen to the p.d. between B and C.

(*b*) Explain what would now happen if, with the diaphragm still pushed in, the switch S were closed.

(*c*) Why is the instrument constructed so that the diaphragm is as close to the first plate as possible?

(*d*) What is the time constant of this circuit?

(*e*) Assuming that switch S is closed, state the changes of p.d. between B and C that you would expect to occur if a compression wave moved the diaphragm inwards in a time:

(i) which was short compared with the time constant of the circuit—say in about 10^{-5} second.

(ii) which was long compared with the time constant of the circuit—say in about 1 second.

(*f*) Two sources of sound, one having a frequency of 10,000 Hz, the other of 50 Hz, are each found to produce the same amplitude of mechanical vibration in the diaphragm.

(i) Why is the amplitude of the resulting variations of p.d. across BC smaller for the 50 Hz vibrations than for the 10,000 Hz vibrations?

(ii) Explain what change you could make in the circuit to bring the amplitude of the electrical output from the microphone, when responding to the 50 Hz note, to approach more closely that produced by the 10,000 Hz note.

[You may find the following formulae useful in dealing with some parts of this question:

$$C = \frac{Q}{V}, \qquad C = \frac{\varepsilon_0 A}{t}.\Big]$$

3. Oxford and Cambridge Schools Examination Board. *Nuffield advanced physics*. Short answer paper, 1971.

'Faraday's Law of Induction says that the electromotive force induced in a circuit is proportional to the rate at which the flux linking the circuit is changing.

A student is investigating the flux in an iron rod standing in a coil carrying alternating current [Fig. 11.71]. (The magnetic field is sketched at an instant when the current is at a maximum.)

He uses a probe coil which surrounds the iron rod, as shown, and connects this coil to an oscilloscope to display the induced voltage in the probe coil. With the coil in the position shown, he obtains the trace illustrated below and measures the height h.

State and explain the effect or lack of effect on the oscilloscope trace of the following changes. Each change is made separately and starts from the original arrangement shown above. Give quantitative information wherever you can.

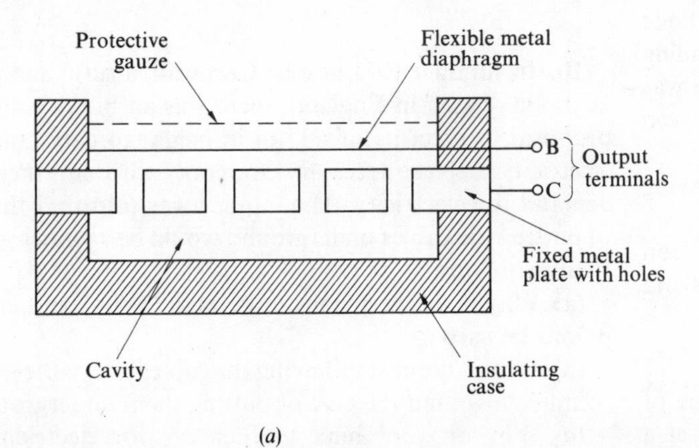

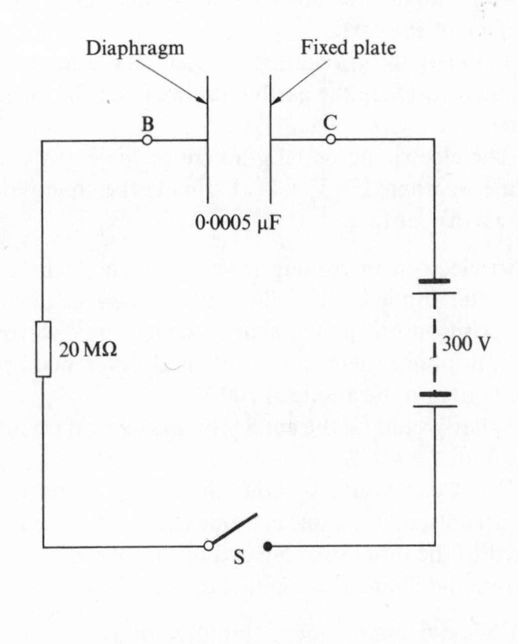

Fig. 11.70

(*a*) Doubling the number of turns in the probe coil.

(*b*) Using, in the same place, a probe coil whose area is twice as big.

(*c*) Raising the probe coil to the top of the rod.

(*d*) Doubling the frequency of the supply to the magnetizing coil, keeping the current the same.'

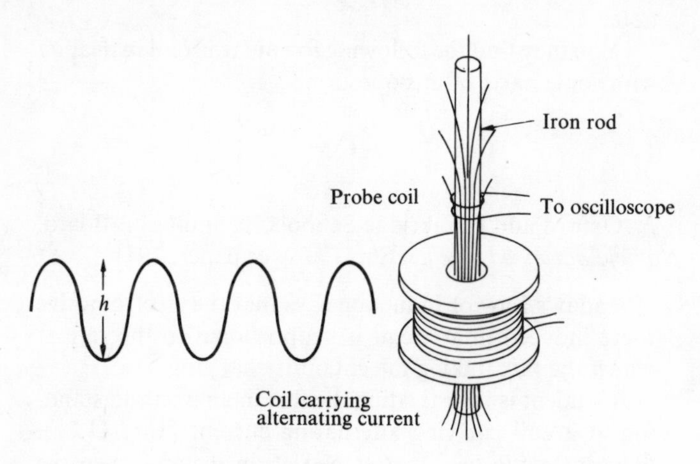

Fig. 11.71

4. 'Physics is mathematical not because we know so much about the physical world, but because we know so little: it is only its mathematical properties that we can discover.'

B. Russell. *An outline of philosophy.*

Discuss this statement and introduce examples of physics which can be used to either support or oppose the statement.

5. (a) Estimate the gravitational potential gradient at the surface of the earth.

(b) Estimate the gravitational potential gradient at a distance equal to twice the earth's radius from the surface of the earth.

(c) The electric potential gradient at the earth's surface is, in fine weather, 130 V m^{-1}. Estimate the charge density on the earth's surface.

6. An electron in passing from the filament in a diode valve to the anode could follow different paths depending on the position on the filament at which the electron was emitted. Suppose there was a potential difference between the filament and the anode of 100 V.

(a) What would be the energy of an electron on arriving at the anode?

(b) What can you say about the energy of an electron that was emitted at say one extreme end of the filament and travelled to the other extreme end of the anode?

Assume the filament is parallel to the anode if this helps.

7. A simple model for a thundercloud is a charge of -40 C at a height of 5 km and a charge of $+40$ C at a height of 10 km.

(a) What would be the electric field strength directly under the cloud due to such charges?

(b) How would you expect the electric field strength to vary with distance along the earth's surface from the point directly under the cloud?

(c) Is 40 C a big charge? Give a comparison with, possibly, a capacitor.

(d) You might care to extend this question and consider how big the charges need to be for lightning to occur.

8. Search coils connected to galvanometers are used to explore magnetic fields. What do you think would be registered if such a coil were placed round a lightning conductor? Remember that lightning flashes are of very short duration.

9. Figure 11.72 shows how the charge on the plate of a capacitor varies with time.

(a) How does the p.d. between the two plates vary with time?

(b) How does the current flowing from the plates vary with time? Sketch a graph.

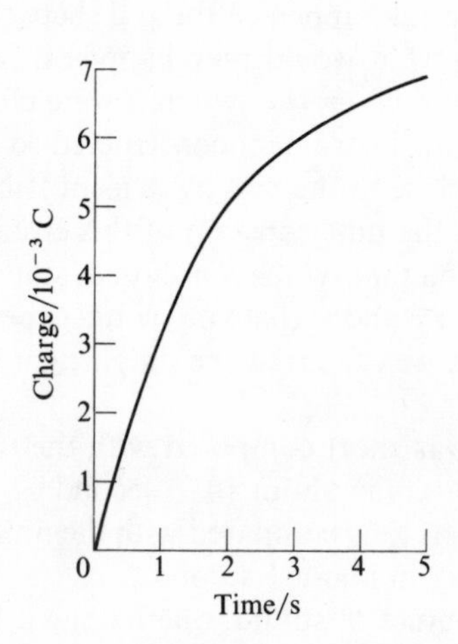

Fig. 11.72

10. In summer 1971 in west Cornwall, a rural and popular holiday area in England, there was an inquiry into the proposal to erect pylons 37 m in height to carry the new electricity cables necessary to cope with an increased demand for electricity. The inquiry was told that the cost of putting the cables underground would be $7\frac{1}{2}$ times greater than by pylons.

(a) Why have the pylons to be high? Why can't smaller pylons be used?

(b) Why is the cost of having the cables above the ground so much less than the cost of putting them underground?

(c) Why do you think the demand for electricity has increased in what is a non-industrial area?

Teaching note Further problems appropriate to this chapter will be found in:

Nuffield advanced physics. Units 3 and 7, student's books, Penguin, 1972.
Physical Science Study Committee. *Physics*, 2nd ed, chapters 27, 28, 30, 31, Heath, 1965.
E. M. Rogers. *Physics for the inquiring mind*, chapters 33, 36, 37, OUP, 1960.

Practical problems

1. When a piece of magnetic material moves into a solenoid, the inductance of that solenoid changes. Consider a balance based on this principle, i.e., a pivoted beam with a piece of magnetic material suspended from one arm and the balance pan from the other—when an object is placed in the pan the piece of magnetic material moves into or out of a solenoid and the change in inductance is used as a measure of the weight of the object.

Produce a balance based on this method. Such a balance will produce an electrical output related to weight and can therefore be used both for a continuous assessment of weight and to actuate other devices such as a mechanism to reject components of incorrect weight.

2. Examine the problems associated with passing an alternating current through cables in air and buried in soil.

3. When a capacitor is specified in a circuit you will generally find a voltage specified. The reason for this is that at higher voltages the dielectric in the capacitor may break down. Investigate this 'electric strength' and determine the factors which affect it.

Suggestions for answers

1. This variation would suggest that the forces between masses were once much greater than they are now. One consequence of this would be that the earth was once much smaller and is expanding over the centuries. There does seem to be evidence that once the continents of the world were all together and have moved apart.

2. (a) (i) The capacitance would decrease.
(ii) The p.d. would decrease.
(b) The p.d. would rise to the original value.
(c) To obtain as large a capacitance as possible.
(d) 0·01 s.
(e) (i) The p.d. would drop.
(ii) The p.d. would remain constant.
(f) (i) The 50 Hz signal is moving the diaphragm inwards and outwards in a time comparable with that of the time constant. The 10 000 Hz signal gives times which are much smaller than the time constant.
(ii) The time constant must be made bigger, perhaps by increasing the resistance in the circuit.

3. (a) Double h. (b) No change.
(c) Some reduction in h.
(d) Doubles the frequency of the probe coil output and doubles h.

4. The field concept, the inverse square law, Newton's laws can be considered to be mathematical properties. Why is there an inverse square law? Do fields really exist? What is meant by the term 'force'?

5. (a) $9{\cdot}8\ \mathrm{V\,m^{-1}}$.
(b) $9{\cdot}8/9\ \mathrm{V\,m^{-1}}$.
(c) $11{\cdot}5 \times 10^{-10}\ \mathrm{C\,m^{-2}}$.

6. (a) $1{\cdot}6 \times 10^{-17}$ J. (b) It is the same.

7. (a) $5{\cdot}7 \times 10^{7}\ \mathrm{V\,m^{-1}}$.
(b) The distance from each charge will vary and also the electric fields due to each charge will be in different directions. At about 10 km away from the point directly under the charges there will be zero electric field at right angles to the earth's surface.
(c) Yes. It would need 40 kV across a 1000 μF capacitor.
(d) The breakdown electric field is about $3 \times 10^{6}\ \mathrm{V\,m^{-1}}$. Lightning could be expected for our cloud.

8. Electromagnetic induction produces a current in the meter. The meter reading will depend on the rate of change of magnetic flux linked by the coil. A steady current down the lightning conductor will not give a reading on the meter. If the pulse of current passing down the lightning conductor is of very short duration, shorter than the time taken for the meter to respond then the swing of the meter pointer is proportional to the charge passed down the conductor. A meter used in this way is often called a ballistic galvanometer.

The rate of change of flux with time is proportional to the rate of change of current in the conductor. But the change of current is equal to the charge that has passed in the time considered. Hence the rate of change of flux is proportional to the charge. The meter reading is proportional to the rate of change of flux and hence to the charge.

9. (a) Similar shape graph to the charge time graph.
(b) Current is $\mathrm{d}Q/\mathrm{d}t$ and thus a graph of the slope of the charge time graph is required against time.

10. (a) Because of the high voltage. You can discuss why high voltages are necessary.
(b) See 'The transmission of power' in this chapter.
(c) More machines, e.g., television, vacuum cleaners, etc.

12 Electromagnetic waves

Teaching note Practical work appropriate to this chapter will be found in:
W. Bolton. *Physics experiments and projects*, Vol. 2, Pergamon, 1968.
Nuffield advanced physics. Units 4 and 8, teachers' guides, Penguin, 1972.

12.1 Radio waves

'Radio is the miracle of the ages. Aladdin's Lamp, the Magic Carpet, the Seven League Boots of fable and every vision that mankind has ever entertained, since the world began, of laying hold upon the attributes of the Almighty, pale into insignificance beside the accomplished fact of radio. By its magic the human voice may be projected around the earth in less time than it takes to pronounce the word "radio".'

Archer. *History of radio to 1926*, American history Soc. Inc., 1938.

In 1888 Hertz produced a spark between the terminals of the secondary of an induction coil (an induction coil is essentially a transformer in which a large e.m.f. is induced in the secondary turns as a result of applying an interrupted d.c. pulse to the primary instead of a.c.) and a short distance away a spark occurred in a gap in another, isolated loop of wire. Something had passed from the spark gap of the induction coil, the transmitter, to the spark gap of the loop, the receiver (Fig. 12.1). Radio was born.

Hertz rapidly determined the properties of this new radiation. The following extracts are taken from his work.

'*Rectilinear propagation*

If a screen of sheet zinc 2 metres high and 1 metre broad is placed on the straight line joining both mirrors [the source spark gap and the detecting (secondary) spark gap were at the focal points of cylindrical mirrors], and at right angles to the direction of the ray, the secondary sparks disappear completely. An equally complete shadow is thrown by a screen of tinfoil or gold paper. If an assistant walks across the path of the ray, the secondary spark-gap becomes dark as soon as he intercepts the ray, and again lights up when he leaves the path clear. Insulators do not stop the ray—it passes right through a wooden partition or door. . . .

There is no sharp geometrical limit to either the ray or the shadows; it is easy to produce phenomena corresponding to diffraction. . . .

Polarization

From the mode in which our ray was produced we can have no doubt whatever that it consists of transverse vibrations and is plane-polarized in the optical sense. We can also prove by experiment that this is the case. If the receiving mirror be rotated about the ray as axis until its focal line, and therefore the secondary conductor also, lies in a horizontal plane, the secondary sparks become more and more feeble, and when the two focal lines are at right angles, no sparks whatever are obtained even if the mirrors are moved close up to one another. . . .

I next had made an octagonal frame, 2 metres high and 2 metres broad; across this were stretched copper wires 1 mm thick, the wires being parallel to each other and 3 cm apart. If the two mirrors were now set up with their focal lines parallel, and the wire screen was interposed perpendicularly to the ray and so that the direction of the wires was perpendicular to the direction of the focal lines, the screen practically did not interfere at all with the secondary sparks. But if the screen was set up in such a way that its wires were parallel to the focal lines, it stopped the ray completely. . . .

Reflection

. . . I allowed the ray to pass parallel to the wall of the room in which there was a doorway. In the neighbouring

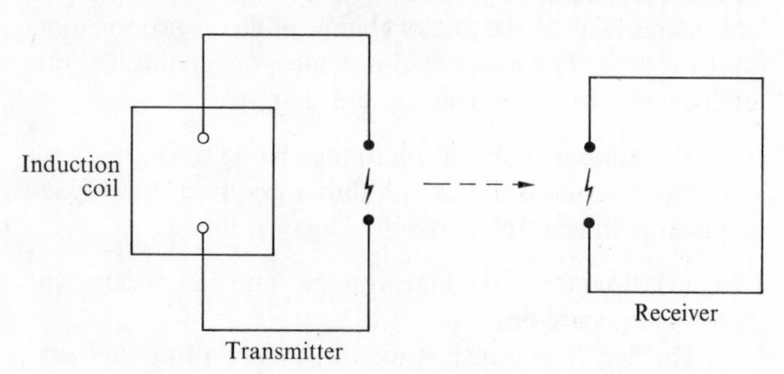

Fig. 12.1

room to which this door led I set up the receiving mirror so that its optic axis passed centrally through the door and intersected the direction of the ray at right angles. If the plane conducting surface was now set up vertically at the point of intersection, and adjusted so as to make angles of 45° with the ray and also with the axis of the receiving mirror, there appeared in the secondary conductor a stream of sparks.... When I turned the reflecting surface about 10° out of the correct position the sparks disappeared. Thus the reflection is regular, and the angles of incidence and reflection are equal....

Refraction

In order to find out whether any refraction of the ray takes place in passing from air into another insulating medium, I had a large prism made of so-called hard pitch, a material like asphalt. The base was an isosceles triangle 1·2 metres in the side, and with a refracting angle of 30°.... The producing mirror was set up at a distance of 2·6 metres from the prism and facing one of the refracting surfaces, so that the axis of the beam was directed as nearly as possible towards the centre of mass of the prism, and met the refracting surface at an angle of incidence of 25° (on the side of the normal towards the base). Near the refracting edge and also at the opposite side of the prism were placed two conducting screens which prevented the ray from passing by any other path than that through the prism.... [The secondary mirror was placed on the emerging side of the prism.]... sparks appeared when the mirror (secondary) was moved towards the base of the prism, beginning when the angular deviation... was about 11°. The sparking increased in intensity until the deviation amounted to about 22°, and then again decreased.... corresponds to a refractive index of 1·69....'

H. Hertz (1888). *Sitzungsber. d. Berl. Akad. d. Wiss.*

Radio waves were thus shown to behave in a similar manner to light or indeed any wave motion. They showed the properties of rectilinear propagation, polarization, reflection, refraction, diffraction, and interference.

The first radio waves were transmitted from an aerial system consisting of two spheres connected by wires to the terminals of an induction coil. It was soon found that

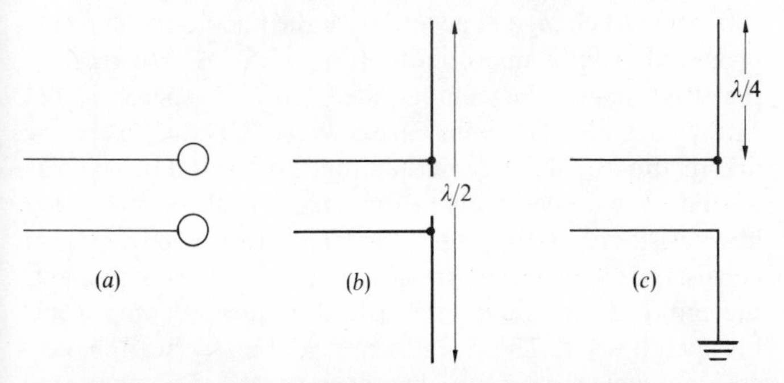

Fig. 12.2 (a) Hertz's aerial system. (b) Dipole. (c) Marconi's quarter wavelength aerial.

the spheres could be replaced by two rods or lengths of wire —the aerial is then known as a dipole. If one of the terminals is connected to the ground only one half of the dipole is needed; the ground acts as the other half (Fig. 12.2).

The production of radio waves

In Hertz's original transmitter radio emission occurred when sparks passed between two spheres. A high potential difference was maintained between the spheres. The system is a capacitor which is charged by the induction coil. Charging continues until the potential difference between the two spheres has risen high enough for the air to break down and a spark be produced. This discharges the capacitor (Fig. 12.3). But because the aerial system has some

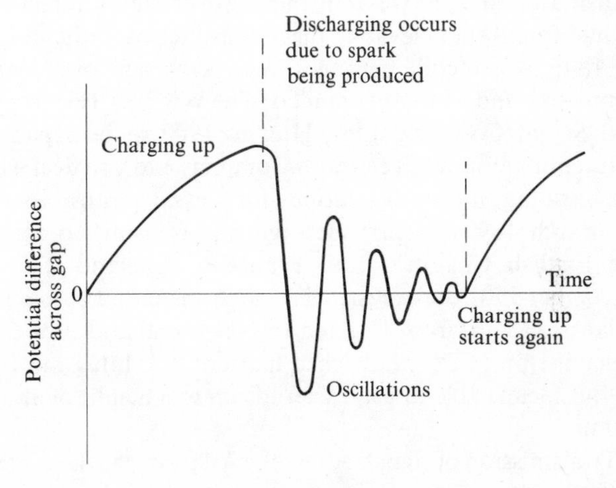

Fig. 12.3

inductance the capacitor is being discharged through an inductive circuit and thus oscillations occur. The oscillations are heavily damped. The spark discharges the capacitor because ions are produced in the air and give a conducting path between the two spheres.

Instead of using an induction coil and relying on a spark to trigger off the oscillations, an oscillating potential difference can be applied to the aerial system. A simple radio transmitter can be a low voltage alternating at say 10^9 Hz and applied to an aerial consisting of two conducting rods a few millimetres apart (Fig. 12.4). The high potential difference was only necessary with Hertz's apparatus as a means of discharging the capacitor through the aerial inductance by means of a spark.

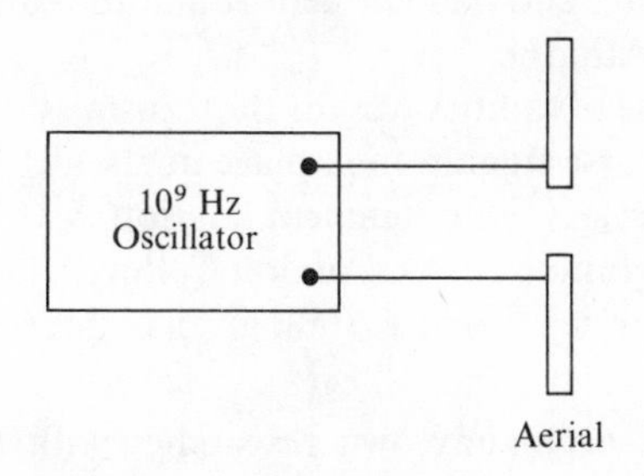

Fig. 12.4

Radio—the miracle of the ages

'It seemed to me at this time that if this radiation could be increased, developed and controlled, it would most certainly be possible to signal across space, for very considerable distances. . . .'

Marconi

Marconi—master of space. B. L. Jacot and D. N. Collier (1935). Hutchinson.

Marconi was referring to his thoughts immediately following Hertz's discovery of radio waves. Hertz had only managed to detect radio radiation a matter of metres from his transmitter. In December 1895 Marconi had a great triumph—he detected radio radiation 10 m away from the transmitter. Soon he was transmitting over a few kilometres; he had found that elevating the aerials increased the range. By 1897 Morse code messages were being sent over many kilometres and Marconi founded 'The Wireless Telegraphy and Signal Company Ltd', later in 1900 to be renamed 'Marconi's Wireless Telegraph Company Ltd', to deal with the installation of radio stations for message transmission. In March 1899 the first message was transmitted across the English Channel, from France to England. On 12 December 1901 the first signals were transmitted across the Atlantic Ocean, from England to Newfoundland. For this transmission Marconi attached his aerials to kites and was by this means able to get his aerials up to a height of about 150 m.

Transmission of signals across the Atlantic was the experiment that should not have worked.

'Despite the opposition which I had received from many quarters, often that of most eminent men, it was still my opinion that electric waves would not be stopped by the curvature of the earth, and therefore could be made to travel any distance, separating any two places on our planet. . . .'

Marconi.

Marconi—master of space. B. L. Jacot and D. N. Collier (1935). Hutchinson.

Two years later Kennelly and Heaviside advanced the suggestion that the radio signals had been reflected from a layer of ionized particles in the earth's upper atmosphere, the ionosphere, and had not bent round the earth's surface to cross the Atlantic.

The first use of radio waves for the transmission of speech, instead of Morse signals, took place in about 1906. In 1922 the first wireless entertainment station was opened in London (the famous 2LO, London Calling). The following are extracts from a broadcast talk given by Marconi in 1931.

'In 1895 I began my own researches with the express intention of utilizing electric waves for telegraphing across considerable distances.

These first tests were soon followed by important improvements which made possible tuning and selectivity and by new discoveries such as that of the enormous distance over which these waves can travel. . . .

The beginnings of telephony as we know it, whether operated by line or radio waves, naturally date from the invention of the electro-magnetic receiver and the carbon microphone. . . .

As is well known (with radio waves), the speech currents (from the microphone) are superimposed on some form of current or high-frequency waves which must be unbroken, not intermittent, and the spark transmission by induction coil and interrupter of that day, although quite satisfactory for telegraph working—I was then effecting radio communication over 36 miles (56 km)—because of the dead intervals between the sparks, was quite unsuitable for telephony. . . . in 1906, a real advance was achieved by Fessenden by employing for the first time the high frequency alternator, which gave him a useful carrier wave of 2000 cycles per second. This enabled him in the following year to transmit speech from Brant Rock, Jamaica, to Long Island (a distance of about 200 miles (320 km)). . . . The invention of the Fleming valve in 1904 and the three electrode valve of Lee Forest in 1907 enabled the disabilities which had delayed the commercial development of wireless telephony to be removed, and the present state of the art realized. In June, 1913, Dr Meissner employed the oscillating valve for the first time as carrier-wave generator for transmitting speech between Berlin and Nuern, a distance of 23 miles (37 km). . . .'

Marconi (1931).

Marconi—master of space. B. L. Jacot and D. N. Collier (1935). Hutchinson.

Hertz's original transmitter had been the intermittent sparks produced between two spheres by an induction coil. This was capable of transmitting Morse as all that was needed was a key to make the pulse of waves transmitted either long or short. The key was thus used to break the transmission circuit. For the continuous transmission needed for the transmission of speech this was unsatisfactory. A continuous radio wave was necessary so that it could be modulated, i.e., the amplitude of the radio wave was made to change in accord with the fluctuating currents produced from a microphone (Fig. 12.5). At the receiver the wave has to be demodulated, i.e., the speech signal extracted from the radio carrier wave. Crystals were the first method used for demodulating. The crystal behaved as a solid state diode—only positive parts of the signal had a low resistance path (Fig. 12.6). The demodulated signal consists of the positive part of a high frequency component, the radio carrier wave, and a low frequency component, the speech wave. These are then picked up by headphones, across which a capacitor has been placed. The capacitor offers a low reactance path for the radio wave and a high

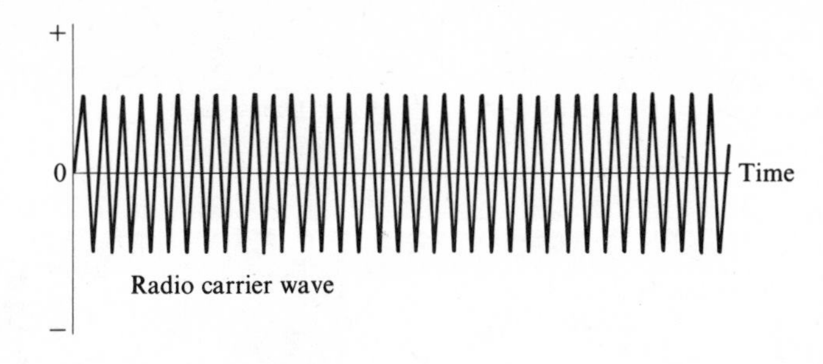

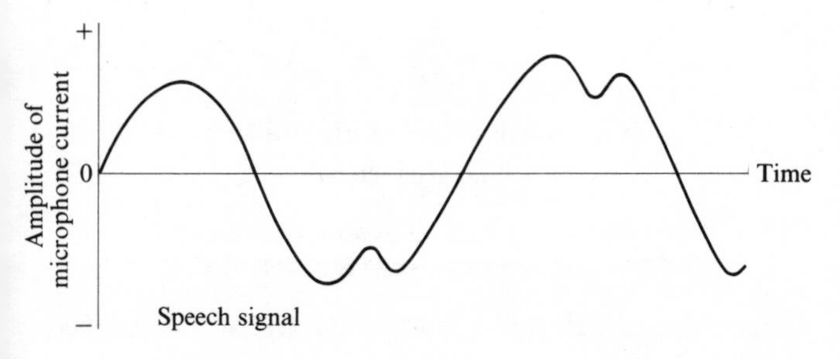

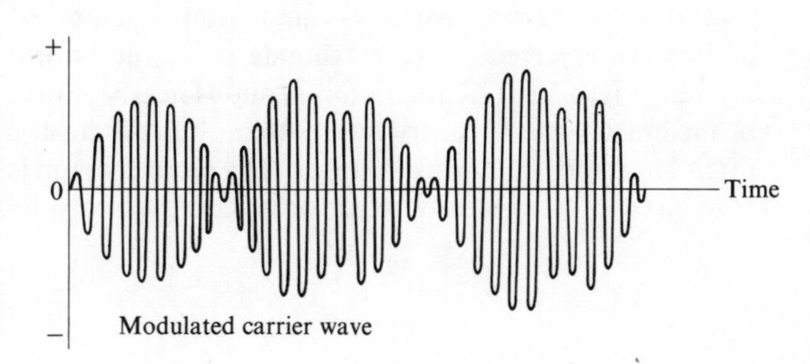

Fig. 12.5

reactance path for the lower frequency speech signal. The headphones offer a high resistance path for the radio wave and a low resistance for the speech signal—the result is that the headphones respond to the speech signal.

The radio signal has a frequency which depends on the capacitance and inductance of the transmitting circuit. By employing an inductor and a variable capacitor in the receiving circuit tuning can be employed to make the receiving circuit respond only to the frequency of the transmitted signal. In fact the receiving circuit would respond to a frequency band which would be centred on the transmission frequency. The frequency the circuit responds to is the resonant frequency of the capacitor–inductor circuit.

The advent of valves considerably improved the performance of receivers. Fleming's valve was a diode and thus could be used in place of the crystal. Lee Forest's three-electrode valve, the triode, enabled the signals to be amplified before being received in the headphones, or loudspeaker. In addition the triode valve could be used with a tuning circuit to give an oscillator for the production of the radio carrier waves.

Figure 12.7 shows a schematic diagram of a simple receiver. The signal is picked up by the aerial, amplified, demodulated, the speech signal is amplified (audio frequency amplifier), and finally the signal is passed to the loudspeaker. Transistors and junction diodes are nowadays often used for the various stages. These are solid-state equivalents to the triode and diode valves.

Figure 12.8 shows a schematic diagram of what is known as a superhet or super-sonic heterodyne receiving set. This has the advantage over the previous receiver of greater selectivity, i.e., a narrower frequency band is selected. The signal is picked up by the aerial, amplified, the carrier frequency changed to a fixed frequency, amplified, demodulated, amplified, and finally passed to the loudspeaker. The advantage of changing the frequency is that all the rest of the receiver circuit has only to be designed to handle

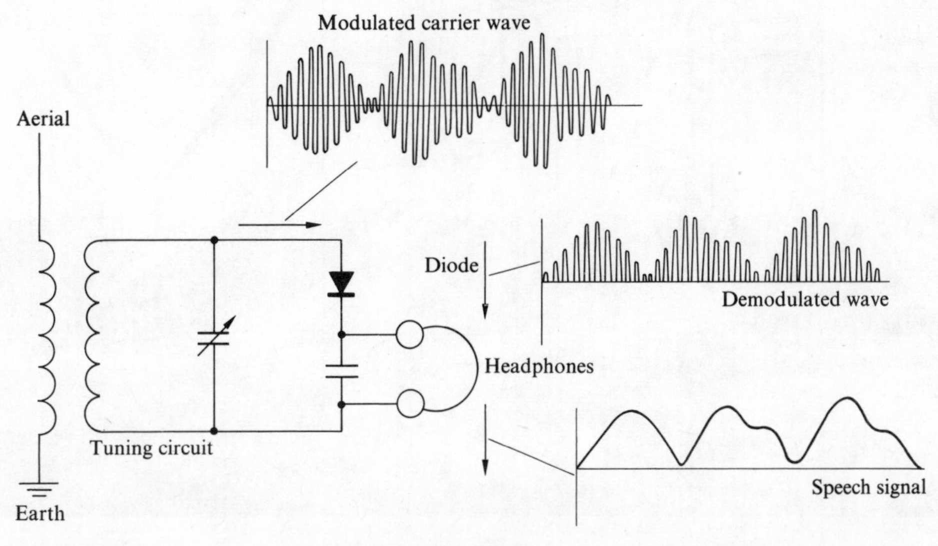

Fig. 12.6

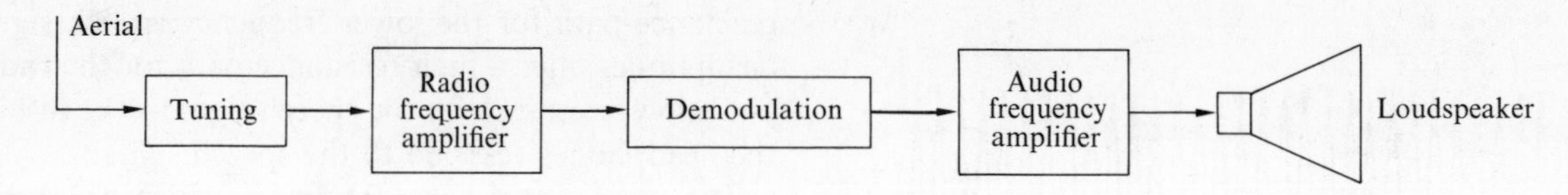

Fig. 12.7 Simple receiver.

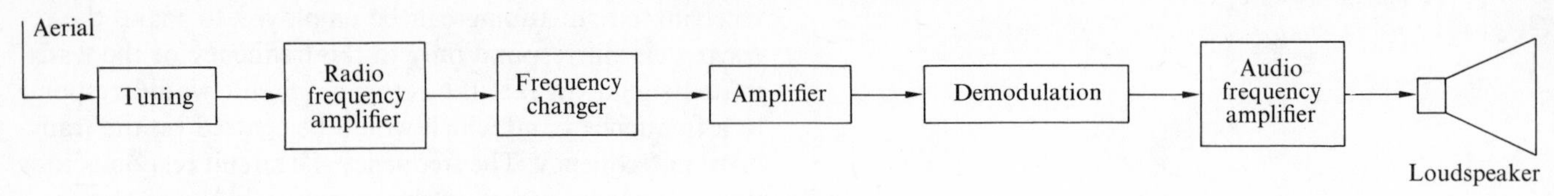

Fig. 12.8

just one constant frequency—not as in the other set, all the frequencies that the set is used for.

Amplitude modulation is not the only way the carrier wave can be made to carry the audio signal. Frequency modulation is used by some stations. Instead of the audio signal being used to control the amplitude of the radio wave it is used to vary the radio wave frequency.

Radioastronomy

Radioastronomy began in 1932 with the work of K. G. Jansky in the United States. He was not looking for radio emissions from non-terrestrial objects but was investigating (Fig. 12.9) the general background noise detected by radio receivers. This noise is known as atmospherics or static.

'Directional studies of atmospherics at high frequencies

By K. G. Jansky
(*Bell Telephone Laboratories, New York City*)

Since the middle of August, 1931, records have been taken . . . of the direction of arrival and the intensity of static on 14·6 meters. . . .

From the data obtained it is found that three distinct groups of static are recorded. The first group is composed of the static received from local thunderstorms and storm centres [Fig. 12.10]. Static in this group is nearly always of the crash type. It is very intermittent, but the crashes often have very high peak voltages. The second group is composed of very steady weak static coming probably

Fig. 12.9 Karl G. Jansky and his 14·6 metre rotatable directional antenna. (Courtesy of Bell Laboratories.)

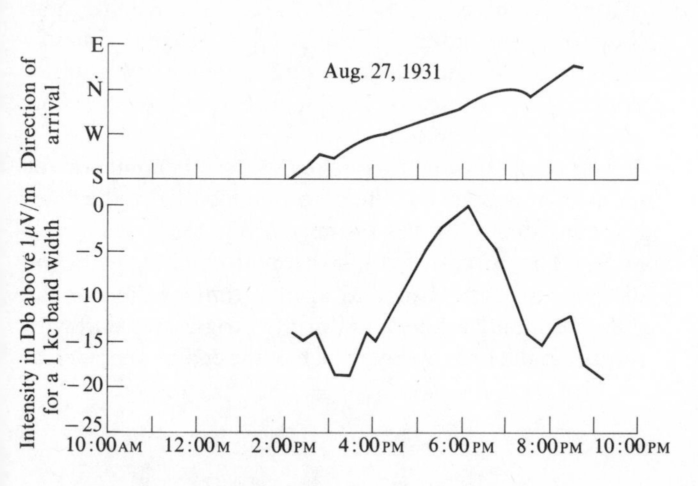

Fig. 12.10 Direction of arrival and intensity of local storm type static on 14·6 metres.

Fig. 12.11 Direction of arrival of hiss type static on 14·6 metres.

Fig. 12.12 Grote Reber's radio-telescope (about 10 m diameter). This is a reconstruction by the United States Government. They have added an azimuth track. (National Radio Astronomy Observatory, Green Bank, Virginia.)

by Heaviside layer refractions from thunderstorms some distance away. The third group is composed of a very steady hiss type static the origin of which is not yet known.

The static of the third group is also very weak. It is, however, very steady, causing a hiss in the phones that can hardly be distinguished from the hiss caused by set noise. It is readily distinguished from ordinary static and probably does not originate in thunderstorm areas. The direction of arrival of this static changes gradually throughout the day going almost completely around the compass in twenty-four hours. It does not quite complete the circuit, but in the middle of the night when it reaches the northwest, it begins to die out and at the same time static from the northeast begins to appear on the record. This new static then gradually shifts in direction throughout the day and dies out in the northwest also and the process is repeated day after day. Fig. 13 [12.11] shows the direction of arrival of this static for three different days plotted against time of day. Curve 1 is for January 2, 1932, curve 2 is for January 26, 1932, and curve 3 is for February 24, 1932. . . .

This type of static was first definitely recognized only this last January. Previous to this time it had been considered merely as interference from some unmodulated carrier. Now, however, that it has been detected it is possible to go back to the old records and trace its position on them.

During the latter part of December and the first part of January the direction of arrival of this static coincided, for most of the daylight hours, with the direction of the sun from the receiver. (See curve 1, Fig. 13.) However, during January and February the direction has gradually shifted so that now (March 1) it precedes in time the direction of the sun by as much as an hour. It will be noticed that the curves 2 and 3 of Fig. 13 have shifted to the left. Since December 21, the sun's rays have been getting more and more perpendicular at the receiving location causing sunrise to occur at the receiver earlier and earlier each day. It would appear that the change in the latitude of the sun is connected with the changing position of the curves. However, the data as yet only cover observations taken over a few months and more observations are necessary before any hard and fast deductions can be drawn.'

Proc. Inst. Radio Eng., **20**, No. 12, December 1932.

After more measurements Jansky was able to report as follows:

'*Radio Waves from Outside the Solar System*

In a recent paper on the direction of arrival of high-frequency atmospherics, curves were given showing the horizontal component of the direction of arrival of an electromagnetic disturbance, which I termed hiss type atmospherics, plotted against time of day. These curves showed that the horizontal component of the direction of arrival changed nearly 360° in 24 hours and, at the time the paper was written, this component was approximately the same as the azimuth of the sun, leading to the assumption that the source of this disturbance was somehow associated with the sun.

Records have now been taken of this phenomenon for more than a year, but the data obtained from them are not consistent with the assumptions made in the above paper. The curves of the horizontal component of the direction of arrival plotted against time of day for the different months show a uniformly progressive shift with respect to the time of day, which at the end of one sidereal

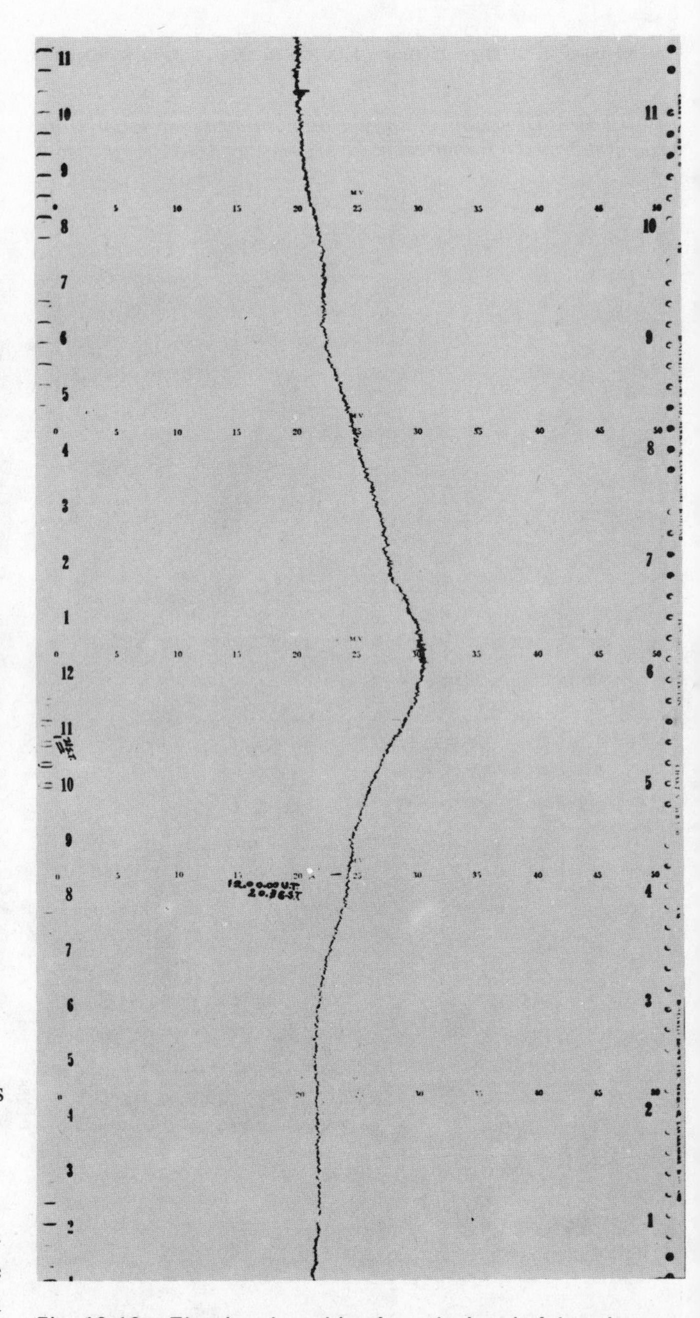

Fig. 12.13 The signal resulting from the bowl of the telescope sweeping across the plane of the Milky Way. (The Director, Nuffield Radio Astronomy Laboratories, University of Manchester, Jodrell Bank, Macclesfield, Cheshire.)

year brings the curve back to its initial position. Consideration of this shift and the shape of the individual curves leads to the conclusion that the direction of arrival of this disturbance remains fixed in space, that is to say, the source of this noise is located in some region that is stationary with respect to the stars.'

K. G. Jansky (1933). *Nature.*

This discovery of radio emissions from outer space was not followed up by the scientific world, with one exception, for a decade. The exception was G. Reber who designed and constructed the first radio telescope (Fig. 12.12). The following quotations are from an article by Reber in 1958 in which he looked back over those early years of radio-astronomy.

'In my estimation it was obvious that K. G. Jansky had made a fundamental and very important discovery. Furthermore he had exploited it to the limit of his equipment facilities. If greater progress were to be made, it would be necessary to construct new and different equipment especially designed to measure cosmic static.... Consideration of the antenna problem showed that any type of wire network would be exceedingly complicated since several hundred minute dipoles would be needed. The only feasible antenna would be a parabolic reflector or mirror....'

G. Reber. *Proc. I.R.E.*, **46**, 16, 1958.

Reber's telescope was a 10 m diameter parabolic dish with a focal length of about 7 m. With this he confirmed Jansky's

Fig. 12.14 Jodrell Bank Mark I Radiotelescope. (The Director, Nuffield Radio Astronomy Laboratories, University of Manchester, Jodrell Bank, Macclesfield, Cheshire.)

results and identified radio emission as coming from the Milky Way. He considered that it might result from ionized hydrogen gas in interstellar matter. Figure 12.13 shows the type of radiotelescope signal obtained by an instrument sweeping across the plane of the Milky Way. This trace was the first signals obtained by the Jodrell Bank telescope on 2 August 1957. The bowl of the telescope was kept fixed and the movement of the earth used to sweep the beam of the telescope across the Milky Way. The Jodrell Bank instrument is much bigger than Reber's instrument, having a bowl diameter of about 80 m (Fig. 12.14).

Our galaxy, known as the Milky Way, is a flattened disc of stars in a spiral configuration. Figures 12.15 and 12.16 show photographs of other galaxies that are, we think, rather like our own galaxy. Figure 12.17 shows a radio map of the sky, at a wavelength of 3·7 m. The Milky Way runs along the centre of the map.

The radio map shows the radiation being emitted over the entire area of the Milky Way without any high degree of localization. Reber's idea that such radiation resulted from ionized hydrogen gas is felt to be the explanation. In addition to this radiation there are several localized sources of radio emission, both within our galaxy and external to it. The first such localized radio source to be discovered was in our galaxy and identified as the Crab Nebula. The Crab Nebula is the remnants of a star that was seen to explode in A.D. 1054, by Chinese astronomers. The star was seen to increase very markedly in brightness before diminishing again—the star had increased in temperature very rapidly and then exploded to leave just a glowing cloud of gas (Fig. 12.18).

> '2nd cyclical day, 5th month, 1st year of Chih-ho of Sung, guest star appearing at South East of T'ien-kuan several ts'un long, lasting more than one year.'

Other such exploding stars, supernovae, are found to act as sources of radio waves. Not all the localized radio sources were found to be within our galaxy. Cygnus A, an exploding galaxy, is one of the most powerful radio sources in the sky—it is more than 500 million light-years away; the Crab nebula within our galaxy is a mere 4000 light-years away. At first it was thought that all distant radio sources were exploding galaxies, but in 1963 small, very powerful, very distant radio sources were discovered. They were called quasars. Quasars have been identified with stars (not galaxies). In 1968 even stranger radio sources were discovered—pulsars, small radio sources which emit flashes of radio energy at regular intervals (intervals of about 1 second).

These latest discoveries have presented astonomers with perplexing problems.

Fig. 12.15 Galaxy NGC 4594, similar to our own galaxy, seen edge on. (Courtesy of the Hale Observatories.)

Fig. 12.16 Galaxy NGC 3031, similar to our own galaxy, seen from an oblique angle. (Courtesy of the Hale Observatories.)

Radar

Essentially radar is a method of measuring distances by sending a pulse of radio waves to a distant object and

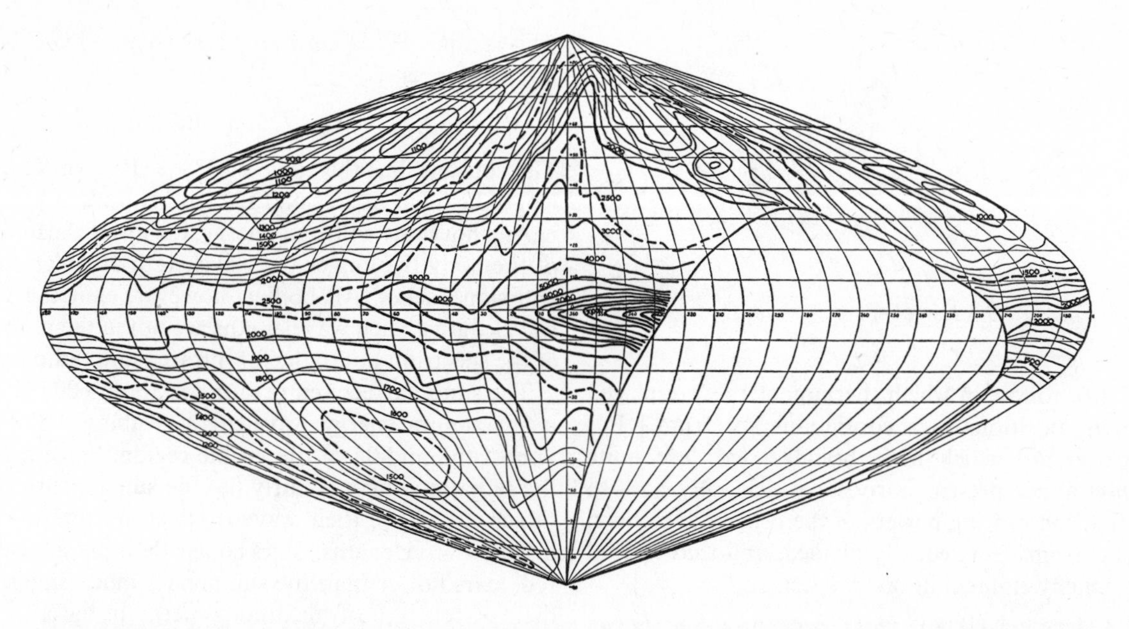

Fig. 12.17 Radio map of the sky at 3·7 metres wavelength. (The blank area is the part of the sky not visible at Cambridge from where the map was made.) (F. G. Graham Smith (1966). Radio Astronomy, *Penguin.)*

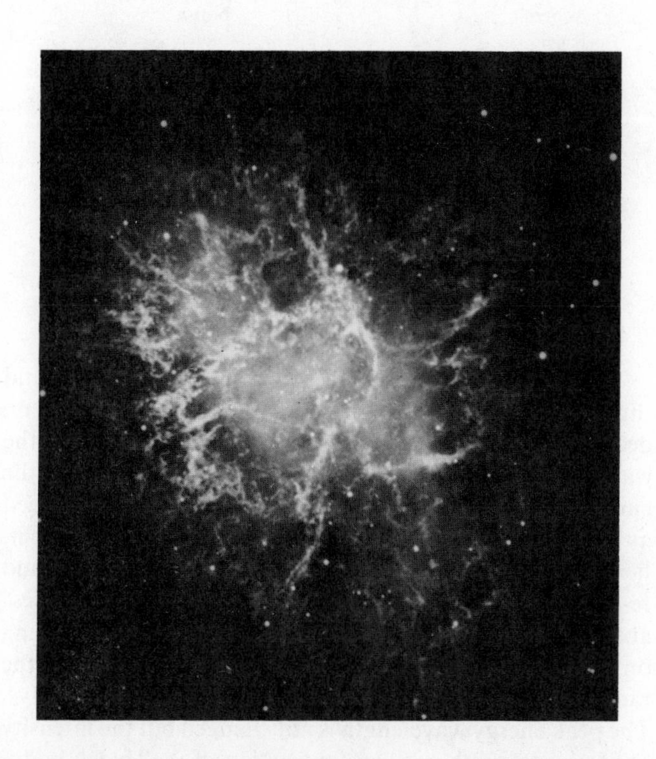

Fig. 12.18 The Crab nebula. Mt. Wilson and Palomar Observatories. (Courtesy of the Hale Observatories.)

measuring the time taken for the reflected pulse to come back to the source, i.e., an echo method. In 1924 Appleton and Barnett used this method to determine the height of the ionosphere (a region in the upper atmosphere which reflects radio signals). The use of radar to detect aircraft, and locate them, emerged during the Second World War. An aircraft reflects only a very minute part of the radio pulse back to the receiver; only about one part in 10^{19} of the emitted energy comes back to the receiver. The early radar transmitted pulses every 1/25 s and used wavelengths in the 6 to 15 m region. The energy transmitted in each pulse was about 8000 J; about 10^{-15} J came back from the aircraft.

In 1946 the first radar echoes from the moon were detected; 1961 saw the first echoes detected from Venus. Since then echoes have been detected from both Mars and Mercury.

12.2 Infrared waves

'It is sometimes of great use in natural philosophy, to doubt of things that are commonly taken for granted; especially as the means of resolving any doubt, when once it is entertained, are often within our reach. We may therefore say, that any experiment which leads us to investigate the truth of what was before admitted upon trust, may become of great utility to natural knowledge. Thus, for instance, when we see the effect of the condensation of the sun's rays in the focus of a burning lens, it seems natural to suppose, that every one of the united rays contributes its proportional share to the intensity of the heat which is produced; and we should probably think it highly absurd, if it were asserted that many of them had but little concern in the combustion, or vitrification, which follows, when an object is put into that focus. It will therefore not be amiss to mention what gave rise to a surmise, that the power of heating and illuminating objects, might not be equally distributed among the variously coloured rays. . . . [Details of an experiment,

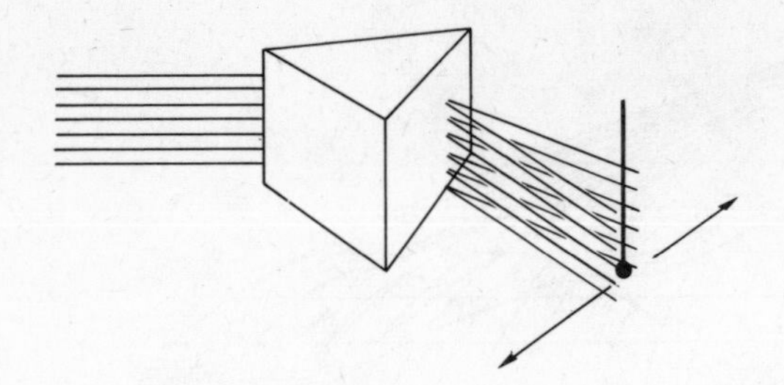

Fig. 12.19 Herschel's experiment.

(Fig. 12.19), follow in which thermometers were placed at different positions in a continuous spectrum.] But the experiments which have been related, are quite sufficient for my present purpose; which only goes to prove, that the heating powers of the prismatic colours, is very far from being equally divided, and that the red rays are chiefly eminent in that respect. . . .'

F. W. Herschel (1800). *Phil. Trans. Roy. Soc. Lond.*, **11**, part II, 255.

Herschel, in 1800, moved thermometers along through the colours of the continuous spectrum produced by the sun's rays passing through a prism. The result was—a greater rise in temperature at the red end of the spectrum than at the violet end. Even stranger, when the thermometers were moved past the red end of the spectrum even larger temperature rises were found—the sun was emitting invisible rays. These rays we now call the infrared. Herschel found that the rays were capable of refraction and reflection.

Measurements of the energy at different wavelengths emitted by a hot solid show a maximum whose wavelength depends on the temperature of the hot body (Fig. 12.20). The wavelength of maximum energy is found to be inversely proportional to the absolute temperature.

$$\lambda_{max} \propto \frac{1}{T}$$

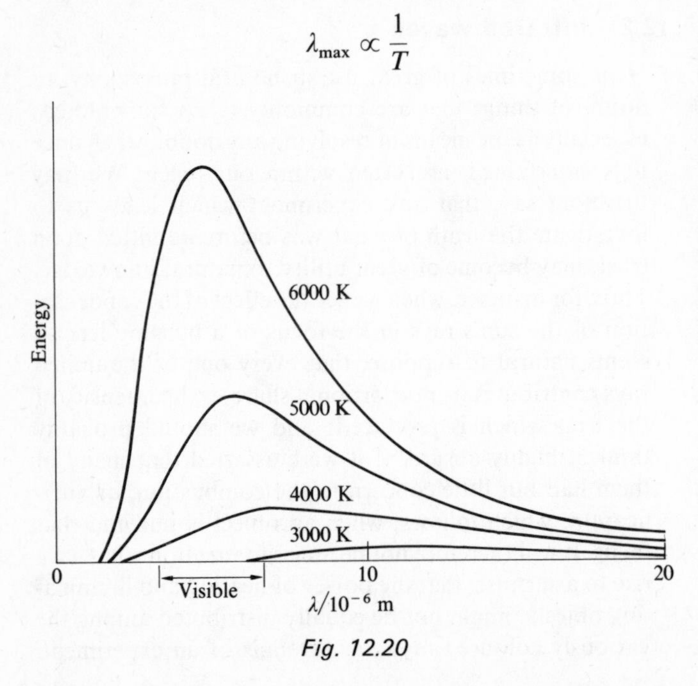

Fig. 12.20

This is called Wien's law, first arrived at in 1894, and generally written as

$$\lambda_{max} T = \text{a constant.}$$

The constant has the value $2{\cdot}8978 \times 10^{-2}$ m K, the wavelength being in metres and the temperature in kelvins.

Any hot body will emit energy, the lower the temperature of the body the longer the wavelength at which maximum emission occurs. All bodies, however, emit the very long wavelengths. Such wavelengths are not in the visible region of the spectrum but in the far infrared and radio regions.

The sun's surface temperature is about 6000 K and thus its maximum energy wavelength is about $4{\cdot}8 \times 10^{-7}$ m, almost the middle of the visible region. Since the peak of the radiation curve is fairly flat the sun appears yellowish. Other stars have their wavelength of maximum energy at different wavelengths. Stars cooler than the sun can appear red, stars hotter than the sun appear blue. Blue stars have their wavelength of maximum energy in the ultraviolet.

Name	*Temperature*, K	*Colour*
Alpha Crucis	23 000	bluish
Spica	20 400	bluish
Acherna	15 500	bluish
Rigel	12 300	bluish
Sirius	10 700	white
Altair	8530	white
Procyon	6800	straw
Sun	6000	yellow
Sigma Eridani A	5360	yellow
Epsilon Eridani	4910	orange
61 Cygni A	3900	orange
Lacaille 9352	3200	red

Adapted from G. Gamow (1964). *A star called the sun.* Reprinted by permission of the Viking Press, Inc, New York, and Macmillan, London and Basingstoke.

The average temperature of the earth is about 281 K and thus its wavelength of maximum energy is about 1×10^{-5} m, deep in the infrared. The intensity of emission at all the wavelengths is not, however, expected to be the same as the curve given by a hot 'black body'. A 'black body' is defined as a body that absorbs all the radiation that falls on it, hence it appears black because no radiation is reflected, and re-emits all the radiation. Most bodies are 'grey bodies' absorbing only a certain percentage of the radiation falling on them and hence emitting only that percentage of the radiation that would have been emitted by a 'black body'. The peak energy wavelength is not changed but the intensity at any wavelength is a certain fraction of the 'black body' intensity. This fraction is called the emissivity. The usefulness of the 'black body' energy distribution graphs is that the shape of the graphs are common to many substances; all that may have to be changed is the scale for the intensity (Fig. 12.21).

If the form of the energy distribution with wavelength relationship is known then measurement of the energy emitted at one wavelength can be used to give a measure of the 'black body' temperature. The following extract is from a paper giving the results of such measurements.

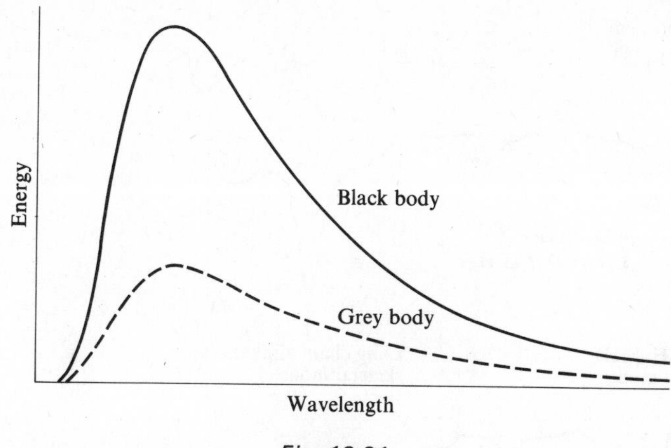

Fig. 12.21

'Observations of thermal radio radiation from the planets are of interest to supplement the infrared observations. This is especially true for the planets with extensive atmospheres, for in general, the much longer wavelength radio radiation would be expected to penetrate the atmospheres more readily than would the infrared radiation. . . .

In May, 1965, a radiometer at a wavelength of 3·15 cm was installed in the Naval Research Laboratory 50 foot [17 m] paraboloid to observe the close approach of Venus. The wavelength near 3 cm was considered a good compromise between the desire to use as short a wavelength as possible, and the limitations in the aiming and focusing ability of the 50-foot antenna. . . . The 3·15-cm radiometer basically was calibrated with thermal noise sources. . . .

Planet	*Infrared blackbody temp.* /K	*Wavelength of observation* /cm	*Antenna temp.* /K	*Radio blackbody temp.* /K
Venus	240	9·4	0·28	580 ± 230
		3·15	1·42	620 ± 110
		3·15	3·54	560 ± 73
Jupiter	130	3·15	0·35	140 ± 56
		3·15	0·51	145 ± 26
Mars	260	3·15	0·24	218 ± 76

Table (some data omitted)

The present measurements are summarized in [the] table along with the blackbody temperatures measured by infrared radiometric methods. The region of emission of the radio radiation is probably not the same as for the infrared radiation for the cases of Venus and Jupiter, and consequently the blackbody temperatures would not necessarily agree. . . .'

C. H. Mayer, T. P. McCullough, R. M. Sloanaker (1958). *Proc. IRE*, **46**, 260.

Essentially the method used in the above paper was to measure the temperature change produced at the antenna by the energy received there from the distant planet. The energy needed to give these temperature rises was determined with a calibration experiment in which known amounts of energy were directly fed to the antenna to give similar temperature changes.

In 1962 the Mariner II satellite passed close to Venus and measured the microwave energy. The temperature indicated was about 700 K. Infrared measurements gave a temperature of 240 K. The higher result is thought to be the ground temperature and the lower reading the cloud temperature. Venus is completely covered in cloud.

In 1879 J. Stefan deduced (a fortuitous deduction, being essentially based on a few very poor results) from an examination of experimental evidence that the total energy radiated per unit surface area per second from a blackbody was proportional to the fourth power of the absolute temperature of the body.

$$E \propto T^4$$

or by introducing a constant, σ, called Stefan's constant,

$$E = \sigma T^4$$

σ has the value $5{\cdot}67 \times 10^{-8}\ \mathrm{J\,s^{-1}\,m^{-2}\,K^{-4}}$. A firm experimental foundation for this equation was provided in 1897 and a theoretical proof by Boltzmann in 1884.

We can use this equation to obtain estimates of the temperature of the sun. The only experimental information necessary is the amount of energy radiated per unit area per second from the sun. We can measure the amount of radiation reaching unit area of the earth per second. As some of the radiation would be absorbed in the earth's atmosphere we need either to estimate the amount not reaching the surface of the earth or make the measurements above the earth's atmosphere. Both types of measurement have been made and we can take the energy per second per square metre as being $1338\ \mathrm{J\,s^{-1}\,m^{-2}}$. This is known as the solar constant. The earth is 149 450 000 km from the sun, so the total radiation emitted by the sun will be

$$4\pi(1{\cdot}495 \times 10^{11})^2 \times 1338$$

The sun has a radius of about 695 300 km so the energy emitted per unit area of the sun's surface per second is

$$\frac{4\pi(1{\cdot}495 \times 11^{11})^2 \times 1338}{4\pi(6{\cdot}953 \times 10^8)^2}$$

or $6{\cdot}176 \times 10^7\ \mathrm{J\,s^{-1}\,m^{-2}}$. This gives a temperature of 5765 K.

Infrared spectrometry

When infrared radiation passes through a substance it may set the molecules in that substance into oscillation. The atoms in a molecule are held together by bonds which can be thought of as springs linking the atoms. If the frequency of the infrared radiation is equal to one of the resonant frequencies of the molecule it will set it in oscillation. The result of this is to take energy from the radiation at the specific frequencies corresponding to the resonant frequencies of the molecule (see chapter 4). Examination of the

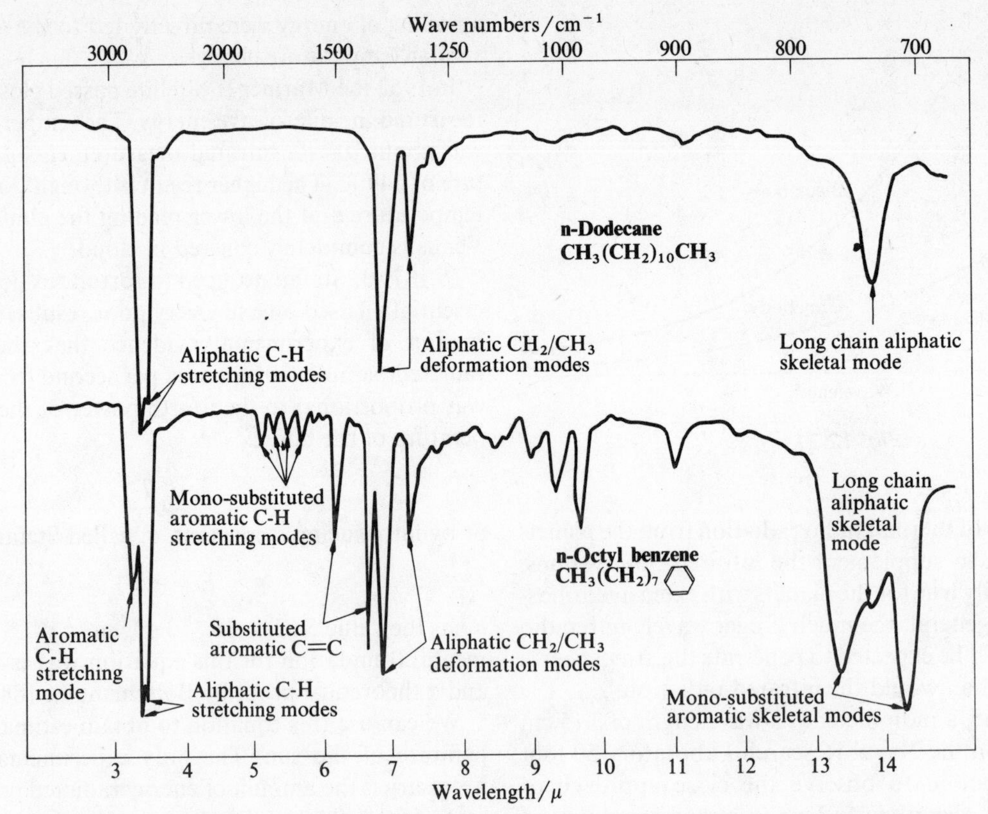

Fig. 12.22 Taken from The physics of chemical structure. *R. J. Taylor (1968), Unilever Educational Publications.*

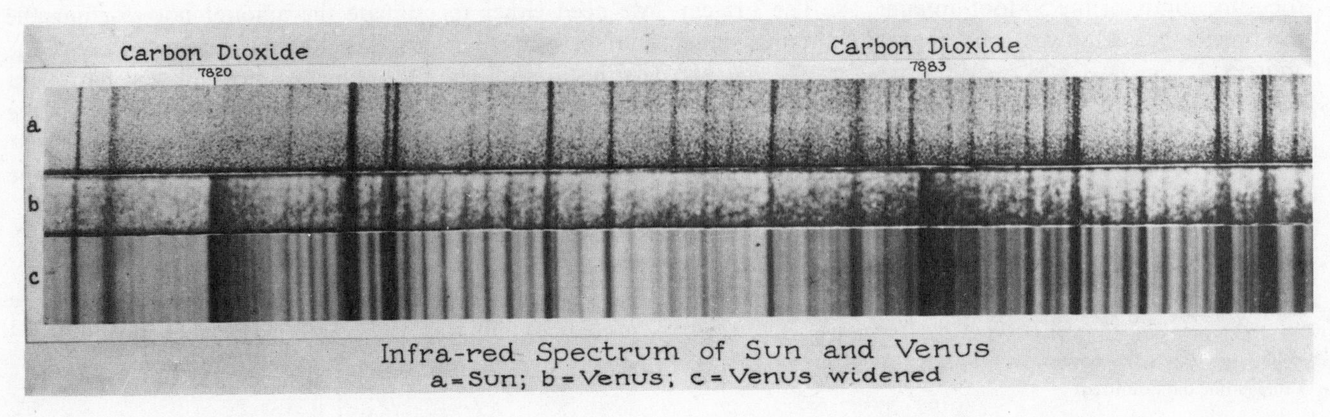

Fig. 12.23 Venus, showing infrared carbon dioxide absorption bands. Photograph from the Mount Wilson and Palomar Observatories. (Royal Astronomical Society.)

infrared absorption spectrum can thus lead to the identification of the types of bonds responsible for the absorption, and thus information about the molecular structure. Figure 12.22 shows two infrared absorption spectra. In addition these bond identifications can serve to 'fingerprint' the substance responsible for them. Thus in addition to determining the types of bonds, infrared absorption spectroscopy can be used to identify substances.

Infrared absorption spectroscopy had led to the identification of some gases in planetary atmospheres (Fig. 12.23). A planet is seen by reflected sunlight which passes through the planet's atmosphere—thus absorption by the gases in the atmosphere can occur. In this way carbon dioxide has been detected in the atmosphere of Venus.

12.3 X-rays

1896 started with the main headlines and stories of the newspapers devoted to events in the Boer war.

'CRISIS IN THE TRANSVAAL
DR. JAMESON CROSSES THE FRONTIER WITH 700 MEN'

The Times, 1 January 1896.

The papers were not, however, without the conventional news for that time of the year.

'THE LEAGUE MATCHES.—FIRST DIVISION

Many thousand people witnessed the defeat of Derby County by Bolton Wanderers yesterday at Bolton by two goals to one.

Everton won their match with Blackburn Rovers at Blackburn yesterday by three goals to two. . . .'

The Times, 1 January 1896.

However, on the sixth of January a report hit the headlines of an event which was to have world-wide significance.

'The noise of the war's alarm should not distract attention from the marvellous triumph of science which is reported from Vienna. It is announced that Prof. Routgen [should be Röntgen] of the Wurzburg University has discovered a light which for the purpose of photography will penetrate wood, flesh, cloth, and most other organic substances. The professor has succeeded in photographing metal weights which were in a closed wooden case, also a man's hand which showed only the bones, the flesh being invisible.

Daily Chronicle, 6 January 1896.

Röntgen's discovery was first announced in a paper in the journal *Sitzungsberichte der Physikalisch–Medizinischen Gesellschaft zu Würzburg* on 28 December 1895. A translation of the paper appeared in *Nature*, 23 January 1896.

'On a New Kind of Rays

(*By W. C. Röntgen. Translated by Arthur Stanton from . . .*)

(1) A discharge from a large induction coil is passed through a Hittorf's tube, or through a well-exhausted Crookes' or Lenard's tube. The tube is surrounded by a fairly close-fitting shield of black paper; it is then possible to see, in a completely darkened room, that paper covered on one side with barium platinocyanide lights up with a brilliant fluorescence when brought into the neighbourhood of the tube, whether the painted side or the other be turned towards the tube. The fluorescence is still visible at two metres distance. It is easy to show that the origin of the fluorescence lies within the vacuum tube.

(2) It is seen, therefore, that some agent is capable of penetrating black cardboard which is quite opaque to ultraviolet light, sunlight, or arc-light. It is therefore of interest to investigate how far other bodies can be penetrated by the same agent. It is readily shown that all bodies possess this same transparency, but in very varying degrees. For example, paper is very transparent; the fluorescent screen will light up when placed behind a book of a thousand pages; printer's ink offers no marked resistance. . . .

. . . A piece of aluminium, 15 mm thick, still allowed the X-rays (as I will call the rays for the sake of brevity) to pass, but greatly reduced the fluorescence. Glass plates of similar thickness behave similarly; lead glass is, however, much more opaque than glass free from lead. Ebonite several centimetres thick is transparent. If the hand be held before the fluorescent screen, the shadow shows the bones darkly, with only faint outlines of the surrounding tissues. . . .

(3) The preceding experiments lead to the conclusion that the density of the bodies is the property whose variation mainly affects the permeability. . . .

(4) Increasing thickness increases the hindrance offered to the rays by all bodies. . . .

(6) The fluorescence of barium platinocyanide is not the only noticeable action of the X-rays. . . .

Of special interest in this connection is the fact that photographic dry plates are sensitive to the X-rays. . . .

(7) After my experiments on the transparency of increasing thicknesses of media, I proceeded to investigate whether the X-rays could be deflected by a prism. Investigations with water and carbon bisulphide in mica prisms of 30° showed no deviation either on the photographic or the fluorescent plate. For comparison, light rays were allowed to fall on the prism as the apparatus was set up for the experiment. They were deviated 10 mm and 20 mm respectively in the case of the two prisms. . . .

(8) The preceding experiments, and others which I pass over, point to the rays being incapable of regular reflection. . . .

(11) A further distinction, and a noteworthy one, results from the action of a magnet. I have not succeeded in observing any deviation of the X-rays even in very strong magnetic fields. . . .

(12) As the result of many researches, it appears that the place of most brilliant phosphorescence of the walls of the discharge tube is the chief seat whence the X-rays originate and spread out in all directions; that is, the X-rays proceed from the front where the kathode rays [electrons] strike the glass. . . .

(14) . . . I have observed and photographed many such shadow pictures. Thus I have an outline of part of a door covered with lead paint; the image was produced by placing the discharge tube on one side of the door, and the sensitive plate on the other. I have also a shadow of the bones of the hand [Fig. 12.24], of a compass card and needle completely enclosed in a metal case, of a piece of metal where the X-rays show the want of homogeneity, and other things. . . .

(15) I have sought for interference effects of the X-rays, but possibly, in consequence of their small intensity, without result.

(17) If one asks, what then are these X-rays; since they are not kathode rays one might suppose, from their power of exciting fluorescence and chemical action, them to be due to ultra-violet light, then that light must possess the following properties.

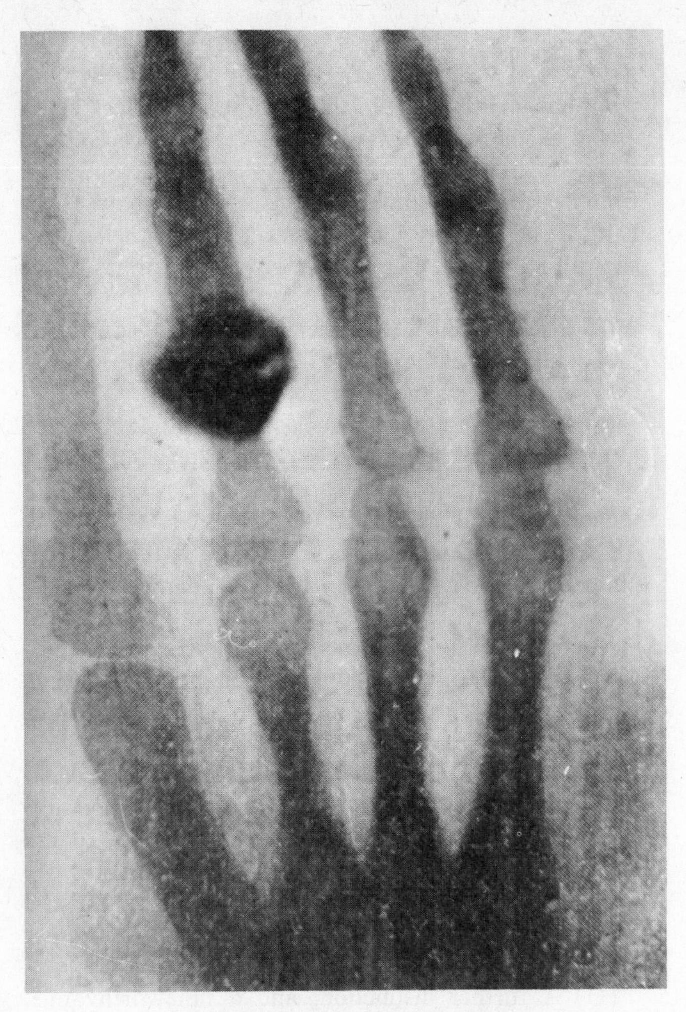

Fig. 12.24 Nature *53*, 274, *1896*.

(a) It is not refracted in passing from air into water, carbon bisulphide, aluminium, rock-salt, glass or zinc.

(b) It is incapable of regular reflection at the surface of the above bodies.

(c) It cannot be polarized by any ordinary polarizing media.

(d) The absorption by various bodies must depend chiefly on their density.

That is to say, these ultra-violet rays must behave quite differently from the visible, infra-red, and hitherto known ultra-violet rays.

These things appear so unlikely that I have sought for another hypothesis.

A kind of relationship between the new rays and light appears to exist; at least the formation of shadows, fluorescence, and the production of chemical action point in this direction. Now it has been known for a long time, that beside the transverse vibrations which account for the phenomena of light, it is possible that longitudinal vibrations should exist in the ether, and according to the view of some physicists, must exist. It is granted that their existence has not yet been made clear, and their properties are not yet experimentally demonstrated. Should not the new rays be ascribed to longitudinal waves in the ether?

I must confess that I have in the course of this research made myself more and more familiar with this thought, and venture to put this opinion forward, while I am quite conscious that the hypothesis advanced still requires a more solid foundation.'

Nature, **53**, 274–6, 1896.

The discovery of X-rays was an accident but the realization that there was a discovery to be investigated when a fluorescent screen glowed some distance from the tube marks this as Röntgen's discovery. Others had almost certainly been involved with X-rays in their experiments but had not recognized that a new radiation was involved. For example in a paper by J. J. Thomson, some two years previous:

'I have been able to detect phosphorescence in pieces of ordinary German-glass tubing held at a distance of some feet from the discharge tube, though in this case the light had to pass through the glass walls of the vacuum tube and a considerable thickness of air before falling on the phosphorescent body. . . .'

Phil. Mag., **37**, 358, 1894.

In the issue of *Nature* in which the translation of Röntgen's paper appeared there were already letters and an article on the topic. In the succeeding months the numbers of letters and articles appearing in journals and newspapers reached alarming numbers.

'The Röntgen Rays

The discovery by Prof. Röntgen of the rays which bear his name has aroused an interest perhaps unparallelled in the history of physical science. Reports of experiments on these rays come daily from laboratories in almost every part of the world. A large part of these relate to the methods of producing Röntgen photographs, and the application of the 'new photography' to medical and other purposes. A considerable amount of work has, however, been done on the physical properties of these rays; this has entirely confirmed the results stated by Röntgen in the paper in which he announced his discovery. . . .'

J. J. Thomson. *Nature,* **53**, 391, 1896.

'Medical science seems likely to benefit much by the application of Prof. Röntgen's discovery. The British Medical Journal thinks, as an aid to diagnosis of obscure fractures and lesions generally, the new photography will be of great value. From our contemporary we note that already a beginning has been made in this direction, and Prof. Mosetig, of Vienna, has taken photographs which showed with the greatest clearness and precision the injuries caused by a revolver-shot in the left hand of a man, and the position of the small projectile. . . .'

Nature, **53**, 324, 1896.

'A New Mode of Motion

. . . What may be called the popular and superficial aspect of his discovery has been seized upon with avidity. Shadow photographs have suddenly become an article of commerce; no illustrated paper is complete without reproductions of pictures showing the transparency of the human hand; every one who can command a vacuum tube and a few sensitive plates is busy repeating the primary experiment; ladies prattle of the new photography, and physicians already dream of unheard-of cures by its agency, and the market price of exhausted tubes—many of them of little value for the purpose in view—is rapidly rising.'

The Times, 4 February 1896.

'Sir,—It may be worthwhile just putting on record that during the past week I have seen fluorescence excited by Röntgen rays after they had penetrated the bodies of two men standing one behind the other in their clothes. . . .'

The Times, 31 March 1896.

Unfortunately for the early experimenters and their photographic subjects the idea of damage to persons due to the X-rays was not known. It was not until nearly the end of the year that the damaging effects of X-rays became a topic for papers.

'Some Effects of the X-Rays on the Hands

At the request of the editor of Nature, I append the following description, compiled from notes, of the effect of repeated exposure of the hands to X-rays. The result, though perhaps interesting from a medical and scientific point of view, has been most unpleasant and inconvenient to myself—the patient—and although my theories may be incorrect, and my conclusions easy to demolish, there is no mistaking the fact that X-rays are quite capable of inflicting such injury upon the hands as to render them almost useless for a time, and to leave in doubt their ultimate condition when entirely freed from frequent daily exposure to their influence.

Now for facts. I commenced demonstrating early in May with a coil capable of giving an 8″ spark [about 20 cm], and have been engaged in the work for several hours per day until the present time. For the first two or three weeks no inconvenience or discomfort were felt, but there shortly appeared on my right hand fingers numerous little blisters of a dark colour under the skin. These gradually became very irritating, the skin itself very red and apparently much inflamed. . . . About the middle of July the tips of my fingers began to swell considerably, and appeared as if they would burst. The tension of the skin was very great, and, to crown all, I noticed for the first time that my nails were beginning to be affected. This was the commencement of a long period of really serious discomfort and pain, which was only partly relieved when, from under the nails, there appeared a somewhat copious and unpleasant-smelling colourless discharge, which continued until the old nails were thrown off. . . .'

Nature, **54**, 621, 1896.

X-rays during 1896 were a topic of conversation both in scientific circles and amongst the general public. X-rays were NEWS.

'O, Röntgen, then the news is true,
  And not a trick of idle rumour,
That bids us each beware of you,
  And of your grim and graveyard humour.

We do not want, like Dr. Swift,
  To take our flesh off and to pose in
Our bones, or show each little rift,
  And joint for you to poke your nose in.

We only crave to contemplate,
  Each other's usual full-dress photo,
Your worse than "altogether" state,
  Of portraiture we bar in toto!
  . . .'

Punch, 25 January 1896.

The Nature of X-Rays

Though only transverse electromagnetic waves had been detected it was not considered impossible for longitudinal forms of electromagnetic waves to exist.

'. . . A series of experiments were made by taking photographs through tourmaline plates, (1) with their axes parallel, (2) with their axes crossed; it was hoped by this method to get some evidence as to whether the rays were longitudinal or transverse. A considerable number of photographs were taken in this way, but no difference could be detected in the obstruction offered to the rays by the tourmaline plates in the two cases.'

Prof. Thomson.

A paper read before the Cambridge Philosophical Society on January 27.

Nature, **53**, 378, 1896.

'Up to the present, however, no phenomena have been observed which enable us to say whether these waves are or are not transverse vibrations of very small wavelength, longitudinal vibrations, or even vibrations at all. Nothing of the nature of polarization or of interference have been described. The absence of polarization can at the present stage of the investigation hardly be pressed as an argument against these rays being transverse vibrations. For, of the three methods of producing polarization in light—reflection, refraction, and absorption—only the latter is available for these rays. . . .'

J. J. Thomson. *Nature*, **53**, 391, 1896.

The principal facts, which any satisfactory theory of the X-rays is called upon to explain, may be summarized as follows:

(1) The production of the rays by electric impulse, at the kathode, in a highly evacuated enclosure.

(2) Propagation in straight lines and absence of interference, reflection, refraction and polarization.

(3) The importance of density of the medium as the determining factor in the transmission of the rays.

(4) The production of fluorescence and actinic effects, and the action on electrified conductors.

Two theories have been proposed to account for these remarkable phenomena; (1) the theory of longitudinal waves; (2) the theory of projected particles.'

A. A. Michelson. From the *American Journal of Science*, April.
Nature, **54**, 21 May 1896.

'The two aspects in which the Röntgen rays differ from light is in the absence of refraction and perhaps polarization . . . the theory of dispersion of light shows that there will be no bending when the frequency of the vibration is very great. . . . Thus the absence of refraction, instead of being in contradiction to the Röntgen rays, being a kind of light, is exactly what we would expect if the wavelength of the light were exceedingly small.'

Prof. J. J. Thomson.

Opening address for the Mathematics and Physics Section of the British Association meeting on 4 September 1896.

Nature, **54**, 473–4, 1896.

The first real evidence for the transverse wave nature of X-rays was given by the experiments of Barkla in 1906 when the polarization of X-rays by reflection was shown.

'A mass of carbon was placed near an excited X-ray tube so as to be subject to a primary beam of considerable intensity. It was then the source of secondary radiation, the total energy of which was quite a large fraction of the energy incident upon it. A beam of this secondary radiation proceeding in a direction perpendicular to that of propagation of the primary falling on the carbon was studied.

In this secondary beam was placed a second mass of carbon, and the intensities of tertiary radiation proceeding in two directions at right angles and perpendicular to the direction of propagation of the secondary beam was observed by means of electroscopes placed in its path. The X-ray tube was turned round the axis of the secondary beam, while everything else was fixed, and the relative

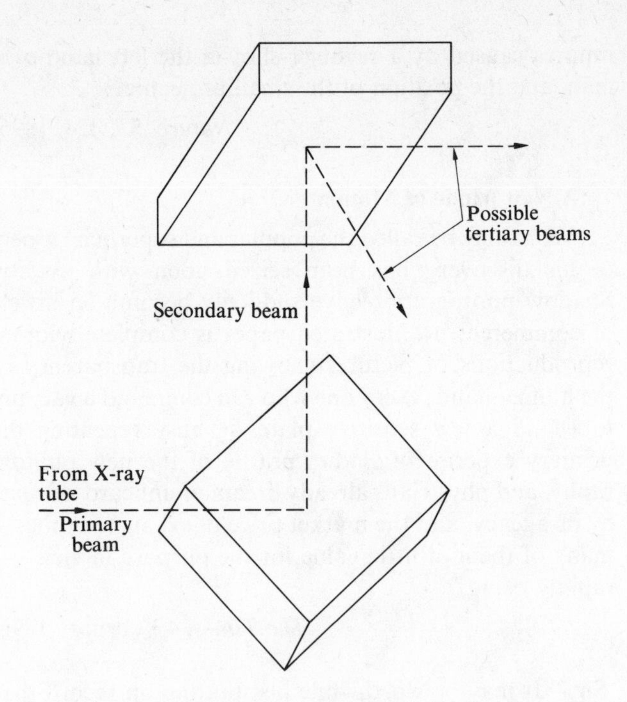

Fig. 12.25 Type of arrangement used.

intensities of the tertiary radiations observed for different positions of the tube [Fig. 12.25].

It was found that the intensity of tertiary radiation reached a maximum when the directions of propagation of the primary and tertiary were parallel, and a minimum when they were at right angles, showing the secondary radiation proceeding from carbon in a direction perpendicular to that of the incident primary to be polarized.'

Direction of primary beam	*Deflection of electroscope* A_1 *receiving horiz. tertiary beam*	*Deflection of electroscope* A_2 *receiving vert. tertiary beam*
Horizontal	6·3	1·9
Vertical	1·95	5·85

Proc. Roy. Soc. A., **77**, 247–255, 1906.

Interference and diffraction effects had been sought by many scientists but without success, until in 1899 Haga and Wind performed a series of experiments in which they passed X-rays through a wedge-shaped slit, at the widest end only about 10^{-6} m wide. The resulting photographs showed a broadening of the slit image at the narrowest end of the slit, consistent with a wavelength of $1{\cdot}3 \times 10^{-10}$ m (Haga and Wind, *Wied. Ann. d. Phys.*, **68**, 884, 1899). This result was confirmed by other experimenters using similar apparatus (Walter and Pohl, *Ann. d. Phys.*, **29**, 331, 1909; Koch, *Ann. d. Phys.*, **38**, 473, 1912). These results led Von Laue to compare the wavelength of X-rays with the spacing between atoms in a crystal and to consider a crystal as a diffraction grating for X-rays.

'The Crystal Space Lattice revealed by Röntgen rays

During a visit to Munich at the beginning of August last the writer was deeply interested in some extraordinary photographs which were shown to him.... They had been obtained by Dr. M. Laue, assisted in the experiments by Herren W. Friedrich and P. Knipping, ... by passing a narrow cylindrical beam of Röntgen rays through a crystal of zinc blende, the cubic form of naturally occurring sulphide of zinc, and receiving the transmitted rays upon a photographic plate. They consisted of black spots arranged in a geometrical pattern, in which a square predominated, exactly in accordance with the holohedral cubic symmetry of the space-lattice attributed by crystallographers to zinc blende. [See Fig. 9.7b]

Prof. von Groth expressed the opinion, in agreement with Herr Laue, that owing to the exceedingly short wavelength of the Röntgen rays (assuming them to be of electromagnetic wave character), they had been able to penetrate the crystal structure and to form an interference (diffraction) photograph of the ... space-lattice. ... The space lattice, in fact, was conceived to play the same function with the short-wave Röntgen rays that the diffraction grating does to the longer electromagnetic waves of light.'

A. E. H. Tutton. *Nature*, **90**, 306–7, 1912.

(The original paper appeared by Friedrich, Knipping, and Laue in *Ber. bayer. Akad. Wiss*, pp. 303 and 363, 1912.)

'In a paper read recently before the Cambridge Philosophical Society my son has given a theory which makes it possible to calculate the positions of the spots for all dispositions of crystal and photographic plate.... It is based on the idea that any plane within the crystal which is "rich" in atoms can be looked on as a reflecting plane; the positions of the spots can then be calculated by the reflection laws in the ordinary way.'

W. H. Bragg. *Nature*, **90**, 360, 1912.

(The paper referred to is Bragg, *Proc. Camb. Phil. Soc.*, **17**, 43, 1912.) Chapter 9 gives details of this theory and X-ray diffraction photographs.

'The Refraction of X-Rays

A. Larson, M. Siegbahn and T. Waller

While Röntgen tried to find a refraction of X-rays in solids, it is well-known that he did not succeed, and up to this time all attempts to show a refraction in amorphous substances have failed. Stenström was the first to give experimental evidence of a deviation from the Bragg formula for the reflection of X-rays at crystals, which indicated the existence of a refraction. On the other hand Compton and Siegbahn with Lundquist have shown that amorphous and crystalline substances act as total reflectors for X-rays when the incident glancing angle is very small. These two phenomenon strongly support the view that a refraction of X-rays at the surface of amorphous and crystalline substances does exist. By means of a suitable experimental arrangement it has been possible for the authors to get measurable deviations of X-ray beams passing through prisms of amorphous or crystalline substances, which permitted the calculation of the refractive index. With beams of non-monochromatic X-rays a real prism spectrum was registered on the photographic plate.'

Phys. Rev., **25**, 235, 1925.

About the same time as refraction by a prism was realized diffraction by a man-made ruled grating was achieved, thus the methods of realizing dispersion in the visible part of the spectrum were now possible with X-rays.

'X-Ray Spectra from a Ruled Reflection Grating

By A. H. Compton and R. L. Dean

We have recently obtained spectra of ordinary X-rays by reflection at very small glancing angles from a grating ruled on speculum metal. Typical spectra thus obtained are shown in the accompanying figures [Fig. 12.26]. From some of these spectra it is possible to measure X-ray wavelengths with considerable precision.

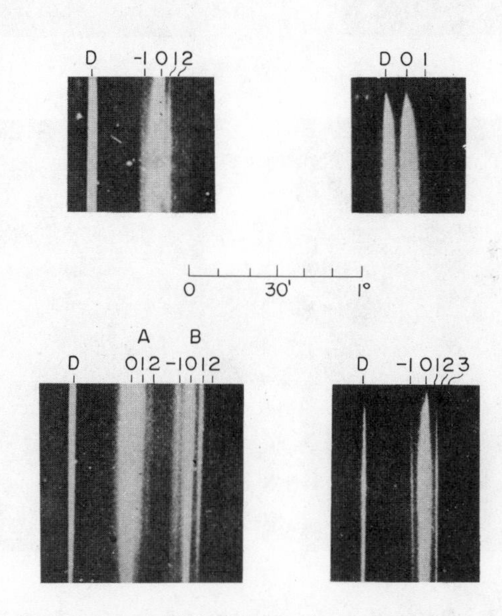

Fig. 12.26 Spectrum of X-rays from copper target. D = image of direct beam. Numbers −1, 0, 1, 2 indicate the order of the spectrum.

In order to reflect any considerable X-ray energy from a speculum surface it is necessary to work at small glancing angles, within the critical angle for total reflection. (See A. H. Compton, Phil. Mag. Vol. 45, p 1121, 1923.) Within this critical angle, which in our experiments, using wavelengths less than 1·6 ångstroms [one ångstrom is 10^{-10}m], was less than 25 minutes of arc, the diffraction

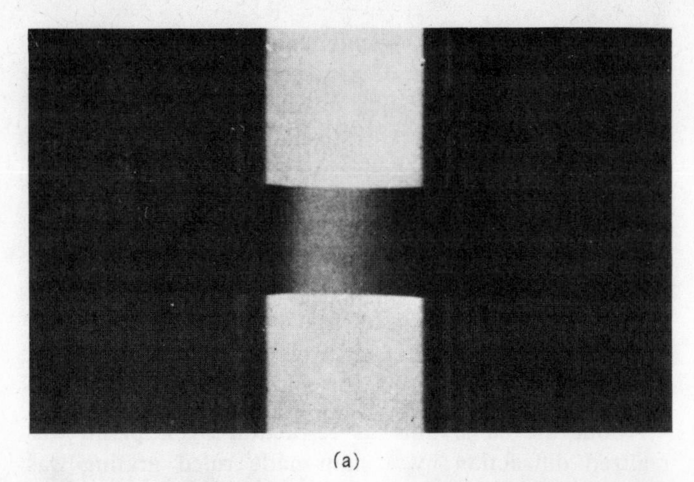

(a)

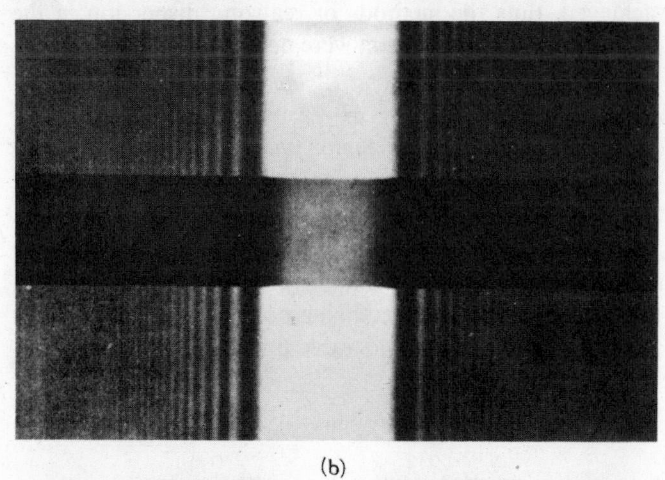

(b)

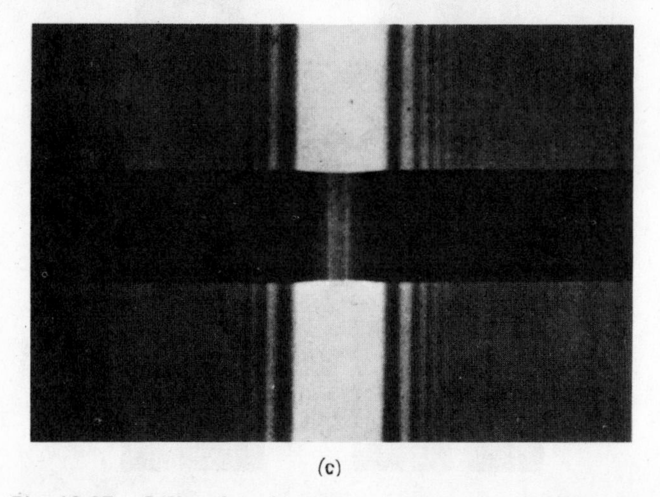

(c)

Fig. 12.27 Diffraction by narrow wires. (a) 0·0436 mm diameter. (b) 0·0379 mm diameter. (c) 0·0188 mm diameter. The central parts of the photographs are longer exposures. (G. Kellstrom. Nova Acta Regiae Societatis Scienturium Upsaliensis, *Ser. IV,* **8**, *61, 1932).*

grating may be used in the same manner as in optical work. The wavelength is given by the usual formula,

$$n\lambda = D(\sin \phi + \sin i)$$

where i is the angle of incidence and ϕ is the angle of diffraction for the nth order. . . .

In order that several orders of the spectrum should appear inside the critical angle, we had a grating ruled with a comparatively large grating space, $D = 2{\cdot}000 \times 10^{-3}$ cm. . . .

We see no reason why measurements of the present type may not be made fully as precise as the absolute measurements by reflection from a crystal, in which the probable error is due chiefly to the uncertainty of the crystalline grating space.'

Proc. Nat. Acad. Sci., **11**, 598–601, 1925.

The remarkable photographs of Kellstrom (*Nova Acta Regiae Societatis Scientiarum Upsaliensis*, Ser. 4, Vol. 8, No. 5) showing the conventional light diffraction and interference experiments with X-rays serve as convincing evidence of the wave nature of X-rays. Kellstrom showed interference effects with Fresnel mirrors and Lloyd's mirror arrangements. In addition diffraction through slits, at straight edges and round a fine wire were shown (Fig. 12.27).

The production of X-rays

In Röntgen's first apparatus the X-rays were produced by applying a high potential difference between two electrodes enclosed in a tube containing air under a low pressure. Stray ions and electrons were present in the tube (they would be produced by cosmic radiation and natural radioactivity) and these became accelerated by the potential difference. The X-rays were noticed as emanating at the anode and thus the source identified as being the 'collision' of electrons with a solid. Modern X-ray tubes produce the electrons by heating the cathode: thermionic emission (Fig. 12.28).

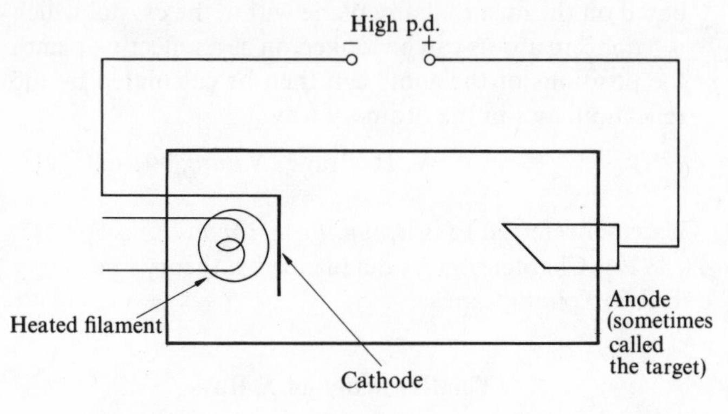

Fig. 12.28

A ruled grating or a crystal can be used to determine how the intensity of the emitted rays varies with wavelength. Figure 12.29 shows a typical result. Superimposed on a continuous background there are a number of sharp wavelength spikes. The continuous background is for the same accelerating p.d. the same for all elements, its shape depending only on the accelerating p.d. The interpretation placed on this is that the continuous X-rays are produced by the slowing down of the electrons in the target. This type of radiation is known

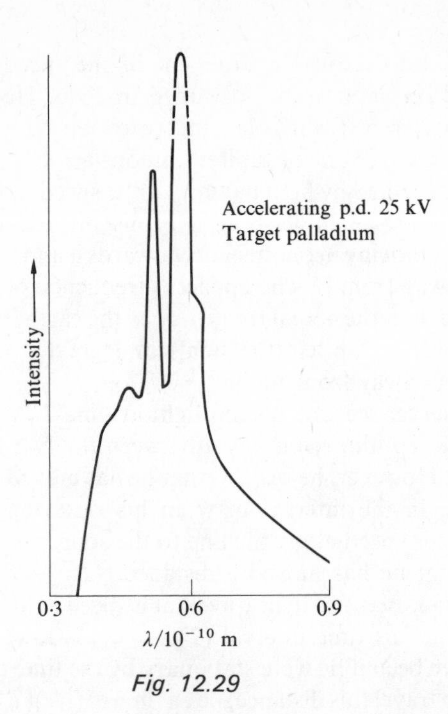

Fig. 12.29

as bremsstrahlung. The sharp spikes are, however, found to occur at the same wavelengths for any one element, regardless of the value of the accelerating p.d., and to be characteristic of the element (see the paragraph on Moseley's work in chapter 13). These emissions are thus from atoms in the target which have been excited by the collisions.

12.4 Electromagnetic waves

The group of waves known as electromagnetic waves (Fig. 12.30) share a number of common properties. They all have the same speed in a vacuum. The following table is adapted from that in *Special relativity* by A. P. French (Nelson) page 12:

Frequency /Hz	*Wavelength* /m	*Speed (with error)* $/10^8$ m s^{-1}
$4{\cdot}7 \times 10^7$	6·4	$2{\cdot}9978 \pm 0{\cdot}0003$
$1{\cdot}7 \times 10^8$	1·8	$2{\cdot}99795 \pm 0{\cdot}00003$
$3{\cdot}0 \times 10^8$	1·0	$2{\cdot}99792 \pm 0{\cdot}00002$
$3{\cdot}0 \times 10^9$	$1{\cdot}0 \times 10^{-1}$	$2{\cdot}99792 \pm 0{\cdot}00009$
$2{\cdot}4 \times 10^{10}$	$1{\cdot}2 \times 10^{-2}$	$2{\cdot}997928 \pm 0{\cdot}000003$
$7{\cdot}2 \times 10^{10}$	$4{\cdot}2 \times 10^{-3}$	$2{\cdot}997925 \pm 0{\cdot}000001$
$5{\cdot}4 \times 10^{14}$	$5{\cdot}6 \times 10^{-7}$	$2{\cdot}997931 \pm 0{\cdot}000003$
$1{\cdot}2 \times 10^{20}$	$2{\cdot}5 \times 10^{-12}$	$2{\cdot}983 \pm 0{\cdot}015$
$4{\cdot}1 \times 10^{22}$	$7{\cdot}3 \times 10^{-15}$	$2{\cdot}97 \pm 0{\cdot}03$

All the waves are transverse waves. All the properties of the waves can be explained by the same basic physics, provided we make adjustments for differences in frequency. There are no firm boundaries between the different types of waves. If you did not know the source no difference could be detected between X-rays produced by slowing down electrons or by excited atoms and gamma rays of the same frequency produced by excited nuclei. A photoelectric cell detects ultraviolet and visible radiation with no apparent difference other than a frequency dependence which applies

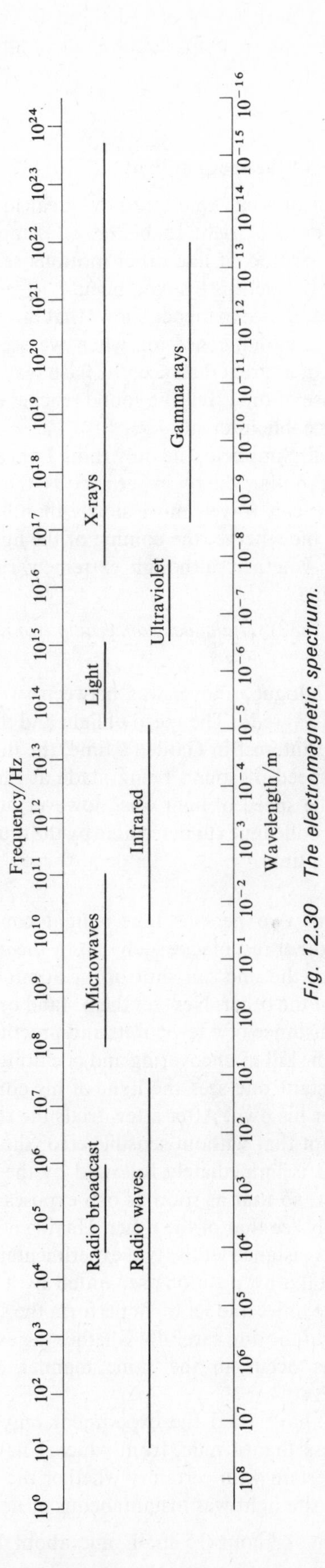

Fig. 12.30 The electromagnetic spectrum.

equally well to both forms of radiation. In the following pages we will look in more detail at the properties mentioned above.

Measurements of the speed of light

'*Sagr*. But of what kind and how great must we consider this speed of light to be? Is it instantaneous or momentary or does it like other motions require time? Can we not decide this by experiment?

Simp. Everyday experience shows that the propagation of light is instantaneous; for when we see a piece of artillery fired, at great distance, the flash reaches our eyes without lapse of time; but the sound reaches our ear only after a noticeable interval.

Sagr. Well, Simplicio, the only thing I am able to infer from this familiar bit of experience is that sound, in reaching our ear, travels more slowly than light; it does not inform me whether the coming of the light is instantaneous or whether, although extremely rapid, it still occupies time.'

Galileo (1638). *Dialogues concerning two new sciences.*

Galileo's dialogue, above, was between two characters Simplicio and Sagredo. The speed of light and that of sound were topics of interest in Galileo's time, the first measurement of the speed of sound being made at that time (see chapter 6). The speed of light was, however, too great for Galileo or his colleagues to determine by the crude methods they were adopting.

'Let each of two persons take a light contained in a lantern, or other receptacle, such that by the interposition of the hand, the one can shut off or admit the light to the vision of the other. Next let them stand opposite each other at a distance of a few cubits and practice until they acquire such skill in uncovering and occulting their lights that the instant one sees the light of his companion he will uncover his own. After a few trials the response will be so prompt that without sensible error the uncovering of one light is immediately followed by the uncovering of the other, so that as soon as one exposes his light he will instantly see that of the other. Having acquired skill at this short distance let the two experimenters, equipped as before, take up positions separated by a distance of two or three miles and let them perform the same experiment at night, noting carefully whether the exposure and occultations occur in the same manner as at short distances. . . .

In fact I have tried the experiment only at a short distance, less than a mile, from which I have not been able to ascertain with certainty whether the appearance of the opposite light was instantaneous or not, . . .'

[1 cubit is about 0·5 m. 1 mile about 1·6 km]

Galileo (1638). *Dialogues concerning two new sciences.*

The first successful measurement of the speed of light was based on data Römer obtained in 1676. He realized that the apparently variable times elapsing between successive passages of one of Jupiter's moons behind the planet could be explained by light having a finite speed. Essentially the problem is—a light flashing at a constant frequency and an observer moving in one instance towards it and in another instance away from it. The apparent frequency of the light will be less than the actual frequency as the earth (observer) moves towards Jupiter (the light) and greater when the earth moves away from Jupiter.

The observer sees the distant light at time T'_1 and again at time T'_2. To him the interval between the two flashes is $T'_2 - T'_1$. However, he realizes that he has moved closer to the source in the interval between his receiving the two flashes. If his speed along the line to the source is v then he reckons that he has moved a distance $v(T'_2 - T'_1)$ in the time interval. Because light travels at a speed c the observer reckons that his time interval $(T'_2 - T'_1)$ is lower than it would have been if he were stationary by the time taken for the light to travel this distance, i.e., a time of $(v/c)(T'_2 - T'_1)$. He therefore reckons that the actual time interval, at the source, must be

$$(T'_2 - T'_1) + \frac{v}{c}(T'_2 - T'_1)$$

or

$$t = t'(1 + v/c)$$

t is the observer's estimate of the time interval at the source, t' is his measured time interval. We will be looking more closely at this equation later in this chapter and arriving at a modified equation

$$t = t'\sqrt{\frac{1 + v/c}{1 - v/c}} \quad \text{or} \quad t' \frac{(1 + v/c)}{\sqrt{(1 - v^2/c^2)}}$$

To a first approximation this is the same as the equation derived above, i.e., when $v \ll c$.

If the source emits the flashes at a regular rate then we can write the equation in terms of frequencies

$$f' = f(1 + v/c)$$

This is known as the Doppler equation. As an observer moves towards a source emitting a constant frequency light (or sound) the apparent frequency rises.

Römer's data gave a value of about 214 000 km s^{-1} for the speed of light. The idea that light had a finite speed and was not instantaneous was not immediately accepted by all. Universal acceptance of the finite speed did not come until about 1727 when Bradley obtained a similar value by a different method, the aberration of light.

Bradley was the third Astronomer Royal in England and over a period of years had noticed that the positions of some stars appeared to change. He first thought that these changes were due to errors of observation but later noticed that the changes were regular and were related to the time

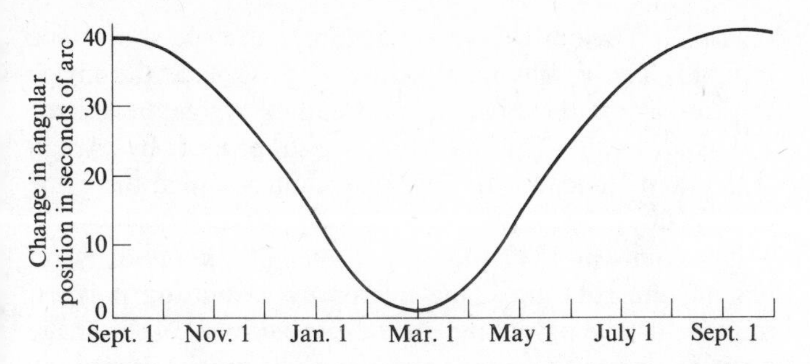

Fig. 12.31 Taken from Bradley. Phil. Trans. Roy. Soc., **35**, *637, 1729.*

of year when the observation was made. Figure 12.31 shows the apparent movement of the star γ Draconis in the north–south direction, taken from Bradley's paper. Bradley's explanation of the effect was that the apparent direction of the light reaching the earth from a star is altered by the velocity of the earth in its orbit. The observer and his telescope are carried along with the earth in its orbital motion. Think of the light from the star as rain falling on you. If the rain falls vertically and you are still you can stop the rain reaching your head by holding an umbrella over your head. If, however, you run then your umbrella has to be tilted to stop the rain falling on your head (Fig. 13.32(a)). If in Fig. 12.32(b) your head moves through a distance AB in the time a drop of rain covers the distance CB then the drop at C will hit your head instead of the drop at D. To stop the rain hitting your head your umbrella must be along a line AC instead of AD. To keep the image of a star lined up with the crosswires in a telescope, when the telescope is moving, it must be tilted. Light entering the objective of the telescope moves with a speed c towards the crosswires; the crosswires, however, are moving with a speed v at right angles (near enough if the star is directly overhead) to the light path. If the light takes a time t to travel from the objective to the crosswires then the crosswires will have moved a distance vt while the light travels a distance ct. The telescope must be tilted at an angle of

$$\tan \alpha = \frac{vt}{ct} = \frac{v}{c} \quad \text{(Fig. 12.32(c))}.$$

Bradley found that the angle changed by about 40 seconds for any two points on the earth's orbit six months apart. Figure 12.33 shows the situation from the point of view of the man running in a circular orbit in the rain. The earth 'runs' round its orbit with a speed of about 30 km s^{-1}. Thus we have

$$\tan 20'' = \frac{30}{c}$$

tan 20″ is about 0·001 and thus c is about 300 000 km s^{-1}.

The first terrestrial method for the speed of light was by Fizeau in 1849 (Fig. 12.34). Essentially his method consisted in sending a pulse of light to a distant mirror and back again and determining the time taken. A rotating toothed wheel

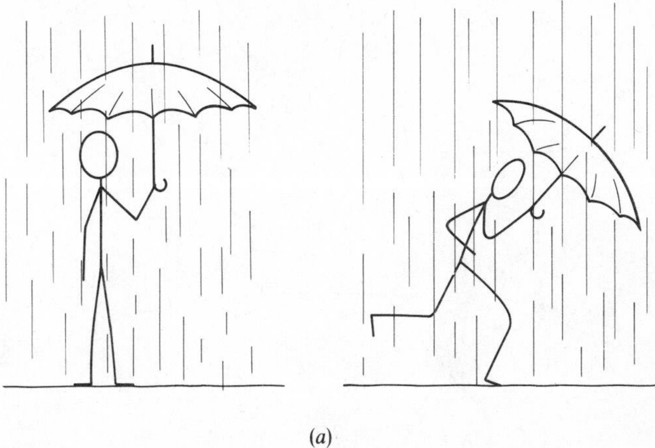

(a)

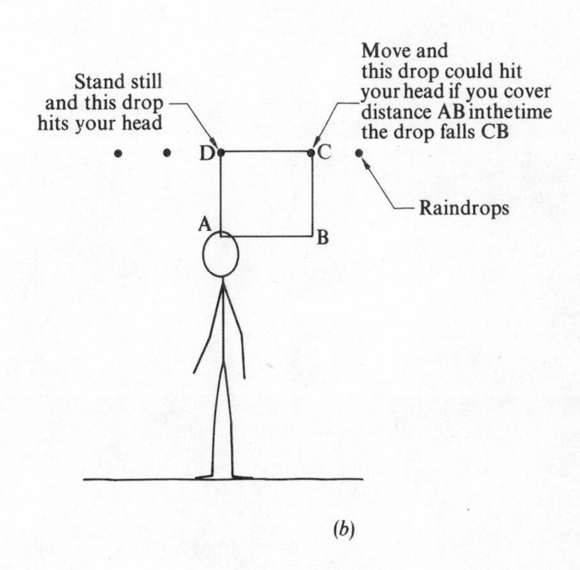

(b)

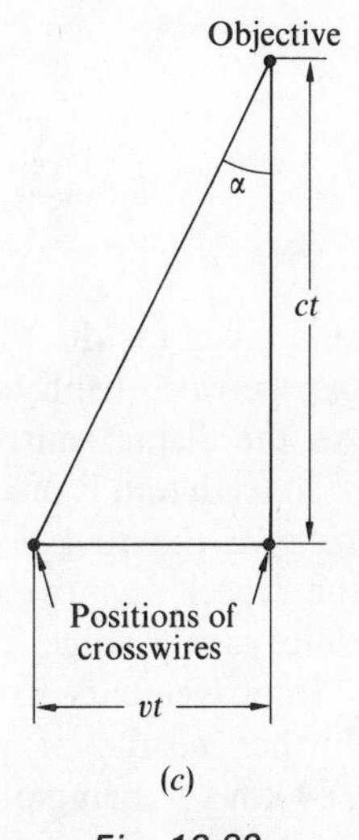

(c)

Fig. 12.32

was used to produce brief pulses of light. The pulses then passed to the distant mirror and back again to the toothed wheel. If the wheel had rotated sufficiently the pulse could arrive back in time to hit the next tooth on the wheel and so no light pass onto the eye of the observer. Thus the observer gradually increases the speed of rotation of the toothed wheel until he sees the light is eclipsed. He then

Fig. 12.33

mirror and reach the rotating mirror after a short time interval. The result of this will be to displace the image position from the position it occupied when the mirror was stationary. The method was improved by Arago (1850) and Foucault. In 1862 Foucault obtained the value of 298 000 $\pm$ 500 km s^{-1}.

Newcomb in 1880 modified Foucault's method by replacing the rotating plane mirror by a rotating polygon mirror. Instead of looking for the displacement of the image position due to the rotation of the plane mirror during the finite time taken for the light to travel to a distant mirror and back, Newcomb looked for no displacement of the image—one face of his polygon mirror having exactly replaced the previous face during the time taken for the light to go to the distant mirror and back. Michelson, together with Newcomb, made many measurements using polygon mirrors. Four, eight, twelve, and sixteen-sided mirrors were used in their various determinations.

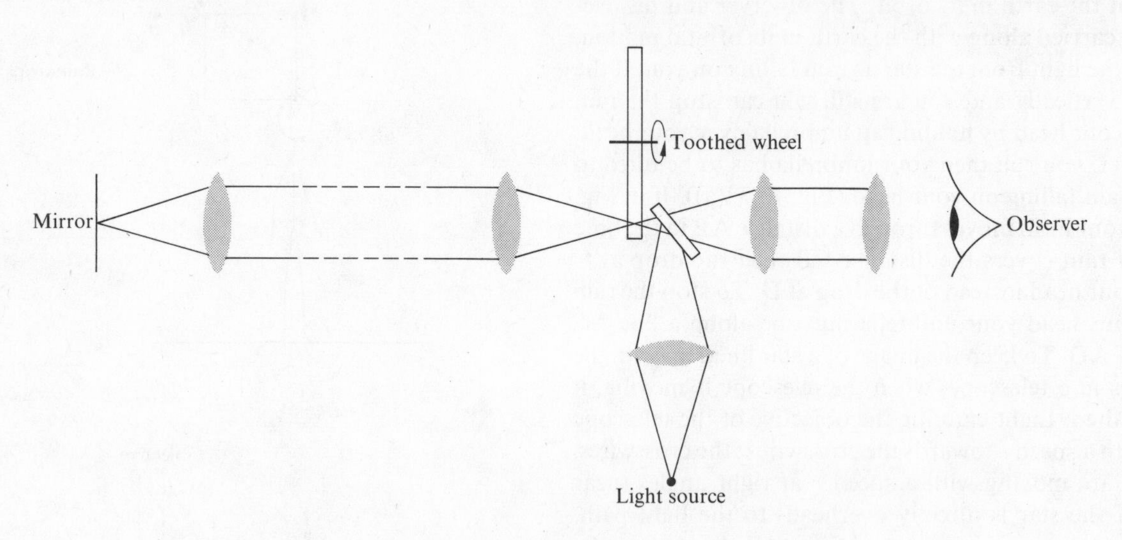

Fig. 12.34 Fizeau's experimental arrangement.

knows that in the time taken for the wheel to rotate by the amount from one gap between teeth to the next tooth the light has travelled to the distant mirror and back again. Fizeau's wheel had 720 teeth and it revolved at 12·6 revolutions per second to give the first eclipse position. The distance between the wheel and the distant mirror was 8·6 km. Fizeau's results gave a value for the speed of light of 312 000 km s^{-1}. Improvements in the apparatus and technique led to further results of greater accuracy, a value of 299 901 $\pm$ 84 km s^{-1} being obtained by Perrotin, using this method, at the beginning of the twentieth century.

In 1834 Wheatstone suggested an alternative method for determining the time taken for a pulse of light to traverse a distance. If a beam of light is reflected from a mirror to a distant mirror and then back again the final image appears at a certain place in the field of view of an eyepiece (Fig. 12.35). If the first mirror rotates then only pulses of light will be reflected at the right angle to go to the distant mirror. These pulses will be reflected back by the distant

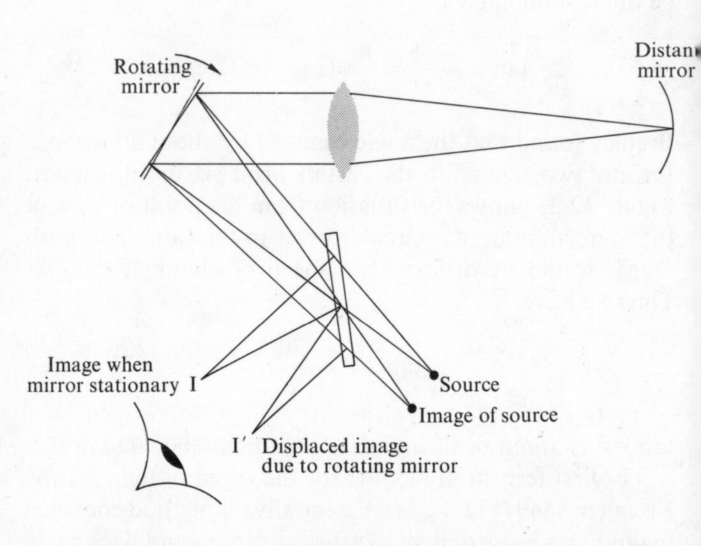

Fig. 12.35 Foucault's experimental arrangement.

'Measurement of the Velocity of Light in a Partial Vacuum
(*Astrophysical J.*, **82**, 1935)
By A. A. Michelson, F. G. Pease and F. Pearson*

ABSTRACT

The observations were made by the rotating-mirror method, the light passing through a steel tube 1 mile long [1·6 km], evacuated to pressures which ranged from 0·5 to 5·5 mm mercury. By multiple reflections the path length was increased to 8 or 10 miles [about 12 to 16 km]. . . .

DESCRIPTION OF APPARATUS

Optical layout.—A diagram of the optical arrangement of the apparatus is shown in Fig. 1.1 [Fig. 12.36]. Light from an arc lamp A was imaged on the slit C by the condensing lens B. For the first 46 series of the 1931 observations it passed above the right-angle prism I to the upper half of the rotating mirror, D, thence through the plane-parallel glass window L into the tube to the diagonal flat E and the concave mirror F. It next passed above the flat mirror H and then, by means of repeated reflections at the flat mirrors G and H until the desired distance had been traversed, formed on the surface of G and H a magnified conjugate image of the slit C. The beam then retraced its path directly below the incoming path and emerged from the tube, striking the lower half of the rotating mirror D; it then passed through the reflecting prism I on to the crosswires J and was observed in the eyepiece K. . . .

* Dr Michelson died on May 9, 1931, when 36 of the 54 series of 1931 observations had been completed.

SYSTEMS OF MEASUREMENT

Measurement of time.— . . . In the null method used . . . the light emerges from one face (of the rotating mirror) and is received on some other face—the adjacent face in . . . [this] experiment. As the mirror starts rotating, the image gradually passes from the field of view and reappears in the other side of the field only when the rotating mirror is approaching its proper speed.

While several methods are available for measuring the velocity of light with this arrangement, the one chosen is as follows: The rotating mirror is brought into synchronism with a tuning fork whose period of vibration is determined; the position of the image is then read with a micrometer for the right- and left-hand rotations of the mirror. The distance remains fixed. The time interval to be measured is therefore that during which the rotating mirror turns 1/32 revolution (it has 32 sides), plus or minus a small angle derived from the readings of the micrometer. The period of the fork is determined by stroboscopic methods in terms of free-pendulum beats. . . .

The light from a 6-volt lamp . . . after striking the small mirror on the tuning fork, was imaged by the small achromatic lens on the one polished face of the nut which clamps the rotating mirror to its shaft. Since the fork stood vertically, the image vibrated up and down on the nut. . . . Since the nut rotated in a horizontal plane, the image, as the mirror speeded up, passed through a series of vibrating and stationary states and finally reached a permanent stationary state when the beats heard between the fork and the rotating mirror ceased. . . .

Measurement of distance.—After a conference with Commander C. L. Garner, assistant chief of the division

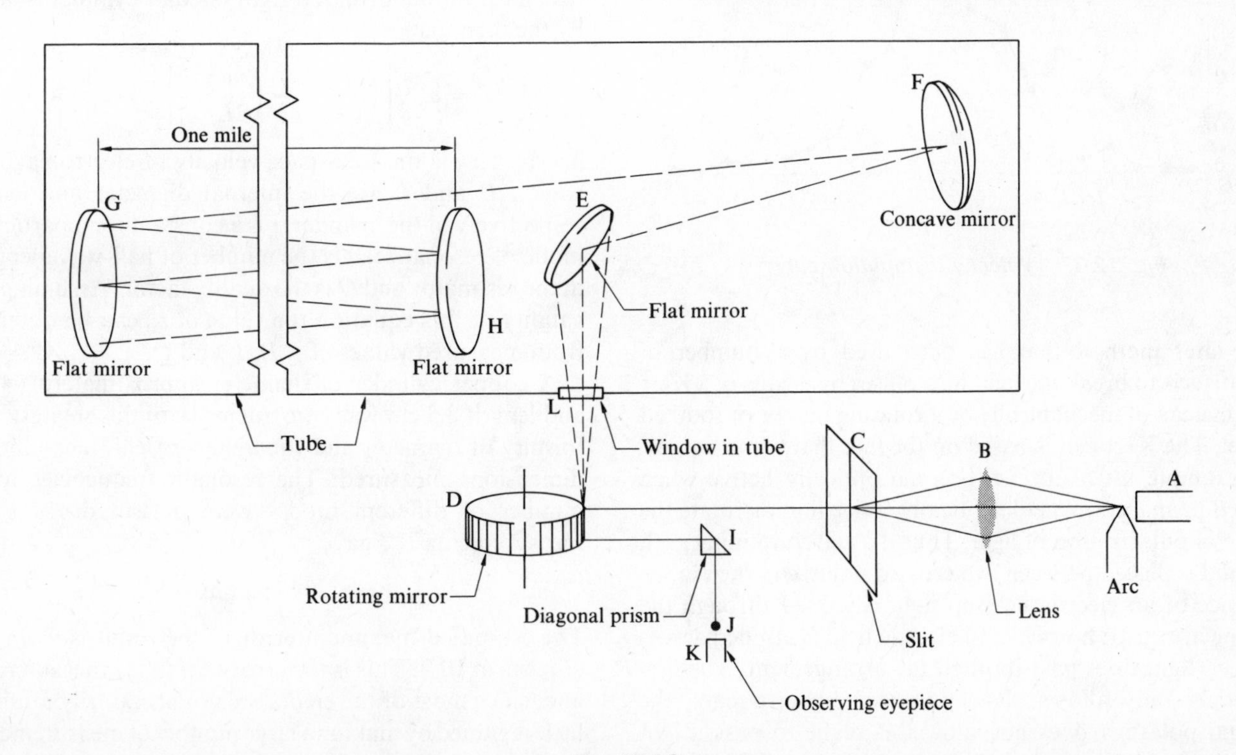

Fig. 12.36

of Geodesy of the United States Coast and Geodetic Survey, it was decided to lay out the base line about 10 feet to the west of the pipe and place six piers along the line.

... The mean length of this base line was found by the United States Coast and Geodetic Survey to be 1594·2658 m. ...

DISCUSSION

Distribution of velocities.—The mean velocities ... have been grouped into divisions covering a range of 5 km s^{-1} and are shown in Table VIII. A plot of these data in Fig. 1.8 [Fig. 12.37] resembles a probability-curve and indicates that the probable value of a constant velocity would be 299 774 km s^{-1}....

Table VIII

FREQUENCY DISTRIBUTION OF MEASURED VELOCITIES

Velocity range	*Number*	*Velocity range*	*Number*
299 000+		299 000+	
726–730	4	776–780	515
731–735	6·5	781–785	270
736–740	3·0	786–790	236
741–745	55	791–795	90
746–750	29	796–800	62
751–755	86	801–805	33
756–760	184	806–810	30
761–765	304	811–815	32·5
766–770	353·5	816–820	0
771–775	580	821–825	12

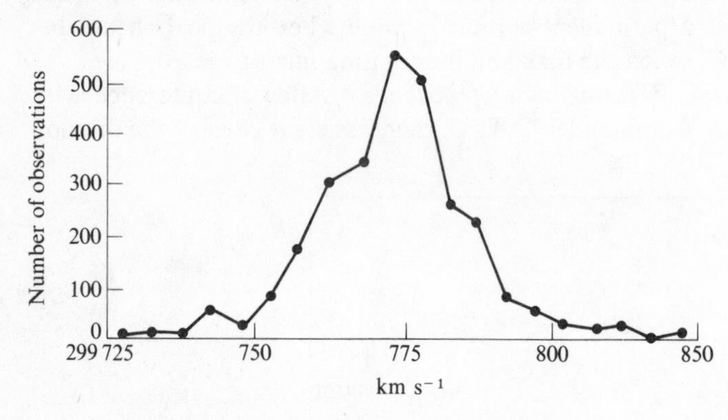

Fig. 12.37 Velocity distribution curve.

A later method that has been used by a number of scientists is to break the light into pulses by means of a Kerr cell, instead of mechanically by a rotating mirror or toothed wheel. The Kerr cell is based on the fact that some liquids, for example nitrobenzene, become optically active when placed in an electric field. Optically active liquids rotate the plane of polarization of light. Thus if a cell containing such a liquid is placed between two crossed polarizers then in the absence of an electric field no light is passed through the arrangement. If, however, an electric field is applied across the cell light does pass through the arrangement. The first polarizer only allows plane polarized light to pass, the second polarizer does not allow this plane to pass. Light can thus only get through the second polarizer if the liquid in the cell rotates the plane of polarization. If the electric field is produced by an alternating potential difference then a train of light pulses are produced. Frequencies in the region 10 to 100 MHz have been used.

The present value (1964) taken for the speed of light is 299 792·5 km s^{-1}, with an uncertainty of about one part in a million.

'Behold the Light emitted from the Sun,
What more familiar, and what more unknown;
While by its spreading Radiance it reveals
All Nature's Face, it still it self conceals...
How soon th' Effulgent Emanations fly
Thro' the blue Gulph of interposing Sky!
How soon their Lustre all the Region fills,
Smiles on the Vallies, and adorns the Hills!
Millions of Miles, so rapid is their Race,
To cheer the Earth, they in few Moments pass.
Amazing Progress! At its utmost Stretch,
What human Mind can this swift Motion reach?'

R. Blackmore (1715), *Creation*, II, p. 386.

Measurement of the speed of microwaves

'The Velocity of Propagation of Electromagnetic Waves Derived from the Resonant Frequencies of a Cylindrical Cavity Resonator'

(*Proc. Roy. Soc.* A, **194**, 1948)

By L. ESSEN, D.Sc., Ph.D., and A. C. GORDON-SMITH.
The National Physical Laboratory

The frequency of resonance of an evacuated cavity resonator in the form of a right circular cylinder is given by the formula

$$f = v_0\sqrt{\left[\left(\frac{r}{\pi D}\right)^2 + \left(\frac{n}{2L}\right)^2\right]}\left[1 - \frac{1}{2Q}\right],$$

in which v_0 is the free-space velocity of electromagnetic waves, D and L are the internal diameter and length respectively of the cylinder, r is a constant for a particular mode of resonance, n is the number of half-wave-lengths in the resonator and Q is the quality factor. Assuming the validity of this equation the value of v_0 can be obtained from measured values of f, D, L and Q.

A copper cylinder of diameter approximately 7·4 cm and length 8·5 cm was constructed with the greatest uniformity of diameter and squareness of end-faces and its dimensions measured. The resonant frequencies for a number of different modes were measured.... Final measurements ... gave

$$v_0 = 299\ 792 \text{ km s}^{-1}$$

The estimated maximum error of the result is 9 km s^{-1} (3 parts in 10^5). This is the error of a single measurement and, since most of the errors are not necessarily random, little is gained by making a large number of measurements. ...'

The frequency of the oscillations used by Essen and Gordon-Smith was of the order of 3×10^9 Hz, a wavelength of 1×10^{-1} m.

Measurement of the speed of gamma radiation

'The Velocity of Gamma-Rays in Air

M. R. CLELAND and P. S. JASTRAM
Physics Department, Washington University, St Louis, Missouri
(*Phys. Rev.*, **84**, 271, 1951)

The velocity of 0·5-Mev gamma-rays resulting from positron annihilation has been measured by means of scintillation counters and delayed-coincidence techniques. The value obtained is $(2{\cdot}983 \pm 0{\cdot}015) \times 10^{10}$ centimeters per second.

INTRODUCTION

Development of a coincidence-counting system capable of resolving events separated by 10^{-9} second suggested the possibility of making a direct measurement of the velocity of gamma-rays. The availability of annihilation radiation, in which two gamma-rays are emitted simultaneously in opposite directions, was essential to the success of the experiment, since reflection techniques could not be used. A suitable source of such radiation consists of a positron emitter such as Cu^{64} or Na^{22} placed in a metal container in which the positrons are stopped and annihilated....

PROCEDURE

Scintillation counters employing a solution of terphenyl in phenylcyclohexane and EMI Type 5311 photomultiplier tubes were placed on opposite sides of a radiator containing initially about 1 curie of Cu^{64}. The source and one of the counters were attached to a suitable frame, and the transit-time difference of the two annihilation quanta was determined for five different path differences. The velocity was obtained from the slope of the best straight line, as determined by the usual least-squares method, through the experimental points representing distance *versus* delay. The general arrangement of the apparatus is shown diagrammatically in Fig. 1 [12.38]....

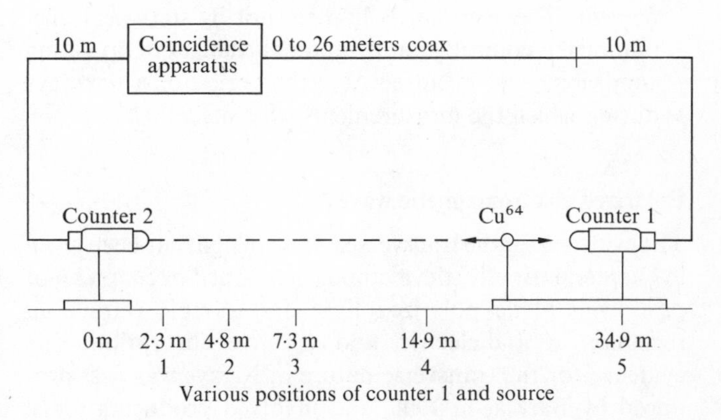

Fig. 12.38 Arrangement of counters, source, and coincidence apparatus.

RESULT

The measured transit times are plotted against distance in Fig. 5 [Fig. 12.39]; the experimental points are indicated by the intersections of the crossed lines. The length of these lines is not related to the statistical errors, which are too small to exhibit.

The velocity is determined by fitting the best straight line to the experimental points, under the assumptions

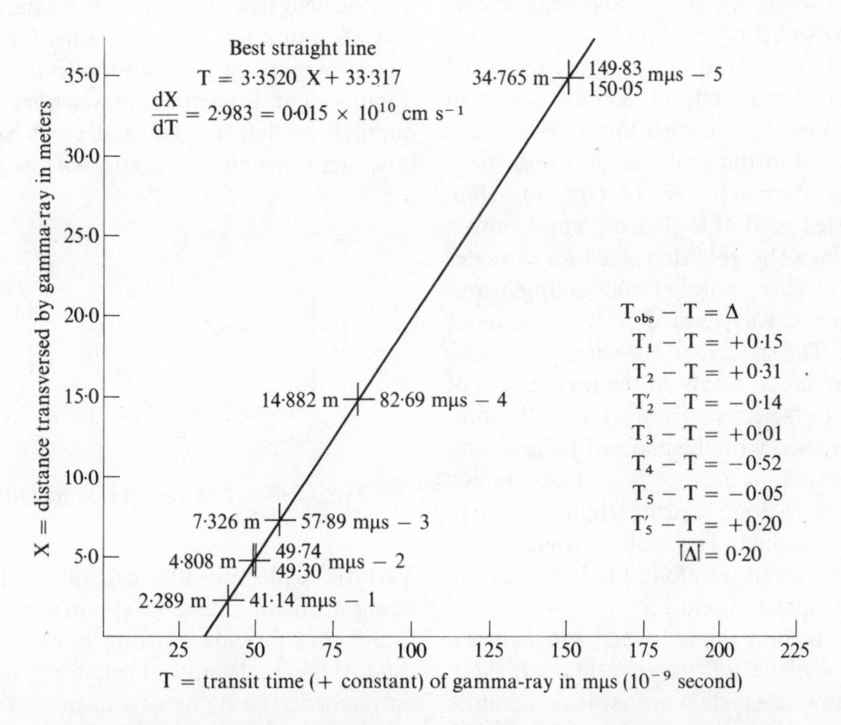

Fig. 12.39 Distance versus time of flight for annihilation radiation.

that only the abscissa values are subject to appreciable error, and that the error is the same for all measurements, including those repeated at positions 2 and 5. The velocity that gives the best least-squares fit is given by

$$c = [n\Sigma x_k^2 - (\Sigma x_k)^2]/[n\Sigma x_k t_k - \Sigma x_k \Sigma t_k],$$

where x_k and t_k are the measured values of the coordinates, and n is the number of points. The result is $2{\cdot}983 \times 10^{10}$ centimeters per second, with an estimated error of 0·5 percent. The accuracy is limited, not by statistical fluctuations in counting but by gradual changes in operating conditions that occurred over the period of a few days during which the measurements were made. . .'

Polarized electromagnetic waves

The evidence for the transverse nature of light has been given in chapter 6. Briefly, the methods mentioned in chapter 6 of producing plane polarized light are—double refraction, reflection at a dielectric, and selective absorption. The evidence for the transverse nature of X-rays was first produced by Barkla, in 1906, and involved producing plane polarized X-rays by reflection from carbon (see earlier in this chapter). The polarization of radio waves can be shown by what we might call selective emission—the waves are restricted to just one plane by the use of linear aerials for transmission. If the detecting aerial is at right angles to the transmitting aerial then no signal is detected, if the detecting aerial is parallel to the transmitting aerial a signal can be detected. A linear aerial emits plane polarized waves. In the case of microwaves selective absorption by a grid of wires can be used to show the transverse nature (see earlier in this chapter). The methods used to show the transverse nature of these different waves are similar and suggest that the waves are all similar—part of one family.

Polarization by selective absorption is not just restricted to the microwave region. Similar effects can be achieved in the radio region with wires much farther apart. Wires very close together can be used in the visible region to achieve the same effect. Bird and Parrish in 1960 (*J. Opt. Soc. Am.*, **50**, 886, 1960) evaporated gold at a glancing angle onto a plastic diffraction grating. The gold deposited on the sides of ruled grating slits to form parallel conducting wires. The spacing of these 'wires' was about 5×10^{-7} m, about the wavelength of light. The spacing of the wires with radio or microwave radiation has to be about the wavelength of the radiation involved for selective absorption to be complete when the grid is crossed with the plane of polarization of the radiation. 'Polaroid' is in fact a grid of 'wires'. 'Polaroid' film consists of long hydrocarbon chains to which iodine has been attached. The iodine provides conduction electrons which can move along the hydrocarbon chains but not perpendicular to them.

Certain crystals and certain liquids when put between 'Polaroids' rotate the plane of polarization of light. Quartz and sodium chlorate crystals are typical examples of what are termed optically active crystals. A few millimetres thickness of such crystals will rotate the plane of polarization many degrees. Quartz or sodium chlorate in solution is not optically active—no rotation of the plane of polarization occurs. The rotating or twisting ability of quartz and sodium chlorate must arise from the arrangement of the atoms within the crystal. Spirals (or helices) of copper wire placed in a beam of plane polarized microwaves will rotate the plane of polarization. The spirals should be about 1 cm long, about 0·5 cm diameter and contain about three turns of wire. Right-handed spirals rotate the plane in a right-handed (clockwise) direction, left-handed spirals in a left-handed (counterclockwise) manner. Sodium chlorate and quartz crystals can be obtained in both left-handed and right-handed rotation forms. Perhaps optically active crystals have spiral shaped conducting paths due to the arrangement of their atoms in the crystal structure.

Certain organic substances show optical rotation when they are in solution. No crystal lattice this time to give rotation but only the molecule; the rotation must come from the arrangement of the atoms within the molecule. Copper spheres arranged in an asymmetric way can give rotation of the plane of polarization of microwaves. An asymmetric arrangement of atoms in a molecule can give rotation of the plane of polarization.

The early work on this was done by Pasteur. He knew that tartaric acid obtained from grapes and other fruits was optically active. Tartaric acid in another form called racemic acid was not optically active even though otherwise it had precisely the same chemical and physical properties. Pasteur found that crystals of the tartaric acid obtained from fruit were all asymmetrical in the same way but crystals prepared from racemic acid contained two forms of asymmetrical crystal (Fig. 12.40), one form the same as that of tartaric acid from fruit. Pasteur separated the two forms of crystals and made two solutions. One of the solutions rotated the plane of polarization to the left, the other rotated it to the right. Racemic acid was normally made up of equal numbers of left-handed and right-handed molecules and thus there was no net rotation of the plane of polarization.

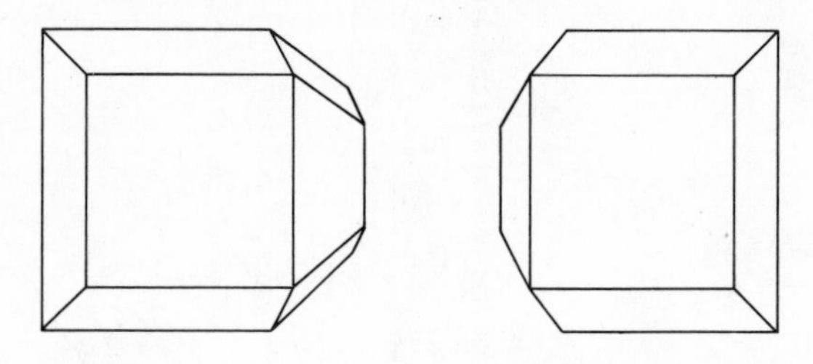

Fig. 12.40 The two forms of tartaric acid crystals.

Tartaric acid molecules can thus exist in two forms, one being a mirror image of the other. In later work Pasteur found that moulds growing in racemic acid destroyed all the tartaric molecules of one form but left the other form untouched. The living organism 'fed' on only one form of asymmetric molecule.

'The asymmetric living organism selects for its nutrient that particular form of tartaric acid which suits its needs—the form, doubtless, which in some way fits its own asymmetry—and leaves the opposite form either wholly, or for the most part, untouched....'

Pasteur.

Most organic substances found in living things are optically active, the same substances prepared in the laboratory are generally not active. Living things seem to like asymmetry. Our bodies seem to prefer the left-hand form of asymmetry. The left-hand form of compounds generally affect us more strongly than the right-handed forms. The left-hand form of nicotine occurs in cigarettes made from tobacco plants and has a much greater effect on man than the artificially produced right-hand form of nicotine.

Further reading

M. Gardner. *The ambidextrous Universe*. Penguin 1967.

In 1845 Faraday found that when he placed a piece of glass in a strong magnetic field it became optically active. The effect became known as the Faraday effect and is not restricted to glass but applies to many solids, liquids, and gases. It is also not restricted to light frequencies but applies to other electromagnetic waves. The earth's ionosphere rotates the plane of polarization of radiowaves by a considerable amount. This has been attributed to the Faraday effect, i.e., the action of a magnetic field on the ionosphere. Evans in 1956 found that with a frequency of 120 MHz the rotation was about 10 radians in daytime. The radiation coming from many stars and parts of the sun is partially plane polarized and it is considered that the explanation might lie in terms of the Faraday effect.

A piece of normally transparent material, such as 'Sellotape', placed between a pair of 'Polaroids' or other polarizers and viewed in white light appears coloured. Materials which show this property are those which show double refraction. Some materials are made doubly refracting when subject to stress. In double refraction two refracted rays at different angles of refraction are produced—there must therefore be two refractive indices. This means that the two refracted components travel at different speeds through the material (see section 6.4). In passing through the material one ray will get ahead of the other by an amount depending on the difference in refractive indices and the thickness of the material. The second 'Polaroid' brings together the components of the two refracted rays that lie in one direction, interference occurs between these two components. Thus if the 'path difference' produced by the different speeds is such as to give destructive interference for blue light then the resulting colour is white minus blue, an orange colour.

'Sellotape' and many of the transparent wrapping materials have their molecules lined up. Plane polarized light incident at an angle to these lines of molecules splits up into two vibrations, along the line of the molecules and at right angles to it. These two rays travel with different speeds. Some of the transparent materials have their molecules all tangled up and do not show this effect until the material is stressed, causing the molecules to line up.

An application of this effect is in polarizing microscopes to heighten the contrast between the different parts of what might otherwise be a colourless specimen. Another application is in stress analysis of structures. Models of the structure are made in a material which becomes doubly refracting under stress. The difference in the two speeds of light in the specimen depend on the amount of stress. Thus the colour seen at a particular point when the specimen is between 'Polaroids' is a measure of the stress at that point. This is known as photoelasticity.

When unpolarized light is reflected from a dielectric, e.g., glass, the reflected ray is completely plane polarized at one particular angle of incidence. This angle is known as the Brewster angle. At this angle the reflected and refracted rays are 90° apart. Thus the angle of incidence plus the angle of refraction equal 90° (Fig. 12.41).

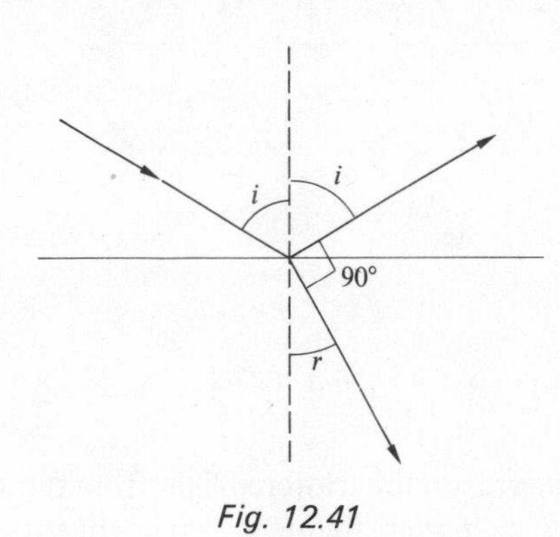

Fig. 12.41

$$\mu = \frac{\sin i}{\sin r} = \frac{\sin i}{\sin (90 - i)} = \frac{\sin i}{\cos i}$$

$$\mu = \tan i$$

The light is being reflected from a pile of atoms—the solid. How is light affected when reflected by small isolated particles? If we examine the light reflected by, say, the suspension of fine particles produced by putting a drop of milk in water or by looking at the sky we find that the reflected light is partially plane polarized. Try looking at the sky through 'Polaroid' sun glasses—try rotating the glasses. The maximum polarization is found in a direction at right angles to the incident light (Fig. 12.42(a)). If multiple scattering did not occur the light in this direction would be completely plane polarized.

Light can only consist of transverse oscillations, thus when an unpolarized wave meets a scattering centre the only planes which can be reflected in a direction at right angles to the incident light are just one plane—the other planes would have to be reflected as longitudinal oscillations and this is not possible (Fig. 12.42(b)).

The degree of scattering of light by small particles is proportional to the fourth power of the frequency of the

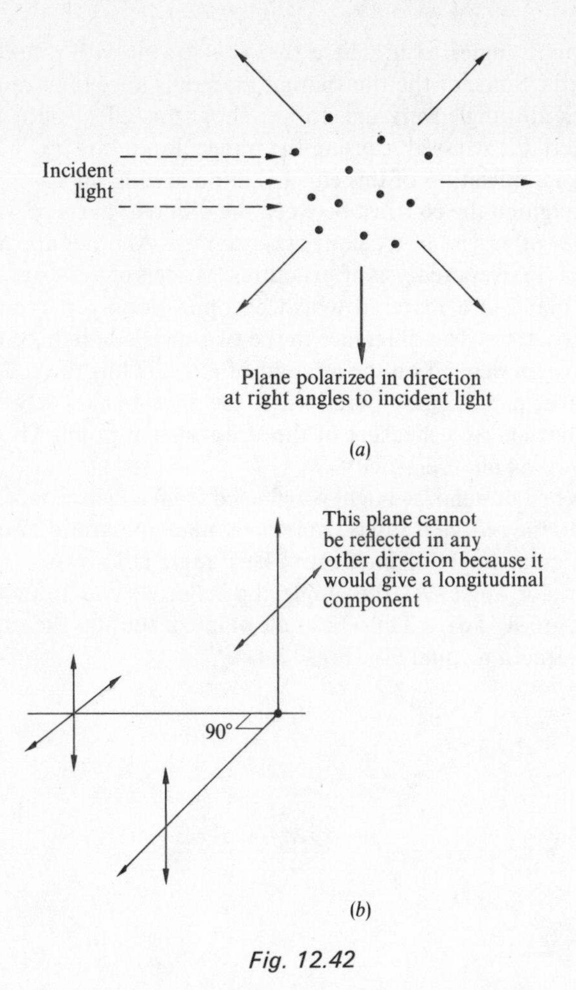

Fig. 12.42

light. For this reason the scattered light from the sky is blue, blue having a higher frequency and thus being more scattered than the red. This is true for particles whose size is less than the wavelength of the incident light. Only under this condition can the scattered light be all in phase when reflected. For bigger particles the different parts of the particle will scatter at different times such that path differences are produced which can give rise to destructive interference. Thus for very small particles whose size is less than the wavelength of blue light the predominant colour scattered is blue; for particles whose size is larger than the wavelength of blue light and less than that of red light some blue and a significant amount of red and other colours are scattered; for particles larger than the wavelength of red light the scattering becomes almost independent of wavelength and the scattered light appears white, hence clouds appear white.

> 'Progress in the observation and theoretical interpretation of polarized reflected sunlight has led to a precise knowledge of the refractive index, particle size and shape, and altitude of the top of the Venus clouds. Hansen concluded that the cloud particles are spherical liquid droplets of extremely uniform size, with mean radii very near 1 μm, with a refractive index of $1{\cdot}45 \pm 0{\cdot}02$ near $\lambda = 0{\cdot}55$ μm. These particles are found near the 50 mbar pressure level.... Theoretical calculations... suggest a temperature possibly as low as 180–210 K at the 50 mbar level.... [1 mbar = 10^2 N m^{-2}]
>
> Several suggestions have been made for the chemical composition of the topmost clouds, including two liquids with refractive indices in the range 1·4–1·5. These are carbon suboxide (C_3O_2) and aqueous HCl solutions.
>
> The refractive index of C_3O_2 has been measured at three temperatures at a wavelength of 5890 Å [the Fraunhofer D line, 1Å = 10^{-10} m]. At 273·2 K, n_D = 1·4538; at 271·9 K, n_D = 1·4596; and at 261·2 K, n_D = 1·4676. A very approximate value for the temperature coefficient of the refractive index is $-0{\cdot}00115$ deg^{-1}. Clearly, even a short extrapolation to 250 K gives a refractive index of 1·480, already outside the error limits given by Hansen (unpublished work) for the Venus clouds.
>
> Detailed data on the concentration dependence of the refractive index of HCl solutions are available for 291 K, 298 K and 303 K, and it is possible to derive approximate temperature coefficients for the refractive index. ... we find n_F (wavelength 4860 Å) for a 27·6 per cent by weight HCl solution at 200 K to be 1·428....
>
> A refractive index of $n_F \geqslant 1{\cdot}42$ corresponds to the cloud formation conditions given by Lewis.... I think the rejection of HCl solution clouds because of the discrepancy between the ~1·420 to 1·428 values which I have calculated, and the $1{\cdot}45 \pm 0{\cdot}02$ found by Hansen, is extremely unwise....'
>
> J. S. Lewis (1971). *Nature*, **230**, 295.

When you gaze up at the sky the blue may seem to be much the same whatever the time of day or the point in the sky you look at. The scattered blue light is, however, partially polarized, the angle of the polarization depending on the time of day and the point in the sky considered. Bees are able to use this partial polarization for direction finding. The bee has compound eyes, each eye consisting of some 2500 separate eyes, with each having its own lens. Each lens can excite eight vision cells, four being sensitive to light with a plane in one direction and the other four being sensitive to light with a plane at right angles to this direction. This sensitivity to a particular plane of polarization is produced by the substance responsible for converting the light into nerve signals being arranged in tubes, stacked in piles. In each pile all the tubes are stacked in the same direction. Just two sorts of pile exist, the tubes in these piles being at right angles to each other.

Travelling fields

> 'Perhaps the most dramatic moment in the development of physics during the 19th century occurred to J. C. Maxwell one day in the 1860's, when he combined the laws of electricity and magnetism with the laws of the behaviour of light. As a result, the properties of light were partly unravelled—that old and subtle stuff that is so important and mysterious that it was felt necessary to

arrange a special creation for it when writing Genesis. Maxwell could say, when he was finished with his discovery, "Let there be electricity and magnetism, and there is light!"'

R. P. Feynman, R. B. Leighton, M. Sands (1963). *The Feynman lectures*, Vol. 1. Addison-Wesley, Reading, Mass.

Electromagnetic waves travel through space at a speed of 3×10^8 m s^{-1}. How fast will electric and magnetic fields travel through space?

Consider a long pair of parallel plates (Fig. 12.43) and a battery connected by means of a switch across one end of the pair. When the switch is closed the plates will become

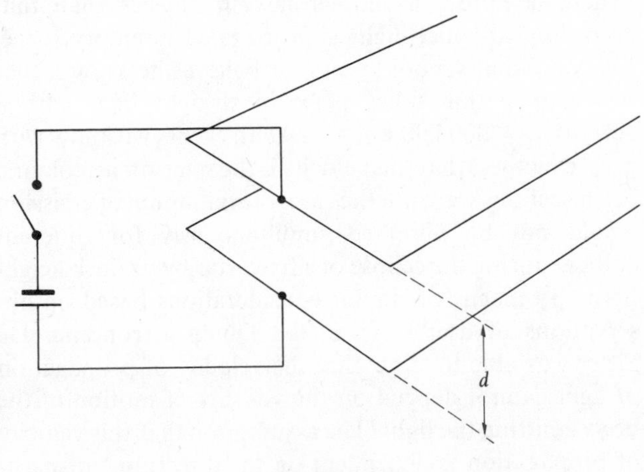

Fig. 12.43

gradually charged, starting at the ends nearest the battery. In order that the far ends of the plates shall become charged the charge must flow along the plates. Each plate must therefore carry a current during the charging process. The charge per unit area of plate, i.e., charge density σ, depends on the electric field E;

$$\sigma = \varepsilon_0 E$$

Thus as the charge moves along the plates so must the electric field. An electric field will thus travel along the space between the plates.

Suppose the charge travels along the plates with a speed v, the electric field will travel down the space between the plates with a speed v. In time Δt the charge will have travelled a distance $v\,\Delta t$ along the plates. If the width of the plates is w then in time Δt the charge will have spread over an area $wv\,\Delta t$ of a plate. The charge that has spread over this area will thus be $\sigma wv\,\Delta t$.

Current is the rate at which charge moves and so the current along the plates must be σwv. If there is a current then there must be a magnetic field. For a single turn solenoid, these plates can be thought of as that, we have $B = \mu_0 I/w$. The number of turns per metre is $1/w$. Thus we have

$$B = \mu_0 \sigma v$$

Figure 12.44 shows the directions of the electric and magnetic fields. They are at right angles to each other. Both are moving down the space between the plates as the charge flows out along the plates.

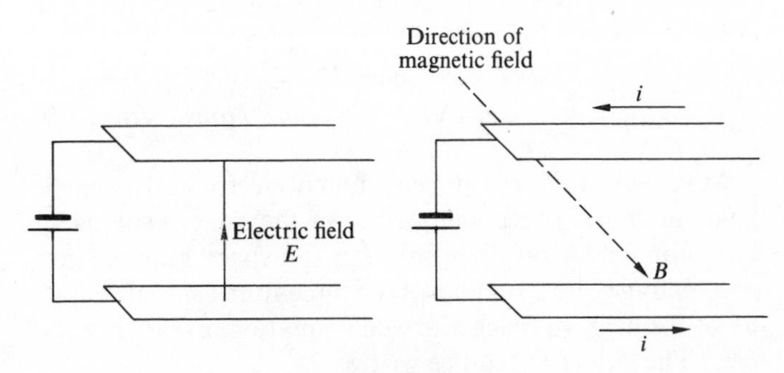

Fig. 12.44

The electric field between the plates, E, can be introduced in the equation, thus

$$B = \mu_0 \varepsilon_0 E v$$

But as the charge moves out along the plates, giving a current which extends down the plates, so the magnetic field travels along the space between the plates. In a time Δt the magnetic field occupies an extra area $dv\,\Delta t$. If the flux density is B then the extra flux produced in time Δt will be $Bdv\,\Delta t$. The rate of change of magnetic flux is thus Bdv. This rate of increase must be maintained by a voltage V between the plates, where $V = Bdv$. But a voltage V between two plates a distance d apart gives an electric field E, where $E = V/d$. Eliminating V between these two equations gives

$$E = Bv$$

The two equations $B = \mu_0 \varepsilon_0 E v$ and $E = Bv$ can only both hold at the same time if

$$v^2 = \frac{1}{\varepsilon_0 \mu_0}$$

ε_0 has the value $8{\cdot}854 \times 10^{-12}$ C^2 N^{-1} m^2 and μ_0 the value $4\pi \times 10^{-7}$ N A^{-2} in air. If we put these values in our equation we find that the speed at which the charge moves down the plates, the speed at which the electric field moves along the gap between the plates, the speed at which the magnetic field moves along between the plates, is 3×10^8 m s^{-1}. This is the speed of light.

The electric and magnetic fields are at all points at right angles to the direction of travel of the disturbance—a transverse disturbance.

The speed and the transverse nature of the disturbance both suggest that light might be an electromagnetic wave. Maxwell arrived at similar conclusions.

'... The agreement of the results [the speed of light and the fields] seem to shew that light and magnetism are affections of the same substance, and that light is an electromagnetic disturbance propagated through the field according to electromagnetic laws.... Hence electromagnetic science leads to exactly the same conclusions as optical science with respect to the direction of the disturbances which can be propagated through the field; both affirm the propagation of transverse vibrations, and both give the same velocity of propagation....'

Maxwell (1864). A dynamical theory of the electromagnetic field, part VI. *Roy. Soc. Trans.*, Vol. CLV.

Maxwell considered the behaviour of electric and magnetic fields in empty space and arrived at the same result as we have for fields travelling through the space between two conducting plates. Because of the three-dimensional character of the fields in space Maxwell's equations are more complex. The equations can be written as

$$\text{div } E = 0$$

$$\text{div } B = 0$$

$$\text{curl } E = -\frac{\partial B}{\partial t}$$

$$\text{curl } B = \varepsilon_0 \mu_0 \frac{\partial E}{\partial t}$$

The div and curl are to specify the directional characteristics of the equations. The first two equations state that there is no source of electric or magnetic flux in a small volume of space, i.e., the flux entering the volume is equal to the flux leaving the volume. The third equation describes how the e.m.f. is proportional to the rate of change of magnetic flux producing it. The fourth equation is Maxwell's postulate that a changing electric field gives rise to a magnetic field.

It is not proposed that these equations and the resulting wave equation should be derived in this book. I would ask that on the evidence of a pulse travelling between plates at the speed of light and the transverse nature of the pulse you accept as a feasible proposition the idea that light and the rest of the electromagnetic spectrum of waves can be represented as moving electric and magnetic fields.

Further reading

Appendix B of the *Nuffield advanced physics, teachers' guide for unit 8* (Penguin) gives a number of ways of tackling the above equations.

12.5 The peculiarity of the speed of light

'"I can't believe that," said Alice.

"Can't you?" the Queen said, in a pitying tone. "Try again: draw a long breath, and shut your eyes."

Alice laughed. "There's no use trying," she said: "One can't believe impossible things."

"I daresay you haven't had much practice," said the Queen. "When I was your age, I always did it for half-an-hour a day. Why, sometimes I've believed as many as six impossible things before breakfast."'

L. Carroll (1871). *Through the looking glass.*

Suppose you walk along a station platform at 3 km h^{-1}. Now you walk along at 3 km h^{-1} in a train carriage moving past the platform at 80 km h^{-1}. To someone sitting on the platform you appear to be moving in one case at 3 km h^{-1} past him and in the other case at 83 km h^{-1}, if you are walking in the direction of the train's motion, past him. He would hardly think that in both cases you were moving past him at 3 km h^{-1}. So far this seems quite believable. But suppose 'you' were a pulse of light travelling at 3×10^8 m s^{-1} emitted on some distant star. Now consider the star moving away from the earth at a speed of 6×10^7 m s^{-1} (the speed for the galaxy Hydra). If we on earth measure the speed of light from that galaxy it is still 3×10^8, not $2{\cdot}4 \times 10^8$ m s^{-1}. Is this believable?

'There is hardly a simpler law in physics than that according to which light is propagated in empty space. Every child at school knows, or believes he knows, that this propagation takes place in straight lines with a velocity c = 300 000 km s^{-1}. At all events we know with great exactness that this velocity is the same for all colours, because if this were not the case, the minimum of emission would not be observed simultaneously for different colours during the eclipse of a fixed star by its dark neighbour. By means of similar considerations based on observations of double stars, the Dutch astronomer De Sitter was able to show that the velocity of propagation of light cannot depend on the velocity of motion of the body emitting the light. The assumption that this velocity of propagation is dependent on the direction 'in space' is in itself improbable.

In short, let us assume that the simple law of the constancy of the velocity of light c (in vacuum) is justifiably believed by the child at school. Who would imagine that this simple law has plunged the conscientiously thoughtful physicist into the greatest intellectual difficulties?...'

A. Einstein (1920). *The theory of relativity*, Methuen, 3rd edition.

Double stars revolve round one another and thus the light emitted from one will be from a body moving at a different velocity relative to us than the other star. If the speed of light depended on the speed of the body emitting it then there would appear to be irregularities in the orbits. If the speed of light increased as the star revolved towards the earth and decreased as the star revolved away from us then a uniform speed orbit would appear non-uniform, the star apparently speeding up as it moves towards us and slowing down as it moves away from us.

A more critical test was carried out by Michelson and Morley and reported in the *Philosophical Magazine* of 1887. We believe the earth is moving so if this motion affected the speed of light we could expect differences in time measurements made in the direction of the earth's motion and at right angles to the direction. Michelson and Morley sent off light in two directions at right angles to each other, reflected the light from mirrors the same distance from the

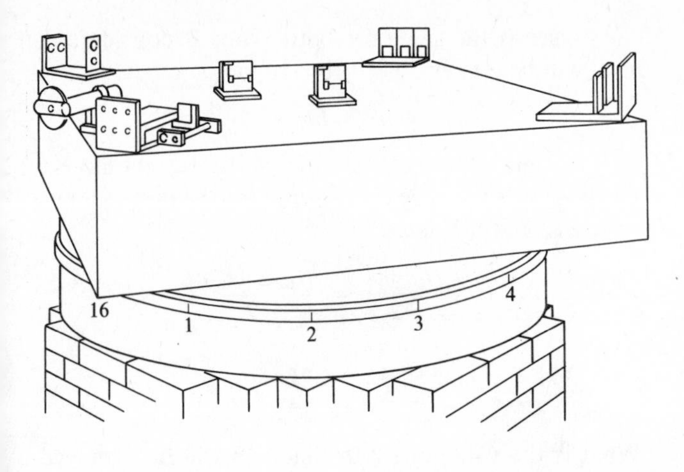

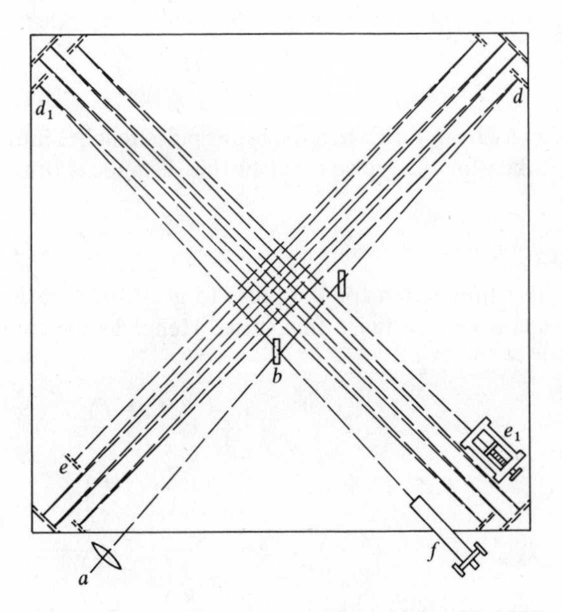

Fig. 12.45 *Michelson and Morley*. Phil. Mag., **190**, *449, 1887.*

starting point, and compared the travel times of the two beams of light when they arrived back together. Figure 12.45 shows their apparatus.

'. . . The stone is about 1·5 metre square and 0·3 metre thick. It rests on an annular wooden float, 1·5 metre outside diameter, 0·7 metre inside diameter, and 0·25 metre thick. The float rests on mercury contained in the cast-iron trough, 1·5 centimetre thick, and of such dimensions as to leave a clearance of about one centimetre around the float. . . .

At each corner of the stone were placed four mirrors *dd ee*. Near the centre of the stone was a plane parallel glass *b*. These were so disposed that light from an argand burner *a*, passing through a lens, fell on *b* so as to be in part reflected to *d*; the two pencils followed the paths indicated in the figure, *bdedbf* and $bd_1e_1d_1bf$ respectively, and were observed by the telescope *f*. Both *f* and *a* revolved with the stone. . . .'

Michelson and Morley (1887). *Phil. Mag.*, **190**, 449.

The light beam from the lamp was split into two parts by a glass block, one part being transmitted and the other reflected. One of the beams then followed one path and the other beam a path at right angles to the first beam. To obtain a large path length reflections from a number of mirrors were used. Thus one beam follows the path abde and back again from e via path edb and then a reflection to the telescope at f. The other beam follows the path abd_1e_1 and back again from e_1 via e_1d_1b and by transmission to the telescope f. As the lengths of the two paths are the same the light will take the same time to cover the two paths, if the speed of light is the same in the two directions. The two beams of light will thus give constructive interference. If the two path lengths are not precisely the same, almost inevitable, there will be a phase difference between the light in the two beams. Rotating the apparatus through 90° should, however, make no difference to this phase difference if the speed of light is the same in all directions—the interference between the two beams should not change. In fact, Michelson and Morley found no change.

The speed of light seems to be the same regardless of the speed at which the source of light is moving.

Time dilation

It is possible for us to do experiments with 'clocks' moving at nearly the speed of light and 'clocks' which are stationary. The 'clocks' are μ-mesons. They are radioactive clocks, decaying with time. Time can be determined by measuring the way the number of mesons changes. They decay in the same manner as other radioactive substances. μ-mesons are produced high in the earth's atmosphere and come shooting down towards the earth with speeds of the order of 0·99 × the speed of light. Decay of the mesons takes place during this high-speed flight. The way in which these high-speed mesons decay can be ascertained by counting the number reaching a particular height above the earth's surface and the number reaching the ground level in the same intervals of time. The following table shows the results from such a measurement.

Run	*Number of μ-mesons counted per hour* On *Mt. Washington,* 1910 m high	*At Cambridge* 3 m high
1	568	412
2	554	403
3	582	436
4	527	395
5	588	393
6	559	. . .
Av. hourly rate	563 ± 10	408 ± 9

(Frisch and Smith. *Am. J. Phys.*, **31**, 348, 1963.)

The speed of the mesons was measured as 0·9952 *c*. At this speed they took $6{\cdot}4 \times 10^{-6}$ s to cover the 1907 m. Figure 12.46 shows the decay curve for stationary mesons, they were stopped inside a block of iron. According to this curve the number of mesons should have changed from 563 to 27 in $6{\cdot}4 \times 10^{-6}$ s. The results with the high-speed mesons

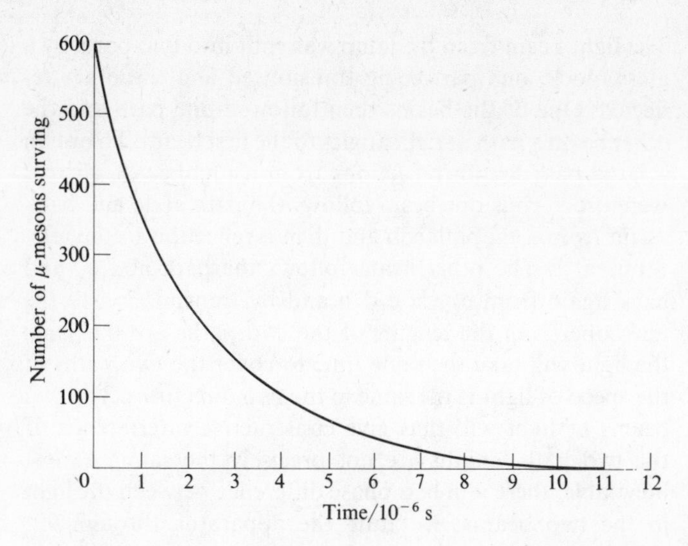

Fig. 12.46 Frisch and Smith. Am. J. Phys., **31**, *347, 1963.*

do not agree with the results for stationary mesons. For the number to change from 563 to 408 only about $0{\cdot}7 \times 10^{-6}$ s appears necessary according to our stationary 'clock'. $0{\cdot}7 \times 10^{-6}$ s of 'stationary time' seems to be equivalent to $6{\cdot}4 \times 10^{-6}$ s of 'time at $0{\cdot}9952\,c$'. The mesons lasted longer at high speed than when stationary—their clocks ran slower. This effect is called time dilation. The high speed clocks seem slower by a factor of about 9. Time for moving things does not seem to be the same as time for something else moving at a different speed.

Suppose we have two space ships A and B and at some instant when they are very close together they both start clocks. They then move apart along a straight line. A time t' after they parted, according to A's clock, A sends a radar pulse to B. Because the radar pulse will travel with a finite speed B will record the arrival of the pulse at a time later than t'. Let us say that the time will be larger than t' by a factor k.

$$t_B = kt'$$

The factor k will obviously depend on the speed of the radar pulse and the speed with which the two spacecraft are moving apart.

The radar pulse is reflected by B back to A. A can calculate what time, when time is measured on A's clock, the pulse must have left B. It should take the same time to go to B from A as to come back to A from B. The pulse left A at time t' and returned at time t''. According to A the time of arrival of the pulse at B must be $(t'' + t')/2$, i.e., the average. Thus the pulse arrival time, according to A, will be

$$t_A = \frac{t'' + t'}{2}$$

According to B a radar pulse left him at time t_B, on his clock, and would arrive some time later at A. The factor by which the time will be larger than t_B will be k, the same as before. They must both consider the other moving away with the same speed and we will assume that the speed of the radar pulses is the same for both. Thus B considers that time t'' will be greater than t_B by the factor k.

$$t'' = kt_B$$

t_A is the time A calculated that the pulse should have reached A, t_B is the time B measured for the pulse's arrival. How are t_A and t_B related?

$$t_A = \frac{t'' + t'}{2} = \frac{kt_B + (t_B/k)}{2}$$

Hence

$$t_B = t_A \frac{2k}{1 + k^2}$$

What is the value of k? In time t' B will have moved a distance vt' away from A, in time t'' B will have moved a distance vt''. According to A, B will be at a distance of

$$\frac{vt'' + vt'}{2}$$

away from him when the radar pulse reaches him. The time taken for this pulse to cover this distance is thus

$$\frac{vt'' + vt'}{2c}$$

The time taken for the pulse to go from A to B and back again will be twice this value. Hence as the time taken is $t'' - t'$:

$$t'' - t' = \frac{v}{c}(t'' + t')$$

$$t''\left(1 - \frac{v}{c}\right) = t'\left(1 + \frac{v}{c}\right)$$

$$kt_B\left(1 - \frac{v}{c}\right) = \frac{t_B}{k}\left(1 + \frac{v}{c}\right)$$

Thus

$$k^2 = \frac{(1 + v/c)}{(1 - v/c)}$$

$$k = \sqrt{\frac{1 + v/c}{1 - v/c}}$$

Multiplying the top and bottom lines by $\sqrt{(1 + v/c)}$

$$k = \frac{1 + v/c}{\sqrt{(1 - v^2/c^2)}}$$

Hence as

$$t_B = t_A \frac{2k}{1 + k^2}$$

$$t_B = t_A \frac{2(1 + v/c)}{\sqrt{(1 - v^2/c^2)}} \cdot \frac{1}{1 + [(1 + v/c)/(1 - v/c)]}$$

$$t_B = t_A \frac{2(1 + v/c)}{\sqrt{(1 - v^2/c^2)}} \cdot \frac{(1 - v/c)}{2}$$

$$t_B = t_A \frac{(1 - v^2/c^2)}{\sqrt{(1 - v^2/c^2)}}$$

$$t_B = t_A\sqrt{(1 - v^2/c^2)}$$

This is the time dilation equation. t_B will always be less than t_A because v^2/c^2 is always positive. t_B is the time recorded by B on his clock, t_A is the 'same' time calculated by A from measurements made with his clock.

For the meson experiment t_B is the time measured by the observers, viz., $0{\cdot}7 \times 10^{-6}$ s. t_A, the calculated time, is $6{\cdot}4 \times 10^{-6}$ s. For the mesons the time dilation factor $\sqrt{(1 - v^2/c^2)}$ is $\sqrt{(1 - 0{\cdot}9952^2)}$, which is 0·0977. The experiment indicated a factor of 0·7/6·4 or 0·109. Within the accuracy of the experiment this is reasonable agreement.

Doppler effect

Suppose A sent out not just one pulse but a series of pulses. The first pulse will leave A a time t' after the spacecraft separate. For B the arrival time will be t_B where

$$t_B = kt'$$

The second pulse leaves A after time $2t'$, i.e., the interval between the pulses is t'. The arrival time for B will be $2kt'$ and thus the time interval between the two signals will be, for B, kt'. If A sends out a train of regularly spaced pulses at a frequency f then the time interval between them will be $1/f$. The time intervals recorded by B will be k/f.

$$\text{Frequency observed by B} = f/k$$

This is for the two spacecraft moving farther apart. For the spacecraft coming together the sign of v becomes reversed. This leads to the frequency observed by B being kf. This apparent change in frequency is known as the Doppler effect.

The frequency observed by an observer viewing a moving source increases as the source moves towards the observer and decreases as the source moves away from the observer.

For the source moving away from the observer the apparent frequency is given by

$$\text{Apparent frequency} = f\,\frac{\sqrt{(1 - v^2/c^2)}}{1 + v/c}$$

For $v \ll c$ the v^2/c^2 term becomes negligible and the equation becomes

$$\text{Apparent frequency} = \frac{f}{1 + v/c}$$

In terms of wavelengths, $c = f\lambda$, we have

$$\text{Apparent wavelength} = (1 + v/c)\lambda$$

The apparent change in wavelength is thus $v\lambda/c$.

The spectral lines of galaxies are found to be all shifted towards the red end of the spectrum—the conclusion, they are moving away from us. Figure 12.47 shows the spectra produced by a number of galaxies. The spectra appear as absorption lines, the central region of each photograph. The outer spectra on each photograph are comparison lines produced by a mixture of helium and hydrogen. In 1919 Hubble suggested a law, now known by his name, that the distance of a galaxy from us is directly proportional to its velocity. Using this law, and the clues given by the relative brightnesses of galaxies, the following table results for the galaxies shown in the photographs.

Galaxy in	*Velocity* $/10^4$ km s^{-1}	*Distance /light-years*
Virgo	0·12	$0{\cdot}4 \times 10^8$
Ursa Major	1·40	$5{\cdot}0 \times 10^8$
Corona Borealis	2·14	$7{\cdot}0 \times 10^8$
Boötes	3·90	13×10^8
Hydra	6·10	20×10^8

Quasars, intense sources of radio emission, have been in many instances identified as originating in visible galaxies. These galaxies are found to have very high 'red shifts' of their wavelengths (Fig. 12.48). If Hubble's law applies then these galaxies must be a very long way off. But their radio emission is very strong—stronger than many near sources. The large amounts of energy radiated could be explained if they were close but if they are very far away they must radiate very very large amounts of energy for such high strength signals to be received on earth. Quasars present problems—as yet without answers.

'Quasi-stellar Objects—A Progress Report

By G. R. Burbidge and E. M. Burbidge
Department of Physics, University of California, San Diego, and Institute of Theoretical Astronomy, University of Cambridge

After nine years, quasars are still puzzling astronomers. This review brings the story up to date.

Since 1960 both optical and radio astronomers have devoted a considerable amount of effort to investigating the properties of the QSOs. The observational milestones along the way have been: (1) The original identification of 3*C* 48 in 1960. (2) The first red-shift determination of 3*C* 273 in 1963. (3) The determination in 1965 of a red-shift as large as $\Delta\lambda/\lambda = 2$; this implied enormous distances according to the obvious interpretation that the red-shifts are due to the expansion of the universe. Many red-shifts as large as this are now known. (4) The demonstration that the objects vary in both radio and optical fluxes made in the period 1960–65. (5) The demonstration, in 1965, that there was probably a very large population of QSOs, both radio and radio-quiet objects; further studies have confirmed this. (6) The identification of the first absorption lines in 1966. (7) The identification of absorption lines displaying multiple red-shifts in 1968. (8) The measurement in the past few years of exceedingly small diameters for radio sources associated with both QSOs and radio galaxies.

The theoretical developments have come as follows. Parallel with the observational discoveries in 1963 came the proposal by Hoyle and Fowler that the very large amounts of energy which are released to give rise to strong radio sources must come from the collapse of massive stars. While this idea was put forward to explain the

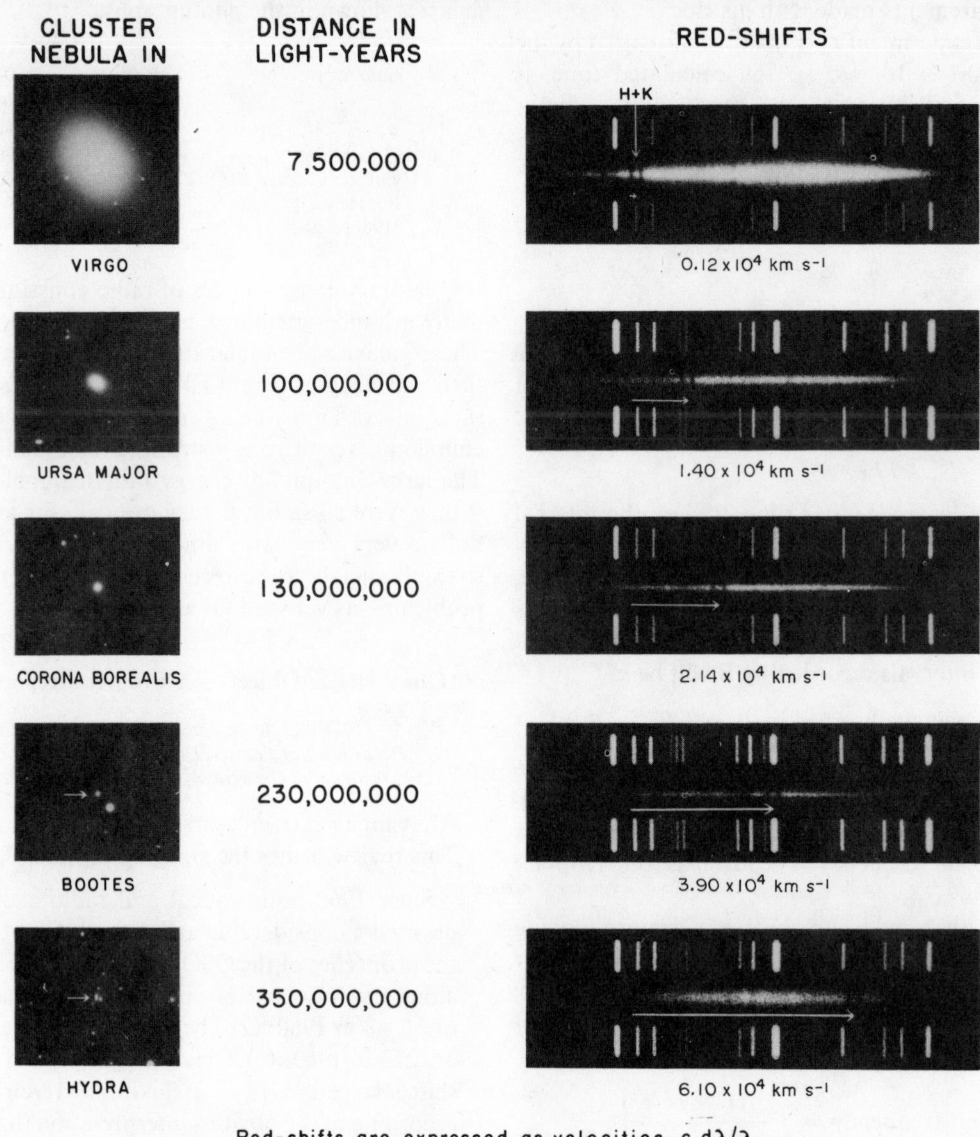

Fig. 12.47 Courtesy of the Hale Observatories. The distances given here are different from those in the table on page 389. The distance scales are constantly being reviewed, the table gives the most recent estimates.

properties of the strong radio galaxies, the proposal did, in a sense, predict the generation of star-like radio sources which are closely related to the radio galaxies. This proposal caused tremendous interest in gravitational collapse, and relativists again began to work on this topic. . . .

Variability

It was the variable nature of the QSOs which first led some to question the nature of the red-shifts. Earlier, Terrell had proposed that QSOs were objects being thrown out from the central part of our Galaxy at relativistic speed, but few found this idea attractive, and to the physicist the principal problem was, and still is, how coherent objects could be accelerated to relativistic speeds. The reasons why variations in radio or optical flux caused some doubts about the cosmological nature of the red-shifts were, in part, simply intuitive—the idea being that very distant objects a hundred times more luminous than galaxies could not be so small that they would vary significantly in times of the order of years or

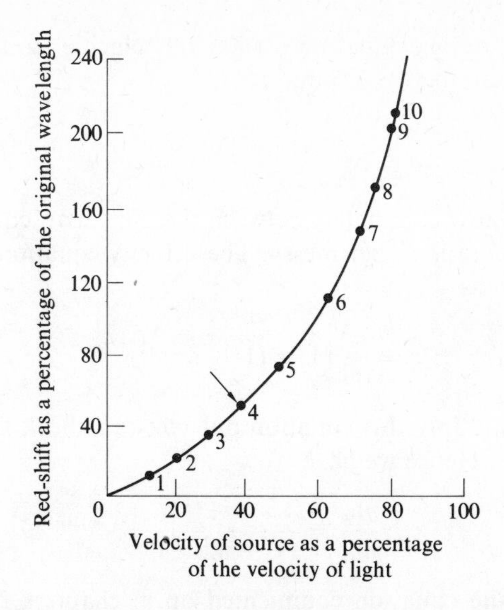

Fig. 12.48 A plot of the red-shift as a percentage of the original wavelength against the velocity of the source as a percentage of light. (After J. L. Greenstein). The top of the curve is uncertain because as yet the large scale geometry of the Universe is unknown. The arrow marks the point of the most distant radiogalaxy and above this all points are for quasars. (1) 3C 273: (2) most distant (measured) ordinary galaxy: (3) 3C 48: (4) Most distant radiogalaxy: (5) 3C 254: (6) 3C 287: (7) 3C 208: (8) 3C 454: (9) 3C 9: (10) 1116 + 12. (F. J. Smith, Nature, *11 March 1967, 967, Fig. 3.)*

months (or days as we now know). These arguments translated into physical terms by Hoyle, Burbidge and Sargent led them to point out that there seemed to be severe difficulties in making self-consistent models if the QSOs were at cosmological distances. These were theoretical arguments, and several attempts were made to rebut them or to make models which would overcome the difficulties. The various ideas have different degrees of plausibility to different people. Behind this discussion is the really fundamental question. If the QSOs are not at cosmological distances, their red-shifts cannot simply be due to the expansion of the universe; there must be large red-shift components which are directly associated with the objects. Now the only alternative explanations are recession at high speeds in a local region, or the presence of strong gravitational fields. A third alternative, that the masses or charges of nuclei or electrons are different in these objects, faces many difficulties, and has not been taken seriously up to the present.

(*Nature*, **224**, 21, 1969.)

Adding velocities

The speed of light is not affected by the speed with which its source is moving or by the speed at which an observer of the light is moving. Consider speeds less than the speed of light—how should we add these?

Consider three spacecraft, A, B, and C. A is moving away from B with a velocity v, B is moving away from C with a velocity v'. How fast is A moving away from C? A emits regularly spaced radar pulses. These are received by B and then immediately relayed on to C. According to B the time between the pulses is

$$t_B = k't'$$

t' is the time between the pulses when they leave A.

$$k' = \frac{1 + v/c}{\sqrt{(1 - v^2/c^2)}}$$

C observes the pulse coming from B and sees the time interval between the pulses as

$$t_C = k''t_B$$

$$k'' = \frac{1 + v'/c}{\sqrt{(1 - v'^2/c^2)}}$$

Hence

$$t_C = k''k't'$$

$$t_C = \left(\frac{1 + v'/c}{\sqrt{(1 - v'^2/c^2)}}\right)\left(\frac{1 + v/c}{\sqrt{(1 - v^2/c^2)}}\right)t'$$

But if C had directly viewed A we would have

$$t_C = kt'$$

where

$$k = \frac{1 + V/c}{\sqrt{(1 - V^2/c^2)}}$$

V is the velocity of A as seen by C.

For both these expressions to agree we must have

$$\frac{1 + V/c}{\sqrt{(1 - V^2/c^2)}} = \left(\frac{1 + v'/c}{\sqrt{(1 - v'^2/c^2)}}\right)\left(\frac{1 + v/c}{\sqrt{(1 - v^2/c^2)}}\right)$$

From which

$$V = \frac{v + v'}{(1 + vv'/c^2)}$$

The value of V cannot exceed c, even if either or both v and v' are equal or close to c.

Mass

Time dilation has a very important consequence in dynamics. Suppose we fire a rocket to the moon and so arrange matters that it crashes into the moon. Because our time measurements will differ from those of some other observer, say one in a freely coasting spaceprobe somewhere in space, our estimate of the velocity with which the rocket hits the moon will differ from that of the spaceprobe observer. Both observers must, however, agree on the effects, the damage, produced on the moon by the impact. For the same effects to be produced by different velocity rockets would suggest that the two observers see the rocket as having different masses—the mass of the rocket appearing as some function of the velocity.

If we assume that the conservation of momentum applies regardless of the velocity of the observer then we can have the situation where one observer sees two identical objects with opposite and equal momenta colliding (Fig. 12.49(a)) and another observer sees the situation as one object hitting another object which is at rest (Fig. 12.49(b)). Suppose A

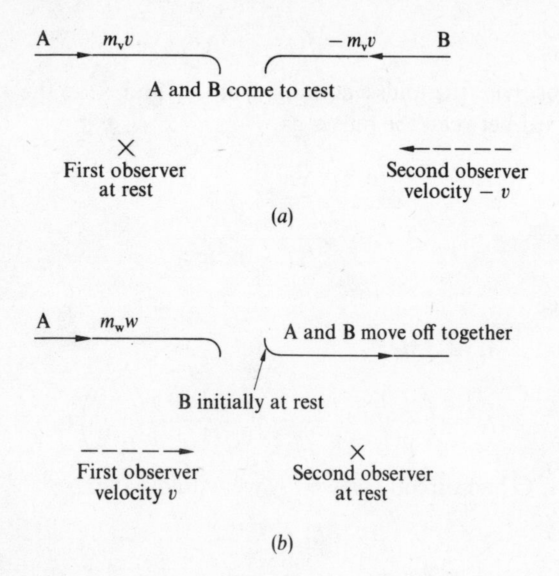

Fig. 12.49 (a) First observer's view. (b) Second observer's view.

and B stick together when they collide, i.e., the collision is completely inelastic. After the collision the first observer will conclude that both A and B have come to rest while the second observer will see them both moving off together.

For the first observer:

Before the collision momentum $= m_V v - m_V v$
After the collision momentum $= 0$

For the second observer:

Before the collision momentum $= m_W w$
After the collision momentum $= Mv$

m_V is the mass at velocity v, m_W is the mass at velocity w, and M is the combined mass after the collision.

What is the combined mass? If we assume that the combined mass is the sum of the masses of the two colliding objects then

$$M = m_W + m_O$$

m_O is the mass of B before the collision, for the second observer. For both observers to see the same effects the velocity of the combined mass must, for the second observer, be v.

Applying the conservation of momentum gives

$$m_W w = Mv$$

Eliminating M

$$m_W w = (m_W + m_O)v$$

$$\frac{m_W}{m_O} = \frac{v}{w - v}$$

We can use our equation for adding velocities to obtain a relationship between w and v.

$$w = \frac{v + v}{(1 + v^2/c^2)}$$

We can now eliminate v between these last two equations. The algebra is rather messy. The velocity equation can be rewritten as

$$v = \frac{c^2}{w}[1 - (1 - w^2/c^2)^{1/2}]$$

If you multiply this equation out you can check that it is the same. Hence we have

$$\frac{m_W}{m_O} = \frac{1}{(1 - w^2/c^2)^{1/2}}$$

This is the equation commented on in chapters 2 and 3. Mass depends on velocity. In this analysis we have considered mass as the coefficient relating momentum and velocity, the coefficient being some function of the velocity.

Appendix 12A Bow waves

A speed boat racing through water generates a bow wave—a V-shaped wave front spreading out behind the boat. This occurs because the speed of the boat is greater than the speed of the water waves. A stationary disturbance would produce circular wavefronts. If we had a row of stationary sources of disturbances and excited them in sequence then we would produce something like a bow wave (Fig. 12.50(a)). As can be seen from Fig. 12.50(b), the bow wave angle θ is given by

$$\frac{wt}{vt} = \sin\theta$$

An object travelling in a medium faster than the speed of light in that medium produces a cone of light, like the bow wave in water. This light is called Cerenkov radiation. An aircraft travelling faster than the speed of sound produces a 'bow' wave, the shock wave which when it passes gives the sonic bang.

Speeds greater than the speed of light in a medium are possible, speeds greater than the speed of light in a vacuum are not possible.

Problems

1. Explain the following extracts from the quotation given at the beginning of this chapter from the work of Hertz.

(a) 'From the mode in which our ray was produced we can have no doubt whatever that it consists of transverse vibrations....'

(b) 'Insulators do not stop the ray.'

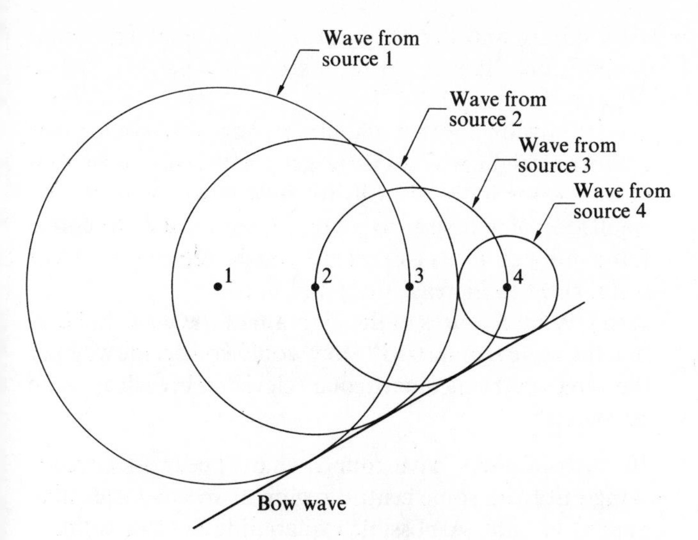

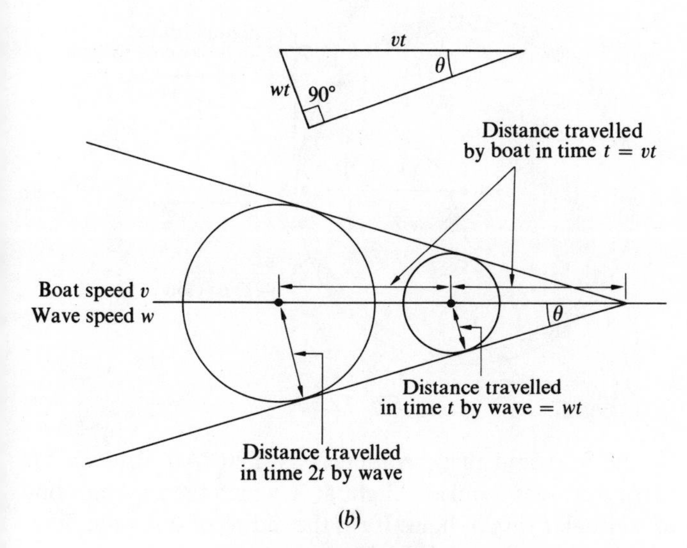

Fig. 12.50

(c) 'If the two mirrors were now set up with their focal lines parallel, and the wire screen was interposed perpendicularly to the ray and so that the direction of the wires was perpendicular to the direction of the focal lines, the screen practically did not interfere at all with the secondary sparks.'

2. The following information refers to a paper by J. H. Trexler in *Proc. IRE*, **46**, 286, 1958: In October 1951 microwave pulses were sent to the moon and the echo detected. The transmitted pulse width was 12×10^{-6} s and the echo pulse width was about 350×10^{-6} s. About 75 per cent of the echo pulse energy was in the first 100×10^{-6} s of the echo.

(a) Why should some of the echo pulse arrive back earlier than other parts?

(b) How great an area of the moon would you estimate that the pulse was reflected from?

(Hint: remember the moon is a sphere of radius 1738 km.)

3. Estimate the total energy that would be received by a radiometer of area 200 m^2, on earth due to the planet Venus being at a temperature of 600 K.

Take the distance of Venus from the Earth to be 5×10^{10} m.

Radius of Venus $= 6 \times 10^6$ m.

Stefan's constant $\sigma = 5{\cdot}67 \times 10^{-8}$ J s^{-1} m^{-2} K^{-4}.

4. Figure 12.51 is a sketch of the spectrum lines produced by the planet Saturn. The outer parts of the spectrum are due to the rings of Saturn, the inner part due to the core of the planet. The planet is viewed in the 'edgewise' position.

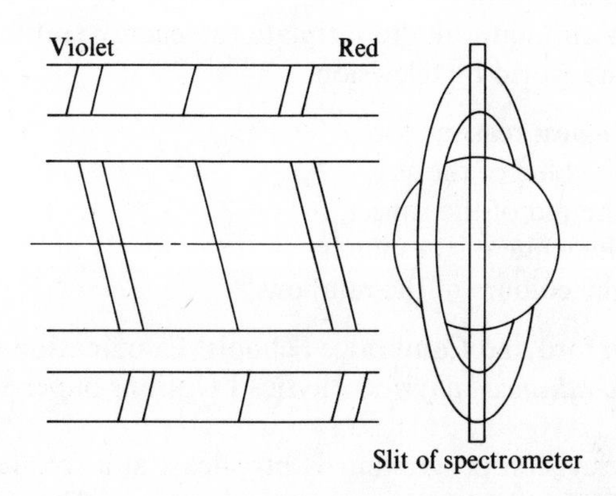

Fig. 12.51 R. H. Baker. Astronomy, *8th Ed., Van Nostrand Reinhold Co. Copyright 1964 by Litton Educational Publishing.*

(a) Why are the spectrum lines slanting?

(b) Which parts of the planet are approaching us and which receding?

5. The following extract is taken from a paper by W. L. Bragg in *Proc. Camb. Phil. Soc.*, **17**, 43, 1912.

'A very narrow pencil of rays from an X-ray bulb is isolated by a series of lead screens pierced with fine holes. In the path of this beam is set a small slip of crystal; and a photographic plate is placed a few centimetres behind the crystal at right angles to the beam. When the plate is developed, there appears on it, as well as the intense spot caused by the undeviated X-rays, a series of fainter spots forming an intricate geometrical pattern.'

(a) W. L. Bragg went on to offer an explanation for these fainter spots. What is the explanation?

(b) The equation $2d \sin \theta = n\lambda$ was developed by Bragg to account for the angles at which interference maxima are produced. Explain how the equation originates.

(c) What information about the crystal is given by the X-ray diffraction picture?

6. A radar signal reflected from a moving object will be received back at the transmitting aerial with a different frequency due to the Doppler effect.

(a) How does the change in frequency depend on the velocity of the moving object?

(b) Such a system is used by police to determine whether a car is exceeding a speed limit. Roughly how big do you think the frequency changes will be for prosecutions to be levelled for the speed limit being exceeded?

7. It has been said that 'people' on some distant planet would be able to detect the emergence of world-wide television on earth by the large amount of energy radiated in the VHF region.

(a) If we were to receive some signal back from the distant planet before the end of the century, in response to our sudden change in VHF radiation, how near would they have to be to our earth?

(b) You might like to estimate the energy emitted over the entire world by television.

8. Suggest reasons for
(a) the blue of the sky,
(b) the red of the sunset,
(c) the white of the clouds,
(d) the colours of the rainbow.

9. Oxford and Cambridge Schools' Examination Board. *Nuffield advanced physics*. Physics I Options paper 1970.

'A television programme is broadcast at a frequency of 600 MHz with the electric vector vertical. The diagram [Fig. 12.52] is meant to illustrate the waves at a considerable distance from the transmitter.

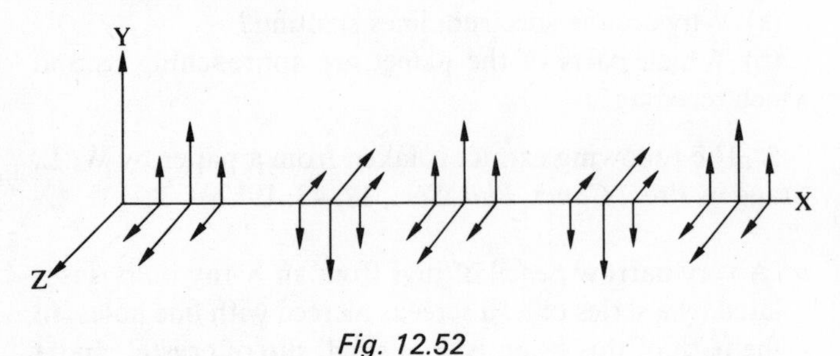

Fig. 12.52

(*a*) Make a rough copy of the diagram, add suitable labelling, and explain how it is meant to represent the waves.

(*b*) Draw another diagram to show what this pulse would look like 5×10^{-10} seconds later. ($c = 3 \times 10^8\ \mathrm{m\ s^{-1}}$)

(*c*) Show on another diagram (i) how you would place a short straight wire so as to get the maximum electromotive force induced in it, (ii) how you would place a small loop of wire so as to get the maximum electromotive force induced in it, (iii) where you would place a large metal sheet to increase the e.m.f. in (i).

(d) What features of the diagram of the wave indicate that the wave is polarized? How would you decide whether the waves carrying a particular television broadcast were polarized?'

10. Astronomers have found what appear to be stars moving out from some central region at speeds faster than the speed of light. A possible explanation for this is that a star surrounded by a spherical shell of dust has exploded (Fig. 12.53). The radiation emitted by the star is absorbed

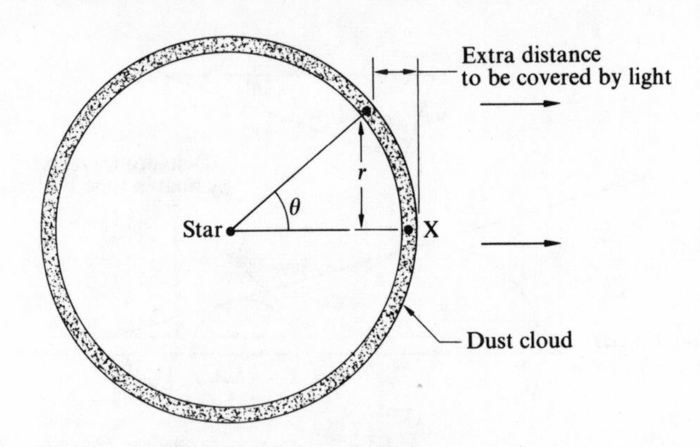

Fig. 12.53

by the dust and then reradiated as light. An observer on earth thus sees a pulse of light at X which then spreads out as a circular ring of light. If r is the radius of this ring, show that the rate of growth of the ring

$$\frac{dr}{dt} = \frac{c}{\tan\theta}$$

What is the condition for the apparent speed to exceed the speed of light?

Teaching note Further problems appropriate to this chapter will be found in:
Nuffield advanced physics. Units 4 and 8, students' books, Penguin, 1972.
Physical Science Study Committee. *Physics*, 2nd ed., chapter 31, Heath, 1965.

Practical problems

1. It is thought that the growth of some organisms near the surface of water is prevented by the ultraviolet radiation from the sun. The same organisms are generally found at greater depths and it is assumed that the ultraviolet radiation has therefore been absorbed in the upper layers of the water. Determine how the transmission of ultraviolet radiation through water varies with the depth of the water.

2. Determine the variation of energy with wavelength for a flash bulb and hence obtain a value for the colour temperature. It is assumed that the flash bulb behaves as a black body and Wien's law holds—

$$\lambda_{max} T = \text{a constant } (0{\cdot}294 \text{ cm K})$$

λ_{max} is the wavelength of maximum energy. T is the absolute temperature and is known as the colour temperature.

3. Investigate the design of television aerials. You might find it useful to scale them down so that you can use microwave apparatus for your investigation.

Suggestions for answers

1. (a) The waves are restricted to one plane by being emitted from linear aerials.

(b) Absorption is due to energy being used to move charges and as no free charges occur in perfect insulators no absorption occurs.

(c) For the screen to absorb, the wires must be along the direction of the electric field.

2. (a) Because the moon is a sphere the outer parts of the radar pulse will be reflected from a different part of the moon to the inner part of the pulse (Fig. 12.54). A time

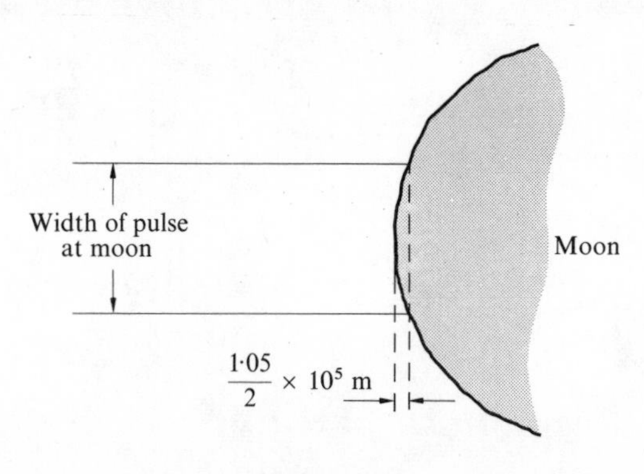

Fig. 12.54

delay of 350×10^{-6} s means a difference in path of $1{\cdot}05 \times 10^5$ m. Using the theorem of intersecting chords (appendix 2A) we can find the radius of the patch on the moon and hence an area of about 6×10^{11} m^2.

3. The radiation emitted per unit area of Venus can be found by $E = \sigma T^4$ and then by considering this spreading out over a sphere the amount received per unit area can be calculated.

$$\text{Total energy} = 5{\cdot}67 \times 10^{-8} \times 600^4 \times \pi \times (6 \times 10^6)^2$$

Energy received on a 200 m^2 surface on earth

$$= \frac{5{\cdot}67 \times 10^{-8} \times 600^4 \times \pi \times (6 \times 10^6)^2 \times 200}{\pi \times (5 \times 10^{10})^2}$$

$$= 0{\cdot}021 \text{ J s}^{-1}$$

4. (a) The different parts of the rings and the core are moving at different speeds towards or away from the earth. Thus because of the Doppler effect the spectrum lines emitted from the different zones occur at different wavelengths.

(b) See Fig. 12.55.

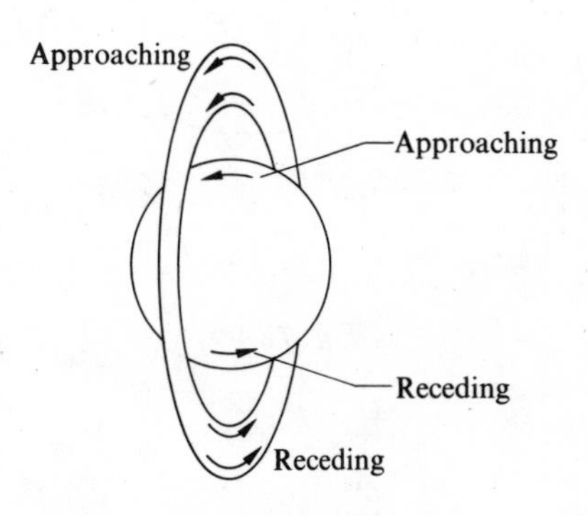

Fig. 12.55

5. (a) Interference occurring between the reflections from the different atomic planes within the crystal.

(b) $2d \sin \theta$ is the path difference between the rays reflected from successive layers in the crystal. When this is equal to a whole number of wavelengths, $n\lambda$, constructive interference occurs.

(c) The spacing between atomic planes.

6. (a) $\Delta f = \dfrac{vf}{c}$

(b) Frequency of microwaves 10^{10} Hz. For a velocity change from 50 km h^{-1} to 55 km h^{-1} the frequency change Δf will be from about 50 to 55 Hz.

7. (a) If we put the date of the emergence of world-wide television as 1950 then there is 50 years for the signal to travel to a distant planet and back. They would have to be within 75×10^8 m, 25 light-years. The nearest star is 4·3 light-years away.

(b) Suppose the average TV station beams out 100 kW and there are 100 stations in Britain. In the world there might be 2000 such stations. The power would be 2×10^8 W. This is pure guesswork on my part—your guesses might be better.

8. (a) See earlier in this chapter and chapter 8. Scattering from non-uniformly distributed air molecules.

(b) At sunset we are seeing the sun through the maximum thickness of the atmosphere, after much of the blue has been scattered out of the light.

(c) The scattering particles, water drops and ice crystals, are bigger than the wavelengths of light.

(d) Light from the sun refracted through raindrops. Figure 12.56 shows a typical path for light in a drop. The path depends on the refractive index which in turn depends on the colour of the light.

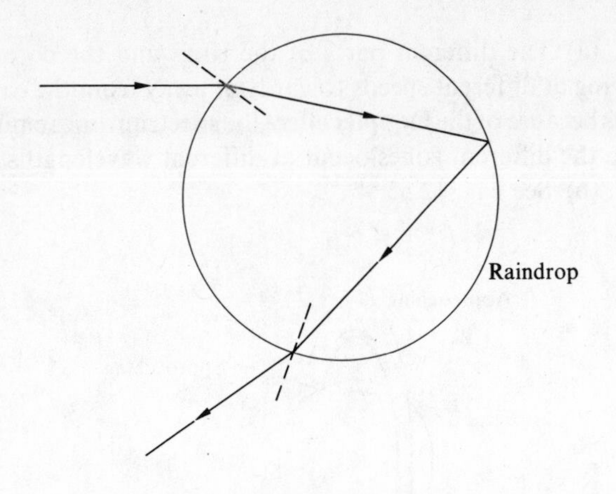

Fig. 12.56

9. (a) The electric vector is vertical, the magnetic vector horizontal.

(b) The pulse would have the vectors appropriate to $\frac{1}{3}\lambda$ further on.

(c) (i) Along the direction of the electric vector.

(ii) With its plane in the plane of the electric vectors.

(iii) At right angles to the plane of the electric vectors and $\lambda/4$ behind the aerial. There is a phase change of $\lambda/2$ at the reflector.

(d) The electric vector is only in one plane. Rotation of the receiving aerial can be used to determine whether the wave is plane polarized.

10. θ greater than 45°.

13 Atoms

Teaching note Practical work appropriate to this chapter will be found in:
W. Bolton. *Physics experiments and projects*, Vol. 3, Pergamon, 1968.
Nuffield advanced physics. Units 5 and 10, teachers' guides, Penguin, 1971.
Project physics course. Handbook, Holt, Reinhart and Winston, 1971.

13.1 The atom subdivided

Towards the end of the nineteenth century there was considerable interest in the various phenomena produced by the passage of electricity through air under low pressure. At the different pressures the air was found to exhibit different patterns of light emission. At low pressures of the order of 0·01 mm of mercury only feebly luminescent bundles of rays are visible. The cathode, the negative electrode, seems to be emitting rays which cause a vivid green fluorescence where they strike the glass container. These rays were called cathode rays.

The following is an account by J. J. Thomson of experiments designed to ascertain the nature of these cathode rays.

'If these rays are negatively electrified particles, then when they enter an enclosure they ought to carry into it a charge of negative electricity. This has been proved to be the case by Perrin, who placed in front of a plane cathode two coaxial metallic cylinders which were insulated from each other: the outer of these cylinders was connected to earth, the inner with a gold leaf electroscope. These cylinders were closed except for two small holes, one in each cylinder, placed so that the cathode rays could pass through them into the inside of the inner cylinder. Perrin found that when the rays passed into the inner cylinder the electroscope received a charge of negative electricity, while no charge went to the electroscope when the rays were deflected by a magnet so as no longer to pass through the hole. . . .

The rays from the cathode C pass through a slit in the anode A, which is a metal plug fitting tightly into the tube and connected with the earth [Fig. 13.1]; after passing through a second slit in another earth-connected metal plug B, they travel between two parallel aluminium plates about 5 cm. long by 2 broad and at a distance of 1·5 cm apart; they then fall on the end of the tube and produce a narrow well-defined phosphorescent patch. A scale pasted on the outside of the tube serves to measure the deflexion of this patch. At high exhaustions the rays were deflected when the two aluminium plates were connected with the terminals of a battery of small storage cells; the rays were depressed when the upper plate was connected with the negative pole of the battery, the lower with the positive, and raised when the upper plate was connected with the positive, the lower with the negative pole. . . .

As the cathode rays carry a charge of negative electricity, are deflected by an electrostatic force as if they were negatively electrified, and are acted on by a magnetic force in just the way in which this force would act on a negatively electrified body moving along the path of these rays, I can see no escape from the conclusion that they are charges of negative electricity carried by particles of matter. The question next arises, What are these particles? are they atoms, or molecules, or matter in a

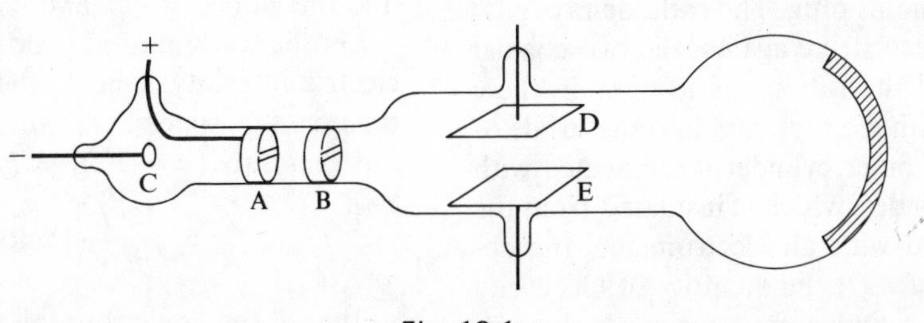

Fig. 13.1

still finer state of subdivision? To throw some light on this point, I have made a series of measurements of the ratio of the mass of these particles to the charge carried by it. To determine this quantity, I have used two independent methods. The first of these is as follows:- Suppose we consider a bundle of homogeneous cathode rays. Let m be the mass of each of the particles, e the charge carried by it. Let N be the number of particles passing across any section of the beam in a given time; then Q the quantity of electricity carried by these particles is given by the equation

$$Ne = Q$$

We can measure Q if we receive the cathode rays in the inside of a vessel connected with an electrometer. When these rays strike against a solid body, the temperature of the body is raised; the kinetic energy of the moving particles being converted into heat; if we suppose that all this energy is converted into heat, then if we measure the increase in the temperature of a body of known thermal capacity caused by the impact of these rays, we can determine W, the kinetic energy of the particles, and if v is the velocity of the particles,

$$\tfrac{1}{2}Nmv^2 = W$$

If [R] is the radius of curvature of the path of these rays in a uniform magnetic field [B], then

$$\left[\frac{mv^2}{R} = Bev, \text{ with modern notation}\right]$$

From these equations we get

$$\frac{1}{2}\frac{m}{e}v^2 = \frac{W}{Q}.$$

$$\left[v = \frac{2W}{QBR}\right]$$

$$\left[\frac{m}{e} = \frac{QB^2R^2}{2W}\right]$$

Thus, if we know the values of Q, W, and BR we can deduce the values of v and m/e.

To measure these quantities, I have used tubes of three different types. The first I tried is like that represented in Fig. [13.1], except that the plates E and D are absent, and two coaxial cylinders are fastened to the end of the tube. The rays from the cathode C fall on the metal plug B, which is connected to earth, and serves for the anode; a horizontal slit is cut in this plug. The cathode rays pass through this slit, and then strike against the two coaxial cylinders at the end of the tube; slits are cut in these cylinders, so that the cathode rays pass into the inside of the inner cylinder. The outer cylinder is connected with the earth, the inner cylinder, which is insulated from the outer one, is connected with an electrometer, the deflexion of which measures Q, the quantity of electricity brought into the inner cylinder by the rays. A thermo-electric couple is placed behind the slit in the inner cylinder; this couple is made of very thin strips of iron and copper fastened to very fine iron and copper wires.... The strips of iron and copper were large enough to ensure that every cathode ray which entered the inner cylinder struck against the junction.... The strips of iron and copper were weighed, and the thermal capacity of the junction calculated....

The value of BR, where R is the curvature of the path of the rays in a magnetic field of strength B was found as follows:- The tube was fixed between two large circular coils placed parallel to each other, and separated by a distance equal to the radius of either; these coils produce a uniform magnetic field, the strength of which is got by measuring with an ammeter the strength of the current passing through them. The cathode rays are thus in a uniform field, so that their path is circular. Suppose that the rays, when deflected by a magnet, strike against the glass of the tube at E [Fig. 13.2], then, if R is the radius of the circular path of the rays,

$$2R = \frac{CE^2}{AC} + AC;$$

thus, if we measure CE and AC we have the means of determining the radius of curvature of the path of the rays....

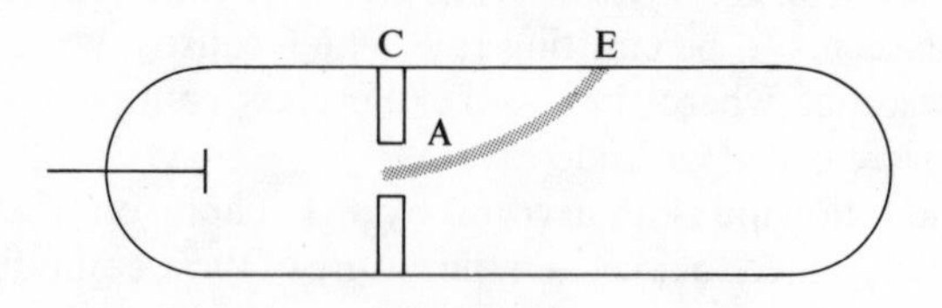

Fig. 13.2

The results of a series of measurements with these tubes are given in the following table:-

Gas	*Value of W/Q*	*BR*	*m/e*	*v*
Air	$4{\cdot}6 \times 10^{11}$	230	$0{\cdot}57 \times 10^{-7}$	4×10^{9}
Air	$1{\cdot}8 \times 10^{12}$	350	$0{\cdot}34 \times 10^{-7}$	1×10^{10}
Air	$6{\cdot}1 \times 10^{11}$	230	$0{\cdot}43 \times 10^{-7}$	$5{\cdot}4 \times 10^{9}$
...				

... If we measure the deflexion experienced by the rays when traversing a given length under a uniform electric intensity, and the deflexion of the rays when they traverse a given distance under a uniform magnetic field, we can find the values of m/e and v in the following way:-

Let the space passed over by the rays under a uniform electric intensity E be L, the time taken for the rays to traverse this space is L/v, the velocity in the direction of E is therefore [$a = F/m = Ee/m$]

$$\frac{Ee}{m}\frac{L}{v}, \text{ [velocity} = at]$$

so that θ, the angle through which the rays are deflected when they leave the electric field and enter a region free from electric force, is given by the equation

$$\theta = \frac{Ee}{m}\frac{L}{v^2}\cdot\text{[see Fig. 13.3]}$$

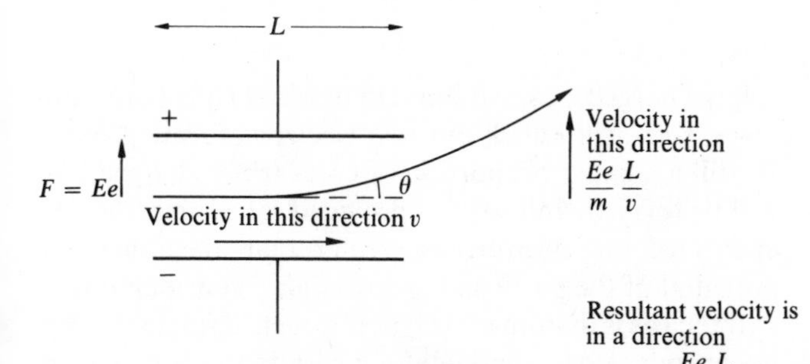

Fig. 13.3

If instead of the electric intensity, the rays are acted on by a magnetic force [$F = Bev$] at right angles to the rays, and extending across the distance L, the velocity at right angles to the original path of the rays is

$$\frac{Bev}{m}\frac{L}{v}, \text{[velocity} = at]$$

so that ϕ, the angle through which the rays are deflected when they leave the magnetic field, is given by the equation

$$\phi = \frac{Be}{m}\frac{L}{v}.$$

From these equations we get

$$v = \frac{\phi}{\theta}\frac{E}{B}$$

and

$$\frac{m}{e} = \frac{B^2\theta L}{E\phi^2}.$$

In the actual experiments B was adjusted so that $\phi = \theta$; in this case the equations become

$$v = \frac{E}{B},$$

$$\frac{m}{e} = \frac{B^2 L}{E\theta}$$

The apparatus used to measure v and m/e by this means is that represented in Fig. [13.1]. . . .

Gas	*m/e*	*v*
Air	$1{\cdot}3 \times 10^{-7}$	$2{\cdot}8 \times 10^{9}$
Air	$1{\cdot}1 \times 10^{-7}$	$2{\cdot}8 \times 10^{9}$
Air	$1{\cdot}2 \times 10^{-7}$	$2{\cdot}3 \times 10^{9}$
Hydrogen	$1{\cdot}5 \times 10^{-7}$	$2{\cdot}5 \times 10^{9}$
Carbonic acid	$1{\cdot}5 \times 10^{-7}$	$2{\cdot}2 \times 10^{9}$
. . .		

From these determinations we see that the value of m/e is independent of the nature of the gas, and that its value 10^{-7} is very small compared with the value 10^{-4}, which is the smallest value of this quantity previously known, and which is the value for the hydrogen ion in electrolysis. . . .'

J. J. Thomson (1897) *Phil. Mag.* S.5, **44**, 293–316

The cathode ray particles were later named electrons. Thomson's experiments showed that they were part of all atoms. The figures quoted by Thomson are in different units to those used in this book; in these units m/e becomes $5{\cdot}7 \times 10^{-11}$ kg C^{-1} (Thomson's result is in grammes per electromagnetic unit) and the velocities about $2{\cdot}5 \times 10^7$ m s^{-1}. Chapter 10 gives an account of Millikan's determination of the charge on the electron, e.

In 1913 Thomson used magnetic deflection to determine the mass-to-charge ratio for these atoms which had lost an electron, or electrons. This was the beginning of mass spectroscopy—the separation of atoms by mass to give a mass spectrum.

'There can . . . be little doubt that what has been called neon is not a simple gas but a mixture of two gases, one of which has an atomic weight about 20 and the other about 22. . . .'

J. J. Thomson (1913). *Proc. Roy. Soc.* Series A, **89**, 20

Neon had been found to have two isotopes; two forms identical in chemical properties, differing only in their atomic mass. Isotopes had already been proposed by Soddy in 1910 to account for radioactive substances having the same chemical properties but different radioactive behaviours (see later in this chapter). Mass spectroscopy was extended by F. W. Aston to other elements, revealing that many elements had isotopes. He found that chlorine had isotopes with masses of 35 and 37. Argon was found to consist mainly of mass 40 with a faint trace of 36. These masses are on a comparative scale, hydrogen atoms having masses of approximately 1. On the scale at present in use the carbon 12 isotope is taken to have a mass of precisely 12 of these atomic mass units.

One of the problems over the years had been to establish regularity amongst atomic masses. Hydrogen has a mass very close to 1, helium very close to 4, oxygen close to 16—on the basis of whole-number atomic masses it was possible to think of a basic building block from which atoms were constructed. However, not all elements had whole, or even close to whole, number atomic masses. Chlorine has an atomic mass of 35·5. It was only with the discovery of isotopes that the whole-number pattern was extended to all elements and the possibility of an atomic building block became feasible. The building blocks were considered to be the electron and the positive charged part of the hydrogen atom, called the proton. This view continued until 1932 when Chadwick discovered the neutron. There were then three building blocks.

Mass spectroscopy

The following extracts are taken from an article 'Mass spectroscopy. An old field in a new world' by A. O. Nier. It first appeared in the December 1966 issue of *American Scientist* (p. 359).

'... It was not until the late 1930's, when high vacuum and electronic techniques became an essential part of physics laboratory work, that mass spectroscopy became a tool which could serve scientists in a variety of disciplines. It was not until approximately 1940 that mass spectrometers were produced commercially and used, principally in the oil industry, for making gas analyses. Even in 1942 there was only one place in the world—a university laboratory—where precise uranium isotope analyses could be made. During World War II, mass spectroscopy was introduced to a large number of scientists and engineers. When the war ended and these individuals returned to their peacetime pursuits, mass spectroscopy was recognized as an extremely useful technique, and under the stimulus of demand a variety of instruments become commercially available.

THE MASS SPECTROSCOPE

A mass spectroscope consists of three essential parts: an ion source which converts neutral atoms or molecules into ions so they may be affected by electric or magnetic fields, an analyzer for separating the ions into a mass spectrum, and an ion detector. Ions may be produced by the electron impact on a gas or vapor, or by heating certain substances to a high temperature. They are also produced in a spark discharge or by field emission.

Mass analyses may be accomplished by using a magnetic field or a combination of electric and magnetic fields, or by employing a time-of-flight or a resonant method such as the cyclotron.

Ion currents can be measured with a sensitive electrical instrument such as an electrometer tube amplifier or electron multiplier, or by having the ions fall on a photographic plate which is subsequently developed. If an electrical detecting system is used the instrument is known as a mass spectrometer. If photographic recording is employed, the instrument is known as a mass spectrograph. The two types together are known as mass spectroscopes.

... Most of what follows will be about mass spectrometers in which ions are produced by the electron impact on a gas or vapor, and the mass analysis is accomplished with a magnetic field. Let us inquire how a simple instrument of this type operates. If a beam of electrons, having energy of say 75 eV, strikes the molecules of a gas at a low pressure, ions will be formed according to the reactions shown in Table 1.

Table 1
Some Common Processes for the Formation of Ions by the Impact of Electrons on Atoms or Molecules

$$Ar + e^- \longrightarrow (Ar^+ + 2e^-), (Ar^{2+} + 3e^-), \text{etc.}$$
$$N_2 + e^- \longrightarrow (N_2^+ + 2e^-), (N^+ + N + 2e^-)$$
$$CH_4 + e^- \longrightarrow (CH_4^+ + 2e^-), (CH_3^+ + H + 2e^-), (CH_2^+ + 2H + 2e^-), \text{etc.}$$

If a monatomic gas such as argon is present, the bombarding electrons will, in a certain proportion of collisions, knock one electron out of the atom, and thus form singly charged positive ions. In a smaller proportion of cases, they will knock out two electrons forming Ar^{2+}. In still a smaller proportion of cases triply charged ions will be formed, and so on—provided, of course, that the energy of the electrons is greater than the ionization potential of the particular species under consideration.

If the gas is diatomic, such as nitrogen, a single electron may produce dissociation along with ionization, and we find N_2^+, N^+, N_2^{2+}, etc. If polyatomic, such as methane, we form CH_4^+, CH_3^+, CH_2^+, etc. As a general statement, one can say that one finds most fragments, which would result if pieces of various sizes were broken from the molecule, plus, in some cases, fragments which come about due to rearrangement of the atoms remaining after the collisions. It is not hard to see that, for a complex molecule such as a heavy hydrocarbon, one will obtain a wide variety of fragments which might be identified by their masses provided one has the proper means for doing so.

The 1930's saw the development of the single-focusing magnetic mass spectrometer to the point where it became an extremely useful tool. Figure 1 [13.4] shows a schematic drawing of such an instrument. In the upper part we see the ion source. A heated filament gives off electrons which are collimated by means of a weak magnetic field not shown, and sent across the apparatus as indicated. On their way they collide with gas molecules, forming positive ions, which are drawn out of the ionizing region by an electric field between plates A and B and accelerated downward between the plates B and C. The ions in the beam have a slight spread in energy since they originate in a region which is not equipotential. Also, the beam diverges as indicated, due to the finite width of the slits employed. It next enters a region between the poles of a magnet where the charged particles follow circular paths, the radii of curvature depending on their speed and their masses, as well as on the magnetic field strength. We see two groups of particles in the figure. One group has the correct mass to continue to the analyzer exit plate where it passes through a slit and on to a collector connected to a sensitive amplifier. The ions in the second group—in this case, the lighter ones—have too small a radius (and hence are bent as shown), hit the conducting walls of the containing vessel and are harmlessly discharged. As the diagram suggests, the device has a focusing property in that it takes a diverging beam and focuses it to a narrow line.

A mass spectrum (Fig. 2) [13.5] may be obtained either by varying the voltage accelerating the ions between

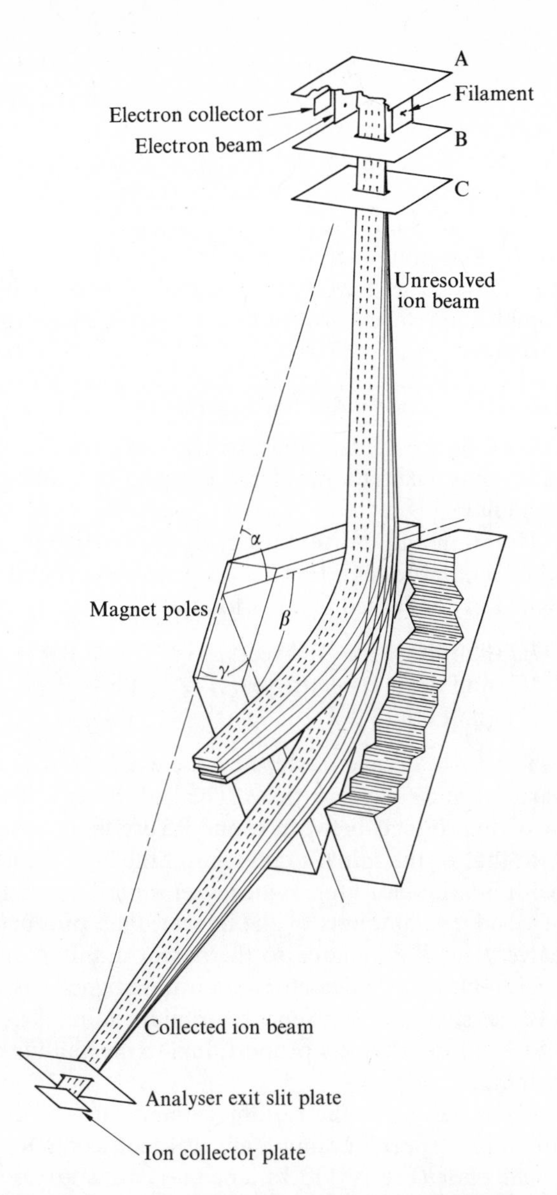

Fig. 13.4 Schematic mass spectrometer employing magnet with wedge-shaped poles to produce separation of ions according to mass. This style instrument focuses charged particles if the source of ions, the apex of magnet pole wedge, and the collector lie on a straight line, i.e., if angles $\alpha + \beta + \gamma = 180°$. Early instruments used $\alpha = \gamma = 0$ and $\beta = 180°$. To conserve magnet material and power and to have ion source and collectors outside of the magnet, where there is more space and opportunity for experimentation or use of electron multipliers as detectors, modern instruments generally use $\beta = 60°$ or $90°$ with $\alpha = \gamma$.

plates B and C (Fig. 1) [13.4] or by varying the magnetic field and thus bringing ions of different masses to the ion collector plate. The focusing property of this type of instrument is simple: if the source slit, the collector slit, and the apex of the wedge-shaped magnetic field all lie in a straight line, the diverging beam leaving the ion source will focus upon the exit slit. Successful instruments employing a wide variety of angles have been built. The radius of curvature of the selected beam in common laboratory instruments lies in the range 15–30 cm. Miniaturized instruments, having radii of curvature of 3·7 cm,

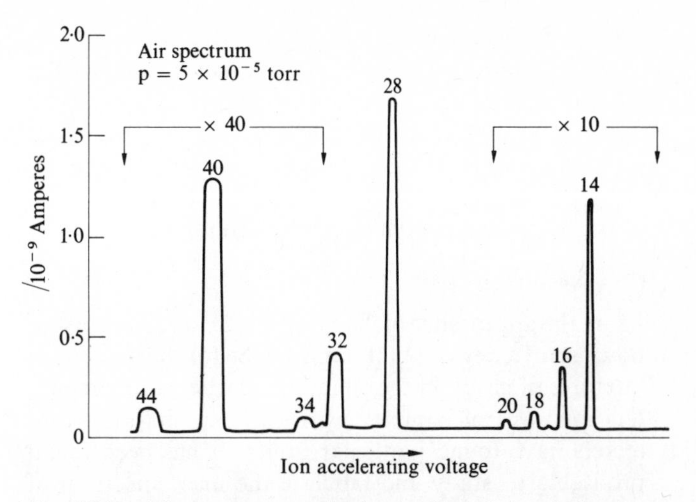

Fig. 13.5 Mass spectrum obtained when air is admitted to a small, low resolving power mass spectrometer such as is used for rocket exploration of the upper atmosphere. In this figure ions which differ in mass by one part in 30 are clearly resolved. Large laboratory instruments have been constructed which have resolving powers in excess of 1 part in 100 000. •

have been carried aloft in rockets and used to determine the composition of the upper atmosphere.

In examining Fig. 1 [13.4], it must, of course, be appreciated that the ion source, ion trajectory, and collector region are enclosed in an evacuated chamber since the device will not be operable unless a relatively good vacuum is present. When a gas is undergoing analysis, the pressure will be of the order of magnitude of 10^{-5} torr* or lower.

Figure 2 [13.5] shows the spectrum obtained if air is admitted to an instrument of this general type. A consideration of the equations governing the paths of the ions shows that, for a constant magnetic field, the mass of the ions collected is inversely proportional to the accelerating voltage applied to the ions. If one increases the accelerating voltage from left to right, ions with decreasing mass will be successively focused into place. Note, first of all, the principal peaks 32 and 28 corresponding to molecular oxygen, $^{16}O^{16}O$, and nitrogen, $^{14}N^{14}N$, respectively. Then, with the amplifier adjusted to 40 times the sensitivity, one sees a peak at mass 44 corresponding to carbon dioxide, $^{12}C^{16}O_2$, an impurity in the apparatus, and a peak at mass 40, ^{40}Ar, an atmospheric constituent having an abundance of approximately 1 %. Note also the small peak at mass 34. This arises from the isotopic molecule $^{18}O^{16}O$ and has an abundance approximately 1/250 of the 32 peak, $^{16}O^{16}O$. Not marked is the 29 peak, $^{14}N^{15}N$, which was traced at the lower sensitivity. Nevertheless, this peak, having an abundance of 1/135 of the 28 peak, $^{14}N^{14}N$, is clearly discernible.

At 20 one sees $^{40}Ar^{2+}$. Since ions are sorted according to their mass-to-charge ratio, a doubly charged ion will appear at half of its mass in the spectrum of singly charged ions. Next one sees peaks at 18 and 17, the former being

* 1 torr = pressure produced by a column of Hg, 1 mm in height, at 0°C.

the molecular ion of water, H_2O, the latter a dissociation product OH formed by the dissociation of H_2O molecules by electron impact. Water vapor appears as a residual impurity in all but the most highly pumped and baked vacuum systems. Finally, one sees peaks at 16 and 14, corresponding to O and N ions formed in the dissociation of the respective molecules by electron impact.

Gas Analyses

From the appearance of the spectrum for air, it is obvious how useful a device of this kind can be for the analysis of mixtures of gases. In the oil industry, where one encounters mixtures of complex hydrocarbons, mass spectrometers have found particular utility. It has been found profitable to study and tabulate the mass spectrum of every possible hydrocarbon which one might encounter. It turns out that the mass spectra of molecules have unique features just as no two individuals have the same fingerprints.

The top spectrum in Fig. 3 [13.6] is a schematic representation showing the spectral pattern obtained when normal butane, C_4H_{10}, is admitted to an instrument having a somewhat higher resolution than that used for obtaining the spectrum shown in Fig. 2 [13.5]. Note how the normal butane spectrum differs from those for isobutane or ethane. Finally, in the lowest spectrum in Fig. 3 [13.6] one sees the pattern found if 25 % normal butane, 25 % isobutane and 50 % ethane are present simultaneously. It is clear from the figure that if one were given an unknown sample and found a spectrum as shown below, and knew the patterns for the individual constituents, one could readily deduce the composition merely by solving simultaneous equations. Those actively working in the field have extended the method to include complex mixtures of a dozen or more heavy hydrocarbons. . . .

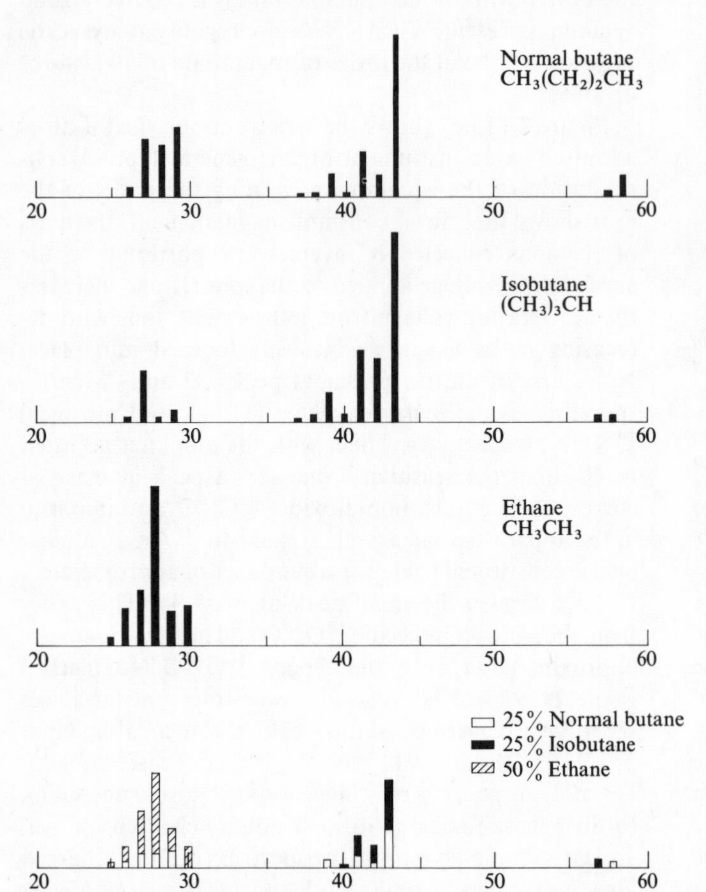

Figure 13.6 Spectral patterns for normal butane, isobutane, ethane, and a mixture of the three.

Geochronology: Age of the Earth

One of the most interesting applications of mass spectroscopy has been to problems relating to geology and cosmology.

Uranium and thorium decay by radioactive processes whose end product is lead. These processes may be represented in condensed form as follows:

$$^{238}\text{U (half life } 4{\cdot}5 \times 10^9 \text{ years)} \rightarrow {}^{206}\text{Pb} + 8\,\alpha + 6\,\beta,$$
$$^{235}\text{U (half life } 0{\cdot}71 \times 10^9 \text{ years)} \rightarrow {}^{207}\text{Pb} + 7\,\alpha + 4\,\beta,$$
$$^{232}\text{Th (half life } 13{\cdot}9 \times 10^9 \text{ years)} \rightarrow {}^{208}\text{Pb} + 6\,\alpha + 4\,\beta.$$

The ^{238}U decays in steps by the emission of 8 alpha particles and 6 beta particles. The half lives of the intermediate products between U and Pb are short compared with that of the initial uranium and the final stable lead (with infinite half life). Hence after a period of millions of years, the amounts of the intermediate products will be very small compared to the parent uranium and the final stable lead. We can for our purposes ignore the intermediate steps and consider the uranium going directly to lead at a rate inversely proportional to the half life of the uranium.

If one examines the isotopic composition of the lead found in a "pure" uranium mineral such as the Katanga pitchblende (Fig. 5) [13.7], one sees two isotopes ^{206}Pb and ^{207}Pb. The ^{206}Pb results from the radioactive decay of the principal isotope of uranium, ^{238}U, ^{207}Pb from the radioactive decay of the less abundant but more famous isotope, ^{235}U.

As an analogy to uranium-lead sequence, one may consider the behavior of sand in an hour glass. The lead corresponds to the sand at the bottom of the glass, the uranium to that at the top. To make the analogy more nearly correct, the number of grains of sand falling per unit time should be proportional to the amount of sand remaining in the upper section of the glass. (This is probably not true for most hour glasses.) If one measures the amount of sand in the lower section and that remaining in the upper section, he can, with his knowledge of the behavior of the sand, compute the time the hour glass has been in operation.

Similarly, if one knows the rate at which the uranium has been transformed to lead over the past ages (as one assumes he does from the present day behavior of uranium), he can compute the age of the mineral from the amounts of uranium and lead now present.

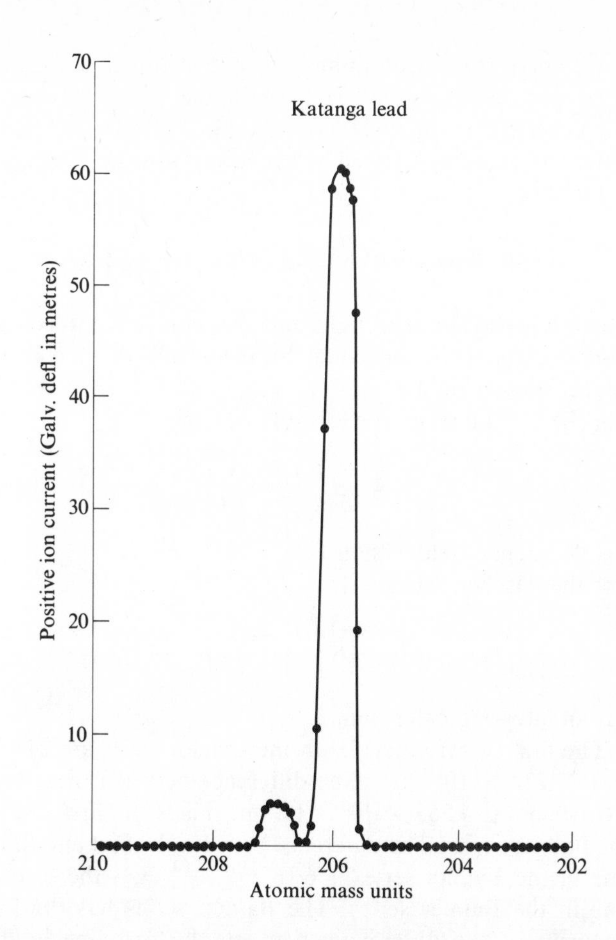

Fig. 13.7 Mass spectrum for a sample of pure lead resulting from the radioactive decay of uranium. Relative heights of peaks depend upon age of mineral, the older the mineral the higher the ^{207}Pb: ^{206}Pb ratio.

This is known as the uranium-lead method of measuring geological age. The "age" determined in this way is the time since the mineral was formed.

Thorium, with one isotope ^{232}Th which decays to ^{208}Pb at a known rate, also furnishes a means of determining the age of minerals.

An isotopic analysis of the lead formed from uranium provides another completely independent method for measuring the age of the mineral since the decay rates of ^{235}U and ^{238}U are quite different.

The parent nuclei, ^{238}U, from which ^{206}Pb is derived, have a half life about six times that of ^{235}U from which ^{207}Pb arises. That of ^{238}U is $4{\cdot}5 \times 10^9$ years, that of ^{235}U is $0{\cdot}7 \times 10^9$ years. Thus the number of nuclei of ^{235}U in the rocks has, during the past ages, changed (decreased) more rapidly than the number of the more abundant ^{238}U. For example, if an age of $4{\cdot}5 \times 10^9$ years is assumed for the earth (approximately the correct age) the initial number of ^{238}U nuclei present was twice the number now present, and the number of ^{235}U initially would have been more than 64 (i.e. 2^6) times the number of ^{235}U now present. (Six half lives for ^{235}U will have passed in $4{\cdot}5 \times 10^9$ years.)

Because of this difference, the rates at which ^{206}Pb and ^{207}Pb have been forming in a given mineral have been changing over the past ages. (The numbers of Pb isotopes formed per unit time at a given time are proportional to the number of parent atoms at that time in the uranium.) The ratio of ^{206}Pb to ^{207}Pb formed from ^{238}U and ^{235}U thus changes with time, and a measurement of this ratio *alone* (the relative heights of the two peaks in Fig. 5 [13.7]) affords a method of determining the age of the mineral containing the lead. It is generally agreed that this method of age determination is more accurate than the more conventional uranium-lead method.

Common lead, such as one finds in solder or other commercial products, is extracted from minerals such as galena (pure lead sulfide). It thus consists of a mixture of primordial lead (that is, lead in existence at the time when the earth was formed) and uranium and thorium lead generated in the magna and rocks prior to the time the lead was separated from these to form galena. The isotopic analysis of a typical sample of common lead is shown in Fig. 6 [13.8]. Fortunately an isotope, ^{204}Pb,

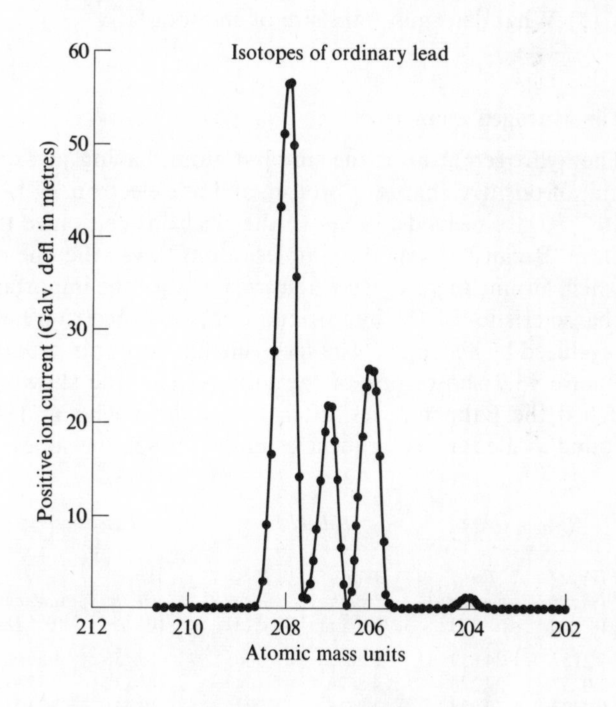

Fig. 13.8 Mass spectrum of a sample of common lead.

occurs which does not arise from any known radioactive disintegration. Should any common lead appear as an impurity in a uranium or thorium mineral whose age is being determined, its presence would be detected by the occurrence of ^{204}Pb. Since the ratios of ^{204}Pb to ^{206}Pb, ^{207}Pb and ^{208}Pb in common lead are reasonably well known, corrections for amount of common ^{206}Pb and ^{207}Pb that may have contaminated the radiogenic lead can be made by measuring the amount of ^{204}Pb present. During the past twenty-five years, many samples of minerals have been analyzed and much has been learned about the early history of the earth. Calculations show that a consistent picture is obtained by setting the maximum age of the earth (as defined earlier) at about 5×10^9 years. . . .'

13.2 Atomic theories

What are the experimental facts that atomic theories have to give reasons for?

(a) Atoms have both positive and negative charge—how is it arranged? Why are combinations of such positive and negative charges stable? Atoms collide frequently in gases yet they still retain their individual nature.

(b) Atoms can combine to form molecules. How are such bonds formed? Why is hydrogen normally available as H_2 and not H or H_3 or H_4?

(c) An element when suitably excited emits radiation, the frequencies of which are characteristic of the element. Why does an atom only emit certain discrete frequencies? What determines the frequencies?

(d) What determines the size of an atom?

The hydrogen atom

The hydrogen atom is the simplest atom, having just one unit of positive charge, a proton, and one electron. $21{\cdot}8 \times 10^{-19}$ J are needed to remove the electron, i.e. ionize the atom. Because it was the simplest atom it was the one on which atomic theories were first tried. One of the important characteristics of the hydrogen atom is the spectrum lines produced by hydrogen. The spectrum lines occur in groups. Figure 13.9 shows one of the groups. The one shown is called the Balmer series, after J. J. Balmer who in 1885 found a rule for the frequencies of the lines in the series.

Lyman series		*Balmer series*		*Paschen series*	
Wave-length $/10^{-7}$ m	*Fre-quency* $/10^{14}$ Hz	*Wave-length* $/10^{-7}$ m	*Fre-quency* $/10^{14}$ Hz	*Wave-length* $/10^{-7}$ m	*Fre-quency* $/10^{14}$ Hz
1·2157	24·659	6·5647	4·5665	18·756	1·5983
1·0257	29·226	4·8626	6·1649	12·822	2·3380
0·9725	30·824	4·3416	6·9044	10·941	2·7399
0·9497	31·564	4·1029	7·3064	10·052	2·9822
0·9378	31·966	3·9712	7·5487	9·5484	3·1395
0·9307	32·208	3·8901	7·7060	. . .	
0·9262	32·365	. . .			
. . .					

The Lyman series gives many more lines, up to a limit of frequency $32{\cdot}881 \times 10^{14}$ Hz, the Balmer series reaches a limit of $8{\cdot}228 \times 10^{14}$ Hz (wavelength $3{\cdot}6456 \times 10^{-7}$ m).

The relationship arrived at by Balmer can be expressed in the form

$$\text{Frequency of line } f = cR\left(\frac{1}{2^2} - \frac{1}{n^2}\right)$$

where c is the speed of light and R a constant, called the Rydberg constant and equal to 10 969 000 m^{-1}. n is an integer greater than 2.
For the Lyman series the formula becomes

$$f = cR\left(\frac{1}{1^2} - \frac{1}{n^2}\right)$$

n is an integer greater than 1.
For the Paschen series

$$f = cR\left(\frac{1}{3^2} - \frac{1}{n^2}\right)$$

n is an integer greater than 3.

The first two frequencies in the Lyman series are 24·659 and $29{\cdot}222 \times 10^{14}$ Hz. The difference between these two frequencies is $4{\cdot}563 \times 10^{14}$ Hz, but this is the first line in the Balmer series. The difference between the first and third line in the Lyman series is $6{\cdot}165 \times 10^{14}$ Hz—the second line in the Balmer series. The Balmer series has the frequencies of the differences between the first line in the Lyman series and each successive line. The Paschen series has the frequencies of the differences between the second line in the Lyman series and each successive line.

In terms of energy levels (see chapter 7), the Lyman series of frequencies are produced by jumps from energy levels to the $n = 1$ level, the Balmer series are produced by jumps to the $n = 2$ level and the Paschen series by jumps to the $n = 3$ level. Thus rewriting the Lyman, Balmer, and Paschen equations we have

$$f = \frac{cR}{1^2} - \frac{cR}{n^2} \quad \text{Lyman series}$$

and as $E = hf$ we have

$$\text{Energy release} = \frac{hcR}{1^2} - \frac{hcR}{n^2}$$

For the Balmer series

$$\text{Energy release} = \frac{hcR}{2^2} - \frac{hcR}{n^2}$$

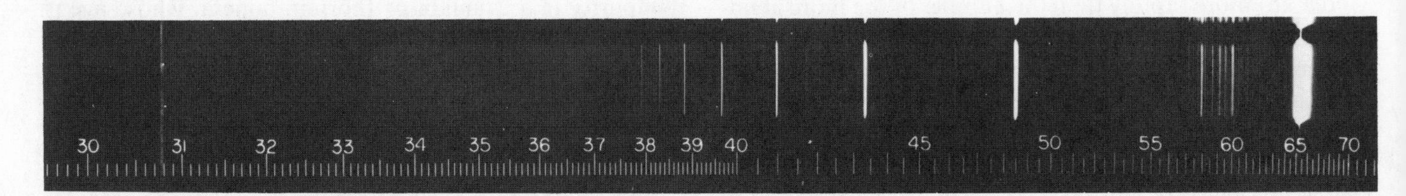

Fig. 13.9 The hydrogen spectrum. (Nuffield Foundation Science Teaching Project. Advanced physics. Unit 10, *Penguin.)*

and for the Paschen series

$$\text{Energy release} = \frac{hcR}{3^2} - \frac{hcR}{n^2}$$

Thus the hydrogen atom has energy levels at energies of

$$-\frac{hcR}{1^2}, -\frac{hcR}{2^2}, -\frac{hcR}{3^2}, -\frac{hcR}{4^2}, -\frac{hcR}{5^2}, \text{ etc.}$$

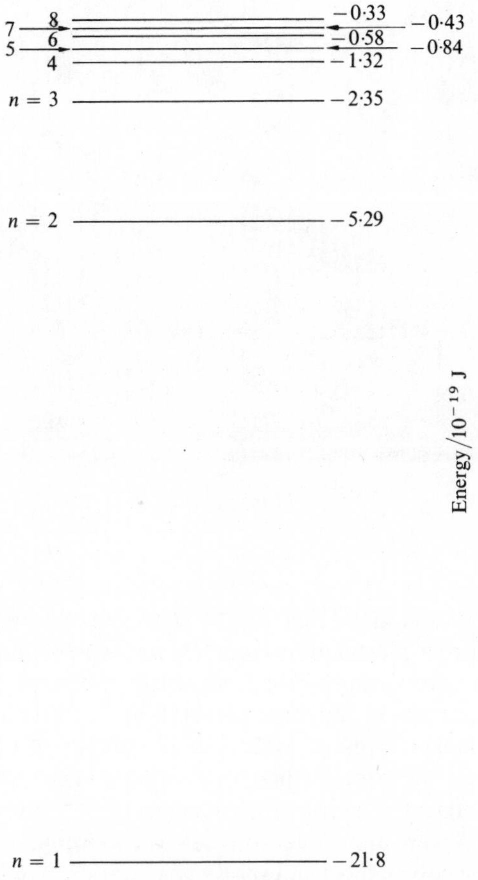

Fig. 13.10 *Energy levels for hydrogen.*

Figure 13.10 shows these energies on an energy level diagram. The energies are proportional to $1/n^2$ and given by $-21{\cdot}8 \times 10^{-19}/n^2$ J. The minus sign appears because we take the view that the atom is an energy well, it needs $21{\cdot}8 \times 10^{-19}$ J to dig out the $n = 1$ electron and bring it to the surface where we take the energy to be zero. The more energy we supply to the atom the higher up the energy level ladder we can progress, i.e., the larger the value of n. Escape is when n is infinity. In chapter 7, the energy levels have been plotted with the $n = 1$ state being taken as zero energy, the $n =$ infinity state then becomes $21{\cdot}8 \times 10^{-19}$ J for hydrogen. This is purely a difference in origin of the energy scale and does not affect the frequencies.

The Rutherford atom

'It is well known that α and β particles suffer deflections from their rectilinear paths by encounters with atoms of matter. This scattering is far more marked for the β than for the α particle on account of the much smaller momentum and energy of the former particle. There seems to be no doubt that such swiftly moving particles pass through the atoms in their path, and that the deflections observed are due to the strong electric field traversed within the atomic system. It has generally been supposed that the scattering of a pencil of α or β rays in passing through a thin plate of matter is the result of a multitude of small scatterings by the atoms of matter traversed. The observations, however, of Geiger and Marsden on the scattering of α rays indicate that some of the α particles must suffer a deflection of more than a right angle at a single encounter. They found, for example, that a small fraction of the incident α particles, about 1 in 20 000, were turned through an average angle of 90° in passing through a layer of gold foil about 0·000 04 cm thick, which was equivalent in stopping power of the α particle to 1·6 mm of air. Geiger showed later that the most probable angle of deflection for a pencil of α particles traversing a gold foil of this thickness was about 0·87°. A simple calculation based on the theory of probability shows that the chance of an α particle being deflected through 90° is vanishingly small. In addition, it will be seen later that the distribution of the α particles for various angles of large deflection does not follow the probability law to be expected if such large deflections are made up of a large number of small deviations. It seems reasonable to suppose that the deflection through a large angle is due to a single atomic encounter, for the chance of a second encounter of a kind to produce a large deflection must in most cases be exceedingly small. A simple calculation shows that the atom must be a seat of an intense electric field in order to produce such a large deflection at a single encounter.'

E.Rutherford (1911). *Phil. Mag.* **21**, 699.

This was the introduction by Rutherford of the idea of atoms having a central minute nucleus surrounded by a diffuse region containing the electrons (see chapter 2).

How big is a nucleus? Geiger and Marsden found that for a metal foil, 4×10^{-7} m thick, 1 in 20 000 particles were turned back through 90° or more. If we take the atoms in the foil as being about $2{\cdot}5 \times 10^{-10}$ m apart then there will be

$$\frac{4 \times 10^{-7}}{2{\cdot}5 \times 10^{-10}} = 1{\cdot}6 \times 10^3$$

layers of atoms. One layer could therefore be expected to turn back about

$$\frac{1}{20000 \times 1{\cdot}6 \times 10^3} = \frac{1}{32 \times 10^6}$$

particles. Thus in the first layer the actual area which turns back alpha particles is

$$\frac{1}{32 \times 10^6}$$

of the surface area of the foil. If each atom occupies an area of about $(2{\cdot}5 \times 10^{-10})^2$ m^2 and each nucleus an area of d^2,

$$\frac{d^2}{(2{\cdot}5 \times 10^{-10})^2} = \frac{1}{32 \times 10^6}$$

$$d \approx 5 \times 10^{-14} \text{ m}$$

The radius of the nucleus is about one five thousandth of the radius of the atom.

What is the charge on the nucleus? We could ask, the same question, what is the number of electrons in an atom? As an atom is neutral there must be equal amounts of negative and positive charge. The deflection of alpha particles by the nuclei of atoms in a foil depends on both the charge carried by the alpha particles and the charge carried by the nuclei of the foil atoms. At a particular angle the number of particles deflected is in fact proportional to the square of the number of charges carried by the nucleus.

Geiger and Marsden in the work which led to the concept of the nucleus deduced that the charge on the nucleus of an atom was roughly $\frac{1}{2}Ae$, where A is the atomic mass and e the charge on the electron. Later (1913) van den Broek suggested that the nuclear charge might be proportional to the atomic number of the element. The atomic number was the number of the element when all the elements were placed in order of increasing atomic mass. J. Chadwick in 1920 repeated the alpha scattering experiment with the intent of finding the charge on the atomic nucleus. The following are his results, taken from *Phil. Mag.*, 6th Ser. **40**, 742, 1920

	Nuclear charge	*Atomic mass*	*Atomic number*
Platinum	77·4e	195	78
Silver	46·3e	108	47
Copper	29·3e	63·5	29

The results clearly show that the charge on the nucleus is equal to the atomic number times the charge on the electron.

Chadwick's results fitted in with the work of H. G. J. Moseley who in 1913 had measured the frequencies of the characteristic X-rays emitted by many elements when they are struck by electrons. The X-rays produced when electrons strike a material are composed of a continuous spectrum and superimposed on that a number of highly intense frequencies characteristic of the substance being bombarded. Moseley was able to use the newly discovered method by Bragg for wavelength measurement of X-rays—diffraction at a crystal and the use of $2d \sin \theta = n\lambda$.

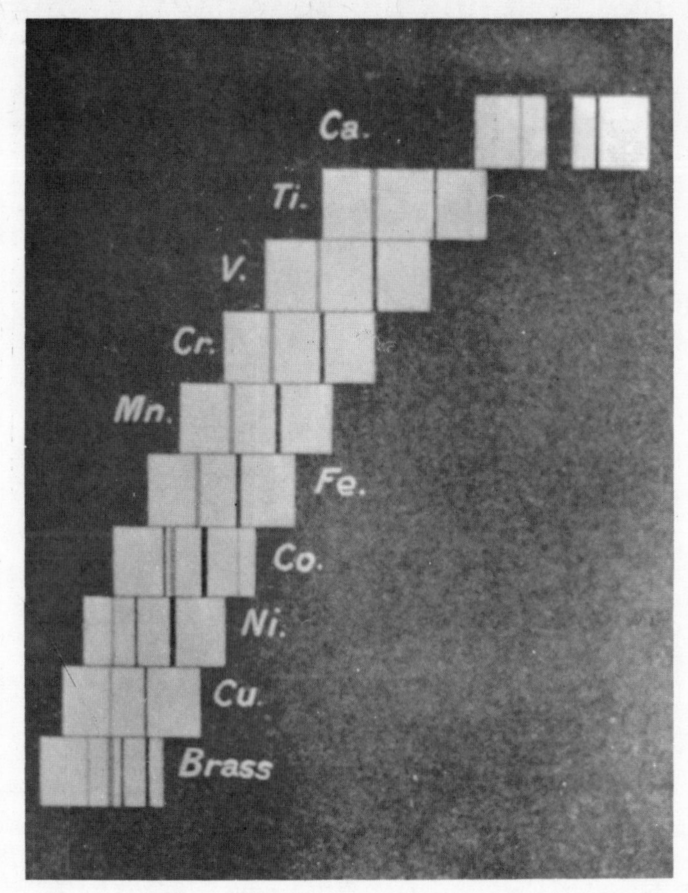

Fig. 13.11

'... Plate XXIII [Fig. 13.11] shows the spectra in the third order placed approximately in register. Those parts of the photographs which represent the same angle of reflection are in the same vertical line.... It will be seen that the spectrum of each element consists of two lines. Of these the stronger has been called α in the table, and the weaker β. The lines found on any of the plates besides α and β were almost certainly all due to impurities. Thus ... the cobalt spectrum shows Niα very strongly and Feα faintly....'

H. G. J. Moseley (1913). *Phil. Mag.*, **26**, 1024.

Moseley found that there was a linear relationship between the square root of the frequencies of the alpha lines and the atomic number, also the square root of the beta lines and the atomic number. In a later paper he concludes:

'1. Every element from aluminium to gold is characterised by an integer N which determines its X-ray spectrum. Every detail in the spectrum of an element can therefore be predicted from the spectra of its neighbours.

2. This integer N, the atomic number of the element, is identified with the number of positive units of electricity contained in the atomic nucleus.

...

6. The frequency of any line in the X-ray spectrum is approximately proportional to $A(N-b)^2$, where A and b are constants.'

H. G. J. Moseley (1914) *Phil. Mag.*, **27**, 713.

Further reading

B. Jaffe. *Moseley and the numbering of the elements.* Heinemann. Science Study Series No. 40, 1972.

The Bohr atom

How are the electrons arranged in the Rutherford atom? Rutherford's atom with its minute nucleus was barely two years old when N. Bohr, a pupil of Rutherford, wrote a paper in which he put forward what was a revolutionary theory of the atom. The idea was that the atomic electrons orbited the nucleus in certain stable orbits or states. When they were in the orbits no emission of energy occurred. Energy was only emitted when the electrons jumped between orbits. The following were the words in Bohr's paper describing his main postulates.

'(1) That the dynamical equilibrium of the systems in the stationary states can be discussed by help of the ordinary mechanics, while the passing of the systems between different stationary states cannot be treated on that basis.

(2) That the latter process is followed by the emission of a homogeneous radiation, for which the relation between the frequency and the amount of energy emitted is given by Planck's theory.'

N. Bohr (1913). *Phil. Mag.*, **26**, 1.

Bohr considered that ordinary mechanics could be applied to the orbit of the electron around the nucleus. Thus the force necessary to keep the electron in orbit was written as mv^2/r. This force was provided by the electric force of attraction between the nucleus and the electron. In the case of hydrogen with just one proton and one electron this force was $e^2/(4\pi\varepsilon_0 r^2)$, Coulomb's equation.

$$\frac{mv^2}{r} = \frac{e^2}{4\pi\varepsilon_0 r^2}$$

Thus the kinetic energy of the electron is

$$\tfrac{1}{2}mv^2 = \frac{e^2}{8\pi\varepsilon_0 r}$$

The potential energy of the electron in its orbit is $-Ve$, where V is the potential at that distance from the nucleus (i.e., the energy needed to remove the electron to infinity).

$$\text{Potential energy} = -\frac{e^2}{4\pi\varepsilon_0 r}$$

Thus the total energy of the electron is

$$\frac{e^2}{8\pi\varepsilon_0 r} - \frac{e^2}{4\pi\varepsilon_0 r} = -\frac{e^2}{8\pi\varepsilon_0 r}$$

Bohr made a third assumption in order to specify the permitted stable orbits.

'For a system consisting of a nucleus and an electron rotating round it, this state [stable orbit state] is... determined by the condition that the angular momentum of the electron round the nucleus is equal to $h/2\pi$.'

N. Bohr (1913). *Phil. Mag.*, **26**, 1.

Thus we have

$$mvr = \text{angular momentum}$$

(see chapter 2)

Bohr's condition stated above specifies the single orbit that would be normally occupied by the single electron. The condition for the other stable orbits was that the angular momentum should be $nh/2\pi$, where n is an integer.

$$mvr = \frac{nh}{2\pi}$$

Using this we can eliminate v in the kinetic energy equation.

$$\frac{m}{2}\left(\frac{nh}{2\pi mr}\right)^2 = \frac{e^2}{8\pi\varepsilon_0 r}$$

Hence

$$r = \frac{n^2h^2\varepsilon_0}{\pi me^2}$$

(This gives a radius of the $n = 1$ orbit of $0{\cdot}5 \times 10^{-10}$ m, something like the size of a hydrogen atom.) Thus we can write the total energy equation as

$$\text{Energy} = -\frac{me^4}{8\varepsilon_0^2h^2n^2}$$

This is the energy associated with the electron in the nth orbit. Thus

$$\text{Energy} \propto -\frac{1}{n^2}$$

The permitted energies of an atomic electron were thus proportional to

$$\tfrac{1}{1}, \tfrac{1}{2}, \tfrac{1}{9}, \tfrac{1}{16}, \text{etc.}$$

Atoms thus had a number of energy levels.

Bohr's second postulate was that radiation, of a single frequency, was emitted when an electron jumped between two orbits. The energy emitted by the jump was equal to hf. Consider the energy changes involved in an electron dropping from the nth orbit to the $n = 1$ orbit.

$$f = \frac{E_n - E_1}{h}$$

$$f = \frac{me^4}{8\varepsilon_0^2h^3}\left(\frac{1}{1^2} - \frac{1}{n^2}\right)$$

$$f = cR\left(\frac{1}{1^2} - \frac{1}{n^2}\right)$$

where R is known as the Rydberg constant, c is the velocity of light. This gives the frequencies of the Lyman series of lines in the hydrogen spectrum.

If the energy changes are taken with reference to the $n = 2$ level the series of spectrum lines produced is the Balmer series. Bohr's atom gave an explanation of the spectrum lines of hydrogen. The cost of the explanation had been a number of arbitrary rules. The justification for these rules was that they gave an explanation of the hydrogen spectrum.

Rutherford had this to say about the Bohr atom:

'Your ideas ... are very ingenious and seem to work out well; but this mixture of Planck ideas with the old mechanics makes it very difficult to form a physical idea of what is the basis of it all. There appears to me one grave difficulty in your hypothesis ... namely, how does an electron decide what frequency it is going to vibrate at when it passes from one stationary state to the other? It seems to me that you would have to assume that the electron knows beforehand where it is going to stop.'

Rutherford. 20 March, 1913 in letter to Bohr
S. Toulmin and J. Goodfield (1962). *The architecture of matter*, p. 320, Penguin.

On purely ordinary mechanics an electron moving in a circular path was accelerating, towards the centre, and accelerating electrons radiate electro-magnetic waves (see chapter 12). Thus the electron is continually losing energy. If it loses energy, at the expense of its potential energy, it must spiral into the nucleus. Bohr 'solved' this problem by just stating that electrons in stable orbits did not radiate and that they only radiated when jumping from one orbit to another. How did they radiate? The only mechanism by which it was known that an electron could radiate was acceleration and this had been disallowed. The Bohr model gives no physical picture of how the radiation is produced.

In general Bohr's model of the atom met with approval.

'Very remarkable! There must be something behind it. I do not believe that the derivation of the absolute value of the Rydberg constant is purely fortuitous.'

Einstein (1913).

'Dr. Bohr has arrived at a most ingenious and suggestive, and I think I must add convincing, explanation of the laws of the spectral series.'

J. H. Jeans (1913).
Report of the 83rd Meeting of the British Association for the Advancement of Science, Sept. 10–17, p. 376.

'This is all nonsense!'

Von Laue (1913).

Is the electron moving in an orbital path around the nucleus? If it does shouldn't the frequency of rotation somehow affect the frequency of the radiation emitted? If the electrons in a stable orbit do not revolve round the nucleus—what keeps them in the stable orbit? Bohr's atomic model was a breakthrough in that quantum ideas were introduced into the atom but it was a model based on a mixture of ordinary mechanics and quantum ideas; it was a step along the path towards an atomic model based only on no ordinary mechanics.

'It is by no means necessary that a theory should be absolutely true in order to be a great help to the progress of science.'

E. Frankland.

Quantum mechanics

Bohr's theory suffered from the difficulties inherent with his mixing quantum ideas with ordinary mechanics. With ordinary mechanics we need orbits and a simple mechanical picture of the atom with electrons leaping between orbits. The quantum idea that Bohr grafted on to this model was that only certain orbits were permitted and a restriction that radiation was only emitted when electrons moved between orbits. The electrons were particles like large-scale particles such as tennis or table-tennis balls.

In 1924 de Broglie put forward his theory that everything, not just photons, had a wave-particle dual nature (see chapter 7).

'The basic idea of quantum theory is, of course, the impossibility of considering an isolated fragment of energy without assigning a certain frequency to it.'

L. de Broglie (1925). *Ann. de Physique*, **3**, 22.

De Broglie considered that all matter had a frequency, a wave nature, associated with it. Quantum ideas had been extended from radiation to all matter.

$$\lambda = \frac{h}{mv}$$

where λ is the wavelength, h Planck's constant, m and v the mass and velocity of the object.

'We can no longer imagine the electron as being just a minute corpuscle of electricity: we must associate a wave with it. And this wave is not just a fiction: its length can be measured and its interferences calculated in advance....'

L. de Broglie (1929). Nobel Prize address. Boorse and Motz, *The world of the atom*, **2**, Basic Books, p. 1059.

These wave properties lead us to a new mechanics. Electrons can no longer be regarded as like tennis balls. The new mechanics is called quantum mechanics.

'"Quantum mechanics" is the description of the behaviour of matter in all its details and, in particular, of the happenings on an atomic scale. Things on a very small scale behave like nothing that you have any direct experience about. They do not behave like waves, they do not behave like particles, they do not behave like clouds, or billiard balls, or weights on springs, or like anything that you have ever seen.'

R. B. Feynman, R. P. Leighton, M. Sands (1969). *The Feynman lectures on physics*, Vol. 1, pp. 37–1. Addison Wesley, Reading, Mass.

In 1925 Heisenberg published a paper 'The interpretation of kinematic and mechanical relationships according to the quantum theory'. This was the beginning of quantum mechanics.

'It is well known that the formal rules that are generally used in the quantum theory for the calculation of observable quantities (e.g. the energy in the hydrogen atom) are often subjected to the serious objection that those rules contain as an essential constituent relationships between quantities that are apparently incapable of observation in principle (such as, for example, the position and the period of rotation of the electron) and that therefore those rules obviously lack any visualizable physical foundation, if we do not desire to continue to retain the hope that those so far unobservable quantities could later perhaps be made experimentally accessible. ... it appears better to give up the hope of an observation of the previously unobservable quantities (such as the position and time of revolution of the electron) completely, and simultaneously, therefore, to grant that the partial agreement of the known quantum rules with experience is more or less fortuitous and to attempt to formulate a quantum-theoretical mechanics analogous to classical mechanics in which only relationships between observable quantities occur....'

W. Heisenberg (1925). *Zeitschrift für Physik*, **33**, 879.

Heisenberg had discerned a basic limitation which exists on our experimental capabilities. Certain quantities are unobservable—not measurable. It is not that we might in the future be able to devise experimental techniques which will let us observe these quantities—they are just unobservable. In the case of a photon or electron passing through a single slit: if the slit is about one wavelength wide then the uncertainty about the position is λ—we know within a wavelength λ where the photon or electron goes; the momentum is $mv = h/\lambda$ and thus the uncertainty about the momentum is h/(uncertainty about position). A single slit of this width gives complete diffraction through 180°—the photon or electron beam spreads out into all directions on emerging from the slit, the sideways momentum the particle obtained at the slit is very indefinite when the particle could go anywhere on passing through the slit. If we make the slit wider we produce more uncertainty about position and less about momentum—the diffraction pattern is much narrower. We have:

$$(\text{Uncertainty about position}) \times (\text{Uncertainty about momentum}) \geqslant h$$

A similar relationship exists for energy of a particle and spread of time over which the energy is determined. For a photon, the momentum is mc and as $E = mc^2$ we have the uncertainty in momentum equal to (uncertainty about energy)/c. The photon travels with a speed c so that position is ct, and so the uncertainty in position means an uncertainty in time. Hence

$$(\text{Uncertainty about energy}) \times (\text{Uncertainty about time}) \geqslant h$$

'... It has been said that the atom consists of a nucleus and electrons moving around the nucleus; it has also been stated that the concept of an electronic orbit is doubtful. One could argue that it should at least in principle be possible to observe the electron in its orbit. One should simply look at the atom through a microscope of a very high resolving power, then one would see the electron moving in its orbit. Such a high resolving power could to be sure not be obtained by a microscope using ordinary light, since the inaccuracy of the measurement of the position can never be smaller than the wave length of the light. But a microscope using γ-rays with a wave length smaller than the size of the atom would do. Such a microscope has not yet been constructed but that should not prevent us from discussing the ideal experiment.

...The position of the electron will be known with an accuracy given by the wave length of the γ-ray. The electron may have been practically at rest before the observation. But in the act of observation at least one light quantum of the γ-ray must have passed the microscope and must first have been deflected by the electron. Therefore, the electron has been pushed by the light quantum, it has changed its momentum and its velocity, and one can show that the uncertainty of this change is just big enough to guarantee the validity of the uncertainty relations. Therefore, there is no difficulty with the first step.

At the same time one can easily see that there is no way of observing the orbit of the electron around the nucleus. The second step shows a wave packet moving not around the nucleus but away from the atom, because the first light quantum will have knocked the electron out from the atom. The momentum of the light quantum of the γ-ray is much bigger than the original momentum of the electron if the wavelength of the γ-ray is much smaller than the size of the atom. Therefore, the first light quantum is sufficient to knock the electron out of the atom and one can never observe more than one point in the orbit of the electron; therefore, there is no orbit in the ordinary sense. The next observation—the third step—will show the electron on its path from the atom. Quite generally there is no way of describing what happens between two consecutive observations. It is of course tempting to say that the electron must have been somewhere between the two observations and that therefore the electron must have described some kind of path or orbit even if it may be impossible to know which path. This would be a reasonable argument in classical physics. But in quantum theory it would be a misuse of the language which ... cannot be justified.... Actually we need not speak of particles at all. For many experiments it is more convenient to speak of matter waves; for instance, of stationary matter waves around the atomic nucleus....'

W. Heisenberg (1959) *Physics and philosophy*, Allen and Unwin.

Electrons in, say, an atom are considerably disrupted by the process of observing them. There is thus no possibility of continuous observation of an electron. If we observe an atom our results are a series of discrete discontinuous events. Sometimes these events may form some chain such that we consider the sequence to represent the observation of a particle moving in some particular direction. It is like watching a television screen—we can get the impression that some object, perhaps a person, is moving across the screen but what we are seeing is a sequence of discrete events.

'... it is better to regard a particle not as a permanent entity but as an instantaneous event.'

E. Schrödinger.

Suppose we think of an atom as being like a box with an electron bouncing backwards and forwards between an opposite pair of walls. If the walls are a distance L apart, then the uncertainty in position of the electron is L and hence the uncertainty in momentum is h/L. Thus we might suggest that the momentum be + or $-h/2L$. Between the walls the electron we will say has only kinetic energy. Thus the energy of the electron in the atom corresponding to this momentum will be

$$\text{energy} = \tfrac{1}{2}mv^2 = \frac{(mv)^2}{2m} = \frac{h^2}{8mL^2}$$

How big is the 'atom' box? We know that about 22×10^{-19} J is needed for an electron to escape from a hydrogen atom. This is the potential energy at a distance of about 1×10^{-10} m from a nucleus having a charge of $1{\cdot}6 \times 10^{-19}$ C.

$$\text{potential energy} = Ve = \frac{e^2}{4\pi\varepsilon_0 r}$$

where r is the radius of the atom. Thus we can think of walls being at this distance. For an electron to escape from the atom it would need to have 22×10^{-19} J to just reach the top of the wall. Taking L as 2×10^{-10} m, m as 9×10^{-31} kg and h as $6{\cdot}6 \times 10^{-34}$ J s gives the energy of the electron as about 14×10^{-19} J. We can have an electron in this size box.

If our atom had been the size of the nucleus, the uncertainty in momentum would be much higher because of the reduction in the uncertainty of the position. The nucleus is about 10 000 times smaller than the atom, and so the kinetic energy would be about 10 000 × 10 000 times larger, of the order of 10^{-11} J or about 100 MeV. This is an immense energy and indicates why electrons are not thought to exist inside nuclei. The above argument thus gives some reason for atoms being the size they are. If they were smaller, the electron would have too high an energy to exist in the atom. The atom is about the smallest size box in which we can contain the electron.

Waves and atoms (wave mechanics)

'In this communication I should first like to show on the basis of the simplest case of the ... hydrogen atom that the usual quantization procedure can be replaced by another condition in which there is no longer any mention of "integers". On the contrary, the occurrence of integers follows in the same natural way as, for instance, in the case of the number of nodes of a vibrating string....'

E. Schrödinger (1926). *Ann. der Physik*, **79**, 361.

Thus Schrödinger began his historic paper in 1926 which introduced wave mechanics. Figure 13.12 shows how a

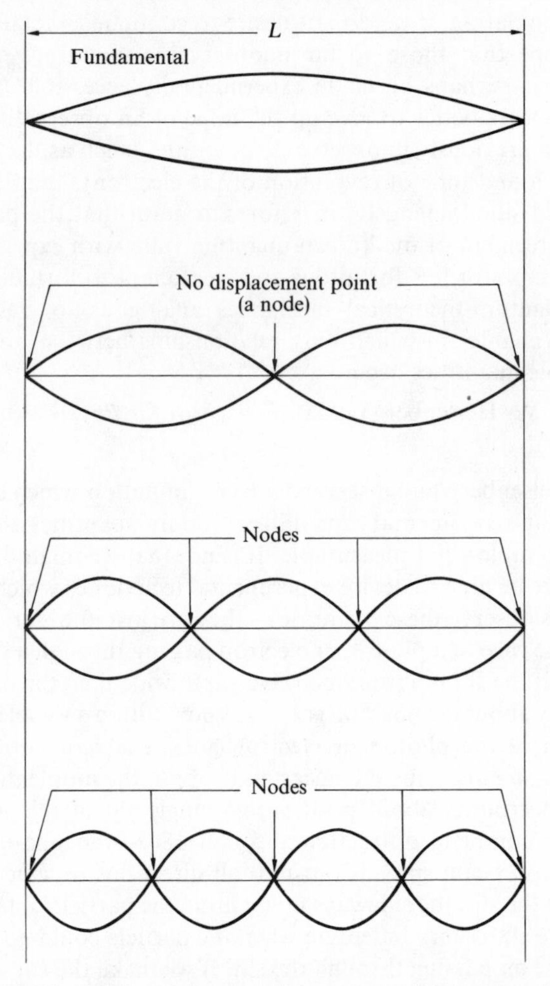

Fig. 13.12 Some of the possible ways a string can vibrate.

string, clamped at the ends, vibrates. In the fundamental oscillation the wavelength of the string is equal to $2L$, in the next form of oscillation $\lambda = L$, in the next $\lambda = \frac{2}{3}L$, in the next $\lambda = L/2$, and so on.

What would the wavelength of an electron have to be if it is to fit like a string inside an atom-sized box? The size of an atom is about 2×10^{-10} m (see previous item), thus the wavelength of the fundamental oscillation will be about 4×10^{-10} m ($2L$).

As

$$\lambda = \frac{h}{mv}$$

the momentum of the electron giving this wavelength will be

$$mv = \frac{6{\cdot}6 \times 10^{-36}}{4 \times 10^{-10}}$$

$$= 1{\cdot}6 \times 10^{-24} \text{ kg m s}^{-1}$$

The kinetic energy of such an electron will be

$$\text{Kinetic energy} = \tfrac{1}{2}mv^2$$

and if $m = 9{\cdot}1 \times 10^{-31}$ m s^{-1},

$$\text{Kinetic energy} = \frac{(1{\cdot}6 \times 10^{-24})^2}{2 \times 9{\cdot}1 \times 10^{-31}}$$

$$= 14 \times 10^{-19} \text{ J}$$

Could an electron with this energy be contained within the atom? Within the box the electron is only considered to have kinetic energy. The energy needed to remove an electron from the hydrogen atom is about 22×10^{-19} J. This is greater than the kinetic energy we calculated for the electron wave and thus the electron can have this energy and remain within the atom.

If the atom were half the size the kinetic energy is increased by a factor of four, to 56×10^{-19} J, and the energy the electron needs to escape is increased by a factor of two, to 46×10^{-19} J, potential energy of the 'wall' is proportional to $1/r$. The electron could not be contained within an atom of this size. The electron could certainly not be contained within the nucleus—its kinetic energy would be much too large, about 14×10^{-13} J for a nucleus 2×10^{-14} m in size. The wavelength of a proton can fit within the nucleus, its wavelength being smaller because of its larger mass.

What about the other forms of oscillation of a string—can electron waves have these forms? The fundamental wave had just one loop, the next wave has two loops, the next three loops, and so on. The wavelength is $2L/n$, where n is the number of loops.

$$\text{Kinetic energy} = \frac{(mv)^2}{2m} = \frac{h^2}{2m\lambda^2}$$

$$= \frac{h^2n^2}{8mL^2}$$

We imagine the electron shuttling back and forth in the box with this constant kinetic energy between walls. If within the box the potential energy is zero, the expression gives us the total energy possible for the electron. Unlike the Bohr atom where the integers had to be specifically introduced by assumption they here spring naturally out of wave ideas.

We can apply this simple idea to the case of a dye molecule where electrons are able to move along the length of a chain of carbon atoms between walls of CH_3 molecules, these being the ends of the carbon chain. A particular dye has a carbon chain length of $8{\cdot}4 \times 10^{-10}$ m, thus the energy levels are $8{\cdot}48 \times 10^{-20}$ J, $33{\cdot}92 \times 10^{-20}$ J, $76{\cdot}32 \times 10^{-20}$ J, $135{\cdot}68 \times 10^{-20}$ J, $212{\cdot}00 \times 10^{-20}$ J, etc (Fig. 13.13). The

212·00 n=5

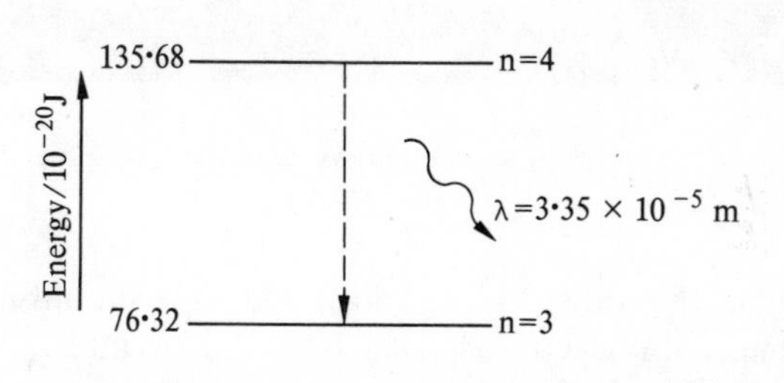

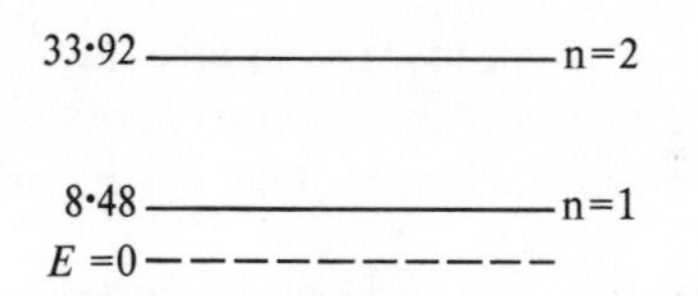

Fig. 13.13 Energy levels for a particle in a box. $E \propto n^2$.

colour of the dye is given by the transition from level 4 to level 3, an energy change of $59{\cdot}36 \times 10^{-20}$ J. This corresponds to a frequency (E/h) of $8{\cdot}94 \times 10^{14}$ Hz, a wavelength of $3{\cdot}35 \times 10^{-5}$ m, and a resulting purple colour.

Our wave model is a crude model, a wave in a rectangular box. We can imagine it as a standing water wave produced between straight barriers (see chapter 6). Figure 13.14 shows a standing water wave produced in a circular ring in a ripple tank. The kinetic energy of the electron is, however, dependent on how far away from the nucleus we consider it to be, the momentum thus depends on the distance and hence the wavelength is not constant over the entire atom. Our box containing the standing wave is not rectangular but is much more complex. A rectangular box is a rectangular potential well, the electron moving unimpeded between the walls and being unable to surmount them because its kinetic energy is not as large as the potential

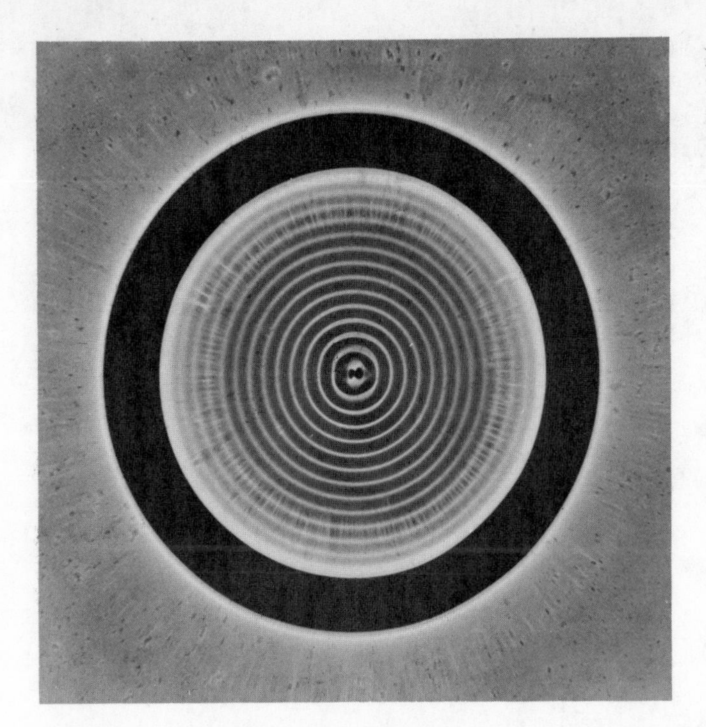

Fig. 13.14 Standing water waves in a circular ring. (PSSC Physics, *2nd Ed., D. C. Heath and Co.*)

energy at the walls (Fig. 13.15(a)). Our box should have a $1/r$ shape, the electron having both kinetic and potential energy within the box

$$V = -\frac{Q}{4\pi\varepsilon_0 r}$$

This is consistent with an inverse square law force. Thus, for a constant total energy, the kinetic energy will decrease as the potential energy increases. This occurs as r becomes larger.

Figure 13.15(b) shows a $1/r$ potential well. What would the fundamental standing wave in such a well look like? Near the centre the kinetic energy would be high, where the

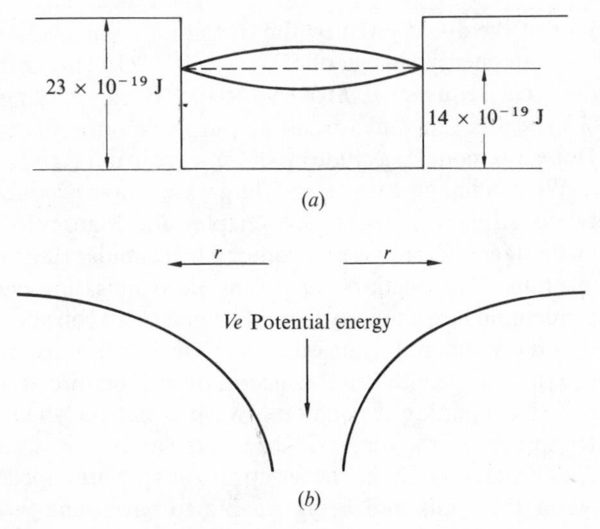

Fig. 13.15

potential energy is small, and thus the wavelength small, near the edge of the atom the kinetic energy would be smaller, the potential energy is higher, and thus the wavelength longer.

$$\text{Kinetic energy} \propto v^2 \propto \frac{1}{\lambda^2}$$

Figure 13.16 shows the type of fundamental wave pattern we might expect for our hypothetical string.

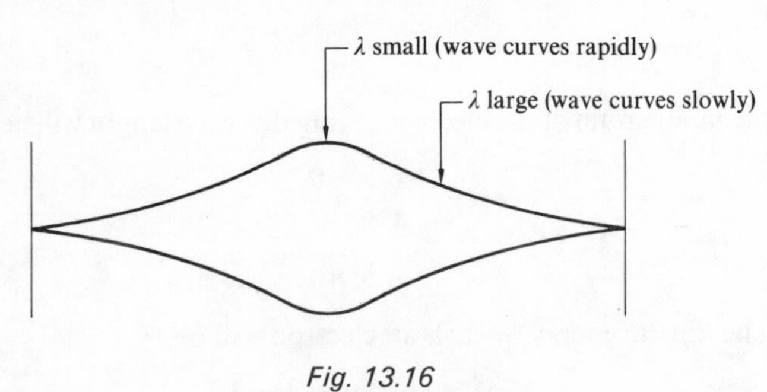

Fig. 13.16

What do the electron waves represent? Schrödinger considered that the electron was smeared out over its wave pattern. Born later developed the interpretation that the wave was a phantom wave which guided the electron to regions of high and low probability. A large amplitude region on a standing wave indicates a region of high probability of finding an electron there. In an interference experiment with light incident on two slits we use the idea of light as a wave to find the regions where the interference fringes will be a maximum—the wave tells us where the greatest chance is of a photon being found (see chapter 7). Thus in our atomic model, the greatest chance of finding an electron is where the amplitude of our standing wave is a maximum. The chance is proportional to the square of the amplitude (as the amplitude could be negative this means we do not have 'negative' chances.)

Our hypothetical string in Fig. 13.16 has a maximum amplitude at the centre. The chance of finding an electron at a point would seem to be highest at the centre. Because atoms are three dimensional it is generally more convenient to give the chance of finding the electron at a given distance

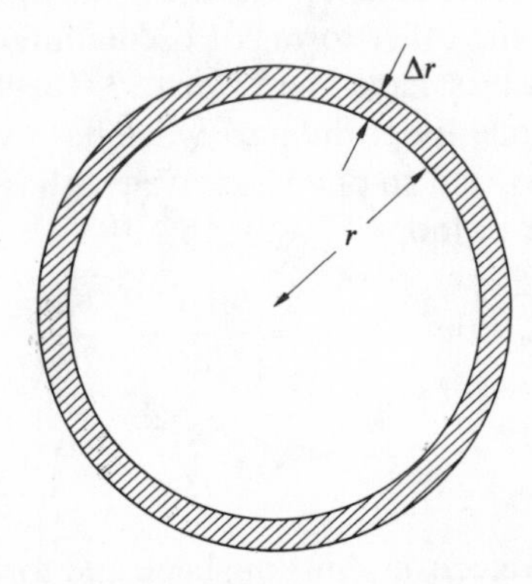

Fig. 13.17 Section through the three-dimensional shell.

from the centre, say at radius r. This radial charge density is $4\pi r^2 \Delta r$ times the chance of finding an electron at a point distance r from the centre. This is the chance of finding the electron in a shell of thickness Δr and radius r (Fig. 13.17). At the centre r is zero and thus the radial charge density is zero. Figure 13.18 shows the result of expressing Fig. 13.16 as a radial distribution. Figure 13.19 shows another way of representing the radial distribution—the blackness of the shading in any region indicates the chance of finding the electron there. Our result looks like a fuzzy Bohr orbit.

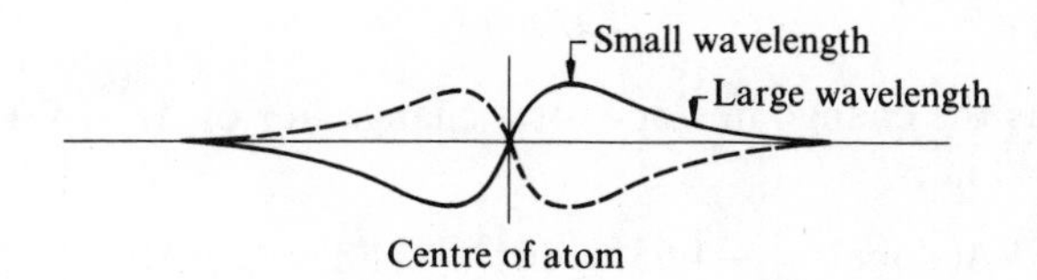

Fig. 13.18 n = *1 radial distribution.*

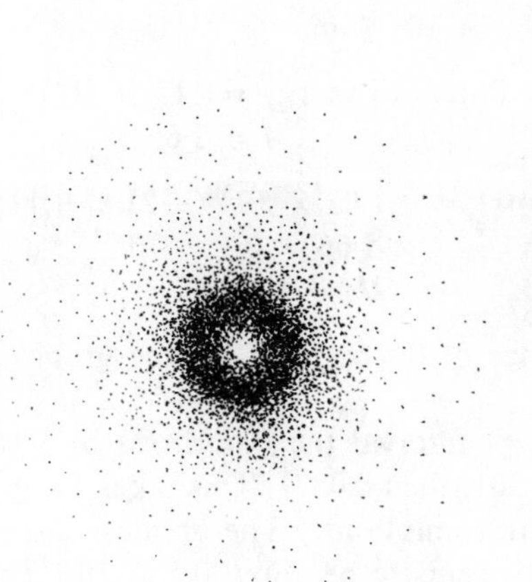

Fig. 13.19 n = *1 radial distribution.*

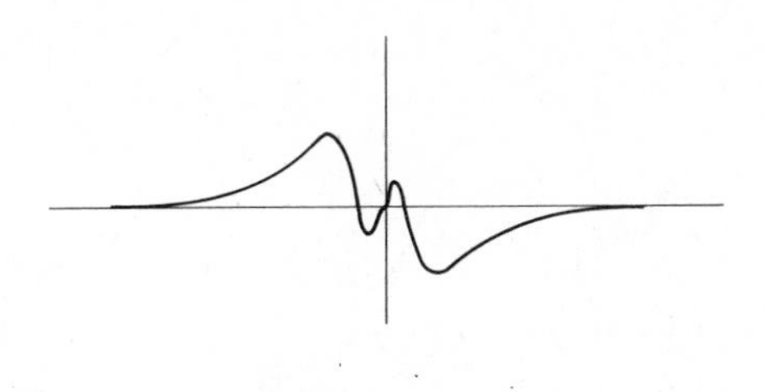

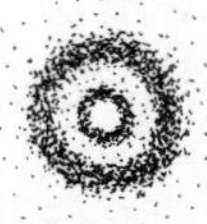

Fig. 13.20 n = *2 radial distribution.*

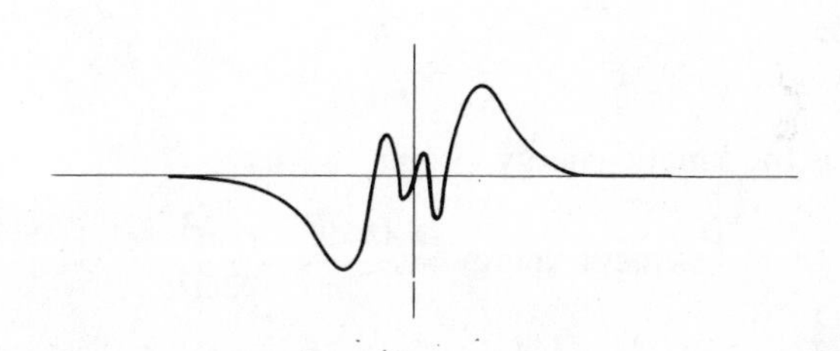

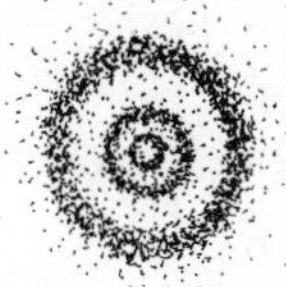

Fig. 13.21 n = *3 radial distribution.*

What would the other modes of oscillation look like? The next mode would have to fit the idea of the wavelength being smaller near the centre than farther out and have an extra maximum (Fig. 13.20). The next mode would have yet one more extra maximum (Fig. 13.21).

The variation of displacement with time for an oscillating object performing simple harmonic motion can be represented by (see chapter 4 for a similar equation)

$$x = A_0 \sin\left(\frac{2\pi t}{T}\right)$$

A_0 is the maximum amplitude, x the displacement at time t, and T the time to complete one oscillation. Figure 13.22 shows a graph of x against t; it is a sine curve. We can by comparison write a similar equation for a wave which varies with distance r, instead of time and which has a wavelength λ instead of a periodic time.

$$A = A_0 \sin\left(\frac{2\pi r}{\lambda}\right)$$

The differential equation which describes this wave can be written as

$$\frac{d^2A}{dr^2} = -\left(\frac{2\pi}{\lambda}\right)^2 A$$

Compare this with

$$\frac{d^2x}{dt^2} = -\omega^2 x = -\left(\frac{2\pi}{T}\right)^2 x$$

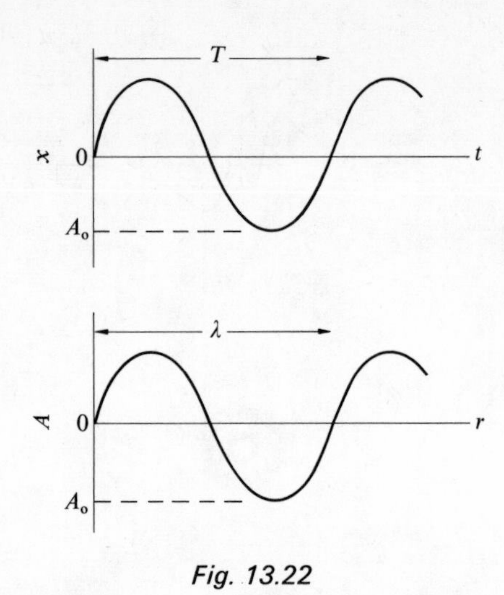

Fig. 13.22

for the simple harmonic motion case. If λ is constant we know that the solution of our differential equation is a sine curve. What is the solution where λ is not a constant?

The de Broglie wave equation gives us for the wavelength

$$\lambda = \frac{h}{mv}$$

and as the kinetic energy $= \frac{1}{2}mv^2$ we have

$$\text{Kinetic energy} = \frac{(mv)^2}{2m} = \frac{h^2}{2m\lambda^2}$$

For any oscillator or wave the total energy, E, is the sum of the potential and kinetic energies at any instant (see chapter 4).
Hence

$$E - (\text{potential energy}) = \frac{h^2}{2m\lambda^2}$$

Thus we have

$$\frac{d^2A}{dr^2} = -\frac{8\pi^2 m}{h^2}[E - (\text{potential energy})]A$$

We can easily determine the potential energy of an electron at different distances from a proton if we assume that it follows a $1/r$ law. The lower half of Fig. 13.23 shows how the potential energy varies with r, for a single electron near a proton. The differential equation can be solved in a similar manner to those in chapters 1, 2, and 4 (in particular see appendix 1A).

$$\frac{d^2A}{dr^2} = \frac{d}{dr}(\text{slope of the } A \text{ against } r \text{ graph})$$

If we only work with finite differences we can write

$$\frac{\Delta(\text{slope})}{\Delta r} = -\frac{8\pi^2 m}{h^2}(E - \text{potential energy})A.$$

Using the numerical values of m and h

$$\frac{\Delta(\text{slope})}{\Delta r} = -1{\cdot}65 \times 10^{38}\,(E - \text{potential energy})A$$

The following is the solution of the differential equation for $E = -21{\cdot}8 \times 10^{-19}$ J. This value is chosen because the experimental evidence of the Lyman spectral series of hydrogen indicates that this is the energy of the $n = 1$ state (see earlier in this section). We will make a starting condition that $A = 0$ at the centre of the atom, we start plotting the graph from there, and that the initial slope of the A against r graph takes A from 0 to 1·0 in a distance of $0{\cdot}1 \times 10^{-10}$ m (see upper part of Fig. 13.23). The units of A are arbitrary; we are only interested in the shape of the graph. At $r = 0{\cdot}1 \times 10^{-10}$ m,

$$\text{Potential energy} = -230 \times 10^{-19}\text{ J}$$
$$A = 1{\cdot}0$$

Thus the change in slope for a change in r of $\Delta r = 0{\cdot}1 \times 10^{-10}$ m is

$$\Delta(\text{slope}) = -1{\cdot}65 \times 10^{38}(-21{\cdot}8 + 230)10^{-19} \times 1{\cdot}0 \times 0{\cdot}1 \times 10^{-10} = -343{\cdot}5 \times 10^{8}$$

In a distance of $0{\cdot}1 \times 10^{-10}$ m the graph line drops by 0·3435.

At $r = 0{\cdot}2 \times 10^{-10}$ m,

$$\text{Potential energy} = 115 \times 10^{-19}\text{ J}$$
$$A = 1{\cdot}66$$

$$\Delta(\text{slope}) = -1{\cdot}65 \times 10^{38}(-21{\cdot}8 + 115)10^{-19} \times 1{\cdot}66 \times 0{\cdot}1 \times 10^{-10} = -256{\cdot}0 \times 10^{8}$$

In a distance of $0{\cdot}1 \times 10^{-10}$ m the graph line drops by 0·256.

In the next interval the line drops by 0·1822, the next interval 0·1201, then 0·0871. This takes the graph line past the maximum amplitude. The greatest chance of finding the electron seems to be about $0{\cdot}5 \times 10^{-10}$ m. At $1{\cdot}05 \times 10^{-10}$ m the potential energy becomes equal to $21{\cdot}8 \times 10^{-19}$ J, the kinetic energy is thus zero. We would expect the electron to be unable to progress beyond this point—all its kinetic energy has been converted into potential energy. A ball rolling up a hill does not progress farther up the hill than the point where all its kinetic energy has been converted into potential energy. This point on our graph might be considered as the 'edge' of the atom. Our equation, however, will operate with negative kinetic energy values. If the graph is plotted further we find that Δ(slope) becomes positive and the curve gradually comes down to the axis, at more than $r = 2 \times 10^{-10}$ m. The electron would seem to have the chance of moving to regions where its kinetic energy is insufficient to get it. This may seem absurd but strangely enough it does happen.

Alpha particles generally have energies of the order of 5 MeV, but the potential barrier they have to climb to get out of the nucleus is higher than this. The alpha particle has a small but finite chance of getting to regions where its kinetic energy is insufficient to get it. The phrase 'it has tunnelled its way out' is often used. The following extract

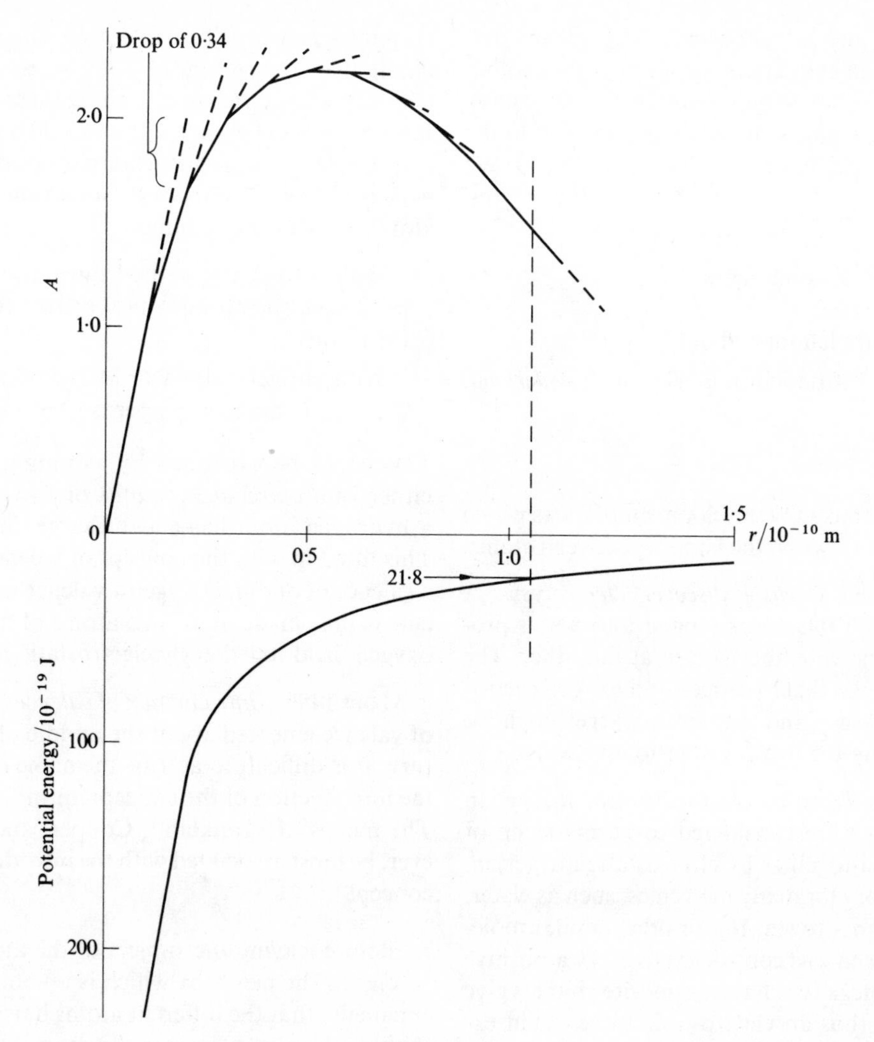

Fig. 13.23

is from the paper by G. Gamow in which he puts forward this argument.

> 'To fly off, the α-particle must overcome a potential barrier of height U_0 [Fig. 13.24]; its energy may not be less than U_0. But the energy of the emitted α-particle, as verified experimentally, is much less ... the α-particles emitted by uranium itself have an energy which represents a distance of $6{\cdot}3 \times 10^{-12}$ cm [r_2 in Fig. 13.24] on the repulsive curve. If an α-particle, coming from the interior of the nucleus, is to fly away, it must pass through the region r_1 and r_2 where its kinetic energy would be negative, which, naturally, is impossible classically.
>
> ... If we consider the problem from the wave mechanical point of view, the above difficulties disappear by themselves. In wave mechanics a particle always has a finite probability, different from zero, of going from one region to another region of the same energy, even though the two regions are separated by an arbitrarily large but finite potential barrier. ...'

G. Gamow (1928). *Zeitschrift für Physik*, **51**, 204.

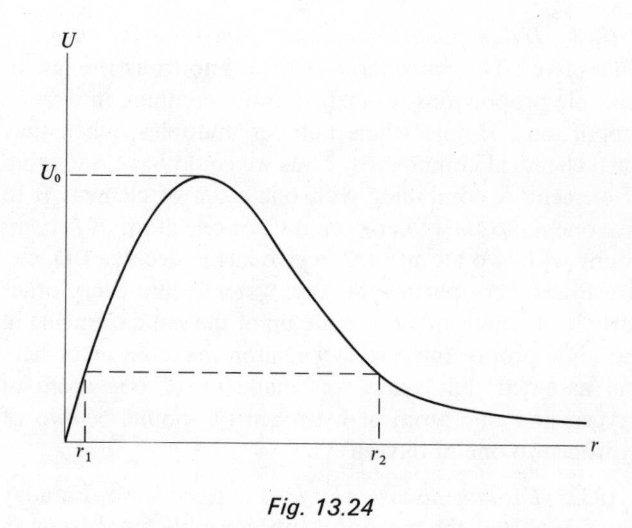

Fig. 13.24

We can solve the differential equation for other energy values. It turns out that only for those energy values given by $(-21{\cdot}8 \times 10^{-19})/n^2$ J, where n is an integer, does the amplitude come down to the r axis. At all other energy values the amplitude goes off to infinity. This means that the chance of finding an electron in the atom at these other energy values is vanishingly small. Only at certain energy values is there a finite chance of finding an electron in the atom. The results are the same as those in Fig. 13.10. We thus have an explanation of energy levels and so the spectral series.

> 'Thus grew the tale of wonderland:
> thus slowly, one by one,
> its quaint events were hammered out.'
>
> L. Carroll. Introduction to *Alice in Wonderland.*

The chemical bond

The following briefly traces the development of ideas which have led to the present views on the bonding between atoms.

1800 *Nicholson and Carlisle discover electrolysis.* A current passed between two wires dipped into water produced hydrogen at one wire and oxygen at the other. The interpretation of this was that hydrogen and oxygen carried opposite electrical charges, and that these charges might be responsible for binding the two together to give water.

1806 *Berzelius produces his electrochemical theory.* In this theory molecules were considered to be made up of atoms or radicals held together by electrostatic attraction. This seemed satisfactory for many molecules, such as water, but not for molecular hydrogen, H_2, or other similar molecules. In water hydrogen was considered to carry a positive charge and oxygen a negative charge, opposite charges give attractive forces and thus an electrostatic force could explain the 'sticking' together of hydrogen and oxygen atoms to give water. But how do you explain two hydrogen atoms 'sticking' together to give molecular hydrogen?

1808 *Dalton publishes his book 'New system of chemical philosophy'.* This introduced what is known as the law of multiple proportions; elements always combine in definite proportions, simple whole number multiples, when they form chemical compounds. Thus we could have one atom of element A combining with one atom of element B to give one molecule of compound C, or one atom of A combining with two atoms of B to give one molecule of D, etc. However, every particle of, say, water is like every other particle of water in being made up of the same elements in the same proportions. In fact Dalton made an error here and assumed that water was made up of one atom of oxygen and one atom of hydrogen (it should be two of hydrogen to one of oxygen).

1832 *Faraday discovers his laws of electrolysis.* Faraday discovered that the mass of a substance liberated from an electrolyte was proportional to the quantity of electricity, i.e., the charge, passed through the solution.

$$\text{Mass} \propto It$$

where I is the constant current and t the time for which the current is passed. The conclusion that can be drawn from this is that every atom of a given substance has the same quantity of electricity, i.e., charge, associated with it. The same quantity of electricity passed through different electrolytes liberates masses which are proportional to the atomic masses of the substances or some simple fraction of their atomic masses. For example:

> with water; 1 g of hydrogen and 8 g of oxygen are produced, the atomic masses are 1 for hydrogen and 16 for oxygen.
>
> with copper sulphate; 31·7 g of copper is produced, the atomic mass of copper is 63·5.

This could be explained by assuming that atoms carried either 1 unit of charge, 2 units of charge, 3 units, etc. Thus a hydrogen atom has a $+1$ charge, an oxygen atom -2. This fitted in with the concept of valence, hydrogen having a valence of one and oxygen a valence of two. A water molecule is thus made up of two atoms of hydrogen and one of oxygen, held together by electrostatic forces.

About 1850 *Introduction of valence concept.* The concept of valence emerged about the middle of the nineteenth century. It is difficult to ascribe the name of any one person to the introduction of the concept, or indeed any specific date. The names of Frankland, Couper, and Kekulé can, however, be most associated with the introduction of the valence concept.

> 'Before leaving the subject of the atom it is desirable to refer to the new idea which is revolutionizing chemistry, namely, that the different atoms have different exchangeable values, or valencies. An atom is the seat of attractive forces, and the valency is the number of centres of such forces.'
>
> From an essay by 'Sigma' in the *English Mechanic* Vol. **10**, issue of 19 Nov. 1869.
>
> (W. G. Palmer. *A history of the concept of valency.* CUP, 1965, p. 27.)

Thus hydrogen has a valence of one and oxygen a valence of two because one oxygen atom provides 'centres of forces' to have two hydrogen atoms attached to it in the case of water, H_2O.

1865 *Newland's law of octaves.* Newland put the elements in order of atomic mass and proposed that similar chemical properties occurred for every eighth element. For example:

Element	*Position in the table*
Lithium	2
Sodium	9
Potassium	16
Copper	23
Rubidium	30
Silver	37

His work, however, received little attention, other than abuse.

1870 *Mendeleev and the periodic table.* When the different elements are listed in order of their atomic masses, certain patterns of chemical properties recurred at regular intervals. Instead of having a recurrence of similar properties every eight elements, Mendeleev had the properties recurring every 2, 8, 18, 18, etc., elements.
As an example, consider the alkali elements:

Element	*Atomic mass*	*Position in table*
Lithium	6·94	3
Sodium	23·00	11
Potassium	39·10	19
Rubidium	85·48	37
Caesium	132·91	55
Francium	223	87

(Not all these elements were known to Mendeleev, he left gaps for 'unknown' elements.)
The alkali elements are all very reactive, all have a valence of one and thus form similar compounds.

1913 *Bohr and his theory of the atom.* Bohr postulated that electrons in atoms can only occupy well-defined orbits around the nucleus. The orbits were considered to be in groups; two giving the first quantum state, eight giving the next quantum state, eight the third quantum state, eighteen the next, eighteen the next, thirty-two the next. Thus lithium would have two electrons in the first quantum state and one lone electron in the next state, sodium would have two electrons in the first state, eight in the next, and one lone electron in the third state. All the alkali elements would have a lone electron in the outer quantum state.

1916 *Kossel, Lewis, and Langmuir and their theory of the electron nature of the chemical bond.* They proposed that the chemical bond was due to the coupling of lone electrons which occupy the outer shells of linked atoms. Hydrogen atoms have just one electron each. Put two hydrogen atoms together and you can have two atoms each contributing an electron to give a molecule with two electrons, and so a completed quantum state. Completed quantum states produce atoms or molecules with low chemical reactivity. For example the inert gases are helium, with two electrons; neon, with ten electrons; argon with eighteen electrons; and so on. The energy needed to remove an electron from an inert element is higher than that for the following alkali element with one more electron. For example, helium needs 24·6 eV while lithium only needs 5·4 eV. If one electron is taken from lithium, to leave it with two—like helium—we need 75·6 eV to remove another electron. 13·6 eV is needed to take an electron from a hydrogen atom; 15·4 eV is needed to take an electron from the hydrogen molecule.

The oxygen molecule is formed by the two oxygen atoms each taking a share of two of the other atom's electrons to give the full quantum state of ten electrons; the oxygen atom has eight electrons. The nitrogen atom has seven electrons and shares three electrons with another nitrogen atom to give full quantum states of ten electrons. The valence of oxygen is two and that of nitrogen three.

Sodium chloride involves the complete transfer of an electron from the sodium atom to the chlorine atom to give each a full quantum state.

In structures consisting of ions there is a complete transfer of an electron to give two oppositely charged ions which are held together by electric forces of attraction. In the case of molecules where no ions are present the electrons are shared. Evidence for these ideas comes from X-ray crystallography. The intensity of scattered X-rays is proportional to the number of electrons the scattering centre has. The scattering from sodium chloride is from ions, i.e., the sodium atom minus one electron and the chlorine atom plus one electron. The scattering from molecules like the hydrogen molecule shows charge spread between the atoms.

1924 *De Broglie's equation.*

1925 *Heisenberg's quantum mechanics.*

1926 *Schrödinger's quantum mechanics.*
The result of these events was that the electron was considered to have a wave characteristic which determined where the electron was likely to be found. Over a period of time the electron presented the appearance of being smeared out.

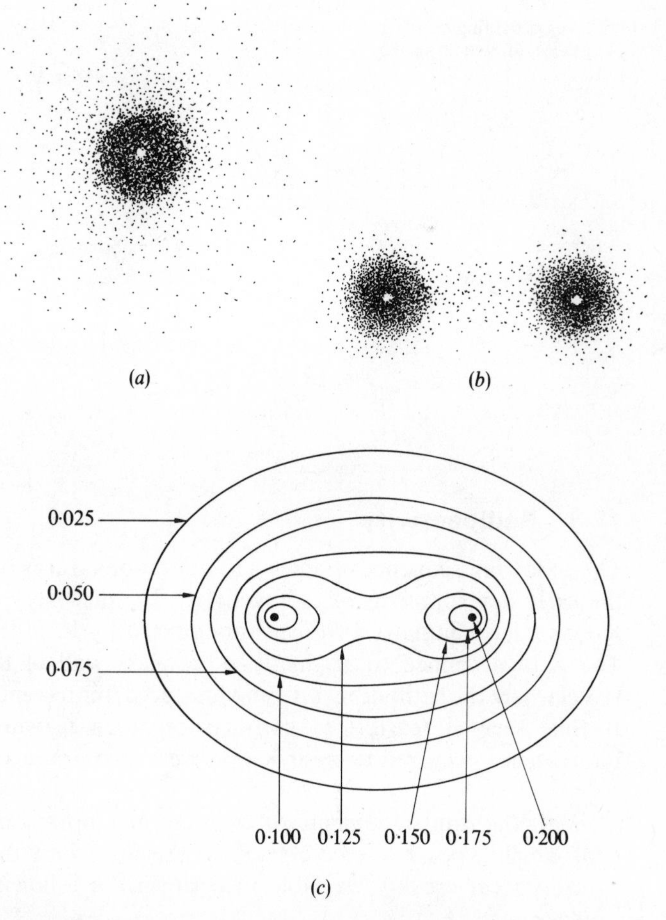

Fig. 13.25 (a) Hydrogen H atom cloud (b) H_2^+ electron cloud. (c) Electron cloud density map for H_2^+. (After C. A. Coulson, Proc. Camb. Phil. Soc., **34**, *210, 1938.)*

1927 *Heitler and London and their theory of molecular bonds*. They used an electron cloud concept of atomic structure to explain the bond between the hydrogen atoms in a molecule. The hydrogen atom was considered to be a nucleus surrounded by a spherical cloud, Fig. 13.25, the density of the cloud at any point representing the chance of finding an electron at the point. In the case of the molecule with just one electron, H_2^+, there is a high density of electron cloud between the atoms. With the H_2 molecule the density is even higher. In the case of the molecule there is a substantial amount of electron cloud between the atoms. There is thus a large chance of finding the electron between the hydrogen atoms. An electron between the atoms will exert attractive forces on both the hydrogen nuclei and so provide a bonding force (Fig. 13.26).

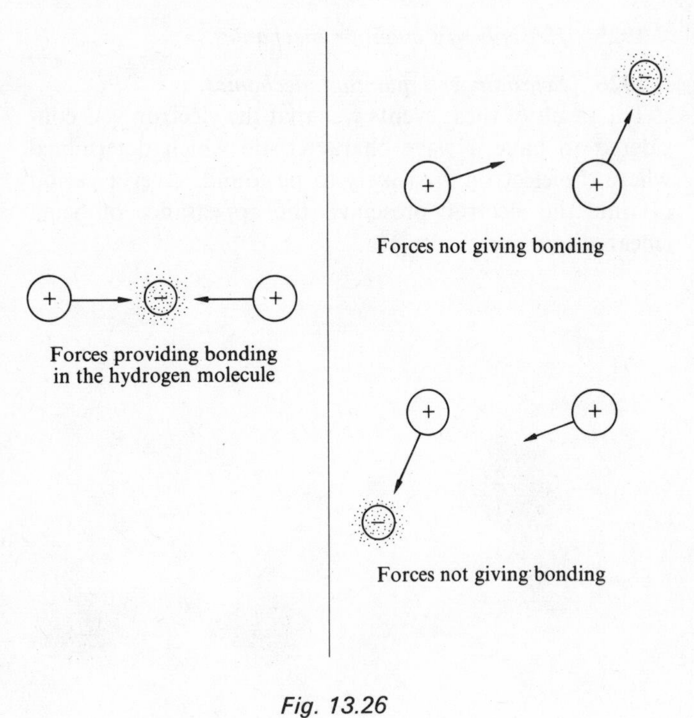

Fig. 13.26

13.3 Radioactivity

The following sequence of notes and quotations traces out the early developments of radioactivity. The discovery of X-rays had in January 1896 been announced by Röntgen. The X-rays seemed to originate in the glass wall of the vessel where the cathode rays hit and produced fluorescence. It thus seemed feasible to consider other cases where fluorescence occurred to see if X-rays were also produced.

'With the double sulphate of uranium and potassium, of which I possess some crystals in the form of a thin transparent crust, I was able to perform the following experiment:

One of Lumière's gelatine-bromide photographic plates is wrapped in two sheets of very heavy black paper, so that the plate does not fog on a day's exposure to sunlight.

A plate of the phosphorescent substance is laid above the paper on the outside and the whole exposed to the sun for several hours. When later the photographic plate is developed, the silhouette of the phosphorescent substance is discovered, appearing in black on the negative. . . .

We may then conclude from these experiments that the phosphorescent substance in question emits radiations which penetrate paper opaque to light and reduce the salts of silver.'

H. Becquerel (1896). *Comptes rendus*. **122**, 420. (Feb. 24)

Becquerel was not the only one investigating the fluorescence of materials and also not the only one reporting that his fluorescent material also emitted a penetrating radiation. The work of the other investigators, however, failed to be confirmed. Becquerel continued with his experiments and within a week gave another report.

'It is very simple to verify that the radiations emitted by this substance (the double sulphate of potassium and uranium) when exposed to the sun or to diffuse daylight will penetrate not only sheets of black paper but even some metals, for example, an aluminium plate and a thin copper foil. . . .

I shall particularly insist on the following fact, which appears to me very important and quite outside the range of the phenomena one might expect to observe. The same crystalline lamellas, placed opposite photographic plates, under the same conditions, separated by the same screens, but shielded from excitation by incident radiation and kept in darkness, still produce the same photographic impressions. . . .'

H. Becquerel (1896). *Comptes rendus*. **122**, 501. (2 March.)

'A few months ago, I showed that uranium salts emit radiations whose existence had not been recognized and that these radiations enjoyed some remarkable properties, some of which are comparable with the properties of the radiation studied by Röntgen. The radiations of the uranium salts are emitted not only when the substances are exposed to light, but even when they are kept in darkness, and for more than two months the same fragments of various salts, shielded from all the exciting radiation known, have continued to emit the new rays, almost without perceptible weakening. . . .

All the salts of uranium I have studied, whether phosphorescent or not with respect to light, crystallized, fused, or in solution, have given comparable results. Thus I have been led to think that the effect was due to the presence in these salts of the element uranium, and that the metal would give more intense effects than its compounds would.

A few weeks ago an experiment with a commercial powder of uranium which had long been in my laboratory confirmed this prediction. . . .'

H. Becquerel (1896). *Comptes rendus*. **122**, 1086.

The rays of Becquerel were thus something emanating from uranium, regardless of the chemical or physical state of that compound. The rays did not appear to have any link with the fluorescence which some of the uranium compounds showed. The fluorescence from the uranium salts used by Becquerel lasted for less than a second after the light was removed.

What was this radiation? In January 1899 Rutherford found that the radiation was made up of more than one component.

'. . . there are present at least two distinct types of radiation—one that is very easily absorbed, which will be termed for convenience α radiation, and the other of a more penetrating character which will be termed the β radiation. . . .'

E. Rutherford (1899). *Phil. Mag.*, (5), **48**, 109.

Later P. Villard found that there was an even more penetrating radiation and called it γ (gamma) radiation.

It was soon found that beta radiation could be deflected by a magnetic field, the deflection being in the direction that negative charges would be deflected. The alpha radiation was much more difficult to deflect by a magnetic field. In 1903 Rutherford was able to show a slight deflection of the particles in a strong magnetic field. The direction of the deviation was in the opposite direction to the deflection of cathode rays (electrons) and thus alpha radiation must consist of positively charged particles.

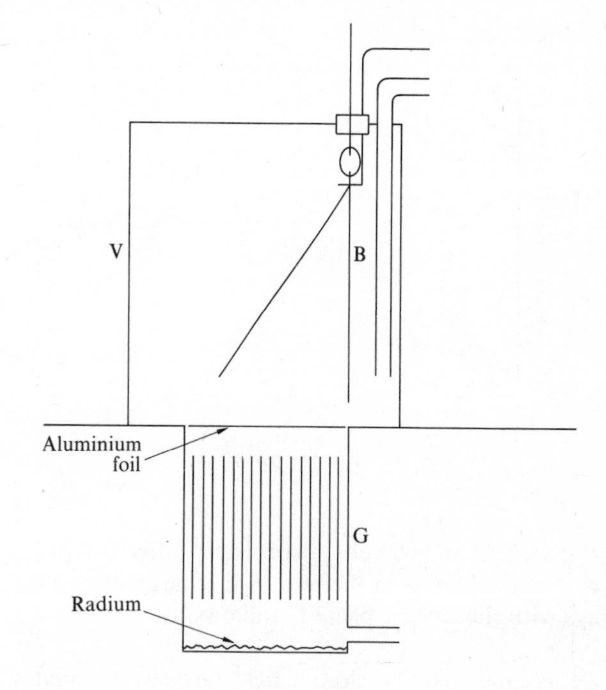

Fig. 13.27

'Figure [13.27] shows the general arrangement of the experiment. The rays from a thin layer of radium passed upwards through a number of narrow slits, G, in parallel, and then through a thin layer of aluminium foil 0·00034 cm thick into the testing vessel V. The ionisation produced by the rays in the testing vessel was measured by the rate of movement of the leaves of a gold-leaf electroscope B. . . .

The magnetic field was applied perpendicular to the plane of the paper and parallel to the plane of the slits. . . . The following is an example of an observation on the magnetic deviation:

	Rate of discharge of electroscope in volts per minute
(1) Without magnetic field	8·33
(2) With magnetic field	1·72
(3) Radium covered with thin layer of mica to absorb all α rays	0·93
(4) Radium covered with mica and magnetic field applied	0·92

The mica plate, 0·01 cm thick, was of sufficient thickness to completely absorb all the α rays, but allowed the β and γ rays to pass through without appreciable absorption. The difference between (1) and (3), 7·40 volts per minute, gives the rate of discharge due to the α rays alone; the difference between (2) and (3), 0·79 volts per minute, that due to the α rays not deviated by the magnetic field employed. . . .'

E. Rutherford (1903). *Phil. Mag.* (6) **5**, 178.

Rutherford found that when the magnetic field was large enough all the alpha rays were deviated to such an extent that none entered the electroscope chamber. When this occurs the maximum radius of curvature of the alpha rays is given by

$$D^2 = (2R - d)d \quad \text{(theorem of intersecting chords)}$$

See Fig. 13.28 for an explanation of the symbols. As D is considerably greater than d, the equation approximates to

$$R = \frac{D^2}{2d}$$

For complete deflection with a magnetic field of flux density 0·84 N A^{-1} m^{-1}, d was 0·055 cm and $2D$ 4·5 cm. Thus R was 0·46 m. For charged particles moving in a circular path in a magnetic field,

$$Bqv = \frac{mv^2}{R}$$

This gives a value for q/mv of 2·6 C s kg^{-1} m^{-1}. Rutherford also attempted with the same apparatus to deflect the alpha particles with an electric field; he connected alternate plates together and produced a potential difference of 600 V between the plates.

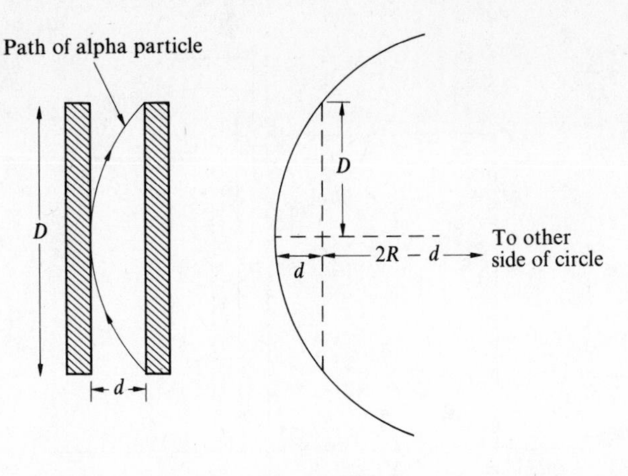

Fig. 13.28

'With a P.D. of 600 volts, a consistent difference of 7 per cent was observed in the rate of discharge due to the α rays with the electric field off and on. . . .'

If we assume that the electric field between the plates is everywhere normal to the rays they will move in a circular path such that we have

$$Eq = \frac{mv^2}{R}$$

$$E = \frac{V}{d}$$

where V is the p.d. between the plates. If we know the p.d. for the rays to describe the same radius R as with the magnetic field, i.e., complete deflection, then

$$v = \frac{E}{B}$$

If we assume that the 600 V is only 7 per cent of the value necessary for complete deflection then the velocity v is about 2×10^7 m s^{-1}. Rutherford estimated that the velocity would be about $2{\cdot}5 \times 10^7$ m s^{-1}. Using this value with the magnetic field result gives for the charge-to-mass ratio for alpha particles about 6×10^7 C kg^{-1}.

Electrons have charge-to-mass ratios of $1{\cdot}76 \times 10^{11}$ C kg^{-1}, protons $9{\cdot}58 \times 10^7$ C kg^{-1}. Alpha rays have charge-to-mass ratios of about half that of protons. Later results gave values much closer to half the proton value.

In a paper in 1906 Rutherford posed a number of questions about alpha particles, and answered them.

'1. Has the α particle expelled from all radioactive products the same mass?

2. Does the value of e/m of the α particle vary in its passage through matter?

3. What is the connection between the velocity of the α particle and its range of ionization in air?

4. What is the connection, if any, between the α particle and the helium atom?

5. Is the heating effect of radium or other radioactive substances due to the bombardment of the radioactive matter by the α particles expelled throughout its own mass?'

E. Rutherford (1906). *Phil. Mag.* (6) **12**, 348.

To the first question Rutherford found the answer—all the alpha particles that he measured had near enough the same value of e/m and thus, he concluded, the same mass. For the second question he found that the value of e/m did not change for alpha particles passing through matter. Alpha particles can traverse a certain thickness of air; this distance travelled is called the range of the alpha particle. Rutherford found that the range and the velocity of an alpha particle were related.

Product	*Range*/cm	*Velocity*/m s^{-1}
Radium	3·50	$1{\cdot}56 \times 10^7$
Emanation	4·36	$1{\cdot}70 \times 10^7$
Radium A	4·83	$1{\cdot}77 \times 10^7$
Radium C	7·06	$2{\cdot}06 \times 10^7$
. . .		

As regards the fourth question and a connection between the α particle and helium.

'The value of e/m for the α particle may be explained on the assumptions that the α particle is (1) a molecule of hydrogen carrying the ionic charge of hydrogen, (2) a helium atom carrying twice the ionic charge of hydrogen, or (3) one half of the helium atom carrying a single ionic charge. . . .

The hypothesis that the α particle is a molecule of hydrogen seems for many reasons improbable. If hydrogen is a constituent of radioactive matter, it is to be expected that it would be expelled in the atomic, and not in the molecular state. . . . If the α particle is hydrogen, we would expect to find a large quantity of hydrogen present in the old radioactive minerals, which are sufficiently compact to prevent its escape. This does not appear to be the case, but, on the other hand, the comparatively large amount of helium present supports the view that the α particle is a helium atom. . . . We are thus reduced to the view that either the α particle is a helium atom carrying twice the ionic charge of hydrogen, or is half of a helium atom carrying a single ionic charge.

The latter assumption involves the conception that helium while consisting of a monovalent atom under ordinary chemical and physical conditions, may exist in a still more elementary state as a component of the atoms of radioactive matter, and that, after expulsion, the parts of the atom lose their charge and recombine to form atoms of helium; while such a view cannot be dismissed as inherently improbable, there is as yet no direct evidence in its favour. On the other hand, the second hypothesis has the merit of greater simplicity and probability. . . .'

E. Rutherford (1906). *Phil. Mag.* (6), **12**, 366.

As regards the fifth question, in 1903 P. Curie and A. Laborde had announced

'We have discovered that the salts of radium constantly release heat.'

P. Curie and A. Laborde (1903). *Comptes rendus*, **136**, 673.

Radioactive materials are at a higher temperature than their surroundings (see chapter 3 for a longer quotation from the above paper). From his measurements of the mass and velocities of alpha particles Rutherford was able to calculate the mean kinetic energy of an alpha particle.

$$\text{Kinetic energy} = \tfrac{1}{2}mv^2$$

If we assume that the charge on the alpha particle is $2e$ then

$$m = \frac{2e}{\text{charge-to-mass ratio}} = \frac{2 \times 1{\cdot}6 \times 10^{-19}}{6 \times 10^7}$$
$$= 5{\cdot}3 \times 10^{-27}\ \text{kg}$$

Hence

$$\text{Kinetic energy} = \tfrac{1}{2} \times 5{\cdot}3 \times 10^{-27} \times (1{\cdot}7 \times 10^7)^2$$
$$= 7{\cdot}7 \times 10^{-13}\ \text{J}$$

This is about 5 MeV.

The energy release from one gramme of radium raises the temperature of the radium to 1·5°C above the surroundings. This is an energy release of 420 J per hour or 0·12 J per second. One gramme of radium emits about $3{\cdot}7 \times 10^{10}$ particles per second (this activity is called one curie). This means an energy release per alpha particle of about

$$\frac{0{\cdot}12}{3{\cdot}7 \times 10^{10}} \approx 3 \times 10^{-13}\ \text{J}$$

This is approximately of the same order as the kinetic energy of an alpha particle and would indicate that the rise in temperature of radium is at least largely accounted for by the collision of alpha particles with the radium atoms.

Conclusive evidence as to the nature of alpha particles was obtained by Rutherford and Royds in 1909.

'We have recently made experiments to test whether helium appears in a vessel into which the α particles have been fired, the active matter itself being enclosed in a vessel sufficiently thin to allow the α particles to escape, but impervious to the passage of helium or other radioactive products.

The experimental arrangement is clearly seen in the figure [Fig. 13.29]. The equilibrium quantity of emanation from about 140 milligrams of radium was purified and compressed by means of a mercury column into a fine glass tube A about 1·5 cms long. This fine tube, which was sealed on a larger capillary tube B, was sufficiently thin to allow the α particles from the emanation and its products to escape, but sufficiently strong to withstand atmospheric pressure.... The thickness of the wall of the tube employed in most of the experiments was less than 1/100 mm, and was equivalent in stopping power of the α particle to about 2 cms of air. Since the ranges of the α particles from the emanation and its products radium A and radium C are 4·3, 4·8, and 7 cms respectively, it is seen that the great majority of the α particles escape through the walls of the tube....

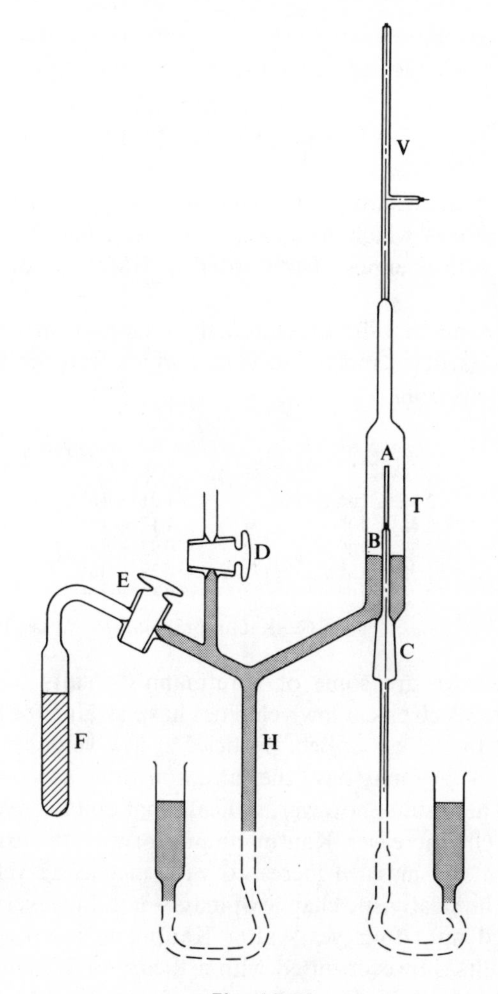

Fig. 13.29

The glass tube A was surrounded by a cylindrical glass tube T.... The outer glass tube T was exhausted by a pump through the stopcock D....

Part of the α particles which escaped through the walls of the fine tube were stopped by the outer glass tube and part by the mercury surface. If the α particle is a helium atom, helium should gradually diffuse from the glass and mercury into the exhausted space, and its presence could then be detected spectroscopically by raising the mercury and compressing the gases into the vacuum tube (V).

...At intervals after the introduction of the emanation the mercury was raised, and the gases in the outer tube spectroscopically examined. After 24 hours no trace of the helium yellow line was seen; after 2 days the helium yellow was faintly visible; after 4 days the helium yellow and green lines were bright; and after 6 days all the stronger lines of the helium spectrum were observed....

... the experiments give a decisive proof that the α particle after losing its charge is an atom of helium.'

E. Rutherford and T. Royds (1909).
Phil. Mag. (6), **17**, 181.

Alpha particles would thus seem to be helium nuclei, i.e., helium atoms which have lost two electrons. They are emitted with energies of the order of 5 MeV, about 8×10^{-13} J.

Kaufmann in 1901 measured the charge-to-mass ratio for beta particles and found values which were similar to those of electrons.

Velocity of β particle /m s^{-1}	*e/m of beta particle* /C kg^{-1}
$2{\cdot}36 \times 10^8$	$1{\cdot}31 \times 10^{11}$
$2{\cdot}48 \times 10^8$	$1{\cdot}17 \times 10^{11}$
$2{\cdot}59 \times 10^8$	$0{\cdot}97 \times 10^{11}$
$2{\cdot}72 \times 10^8$	$0{\cdot}77 \times 10^{11}$
$2{\cdot}85 \times 10^8$	$0{\cdot}63 \times 10^{11}$

(W. Kaufmann. *Göttinger Nachr.*, 1901–3.)

The above are some of Kaufmann's results for beta particles; electrons at low velocities have a value for e/m of $1{\cdot}76 \times 10^{11}$ C kg^{-1}. Beta particles at low velocities look as though they may have the same e/m value as slow electrons. The results, however, indicate that e/m decreases as the velocity increases. Kaufmann put forward the explanation that the mass m increased with increasing velocity. This is the relativistic change of mass—not, however, to be predicted until four years after Kaufmann's experiment. The results, however, fitted with a theory of Lorentz that the dimensions of electrons decreased as their velocity increased. At the time there were two opposing theories—electrons were rigid, electrons could contract.

Beta particles are considered to be high-speed electrons.

Gamma radiation was not deflected by magnetic or electric fields. In all its properties it was similar to X-rays. Final confirmation was the use of crystals as diffraction gratings in a measurement of the wavelengths of gamma rays. Gamma radiation is an electromagnetic wave (see chapter 12 for a velocity measurement).

Alpha, beta, and gamma radiations ionize air, the alpha radiation producing considerably more ions than the beta and the beta more than the gamma. Becquerel was able to use a charged electroscope to determine the presence of the radiation: the radiation caused the electroscope to discharge. The rate of discharge of the electroscope was taken as a measure of the intensity of the radiation.

An alpha particle of energy 5 MeV has a range in air of about 6·5 cm. Along its track it will ionize the air (see cloud chamber photographs in chapter 2). To ionize oxygen or nitrogen takes about 30 eV (see chapter 7). As in a distance of 6·5 cm an alpha particle loses 5 MeV it must, if we assume it only loses energy by ionization and not just by transfer to molecules, produce

$$\frac{5 \times 10^6}{30} \text{ ion pairs.}$$

This is $1{\cdot}7 \times 10^5$ ion pairs. This is an average of 2600 ion pairs per millimetre of path.

In fact the number of ion pairs produced per millimetre of path is greater near the end of the alpha particle's range than near the beginning because it is moving slower and spends more time in those regions. Figure 13.30 shows the ionization curve.

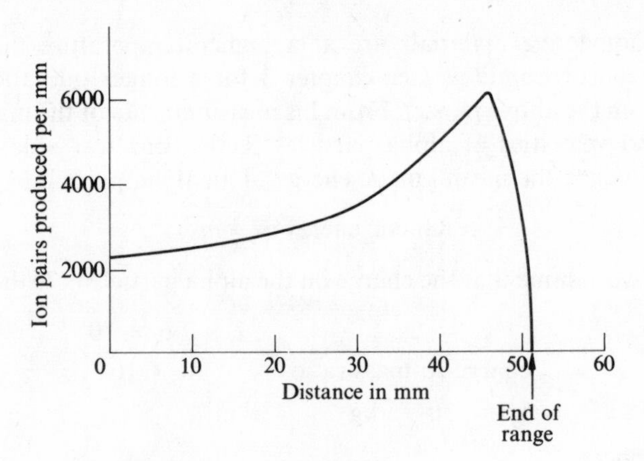

Fig. 13.30

Alpha particles are massive by comparison with electrons and thus little energy is lost by an alpha particle in collisions with electrons. This is, however, not the case with beta particles—they lose a lot of their energy by collisions with electrons, and are more easily deflected from their path.

Alpha and beta particles lose their energy in a number of collisions. Gamma radiation being composed of photons generally loses all its energy at a single encounter. Gamma radiation is extremely penetrating only because encounters between photons and atoms are not too frequent.

A charged gold-leaf electroscope discharges when ions produced inside the electroscope chamber neutralize the charge on the deflected gold leaf. The rate at which the leaf's deflection changes is a measure of the number of ions produced in the chamber. A more convenient way of detecting this is to produce the ions between a pair of electrodes. If the electrodes have a potential difference applied between them they will have become charged; they are a capacitor. If the charges on the two electrodes are partially neutralized by the ions produced in the chamber a current will flow in the electrode circuit as the p.d. again charges up the electrodes (Fig. 13.31). This arrangement is called an ionization chamber.

If the air pressure inside the ionization chamber is reduced below atmospheric pressure the ions are able to reach high velocities, and hence high kinetic energies, in their passage to the electrodes.They are being produced in the electric field given by the p.d. applied between the two electrodes. Reducing the pressure increases the mean free path of the ions. If the ions have high enough kinetic energy they can produce further ions when they collide with gas atoms in the chamber. With a suitable combination of low pressure and high voltage the ion current may be multiplied by many thousands by these further ionizations. With a high enough

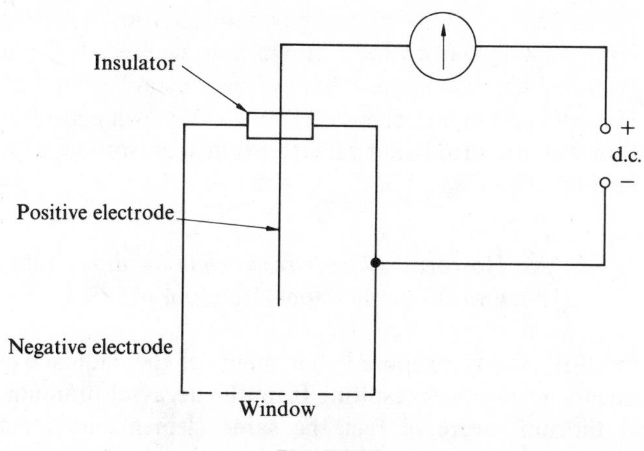

Fig. 13.31

voltage an avalanche of ions is produced as each ion produces more ions which in turn produce more ions which in turn ... and so on. Each primary ion may give rise to 10^9 further ion pairs before reaching an electrode. A chamber operating under these conditions is called a Geiger–Müller tube (first produced in 1928). This is a simplified account.

Radioactive elements

Uranium and thorium were both found to be radioactive, uranium by Becquerel and thorium by Schmidt in 1898. Were other substances radioactive? In July 1898 the Curies reported the possible existence of another radioactive element, polonium, and in December 1898 another element, radium.

'Certain minerals containing uranium and thorium (pitchblende, chalcolite, uranite) are very active from the point of view of the emission of Becquerel rays. In a previous paper, one of us (Mme Curie) has shown that their activity is even greater than that of uranium and thorium, and has expressed the opinion that this effect was attributable to some other very active substance included in small amounts in these minerals....

The pitchblende which we have analysed was approximately two and a half times more active than uranium.... We have treated it with acids and have treated the solutions obtained with hydrogen sulphide. Uranium and thorium remain in solution. We have verified the following facts:

The precipitated sulphides contain a very active substance together with lead, bismuth, copper, arsenic, and antimony. This substance is completely insoluble in the ammonium sulphide which separates it from arsenic and antimony. The sulphides insoluble in ammonium sulphide being dissolved in nitric acid, the active substance may be partially separated from lead by sulphuric acid. On washing lead sulphate with dilute sulphuric acid, most of the active substance entrained with the lead sulphate is dissolved.

The active substance present in solution with bismuth and copper is precipitated completely by ammonia [along with bismuth] which separates it from copper. Finally the active substance remained with bismuth....

We believe therefore that the substance which we have removed from pitchblende contains a metal not yet reported [and] close to bismuth in its analytical properties. If the existence of this new metal is confirmed, we propose to call it polonium from the name of the country of origin of one of us....'

P. and M. Curie (1898). *Comptes rendus*, **127**, 175.

'...The investigations which we are now following are in accord with the first results obtained, but in the course of these researches we have found a second substance strongly radioactive and entirely different in its chemical properties from the first.... The new radioactive substance we have just found has all the chemical aspects of nearly pure barium ... we propose to give the name radium....'

P. Curie, M. Curie, and G. Bemont (1898). *Comptes rendus*, **127**, 1215.

The end of the nineteenth and the beginning of the twentieth centuries saw the discovery of a multitude of other radioactive substances in uranium and thorium compounds. These substances were separated chemically from the uranium and thorium.

'On the discovery by M. and Mme Curie of polonium and radium, bodies of enormous radio-active powers, it was suggested that uranium might possibly owe its power to the presence of a small quantity of one of these bodies....'

W. Crookes (1899–1900). *Proc. Roy. Soc.*, **66**, 409.

Crookes proceeded to try a multitude of chemical reactions with uranium salts in an endeavour to separate other radioactive bodies from uranium. He succeeded and writes

'For the sake of lucidity the new body must have a name. Until it is more tractable I will call it provisionally UrX —the unknown substance in uranium.'

Then Rutherford investigated thorium.

'Thorium compounds under certain conditions possess the property of producing temporary radioactivity in all solid substances in their neighbourhood.... Unlike the radiations from thorium and uranium, which are given out uniformly for long periods of time, the intensity of the excited radiation is not constant, but gradually diminishes....'

E. Rutherford (1900). *Phil. Mag.* (5), **49**, 161.

The thorium compounds were emitting a radioactive gas

whose radioactivity decayed with time (see chapter 5 for a discussion of radioactive decay).

'In previous papers it has been shown that the radioactivity of the elements radium, thorium and uranium is maintained by the continuous production of new kinds of matter which possess temporary activity. In some cases the new product exhibits well-defined chemical differences from the element producing it, and can be separated by chemical processes. Examples of this are to be found in the removal of thorium X from thorium and uranium X from uranium. In some other cases the new products are gaseous in character, and so separate themselves by the mere process of diffusion....'

E. Rutherford and F. Soddy (1903). *Phil. Mag.* (6), **5**, 576.

'A close examination of the origin of these products shows that they are not produced simultaneously, but arise in consequence of a succession of changes originating in the radio-element. Thorium first of all gives rise to the product ThX. The ThX produces from itself the thorium emanation (gas), and this itself is transformed into a non-volatile substance. A similar series of changes is observed in radium....'

E. Rutherford (1905). *Phil. Trans. Roy. Soc* A, **204**, 169.

In this last paper Rutherford gives a table in which he gives the various radioactive series. That part of the table concerned with thorium is given below.

Product	*Half life*	*Some physical and chemical properties*
Thorium	3×10^9 years	Insoluble in ammonia
↓		
Thorium X	4 days	Soluble in ammonia
↓		
Thorium emanation	1 minute	Chemically inert gas; condenses about $-120°C$
↓		
Thorium A	11 hours	Behaves as solid; insoluble in ammonia; volatilized at a white heat; soluble in strong acids. Thorium A can be separated from B by electrolysis.
↓		
Thorium B	55 minutes	
↓		
Thorium C (final product)		

It was soon found that many of the radioactive substances could not be separated from each other or from some common element by chemical methods.

'From very early—around 1905—cases of radio-elements inseparable by chemical analysis from common elements began to accumulate.... The first was 'radiolead' found in pitchblende, and extracted with the relatively large amount of lead therein, first noted as early as 1900, but the early experiments were very inaccurate and conflicting. By 1905, Rutherford had practically shown it to be his Ra D.... All this distracted attention from its chemistry—that it was entirely inseparable chemically from lead—which was indeed finally put beyond doubt by Hevesy and Paneth in about 1910.... My own contribution was the proof that mesothorium was isotopic with radium....'

F. Soddy

M. Howorth. *Pioneer research in the atom*, 1958, New world publications, London, p. 180.

In 1910 Soddy proposed that many of the radioactive elements in the series resulting from the decay of uranium, and thorium, were in fact the same elements as some naturally occurring non-radioactive elements. The term isotope was used to describe the different forms of an element. This leads to the following version of the thorium series.

Element	*Old name*	*Atomic number*	*Atomic mass*
Thorium	Thorium	90	232
Radium	Mesothorium I	88	228
Actinium	Mesothorium II	89	228
Thorium	Radiothorium	90	228
Radium	Thorium X	88	224
Radon	Thoron (emanation)	86	220
Polonium	Thorium A	84	216
Lead	Thorium B	82	212
Bismuth	Thorium C	83	212
Thallium	Thorium C′	81	208
Lead	Thorium D	82	208

The last element, lead 208, is stable.

When an alpha particle is emitted the element loses two units of positive charge and four units of mass from the nucleus, the atomic number thus decreases by two and the atomic mass by four. When a beta particle is emitted the element loses one unit of negative charge and virtually no mass from the nucleus, the atomic number increases by one and the atomic mass remains unchanged. Taking a negative charge from the nucleus increases the positive charge by one unit—an explanation is that a neutron in the nucleus has changed to a proton.

This is the dream of the ancient chemists—the transmutation of elements, one element changing to another element.

Further reading

A. Romer. *The restless atom*, Heinemann Science Study Series, No. 10, 1961.

Bombarding nuclei

The first artificial transmutation was produced by Rutherford in 1919. He bombarded nitrogen atoms with alpha particles and found that a long-range radiation was produced which he identified as protons. Nitrogen had been changed into oxygen.

$$^{14}_{7}N + ^{4}_{2}He \rightarrow ^{17}_{8}O + ^{1}_{1}H$$

Rutherford and Chadwick later showed that many other

light elements could be similarly transmuted. In all cases the bombardment was provided by the alpha particles.

It was the results of bombarding beryllium by alpha particles which led to the discovery of the neutron.

> 'It has been shown by Bothe and others that beryllium when bombarded by α particles of polonium emits a radiation of great penetrating power.... Recently Mme Curie-Joliot and M. Joliot found, when measuring the ionization produced by this beryllium radiation in a vessel with a thin window, that the ionization increased when matter containing hydrogen was placed in front of the window. The effect appeared to be due to the ejection of protons with velocities up to a maximum of nearly 3×10^9 cm per sec.... These results, and others I have obtained in the course of the work, are very difficult to explain on the assumption that the radiation from beryllium is a quantum radiation, if energy and momentum are to be conserved in the collisions. The difficulties disappear, however, if it be assumed that the radiation consists of particles of mass 1 and charge 0, or neutrons. The capture of the α particle by the Be 9 nucleus may be supposed to result in the formation of a C 12 nucleus and the emission of the neutron....'
>
> J. Chadwick (1932). *Nature*, No. 3252, **1291**, 312.

(Chapter 2 gives details of the energy and momentum calculations which led Chadwick to the neutron having a mass of 1.)

The nuclear reaction for the neutron production obtained by Chadwick was

$$^{9}_{4}\mathrm{Be} + {}^{4}_{2}\mathrm{He} \rightarrow {}^{12}_{6}\mathrm{C} + {}^{1}_{0}\mathrm{n}$$

When matter containing hydrogen was placed in the path of the neutrons, protons were ejected.

It was in the years round 1930 that the first accelerators were being built. These were for the production of high-speed particles with which to bombard matter, and find more transmutations. In 1932 Cockcroft and Walton bombarded lithium with protons, of energy about 250 keV. The lithium nucleus broke up to give two helium particles.

$$^{7}_{3}\mathrm{Li} + {}^{1}_{1}\mathrm{H} \rightarrow {}^{4}_{2}\mathrm{He} + {}^{4}_{2}\mathrm{He}$$

The high velocity protons were produced by accelerating them through a large potential difference, this being produced by an arrangement of rectifiers and capacitors.

Other means of accelerating charged particles were soon developed. In 1931 Van de Graaff invented the electrostatic generator now known by his name.

> 'The principle of the generator can be described by making reference to the diagram shown in figure [13.32].

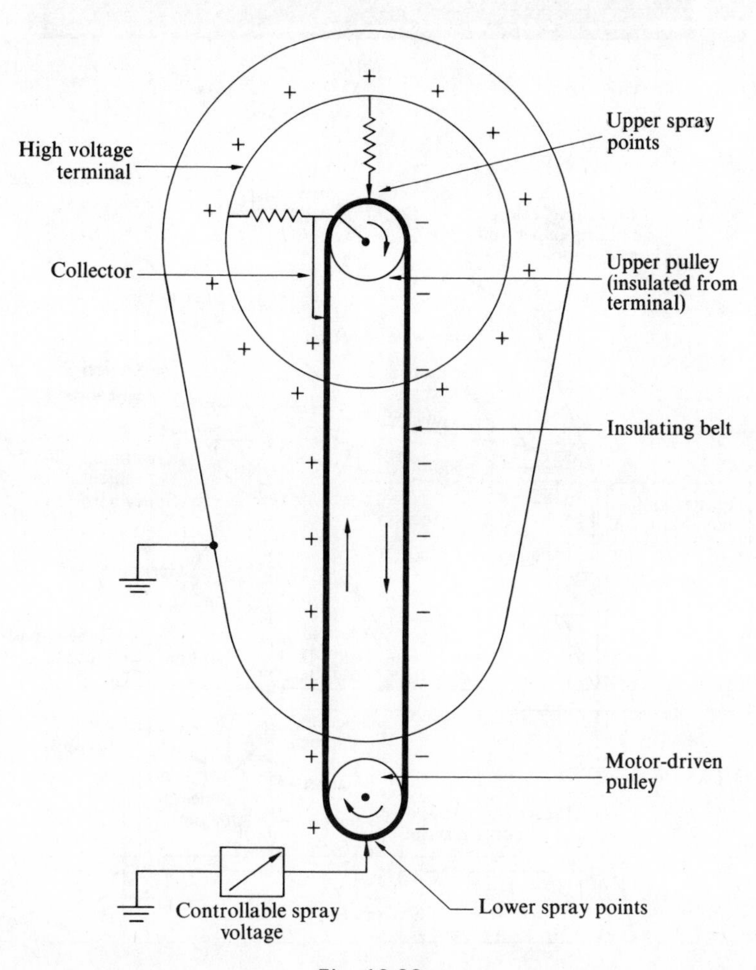

Fig. 13.32

(a)

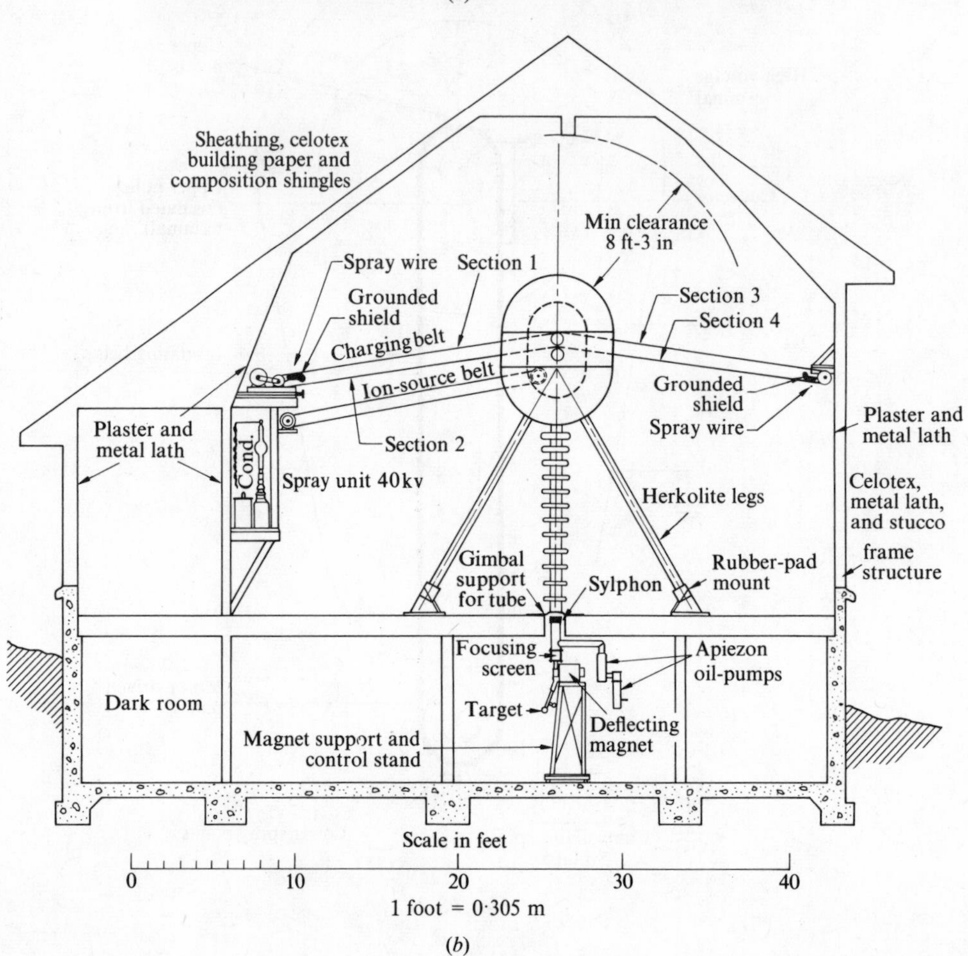
Sheathing, celotex
building paper and
composition shingles
Min clearance
8 ft-3 in
Spray wire
Section 1
Grounded
shield
Charging belt
Ion-source belt
Section 3
Section 4
Grounded
shield
Spray wire
Plaster and
metal lath
Cond.
Section 2
Spray unit 40kv
Plaster and
metal lath
Celotex,
metal lath,
and stucco
Herkolite legs
frame
structure
Gimbal
support
for tube
Sylphon
Rubber-pad
mount
Focusing
screen
Apiezon
oil-pumps
Dark room
Target
Deflecting
magnet
Magnet support and
control stand
Scale in feet
0
10
20
30
40
1 foot = 0·305 m

(b)

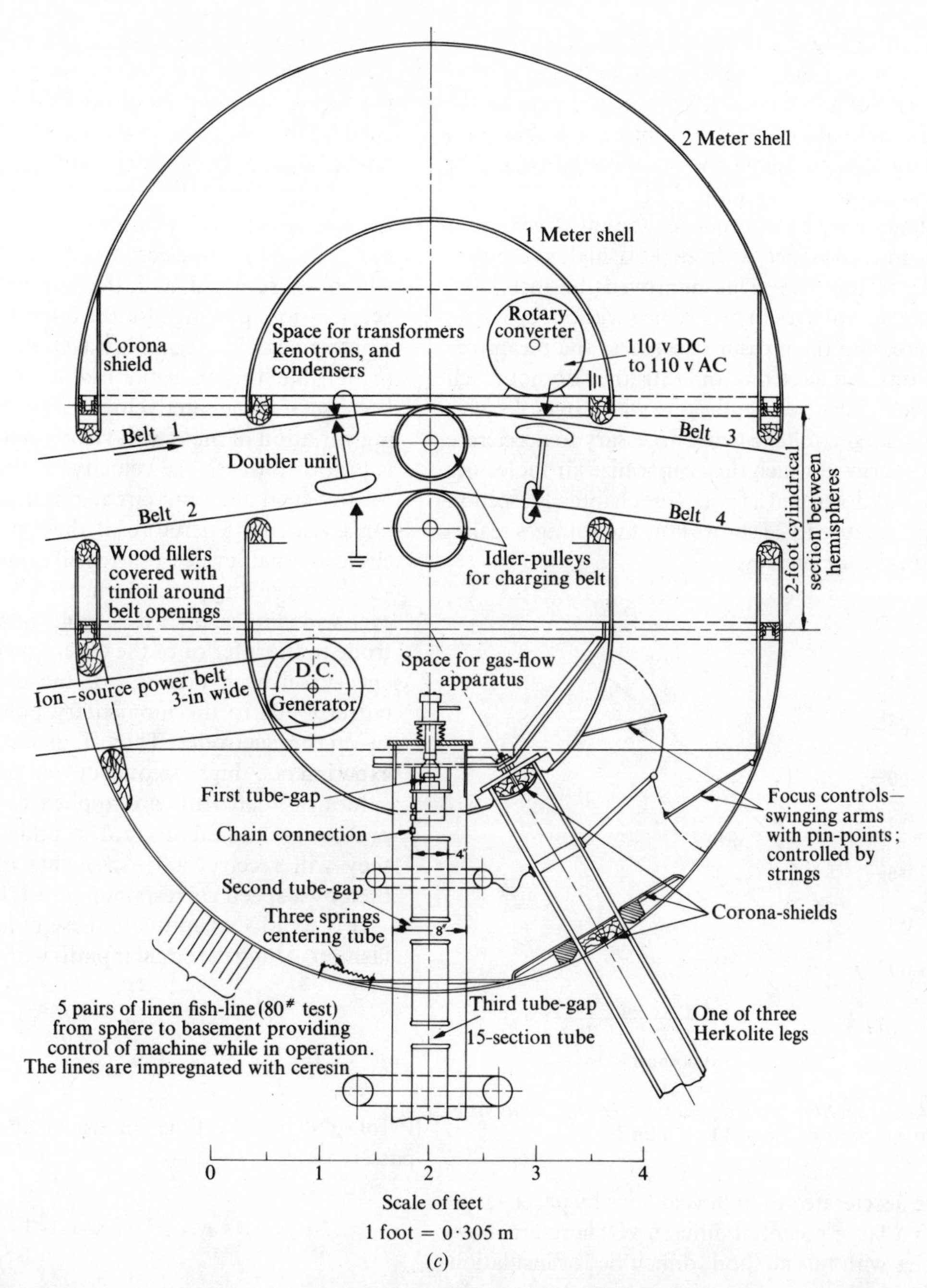

Fig. 13.33 M. A. Tuve, L. R. Hafstad and O. Dahl. Phys. Rev. **48**, *1935. (a) View of generator and cascade tube. (b) Section through high voltage laboratory. (c) Details of internal arrangements in the generator.*

The generator consists of a well-rounded high-voltage terminal supported from ground on an insulating column, and of a charge-conveying system consisting of one or more belts of an insulating material running between this terminal and ground. Electric charge of one polarity is sprayed by corona on the belt at its grounded end and is transported by the belt into the hollow terminal, where it is removed. Within the terminal, charge of the opposite polarity may be deposited on the belt and transferred to the ground. At any instant the terminal potential is $V = Q/C$, where Q is the stored charge and C the capacitance of the terminal to ground. The terminal potential rises initially at the rate $dV/dt = I/C$, where I is the net charging current to the terminal. This rate is commonly of the order of one million volts per second. . . .'

R. J. Van de Graaff, J. G. Trump, W. W. Buechner. (1948). *Reports on progress in physics*, **11**, p. 1.

See chapter 11 and the section on 'lightning' for the mechanism by which the charge is sprayed onto the belt.

Figures 13.33(a), (b), and (c) show details of the Van de Graaff generator used by Tuve, Hafstad and Dahl at the Carnegie Institution in the USA in 1935. The charging belt is not vertical as in Fig. 13.32 but almost horizontal. Another variation is that both the pulley wheels are located outside the high voltage terminal. The vertical tube in the diagrams is the tube along which the ions are accelerated, the target being located below the floor of the generator room. The generator produced a p.d. of 1300 kV.

Higher voltages were produced when the entire generator was mounted in a chamber with air at a higher pressure than atmospheric pressure. This improved the insulation and enabled higher voltages to be reached without sparking occurring. Increasing the pressure decreases the mean free path of stray ions and electrons and thus they do not reach such high velocities between collisions with air molecules. This results in a larger voltage being necessary to accelerate the ions to an energy at which they can ionize air molecules and so cause breakdown and a spark (see chapter 11, section on 'Lightning'). Figure 13.34 shows how the voltages realizable depend on the air pressure.

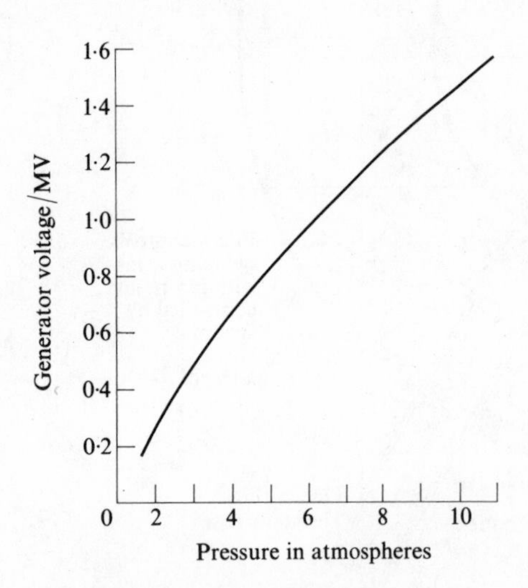

Fig. 13.34 Adapted from van de Graaff, Trump, Buechner. Reports on Progress in Physics, **11**, *6, 1948.*

Ions can be accelerated to high velocities by accelerating them through a large potential difference. There are, however, difficulties with this method; difficulties of insulation, discharges, the production of the high voltage. In the cyclotron the ions are accelerated in a number of small spurts and only relatively low potential differences are used. The following extract is taken from the paper by Lawrence and Livingston in which they describe the first cyclotron.

'Two electrodes A, B in the form of semi-circular hollow plates are mounted in a vacuum tube in coplanar fashion with their diametral edges adjacent [Fig. 13.35]. By placing the system between the poles of a magnet, a magnetic field is introduced that is normal to the plane of the plates. High frequency electric oscillations are applied to the plates so that there results an oscillating electric field in the diametral region between them.

With this arrangement it is evident that, if at one moment there is an ion in the region between the electrodes, and electrode A is negative with respect to electrode B, then the ion will be accelerated to the interior of the former. Within the electrode the ion traverses a circular path because of the magnetic field, and ultimately emerges again between the electrodes; this is indicated in the diagram by the arc $a..b$. If the time consumed by the ion in making the semi-circular path is equal to the half period of the electric oscillations, the electric field will have reversed and the ion will receive a second acceleration, passing into the interior of electrode B with a higher velocity. Again it travels on a semi-circular path ($b..c$), but this time the radius of curvature is greater because of the greater velocity. For all velocities (neglecting variation of mass with velocity) the radius of the path is proportional to the velocity, so that the time required for traversal of a semi-circular path is independent of the ion's velocity. Therefore, if the ion travels its first half circle in a half cycle of the oscillations, it will do likewise on all succeeding paths. Hence it will circulate around on ever widening semi-circles from the interior of one electrode to the interior of the other, gaining an increment of energy on each crossing of the diametral region that corresponds to the momentary potential difference between the electrodes. Thus, if, as was done in the present experiments, high frequency oscillations having peak values of 4000 volts are applied to the electrodes, and protons are caused to spiral around in this way 150 times, they will receive 300 increments of energy, acquiring thereby a speed corresponding to 1,200,000 volts.

It is well to recapitulate these remarks in quantitative fashion. Along the circular paths within the electrodes . . .

$$\left[\frac{mv^2}{r} = Bev\right] \tag{1}$$

It follows that the time for traversal of a semi-circular path is

$$t = \frac{\pi r}{v} = \cdots = \left[\frac{\pi m}{Be}\right] \tag{2}$$

which is independent of the radius r of the path and the velocity v of the ion. The particle of mass m and charge e thus may be caused to travel in phase with the oscillating electric field by suitable adjustment of the magnetic field.'

E. O. Lawrence and M. S. Livingston (1932). *Phys. Rev.*, **40**, 23.

Cyclotrons can accelerate particles up to energies of about 20 MeV. This limit is set by the velocity of the ions becoming appreciable when compared with the speed of light. A changing value of m in the above equations means that ions will not arrive at the gap between the electrodes at the right time to be accelerated. This has been overcome by either varying the magnetic field or the frequency of the electric field during the acceleration of the ions. Such a modified

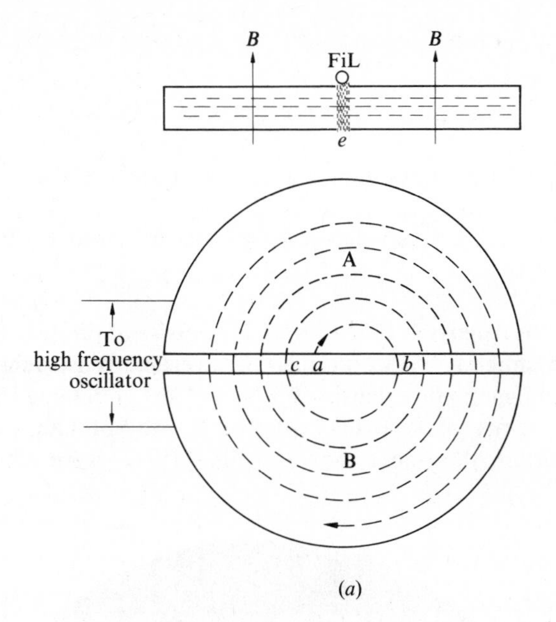

(a)

(c)

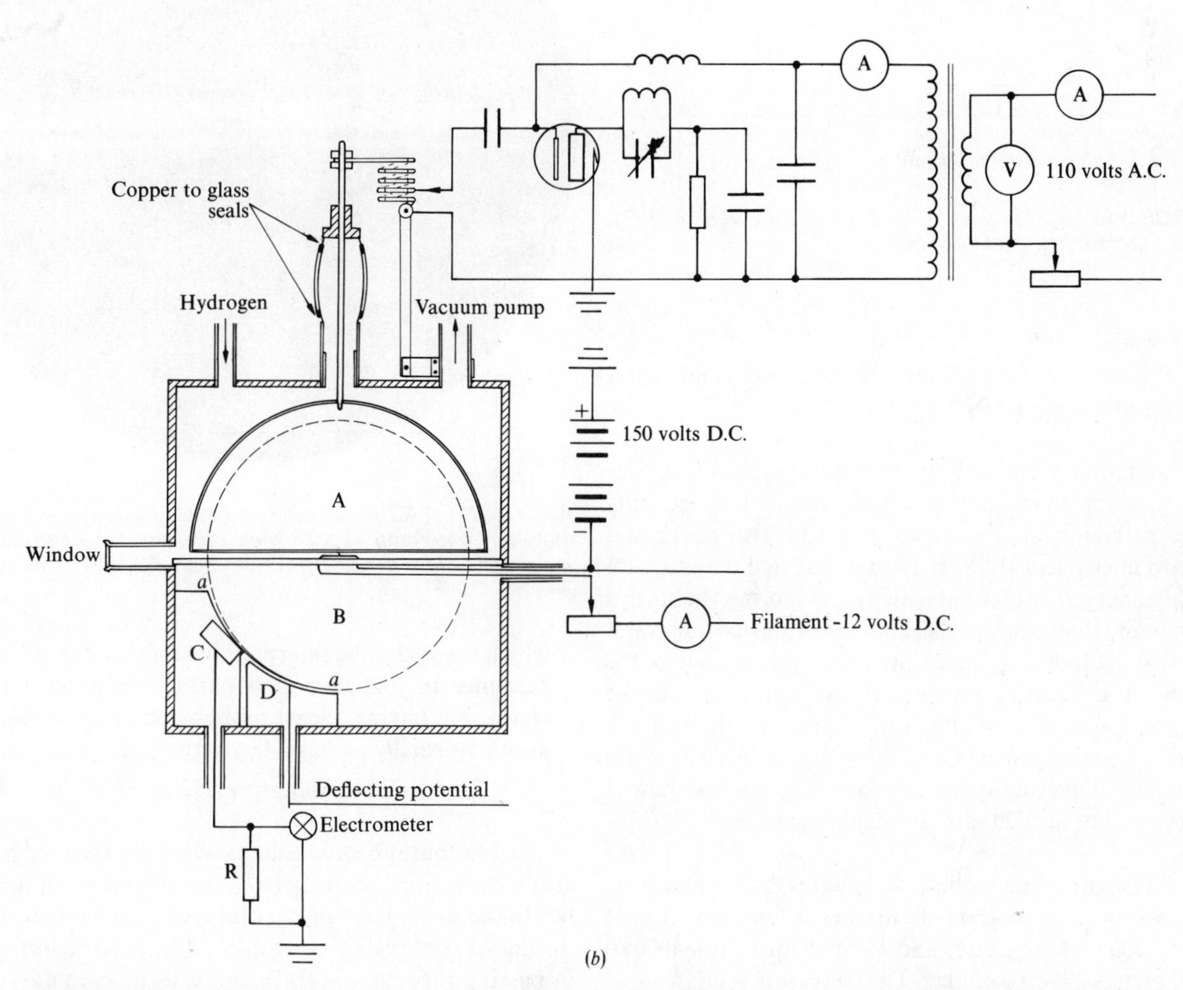

(b)

Fig. 13.35 (a) Diagram of experimental method for multiple acceleration of ions. (b) Diagram of apparatus for the multiple acceleration of ions. (c) Tube for the multiple acceleration of light ions—with cover removed. (Lent to Science Museum, London by Professor E. O. Lawrence, Radiation Laboratory, University of California. Photo: Science Museum, London.)

cyclotron is known as a synchrocyclotron. Such instruments were first produced in about 1946.

Figure 13.36 shows how the energies realizable by different accelerators have changed over the years. Since 1930 the particle energies produced by accelerators have increased from about 1 MeV to almost 100 000 MeV.

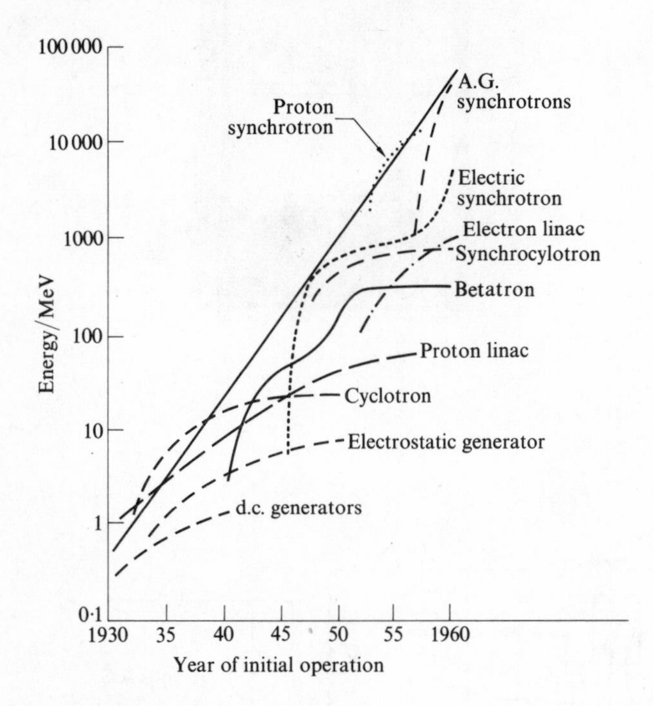

Fig. 13.36 M. S. Livingston and J. P. Blewett (1962). Particle accelerators, McGraw-Hill.

Further reading

R. R. Wilson and P. Littauer. *Accelerators*, Heinemann Science Study Series, No. 15, 1960.

The radiations emitted by naturally radioactive materials and the man-made accelerators are not the only sources of high-energy particles with which to probe and bombard nuclei. In 1912 V. F. Hess discovered the existence of highly energetic radiation which was entering the earth's atmosphere; the radiation became known as cosmic rays. Hess flew balloons to high altitudes and measured the amount of ionization present. If the ionization in the atmosphere was due to the 'radioactive' earth then the amount of ionization in the atmosphere should decrease with height above the earth's surface. The results showed no decrease but at high altitudes an increase.

'. . . At heights of more than 2000 metres there is a marked increase in the radiation. It reaches 4 ions (cm^{-3} s^{-1}) from 3000 to 4000 metres and 16 to 18 ions from 4000 to 5200 metres in two counters. The increase is even stronger in the thin-walled counter No. 3. . . .

The discoveries revealed by the observations here given are best explained by assuming that radiation of great penetrating power enters our atmosphere from the outside. . . .'

V. F. Hess (1912). *Phys. Zeitschrift*, **13**, 1084.

Cosmic radiation is of very high energy and is able to produce nuclear reactions. Examination of cosmic ray tracks in a cloud chamber led to the discovery of the positron.

'On August 2, 1932, during the course of photographing cosmic-ray tracks produced in a vertical Wilson chamber (magnetic flux density 1·5 N A^{-1} m^{-1}) designed in the summer of 1930 by Professor R. A. Millikan and the writer, the tracks shown in Fig. [13.37] were obtained,

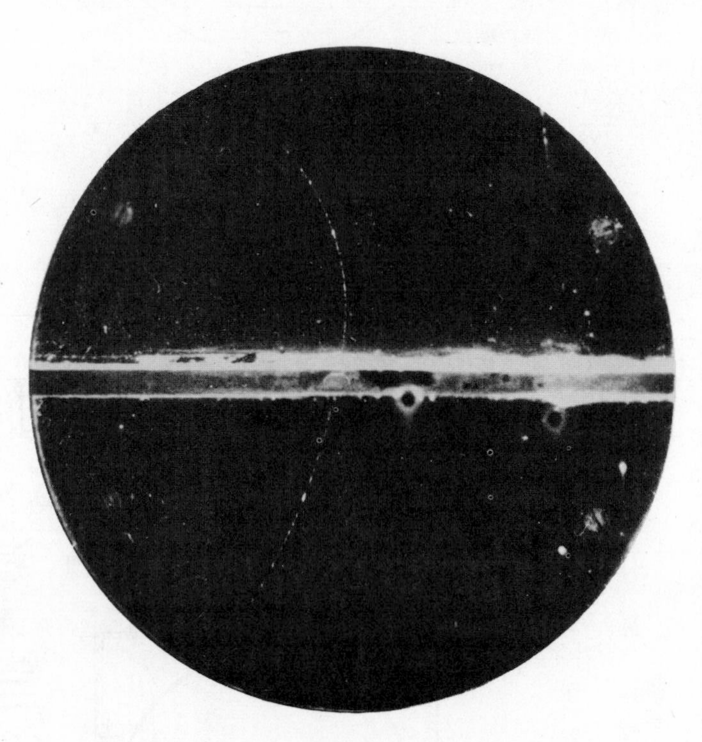

Fig. 13.37 A 63 MeV positron passing through a 6 mm lead plate and emerging as a 23 MeV positron. The chamber is in a magnetic field. (Anderson. Phys. Rev., **43**, *491, 1933.)*

which seemed to be interpretable only on the basis of the existence in this case of a particle carrying a positive charge but having a mass of the same order of magnitude as that normally possessed by a free negative electron. . . .'

C. D. Anderson (1933). *Phys. Rev.*, **43**, 491.

The photograph shows the positive electron, or positron as it is now called, moving from the upper half of the chamber to the lower half. In the centre of the chamber it passes through a lead plate, 6 mm thick. The particle loses energy in passing through the plate and thus its path curvature is less below the plate than above. Measurements of the curvature of the tracks enables a value of mv/q to be determined.

$$\frac{mv^2}{R} = Bqv$$

$$\frac{mv}{q} = BR$$

The effect of the lead on the track enables an estimate to be made of the absorption of the radiation in lead.

Radiocarbon dating

The earth's atmosphere is bombarded by cosmic rays. These produce neutrons which on hitting nitrogen-14 atoms produce carbon-14.

$${}^{14}_{7}N + {}^{1}_{0}n \rightarrow {}^{14}_{6}C + {}^{1}_{1}H$$

Carbon-14 has a half-life of 5568 years. This radioactive carbon produced in the upper atmosphere combines with oxygen to give radioactive carbon dioxide. The radioactive carbon dioxide forms only a very small percentage of the atmospheric carbon dioxide. As plants live off carbon dioxide this means that all plants will show some radioactivity due to carbon-14. During the lifetime of the plant it will take in carbon dioxide and the percentage of its carbon which is radioactive will remain constant, the same as that of the atmosphere. When, however, the plant dies it ceases to take in further supplies of carbon and the percentage of radioactive carbon within the plant decreases as the carbon-14 decays. Figure 13.38 shows how the number of disintegrations per minute per gramme of carbon changes with time. The results are taken from materials whose date could be established by other means. Thus by measuring the activity of the carbon-14 in a specimen it becomes possible to establish its age.

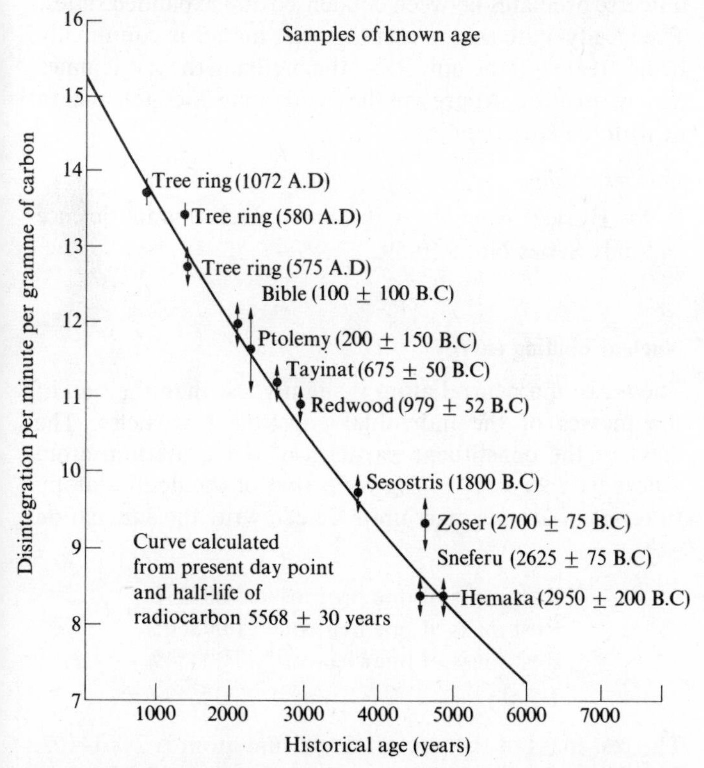

Fig. 13.38 Radiocarbon dating, *2nd Ed., Libby, University of Chicago Press, 1955.*

The age of the earth

How old is the earth? Ten million years? A hundred million years? A thousand million years? Five thousand million years? Figures like these were the subject of great controversy in the latter half of the nineteenth century. In 1858 Darwin published his book *The Origin of Species.*

> '... if the variations useful to any organic being ever do occur, assuredly individuals thus characterized will have the best chance of being preserved for the struggle for life; and from the strong principle of inheritance, these will tend to produce offspring similarly characterized. This principle of preservation, or the survival of the fittest, I have called Natural Selection....'
>
> C. Darwin (1858). *The Origin of Species.*

A vital point in this theory was the length of time the earth had supported life in order to allow for natural selection to evolve the various species. This required more time than if species had been separately created. The time necessary for man to evolve via apes would be much greater than the time in which man was somehow directly created as man.

W. Thomson (later Lord Kelvin) opposed Darwin's views because he considered that the earth was not old enough for evolution by natural selection to have occurred. His argument was based on a calculation of the rate at which the earth was cooling. Kelvin measured the temperature gradient of the earth's crust near the surface and the thermal conductivity of the surface material.

The rate of flow of heat from a cooling object, dQ/dt, is related to the temperature gradient, $d\theta/dx$, by

$$\frac{dQ}{dt} = KA\frac{d\theta}{dt}$$

where K is called the thermal conductivity, and A is the surface area of the object.

$K = 3\ J\ m^{-1}\ K^{-1}\ s^{-1}$
$d\theta/dt = 17 \times 10^{-3}\ K\ m^{-1}$ (Kelvin used 1/2776)
$A = 4\pi r^2 = 4\pi\ (6{\cdot}4 \times 10^6)^2 = 5{\cdot}2 \times 10^{14}\ m^2$

$$\text{Thus}\ \frac{dQ}{dt} = 2{\cdot}7 \times 10^{12}\ J\ s^{-1}$$

Kelvin assumed that the beginning of time for the earth could be taken as the time when the earth was just solidifying. He took the temperature of solidification as 3900°C. The heat flow from a body cooling from 3900°C to about 0°C (assumed to be the average temperature of the earth) is

Q = mass × specific thermal capacity × 3900
$= 6 \times 10^{24} \times 800 \times 3900$
$= 1{\cdot}9 \times 10^{31}\ J$

If dQ/dt remained constant then it would need about 7×10^{18} s or about 10^{11} years. The rate of flow of heat at 3900°C might, however, be expected to be greater than that at 0°C; this would mean less years would be necessary. Kelvin suggested that the value would be about 10^8 years (a hundred million years).

> 'The limitation of geological periods imposed by physical science cannot, of course, disprove the hypothesis of transmutation of species; but it does seem sufficient to disprove the doctrine that transmutation has taken place through 'descent with modification by natural selection'.'
>
> Kelvin. *Popular lectures and addresses*, London, 1894, p. 89.
>
> (*Darwin's century*. L. Eiseley, Gollancz, 1959, p. 240.)

It was not many years before the error of Kelvin's calculation emerged—Kelvin had assumed that the earth was cooling without itself containing any source of energy which could replenish in any way the energy lost. A source was discovered—radioactivity. The energy released by 1 g of uranium is about 10^{-6} J s^{-1}. If we assume that uranium occurs throughout the earth's volume in the same proportions as the surface rocks the total amount of uranium in the earth will be of the order of 10^{19} kg and the total energy released about 10^{13} J s^{-1}. This is about the same as the rate at which the earth is losing energy.

Our estimates are very rough; we have neglected the thorium in the earth and probably overestimated the uranium content (it is doubtful that there is the same proportion in the earth's interior as in the crust). The result is, however, sufficient to indicate that the age of the earth as estimated by Kelvin is probably a considerable underestimate.

Estimates of the ages of rocks can be made from measurements of their helium content. Radioactive materials, which decay by alpha emission, produce helium. If this is trapped in the rocks we can from a measurement of the amount of helium estimate the number of alpha particles produced over the lifetime of the rock. The mineral thorianite from Ceylon was found to contain 68 per cent of thorium and 11 per cent of uranium and yielded 8·9 cm^3 of helium gas (at 273 K and 76 cm pressure) per gramme of mineral. Knowing the rates at which uranium and thorium produce alpha particles, and hence helium, the age of the mineral can be estimated. The thorianite has been aged at 270 million years.

Uranium and thorium decay to give lead; uranium 238 decays to give lead 206, thorium 232 decays to give lead 208. Thus lead in uranium of thorium-bearing rocks would be expected to have a higher concentration of these particular lead isotopes than common lead. Measurement, using a mass spectroscope, of the relative amounts of these isotopes enables the rock to be aged. (See Mass spectroscopy, earlier in this chapter.)

Naturally occurring potassium contains a radioactive isotope, potassium 40. On decay this gives argon 40. The amount of argon in potassium bearing minerals can therefore be used to age those minerals. Rubidium 87 occurs naturally in rocks and decays to give strontium 87. Measurements of the strontium-to-rubidium ratio can thus yield an estimate of the age of the rocks.

The oldest rocks on earth appear to have an age of about $4{\cdot}7 \times 10^9$ years, meteorites an age of about $4{\cdot}5 \times 10^9$ years, soil from the moon about $4{\cdot}7 \times 10^9$ years. All the indications are that the earth, and the moon, solidified about $4{\cdot}7 \times 10^9$ years ago.

How old is the universe? The red shift (see chapter 12) indicates that galaxies are all receding. Measurements of the red shift enable us to calculate the velocity with which they recede. If we couple these with estimates of their distance from us we can make an estimate of how long it has taken for the galaxies to reach their present positions from an initial big-bang. The Virgo cluster is receding at about 1200 km s^{-1} and has been estimated as being 40 million light years from the earth. The time taken to cover 40 million light years at 1200 km s^{-1} is

$$\frac{40 \times 10^6 \times 3 \times 10^8 \times 3{\cdot}2 \times 10^7}{1200 \times 10^3}$$

$$= 3{\cdot}2 \times 10^{17} \text{ s}$$

or about 10^{10} years. More careful estimates put the figure at about $1{\cdot}3 \times 10^{10}$ years.

There are basically (1961) three theories concerning the origin of the universe. One considers that the universe started with a 'bang', originating with some giant primeval atom. A modification of this theory is to suggest that the universe oscillates between condensed and expanded states. The steady-state theory assumes that matter is continually being created in the universe—the big-bang theory assumes no new creation. At present the steady state does not seem to fit with the latest evidence.

Further reading

P. M. Hurley. *How old is the earth?* Heinemann, Science Study Series No. 5, 1959.

Nuclear binding energy

The mass of a neutral atom is slightly less than the sum of the masses of the individual constituent particles. The mass of the constituent particles of the deuterium atom differs by $2{\cdot}96 \times 10^{-30}$ kg from that of the neutral atom. In terms of atomic mass units we can write the summation as

Rest mass of one proton	1·007276
Rest mass of one neutron	1·008665
Rest mass of one electron	0·000549
Total	2·016490

The rest mass of the neutral deuterium atom is 2·0014102, an amount 0·002388 less than that of the total of the constituents' masses.

If we use Einstein's equation relating energy and mass (see chapter 3),

$$E = mc^2$$

then the above mass difference corresponds to an energy change of $3{\cdot}55 \times 10^{-13}$ J or 2·22 MeV. Combining the particles to form deuterium means that this energy must be released.

When hydrogen is bombarded with neutrons deuterium can be produced. In the process gamma rays are produced; they have energy 2·22 MeV. This is how the surplus energy is carried off. If we want to break a deuterium atom down into hydrogen we have to bombard the hydrogen with gamma rays of energy at least 2·22 MeV before the change can occur. This 2·22 MeV can be considered as the binding energy of the deuterium nucleus.

The binding energy increases as the number of particles in the nucleus increases. A convenient way of looking at binding energies is to consider the value of the average binding energy per nuclear particle (the term nucleon is often used). Figure 13.39 shows how this binding energy per nucleon varies with the atomic mass. Some nuclei have

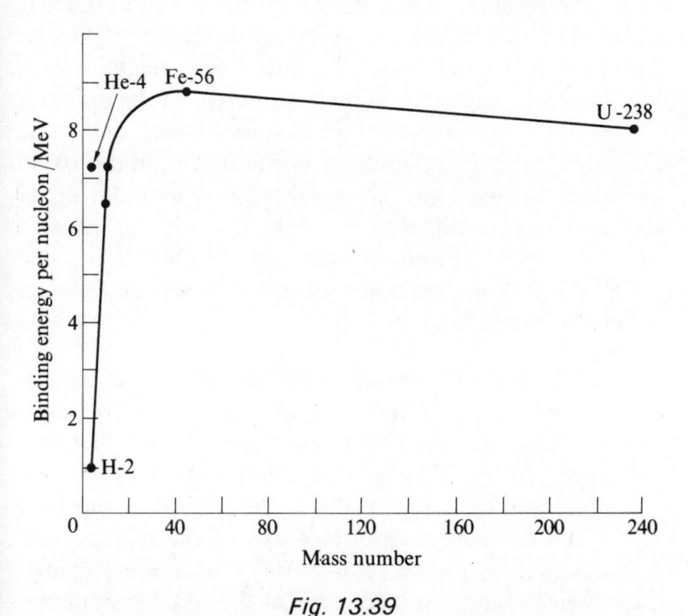

Fig. 13.39

binding energies above the graph curve which passes through the majority of points. Helium, mass 4, is one such example. It has a higher binding energy than neighbouring atoms with mass 3 or 5. Helium is a particularly stable nucleus. Other particularly stable nuclei are those of carbon 12 and oxygen 16. The binding energy per nucleon reaches a maximum at iron 56, a particularly stable nucleus. The elements that are particularly stable are the ones found in greatest abundance on the earth, and on the sun.

Nuclear fission

Neutrons were discovered by Chadwick in 1932 and they were soon used as projectiles to bombard nuclei. In 1934 E. Fermi bombarded uranium with slow neutrons (about the speed of molecules in a room-temperature gas) hoping that the addition of the neutron to the uranium nucleus would give a new element, atomic number 93. His results were, however, puzzling.

> 'Neutron bombardment of uranium resulted in a beta radiation of great intensity, and various periods were observed; one of about a minute, another of 13 minutes, and several that were longer but which have not yet been measured with any precision. The beta rays corresponding with the 13-minute period were particularly intense.
>
> Attempts were made by chemical means to identify the element giving a period of 13 minutes as an isotope of some very heavy element. . . .'
>
> E. Fermi. May 10 1934.
>
> P. de Latil. *E. Fermi*, Souvenir Press, 1965.

Fermi failed to identify the elements produced.

Early in January 1939 a clue emerged as to the nature of the elements produced by the bombardment of uranium with slow neutrons. O. Hahn and F. Strassmann, in Germany, found that one of the products of the bombardment was barium. Barium has an atomic number of 56, nowhere near that of uranium at 92. An explanation for this strange result was soon forthcoming, from L. Meitner and O. R. Frisch—the uranium nucleus had been split; nuclear fission was occurring.

> 'It seems . . . possible that the uranium nucleus has only small stability of form and may, after neutron capture, divide itself into two nuclei of roughly equal size (the precise ratio of sizes depending on finer structural features and perhaps partly on chance). These two nuclei will repel each other and should gain a total kinetic energy of c. 200 MeV. . . .'
>
> L. Meitner and O. R. Frisch (1939). *Nature*, **143**, 239.

200 MeV is a lot of energy per atom. The implications of this large energy release were considerable.

Lise Meitner had been working under Hahn in Germany until 1938 when she fled from Nazi Germany. Hahn advised her of his discovery of barium in a letter which she received just before Christmas, 1938. Her nephew Otto Frisch was on holiday with her in Sweden and in discussing the implications of Hahn's letter they arrived at the idea of fission. On returning from holiday to Copenhagen Frisch mentioned the matter to Bohr who was just departing for a conference in the USA. Bohr acclaimed the idea and advised speedy publication. The news of fission set the conference alight and scientists rushed off to their laboratories to repeat the experiments. Within days of the announcement nuclear fission had been realized at many USA laboratories. The following is part of the account of those days by Bohr's biographer.

'Bohr asked if he might make an announcement of the utmost importance. The fairly humdrum conference exploded in its turn.

As Bohr spoke, Tuve whispered to Hafsted that he had "better go out" and put a new filament in the new Carnegie atom accelerator.

No sooner had Bohr finished than some of the physicists rushed to the door. They were on the way to phone their laboratories. Long distance calls went out to universities all across the country....

...The Carnegie's accelerator was quickly supplied with the new filament Tuve had called for and was ready to go into operation. Tuve invited Bohr, Rosenfeld, and Teller to watch the history-making test that night. The guests arrived about midnight and were led through the deserted library and halls to the laboratory.

The control board of the apparatus was a maze of dials, windows and buttons, but all eyes were turned to the window of the oscilloscope. As Bohr and Rosenfeld walked in they could see great green pulses shooting to its very top. The shining, glowing line of green recorded the energy given off each time a uranium nucleus divided. The atom was being split in their presence, before their very eyes. "The state of excitement challenged description," said Rosenfeld. "I remember it as though it were yesterday."...'

From *Niels Bohr*, by R. Moore and used by permission of the publishers Hodder and Stoughton and Alfred A. Knopf, Inc. Copyright 1967.

Uranium has three naturally occurring isotopes; about 0·006 per cent U-234, about 0·7 per cent U-235, and the remainder as U-238. Do all three undergo fission by neutrons? A mass spectrometer arrangement was used to separate the three isotopes and the minute samples so prepared were bombarded by neutrons. Only U-235 showed fission. U-238 captured neutrons which hit it while moving fast but failed to fission. U-235 was found to capture slow neutrons more easily than fast ones. When fission occurred the U-235 nucleus split to liberate more neutrons—fast neutrons. A chain reaction was possible if the neutrons emerging from the fissioned U-235 could be slowed down and hit other U-235 nuclei; they must not, however, hit U-238 nuclei and be absorbed. Because the greater bulk of natural uranium was U-238 a chain reaction in natural uranium was not possible.

What happens to the U-238 when it absorbs fast neutrons? It changes via U-239 to a new element, neptunium (Np).

$$^{238}_{92}\text{U} + ^{1}_{0}\text{n} \rightarrow ^{239}_{93}\text{Np} + ^{0}_{-1}\text{e}$$

Neptunium is radioactive and decays to give another new element, plutonium (Pu).

$$^{239}_{93}\text{Np} \rightarrow ^{239}_{94}\text{Pu} + ^{0}_{-1}\text{e}$$

The half-life of neptunium-239 is 2·34 days, the half-life of plutonium-239 is $2{\cdot}44 \times 10^4$ years. Plutonium is, however, fissionable by slow neutrons.

Because plutonium can be produced from the main bulk of natural uranium it provided a more readily available source of fissionable material than U-235. A chain reaction in plutonium seemed feasible.

For a chain reaction to be sustained the neutrons produced by fission must be slowed down and sufficient of them must find other fissionable nuclei. The neutrons must not be wasted by being absorbed or leaving the block of material. The neutrons can be slowed down by using carbon, the chance of the neutrons finding other fissionable nuclei can be increased if the size of the piece of fissionable material is increased. For a spherical piece of material the volume and hence the number of accessible fissionable nuclei goes up as the radius cubed, the number of neutrons lost, i.e., the surface area, only increases as the square of the radius. Hence the idea of a 'critical' size.

In 1942 Fermi produced the first 'divergent chain reaction'. (See chapter 5 for an eyewitness account.) The vital factor in a chain reaction is what is called the reproduction factor—the number of neutrons produced in the 'second generation' of fissions by a single neutron producing the 'first generation' fission. If the reproduction factor is less than 1 the reaction is not self sustaining. Fermi in the first chain reaction obtained a reproduction factor of 1·0006. Cadmium strips were used to control the reaction—cadmium absorbs neutrons and when inserted in the uranium pile rapidly reduces the reproduction factor. The pile consisted of lumps of uranium metal and uranium oxide embedded in graphite. About 6000 kg of uranium metal was used and the pile was near spherical with an effective radius of 3·55 m. The power produced was about 200 watts.

On 16 July 1945, the first nuclear bomb was exploded, a test run in New Mexico.

'The effects could well be called unprecedented, magnificent, beautiful, stupendous and terrifying. No man-made phenomenon of such tremendous power had ever occurred before. The lighting effects beggared description. The whole country was lighted by a searing light with the intensity many times that of the midday sun. It was golden, purple, violet, gray and blue. It lighted every peak, crevasse and ridge of the nearby mountain range with a clarity and beauty that cannot be described but must be seen to be imagined. It was that beauty the great poets dream about but describe most poorly and inadequately. Thirty seconds after the explosion came first, the air blast pressing hard against the people and things, to be followed almost immediately by the strong, sustained, awesome roar which warned of doomsday and made us feel that we puny things were blasphemous to dare tamper with the forces heretofore reserved to the Almighty....'

The war department release on the New Mexico test. 16 July 1945.

The explosion released about 10^{13} J of energy, the equivalent of about 20 000 tons of TNT or the energy needed to roast ten million chickens or the energy released by about a million lightning flashes.

On 6 August 1945, a nuclear bomb was dropped on the city of Hiroshima. The bomb involved the fission of U-235. Later that month another bomb, plutonium fission, was dropped on Nagasaki. The war against Japan came to an end.

> '. . . it is evident that the formidable power of destruction which has come within the reach of man may become a mortal menace unless human society can adjust itself to the exigencies of the situation. Civilization is presented with a challenge more serious perhaps than ever before, and the fate of humanity will depend on its ability to unite in averting common dangers and jointly to reap the benefit from the immense opportunity which the progress of science offers. . . .'
>
> N. Bohr (1945). *The Times*, 11 August.

Nuclear fusion

Fusion reactions are those in which the resulting nuclei have higher masses than those of initial particles. For example, when two protons combine to produce deuterium.

$$ {}^1_1\text{H} + {}^1_1\text{H} \rightarrow {}^2_1\text{H} + \underset{\text{positron}}{{}^0_1\text{e}} + \underset{\text{neutrino}}{\nu} + 1{\cdot}44\ \text{MeV} $$

This reaction between the two protons can only occur when the kinetic energies of the two are high enough for them to come close enough to react. The high kinetic energies can be produced by raising hydrogen to very high temperatures, millions of degrees. This reaction is thought to occur in the sun and be partially responsible for the energy released by the sun. The reaction forms part of a sequence of reactions involving hydrogen, deuterium, and tritium—the end product being helium and a considerable release of energy.

$$ 4\,{}^1_1\text{H} \rightarrow {}^4_2\text{He} + 2\,{}^0_1\text{e} + 26\ \text{MeV} $$

Four protons are effectively fused to produce helium.

Nuclear power

In a conventional power station coal or oil is burnt and the hot gases produced used to heat banks of tubes in the boiler and produce steam. The steam then passes to the turbine where it produces rotation and generates electricity. In a nuclear power station fission occurs in the reactor and the energy released heats a fluid which then circulates through a heat exchanger. Here the hot fluid produces steam which then drives a turbine and so produces electricity. Figure 13.40 shows a simple diagram of such a reactor. A typical reactor produces about 600 MW.

What nuclear fuel resources have we? Our fossil fuel resources are dwindling at an alarming rate; can the future energy requirements of the world be met with nuclear fuels? The world reserves of uranium, which are not uneconomic

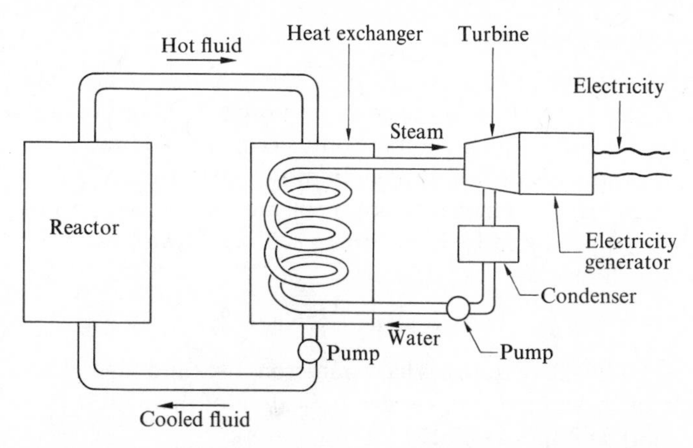

Fig. 13.40 Simple diagram of a power producing nuclear reactor.

to extract, have been estimated as being able to supply between 2·5 and 4·5 × 10^{22} J (ores, at present uneconomic, raise this figure to about 10^{27} J). This, however, does not include the energy that can be produced from the plutonium which is a product of the uranium reaction. The plutonium can be used as a fuel. Taking this into account raises the energy that can be realized from uranium to a maximum of about 2500 × 10^{22} J. In 1970 the world used about 2 × 10^{20} J and the fossil fuel reserves were only about 11 × 10^{22} J. Using only fossil fuels and assuming no increase in demand, extremely unlikely, the fossil fuels can only supply energy for about 550 years. The demand is, however, thought to be more like 8 × 10^{20} J in the year 2000; at this rate the fossil fuels can only last about 140 years. Uranium as a fuel can even at the year 2000 rate increase this time by up to 30 000 years. Reserves of thorium have not been included in this estimate. These amount to about 1000 × 10^{22} J if suitable techniques can be established for the use of the fuel. The future looks as though there will be sufficient energy resources because, and only because, of nuclear energy (Fig. 13.41).

In the above we have only considered nuclear fission. Nuclear fusion in which low atomic number elements combine and release energy in the process will, if suitable methods can be devised, offer much greater energy resources.

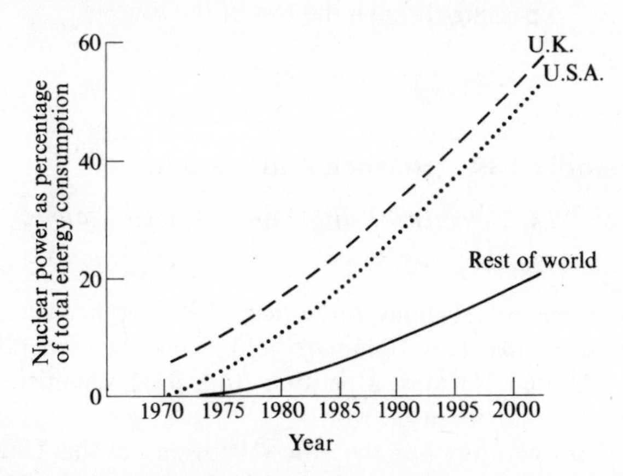

Fig. 13.41 Estimated trend in nuclear power requirements. (P. Searby, New Scientist, *13 Nov. 1969, 22.)*

Radioactive isotopes

Bombardment of nuclei by high energy particles can lead to the production of radioactive isotopes. Thus sodium, for example, which naturally only occurs as sodium 23 can be produced artificially as sodium isotopes of mass 20, 21, 22, 24, and 25. These isotopes are all radioactive. Sodium-24 can be produced by bombarding sodium-23 with neutrons from a reactor.

$$^{23}_{11}\text{Na} + ^{1}_{0}\text{n} \rightarrow ^{24}_{11}\text{Na} + ^{0}_{0}\gamma$$

Radioactive isotopes have numerous uses and the following list can only give a rough indication of these uses.

(a) Isotopes as labels.
- (i) The investigation of the wear of a bearing. The bearing is made radioactive and thus the movement of the radioactive particles of bearing which wear off the bearing can be traced.
- (ii) Fluid level gauge. A float has a radioactive source attached to it. The movement of the fluid surface can be followed by a Geiger counter following the movements of the float. This has application where the fluid is contained in a non-transparent container.
- (iii) The movement of a particular element in a chemical reaction can be followed by making some of the atoms of that element radioactive.

(b) Absorption of radiation.
- (i) Thickness gauges. The thickness of a sample can be determined by determining the amount of radiation absorbed by the sample. A radioactive source on one side of the sample and a Geiger tube on the other is an arrangement in which the output from the Geiger tube is a measure of the thickness of the sample passing between them.
- (ii) A similar arrangement to the thickness gauge can be used to indicate whether a container is completely full.
- (iii) Radiography. The presence of cracks or flaws in a sample can be shown by the difference in absorption of the radiation passing through such parts when compared with the radiation passing through the rest of the sample.

Appendix 13A Science and mankind

'The Scientists' Petition to the United Nations

By Linus Pauling (1958)

At noon on Monday 15 January 1958 I placed in the hands of Mr. Dag Hammarskjold, Secretary-General of the United Nations, a petition from 9235 scientists, of many countries in the world.

This petition has the title "Petition to the United Nations Urging that an International Agreement to Stop the Testing of Nuclear Bombs be Made Now."

The petition consists of five paragraphs, as follows:

We, the scientists whose names are signed below, urge that an international agreement to stop the testing of nuclear bombs be made now.

Each nuclear bomb test spreads an added burden of radioactive elements over every part of the world. Each added amount of radiation causes damage to the health of human beings all over the world and causes damage to the pool of human germ plasm such as to lead to an increase in the number of seriously defective children that will be born in future generations.

So long as these weapons are in the hands of only three powers an agreement for their control is feasible. If testing continues, and the possession of these weapons spreads to additional governments, the danger of outbreak of a cataclysmic nuclear war through the reckless action of some irresponsible national leader will be greatly increased.

An international agreement to stop the testing of nuclear bombs now could serve as a first step towards a more general disarmament and the ultimate effective abolition of nuclear weapons, averting the possibility of a nuclear war that would be a catastrophe to all humanity.

We have in common with our fellow men a deep concern for the welfare of all human beings. As scientists we have knowledge of the dangers involved and therefore a special responsibility to make those dangers known. We deem it imperative that immediate action be taken to effect an international agreement to stop the testing of all nuclear weapons.'

Linus Pauling. *No more war* Dodd, Mead & Co., 1962.

'The compelling need for nuclear tests

By Edward Teller and Albert Latter (1958)

Last month the U.N. received a petition signed by more than 9000 scientists urging an immediate halt to nuclear bomb tests. The petition said that the tests were endangering both the present population of the world and generations yet unborn and declared that an agreement to stop such testing is now feasible. Dr. Linus Pauling, the American scientist who presented the document to the U.N., said further that in his opinion the U.S. will never achieve one of the principal objectives of its nuclear tests: the production of a "clean" bomb—i.e., one with little or no radioactivity. These are statements of tremendous importance. If true, they strongly support the position that we must stop nuclear tests at once.

But they are not true. The statements are at best half-truths, and they are misleading and dangerous. If acted upon, they could bring disaster to the free world. Sober consideration of the facts makes it perfectly clear that we must continue testing nuclear weapons. Such tests do not seriously endanger either present or future generations. It is true that they might be halted by agreement—but there would be no way to prevent the Soviet Union, a notorious violator of agreements, from starting them up again in secret. And finally, American tests are leading to the development of a clean bomb. . . .'

E. Teller and A. Latter. *Life* magazine, 10 February 1958, S. G. Phillips, Inc.

Should the bomb be banned? Would the abolition of nuclear bombs lead to peace or another war? Why was the bomb ever produced? The following quotations are relevant to this last issue.

Letter to President Roosevelt

By ALBERT EINSTEIN

August 2, 1939

Sir:

Some recent work by E. Fermi and L. Szilard, which has been communicated to me in manuscript leads me to expect that the element uranium may be turned into a new and important source of energy in the immediate future. Certain aspects of the situation seem to call for watchfulness and, if necessary, quick action on the part of the Administration. I believe, therefore, that it is my duty to bring to your attention the following facts and recommendations.

In the course of the last four months it has been made probable—through the work of Joliot in France as well as Fermi and Szilard in America—that it may become possible to set up nuclear chain reactions in a large piece of uranium, by which vast amounts of power and large quantities of new radium-like elements would be generated. Now it appears almost certain that this could be achieved in the immediate future.

This new phenomenon would also lead to the construction of bombs, and it is conceivable—though much less certain—that extremely powerful bombs of a new type may thus be constructed. A single bomb of this type, carried by boat or exploded in a port, might very well destroy the whole port together with some of the surrounding territory. . . .'

O. Nathan and H. Norden (1960). *Einstein on peace.* Simon and Schuster, Inc. (1963), Methuen.

'My participation in the production of the atomic bomb consisted of one single act: I signed a letter to President Roosevelt, in which I emphasized the necessity of conducting large-scale experimentation with regard to the feasibility of producing an atom bomb.

I was well aware of the dreadful danger which would threaten mankind were the experiments to prove successful. Yet I felt impelled to take the step because it seemed probable that the Germans might be working on the same problem with every prospect of success. I saw no alternative but to act as I did, although I have always been a convinced pacifist. . . .'

A. Einstein. 20 September 1952 (as letter).

'I have never said I would have approved the use of the atomic bomb against the Germans. I did believe that we had to avoid the contingency of Germany under Hitler being in sole possession of this weapon. This was the real danger at the time. . . .'

A. Einstein. 23 June 1953 (as letter).

Wars are between nations, between groups of people. The causes of war are much the same now as they were centuries ago. Science has not changed any of this. What science has changed is the method of war. We should not fear science—we should fear war.

Problems

1. Use the idea of standing electron waves in a rectangular box to forecast the absorption frequencies of a dye molecule where the electrons can move along a carbon chain $8{\cdot}4 \times 10^{-10}$ m long.

2. Consider a wave incident on a row of scattering centres, Fig. 13.42.

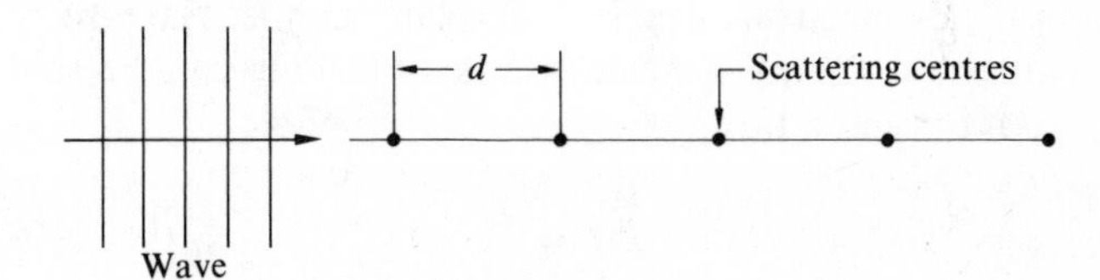

Fig. 13.42

(a) What is the relationship between the wavelength of the radiation and the distance d between the scattering centres for a standing wave pattern to be produced?

(b) The wavelength of electrons is given by

$$\lambda = \frac{h}{mv}$$

What is the condition for electrons travelling along a row of atoms to give a standing wave?

(c) What can be said about the energies of the electrons?

(d) What would be the effects of compressing the 'solid' and so moving the atoms closer together?

3. 'The chemical bond between two hydrogen atoms, which results in the hydrogen molecule, is a result of an electron tunnelling through the potential hill between the two atoms.'

(a) Sketch the potential well/hill of a single hydrogen atom. (b) Sketch the electron standing wave for the atom.

(c) What happens if the two atoms are brought very close together?

(d) Put the above statement in your own words. In particular explain what is meant by the term 'tunnelling'.

4. A beam of electrons is accelerated down a tube by a potential difference of 3 kV applied between their source, a filament, and a target.

(a) What is the kinetic energy of an electron when it hits the target?

(b) What is the velocity of the electrons when they hit the target?

(c) If the beam current of electrons along the tube is 5 μA what is the number of electrons arriving per second at the target?

(d) What is the total energy delivered to the target per second?

(e) If the target has a mass of 20 g and is of steel having a specific heat of 500 J kg^{-1}, calculate the temperature rise that might be expected per second.

mass of the electron = $9{\cdot}1 \times 10^{-31}$ kg
charge on the electron = $1{\cdot}6 \times 10^{-19}$ C

5. What do you think might be the outcome of two systems like our solar system colliding? What happens when atoms collide?

6. Oxford and Cambridge Schools' Examination Board. Nuffield Advanced Physics. Physics I Options paper 1970.

'(*a*) Each of the diagrams [Fig. 13.43] is said to represent an electron in a box. Explain why a diagram of potential energy against distance like this can be said to represent a box.

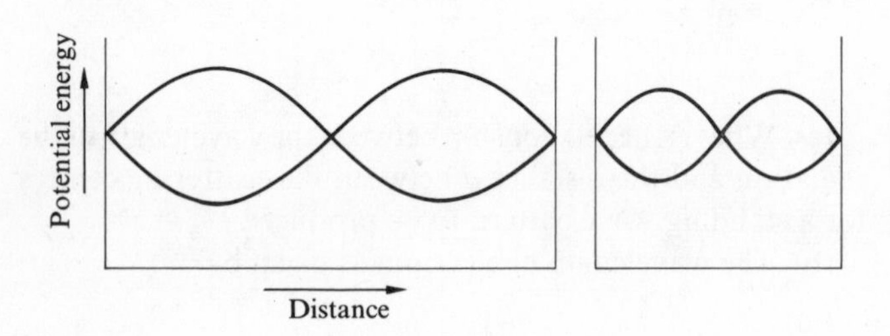

Fig. 13.43

Also explain how the curved lines in each box can be said to represent electrons.

(*b*) "With a single particle in a box, the smaller the box the faster the particle will rattle around." Use the diagrams above, and the relationship $p = h/\lambda$, to explain this statement.

(*c*) The electron in a hydrogen atom cannot be represented by diagrams like these. Draw a rough diagram that better represents an electron in a hydrogen atom "box" with the electron in the lowest energy level. How and why does your diagram differ from those given above?'

7. The first cyclotron (see page 408) produced protons with energies of about 1·2 MeV by the use of an alternating potential of 4000 V (peak value) applied at a frequency of $15{\cdot}6 \times 10^6$ Hz between the two electrodes. The protons were kept in circular paths by a magnetic field.

(a) If the moving protons kept in phase with the alternating electric field how long was taken for protons to traverse a semicircular path within one of the electrodes?

(b) What magnetic field was necessary to keep the protons in suitable circular paths?

(c) How many circuits were used to get the protons to the 1·2 MeV energy?

Mass of proton = $1{\cdot}68 \times 10^{-27}$ kg,
charge = $1{\cdot}6 \times 10^{-19}$ C.

8. Figure 13.44 is taken from a paper by Kaufman and Libby (*Phys. Rev.*, **93**, 1337, 1954) and shows how the amount of tritium in vintage wines depends on the age of the wines. Tritium is an isotope of hydrogen produced in the

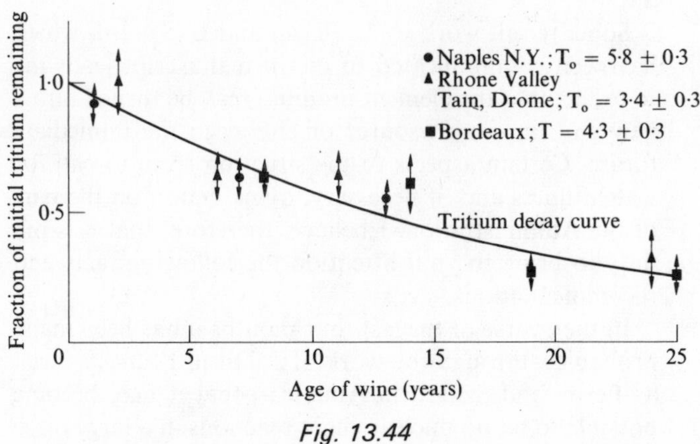

Fig. 13.44

earth's atmosphere by cosmic ray bombardment. It is thus a natural constituent of all atmospheric water vapour and finds its way into all items which use surface water.

(a) Determine from the graph the half-life of tritium.

(b) How does the activity of a fresh wine compare with that of a similar sample which is 25 years old?

9. The world energy demand is increasing, the stock of fossil fuels is dwindling at an alarming rate. The future energy requirements would seem to depend on nuclear reactions. What consequences would you forecast?

10. Oxford and Cambridge Board. *Nuffield advanced physics*. Special paper, 1971.

'The preface to one well-known text-book on Physics begins as follows:—

"Physics is a science which deals with the behaviour of the inanimate world. This world is classified by contemporary physicists into two categories: matter and electromagnetic radiation."

Do you think this is a good summary of the science you understand as physics? Discuss the extent to which you think it is and the extent to which you feel it fails to do justice to the science.

Finally, produce your own short paragraph summarizing what *you* understand "physics" to be all about.'

Teaching note Further problems appropriate to this chapter will be found in:
Nuffield advanced physics. Units 5 and 10, students' books, Penguin, 1972.
The project physics course. Units 5 and 6 texts, Holt, Reinhart and Winston, 1971.
Physical Science Study Committee. *Physics*, 2nd ed. chapters 32 and 34, Heath, 1965.

Practical problems

1. Beta particles are considerably scattered by matter and the multiple scattering experienced by beta particles can result in a net deflection of more than 180°, i.e., beta particles can be scattered back towards the source. How does the amount of back scattering depend on the thickness of the absorber? How does the amount of back scattering depend on the atomic number of the absorber? Consider possible uses for back scattering in the light of your results.

2. Produce and calibrate a device for the indirect measurement of the liquid level in a container.

3. Whenever a Geiger tube is used, allowance has to be made for the background radiation. If a cloud chamber is used background radiation can be clearly seen in the absence of any source placed in the chamber. Background radiation is due to natural activity of the earth, the air, and the materials used in the construction of the tube and shielding. In addition cosmic radiation is a significant factor.

Investigate the background radiation and determine as far as is possible the constituents of the radiation and their origins.

Suggestions for answers

1. See page 391 where the energies permitted to the electron wave are found to be proportional to n^2. This gives the rule for the energy levels and hence the spectral lines.

2. (a) $n\lambda = 2d$

(b) $\dfrac{h}{mv} = \dfrac{2d}{n}$

(c) $E = \dfrac{1}{2m}\left(\dfrac{2d}{n\lambda}\right)^2$

Only energies given by the above equation are possible. There are energy levels.

$$E \propto \frac{1}{n^2}$$

(d) Making d smaller makes E smaller and moves the energy levels closer together.

3. (a) Fig. 13.15.

(b) Fig. 13.18, 13.20, 13.21.

(c) Fig. 13.45 shows how the potential between the atoms becomes modified.

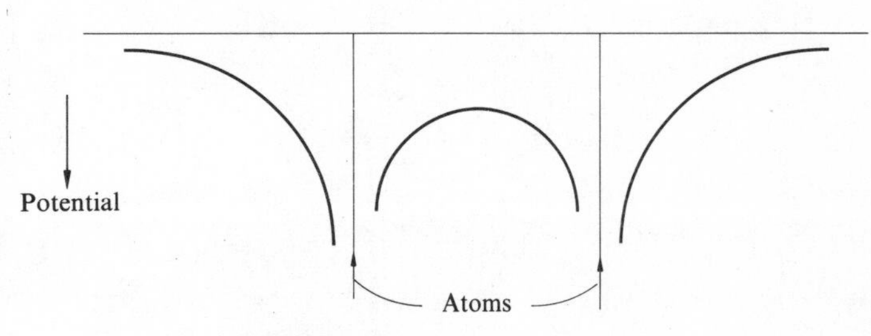

Fig. 13.45

(d) You could talk about the chance of an electron being 'outside' the atom—of being in a place where its kinetic energy is insufficient to get it.

4. (a) $4{\cdot}8 \times 10^{-13}$ J.

(b) Taking m as $9{\cdot}1 \times 10^{-31}$ kg gives a velocity of $1{\cdot}0 \times 10^{9}$ m s^{-1}—faster than the speed of light! The variation of mass with velocity has to be taken into account to obtain the actual velocity—a reasonable guess is to say almost the speed of light.

(c) $3{\cdot}1 \times 10^{13}$.

(d) 14·9 J.

(e) 1·49°C.

5. The solar systems would considerably disrupt each other and the orbits of planets would be considerably changed if not completely destroyed. Except at a few distinct energy values atoms bounce off each other.

6. (a) An abrupt change in potential energy signifies a force which suddenly acts on a particle. This is the type of force we consider, on the large scale, to be characteristic of an object meeting a wall. Your discussion of the curved lines should talk of the chance of finding an electron in a region.

(b) The wavelength equals the length of the box L.

$$E = \frac{h^2}{2mL^2}$$

The smaller L the larger E.

(c) Figure 13.15 shows a 'better' box. Figures 13.18, 13.20, and 13.21 show electron waves in different energy levels. Figure 13.18 is the lowest energy level.

7. (a) $\dfrac{1}{2 \times 1{\cdot}2 \times 10^6}$ s.

(b) $t = \dfrac{\pi m}{Be}$.

$B = 7{\cdot}9 \times 10^{-2}$ N A^{-1} m^{-1}.

(c) 300.

8. (a) 12·3 years.

(b) About $\frac{1}{4}$.

9. Your answer might include discussion of the disposal of radioactive wastes, disposal of 'surplus heat' from power stations, possible hazards due to breakdown in a nuclear power station.

10. I leave this to you.

Index

Pages set in italic type indicate the first page of a sequence. Where a quotation has more than one author it is listed only under the first author.

Printed in Great Britain by William Clowes & Sons, Limited
London, Beccles and Colchester